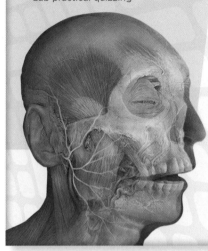

Anatomy & Physiology | REVEALED® 3.0

An Interactive Cadaver Dissection Experience

This unique multimedia tool is designed to help you master human anatomy and physiology with:

- Content customized to your course
- Stunning cadaver specimens
- Vivid animations
- Lab practical quizzing

my Course Content
- Maximize efficiency by studying exactly what's required.
- Your instructor selects the content that's relevant to your course.

Dissection
- Peel layers of the body to reveal structures beneath the surface.

Animation
- Over 150 animations make anatomy and physiology easier to visualize and understand.

Histology
- Study interactive slides that simulate what you see in lab.

Imaging
- Correlate dissected anatomy with X-ray, MRI, and CT scans.

Quiz
- Gauge proficiency with customized quizzes and lab practicals that cover only what you need for your course.

WWW.APREVEALED.COM

Full Textbook Integration!

Icons throughout the book indicate specific **McGraw-Hill Anatomy & Physiology | REVEALED® 3.0** content that corresponds to the text and figures.

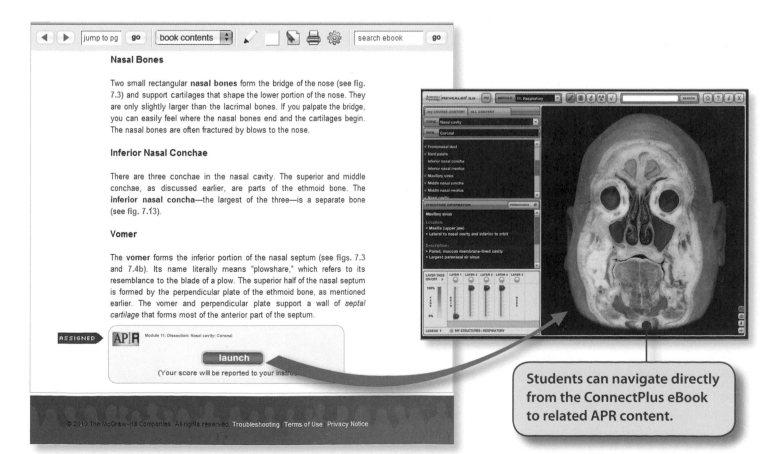

Nasal Bones

Two small rectangular **nasal bones** form the bridge of the nose (see fig. 7.3) and support cartilages that shape the lower portion of the nose. They are only slightly larger than the lacrimal bones. If you palpate the bridge, you can easily feel where the nasal bones end and the cartilages begin. The nasal bones are often fractured by blows to the nose.

Inferior Nasal Conchae

There are three conchae in the nasal cavity. The superior and middle conchae, as discussed earlier, are parts of the ethmoid bone. The **inferior nasal concha**—the largest of the three—is a separate bone (see fig. 7.13).

Vomer

The **vomer** forms the inferior portion of the nasal septum (see figs. 7.3 and 7.4b). Its name literally means "plowshare," which refers to its resemblance to the blade of a plow. The superior half of the nasal septum is formed by the perpendicular plate of the ethmoid bone, as mentioned earlier. The vomer and perpendicular plate support a wall of *septal cartilage* that forms most of the anterior part of the septum.

ASSIGNED — APR Module 11: Dissection: Nasal cavity: Coronal

launch
(Your score will be reported to your instructor.)

Students can navigate directly from the ConnectPlus eBook to related APR content.

Essentials of Anatomy & Physiology

Kenneth S. Saladin
Georgia College & State University

Robin K. McFarland
Cabrillo College

Digital Author

Stephen J. Sullivan
Bucks County Community College

Mc Graw Hill

Connect Learn Succeed™

ESSENTIALS OF ANATOMY & PHYSIOLOGY

Published by McGraw-Hill, a business unit of The McGraw-Hill Companies, Inc., 1221 Avenue of the Americas, New York, NY 10020. Copyright © 2014 by The McGraw-Hill Companies, Inc. All rights reserved. Printed in the United States of America. No part of this publication may be reproduced or distributed in any form or by any means, or stored in a database or retrieval system, without the prior written consent of The McGraw-Hill Companies, Inc., including, but not limited to, in any network or other electronic storage or transmission, or broadcast for distance learning.

Some ancillaries, including electronic and print components, may not be available to customers outside the United States.

This book is printed on acid-free paper.

1 2 3 4 5 6 7 8 9 0 DOW/DOW 1 0 9 8 7 6 5 4 3

ISBN 978-0-07-245828-2
MHID 0-07-245828-3

Senior Vice President, Products & Markets: *Kurt L. Strand*
Vice President, General Manager, Products & Markets: *Marty Lange*
Vice President, Content Production & Technology Services: *Kimberly Meriwether David*
Manager Director: *Michael S. Hackett*
Director: *James F. Connely*
Brand Manager: *Marija Magner*
Director of Development: *Rose Koos*
Senior Developmental Editor: *Donna Nemmers*
Director of Digital Content Development: *Barbekka Hurtt, Ph.D.*
Senior Project Manager: *April R. Southwood*
Senior Buyer: *Laura Fuller*
Senior Designer: *David W. Hash*
Interior Designer: *Greg Nettles*
Cover Image: *A woman climber on top of a mountain,* © *OJO Images/Getty Images*
Senior Content Licensing Specialist: *John C. Leland*
Photo Research: *Mary Reeg*
Compositor: *Electronic Publishing Services Inc., NYC*
Typeface: *10/12 Utopia*
Printer: *R. R. Donnelley*

All credits appearing on page or at the end of the book are considered to be an extension of the copyright page.

Library of Congress Cataloging-in-Publication Data

Cataloging-in-Publication Data has been requested from the Library of Congress.

The Internet addresses listed in the text were accurate at the time of publication. The inclusion of a website does not indicate an endorsement by the authors or McGraw-Hill, and McGraw-Hill does not guarantee the accuracy of the information presented at these sites.

www.mhhe.com

Brief Contents

About the Authors

KENNETH S. SALADIN is Professor of Biology at Georgia College & State University in Milledgeville, Georgia, where he has taught since 1977. Ken teaches human anatomy and physiology, introductory medical physiology, histology, animal behavior, and natural history of the Galápagos Islands. He has also previously taught introductory biology, general zoology, sociobiology, parasitology, and biomedical etymology. Ken is a member of the Human Anatomy and Physiology Society, American Association of Anatomists, American Physiological Society, Society for Integrative and Comparative Biology, and American Association for the Advancement of Science. He is the author of the best-selling textbooks *Anatomy & Physiology: The Unity of Form and Function* and *Human Anatomy*. Ken and his wife Diane have two adult children.

ROBIN MCFARLAND has taught anatomy and physiology at Cabrillo College in Aptos, California, since 1998. She earned a Ph.D. in physical (biological) anthropology from the University of Washington, where she studied the relationship between body fat and reproduction in primates. Robin collaborates on studies of ape anatomy with Adrienne Zihlman at the University of California, Santa Cruz. Robin is a member of the Human Anatomy and Physiology Society, American Association of Anatomists, and American Association of Physical Anthropologists. She was also contributing author to *Human Anatomy,* second edition, by Ken Saladin. She and her husband Jeff have two children, Reid and Madeleine. Robin enjoys hiking and climbing mountains with her family.

STEPHEN J. SULLIVAN has been teaching anatomy and physiology at Bucks County Community College near Philadelphia since 2002. Steve started modifying existing digital tools to fit his classroom needs in 2005 and began collaborating with McGraw-Hill on developing digital tools in 2009. As the Digital Author for this text, Steve has prepared digital assessments that directly reflect the content and style of the textbook authors to help foster student access and success.

Dedicated to my students, who are to my spirits what ATP is to my cells.—K.S.S.

This book is dedicated to my former, present, and future students and to Adrienne Zihlman—mentor, colleague, friend.—R.K.M.

To the students who every day teach me what works; and to Tabitha and Ben, for when the time I sacrifice is theirs.—S.J.S.

Table of Contents

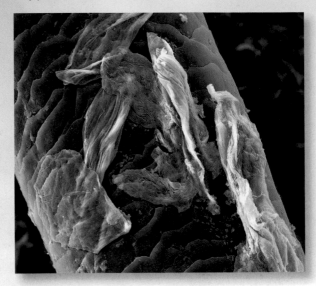

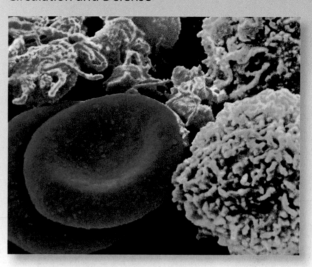

PART 6
Human Life Cycle

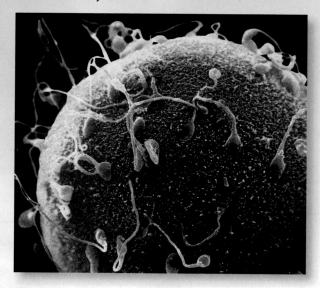

Chapter 19
The Reproductive System 631

Chapter 20
Human Development and Aging 675

Preface

Essentials of Anatomy & Physiology is the newest product of a long collaboration between Ken Saladin and Robin McFarland. Robin was a contributor to the second edition of Ken's *Human Anatomy* text and provided valuable input over the years, as well, to his "flagship" book *Anatomy & Physiology: The Unity of Form and Function.* For several years, the authors met at professional conferences and discussed the challenges and fulfillment of their classroom experiences—Ken's at a four-year state university in Georgia, and Robin's at a community college in California. They developed a shared vision of a book that would extend their enjoyment of teaching beyond the classroom—a book that would meet the needs of students in a one-semester course better than any other book on the market.

Foundation for This Text

In the course of her community college teaching, Robin long felt that no existing textbook fully satisfied her wish for more up-to-date science, fewer factual errors, and truly stimulating writing and art. At the same time, Ken's *Anatomy & Physiology: The Unity of Form and Function* had evolved into an international best seller, drawing letters and e-mails from students all over the world expressing thanks for a book that was so enjoyable to read and so vividly illustrated.

With their harmonious writing styles and shared work ethic, the authors set out to combine Ken's writing and publishing experience with Robin's community college teaching perspective. This book is the result—an outreach to students worldwide through vivid art, effective pedagogy, and a storytelling, reader-friendly writing style.

This is truly a fresh book, not a cut-and-paste reduction of an existing text. The text has been written in entirely new, simple, straightforward prose; and many new illustration ideas were conceived for this audience. But this is no typical first edition, either—first editions often have to go through a revision or two to mature and weed out errors, but the information in this book is from *Anatomy & Physiology* and reflects more than 20 years of peer review, refinement, and correction. It is time-tested content, expressed in a new way, for a new audience, through the authors' joint writing style and new art concepts.

Audience

Essentials of Anatomy & Physiology is intended for students in associate degree, certification, and career-training programs; students in high-school advanced placement classes; students who are seeking a general education science class; and persons who may not have set foot in a college classroom for many years. The prose and vocabulary in *Essentials of Anatomy & Physiology* are appropriate to serve this broad spectrum of readers.

Keeping in mind students who are highly pragmatic and career-oriented, a "Career Spotlight" feature has been included in every chapter, and references to further career information are found in appendix B.

Structure of Chapters

Chapters and pedagogy are structured to provide students a clear sense of direction, the means of evaluating one's own progress and mastery, and a sense of accomplishment at frequent intervals. From the perspectives of Robin's passion for hiking and climbing in the Cascade Range of the Northwest, and Ken's frequent trips with his classes to the volcanoes of the Andes and Galápagos Islands, they used mountain hiking and climbing as the metaphor for a student's progress through the chapters.

Starting in chapter 2, every chapter begins with a content outline and a box called "Base Camp." Like a climber's base camp, this feature gives one the provisions needed for ascending to the next level—a list of vital concepts in earlier chapters that the  student should understand before moving on to new heights, and reference to the pages where those concepts are first introduced.

Just as a mountain climber must have a clear sense of the path ahead, and what is to be achieved in the next stretch of the journey, so should the textbook reader. Each chapter is broken into short, numbered modules, beginning with a list called "Expected Learning Outcomes." This is where students will find the answers to such natural questions as, What am I going to gain by reading these next few pages? What's in it for me? What will my instructor want me to know?

Just as every climber needs places to stop and take stock of what he or she has achieved so far, the feature "Before You Go On" provides this perspective. Answering these questions, based on no more than three or four pages of reading,

Before You Go On

enables students to judge whether they have met the expectations of that section and understand its most important concepts before moving on to scale the next precipice.

Each journey is completed by that moment of triumph at the top of the mountain to look back and take pride in what has been achieved. At the end of every chapter the feature "Assess Your Learning Outcomes" provides a study tool to review the most important concepts in the chapter as a whole, and an opportunity to evaluate how well one has met the initial objectives.

From "Base Camp" to "Assess Your Learning Outcomes," the student should get a clear sense of the path ahead, a convenient means of charting one's progress, and a satisfying sense of accomplishment at the end.

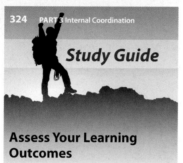

Benefits for Instructors

Many instructors are under increasing pressure to spell out specific learning objectives for a course, and to define how student achievement of these objectives will be measured. Instructors come into A&P from a broad range of academic fields, and it can be especially difficult for adjunct instructors and newer instructors trained in other areas to know what's important and what the key objectives and concepts of an A&P course should be.

This book provides a ready-made course outline of course objectives and means of assessment. The "Expected Learning Outcomes" at the start of each chapter section, the parallel "Assess Your Learning Outcomes" at the end of each chapter, and the intermediate aids such as "Before You Go On" provide an easy means for meeting the requirements of an outcome-driven curriculum.

The "Study Guide" at the end of each chapter also provides an overview of key points, as well as a variety of self-testing question formats, for students who wish to have a study guide for their next exam. A student who masters these study guides should do well on an exam.

Instructors with diverse backgrounds may wonder how they can know whether the goals set in their own classrooms correspond with what is expected of A&P students nationwide. The professional society of A&P instructors, the Human Anatomy and Physiology Society (HAPS), has comprehensively spelled out the desirable goals of an A&P course in its "A&P Learning Outcomes," available to members under Teaching Resources at www.hapsweb.org. The authors have written *Essentials of Anatomy & Physiology* with an eye to fulfilling those goals at a level commensurate with a one-semester essentials course. Instructors using this book can therefore feel confident that they have presented a course on a par with the largest body of A&P instructors in the United States and Canada.

The authors would enjoy hearing from colleagues and students alike who use this book and may wish to offer suggestions for our next edition, or encouragement to continue doing certain things the way we have on this maiden voyage. Such feedback is invaluable for improving a textbook, and the authors will endeavor to answer all correspondence.

Kenneth S. Saladin, PhD
Georgia College, Milledgeville, GA
ken.saladin@gcsu.edu

Robin K. McFarland, PhD
Cabrillo College, Aptos, CA
romcfarl@cabrillo.edu

Guided Tour

Making Essentials of Anatomy & Physiology Intriguing and Inspiring

Essentials of Anatomy & Physiology crafts the facts of A&P into art and prose in a way that makes the book exciting and rewarding to read.

Captivating Art and Photography

A&P is a highly visual subject; beautiful illustrations pique the curiosity and desire to learn. *Essentials of Anatomy & Physiology*'s illustrations set a new standard in the A&P Essentials market, where many students regard themselves as visual learners.

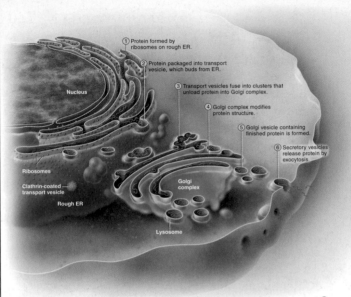

Figure 3.12 **Organelles of Protein Synthesis and Secretion.** The steps in protein synthesis and secretion are numbered ① through ⑥.

Building Cognitive Skill

Essentials of Anatomy & Physiology asks questions that not only test memory, but also exercise and expand the student's thinking skills at multiple levels of Bloom's Taxonomy of Learning Objectives.

Apply What You Know

Interjections in each chapter prompt students to apply what they have just read to a new thought-provoking problem or context, and encourage new insights.

> **Apply What You Know**
> Physical exercise obviously increases cardiac output. Do you think it achieves this through heart rate, contraction strength, or both? Explain.

area of plasma membrane. However, it triggers another action potential in the membrane immediately ahead of it, and that action potential triggers another, and so forth. Thus, we get a chain reaction of one action potential after another along the length of a nerve fiber. This chain reaction constitutes the **nerve signal (nerve impulse).** An illuminating analogy to this is standing up a long row of dominoes and pushing the first one over. When that domino falls, it pushes over the second, and so forth—and the chain reaction produces a wave of energy traveling to the end of the line. No one domino moves to the other end of the line; a falling domino is a local event. Similarly, an action potential is a local event, but it triggers the next one and, like the row of falling dominoes, we get a wave of energy traveling from one end of the axon to the other. That traveling wave is the nerve signal (fig. 8.10). Action potentials do not travel; nerve signals do.

Stimulating Prose

Far more than "just the facts," *Essentials of Anatomy & Physiology*'s narrative style weaves the facts into an absorbing story of human form and function. Vivid analogies that captivate the imagination make complex concepts easy to understand.

Figure Legend Questions

Thought questions in many figure legends encourage students to think analytically about the art, not merely view it.

Figure 5.5 Structure of a Hair and its Follicle. (a) Anatomy of the follicle and associated structures. (b) Light micrograph of the base of a hair follicle. (c) Electron micrograph of a hair emerging from its follicle. Note the exfoliating epidermal cells encircling the follicle like rose petals. AP|R

- *In light of your knowledge of hair, discuss the validity of an advertising claim that a shampoo will "nourish your hair." Where and how does a hair get its sole nourishment?*

Before You Go On

Answer these questions from memory. Reread the preceding section if there are too many you don't know.

1. *Which term refers to all the cell contents between the plasma membrane and nucleus: cytosol, cytoplasm, nucleoplasm, or extracellular fluid?*

2. *About how big would a cell have to be for you to see it without a microscope? Are any cells actually this big? If so, what ones?*

3. *Explain why cells cannot grow to an indefinitely large size.*

Before You Go On

Each chapter is divided into brief modules conveniently suited to short study periods. "Before You Go On" questions let students know whether they are "getting it" before moving ahead.

Analyzing Medical Terms

The task of learning medical terminology seems overwhelming at first, but there is a simple trick to becoming more comfortable with the technical language of medicine. Those who, at first, find scientific terms confusing and difficult to pronounce, spell, and remember usually feel more confident once they realize the logic of how such terms are composed. A term such as *hyponatremia* is less forbidding once we recognize that it is composed of three common word elements: *hypo-* (below normal), *natr-* (sodium), and *-emia* (blood condition). Thus, hyponatremia is a deficiency of sodium in the blood. Those three word elements appear over and over in many other medical terms: *hypothermia, natriuretic, anemia,* and so on. Once you learn the meanings of *hypo-, natri-,* and *-emia,* you already have the tools at least to partially understand hundreds of other biomedical terms.

Coccygeal (coc-SIDJ-ee-ul) **vertebrae** (Co1–Co4) fuse by the age of 20 to 30 into a single small triangular bone, the **coccyx**[42] (COC-six) (fig. 6.22). The Roman anatomist Claudius Galen (c. 130–c. 200) named it this because he thought it resembled the beak of a cuckoo. The coccyx provides attachment for muscles of the pelvic floor. It is a vestige of the bones which, in other mammals, continue as the vertebrae of the tail; hence it is colloquially known as our "tailbone."

- *Pronunciation guides* that appear throughout chapters make it easier to pronounce key terms, and make these words more likely to be remembered and understood.

Building Vocabulary

The plethora of medical terms in A&P is one of a student's most daunting challenges. Chapter 1 teaches core principles of how to break words down into familiar roots, prefixes, and suffixes, making medical terminology less intimidating while teaching the importance of precision in spelling *(ilium/ileum, malleus/malleolus).*

- An end-of-book *Glossary* provides clear definitions of the most important or frequently used terms, and *Appendix D: Biomedical Word Roots, Prefixes, and Suffixes* defines nearly 400 Greek and Latin roots, which make up about 90% of today's medical terms.
- *Footnoted word origins* show how new terms are composed of familiar wood roots.

[43] *aur* = ear; *icul* = little

[44] *coccyx* = cuckoo

Study Guide

Each chapter ends with a self-assessment tool for the student.

- *Assess Your Learning Outcomes* provides a comprehensive overview of the key points in the chapter, and requires the student to reexamine the text to get information, rather than simply handing it to them. This format encourages active learning over passive reading.

Study Guide

Assess Your Learning Outcomes

To test your knowledge, discuss the following topics with a study partner or in writing, ideally from memory.

3.1 The General Structure of Cells (p. 64)

1. Fundamental components of a cell
2. Intracellular and extracellular fluids
3. The typical size range of human cells and what factors limit cell size

3.2 The Cell Surface (p. 66)

1. Molecular components and organization of the plasma membrane
2. Varieties and functions of the plasma membrane proteins
3. The composition, location, and functions of a cell's glycocalyx
4. Structural and functional distinctions between microvilli, cilia, and flagella
5. Structural distinctions and respective advantages of three types of cell junctions—tight junctions, desmosomes, and gap junctions
6. The eight modes of transport through a plasma membrane and how they differ with respect to the use of carrier proteins, direction of movement of the transported substances, and demand for ATP

3.3 The Cell Interior (p. 78)

1. Components and functions of the cytoskeleton
2. Types of cell inclusions and how inclusions differ from organelles
3. What organelles have in common and how they differ, as a class, from other cellular components
4. Structure of the nucleus, particularly of its nuclear envelope, chromatin, and nucleoli
5. Two forms of endoplasmic reticulum, their spatial relationship, their structural similarities and differences, and their functional differences
6. The composition, appearance, locations, and function of ribosomes
7. Structure of the Golgi complex and its role in the synthesis, packaging, and secretion of cell products
8. Similarities and differences between lysosomes and peroxisomes in structure, contents, and functions
9. Structure, function, and evolutionary origin of mitochondria, and the significance of mitochondrial DNA
10. Structure, locations, and functions of centrioles
11. The processes of genetic transcription and translation, including the roles of mRNA, rRNA, and tRNA
12. How the amino acid sequence of a protein is represented by the codons of mRNA
13. How proteins are processed and secreted after their assembly on a ribosome

3.4 The Life Cycle of Cells (p. 87)

1. Four phases of the cell cycle and the main events in each phase
2. How DNA is replicated in preparation for mitosis
3. Functions of mitosis
4. Four stages of mitosis; changes in chromosome structure and distribution that occur in each stage; and the role of centrioles and the mitotic spindle
5. The mechanism and result of cytokinesis

Testing Your Recall

1. The clear, structureless gel in a cell is its
   a. nucleoplasm.
   b. endoplasm.
   c. cytoplasm.
   d. neoplasm.
   e. cytosol.

2. New nuclei form and a cell begins to pinch in two during
   a. prophase.
   b. metaphase.
   c. interphase.
   d. telophase.
   e. anaphase.

3. The amount of ___ in a plasma membrane affects its fluidity.
   a. phospholipid
   b. cholesterol
   c. glycolipid
   d. glycoprotein
   e. integral protein

4. Cells specialized for absorption of matter from the extracellular fluid are likely to show an abundance of
   a. lysosomes.
   b. microvilli.

True or False

Determine which five of the following statements are false, and briefly explain why.

1. The monomers of a polysaccharide are called amino acids.

2. An emulsion is a mixture of two liquids that separate from each other on standing.

3. ATP provides better long-term storage of energy than triglycerides do.

4. If a pair of shared electrons is more attracted to one nucleus than to the other, the electrons form a polar covalent bond.

5. Amino acids are joined by a unique type of bond called a peptide bond.

6. A saturated fat is defined as a fat to which no more carbon can be added.

7. Catabolism is an energy-releasing oxidation process.

8. Cellulose is an important source of dietary calories.

9. Two isomers have identical chemical properties, because their chemical behavior depends on the number and types of atoms present, not on how the atoms are arranged.

10. A solution of pH 8 has one-tenth the hydrogen ion concentration of a solution with pH 7.

Answers in appendix A

Testing Your Comprehension

1. How would the important life-sustaining properties of water change if it had nonpolar covalent bonds instead of polar covalent? Explain.

2. In one form of radioactive decay, a neutron breaks down into a proton and electron and emits a gamma ray. Is this an endergonic or exergonic reaction, or neither? Is it an anabolic or catabolic reaction, or neither? Explain both answers.

3. Some metabolic conditions such as diabetes mellitus cause disturbances in the acid–base balance of the body, which gives the body fluids an abnormally low pH. Explain how this could affect the ability of enzymes to control biochemical reactions in the body.

Multiple Question Types

- *Testing Your Recall* questions check for simple memory of terms and facts.
- *True or False* questions require more than a lucky guess, and ask students to analyze the validity of ideas and to explain or rephrase false statements.
- *Testing Your Comprehension* questions necessitate insight and application to clinical and other scenarios.

Tying It All Together

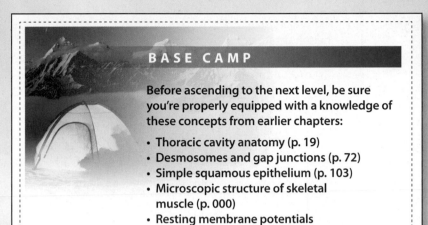

BASE CAMP

Before ascending to the next level, be sure you're properly equipped with a knowledge of these concepts from earlier chapters:

- Thoracic cavity anatomy (p. 19)
- Desmosomes and gap junctions (p. 72)
- Simple squamous epithelium (p. 103)
- Microscopic structure of skeletal muscle (p. 000)
- Resting membrane potentials and action potentials (p. 264)

- *Base Camp* lists key concepts from earlier chapters that a student should know before embarking on the new one, and effectively ties all chapters together into an integrated whole.

- No organ system functions in isolation. The *Connective Issues* tool shows how every organ system affects all other body systems, and generates a more holistic understanding of human function.

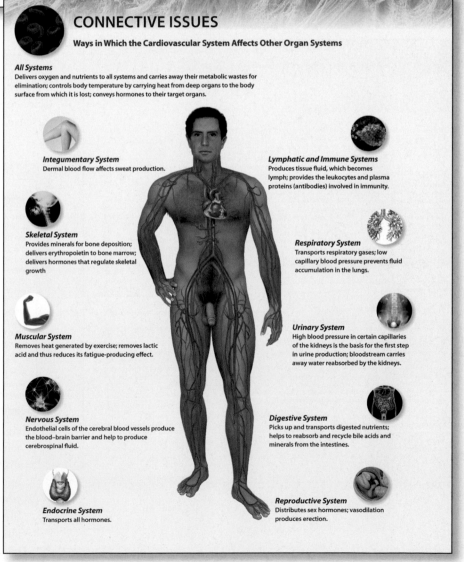

CONNECTIVE ISSUES

Ways in Which the Cardiovascular System Affects Other Organ Systems

All Systems
Delivers oxygen and nutrients to all systems and carries away their metabolic wastes for elimination; controls body temperature by carrying heat from deep organs to the body surface from which it is lost; conveys hormones to their target organs.

Integumentary System
Dermal blood flow affects sweat production.

Skeletal System
Provides minerals for bone deposition; delivers erythropoietin to bone marrow; delivers hormones that regulate skeletal growth

Muscular System
Removes heat generated by exercise; removes lactic acid and thus reduces its fatigue-producing effect.

Nervous System
Endothelial cells of the cerebral blood vessels produce the blood–brain barrier and help to produce cerebrospinal fluid.

Endocrine System
Transports all hormones.

Lymphatic and Immune Systems
Produces tissue fluid, which becomes lymph; provides the leukocytes and plasma proteins (antibodies) involved in immunity.

Respiratory System
Transports respiratory gases; low capillary blood pressure prevents fluid accumulation in the lungs.

Urinary System
High blood pressure in certain capillaries of the kidneys is the basis for the first step in urine production; bloodstream carries away water reabsorbed by the kidneys.

Digestive System
Picks up and transports digested nutrients; helps to reabsorb and recycle bile acids and minerals from the intestines.

Reproductive System
Distributes sex hormones; vasodilation produces erection.

CAREER SPOTLIGHT

Electrocardiographic Technician

An *electrocardiographic (ECG or EKG) technician* prepares electrocardiograms (ECGs) for diagnostic, exercise testing, and other purposes. The ECG technician prepares the patient for the test by attaching electrodes to specific sites on the chest and limbs and monitors the equipment while results are recorded. One can become a certified ECG technician through programs at community colleges or vocational colleges. A typical course of training entails 4 months beyond high school and includes anatomy and physiology, medical terminology, interpretation of cardiac rhythms, patient-care techniques, cardiovascular medication, and medical ethics. Many people, however, become ECG technicians through on-the-job training rather than formal programs. Most employers prefer to train people who are already in a health-care profession, such as nurses' aides. With more advanced training, one may become a cardiovascular technologist and assist physicians in diagnosis, cardiac catheterization, echocardiography, and other more specialized skills and for correspondingly better salaries. For further information on a career as an ECG technician or cardiovascular technologist, see appendix B.

- *Career Spotlight* features provide a relevant career idea in every chapter to provide basic information on educational requirements and entry into a career, and expand student awareness of opportunities in allied health professions. *Appendix B* refers students to online sources of further information on 20 career fields and a list of 83 more health-care career ideas.

Clinical Application 3.2

CALCIUM CHANNEL BLOCKERS

Membrane channels may seem only an abstract concept until we see how they relate to disease and drug design. For example, drugs called *calcium channel blockers* are often used to treat high blood pressure (hypertension). How do they work? The walls of the arteries contain smooth muscle that constricts to narrow the vessels and raise blood pressure, or relaxes to let them widen and reduce blood pressure. Excessive, widespread vasoconstriction (vessel narrowing) can cause hypertension, so one approach to the prevention of hypertension is to inhibit this. In order to constrict, smooth muscle cells open calcium channels in the plasma membrane. The inflow of calcium activates the proteins of muscle contraction. Calcium channel blockers act, as their name says, by preventing calcium channels from opening and thereby preventing arterial constriction.

- *Clinical Applications* are short essays that apply basic science to interesting issues of health and disease.

PERSPECTIVES ON HEALTH

Methods of Contraception

Contraception means any procedure or device intended to prevent pregnancy (the presence of an implanted conceptus in the uterus). This essay summarizes the most popular methods and some issues involved in choosing among them.

Behavioral Methods

Abstinence (refraining from intercourse) is, obviously, a completely reliable method if used consistently. The rhythm method (periodic abstinence) is based on avoiding intercourse near the time of expected ovulation. Among typical users, it has a 25% failure rate, partly due to lack of restraint and partly because it is difficult to predict the exact date of ovulation. Intercourse must be avoided for at least 7 days before ovulation so there will be no surviving sperm in the reproductive tract when the egg is ovulated, and for at least 2 days after ovulation so there will be no fertile egg present when sperm are introduced.

Withdrawal (coitus interruptus) requires the male to withdraw the penis before ejaculation. This often fails because of lack of willpower, because some sperm are present in the preejaculatory fluid, and because sperm ejaculated anywhere in the vulva can potentially get into the reproductive tract.

Barrier and Spermicidal Methods

Barrier methods are designed to prevent sperm from getting into or beyond the vagina. They are most effective when used with chemical spermicides, available as nonprescription foams, creams, and jellies. Second only to birth-control pills in popularity is the male *condom*, a sheath usually made of latex, worn over the penis. Condoms are the only contraceptive methods that also protect against disease transmission, although animal-membrane condoms do not protect against HIV or hepatitis B viruses. Condoms have the advantages of being inexpensive and requiring no medical examination or prescription.

The *diaphragm* is a latex dome worn over the cervix to block sperm migration. It requires a physical examination and prescription to ensure proper fit, but is otherwise comparable to the condom in convenience and reliability, provided it is used with a spermicide. Without a spermicide, it is not very effective.

The *sponge* is a concave foam disc inserted before intercourse to cover the cervix. It is impregnated with spermicidal gel and acts by absorbing semen and killing the sperm. It requires no prescription or fitting. The sponge provides protection for up to 12 hours, and must be left in place for 6 hours after intercourse.

Hormonal Methods

Most hormonal methods of contraception are aimed at preventing ovulation. They mimic the negative feedback effect of ovarian hormones on the pituitary, inhibiting FSH and LH secretion so follicles do not mature. For most women, they are highly effective and present minimal complications.

The oldest and still the most widely used hormonal method in the United States is the *combined oral contraceptive (birth-control pill)*. It is composed of estrogen and progestin, a synthetic progesterone. It must be taken daily, at the same time of day, for 21 days each cycle. The 7-day withdrawal allows for menstruation. Side effects include an elevated risk of heart attack or stroke in smokers and in women with a history of diabetes, hypertension, or clotting disorders.

Other hormonal methods avoid the need to remember a daily pill. One option is a skin patch that releases estrogen and progestin transdermally. It is changed at 7-day intervals (three patches per month and one week without). The NuvaRing is a soft flexible vaginal ring that releases estrogen and progestin for absorption through the vaginal mucosa. It must be worn continually for 3 weeks and removed for the fourth week of each cycle. Medroxyprogesterone (trade name Depo-Provera) is a progestin administered by injection two to four times per year. It provides highly reliable, long-term contraception, although in some women it causes headaches, nausea, or weight gain.

Some drugs can be taken orally after intercourse to prevent implantation of a conceptus. These are called emergency contraceptive pills (ECPs), or "morning-after pills" (trade names Plan B, Levonelle). An ECP is a high dose of estrogen and progestin or a progestin alone. It can be taken within 72 hours after intercourse and induces menstruation within 2 weeks. ECPs inhibit ovulation, inhibit sperm or egg transport in the uterine tube, and prevent implantation. They do not work if a blastocyst is already implanted.

Intrauterine Devices

Intrauterine devices (IUDs) are springy, often T-shaped devices inserted through the cervical canal into the uterus. Some IUDs act by releasing a synthetic progesterone, but most have a copper wire wrapping or copper sleeve. IUDs irritate the uterine lining and interfere with blastocyst implantation, and copper IUDs also inhibit sperm motility. An IUD can be left in place for 5 to 12 years.

- *Perspectives on Health* essays that make the basic science relevant to the student's interest in health and disease.

Aging of the Muscular System

One of the most common changes in old age is the replacement of lean body mass (muscle) with fat, accompanied by loss of muscular strength. Muscular strength and mass peak in the 20s, and by the age of 80, most people have only half as much strength and endurance. Many people over age 75 cannot lift a 4.5 kg (10 lb) weight, making such simple tasks as carrying a bag of groceries very difficult. Tasks such as buttoning the clothes also take more time and effort. The loss of strength is a major contributor to falls, fractures, and dependence on others for living assistance. Fast-twitch muscle fibers show the earliest and greatest atrophy, thus increasing reaction time, slowing the reflexes, and reducing coordination.

There are multiple reasons for the loss of strength. Aged muscle has fewer myofibrils; more disorganized sarcomeres; smaller mitochondria; and reduced amounts of ATP, myoglobin, glycogen, and creatine phosphate. Increased adipose and fibrous tissue in the muscles limits their movement and blood circulation. In addition, there are fewer motor neurons in the spinal cord, so some muscle atrophy may result from reduced innervation. Even the neurons that do remain produce less acetylcholine and stimulate the muscles less effectively.

Even though people typically lose muscle mass and function as they age, these effects are noticeably less in people who continue to exercise throughout life. For example, studies show that even moderate exercise can help elderly people maintain muscle mass and improve balance; it even seems to improve mental agility.

- *Aging of Body Systems* is discussed in a section within each systems chapter to describe how each organ system changes over time, especially in old age. This discussion expands anatomical and physiological understanding beyond the prime of life, and is highly relevant to patient treatment, since older patients constitute most of the health-care market.

Elevate
Studying

ConnectPlus™ Anatomy & Physiology

McGraw-Hill ConnectPlus™ interactive learning platform provides

- auto-graded assignments
- adaptive diagnostic tools
- powerful reporting against learning outcomes and level of difficulty
- an easy-to-use interface
- McGraw-Hill Tegrity Campus™, which digitally records and distributes your lectures with a click of a button
- Learning Objectives from the textbook that are tied to interactive questions in Connect, to ensure that all parts of the chapters have adequate coverage within Connect assignments

ConnectPlus includes the full textbook as an integrated, dynamic eBook, which you can also assign. Everything you need—in one place!

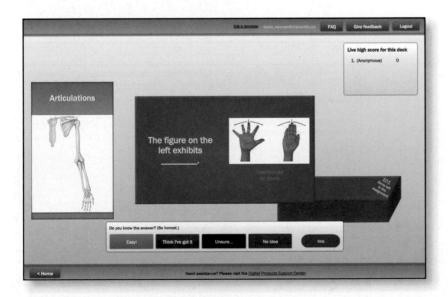

McGraw-Hill LearnSmart™

Unlike static flashcards or rote memorization, LearnSmart ensures your students have mastered course concepts before taking the exam, thereby saving you time and increasing student success.

- The only truly adaptive learning system
- Intelligently identifies course content students have not yet mastered
- Maps out personalized study plans for student success

Anatomy & Physiology REVEALED®

APR 3.0 is an interactive cadaver dissection tool to enhance lecture and lab that students can use anytime, anywhere. Instructors may customize APR 3.0 to their course by selecting the specific structures they require in their course, and APR 3.0 does the rest. Once the structure list is generated, APR highlights these selected structures for students.

APR contains all the material covered in an A&P course, including these three new modules:

- Body Orientation
- Cells and Chemistry
- Tissues

Presentation Center

Accessed from the *Essentials of Anatomy & Physiology* Connect website, **Presentation Center** is an online digital library containing photos, artwork, animations, and other media types that can be used to create customized lectures, visually enhanced tests and quizzes, compelling course websites, or attractive printed support materials. All assets are copyrighted by McGraw-Hill Higher Education, but can be used by instructors for classroom purposes. The visual resources in this collection include:

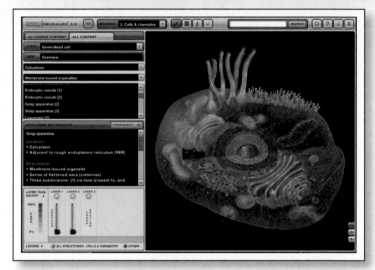

- **Art** Full-color digital files of all illustrations in the book can be readily incorporated into lecture presentations, exams, or custom-made classroom materials. In addition, all files are pre-inserted into PowerPoint slides for ease of lecture preparation.
- **Photos** The photo collection contains digital files of photographs from the text, which can be reproduced for multiple classroom uses.
- **Tables** Every table that appears in the text has been saved in electronic form for use in classroom presentations and/or quizzes.
- **Animations** Numerous full-color animations illustrating important processes are also provided. Harness the visual impact of concepts in motion by importing these files into classroom presentations or online course materials.

Also accessed through the *Essentials of Anatomy & Physiology* Connect website are:

- **PowerPoint Lecture Outlines** Ready-made presentations that combine art and lecture notes are provided for each chapter of the text.

- **PowerPoint Slides** For instructors who prefer to create their lectures from scratch, all illustrations, photos, tables, animations, and relevant *Anatomy & Physiology REVEALED* images are pre-inserted by chapter into blank PowerPoint slides.

Instructors: To access Connect, request registration information from your McGraw-Hill sales representative.

- **Digital Lecture Capture: Tegrity** McGraw-Hill Tegrity Campus records and distributes your lecture with just a click of a button. Students can view anytime/anywhere via computer, iPod, or mobile device. Tegrity indexes as it records your slideshow presentations and anything shown on your computer, so students can use keywords to find exactly what they want to study.
- **Computerized Test Bank** Test questions are served up utilizing EZ Test software to accompany *Essentials of Anatomy & Physiology*. These questions are also available to instructors in Word format.
- **Content Delivery Flexibility** *Essentials of Anatomy & Physiology* is available in many formats in addition to the traditional textbook to give instructors and students more choices when deciding on the format of their A&P text. Choices include
  - **Customizable Textbooks: Create** Introducing McGraw-Hill Create™—a self-service website that allows you to create custom course materials—print and eBooks—by drawing upon McGraw-Hill's comprehensive, cross-disciplinary content. Add your own content quickly and easily. Tap into other rights-secured, third-party sources as well. Then, arrange the content in a way that makes the most sense for your course. Even personalize your book with your course name and information. Choose the best format for your course: color print, black and white print, or eBook. The eBook is now viewable on an iPad! When you are finished customizing, you will receive a free PDF review copy in just minutes. Visit McGraw-Hill Create—www.mcgrawhillcreate.com—today and begin building your perfect book.
- **ConnectPlus™ eBook** McGraw-Hill's ConnectPlus eBook takes digital texts beyond a simple PDF. With the same content as the printed book, but optimized for the screen, ConnectPlus has embedded media, including animations and videos, which bring concepts to life and provide "just-in-time" learning for students. Additionally, fully integrated homework allows students to interact with the questions in the text to determine if they're gaining mastery of the content, and can also be assigned by the instructor.

McGraw-Hill and Blackboard©

McGraw-Hill Higher Education and Blackboard have teamed up. What does this partnership mean for you? Blackboard users will find the single sign-on and deep integration of ConnectPlus within their Blackboard course invaluable benefits. Even if your school is not using Blackboard, we have a solution for you. Learn more at www.domorenow.com.

Acknowledgments

I gratefully acknowledge the team at McGraw-Hill who have provided excellent ideas and unfailing encouragement throughout this project. I am immensely grateful to my coauthor Ken Saladin for a rewarding collaboration and firm friendship. I appreciate my colleagues in the biology department at Cabrillo College who inspire me every day with their dedication to student success. Finally, I wish to thank my husband Jeff and my children, Reid and Madeleine, for their support and patience.

Robin McFarland

My heartfelt appreciation goes to our team at McGraw-Hill who have provided such friendship, collegiality, and support over my 20-year history in textbooks; to Robin for adding this new dimension and stimulating collaboration to my writing career; to my colleagues at Georgia College for an atmosphere that supports and rewards such work; and to Diane for her steadfast love and encouragement.

Ken Saladin

Our grateful thanks are extended to these reviewers, who read early drafts of these chapters and provided instructive comments to help shape the content within these pages.

Peter Allen, *Rockingham Community College*

Anna Marie Avola, *Hodges University*

Charles Benton, Jr., *Madison Area Technical College*

Tracey Bergeron, *McIntosh College*

Dan Bickerton, *Ogeechee Technical College*

Russell Blalock, *Central Georgia Technical College*

Ginger L. Bohlen, *Alpena Community College*

Glen Borchert, *Heartland Community College*

Michael Broughton, *Bohecker College—Cincinnati*

Jocelyn Cash, *Central Piedmont Community College*

Weiru Chang, *Estrella Mountain Community College*

Reggie Cobb, *Nash Community College*

J. Mark Danley, *Central New Mexico Community College*

Donald Galen DeHay, *Tri-County Technical College*

Bill Ebener, *College of Southern Idaho*

Amy Fenech Sandy, *Columbus Technical College*

Paul Fierimonte, *University of Massachusetts Lowell & Merrimack College*

Maria Florez, *Lone Star College—CyFair*

Deborah Furbish, *Wake Technical Community College*

Linda Gerlock, *Middle Georgia Technical College*

Brent Graves, *Northern Michigan University*

Dawn Hillard, *Northeast Mississippi Community College*

Scott Hobson, *University of Arkansas Community College—Hope*

Karen Huffman-Kelly, *Erie Community College—North*

Branko Jablanovic, *College of Lake County*

Thomas Johnson, *Tidewater Community College—Chesapeake*

Christi A. Laurent, *Gwinnett Technical College*

Dean Lauritzen, *City College of San Francisco*

Shari J. Litch Gray, *Chester College of New England*

Leontine M. Lowery, *Delaware Technical Community College—Dover*

Pam McNamara, *Beckfield College*

Leslie R. Miller, *Iowa State University*

Robert Moldenhauer, *St. Clair County Community College*

Norma C. Moore, *Laredo Community College*

April Murphy, *Columbus Technical College*

Cathleen A. Murphy, *Sanford-Brown Nassau Community College—SBI Campus*

Kelly Neary, *Mission College*

Jeanine L. Page, *Lock Haven University*

Tami Panhuis, *Ohio Wesleyan University*

Diane Pelletier, *Green River Community College*

Jeffrey Penton, *Fortis Institute*

Mary Prorok, *South Hills School of Business and Technology*

Susan Rohde, *Triton College*

Robin Robison, *Northwest Mississippi Community College*

Jimmy Rozell, *Tyler Junior College*

Fadi N. Salloum, *Virginia Commonwealth University*

Amy Sandy, *Columbus Technical College*

Leba Sarkis, *Aims Community College*

Gaynelle Schmieder, *Pennsylvania Highlands Community College*

Daniel A. Slutsky, *American River College*

Kelly M. Smith, *Pima Medical Institute*

Majella Smith, *Los Medanos College*

Sherry Stewart, *Navarro College*

Jan Stone, *Bristol Community College*

Mary Torrano, *American River College*

Marilyn M. Turner, *Ogeechee Technical College*

Vivian E. Turner, *Rochester College*

Padmaja Vedartham, *Lone Star College—CyFair*

Christa Voss, *Tulsa Community College*

Leesa G. Whicker, *Central Piedmont Community College*

Rosann Wilcox, *SUNY Empire State College*

Edward Wolfe, *Central Piedmont Community College*

Yu Zhao, *American River College*

Donald Zakutansky, *Gateway Technical College*

The Study of Anatomy and Physiology

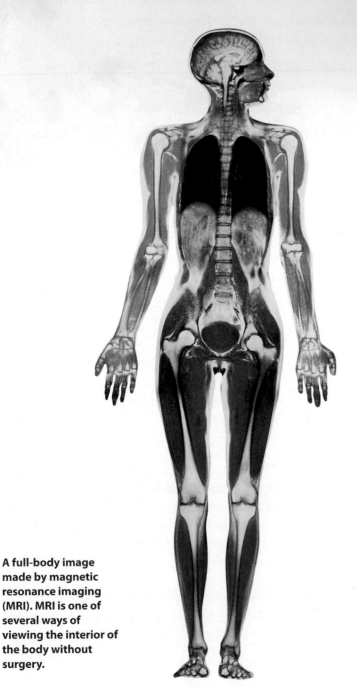

A full-body image made by magnetic resonance imaging (MRI). MRI is one of several ways of viewing the interior of the body without surgery.

Anatomy & Physiology REVEALED®
aprevealed.com

Module 1: Body Orientation

In recent years, people have flocked with eager curiosity to exhibitions of human cadavers dissected and preserved by a dry *plastination* process. The cost of admission is not cheap, but the willingness of so many to pay it attests to the great interest held not just by scientists and medical professionals, but by laypersons, regarding the internal structure of the body and its intricate functions.

Anatomy and physiology are the basic sciences that provide the foundation for advanced study and professions in health care, fitness training, nutrition, and other fields concerned with human performance and well-being. **Human anatomy** is the study of the structural basis of body function, and **human physiology** is the complementary study of the functional relevance of human structure. The two sciences are highly interdependent. All structure results from cell physiology, and it is the anatomy of organs that makes their physiology possible.

This chapter provides a context and vocabulary for the ones to follow. It introduces the sciences of anatomy and physiology and describes some of their methods and subdisciplines. It discusses the major criteria for considering the body alive, and deals especially with *homeostasis,* one of the most vital life processes. It describes the levels of structural detail in the human body and outlines its general structural plan. Because medical vocabulary is one of the greatest challenges to beginning students, this chapter also gives some suggestions for how to become at ease with it.

1.1 Anatomy—The Structural Basis of Human Function

Expected Learning Outcomes
When you have completed this section, you should be able to:

   a. define some subdisciplines of anatomy;
   b. explain the importance of dissection;
   c. describe some methods of examining a living patient;
   d. discuss the principles and applications of some medical imaging methods; and
   e. discuss the significance of variations in human anatomy.

The Anatomical Sciences

There are many approaches to the study of human anatomy, both in research for the purposes of discovery and understanding, and in clinical settings for the purposes of diagnosis and treatment. **Gross anatomy** is structure visible to the naked eye, either by surface observation or dissection. Ultimately, though, body functions result from individual cells. To see those, we usually take tissue specimens, thinly slice and stain them, and observe them under the microscope. This approach is called **histology**[1] (**microscopic anatomy**). **Histopathology** is the microscopic examination of tissues for signs of disease.

Surface anatomy is the external structure of the body, and is especially important in conducting a physical examination of a patient. **Systemic anatomy** is the study of one organ system at a time; this is the approach taken by introductory textbooks such as this one. **Regional anatomy** is the study of multiple organ systems at the same time in a given region of the body, such as the head or chest. Medical schools and anatomical atlases typically teach anatomy from this perspective, because it is more logical to dissect all structures of the head and neck, the chest, or a limb, than to try to dissect the entire digestive system, then the cardiovascular system, and so forth. Dissecting one system almost inevitably destroys organs of another system that stand in the way.

> ### Apply What You Know
>
> Do you think that a surgeon thinks more in terms of systemic anatomy or regional anatomy? Explain your answer.

You can study human anatomy from an atlas; yet as fascinating and valuable as anatomy atlases are, they teach almost nothing but the locations, appearances, and names of structures. This book is much different; it deals with what biologists call **functional morphology**[2]—not just the structure of organs, but the functional reasons behind that structure.

Functional morphology draws heavily on **comparative anatomy,** the study of more than one species in order to learn generalizations, evolutionary trends, and structure–function relationships. Many of the reasons for human structure become apparent only when we compare it to the structure of other animals. The human pelvis, for example, has a unique bowl-shaped configuration that can be best understood by comparison with animals such as a chimpanzee, whose pelvis is adapted to walking on four legs rather than two.

Examination of the Body

The simplest method of examining the body is **inspection** of surface structure, such as physicians perform during a physical examination. A deeper understanding depends on **dissection**[3]—the careful cutting and separation of tissues to reveal their relationships. The word *anatomy*[4] literally means "cutting apart," and dissection was called "anatomizing" until the nineteenth century. The dissection of a dead human body, or **cadaver,**[5] was crucial historically for

---

[1] *histo* = tissue; *logy* = study of

[2] *morpho* = form; *logy* = study of

[3] *dis* = apart; *sect* = cut

[4] *ana* = apart; *tom* = cut

[5] *cadere* = to fall or die

accurately mapping the human body, and remains an essential part of the training of many health science students.

Dissection, of course, is not the method of choice when examining a living patient! Some additional methods of clinical examination include the following.

- **Palpation**[6] is feeling structures with the fingertips, such as palpating a swollen lymph node or taking a pulse.
- **Auscultation**[7] (AWS-cul-TAY-shun) is listening to the natural sounds made by the body, such as heart and lung sounds.
- **Percussion** is tapping on the body and listening to the sound for signs of abnormalities such as pockets of fluid or air.
- **Medical imaging** includes methods of viewing the inside of the body without surgery. The branch of medicine concerned with imaging is called **radiology.**

Techniques of Medical Imaging

It was once common to diagnose disorders through *exploratory surgery*—opening the body and taking a look inside to see what was wrong and what could be done about it. Any breach of the body cavities is risky, however, and most exploratory surgery has been replaced by imaging techniques that allow physicians to see inside the body without cutting. These methods are called *noninvasive* if they involve no penetration of the skin or body orifices. *Invasive* techniques may entail inserting ultrasound probes into the esophagus, vagina, or rectum to get closer to the organ to be imaged, or injecting substances into the bloodstream or body passages to enhance image formation.

Anatomy students today must be acquainted with the basic methods of imaging and their advantages and limitations. Many images in this book have been produced by the following techniques. Most of these methods produce black and white images; those in the book are colorized to enhance detail or for esthetic appeal.

---

[6] *palp* = touch, feel
[7] *auscult* = listen

Figure 1.1 Radiologic Images of the Head. (a) An X-ray (radiograph) of the head. (b) A colorized cerebral angiogram, made by injecting a substance opaque to X-rays into the circulation and then taking an X-ray of the head to visualize the blood vessels. (c) A CT scan of the head at the level of the eyes. The eyes and skin are shown in blue, bone in pink, and the brain in green. (d) An MRI scan of the head at the level of the eyes. The optic nerves appear in red and the muscles that move the eyes in green. (e) A PET scan of the brain of an unmedicated schizophrenic patient. Red areas indicate regions of high metabolic rate. In this patient, the visual center of the brain (at bottom of photo) was especially active.

- *Why is a PET scan considered invasive whereas MRI is noninvasive?*

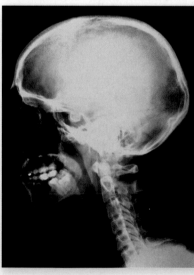

(a) X-ray (radiograph)

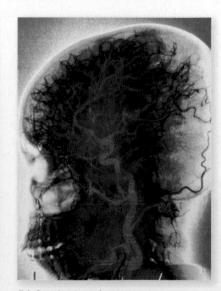

(b) Cerebral angiogram

Radiography (fig. 1.1a, b) is the process of photographing internal structures with X-rays, a form of high-energy radiation. The term *X-ray* also applies to a photograph *(radiograph)* made by this method. X-rays are absorbed by dense tissues such as bone, teeth, and tumors, which produce a lighter image than soft tissues. Radiography is commonly used in dentistry; mammography; diagnosis of fractures; and examination of the digestive, respiratory, and urinary tracts. Some disadvantages of radiography are that images of overlapping organs can be confusing, slight differences in tissue density are not detected well, and X-rays can cause mutations and cancer.

Computed tomography[8] (the **CT scan**) (fig. 1.1c) is a more sophisticated application of X-rays. The patient is moved through a ring-shaped machine that emits low-intensity X-rays on one side and receives them with a detector on the opposite side. A computer analyzes signals from the detector and produces an image of a "slice" of the body about as thin as a coin. CT scanning has the advantage of imaging thin sections of the body, so there is little organ overlap and the image is much sharper than a conventional X-ray. CT scanning is useful for identifying tumors, aneurysms, cerebral hemorrhages, kidney stones, and other abnormalities.

Magnetic resonance imaging (MRI) (fig. 1.1d) is even better than CT for visualizing soft tissues. The patient lies within a tunnel surrounded by a large electromagnet that creates a very strong magnetic field. An image is generated by the responses of tissues to the magnetic field and radio waves. A computer analyzes the signal to produce an image of the body. MRI can "see" clearly through the skull and vertebral column to produce images of the nervous tissue within, and it is better than CT for distinguishing between soft tissues such as the white and gray matter of the brain. *Functional MRI (fMRI)* is a form of MRI that visualizes moment-to-moment changes in tissue function; fMRI scans of the brain, for example, show shifting patterns of activity as the brain applies itself to a specific task. This method is very useful for identifying which parts of the brain perform various sensory, mental, and motor tasks.

Positron emission tomography (the **PET scan**) (fig. 1.1e) is used to assess the metabolic state of a tissue and to distinguish which tissues are most active.

---

[8] *tomo* = section, cut, slice; *graphy* = recording process

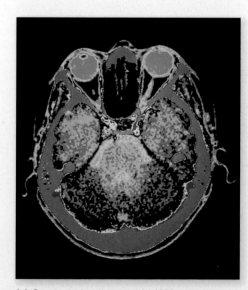

(c) Computed tomographic (CT) scan

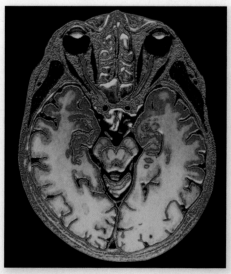

(d) Magnetic resonance image (MRI)

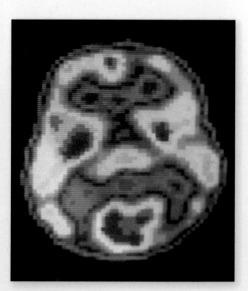

(e) Positron emission tomographic (PET) scan

It uses an injection of radioactively labeled glucose to highlight which tissues are most actively consuming energy at the moment of the scan. In cardiology, PET scans can show the extent of damaged heart tissue. Since damaged tissue consumes little or no glucose, it appears dark. In neuroscience, PET scans are used, like fMRI, to show which regions of the brain are most active when a person performs a specific task. The PET scan is an example of **nuclear medicine**—the use of radioisotopes to treat disease or to form diagnostic images of the body.

Sonography[9] (fig. 1.2) uses a handheld device placed firmly against the skin; it emits high-frequency ultrasound and receives signals reflected back from internal organs. Although sonography was first used medically in the 1950s, images of significant clinical value had to wait until computer technology had developed enough to analyze differences in the way tissues reflect ultrasound. Sonography avoids the harmful effects of X-rays, and the equipment is relatively inexpensive and portable. Its primary disadvantage is that it does not produce a very sharp image. Sonography is the method of choice in obstetrics, where the image *(sonogram)* can be used to locate the placenta and evaluate fetal age, position, and development. *Echocardiography* is the sonographic examination of the beating heart.

Anatomical Variation

A quick look around any classroom is enough to show that no two humans look exactly alike; on close inspection, even identical twins exhibit differences. Anatomy atlases and textbooks can easily give you the impression that everyone's internal anatomy is the same, but this simply is not true. Someone who thinks that all human bodies are the same internally would be a very confused medical student or an incompetent surgeon. Books such as this one

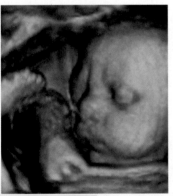

teach only the most common structural patterns—the anatomy seen in approximately 70% or more of people.

Some people completely lack certain organs. For example, most of us have a *palmaris longus muscle* in the forearm and a *plantaris muscle* in the leg, but not everyone does. Most of us have one spleen, but some people have two. Most have two kidneys, but some have only one. Most kidneys are supplied by a single *renal artery* and drained by one *ureter,* but in some people a single kidney has two renal arteries or ureters. Figure 1.3 shows some common variations in human anatomy, and Perspectives on Health (p. 9) describes a particularly dramatic variation.

(b)

[9] *sono* = sound; *graphy* = recording process

Figure 1.2 Sonography. (a) Producing a sonogram for an expectant family. (b) Three-dimensional sonogram of a fetus at 32 weeks of gestation.

- *Why is this procedure safer than radiography for fetal assessment?*

(a)

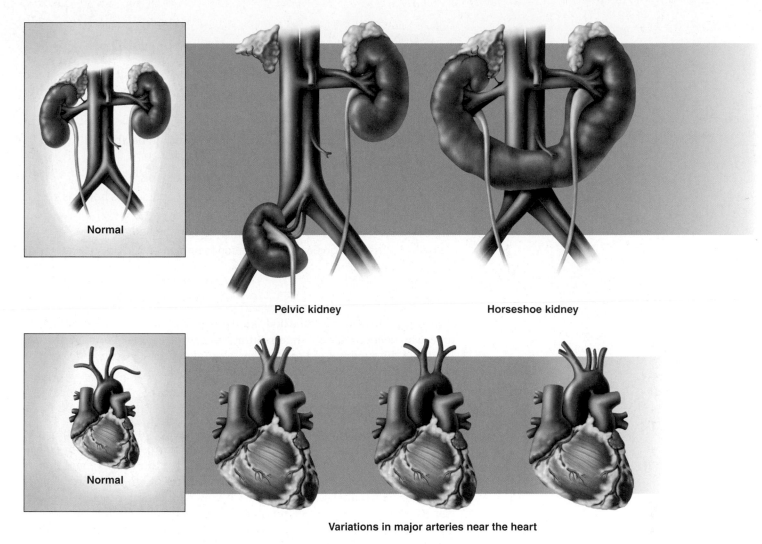

Normal

Pelvic kidney   Horseshoe kidney

Normal

Variations in major arteries near the heart

Figure 1.3 Variations in Human Anatomy. Not all humans have the usual "textbook structure."

Before You Go On

Answer these questions from memory. Reread the preceding section if there are too many you don't know.

1. *What is the difference between gross anatomy and histology?*
2. *In a routine physical examination, a physician may inspect you by palpation and auscultation. What is the difference between these procedures?*
3. *What are the advantages of CT over sonography? Conversely, what are the advantages of sonography over CT?*

1.2 Physiology—The Functional Relevance of Human Structure

Expected Learning Outcomes
When you have completed this section, you should be able to:
a. identify some subdisciplines of physiology;
b. describe the characteristics that define an organism as alive;
c. define *homeostasis,* explain its significance, and discuss how it is maintained by negative feedback;
d. discuss positive feedback and its effects on the body; and
e. discuss the significance of variation in human physiology.

Physiology[10] is the study of the body's life processes. The term comes from Aristotle, who believed in both supernatural and natural causes of human disease. He called the supernatural causes *theologi* and natural causes *physiologi.* For centuries, physicians were called "doctors of physick."

The Physiological Sciences

Physiology uses the methods of experimental science to determine how the body functions. It has many subdisciplines such as *neurophysiology* (physiology of the nervous system), *endocrinology* (physiology of hormones), and *pathophysiology* (mechanisms of disease). Partly because of limitations on experimentation with humans, much of what we know about bodily function has been gained through **comparative physiology,** the study of how different species have solved problems of life such as water balance, respiration, and reproduction. Comparative physiology is also the basis for the development of most new medications and procedures. For example, a new drug is tested for safety in laboratory mammals such as rats before it proceeds to trials with human subjects.

Essential Life Functions

Whereas anatomy views the body as a set of interconnected structures, physiology views it as a set of interconnected processes. Collectively, we call these processes *life.* But what exactly is life? Why do we consider a growing child to be alive, but not a growing crystal? Is abortion the taking of a human life? If so, what about a contraceptive foam that kills only sperm? As a patient is dying, at what point does it become ethical to disconnect life-support equipment and remove organs for donation? If these organs are alive, as they must be to serve someone else, then why isn't the donor considered alive? Such questions have no easy answers, but they demand a concept of what life is—a concept that may differ with one's biological, medical, religious, or legal perspective.

From a biological viewpoint, life is not a single property. It is a collection of qualities that help to distinguish living from nonliving things:

- **Organization.** Living things exhibit a far higher level of organization than the nonliving world around them. They expend a great deal of energy to maintain order, and disease and death result from a breakdown in this order.

---

[10] *physio* = nature; *logy* = study of

PERSPECTIVES ON HEALTH

Situs Inversus and Other Unusual Anatomy

Earlier in this chapter, we saw that not all human bodies are alike, and introductory textbooks can do no more than describe the most common patterns of human anatomy and physiology. Some organs are lacking from some individuals or are in different locations from usual. Some particularly striking examples of anatomical variation are situs (SITE-us) perversus and situs inversus. In *situs perversus*, an organ occupies an atypical locality; for example, a kidney may be located low in the pelvic cavity instead of high in the abdominal cavity (see fig. 1.3), or a parathyroid gland may be found in the root of the tongue instead of on the posterior surface of the thyroid gland.

In most people, the heart tilts toward the left, the spleen and sigmoid colon are on the left, the liver and gallbladder lie mainly on the right, the appendix is on the right, and so forth. But in *situs inversus*, occurring in about 1 out of 8,000 people, the organs of the thoracic and abdominal cavities are reversed between right and left. Selective right–left reversal of the heart is called *dextrocardia*. Some conditions, such as dextrocardia in the absence of complete situs inversus, can cause serious medical problems. Complete situs inversus, however, usually causes no functional problems because all of the viscera, though reversed, maintain their normal relationships to each other.

Defining the End of Life

We have seen in this chapter that *life* is a difficult property to define. That being the case, so is defining the end of life—and yet we're often forced to make decisions on that issue. How do we decide when to "let go" of a terminally ill loved one, perhaps to disconnect life-support equipment?

There is no easily defined instant of biological death. Some organs function for an hour or longer after the heart stops beating. During this time, even if a person is declared legally dead, living organs may be removed for transplantation. For legal purposes, death was once defined as the loss of a spontaneous heartbeat and respiration. Now that cardiopulmonary functions can be restarted and artificially maintained for years, this criterion is less distinct. Clinical death is now widely defined in terms of *brain death*—a lack of any detectable electrical activity in the brain, including the brainstem, accompanied by coma, lack of unassisted respiration, and lack of brainstem reflexes (such as pupillary, blinking, or coughing reflexes). A judgment of death is generally accepted only upon finding a complete lack of brain activity for a period ranging from 2 to 24 hours, depending on state laws. The permanent lack of cerebral activity is called a *persistent vegetative state*. Controversy has lingered, however, over the question of whether death of the entire brain (including the brainstem) should be required as a criterion of clinical death, or whether death may be declared upon lack of activity in only the cerebrum (the upper level of the brain that houses consciousness and thought).

Medical educators, ethicists, philosophers, and theologians struggle continually with the difficulty of defining life and the moment of its cessation. The demand for organs for transplant pressures physicians to make delicate decisions as to when the life of the whole person is irretrievable, yet individual organs are still in sufficiently healthy condition to be useful to a recipient. Theologians, on the other hand, may wish for moral certainty that death has overtaken the whole person, and may see the "culture of organ donation" as incompatible with some religious values.

- **Cells.** Living matter is always compartmentalized into one or more cells.
- **Metabolism.**[11] Living things take in molecules from the environment and chemically change them into molecules that form their own structures, control their physiology, or provide energy. Metabolism is the sum of all this internal chemical change. There is a constant turnover of molecules in the body; although you sense a continuity of personality and experience from your childhood to the present, nearly every molecule of your body has been replaced within the past year.

---

[11] *metabol* = change; *ism* = process

- **Growth.** Some nonliving things grow, but not in the way your body does. When a saturated sugar solution evaporates, crystals grow from it, but not through a change in the composition of the sugar. They merely add more sugar molecules from the solution to the crystal surface. The growth of the body, by contrast, occurs through metabolic change; for the most part, the body is not composed of the molecules one eats, but of molecules made by chemically altering the food.

- **Development.** Development is any change in form or function over the lifetime of the organism. It includes not only growth but also *differentiation*—the transformation of cells and tissues with no specialized function to ones that are committed to a particular task. For example, a single embryonic, unspecialized tissue called *mesoderm* differentiates into muscle, bone, cartilage, blood, and several other specialized tissues.

- **Excitability.** The ability of organisms to sense and react to *stimuli* (changes in their environment) is called *excitability* or *irritability*. It occurs at all levels from the single cell to the entire body, and it characterizes all living things from bacteria to humans. Excitability is especially obvious in animals because of nerve and muscle cells that exhibit high sensitivity to environmental stimuli, rapid transmission of information, and quick reactions.

- **Homeostasis.**[12] Although the environment around an organism changes, the organism maintains relatively stable internal conditions—for example, a stable temperature, blood pressure, and body weight. In his book *The Wisdom of the Body* (1932), American physiologist Walter Cannon coined the term *homeostasis* for this ability to maintain internal stability.

- **Reproduction.** All living organisms can produce copies of themselves, thus passing their genes on to new, younger "containers"—their offspring.

- **Evolution.** All living species exhibit genetic change from generation to generation, and therefore evolve. This occurs because new variations are inevitably introduced by *mutations* (changes in the genes), and environmental conditions favor some variations over others, thus perpetuating some genes and eliminating others. Evolution simply means genetic change in the population over time. Unlike the other characteristics of life, evolution is a characteristic seen only in the population as a whole. No single individual evolves over the course of its life. Evolution, however, holds the explanation for why human structure and function are as they are. *Evolutionary medicine* is a science that interprets human disease and dysfunction in terms of the biological history of the species.

Clinical and legal criteria of life differ from these biological criteria (see Perspectives on Health, p. 9).

Homeostasis and Feedback

Of the foregoing properties of life, the one that arises most frequently in this book is **homeostasis** (ho-me-oh-STAY-sis)—the ability to maintain internal stability. For example, whether it is cold or hot outdoors, your body temperature stays within a range of about 36° to 37°C (97° to 99°F). Homeostatic

---

[12] *homeo* = the same; *stasis* = stability

mechanisms also stabilize such variables as your blood pressure, body weight, electrolyte balance, and pH.

Homeostasis has been one of the most enlightening concepts in physiology. Physiology is mostly a group of mechanisms for maintaining this stability, and the loss of homeostatic control tends to cause illness or death. Pathophysiology is essentially the study of unstable conditions that result when our homeostatic controls go awry.

Negative Feedback and Stability

The fundamental mechanism that maintains homeostasis is **negative feedback**—a process in which the body senses a change and activates mechanisms that negate (reverse) it. Negative feedback does not produce absolute constancy in the body, but maintains physiological values within a narrow range of a certain **set point**—an average value such as 37°C for body temperature. Conditions fluctuate slightly around the set point. Thus, negative feedback is said to maintain a *dynamic equilibrium,* a state of ever-changing balance. By maintaining physiological equilibrium, negative feedback is the key mechanism for maintaining health.

These principles can be understood by comparison to a home heating system (fig. 1.4a). Suppose it is a cold winter day and you have set your thermostat for 20°C (68°F)—the set point. If the room becomes too cold, a temperature-sensitive switch in the thermostat turns on the furnace. The temperature rises to your set point, then the switch breaks the circuit and turns off the furnace. This is a negative feedback process that reverses the falling temperature and restores it to the set point. When the furnace turns off, the temperature slowly drops again until the switch is reactivated—thus, the furnace cycles on and off all day. The room temperature does not stay at exactly 20°C but fluctuates to a small degree—the system maintains a state of dynamic equilibrium in which the temperature deviates from the set point by one degree or less. Because feedback mechanisms alter the original changes that triggered them (temperature, for example), they are often called **feedback loops**.

Body temperature is also regulated by a "thermostat"—a group of nerve cells in the base of the brain. These cells monitor your blood temperature and receive input from nerve endings in the skin that report on body surface temperature. If you become overheated, the thermostat triggers heat-losing mechanisms (fig. 1.4b). One of these is **vasodilation** (VAY-zo-dy-LAY-shun), the widening of blood vessels. When vessels of the skin dilate, warm blood flows closer to the body surface and loses heat to the surrounding air. If this is not enough to return your temperature to normal, sweating occurs; the evaporation

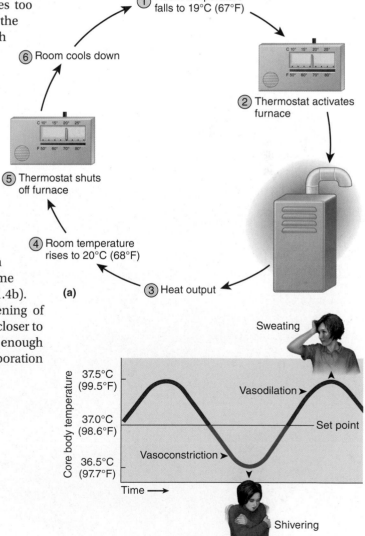

Figure 1.4 Negative Feedback in Thermoregulation. (a) The negative feedback loop that maintains room temperature. (b) Negative feedback usually keeps the human body temperature within about ±0.5°C of a 37.0°C set point. Cutaneous vasoconstriction and shivering raise the body temperature when it is cold, and cutaneous vasodilation and sweating lower it when the body is overheated.

• *Why does vasodilation reduce body temperature?*

Clinical Application 1.1

MEN IN THE OVEN

English physician Charles Blagden (1748–1820) staged a rather theatrical demonstration of homeostasis long before Cannon coined the word. In 1775, he spent 45 minutes in a chamber heated to 127°C (260°F)—along with a beefsteak, a dog, and some research associates. Being dead and unable to maintain homeostasis, the beefsteak was cooked. But being alive and capable of evaporative cooling, the dog panted, the men sweated, and all of them survived. History does not record whether the men ate the beefsteak in celebration or shared it with the dog.

Figure 1.5 Positive Feedback in Childbirth. Repetition of this cycle of events has a self-amplifying effect, intensifying labor contractions until the infant is born. This is one case in which positive feedback has a beneficial outcome.

- *Could childbirth as a whole be considered a negative feedback event? Discuss.*

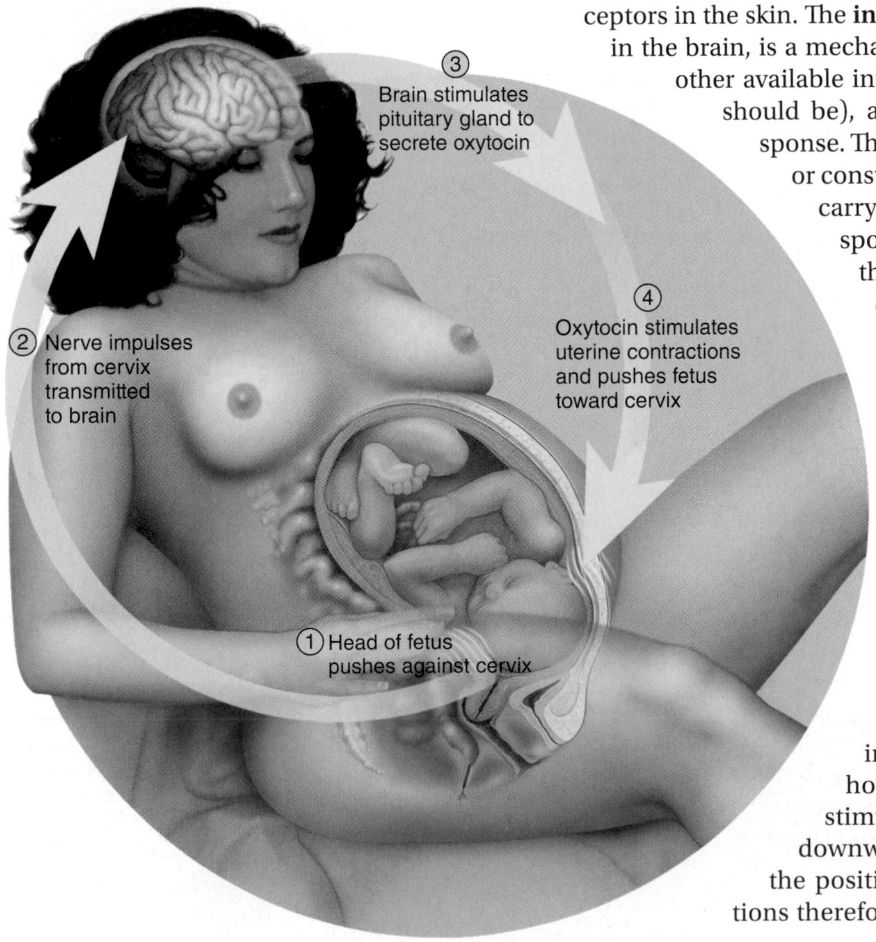

③ Brain stimulates pituitary gland to secrete oxytocin

② Nerve impulses from cervix transmitted to brain

④ Oxytocin stimulates uterine contractions and pushes fetus toward cervix

① Head of fetus pushes against cervix

of water from the skin has a powerful cooling effect (see Clinical Application 1.1). Conversely, if it is cold outside and your body temperature drops a little below 37°C, the brain's thermostat activates heat-conserving mechanisms. The first of these is cutaneous **vasoconstriction,** a narrowing of the blood vessels in the skin, which serves to retain warm blood deeper in your body and reduce heat loss. If this is not enough, shivering follows—muscle tremors that generate heat.

It is common, although not universal, for feedback loops to include three components: a receptor, an integrating center, and an effector. The **receptor** is a structure that senses a change in the body, such as the temperature receptors in the skin. The **integrating (control) center,** such as the thermostat in the brain, is a mechanism that processes this information, relates it to other available information (such as its sense of what the set point should be), and "makes a decision" as to an appropriate response. The **effectors,** in this case the blood vessels that dilate or constrict and the muscles that shiver, are structures that carry out the response that restores homeostasis. The response, such as a lowering of the body temperature, is then sensed by the receptor, and the feedback loop is complete.

Positive Feedback and Rapid Change

Positive feedback is a self-amplifying cycle in which a physiological change leads to even greater change in the same direction, rather than producing the self-corrective effects of negative feedback. Positive feedback is sometimes a normal way of producing rapid change. During childbirth, for example, the head of the fetus pushes against a woman's cervix (the neck of the uterus) and stimulates its nerve endings (fig. 1.5). Nerve signals travel to the brain, which in turn stimulates the pituitary gland to secrete the hormone *oxytocin*. Oxytocin travels in the blood and stimulates the uterus to contract. This pushes the fetus downward, stimulating the cervix still more and causing the positive feedback loop to be repeated. Labor contractions therefore become more and more intense until the fetus is expelled. Other cases of beneficial positive feedback occur in blood clotting, protein digestion, and the generation of nerve signals.

More often, however, positive feedback is a harmful or even life-threatening process. This is because its self-amplifying nature can quickly change the internal state of the body to something far from its homeostatic set point. Consider a high fever, for example. A fever triggered by infection is beneficial up to a point, but if the body temperature rises much above 42°C (108°F), it may create a dangerous positive feedback loop: The high temperature raises the metabolic rate, which makes the body produce heat faster than it can get rid of it. Thus, temperature rises still further, increasing the metabolic rate and heat production still more. This "vicious circle" becomes fatal at approximately 45°C (113°F). Positive feedback loops often create dangerously out-of-control situations that require emergency medical treatment.

Apply What You Know

In a heart attack, the death of cardiac muscle reduces the heart's pumping effectiveness. Thus, blood flow throughout the body slows down. This leads to widespread blood clotting. Blood clots block the flow of blood to the cardiac muscle, so the muscle is less nourished and still more of it dies. Is this positive feedback, negative feedback, or neither? Explain.

Physiological Variation

Earlier we considered the clinical importance of variations in human anatomy, but physiology is even more variable. Physiological variables differ with sex, age, weight, diet, degree of physical activity, and environment, among other things. If an introductory textbook states a typical human heart rate, blood pressure, red blood cell count, or body temperature, it is generally assumed that such values are for a healthy young adult unless otherwise stated. Adjustments must be made for sex, infants, children, the elderly, and ethnicity and are crucial for such purposes as clinical assessment and drug dosages. Failure to consider such variation leads to medical mistakes such as overmedication of the elderly or medicating women on the basis of research that was done on men.

Before You Go On

Answer these questions from memory. Reread the preceding section if there are too many you don't know.

4. What is the difference between growth and development?
5. Explain why positive feedback is likely to cause a loss of homeostasis, and why negative feedback can restore homeostasis.
6. Why is it better to define homeostasis as a dynamic equilibrium than to define it as a state of internal constancy?

1.3   The Human Body Plan

Expected Learning Outcomes

When you have completed this section, you should be able to:

a. list the levels of human complexity in order from the whole organism down to atoms;

b. define or demonstrate the anatomical position and explain its importance in descriptive anatomy;

c. define the three major anatomical planes of the body;

d. identify the major anatomical regions of the body;

e. describe the body's cavities and the membranes that line them; and

f. name the 11 organ systems, their principal organs, and their functions.

This section gives a broad overview of the structural organization of the human body, providing a vocabulary and a context for the study of specific regions and organ systems in the chapters to follow. We also preview the major organs and functions of the body's 11 organ systems.

Figure 1.6 The Structural Hierarchy of the Human Body.

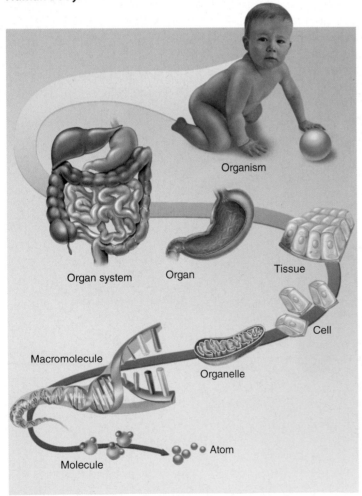

Organism

Organ system   Organ   Tissue

Cell

Macromolecule

Organelle

Molecule   Atom

Levels of Human Structure

The human body can be studied at various levels, from gross anatomy to molecular ultrastructure. Consider for a moment an analogy to human structure: The English language, like the human body, is very complex, yet an endless array of ideas can be conveyed with a limited number of words. All words in the English language are, in turn, composed of various combinations of just 26 letters. Between an essay and the alphabet are successively simpler levels of organization: paragraphs, sentences, words, and syllables. Humans have an analogous hierarchy of complexity (fig. 1.6), as follows:

> The organism is composed of organ systems,
>   organ systems are composed of organs,
>   organs are composed of tissues,
>   tissues are composed of cells,
>   cells are composed (in part) of organelles,
>   organelles are composed of molecules, and
>   molecules are composed of atoms.

The **organism** is a single, complete individual.

An **organ system** is a group of organs that carry out a basic function of the organism such as circulation, respiration, or digestion. The human body has 11 organ systems, defined and illustrated later in this chapter. Usually, the organs of a system are physically interconnected, such as the kidneys, ureters, urinary bladder, and urethra that compose the urinary system.

An **organ** is a structure composed of two or more tissue types that work together to carry out a particular function. Organs have definite anatomical boundaries and are visibly distinguishable from adjacent structures. They include not only what people traditionally think of as the "internal organs," such as the heart and kidneys, but less obvious examples such as the skin, muscles, and bones. Most familiar organs are within the domain of gross anatomy. However, there are organs within organs—the large organs visible to the naked eye contain smaller organs, some of which are visible only with the microscope. The skin, for example, is the body's largest organ. Included within it are thousands of smaller organs: each hair follicle, nail, sweat gland, nerve, and blood vessel of the skin is an organ in itself.

A **tissue** is a mass of similar cells and cell products that forms a discrete region of an organ and performs a specific function. The body is composed of only four primary classes of tissue—epithelial, connective, nervous, and muscular tissues. *Histology,* the study of tissues, is the subject of chapter 4.

Cells are the smallest units of an organism that carry out all the basic functions of life; nothing simpler than a cell is considered alive. A cell is a microscopic compartment enclosed in a film called the *plasma membrane.* It usually contains one nucleus and a variety of other organelles. *Cytology,* the study of cells and organelles, is the subject of chapter 3.

Organelles[13] are microscopic structures that carry out a cell's individual functions. Examples include nuclei, mitochondria, centrioles, and lysosomes.

Organelles and other cellular components are composed of **molecules.** The largest molecules, such as proteins, fats, and DNA, are called *macromolecules.* These are described in chapter 2. A molecule is a particle composed of at least two **atoms.**

Anatomical Position

In describing the human body, anatomists assume that it is in **anatomical position** (fig. 1.7)—that of a person standing upright with the feet flat on the floor, arms at the sides, and the palms and face directed toward the observer. Without such a frame of reference, to say that a structure such as the sternum, thymus, or aorta is "above the heart" would be vague, since it would depend on whether the subject was standing, lying face down *(prone),* or lying face up *(supine).* From the perspective of anatomical position, however, we can describe the thymus as *superior* to the heart, the sternum as *anterior (ventral)* to it, and the aorta as *posterior (dorsal)* to it. These descriptions remain valid regardless of the subject's position.

Unless stated otherwise, assume that all anatomical descriptions refer to anatomical position. Bear in mind that if a subject is facing you in anatomical position, the subject's left will be on your right and vice versa. In most anatomical illustrations, for example, the appendix appears on the *left* side of the page, though it is located in the *right* lower quadrant of the abdomen.

Anatomical Planes

Many views of the body are based on real or imaginary "slices" called sections or planes. *Section* implies an actual cut or slice to reveal internal anatomy, whereas *plane* implies an imaginary flat surface passing through the body. The three major anatomical planes are *sagittal, frontal,* and *transverse* (fig. 1.7).

A **sagittal**[14] (SADJ-ih-tul) **plane** extends vertically and divides the body or an organ into right and left portions. The **median (midsagittal) plane** passes through the midline of the body and divides it into *equal* right and left halves. Other planes parallel to this but off

Figure 1.7 Anatomical Position and Planes of Reference.

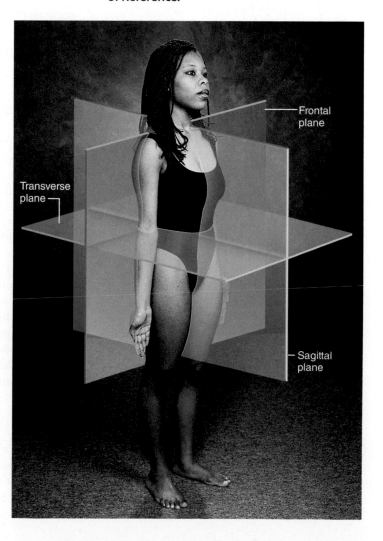

Frontal plane

Transverse plane

Sagittal plane

---

[13] *elle* = little

[14] *sagitta* = arrow

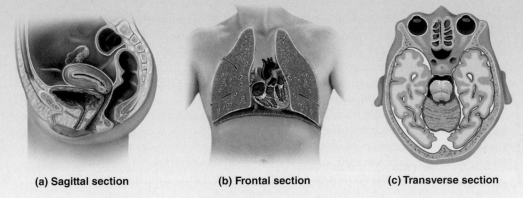

Figure 1.8 Sections of the Body in the Three Primary Anatomical Planes.
(a) Sagittal section of the pelvic region.
(b) Frontal section of the thoracic region.
(c) Transverse section of the head at the level of the eyes.

(a) Sagittal section   **(b) Frontal section**   **(c) Transverse section**

center are called *parasagittal*[15] *planes* and divide the body into unequal right and left portions. The head and pelvic organs are commonly illustrated on the median plane (fig. 1.8a).

A **frontal (coronal**[16]**) plane** also extends vertically, but it is perpendicular to the sagittal plane and divides the body into anterior (front) and posterior (back) portions. A frontal section of the head, for example, would divide it into one portion bearing the face and another bearing the back of the head. Contents of the thoracic and abdominal cavities are commonly shown in frontal section (fig. 1.8b).

A **transverse (horizontal) plane** passes across the body or an organ perpendicular to its long axis; therefore, it divides the body or organ into superior (upper) and inferior (lower) portions. Many CT and MRI scans are transverse sections (fig. 1.8c; see also fig. 1.1c–e).

Major Body Regions

Knowledge of the external anatomy and landmarks of the body is important in performing a physical examination, reporting patient complaints, and many other clinical procedures. For purposes of study, the body is divided into two major regions called the *axial* and *appendicular regions,* each with many smaller regions (fig. 1.9).

The **axial region** forms the central axis of the body—that is, everything but the limbs. It consists of the **head, neck** (**cervical**[17] **region**), and **trunk.** The trunk is further divided into the **thoracic region** above the diaphragm and the **abdominal** and **pelvic regions** below it.

One way of referring to the locations of abdominopelvic structures is to divide the region into quadrants. Two perpendicular lines intersecting at the umbilicus (navel) divide the area into a **right upper quadrant (RUQ), right lower quadrant (RLQ), left upper quadrant (LUQ),** and **left lower quadrant (LLQ)** (fig. 1.10a, b). The quadrant scheme is often used to describe the site of an abdominal pain or abnormality.

The abdomen also can be divided into nine regions defined by four lines that intersect like a tic-tac-toe grid (fig. 1.10c, d). Each vertical line is called a *midclavicular line* because it passes through the midpoint of the clavicle (collarbone). The upper horizontal line is called the *subcostal*[18] *line* because it connects the inferior borders of the lowest *costal cartilages* (cartilage connecting the tenth rib on each side to the inferior end of the sternum). The lower horizontal line is called the *intertubercular*[19] *line* because it passes from

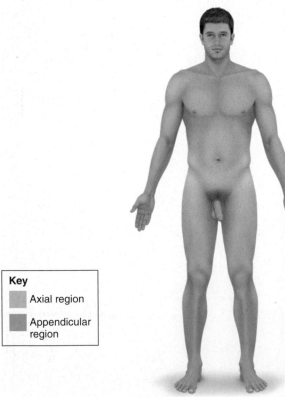

Key

Axial region

Appendicular region

Figure 1.9 The Axial and Appendicular Regions.

[15] *para* = next to

[16] *corona* = crown; *al* = like

[17] *cervic* = neck

[18] *sub* = below; *cost* = rib

[19] *inter* = between; *tubercul* = little swelling

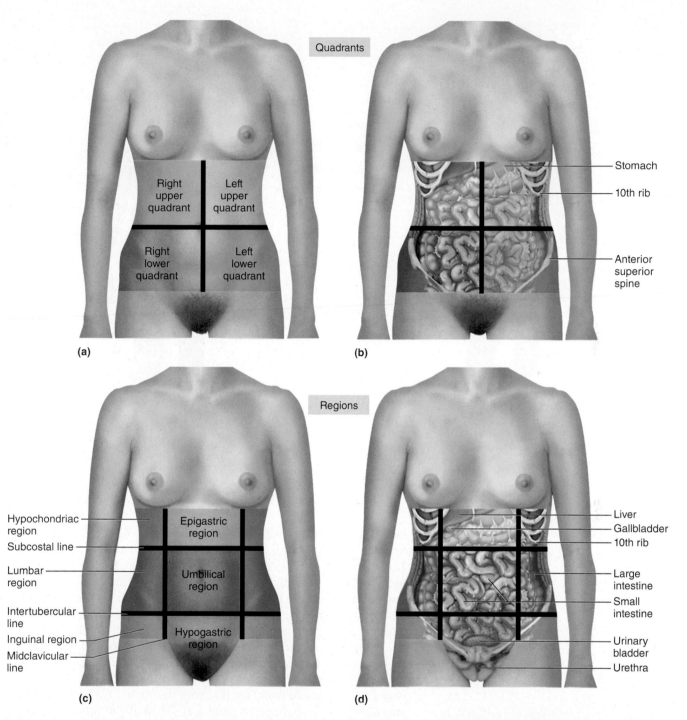

Figure 1.10 Quadrants and Regions of the Abdomen. (a) External division into four quadrants. (b) Internal anatomy correlated with the four quadrants. (c) External division into nine regions. (d) Internal anatomy correlated with the nine regions. **AP|R**

left to right between the tubercles *(anterior superior spines)* of the pelvis—two points of bone located about where the front pockets open on most pants. The three lateral regions of this grid on each side, from upper to lower, are the **hypochondriac,**[20] **lumbar,** and **inguinal**[21] **(iliac) regions.** The three median regions from upper to lower are the **epigastric,**[22] **umbilical,** and **hypogastric (pubic) regions.**

The **appendicular** (AP-en-DIC-you-lur) **region** of the body consists of the upper and lower limbs (also called *appendages* or *extremities*). The **upper limb**

---

[20] *hypo* = below; *chondr* = cartilage

[21] *inguin* = groin

[22] *epi* = above, over; *gastr* = stomach

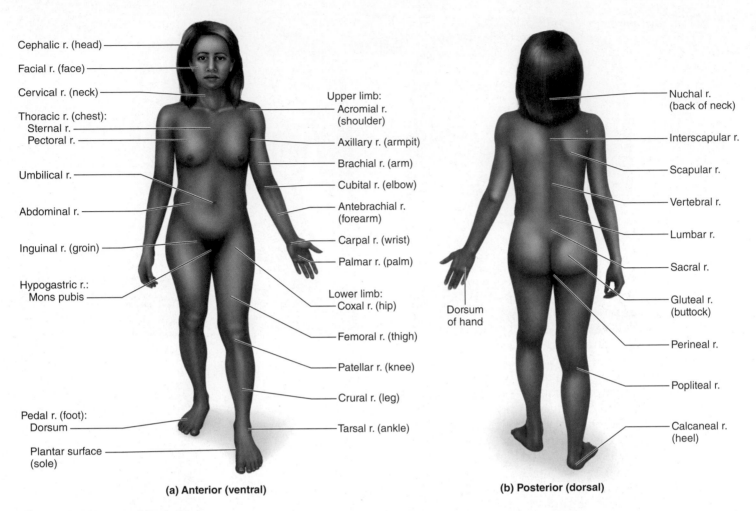

Cephalic r. (head)
Facial r. (face)
Cervical r. (neck)
Thoracic r. (chest):
    Sternal r.
    Pectoral r.
Umbilical r.
Abdominal r.
Inguinal r. (groin)
Hypogastric r.:
    Mons pubis
Pedal r. (foot):
    Dorsum
Plantar surface
(sole)

Upper limb:
    Acromial r.
    (shoulder)
Axillary r. (armpit)
Brachial r. (arm)
Cubital r. (elbow)
Antebrachial r.
(forearm)
Carpal r. (wrist)
Palmar r. (palm)
Lower limb:
    Coxal r. (hip)
Femoral r. (thigh)
Patellar r. (knee)
Crural r. (leg)
Tarsal r. (ankle)

(a) Anterior (ventral)

Nuchal r.
(back of neck)
Interscapular r.
Scapular r.
Vertebral r.
Lumbar r.
Sacral r.
Gluteal r.
(buttock)
Perineal r.
Popliteal r.
Calcaneal r.
(heel)

Dorsum
of hand

(b) Posterior (dorsal)

Figure 1.11 External Body Regions
(r. = region). AP|R

includes the **arm (brachium)** (BRAY-kee-um), **forearm (antebrachium**[23]**)** (AN-teh-BRAY-kee-um), **wrist (carpus), hand (manus),** and **fingers (digits).** The lower limb includes the **thigh (femoral region), leg (crus), ankle (tarsus), foot (pes),** and **toes (digits).** In strict anatomical terms, *arm* refers only to that part of the upper limb between the shoulder and elbow. *Leg* refers only to that part of the lower limb between the knee and ankle. Figure 1.11 identifies several smaller body regions commonly referred to in medical records and literature.

A **segment** of a limb is a region between one joint and the next. The arm, for example, is the segment between the shoulder and elbow joints, and the forearm is the segment between the elbow and wrist joints. Slightly flexing your fingers, you can easily see that your thumb has two segments (proximal and distal—terms defined in table 1.3), whereas the other four digits have three segments (proximal, middle, and distal). The segment concept is especially useful in describing the locations of bones and muscles and the movements of the joints.

Body Cavities and Membranes

The axial region of the body contains a few major cavities containing the "internal organs" or **viscera** (VISS-er-uh) (singular, *viscus*[24]) (table 1.1). These cavities are lined by thin **serous membranes,** which secrete a lubricating film of

---

[23] *ante* = fore, before; *brachi* = arm

[24] *viscus* = body organ

Table 1.1 Body Cavities and Membranes

Name of Cavity	Principal Viscera	Serous Membrane
Cranial cavity	Brain	Meninges
Vertebral canal	Spinal cord	Meninges
Thoracic cavity		
Pleural cavities (2)	Lungs	Pleurae
Pericardial cavity	Heart	Pericardium
Abdominopelvic cavity		
Abdominal cavity	Digestive organs, spleen, kidneys	Peritoneum
Pelvic cavity	Bladder, rectum, reproductive organs	Peritoneum

moisture similar to blood serum (hence the name *serous*). These membranes are distinct from the **mucous membranes** that line the digestive, respiratory, urinary, and reproductive tracts—passages open to the exterior. Serous and mucous membranes are further described in chapter 4.

The **cranial** (CRAY-nee-ul) **cavity** is enclosed by the skull and contains the brain. The **vertebral canal,** continuous with the cranial cavity, is a space about as wide as your finger that passes down the vertebral column (spine) (fig. 1.12a). Both of these cavities are lined by three membranes called **meninges** (meh-NIN-jeez). Among other functions, they protect the delicate nervous tissue from the hard protective bone that encloses it. The meninges are discussed in more detail in chapter 9.

Early in its development, the human embryo exhibits an internal space called the *coelom* (SEE-loam). This space soon becomes divided by a muscular sheet, the **diaphragm,** into the future **thoracic cavity** above and **abdominopelvic cavity** below.

The thoracic cavity is subsequently divided into right, left, and median portions by a partition called the **mediastinum**[25] (ME-dee-ah-STY-num) (fig. 1.12b). The mediastinum is occupied by the esophagus and trachea, a gland called the *thymus,* and the heart and major blood vessels connected to it. The heart is enveloped by a two-layered serous membrane called the **pericardium.**[26] The anatomy of this pericardium and its relationship with the heart are further described in chapter 13. The right and left sides of the thoracic cavity contain the lungs, which are each enfolded in another two-layered serous membrane, the **pleura**[27] (PLOOR-uh) (plural, *pleurae*). The anatomy of the pleurae and their relationship with the lungs are detailed in chapter 15.

The abdominopelvic cavity can be subdivided into the **abdominal cavity** and **pelvic cavity,** although the two form one continuous space; they are not separated by a wall the way the abdominal and thoracic cavities are separated by the diaphragm. The dividing line between them is the margin of the pelvic inlet (see fig. 6.29a, p. 180). The abdominal cavity contains most of the digestive organs as well as the spleen, kidneys, and ureters. The pelvic cavity is markedly narrower and its lower end tilts posteriorly (fig. 1.12a). It contains the lowermost portion of the large intestine, the urinary bladder and urethra, and the reproductive organs.

---

[25] *mediastinum* = in the middle

[26] *peri* = around; *cardi* = heart

[27] *pleur* = rib, side

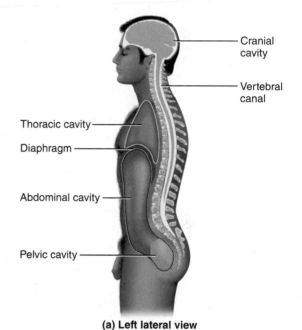

(a) Left lateral view

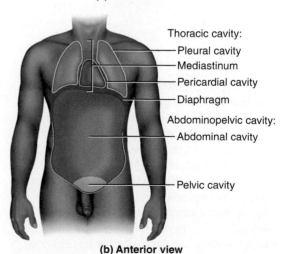

(b) Anterior view

Figure 1.12 The Major Body Cavities. AP|R

The abdominopelvic cavity is lined by a two-layered serous membrane called the **peritoneum**[28] (PERR-ih-toe-NEE-um). The outer layer, lining the abdominal wall, is called the *parietal peritoneum.* Along the posterior midline of the abdominal wall, it turns inward and becomes another layer, the *visceral peritoneum,* suspending certain abdominal viscera from the body wall and covering their outer surfaces (fig. 1.13). The visceral peritoneum is also called a **mesentery**[29] (MEZ-en-tare-ee) at points where it forms a

[28] *peri* = around; *tone* = stretched
[29] *mes* = in the middle; *enter* = intestine

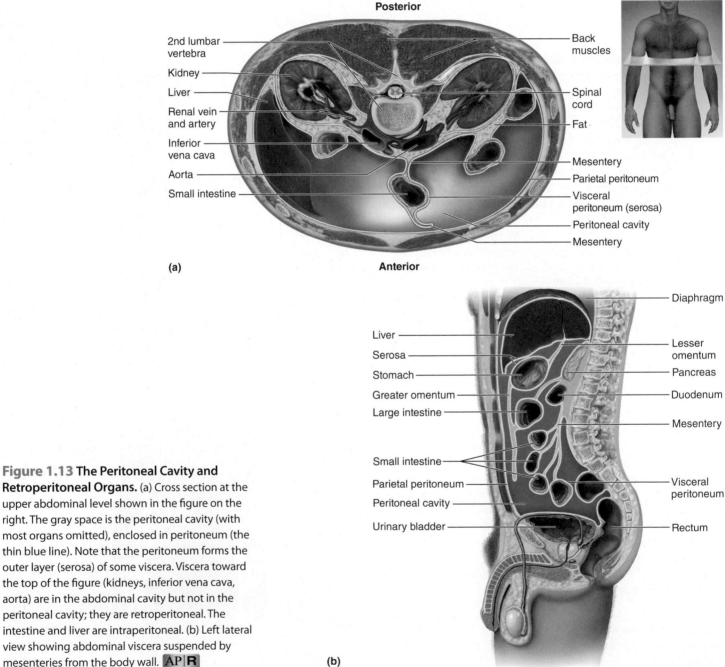

Posterior

2nd lumbar vertebra

Kidney

Liver

Renal vein and artery

Inferior vena cava

Aorta

Small intestine

Back muscles

Spinal cord

Fat

Mesentery

Parietal peritoneum

Visceral peritoneum (serosa)

Peritoneal cavity

Mesentery

(a)

Anterior

Liver

Serosa

Stomach

Greater omentum

Large intestine

Small intestine

Parietal peritoneum

Peritoneal cavity

Urinary bladder

Diaphragm

Lesser omentum

Pancreas

Duodenum

Mesentery

Visceral peritoneum

Rectum

Figure 1.13 The Peritoneal Cavity and Retroperitoneal Organs. (a) Cross section at the upper abdominal level shown in the figure on the right. The gray space is the peritoneal cavity (with most organs omitted), enclosed in peritoneum (the thin blue line). Note that the peritoneum forms the outer layer (serosa) of some viscera. Viscera toward the top of the figure (kidneys, inferior vena cava, aorta) are in the abdominal cavity but not in the peritoneal cavity; they are retroperitoneal. The intestine and liver are intraperitoneal. (b) Left lateral view showing abdominal viscera suspended by mesenteries from the body wall. AP|R

(b)

Clinical Application 1.2

PERITONITIS

Peritonitis is inflammation of the peritoneum. It is a critical, life-threatening condition necessitating prompt treatment. The most serious cause of peritonitis is a perforation in the digestive tract, such as a ruptured appendix. Digestive juices cause immediate chemical inflammation of the peritoneum, followed by microbial inflammation as intestinal bacteria invade the body cavity. Anything that perforates the abdominal wall can also lead to peritonitis, such as abdominal trauma or surgery. So, too, can free blood in the abdominal cavity, as from a ruptured aneurysm (a weak point in a blood vessel) or ectopic pregnancy (implantation of an embryo anywhere other than the uterus); blood itself is a chemical irritant to the peritoneum. Peritonitis tends to shift fluid from the circulation into the abdominal cavity. Death can follow within a few days from severe electrolyte imbalance, respiratory distress, kidney failure, and a widespread blood clotting called disseminated intravascular coagulation.

membranous curtain suspending and anchoring the viscera, and a **serosa** (seer-OH-sa) at points where it enfolds and covers the outer surfaces of organs such as the stomach and small intestine (fig. 1.14).

The space between the parietal and visceral peritoneum is called the **peritoneal cavity** and is lubricated by **peritoneal fluid.** These relationships of the peritoneum, serosa, and peritoneal cavity with the abdominal digestive organs are detailed in chapter 17.

Some organs of the abdominal cavity lie against the posterior body wall and are covered by peritoneum only on the side facing the peritoneal cavity. They are said to have a **retroperitoneal**[30] position. Examples include the kidneys, aorta, and inferior vena cava as shown in figure 1.13a and the duodenum and pancreas as shown in figure 1.13b. Organs that are encircled by peritoneum and suspended from the posterior body wall by mesenteries, such as the loops of small intestine shown in those figures, are designated **intraperitoneal.**[31]

Figure 1.14 Mesentery. This is a translucent serous membrane associated with the small intestine and other abdominal organs. Mesenteries contain blood vessels, lymphatic vessels, and nerves supplying the viscera.

Organ Systems

The human body has 11 organ systems (fig. 1.15) and an immune system, which is better described as a population of cells that inhabit multiple organs rather than as an organ system. These systems are classified in the following list by their principal functions, but this is an unavoidably flawed classification. Some organs belong to two or more systems—for example, the male urethra is part of both the urinary and reproductive systems; the pharynx is part of the respiratory and digestive systems; and the mammary glands can be considered part of the integumentary and female reproductive systems.

---

[30] *retro* = behind

[31] *intra* = within

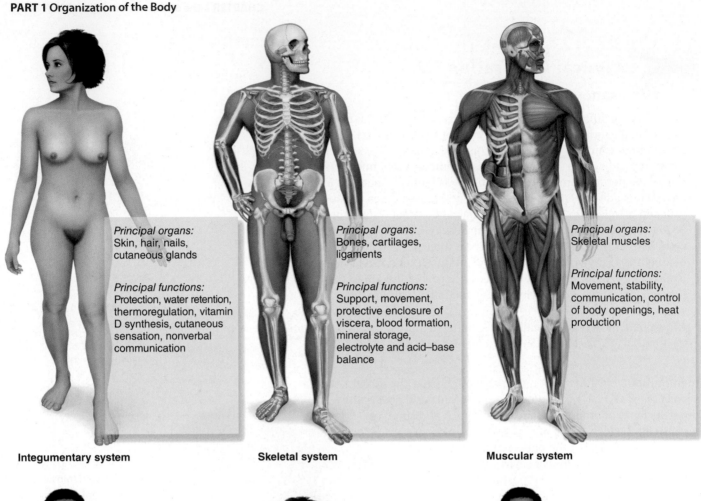

Principal organs:
Skin, hair, nails, cutaneous glands

Principal functions:
Protection, water retention, thermoregulation, vitamin D synthesis, cutaneous sensation, nonverbal communication

Integumentary system

Principal organs:
Bones, cartilages, ligaments

Principal functions:
Support, movement, protective enclosure of viscera, blood formation, mineral storage, electrolyte and acid–base balance

Skeletal system

Principal organs:
Skeletal muscles

Principal functions:
Movement, stability, communication, control of body openings, heat production

Muscular system

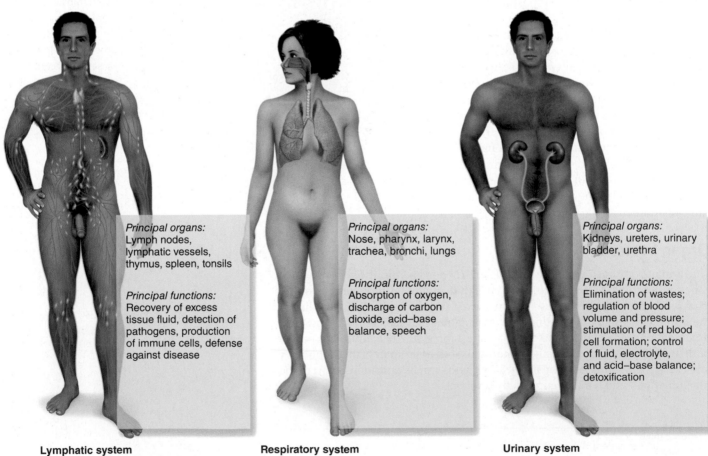

Principal organs:
Lymph nodes, lymphatic vessels, thymus, spleen, tonsils

Principal functions:
Recovery of excess tissue fluid, detection of pathogens, production of immune cells, defense against disease

Lymphatic system

Principal organs:
Nose, pharynx, larynx, trachea, bronchi, lungs

Principal functions:
Absorption of oxygen, discharge of carbon dioxide, acid–base balance, speech

Respiratory system

Principal organs:
Kidneys, ureters, urinary bladder, urethra

Principal functions:
Elimination of wastes; regulation of blood volume and pressure; stimulation of red blood cell formation; control of fluid, electrolyte, and acid–base balance; detoxification

Urinary system

Figure 1.15 The Human Organ Systems.

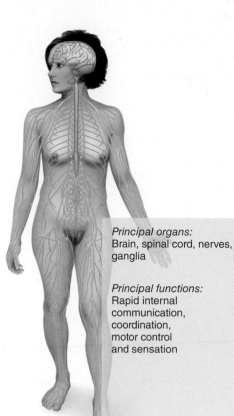

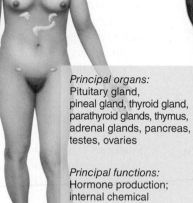

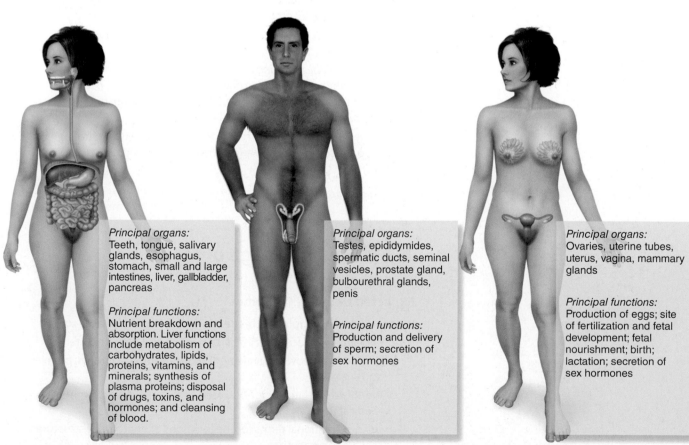

Principal organs:
Brain, spinal cord, nerves, ganglia

Principal functions:
Rapid internal communication, coordination, motor control and sensation

Nervous system

Principal organs:
Pituitary gland, pineal gland, thyroid gland, parathyroid glands, thymus, adrenal glands, pancreas, testes, ovaries

Principal functions:
Hormone production; internal chemical communication and coordination

Endocrine system

Principal organs:
Heart, blood vessels

Principal functions:
Distribution of nutrients, oxygen, wastes, hormones, electrolytes, heat, immune cells, and antibodies; fluid, electrolyte, and acid–base balance

Circulatory system

Principal organs:
Teeth, tongue, salivary glands, esophagus, stomach, small and large intestines, liver, gallbladder, pancreas

Principal functions:
Nutrient breakdown and absorption. Liver functions include metabolism of carbohydrates, lipids, proteins, vitamins, and minerals; synthesis of plasma proteins; disposal of drugs, toxins, and hormones; and cleansing of blood.

Digestive system

Principal organs:
Testes, epididymides, spermatic ducts, seminal vesicles, prostate gland, bulbourethral glands, penis

Principal functions:
Production and delivery of sperm; secretion of sex hormones

Male reproductive system

Principal organs:
Ovaries, uterine tubes, uterus, vagina, mammary glands

Principal functions:
Production of eggs; site of fertilization and fetal development; fetal nourishment; birth; lactation; secretion of sex hormones

Female reproductive system

The human organ systems

Systems of protection, support, and movement
Integumentary system
Skeletal system
Muscular system
Systems of internal communication and integration
Nervous system
Endocrine system
Systems of fluid transport
Circulatory system
Lymphatic system
Systems of intake and output
Respiratory system
Urinary system
Digestive system
Systems of reproduction
Male reproductive system
Female reproductive system

Before You Go On

Answer these questions from memory. Reread the preceding section if there are too many you don't know.

7. *Rearrange the following alphabetical list in order from the largest, most complex components of the body to the smallest, simplest ones: cells, molecules, organelles, organs, organ systems, tissues.*

8. *Examine figures 1.1 and 1.13 and identify whether each figure shows the body on a frontal, sagittal, or transverse plane.*

9. *Identify each of the following regions as belonging to the upper limb, lower limb, or axial region of the body: crural, cubital, pectoral, gluteal, antebrachial, calcaneal, patellar, nuchal, hypogastric, and hypochondriac.*

10. *Name the body cavity in which each of the following viscera are found: spinal cord, liver, lung, spleen, heart, pancreas, gallbladder, and kidney.*

1.4 The Language of Medicine

Expected Learning Outcomes

When you have completed this section, you should be able to:

a. explain why precision is important in the use of medical terms;

b. demonstrate how to break medical terms into their roots, prefixes, and suffixes;

c. identify the relationships between singular and plural forms of a medical term; and

d. define directional terms for the locations of anatomical structures relative to each other.

One of the greatest challenges faced by students of anatomy and physiology is the vocabulary. In this book, you will encounter such Latin terms as *corpus callosum* (a brain structure) and *extensor carpi radialis longus* (a forearm muscle). You may wonder why structures aren't named in "just plain English," and how you will ever remember such formidable names. This section will give you some answers to these questions and some useful tips on mastering anatomical terminology.

It is of greatest importance to use medical terms precisely. It may seem trivial if you misspell *trapezius* as *trapezium,* but in doing so, you would be changing the name of a back muscle to the name of a wrist bone. A "little" error such as misspelling *malleus* as *malleolus* changes the name of a middle-ear bone to the name of a protuberance of the ankle. A difference of only one letter distinguishes *gustation* (the sense of taste) from *gestation* (pregnancy). The health professions demand the utmost attention to detail and precision. People's well-being may one day be in your hands, and indeed many patients die simply because of written and oral miscommunication among hospital staff. The habit of precision must be cultivated early.

> ### *Apply What You Know*
>
> A student means to write about part of the small intestine called the ileum, but misspells it *ilium*. In complaining about the points lost, the student says, "I was only one letter off!" The instructor says, "But you changed the entire meaning of the word." With the help of a dictionary, explain the instructor's reasoning.

Analyzing Medical Terms

The task of learning medical terminology seems overwhelming at first, but there is a simple trick to becoming more comfortable with the technical language of medicine. Those who, at first, find scientific terms confusing and difficult to pronounce, spell, and remember usually feel more confident once they realize the logic of how such terms are composed. A term such as *hyponatremia* is less forbidding once we recognize that it is composed of three common word elements: *hypo-* (below normal), *natr-* (sodium), and *-emia* (blood condition). Thus, hyponatremia is a deficiency of sodium in the blood. Those three word elements appear over and over in many other medical terms: *hypothermia, natriuretic, anemia,* and so on. Once you learn the meanings of *hypo-, natri-,* and *-emia,* you already have the tools at least to partially understand hundreds of other biomedical terms.

Scientific terms are typically composed of one or more of the following elements:

- At least one *root (stem)* that bears the core meaning of the word. In *cardiology,* for example, the root is *cardi-* (heart). Many words have two or more roots. In *adipocyte,* the roots are *adip-* (fat) and *cyte* (cell). Word roots are often linked through an *o* or other vowel to make the word more pronounceable.
- A *prefix* may be present at the beginning of a word to modify its core meaning. For example, *gastric* (pertaining to the stomach or to the belly of a muscle) takes on a variety of new meanings when prefixes are added to it: *epigastric* (above the stomach), *hypogastric* (below the stomach), *endogastric* (within the stomach), and *digastric* (a muscle with two bellies).
- A *suffix* may be added to the end of a word to modify its core meaning. For example, *microscope, microscopy, microscopic,* and *microscopist* have different meanings because of their suffixes alone.

Table 1.2 Singular and Plural Forms of Some Noun Terminals

Singular Ending	Plural Ending	Examples
-a	-ae	pleura, pleurae
-ax	-aces	thorax, thoraces
-en	-ina	lumen, lumina
-ex	-ices	cortex, cortices
-is	-es	testis, testes
-is	-ides	epididymis, epididymides
-ix	-ices	appendix, appendices
-ma	-mata	carcinoma, carcinomata
-on	-a	ganglion, ganglia
-um	-a	septum, septa
-us	-era	viscus, viscera
-us	-i	villus, villi
-us	-ora	corpus, corpora
-x	-ges	phalanx, phalanges
-y	-ies	ovary, ovaries
-yx	-yces	calyx, calyces

Consider another word, *gastroenterology,* a branch of medicine dealing with the stomach and small intestine. It breaks down into

gastro/entero/logy
   gastro = "stomach"
   entero = "small intestine"
   logy = "the study of"

"Dissecting" words in this way and paying attention to the word-origin footnotes throughout this book will help make you more comfortable with the language of anatomy. Breaking a word down and knowing the meaning of its elements make it far easier to pronounce it, spell it, and remember its definition. In appendix D, you will find a lexicon of the word roots, prefixes, and suffixes most frequently used in this book.

Singular and Plural Forms

A point of confusion for many beginning students is how to recognize the plural forms of medical terms. Few people would fail to recognize that *ovaries* is the plural of *ovary,* but the connection is harder to make in other cases: for example, the plural of *cortex* is *cortices* (COR-ti-sees), and the plural of *corpus* is *corpora.* Table 1.2 will help you make the connection between common singular and plural noun terminals.

Directional Terminology

In "navigating" the human body and describing the locations of structures, anatomists use a set of standard **directional terms** (table 1.3). You will need to be very familiar with these in order to understand anatomical descriptions later in this book. The terms assume that the body is in anatomical position.

Intermediate directions are often indicated by combinations of these terms. For example, a structure that is *superomedial* to another is above and medial to it; the bridge of the nose is superomedial to flare of the nostrils.

Because of the bipedal (two-legged), upright stance of humans, some directional terms have different meanings for humans than they do for other

Table 1.3 Directional Terms in Human Anatomy AP|R

Term	Meaning	Examples of Usage
Ventral	Toward the front* or belly	The aorta is *ventral* to the vertebral column.
Dorsal	Toward the back or spine	The vertebral column is *dorsal* to the aorta.
Anterior	Toward the ventral side*	The sternum is *anterior* to the heart.
Posterior	Toward the dorsal side*	The esophagus is *posterior* to the trachea.
Superior	Above	The heart is *superior* to the diaphragm.
Inferior	Below	The liver is *inferior* to the diaphragm.
Medial	Toward the midsagittal plane	The heart is *medial* to the lungs.
Lateral	Away from the midsagittal plane	The eyes are *lateral* to the nose.
Proximal	Closer to the point of attachment or origin	The elbow is *proximal* to the wrist.
Distal	Farther from the point of attachment or origin	The fingernails are at the *distal* ends of the fingers.
Superficial	Closer to the body surface	The skin is *superficial* to the muscles.
Deep	Farther from the body surface	The bones are *deep* to the muscles.

* In humans only; definition differs for other animals.

CAREER SPOTLIGHT

Radiologic Technologist

A radiologic technologist is a person who produces medical images for the purposes of diagnosing and treating illnesses and injuries, or who administers radiation therapy. (The profession is not to be confused with *radiologic technician*—one who sets up, maintains, and repairs radiologic equipment.) Radiologic technologists work closely with radiologists—physicians who interpret the images, make diagnoses, and prescribe courses of treatment. Programs in radiologic technology range from a 2-year associate degree to bachelor's and master's degrees. The training of a radiologic technologist is a mixture of medical and physical sciences, including anatomy, physiology, genetics, pathology, chemistry, general and nuclear physics, medical terminology, patient examination and positioning, radiologic instrumentation, and medical imaging sciences. Writing, speech, and figure drawing also provide useful experience, and one must have good analytical thinking skills, attention to detail, compassion, and patience. Entry into the profession requires passing a board examination; earning certification, although optional, is a further step that enhances one's career mobility. Many employers prefer to hire only certified radiologic technologists. Career-long continuing education courses are necessary to keep one's certification current. Specialties in radiologic technology include radiography (X-ray technology), mammography, sonography, fluoroscopy, CT, MRI, PET, nuclear medicine, medical dosimetry, and bone densitometry. Some radiologic technologists go into clinical administration or radiologic education, or work for equipment manufacturers. See appendix B for additional career information.

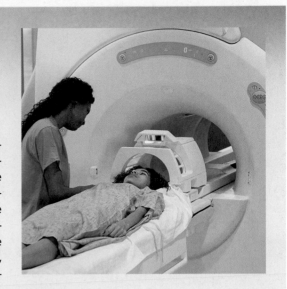

animals. *Anterior,* for example, denotes the region of the body that leads the way in normal locomotion. For a four-legged animal such as a cat, this is the head end of the body; for a human, however, it is the front of the chest and abdomen. Thus, *anterior* has the same meaning as *ventral* for a human but not for a cat. *Posterior* denotes the region of the body that comes last in normal locomotion—the tail end of a cat but the dorsal side of a human. These differences must be kept in mind when dissecting other animals for comparison to human anatomy.

Before You Go On

Answer these questions from memory. Reread the preceding section if there are too many you don't know.

11. *Following the example given for* gastroenterology, *break each of the following words down into their roots, prefixes, and suffixes:* electrocardiography, brachiocephalic, hyperkalemia, substernal, periodontal.

12. *Write the plural form of each of the following terms:* stoma, lacuna, nucleus, epithelium, diagnosis. *Write the singular form of each of the following:* larynges, carpi, ampullae, matrices, ova.

13. *Examine figure 1.11 and use the proper directional term from table 1.3 to describe (a) the location of the axillary region relative to the cubital region; (b) the location of the sacral region relative to the lumbar region; (c) the location of the sacral region relative to the hypogastric region; and (d) the location of the umbilical region relative to the lumbar region.*

Study Guide

Assess Your Learning Outcomes

To test your knowledge, discuss the following topics with a study partner or in writing, ideally from memory.

1.1 Anatomy—The Structural Basis of Human Function (p. 2)

1. The distinction between gross anatomy and histology, and the relationship of histology to histopathology
2. How surface anatomy, systemic anatomy, regional anatomy, and functional morphology differ in their perspectives
3. Why the study of other animal species is important for understanding human functional morphology
4. Distinctions between dissection, palpation, auscultation, percussion, and medical imaging as methods of studying human structure
5. How each of the following methods of medical imaging is performed, and why one might be chosen over the others for specific purposes: radiography, computed tomography (CT), magnetic resonance imaging (MRI), positron emission tomography (PET), and sonography
6. Examples of individual anatomical variation and why such variations are important in the practice of medicine

1.2 Physiology—The Functional Relevance of Human Structure (p. 8)

1. The meaning of *physiology,* and the importance of studying the physiology of other species for understanding humans
2. The properties that define something as alive and the difficulty of defining *life* or the moment of death
3. The meaning of *homeostasis,* the role of negative feedback loops in maintaining homeostasis, and the contrast between the set point and a dynamic equilibrium
4. Examples of negative feedback and homeostasis
5. The fundamental components seen in many negative feedback loops
6. How positive feedback differs from negative feedback, and examples of beneficial and harmful effects of positive feedback
7. Why it is clinically important to be aware of individual physiological variation

1.3 The Human Body Plan (p. 14)

1. The eight levels of structural complexity from organism to atoms, and the definitions of such levels as organ, tissue, cell, organelle, and others
2. How a body is positioned when it is said to be in anatomical position, and why anatomical position is such an important frame of reference
3. Definition of the three principal anatomical planes of the human body
4. Components of the axial and appendicular regions of the human body
5. Landmarks that define the four quadrants of the abdomen; names of the quadrants; and some organs that lie within each quadrant
6. Landmarks that divide the abdomen into nine regions; names of those regions; and some organs that lie within each
7. Terminology of the surface regions of human anatomy
8. Meaning of the *segments* of an appendage, and examples
9. Differences in the locations of the serous and mucous membranes of the body
10. The principal body cavities; names of the membranes that line them; and the most important viscera found in each
11. The meaning of *parietal* and *visceral peritoneum*
12. The location, appearance, and function of the mesenteries, and how the mesenteries are related to the peritoneum and the serosa of the viscera
13. The distinction between intraperitoneal and retroperitoneal organs, and some examples of each
14. The 11 organ systems of the human body, functions of each system, and principal organs of each

1.4 The Language of Medicine (p. 24)

1. Examples of why precision is important in anatomical terminology, and why spelling errors can be more significant than they seem
2. How to break down biomedical terms into roots, prefixes, and suffixes, and why it is helpful to make a habit of seeing terminology in this way
3. Relationships between the singular and plural forms of the same biomedical nouns
4. The distinctions between *dorsal* and *ventral; anterior* and *posterior; superior* and *inferior; medial* and *lateral; proximal* and *distal;* and *superficial* and *deep*
5. Why *anterior* and *posterior, dorsal* and *ventral* have different meanings for a human than for a cat

Testing Your Recall

1. Structure that can be observed with the naked eye is called
   a. gross anatomy.
   b. histology.
   c. ultrastructure.
   d. comparative anatomy.
   e. cytology.

2. The method of medical imaging that exposes a person to radio waves is
   a. a PET scan.
   b. an MRI scan.
   c. radiology.
   d. sonography.
   e. a CT scan.

3. The tarsal region is ___ to the popliteal region.
   a. dorsal
   b. distal
   c. superior
   d. proximal
   e. appendicular

4. Which of the following regions is *not* part of the upper limb?
   a. palmar
   b. antebrachial
   c. cubital
   d. carpal
   e. popliteal

5. Which of the following is *not* an organ system?
   a. muscular system
   b. endocrine system
   c. immune system
   d. lymphatic system
   e. integumentary system

6. In which area do you think pain from the gallbladder would be felt?
   a. the right upper quadrant
   b. the umbilical region
   c. the hypogastric region
   d. the left hypochondriac region
   e. the left lower quadrant

7. Which of these organs is intraperitoneal?
   a. the urinary bladder
   b. a kidney
   c. the heart
   d. the stomach
   e. the brain

8. The lining of the abdominal cavity is
   a. a mucous membrane.
   b. the peritoneum.
   c. the meninges.
   d. the pleura.
   e. the serosa.

9. The word root *patho-* means
   a. doctor.
   b. medicine.
   c. organ.
   d. health.
   e. disease.

10. The prefix *epi-* means
    a. next to.
    b. below.
    c. above.
    d. within.
    e. behind.

11. When a doctor presses on the abdomen to feel the size and texture of the liver, he or she is using a technique of physical examination called _____.

12. A method of medical imaging that uses X-rays and a computer to generate an image of a thin slice of the body is called _____.

13. Most physiological mechanisms serve the purpose of _____, maintaining a stable internal environment in the body.

14. A/an ___ is the simplest body structure to be composed of two or more types of tissue.

15. The carpal region is more commonly known as the _____, and the tarsal region is more commonly known as the _____.

16. In standard directional terms, the sternal region is _____ to the pectoral region.

17. The layer of peritoneum facing the body wall is called the _____ layer, and the layer on the surface of an internal organ is called the _____ layer.

18. Homeostasis is maintained by a cycle of events called a _____, in which the body senses a change and activates mechanisms to minimize or reverse it.

19. The directional terms of human anatomy assume that a person is in _____, which means standing upright with the feet together and the palms, face, and eyes forward.

20. The elbow is said to be _____ to the wrist because it is closer to the upper limb's point of origin (the shoulder).

Answers in appendix A

True or False

Determine which five of the following statements are false, and briefly explain why.

1. It is possible to see both eyes in one frontal section of the head.

2. The diaphragm is ventral to the lungs.

3. An MRI scan is a noninvasive method of medical imaging.

4. The pleural and pericardial cavities are lined by mucous membranes.

5. Abnormal skin color or dryness is one piece of information that could be obtained by auscultation.

6. Sonography is a simpler and safer way than a CT scan to monitor fetal development.

7. Histopathology is a subdiscipline of microscopic anatomy.

8. Negative feedback is more often harmful than beneficial to the body.

9. There are more cells than organelles in the human body.

10. The pericardial sac is superficial to the heart.

Answers in appendix A

Testing Your Comprehension

1. Identify which anatomical plane—sagittal, frontal, or transverse—is the only one that could *not* show (a) both the cerebrum and tongue, (b) both eyes, (c) both the hypogastric and gluteal regions, (d) both the sternum and vertebral column, and (f) both the heart and uterus.

2. Name one structure or anatomical feature that could be found in each of the following locations relative to the ribs: medial, lateral, superior, inferior, deep, superficial, posterior, and anterior. Try not to use the same example twice.

3. For each of the following nonbiological processes, state whether you think it is analogous to physiological positive feedback, negative feedback, or neither of these, and justify each answer: (a) the cruise control of a car maintaining a preset highway speed; (b) a flushed toilet tank refilling to its original resting level; (c) a magnifying glass focusing the sun's rays and catching a piece of paper on fire; (d) a house fire in which heat from the flames ignites adjacent flammable material until, if unchecked, the whole house is consumed; (e) the increasingly loud howl of a loudspeaker as a band is setting up the sound stage for a concert and gets a microphone too close to the speaker.

Life, Matter, and Energy

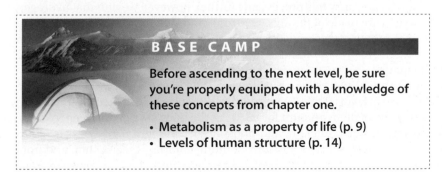

Urea crystals seen with a polarizing microscope. Urea, a product of protein metabolism, is the principal organic waste product of urine.

BASE CAMP

Before ascending to the next level, be sure you're properly equipped with a knowledge of these concepts from chapter one.

- Metabolism as a property of life (p. 9)
- Levels of human structure (p. 14)

Anatomy & Physiology | **REVEALED**®
aprevealed.com

Module 2: Cells and Chemistry

Why is too much sodium or cholesterol harmful? Why does an iron deficiency cause anemia and an iodine deficiency cause a goiter? How can a pH imbalance make some drugs less effective? Why do some pregnant women suffer convulsions after several days of vomiting? How can radiation cause cancer as well as cure it?

None of these questions can be answered, nor would the rest of this book be intelligible, without understanding the chemistry of life. A little knowledge of chemistry can help you choose a healthy diet, use medications more wisely, avoid worthless health fads and frauds, and explain treatments and procedures to your patients or clients. Thus, we begin our study of the human body with basic chemistry, the simplest level of the body's structural organization.

2.1 Atoms, Ions, and Molecules

Expected Learning Outcomes

When you have completed this section, you should be able to:

a. recognize elements of the human body from their chemical symbols;

b. distinguish between chemical elements and compounds;

c. state the functions of minerals in the body;

d. explain the basis for radioactivity and the uses and hazards of ionizing radiation;

e. distinguish between ions, electrolytes, and free radicals; and

f. define the types of chemical bonds.

Chemical Elements

A chemical **element** is the simplest form of matter to have unique chemical properties. Water, for example, has unique properties, but it can be broken down into two elements, hydrogen and oxygen, that have unique properties of their own. If we carry this process any further, however, we find that hydrogen and oxygen are made of protons, neutrons, and electrons—none of which are unique. A proton of gold is identical to a proton of oxygen. Therefore, hydrogen and oxygen are the simplest chemically unique components of water and are its elements.

There are 91 naturally occurring elements on earth, 24 of which play normal roles in humans. Table 2.1 groups these 24 according to their abundance in the body. Six of them account for 98.5% of the body's weight: oxygen, carbon, hydrogen, nitrogen, calcium, and phosphorus. The next 0.8% consists of another 6 elements: sulfur, potassium, sodium, chlorine, magnesium, and iron. The remaining 12 total only 0.7% of body weight; thus, they are known as **trace**

Table 2.1 Elements of the Human Body

Name	Symbol		Percentage of Body Weight
Major Elements (Total 98.5%)			
Oxygen	O		65.0
Carbon	C		18.0
Hydrogen	H		10.0
Nitrogen	N		3.0
Calcium	Ca		1.5
Phosphorus	P		1.0
Lesser Elements (Total 0.8%)			
Sulfur	S		0.25
Potassium	K		0.20
Sodium	Na		0.15
Chlorine	Cl		0.15
Magnesium	Mg		0.05
Iron	Fe		0.006
Trace Elements (Total 0.7%)			
Chromium	Cr	Molybdenum	Mo
Cobalt	Co	Selenium	Se
Copper	Cu	Silicon	Si
Fluorine	F	Tin	Sn
Iodine	I	Vanadium	V
Manganese	Mn	Zinc	Zn

elements. Despite their minute quantities, trace elements play vital roles in physiology. Iodine, for example, is an essential component of thyroid hormone. Other elements without natural physiological roles can contaminate the body and severely disrupt its functions, as in heavy metal poisoning with lead or mercury.

The elements are represented by one- or two-letter symbols, usually based on their English names: C for carbon, Mg for magnesium, Cl for chlorine, and so forth. A few symbols are based on Latin names, such as K for potassium *(kalium),* Na for sodium *(natrium),* and Fe for iron *(ferrum).* The periodic table of the elements (inside back cover) summarizes information on all the natural chemical elements and their relative importance in human physiology.

Several of these elements are classified as **minerals**—inorganic elements that are extracted from the soil by plants and passed up the food chain to humans and other organisms. Minerals constitute about 4% of the human body by weight. Nearly three-quarters of this is Ca and P; the rest is mainly Cl, Mg, K, Na, and S. Minerals contribute significantly to body structure. The bones and teeth consist partly of crystals of Ca, P, Mg, Fl, and sulfate ions. To name but a few additional examples of the biological roles of minerals, sulfur is a component of many proteins; phosphorus is a major component of nucleic acids, ATP, and cell membranes; and iron is a component of hemoglobin.

Atomic Structure

Each chemical element is composed of a unique type of **atom**. At the center of an atom is the **nucleus**, composed of positively charged **protons (p^+)** and uncharged **neutrons (n^0)**. The nucleus is orbited by **electrons (e^-)**, tiny particles with a single negative charge and very low mass (fig. 2.1). A person who weighs

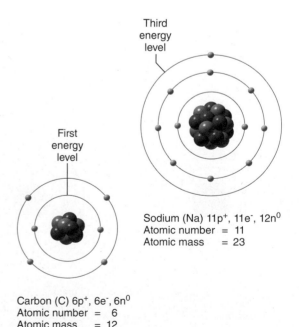

Carbon (C) $6p^+$, $6e^-$, $6n^0$
Atomic number = 6
Atomic mass   = 12

Sodium (Na) $11p^+$, $11e^-$, $12n^0$
Atomic number = 11
Atomic mass   = 23

Figure 2.1 Two Representative Elements.
Protons are represented as p^+, neutrons are represented as n^0, and electrons as e^-. **AP|R**

64 kg (140 lb) contains less than 24 g (1 oz) of electrons, yet this hardly means that we can ignore them. They determine the chemical properties of an atom, governing what molecules can exist and what chemical reactions can occur. The number of electrons equals the number of protons, so their charges cancel each other and an atom is electrically neutral.

Electrons swarm about the nucleus in concentric regions called *energy levels (electron shells)*. Those of the outermost shell, called **valence electrons**, determine the chemical bonding properties of an atom.

Isotopes and Radioactivity

Not every atom of an element is identical; all elements have two or more varieties called **isotopes**,[1] which differ from each other only in number of neutrons. Most hydrogen atoms, for example, have a nucleus composed of only one proton; this isotope is symbolized ^1H. Hydrogen has two other isotopes: *deuterium* (^2H) with one proton and one neutron, and *tritium* (^3H) with one proton and two neutrons. Over 99% of carbon atoms have a nucleus of six protons and six neutrons, and are called carbon-12 (^{12}C), but a small percentage of carbon atoms are ^{13}C, with seven neutrons, and ^{14}C, with eight.

Many isotopes are unstable and *decay* (break down) to more stable isotopes by giving off radiation. This process of decay is called **radioactivity**, and unstable isotopes are therefore called **radioisotopes**. Every element has at least one radioisotope.

Radioactivity is one form of **ionizing radiation** (ultraviolet radiation and X-rays are others). Ionizing radiation can be very damaging to tissues and is capable of causing cancer, birth defects, or in high doses, immediate death. In controlled, targeted doses, however, it is useful for such purposes as radiography, PET scans, and cancer therapy (radiotherapy).

Ions, Electrolytes, and Free Radicals

Ions are charged particles with unequal numbers of protons and electrons. Elements with one to three valence electrons tend to give them up, and those with four to seven electrons tend to gain more. If an atom of the first kind is exposed to an atom of the second, electrons may transfer from one to the other and turn both of them into ions. This process is called *ionization*. The particle that gains electrons acquires a surplus negative charge and is called an **anion** (AN-eye-on). The one that loses electrons is left with an excess positive charge (from a surplus of protons) and is called a **cation** (CAT-eye-on).

Consider, for example, what happens when sodium and chlorine meet (fig. 2.2). Sodium has a total of 11 electrons: 2 in its first (inner) shell, 8 in the second, and 1 in the third (outer) shell. If it gives up the electron in the third shell, its second shell becomes the valence shell and has a stable configuration of 8 electrons. Chlorine has 7 electrons in its valence shell (17 in all). If it can gain one more electron, it can fill the third shell with 8 electrons and become stable. Sodium and chlorine seem "made for each other"—one needs to lose an electron and the other needs to gain one. This is just what they do. When an electron transfers from sodium to chlorine, sodium is left

Figure 2.2 Ionization. A sodium atom donates an electron to a chlorine atom. This electron transfer converts the atoms to a positive sodium ion (Na⁺) and a negative chloride ion (Cl⁻). Attraction of these two oppositely charged ions to each other then constitutes an ionic bond.

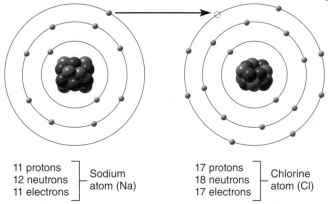

11 protons		17 protons	
12 neutrons	Sodium	18 neutrons	Chlorine
11 electrons	atom (Na)	17 electrons	atom (Cl)

① Transfer of an electron from a sodium atom to a chlorine atom

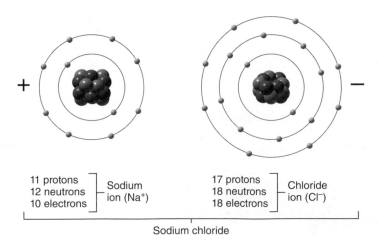

11 protons		17 protons	
12 neutrons	Sodium	18 neutrons	Chloride
10 electrons	ion (Na⁺)	18 electrons	ion (Cl⁻)

Sodium chloride

② The charged sodium ion (Na⁺) and chloride ion (Cl⁻) that result

---

[1] *iso* = same; *top* = place (same position in the periodic table)

with 11 protons but only 10 electrons; thus, it has a positive charge, and we symbolize the sodium ion Na^+. Chlorine is changed to the chloride ion with a surplus negative charge, symbolized Cl^-. Ions are not always single atoms that have become charged; some are groups of atoms—phosphate (PO_4^{3-}) and bicarbonate (HCO_3^-) ions, for example.

Ions with opposite charges are strongly attracted to each other. They often unite in *ionic bonds,* described shortly, and they tend to follow each other through the body. Thus, when Na^+ is excreted in the urine, Cl^- tends to follow it. The attraction of cations and anions to each other is important in maintaining the excitability of muscle and nerve cells, as we shall see in later chapters.

Electrolytes are salts that ionize in water and form solutions capable of conducting electricity (table 2.2). Electrolytes are important for their chemical reactivity (as when calcium phosphate becomes incorporated into bone), osmotic effects (influence on water content and distribution in the body), and electrical effects (which are essential to nerve and muscle function). Electrolyte balance is one of the most important considerations in patient care. Electrolyte imbalances, discussed in chapter 16, have effects ranging from muscle cramps and brittle bones to coma and cardiac arrest.

Free radicals are chemical particles with an odd number of electrons. For example, oxygen normally exists as a stable molecule composed of two oxygen atoms, O_2—but if an additional electron is added, it becomes a free radical called the *superoxide anion,* symbolized $O_2^-\bullet$. The dot on the right symbolizes the odd electron.

Free radicals are produced by some normal metabolic reactions of the body (such as the ATP-producing oxidation reactions in mitochondria), and by external agents such as ionizing radiation and some chemicals (for example, the solvent carbon tetrachloride and food preservatives called nitrites). Free radicals are unstable and short-lived. They combine quickly with molecules such as fats, proteins, and DNA, converting them into free radicals and triggering chain reactions that destroy still more molecules. The damage caused by free radicals includes myocardial infarction (the death of heart tissue), spinal cord death in spinal injuries, and some forms of cancer. One theory of aging is that it results in part from lifelong cellular damage by free radicals.

Because free radicals are so common and destructive, we have multiple mechanisms for counteracting them. **Antioxidants** are chemicals that neutralize free radicals, including selenium, vitamin E (α-tocopherol), vitamin C (ascorbic acid), and carotenoids (such as β-carotene). Dietary deficiencies of antioxidants are associated with increased incidence of heart attacks, sterility, muscular dystrophy, and other disorders.

Table 2.2 **Major Electrolytes and the Ions Released by Their Dissociation** AP|R

Electrolyte		Cations	Anions
Calcium chloride (CaCl)	→	Ca^{2+}	$2\ Cl^-$
Disodium phosphate (Na_2HPO_4)	→	$2\ Na^+$	HPO_4^{2-}
Magnesium chloride ($MgCl_2$)	→	Mg^{2+}	$2\ Cl^-$
Potassium chloride (KCl)	→	K^+	Cl^-
Sodium bicarbonate ($NaHCO_3$)	→	Na^+	HCO_3^-
Sodium chloride (NaCl)	→	Na^+	Cl^-

	Molecular formulae	Structural formulae	Condensed structural formulae
Ethanol	C_2H_6O	H H H—C—C—OH H H	CH_3CH_2OH
Ethyl ether	C_2H_6O	H H H—C—O—C—H H H	CH_3OCH_3

Figure 2.3 Structural Isomers, Ethanol and Ethyl Ether. The molecular formulae are identical, but the structures and chemical properties are different.

Molecules and Chemical Bonds

Molecules are chemical particles composed of two or more atoms united by a covalent chemical bond (the sharing of electrons). The atoms may be identical, as in nitrogen (N_2), or different, as in glucose ($C_6H_{12}O_6$). Molecules composed of two or more *different* elements are called **compounds**. For example, oxygen (O_2) and carbon dioxide (CO_2) are both molecules because both consist of at least two atoms, but only CO_2 is a compound.

Molecules are represented by *molecular formulae* that list each element only once and show how many atoms of each are present. Molecules with identical formulae but different arrangements of their atoms are called **isomers**[2] of each other. For example, both ethanol (grain alcohol) and ethyl ether have the molecular formula C_2H_6O, but they are certainly not interchangeable! To show the difference between them, we use *structural formulae* that show the location of each atom, or *condensed structural formulae* that show all the elements on one line but in groups reflecting the unique structural formula of each isomer (fig. 2.3).

A molecule is held together, and molecules are attracted to each other, by forces called **chemical bonds**. The three bonds of greatest physiological interest are *ionic bonds, covalent bonds,* and *hydrogen bonds*.

An **ionic bond** is the attraction of a cation to an anion. Sodium and chloride ions (Na^+ and Cl^-), for example, are attracted to each other and form the compound sodium chloride (NaCl), common table salt. Ionic bonds are weak and easily dissociate (break up) in the presence of something more attractive, such as water. The ionic bonds of NaCl break down easily as salt dissolves in water, because both Na^+ and Cl^- are more attracted to water molecules than they are to each other.

> **Apply What You Know**
>
> Do you think ionic bonds are common in the human body? Why or why not?

Covalent bonds form by the sharing of electrons. For example, two hydrogen atoms share valence electrons to form a hydrogen molecule, H_2 (fig. 2.4a).

---

[2] *iso* = same; *mers* = parts

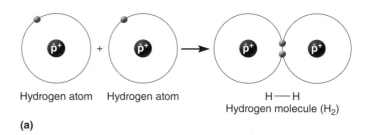

Hydrogen atom   Hydrogen atom   H——H
Hydrogen molecule (H_2)

(a)

Figure 2.4 Covalent Bonding. (a) Two hydrogen atoms share a single pair of electrons to form a hydrogen molecule. (b) A carbon dioxide molecule, in which a carbon atom shares two pairs of electrons with each oxygen atom, forming double covalent bonds.

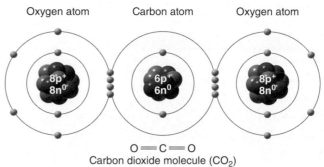

Oxygen atom   Carbon atom   Oxygen atom

O═C═O
Carbon dioxide molecule (CO_2)

(b)

The two electrons, one donated by each atom, swarm around both nuclei in a dumbbell-shaped cloud. A *single covalent bond,* the sharing of a single pair of electrons, is symbolized by a single line between atomic symbols, for example H—H. A *double covalent bond,* the sharing of two pairs of electrons, is symbolized by two lines—for example, in carbon dioxide, O=C=O, where each oxygen shares two electrons with the central carbon (fig. 2.4b).

When shared electrons spend approximately equal time around each nucleus, they form a *nonpolar covalent bond* (fig. 2.5a), the strongest of all chemical bonds. Carbon atoms bond to each other with nonpolar covalent bonds. If shared electrons spend significantly more time orbiting one nucleus than they do the other, they lend their negative charge to the region where they spend the most time, and they form a *polar covalent bond* (fig. 2.5b). For example, in water, H—O—H, the electrons are more attracted to the oxygen nucleus and orbit it more than they do the hydrogen. This makes the oxygen slightly negative (symbolized δ–) and the hydrogen slightly positive (δ+).

A **hydrogen bond** is a weak attraction between a slightly positive hydrogen atom ($H^{\delta+}$) in one molecule and a slightly negative oxygen ($O^{\delta-}$) or nitrogen ($N^{\delta-}$) in another. Water molecules are weakly attracted to each other by hydrogen bonds (fig. 2.6). Hydrogen bonds also form between different regions of the same molecule, especially in very large molecules such as proteins and DNA that can bend back on themselves. Hydrogen bonding causes such molecules to fold or coil into

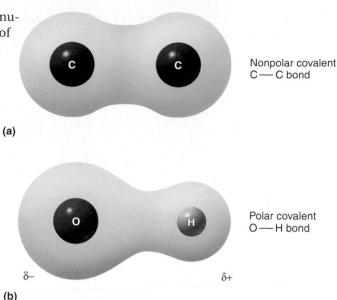

(a)

(b)

Figure 2.5 Nonpolar and Polar Covalent Bonds. (a) A nonpolar covalent bond between two carbon atoms, formed by electrons that spend an equal amount of time around each nucleus, as represented by the symmetric blue cloud. (b) A polar covalent bond, in which electrons orbit one nucleus significantly more than the other, as represented by the asymmetric cloud. This results in a slight negative charge (δ–) in the region where the electrons spend most of their time, and a slight positive charge (δ+) at the other pole.

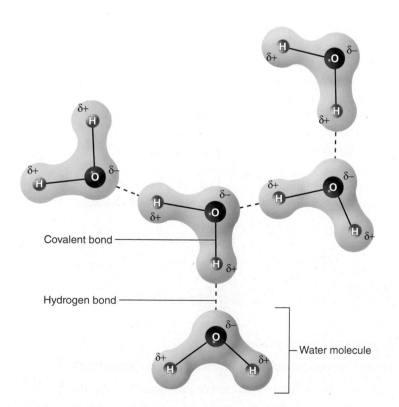

Figure 2.6 Hydrogen Bonding of Water. The polar covalent bonds of water molecules enable each oxygen to form a hydrogen bond with a hydrogen of a neighboring molecule. Thus, the water molecules are weakly attracted to each other.

• *Why would a liquid without hydrogen bonding (such as liquid nitrogen) have a lower boiling point than water?*

precise three-dimensional shapes. Hydrogen bonds are represented by dotted or broken lines between atoms: —C= O···H—N—.

Table 2.3 summarizes the types of chemical bonds.

Table 2.3 Types of Chemical Bonds AP|R

Bond Type	Definition and Characteristics
Ionic bond	Relatively weak attraction between an anion and a cation. Easily disrupted in water, as when salts dissolve.
Covalent bond	Sharing of one or more pairs of electrons between nuclei.
Single covalent	Sharing of one electron pair.
Double covalent	Sharing of two electron pairs. Often occurs between carbon atoms, between carbon and oxygen, and between carbon and nitrogen.
Nonpolar covalent	Covalent bond in which electrons are equally attracted to both nuclei. May be single or double. Strongest type of chemical bond.
Polar covalent	Covalent bond in which electrons are more attracted to one nucleus than to the other, resulting in slightly positive and negative regions in one molecule. May be single or double.
Hydrogen bond	Weak attraction between polarized molecules or between polarized regions of the same molecule. Important in the three-dimensional folding and coiling of large molecules. Weakest of all bonds; easily disrupted by temperature and pH changes.

Before You Go On

Answer these questions from memory. Reread the preceding section if there are too many you don't know.

1. *Consider iron (Fe), hydrogen gas (H_2), and ammonia (NH_3). Which of these is/are atoms? Which of them is/are molecules? Which of them is/are compounds? Explain.*

2. *Where do free radicals come from? What harm do they do? What protections do we have against free radicals?*

3. *How does an ionic bond differ from a covalent bond?*

4. *What is a hydrogen bond? Why do hydrogen bonds depend on the existence of polar covalent bonds? What role do hydrogen bonds play in the three-dimensional structure of large molecules?*

2.2 Water, Mixtures, and pH

Expected Learning Outcomes

When you have completed this section, you should be able to:

a. describe the biologically important properties of water;

b. define *mixture* and distinguish between three types of mixtures; and

c. define *acid* and *base* and interpret the pH scale.

Water

We all know that we need water to live, but why? Our bodies are 50% to 75% water, depending on age, sex, body fat, and other factors—but again, why? Why not some other liquid, or less water? Its structure, simple as it is, has profound biological effects. It is a polar molecule, with a slight negative charge ($\delta-$) on the oxygen and a slight positive charge ($\delta+$) on each hydrogen. Because of this polarity, water molecules are attracted to each other by hydrogen bonds. This gives water a set of properties that account for its ability to support life:

- Water is nature's most versatile **solvent**. More substances (**solutes**) dissolve in water than in any other liquid. Thus, it is the ideal medium for the chemical reactions that must occur in the body, and it is ideally suited for transporting substances from place to place, as in the bloodstream. Substances that dissolve in water, such as sugars and salts, are said to be **hydrophilic**[3] (HY-dro-FILL-ic); the relatively few that do not, such as fats, are **hydrophobic**[4] (HY-dro-FOE-bic).

- Water has the quality of **adhesion**, the tendency to cling to surfaces such as tissue membranes. This makes it a good lubricant in such places as the joints and the pericardial sac around the heart; it reduces friction as surfaces rub against each other during joint movement and the heartbeat.

- Water also exhibits **cohesion**, an attraction of its molecules to each other. Its hydrogen bonds make water molecules cling together like little magnets. This is why water forms puddles and hangs from a faucet in drops, and it accounts for the next property of water.

- The cohesion of water gives it relatively great **thermal stability**, or resistance to temperature changes. To increase in temperature, the molecules of a substance must move around more rapidly. The cohesion of water molecules inhibits their movement, so water can absorb a given amount of heat without changing temperature (molecular motion) as much. Water thus helps to stabilize the body temperature and serves as a very effective coolant. When water evaporates, it carries a large amount of heat with it—an effect that is very apparent when you are sweaty and stand in front of a fan. Indeed, the base unit of heat is defined in reference to water: one **calorie**[5] (**cal**) is the amount of heat that raises the temperature of 1 g of water 1°C. One thousand calories is a **kilocalorie (kcal)**, also known as 1 dietary Calorie (with a capital C).

- The **chemical reactivity** of water is its ability to participate in chemical reactions. Not only does water ionize many other chemicals such as acids and salts, but water itself ionizes into hydrogen and hydroxide ions, H^+ and OH^-. These ions can be incorporated into other molecules, or released from them, in the course of chemical reactions such as *hydrolysis* and *dehydration synthesis,* described later in this chapter.

Apply What You Know

Liquid nitrogen, N_2, is commonly used for frozen storage of blood, sperm, bacterial and tissue cultures, and other biological specimens. It boils at −196°C, in contrast to distilled water, which boils at +100°C. Predict what type of covalent bond—polar or nonpolar—liquid nitrogen has. State the reason for your prediction and then explain how that relates to the boiling point of liquid nitrogen being almost 300° lower than that of water.

---

[3] *hydro* = water; *philic* = loving, attracted to

[4] *hydro* = water; *phobic* = fearing, avoiding

[5] *calor* = heat

Mixtures

Body fluids are complex mixtures of water and other chemicals. A **mixture** consists of substances that are physically blended but not chemically combined; each substance retains its own chemical properties. For example, sucrose (table sugar) is a compound. If you stir it into your coffee, it becomes part of a mixture of sugar, water, and other dissolved matter, but it still retains the molecular structure and properties of sucrose; it is not chemically altered.

The aqueous (water-based) mixtures in the body can be classified as *solutions, colloids,* and *suspensions* (fig. 2.7). A **solution** is transparent, because its dissolved particles are very small (under 1 nanometer, nm) and do not scatter light significantly. The solute and solvent cannot be visually distinguished from each other, even with a microscope. The salts in your blood are in solution.

A **colloid**[6] is a mixture of larger particles (1–100 nm) in water. Colloids are usually cloudy, because particles this large scatter light. In both solutions and colloids, however, the particles are small enough to remain permanently mixed when the mixture stands. Many colloids can change from liquid to gel states—gelatin desserts, agar culture media, and the fluids within and between our cells, for example. The most abundant colloidal material in the body is protein, such as the albumin in the blood plasma.

A **suspension** is a mixture with particles larger than 100 nm. Such particles not only render a suspension cloudy or opaque, but also settle out of the mixture by gravity if the mixture is not constantly agitated. Blood cells form a suspension in the blood plasma and settle to the bottom of a tube when blood is allowed to stand without mixing. An **emulsion** is a suspension of one liquid in another, such as the fat in breast milk, and medications such as Kaopectate and milk of magnesia.

A single mixture can fit into more than one of these categories. Blood is a perfect example—it is a solution of sodium chlo-

---

[6] *collo* = glue; *oid* = like, resembling

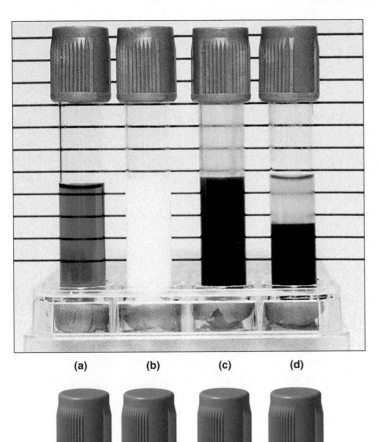

(a)   (b)   (c)   (d)

Solution   Colloid   Suspension

Figure 2.7 A Solution, a Colloid, and a Suspension. Top row: Photographs of a representative solution, colloid, and suspension. Bottom row: Symbolic representation of the particle sizes in each mixture. (a) In this copper sulfate solution, the solute particles are so small that they remain permanently mixed and the solution is transparent. (b) In colloids such as this milk, the particles are still small enough to remain permanently mixed, but they are large enough to scatter light, so we cannot see through the colloid. (c) In suspensions such as this freshly mixed blood, the particles (blood cells) also scatter light and make the mixture opaque. Furthermore (d), they are too large to remain permanently mixed, so they settle out of the mixture, as in this blood sample that stood overnight.

• *How would you classify the yellowish plasma at the top of tube d—as a solution, colloid, suspension, or more than one of these?*

ride, a colloid of protein, and a suspension of cells. Milk is a solution of calcium, a colloid of protein, and an emulsion of fat.

It is often important to specify the **concentration** of a solution—how much solute is present in a given volume of solvent. There are several ways of doing so, suitable for different purposes. Appendix C defines and explains the most commonly used measures of concentration: weight per volume, percentage, molarity, osmolarity, milliequivalents per liter, and pH. Much of our physiology is aimed at maintaining the proper concentration of dissolved matter in the body fluids because abnormalities of concentration can have profound effects on body function, with consequences as severe as seizures, cardiac arrest, and death.

Acids, Bases, and pH

Most people have some sense of what acids and bases are. Advertisements are full of references to excess stomach acid and pH-balanced shampoo. We know that drain cleaner (a strong base) and battery acid can cause serious chemical burns. But what exactly do *acid* and *base* mean, and how can they be quantified?

An **acid** is any *proton donor,* a molecule that releases a proton (H^+) in water. A **base** is a proton acceptor. Since hydroxide ions (OH^-) accept H^+, many bases are substances that release hydroxide ions—sodium hydroxide (NaOH), for example. A base does not have to be a hydroxide donor, however. Ammonia (NH_3) is also a base. It does not release hydroxide ions, but it readily accepts hydrogen ions to become the ammonium ion (NH_4^+).

Acidity is expressed in terms of **pH**, a measure derived from the concentration of H^+ (see appendix C). The pH scale (fig. 2.8) was invented in 1909 by Danish biochemist and brewer Sören Sörensen to measure the acidity of beer. The scale extends from 0.0 to 14.0. A solution with a pH of 7.0 is **neutral**; solutions with pH below 7 are **acidic**; and solutions with pH above 7 are **basic**

Figure 2.8 The pH scale. The pH is shown within the colored bar.

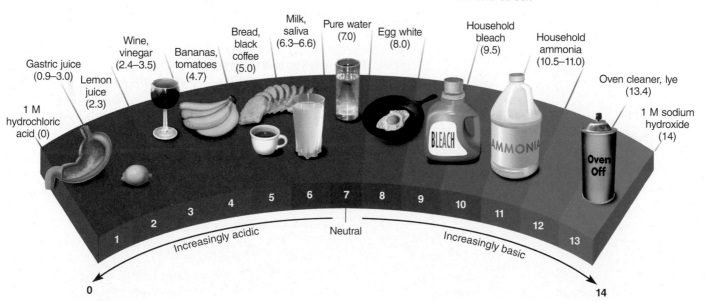

Clinical Application 2.1

pH AND DRUG ACTION

The pH of our body fluids has a direct bearing on how we react to drugs. Depending on pH, drugs such as aspirin, phenobarbital, and penicillin can exist in charged (ionized) or uncharged forms. Whether a drug is charged or not can determine whether it will pass through cell membranes. When aspirin is in the acidic environment of the stomach, for example, it is uncharged and passes easily through the stomach lining into the bloodstream. Here it encounters a basic pH, whereupon it ionizes. In this state, it is unable to pass back through the membrane, so it accumulates in the blood. This effect, called *ion trapping* or *pH partitioning*, can be controlled to help clear poisons from the body. The pH of the urine, for example, can be clinically manipulated so that poisons become trapped there and are more rapidly excreted from the body.

(**alkaline**). The lower the pH value, the more hydrogen ions a solution has and the more acidic it is. Each step of the scale represents a 10-fold change in H^+ concentration. For example, a solution of pH 5.0 has 10 times the H^+ concentration as one with a pH of 6.0; a pH of 4.0 represents 100 times the H^+ concentration at 6.0.

Slight disturbances of pH can seriously disrupt physiological functions and alter drug actions (see Clinical Application 2.1), so it is important that the body carefully controls its pH. Blood, for example, normally has a pH ranging from 7.35 to 7.45. Deviations from this range cause tremors, fainting, paralysis, or even death. Chemical solutions that resist changes in pH are called **buffers**. Buffers and pH regulation are considered in detail in chapter 16.

Before You Go On

Answer these questions from memory. Reread the preceding section if there are too many you don't know.

5. What is meant by the terms hydrophilic *and* hydrophobic?
6. What physical property of water makes it resistant to temperature changes? What specific type of chemical bond is responsible for that physical property?
7. What are the differences between a solution, a colloid, and a suspension? Give an example of each.
8. Define acid *and* base, *and state what ranges of pH values are considered acidic and basic.*

2.3 Organic Compounds

Expected Learning Outcomes

When you have completed this section, you should be able to:

a. discuss the relevance of polymers and macromolecules to biology, and explain how they are formed and broken by dehydration synthesis and hydrolysis;

b. describe the structural properties that distinguish carbohydrates, lipids, proteins, and nucleic acids from each other;

c. describe or define the subclasses of each of those categories of biomolecules;

d. discuss the roles that each of these categories of molecules play in the body;

e. explain how enzymes function; and

f. describe the structure, production, and function of ATP.

Organic chemistry is the study of compounds of carbon, and *biochemistry* is the field that relates these compounds to the processes of life. Carbon is a uniquely suitable element to serve as the basis for a broad variety of biological molecules. Carbon atoms can bond with each other and thus form long chains, branched molecules, and rings. They also bind readily to hydrogen, oxygen, nitrogen, sulfur, and other elements.

Biochemists classify the larger organic molecules of life into four main categories: *carbohydrates, lipids, proteins,* and *nucleic acids.* Some of these molecules attain enormous sizes and are called *macromolecules.* Most macromolecules are **polymers**[7]—molecules made of a repetitive series of identical or similar subunits called **monomers.** Starch, for example, is a polymer of about 3,000 glucose monomers.

Living cells use a process called **dehydration synthesis (condensation)** to join monomers together to form polymers (fig. 2.9a). They remove a hydrogen (—H) from one monomer and a hydroxyl (—OH) group from another, producing water as a by-product. The two monomers become joined by a covalent bond. This is repeated for each monomer added to the chain.

The opposite of dehydration synthesis is **hydrolysis**[8] (fig. 2.9b). In hydrolysis, a water molecule ionizes into OH^- and H^+. The cell breaks a covalent bond linking one monomer to another and adds the H^+ to one monomer and the OH^- to the other one. All chemical digestion consists of hydrolysis reactions.

Figure 2.9 Synthesis and Hydrolysis Reactions. (a) In dehydration synthesis, a hydrogen atom is removed from one monomer and a hydroxyl group is removed from another. These combine to form water as a by-product. The monomers become joined by a covalent bond. Repetition of this process can produce long molecular chains such as polymers. (b) In hydrolysis, a covalent bond between two monomers is broken. Water donates a hydrogen atom to one monomer and a hydroxyl group to the other. Polymers such as starch are digested in this manner.

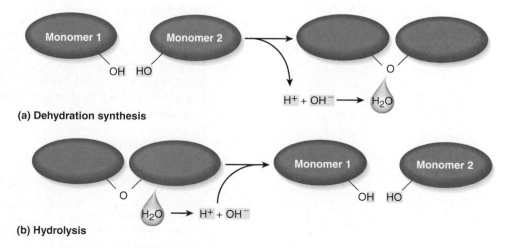

(a) Dehydration synthesis

(b) Hydrolysis

---

[7] *poly* = many; *mer* = part

[8] *hydro* = water; *lysis* = splitting apart

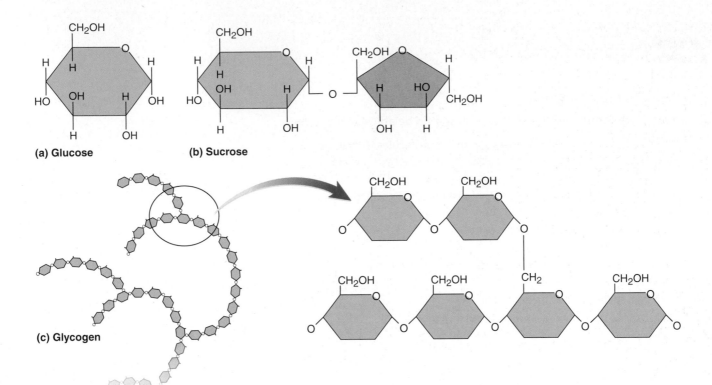

Figure 2.10 Carbohydrate Structures.
(a) Glucose (blood sugar), our most abundant monosaccharide. (b) Sucrose (table or cane sugar), a disaccharide, composed of glucose and fructose (the tan sugar). (c) Glycogen, a polysaccharide, composed of long branched chains of glucose. Each angle in the rings represents a carbon atom except the one where oxygen is shown. This is a conventional way of representing carbon in the structural formulae of organic compounds.

Carbohydrates

Carbohydrates[9] include sugars, starch, and glycogen. They all consist of carbon and a 2:1 ratio of hydrogen to oxygen; their very name means "carbon plus water." For example, glucose ("blood sugar") has the formula $C_6H_{12}O_6$—essentially six carbons and six waters. Carbohydrates are hydrophilic—they readily absorb or mix with water. The names of individual carbohydrates are often built on the word root *sacchar-* or the suffix *-ose,* both of which mean "sugar" or "sweet."

Monosaccharides

The simplest carbohydrates are called **monosaccharides**[10] (MON-oh-SAC-uh-rides), or simple sugars. The three of primary importance are **glucose, fructose,** and **galactose,** all with the molecular formula $C_6H_{12}O_6$; they are isomers of each other. We obtain these sugars mainly by digesting more complex carbohydrates, but many processed foods contain large amounts of high-fructose corn syrup. Glucose (fig. 2.10a) is commonly called *blood sugar.*

Disaccharides

Disaccharides are sugars composed of two monosaccharides bonded to each other. The most important ones are **sucrose** (made of glucose + fructose), **lactose** (glucose + galactose), and **maltose** (glucose + glucose). Sucrose (fig. 2.10b) is common table sugar, lactose is milk sugar, and maltose is mainly a product of starch digestion, but is present in a few foods such as malt beverages and germinating wheat.

Polysaccharides

Polysaccharides (POL-ee-SAC-uh-rides) are polymers of glucose. The most important of these are glycogen, starch, and cellulose. **Glycogen**[11] is an energy-

[9] *carbo* = carbon; *hydr* = water
[10] *mono* = one; *sacchar* = sugar
[11] *glyco* = sugar; *gen* = producing

storage polysaccharide made by the liver, muscles, uterus, and a few other organs. It is a long branched polymer (fig. 2.10c). The liver stores glycogen after a meal, when the blood glucose level is high, and then breaks it down between meals to maintain blood glucose levels when there is no food intake. Muscle stores glycogen for its own energy needs, and the uterus uses it in pregnancy to nourish the embryo. **Starch** is the corresponding energy-storage polysaccharide of plants. It is the only significant digestible polysaccharide in the human diet. **Cellulose** is a structural polysaccharide of plants, and is the most abundant organic compound on earth. It is the principal component of wood, cotton, and paper. Cellulose is indigestible to us, so we derive no energy or nutrition from it; yet it is important as dietary fiber (*bulk* or *roughage*). It swells with water in the digestive tract and promotes movement of other materials through the intestines.

Carbohydrates are, above all, a source of energy that can be quickly mobilized, but they have other functions as well (table 2.4). They are often covalently bonded to lipids and proteins to form components of mucus, tissue gel, cell membranes, the tough matrix of cartilage, the lubricating fluid in joints, and the gelatinous filler in the eyeball.

Lipids

Lipids include triglycerides (fats and oils), phospholipids, steroids (such as cholesterol, estrogen, and testosterone), and others (table 2.5). They are hydrophobic (it is well known that oil and water do not mix, for example). Like carbohydrates, lipids are usually composed only of carbon, hydrogen, and oxygen, but they have a higher ratio of hydrogen to oxygen. A fat called *tristearin* (tri-STEE-uh-rin), for

Table 2.4 Carbohydrate Functions

Type	Function
Monosaccharides	
Glucose	Blood sugar—energy source for most cells
Galactose	Converted to glucose and metabolized
Fructose	Fruit sugar—converted to glucose and metabolized
Disaccharides	
Sucrose	Cane sugar—digested to glucose and fructose
Lactose	Milk sugar—digested to glucose and galactose; important in infant nutrition
Maltose	Malt sugar—product of starch digestion, further digested to glucose
Polysaccharides	
Cellulose	Structural polysaccharide of plants; dietary fiber
Starch	Energy storage in plant cells; dietary source of energy for humans
Glycogen	Energy storage in animal cells (liver, muscle, uterus, vagina)

Table 2.5 Some Lipid Functions

Type	Function
Triglycerides	Energy storage; thermal insulation; filling space; binding organs together; cushioning organs
Fatty acids	Precursor of triglycerides; source of energy
Phospholipids	Major component of cell membranes
Cholesterol	Component of cell membranes; precursor of other steroids
Steroid hormones	Chemical messengers between cells
Bile acids	Steroids that aid in fat digestion and nutrient absorption

example, has the molecular formula $C_{57}H_{110}O_6$—more than 18 hydrogens per oxygen. Lipids are less oxidized than carbohydrates, and thus have more calories per gram. Beyond these criteria, it is difficult to generalize about lipids; they are much more variable in structure than the other macromolecules we are considering.

Triglycerides and Fatty Acids

Triglycerides (try-GLISS-ur-ides) are commonly called **fats** if they are solid at room temperature and **oils** if they are liquid. A triglyceride is built by dehydration synthesis reactions linking a three-carbon molecule called **glycerol**

Figure 2.11 Fatty Acid and Triglyceride (Fat) Structure. (a) Glycerol and three fatty acids, the building blocks of a triglyceride. (b) The triglyceride formed by dehydration synthesis, with a glycerol "backbone" and three fatty acid "tails." This triglyceride is shown with two saturated fatty acids and one unsaturated fatty acid.

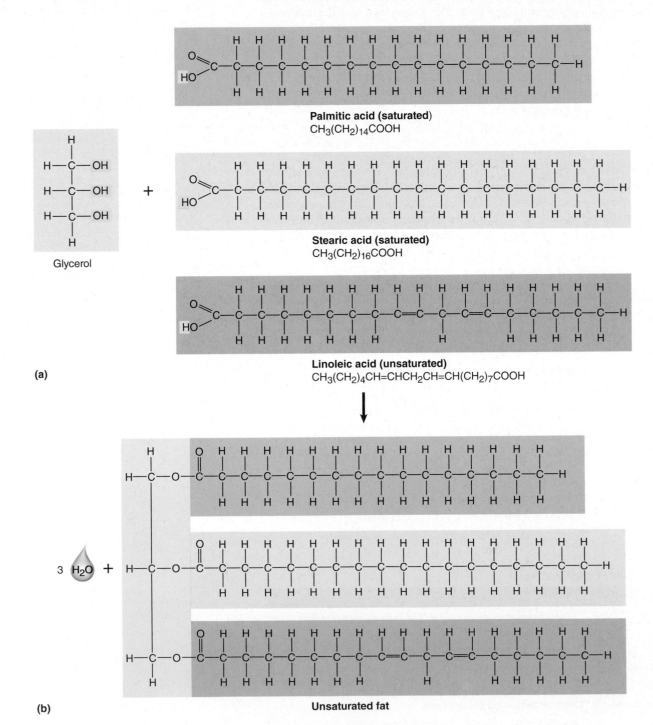

Palmitic acid (saturated)
$CH_3(CH_2)_{14}COOH$

Glycerol

Stearic acid (saturated)
$CH_3(CH_2)_{16}COOH$

Linoleic acid (unsaturated)
$CH_3(CH_2)_4CH=CHCH_2CH=CH(CH_2)_7COOH$

(a)

$3\ H_2O$ +

Unsaturated fat

(b)

Clinical Application 2.2

TRANS FATS AND CARDIOVASCULAR HEALTH

There has been a great deal of public interest lately in *trans* fats and cardiovascular health, with some U.S. states even regulating or banning the use of *trans* fats in restaurants. What, then, are *trans* fats and why have they gotten such a bad reputation?

A *trans* fat is a triglyceride containing one or more *trans*-fatty acids. Ordinarily, when a fatty acid has a C=C double bond, the bond creates a "kink" in the chain (fig. 2.12a). These molecules are called *cis*-fatty acids. The kinks prevent triglyceride molecules from packing too closely together, making them liquid (oily) at room temperature. In the early 1900s, the food industry developed a way to modify these into *trans*-fatty acids. These have straight fatty acid chains (fig. 2.12b) that allow the triglycerides to pack closely together, making them solid at room temperature. Crisco vegetable shortening was the first commercial product composed of *trans* fats. *Trans* fats are much easier to work with in making pie crusts, biscuits, and other baked goods. They are now used abundantly in snack foods, baked goods, fast foods such as french fries, and many other foods. *Trans* fats constitute about 30% of the fat in shortening, but only 4% of the animal fat in butter.

The disadvantage of *trans* fats is that they resist digestion, remain in circulation longer, and have more tendency to deposit in the arteries than saturated fats and *cis* fats do. Therefore, they raise the risk of coronary heart disease (CHD). From 1980 to 1994, medical scientists tracked a cohort of 80,082 female nurses (the Nurses' Health Study II) and, among other things, correlated their incidence of CHD with their self-reported diets. They concluded that for every 2% increase in calories from *trans* fats in place of carbohydrate calories, the women had a 93% higher risk of CHD.

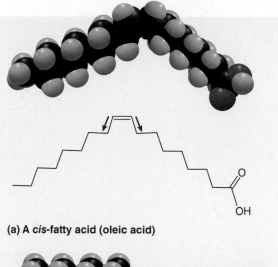

(a) A *cis*-fatty acid (oleic acid)

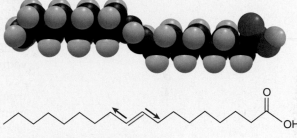

(b) A *trans*-fatty acid (elaidic acid)

Figure 2.12 *Cis*- and *Trans*-Fatty Acids. Each example is unsaturated at the C=C double bond in red. (a) In a *cis*-fatty acid, the bonds angle in the same direction. (b) In a *trans*-fatty acid, the bonds on opposite sides of the C=C bond angle in opposite directions (arrows). The straight chains of *trans*-fatty acids allow fat molecules to pack together more tightly, thus to remain solid (greasy) at room temperature.

with three long zigzag carbon chains called **fatty acids** (fig. 2.11). Fatty acids and the fats made from them are classified as *saturated* or *unsaturated*. A **saturated fatty acid** has as much hydrogen as it can carry, such as the palmitic and stearic acids in the figure. Carbon atoms are limited to forming four covalent bonds, and in saturated fatty acids, no more hydrogen could be added without exceeding that four-bond limit; that is, the molecule is "saturated" with hydrogen. In **unsaturated fatty acids**, such as the linoleic acid shown, some carbon atoms are joined by double covalent (C=C) bonds. Each of these could potentially share one pair of electrons with another hydrogen atom instead of the adjacent carbon, so hydrogen could be added to this molecule. **Polyunsaturated fatty acids** are those with many C=C bonds. A *saturated fat* is a triglyceride made with all saturated fatty acids; an *unsaturated* or *polyunsaturated fat* has at least one unsaturated fatty acid. Saturated fats and *trans* fats are major contributors to cardiovascular disease (see Clinical Application 2.2).

The primary function of triglycerides is energy storage, but when concentrated in *adipose (fat) tissue,* they also provide thermal insulation and act as a shock-absorbing cushion for vital organs.

Phospholipids

Phospholipids are similar to triglycerides except that in place of one fatty acid, they have a phosphate group that, in turn, is linked to another organic group (fig. 2.13). The glycerol and phosphate-containing group constitute a hydrophilic "head," and the two fatty acids constitute hydrophobic zigzag "tails" of the molecule. Together, the head and two tails of a phospholipid give it a shape like a clothespin. A phospholipid is said to be **amphipathic**[12] (AM-fih-PATH-ic)—part of it is attracted to water and part of it is repelled. The most important function of phospholipids is to serve as the structural foundation of cell membranes. (See chapter 3, where the significance of this amphipathic property will be more fully explained.)

Steroids

Steroids are lipids with 17 of their carbon atoms arranged in four rings (fig. 2.14). **Cholesterol** is the "parent" steroid from which the others are synthesized. Others include cortisol, progesterone, estrogen, testosterone, and bile acids (table 2.5). These differ from each other in the location of C=C bonds within the rings and in the functional groups attached to the rings.

The average adult contains over 200 g (half a pound) of cholesterol, of which about 85% is synthesized in one's own body and 15% comes from animal products in the diet. (Plants contain no significant amounts of cholesterol.) Cholesterol has a bad reputation as a factor in cardiovascular disease, and it is true that hereditary and dietary factors can elevate blood

---

[12] *amphi* = both; *pathic* = feeling

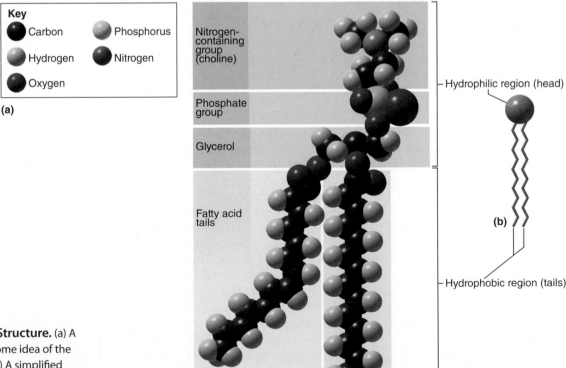

Figure 2.13 Phospholipid Structure. (a) A space-filling model that gives some idea of the actual shape of the molecule. (b) A simplified representation of a phospholipid used in diagrams of cell membranes.

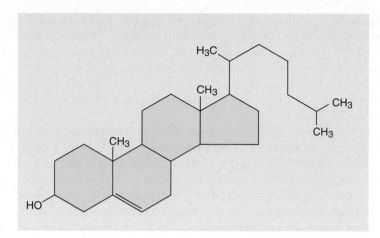

Figure 2.14 Cholesterol. All steroids have this basic four-ringed structure, with variations in the side groups and locations of double bonds within the rings.

cholesterol to dangerously high levels. Nevertheless, cholesterol is a natural product of the body and plays vital roles as the raw material for producing other steroids necessary to life.

Proteins

The word *protein* is derived from the Greek word *proteios,* "of first importance." Proteins are the most versatile molecules in the body, and many discussions in this book will draw on your understanding of protein structure and behavior. Table 2.6 summarizes several of the very diverse functions of proteins.

Table 2.6 Protein Functions

Function	Examples
Structure	*Keratin* is a tough protein that strengthens nails, hair, and the skin surface. *Collagen* is a structural foundation of the bones, cartilage, teeth, and deeper layer (dermis) of the skin.
Communication	Some hormones and other cell-to-cell signals are proteins, as are the receptors to which the signal molecules bind in the receiving cell.
Membrane transport	Some proteins form channels in cell membranes that govern what can pass through the membrane and when. Others act as carriers that briefly bind to solute particles and move them to the other side of the membrane. Among other roles, membrane transport proteins turn nerve and muscle activity on and off.
Catalysis	Most metabolic pathways of the body are controlled by enzymes, which are globular proteins that function as biological catalysts.
Recognition and protection	Cell surface proteins function in immune recognition; antibodies attack and neutralize organisms that invade the body; clotting proteins seal broken blood vessels and reduce blood loss.
Movement	Proteins are the basis for all movement, ranging from transport of molecules and organelles from place to place within a cell to the contractions of muscles and beating of cilia. Some proteins are called *molecular motors* for their ability to repeatedly change shape and produce movement.
Cell adhesion	Proteins bind cells to each other and to the extracellular material, prevent tissues from falling apart, enable sperm to bind to an egg and fertilize it, and enable immune cells to bind to enemy cells such as microbes and cancer cells.

A **protein** is a polymer of amino acids. An **amino acid** is a small organic molecule with a central carbon, an **amino group** (—NH$_2$), and a **carboxyl group** (—COOH) (fig. 2.15a). There are 20 amino acids employed in protein structure, differing from each other in a third functional group, generically symbolized R, attached to the central carbon. In the simplest amino acid, glycine, R is merely a hydrogen atom, while in the largest amino acids it includes rings of carbon atoms.

With the aid of the right enzymes, amino acids can be linked together by dehydration synthesis reactions to form **peptides** (a dipeptide for a chain of two amino acids, tripeptide for a chain of three, and so forth) (fig. 2.15b). The link formed between the C=O of one amino acid and —NH of the next is called a **peptide bond**. Chains larger than 10 or 15 amino acids are called **polypeptides**, and most authorities regard proteins as any polypeptides of 50 or more amino acids. Some hormones are small peptides, such as the one that causes labor contractions, *oxytocin*. Some others, such as growth hormone, are large polypeptides.

You could think of a protein as loosely analogous to a bead necklace with 20 different-colored beads corresponding to the 20 amino acids from which proteins may be made. Proteins differ from one another in which amino acids occur and in what order—their *amino acid sequence*. This sequence is specified by the *genetic code* in one's DNA.

Some proteins can function only when a non–amino acid component is bound to them. Hemoglobin, for example, requires an iron-containing ring called a *heme* group (fig. 2.16b), without which it cannot transport oxygen; this is one reason you must have iron in your diet. Many proteins must have zinc, magnesium, or other ions bound to them in order to function.

Protein Shapes

As a protein is assembled, interactions between the amino acids and water, and between one amino acid and another, cause the chain to coil and fold

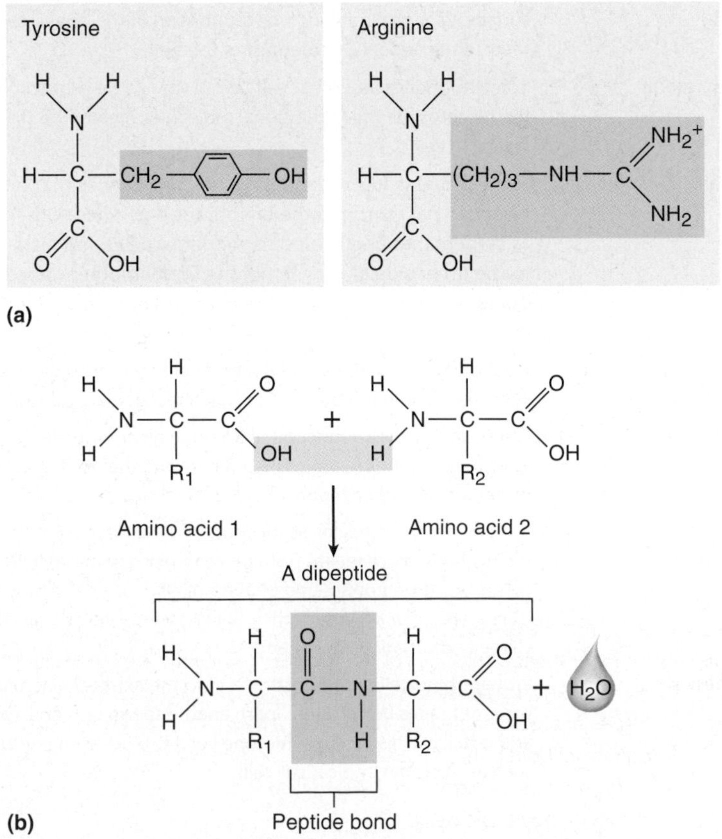

Figure 2.15 Amino Acids and Peptide Bondings. (a) Two representative amino acids. Note that they differ only in the R group, shaded in pink. (b) The joining of two amino acids by a peptide bond, forming a dipeptide. Side groups R$_1$ and R$_2$ could be the groups indicated in pink in part (a), among other possibilities.

into complex three-dimensional structures. One of the most common shapes is a springlike *alpha helix* (fig. 2.16a), which is stabilized by hydrogen bonds between the amino acids of the chain. One protein often has multiple regions of alpha helix connected by nonhelical segments. Other proteins are folded into rounded *globular* shapes like a wadded ball of yarn. This is especially characteristic of proteins such as antibodies, enzymes, hormones, and hemoglobin (fig. 2.16b)—compact proteins that must be able to dissolve and move around freely in the body fluids. On a larger scale, some take on ropelike *fibrous* forms (fig. 2.16c)—especially proteins that provide structural strength to tissues, such as the *collagen* that makes up tendons, ligaments, the dermis of the skin, and part of the bone tissue. The shapes of globular and fibrous proteins are often stabilized by *disulfide bridges,* links between two proteins, or between two parts of the same protein, through two sulfur atoms. These bridges are symbolized —S—S— in some of the art in this book.

The three-dimensional shape *(conformation)* of a protein is critical to its function. Two of the most important properties of proteins are their ability to bind and release other chemicals and their ability to change shape. Conformation changes can be triggered by such influences as voltage changes on a cell membrane or the binding of a hormone to a protein. Subtle, reversible changes in conformation are important to processes such as enzyme function, muscle contraction, nerve signaling, and the opening and closing of pores in cell membranes. **Denaturation** is a more drastic conformation change, with loss of function, in response to conditions such as extreme heat or pH. It is seen, for example, when you cook an egg and the albumen (egg white) turns from clear to opaque. Denaturation is sometimes reversible, but often it permanently destroys protein function.

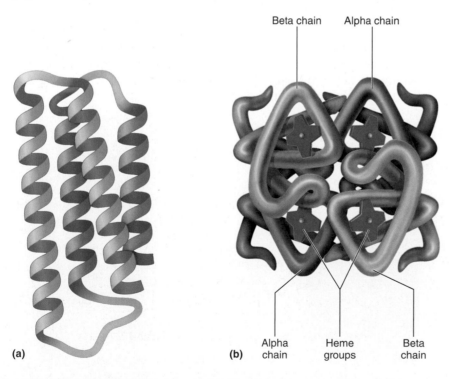

Beta chain   Alpha chain

Alpha chain   Heme groups   Beta chain

(a)   (b)   (c)

Figure 2.16 Protein Shapes. (a) A protein composed of four regions of alpha helix connected by short nonhelical regions. (b) Hemoglobin, the red blood pigment. This is a compact globular protein composed of two identical alpha chains and two identical beta chains, all four of which are alpha helices. Each chain possesses a nonprotein oxygen-carrying *heme* group. (c) Collagen, a fibrous protein, forming much of the substance of bones, tendons, ligaments, and dermis of the skin. This is composed of three alpha helices intertwined into a long ropy triple helix.

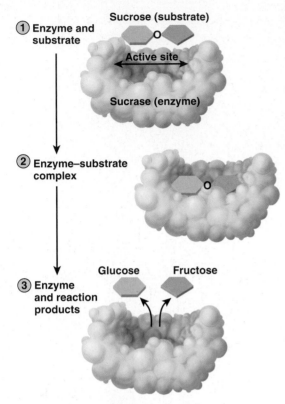

Figure 2.17 The Three Steps of an Enzymatic Reaction. One or more substrate molecules bind to the enzyme's active sites. The substrates and enzyme form a temporary enzyme–substrate complex and the substrates react chemically with each other. The enzyme releases the reaction products and is available to catalyze the same reaction again. **AP|R**

Enzymes

Enzymes are proteins that function as biological catalysts; a catalyst is any chemical that enables a chemical reaction to go faster or at lower temperature, but that is not consumed by that reaction. As catalysts, enzymes permit biochemical reactions to occur rapidly at normal body temperatures.

Enzymes were initially given somewhat arbitrary names, still with us, such as *pepsin* and *trypsin*. The modern system of naming enzymes, however, is more uniform and informative. It identifies the substance the enzyme acts upon, called its **substrate**; it sometimes refers to the enzyme's action; and it adds the suffix *-ase*. Thus, *amylase* digests starch (*amyl-* = starch) and *carbonic anhydrase* removes water (*anhydr-*) from carbonic acid.

An enzyme has surface pockets called **active sites** that bind specific substrates. A substrate fits its active site somewhat like a key fitting a lock. Just as no other key will fit a given lock, no other substrates will fit a given enzyme. An enzyme that acts on glucose, for example, will not act on the similar sugar fructose. Once a substrate is bound, the enzyme chemically changes it to one or more reaction products, releases the products, and is then available to act on another molecule of the same substrate (fig. 2.17). Since enzymes are not consumed by the reactions they catalyze, one enzyme molecule can convert millions of substrate molecules—and at astonishing speeds. A single molecule of carbonic anhydrase, for example, breaks carbonic acid (H_2CO_3) down to H_2O and CO_2 at a rate of 36 million molecules per minute.

Factors that change the shape of an enzyme—notably temperature and pH—can alter or destroy the ability of the enzyme to bind its substrate. They disrupt the weak forces that hold the enzyme in its proper conformation, essentially changing the shape of the "lock" so that the "key" no longer fits. Enzymes vary in optimum pH according to where in the body they normally function. For example, a digestive enzyme that works in the mouth functions best at a neutral pH of 7, whereas one that works in the more acidic environment of the stomach functions best at a pH around 2. Our internal body temperature is nearly the same everywhere, however, and all human enzymes function best at a temperature of about 37° to 40°C.

Figure 2.18 Nucleotides and Nitrogenous Bases. (a) The structure of adenosine, a nucleotide found in DNA and RNA. (b) The five nitrogenous bases found in DNA and RNA nucleotides. DNA has the bases A, T, C, and G. In RNA, U takes the place of T.

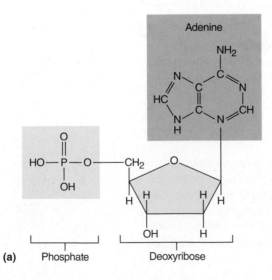

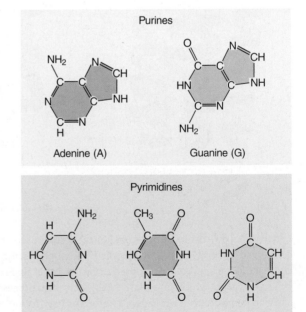

Nucleic Acids (DNA and RNA)

Nucleic acids are organic polymers that serve as the cell's genetic machinery—**deoxyribonucleic acid (DNA),** which resides in the nucleus (hence the term *nucleic*) and contains the individual's genes, and a few forms of **ribonucleic acid (RNA),** which carry out the "orders" given by the DNA. Both of these are composed of monomers called nucleotides.

A **nucleotide** consists of a monosaccharide, a phosphate group, and a **nitrogenous base,** which is a single or double ring of carbon and nitrogen atoms (fig. 2.18a). There are five main nitrogenous bases. Three of them have a single carbon–nitrogen ring and are named **cytosine (C), thymine (T),** and **uracil (U).** The other two have double rings and are named **adenine (A)** and **guanine (G)** (fig. 2.18b).

DNA consists of two nucleotide chains twined together in a *double helix,* shaped like a twisted ladder (fig. 2.19). It is the body's largest polymer, a chain of typically 100 million to 1 billion nucleotides. The "backbone" of each helix, like the vertical rails of a ladder, consists of phosphate groups alternating with a sugar called *deoxyribose.* The rungs of the ladder are pairs of nitrogenous bases. Wherever one chain has an A, the one across from it has a T; where one has C, the other has G. Thus, a single-ringed base is always paired with a double-ringed one, giving the DNA molecule a uniform diameter throughout. A–T and C–G are called the **base pairs** of DNA, and the fact that A pairs only with T and C pairs only with G is called the **law of complementary base pairing.** It is a foundation for the genetic function of DNA and for its exact replication when cells divide.

DNA directs the synthesis of proteins by means of its smaller cousins, the ribonucleic acids. There are three types of RNA involved in carrying out the cell's genetic instructions and making protein. All of them differ from DNA in that they are much smaller; they are composed of just one nucleotide chain rather than two chains twined around each other; and they employ the sugar **ribose** instead of deoxyribose and contain uracil in place of thymine. DNA normally stays in the nucleus, giving orders from there, whereas RNA functions mainly in the cytoplasm, carrying out the instructions encoded in the DNA. The actions of DNA and RNA in protein synthesis and heredity will be described in the next chapter.

Figure 2.19 DNA Structure. (a) A space-filling molecular model of DNA giving some impression of its actual geometry. (b) The "twisted ladder" structure of DNA. The two sugar–phosphate backbones (blue) twine around each other to form the double helix (lower half of figure), and the nitrogenous bases (colored bars) face each other in the center. (c) A small segment of DNA showing the composition of the backbone and complementary pairing of the nitrogenous bases. **AP|R**

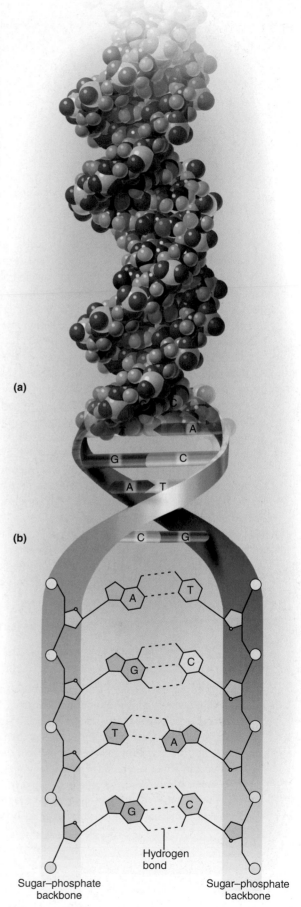

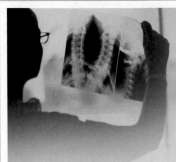

PERSPECTIVES ON HEALTH

Anabolic Steroids

Anabolic (or anabolic–androgenic) steroids are compounds related to testosterone that are used to promote tissue growth and repair. They were first developed in the 1950s to treat anemia, breast cancer, osteoporosis, and some muscle diseases. By the 1960s, athletes were using them to stimulate muscle growth, accelerate the repair of tissues damaged in training or competition, and accentuate the aggressiveness needed in such sports as football and boxing.

The doses used by athletes, however, are 10 to 1,000 times higher than those prescribed for medical purposes. At such concentrations, anabolic steroids raise cholesterol levels and thus promote coronary artery disease, heart and kidney failure, and stroke. Deteriorating blood circulation often causes gangrene and has necessitated many limb amputations among steroid abusers. Liver disease, immune suppression, behavioral disorders, atrophy of the testes, impotence, and overgrowth of the male breasts (gynecomastia) are other common effects. In women, anabolic steroids tend to cause atrophy of the breasts and uterus; menstrual irregularity; and overgrowth of facial hair, body hair, and the clitoris.

Because of these severe effects, anabolic steroid use for performance enhancement is condemned by the American Medical Association, American College of Sports Medicine, International Olympic Committee, National Football League, and National Collegiate Athletic Association.

Collagen Diseases

The fibrous protein collagen is the most abundant protein in the body and gives strength to tendons, ligaments, bone, and cartilage. Defects in collagen synthesis therefore cause a wide range of disorders collectively called collagen diseases or collagenopathies. Humans have about 28 types of collagen subject to more than 1,000 mutations that can affect its chemical structure (amino acid sequence) or the quantity synthesized. Such defects can have severe effects on the many tissues and organs that depend on collagen for strength— tendons, ligaments, bones, blood vessels, eyes, and viscera.

Ehlers–Danlos[13] syndrome (EDS), for example, has six main types that vary in effect from mild to life-threatening. Its most common signs are loose ligaments and hyperextensible joints that look like one is "double-jointed" to an extreme degree, and highly elastic skin that can stretch several centimeters but is very fragile. The skin bruises easily and even minor trauma can cause wide, gaping wounds. It is very difficult for surgeons to successfully suture the skin of such patients, since the sutures tear out so easily. Depending on the mutational type, people with EDS can also show spinal deformation, frequent hip dislocations, fragile eyeballs, extreme muscle weakness, and fragile blood vessels that may spontaneously rupture. Virtuoso violinist Niccolò Paganini (1782–1840) wrote compositions that many other accomplished violinists find nearly impossible to play because they require such extraordinarily wide finger placements. It is thought that Paganini was able to play such pieces because of EDS, exceptionally long fingers, and the associated hypermobility of his finger and wrist joints.

Another collagen disease is osteogenesis imperfecta (OI), or brittle bone disease. The colloquial name comes from the fact that collagen-deficient bones fracture very easily. Infants with the most severe forms of OI often die before, during, or shortly after birth, and are born with severely deformed limbs and multiple fractures from birth trauma. Various forms of OI can be characterized by fragile and misplaced teeth, spinal deformity, short stature, barrel chest, and hearing loss. The white capsule of the eyeball often is so thin that internal blood vessels show through and the "whites of the eyes" have a blue-gray to purple tint. Many persons with OI, however, can enjoy a normal life span and even high professional accomplishment. Among them, British actor, playwright, and novelist Nabil Shaban (1935–) has made a fascinating documentary, "The Strangest Viking" (see YouTube), about speculation that the dwarfish but fearsome Viking known as Ivar the Boneless had OI.

Both EDS and OI appear in infancy and early childhood and are sufficiently rare that they are often unrecognized or misdiagnosed. There is an added tragedy stemming from the frequent bruising of EDS and bone fractures of OI: Parents have been falsely accused of child abuse and taken before child protective agencies and family court before being exonerated by a correct medical diagnosis.

---

[13] Edvard Ehlers (1863–1937), Danish dermatologist; Henri-Alexandre Danlos (1844–1912), French dermatologist

Adenosine Triphosphate (ATP)

Adenosine triphosphate (ATP) is a nucleotide built from the nitrogenous base *adenine*, the sugar ribose, and three phosphate groups (fig. 2.20). It functions as the body's most important energy-transfer molecule. It briefly stores energy gained from reactions such as glucose oxidation, and releases it within seconds for physiological work such as synthesis reactions, muscle contraction, and pumping ions through cell membranes.

The second and third phosphate groups of ATP are attached to the rest of the molecule by high-energy covalent bonds traditionally indicated by a wavy line (P~O) in the structural formula. Since phosphate groups are negatively charged, they repel each other. It requires a high-energy bond to overcome that repulsive force and hold them together—especially to add the third phosphate group to a chain that already has two negatively charged phosphates. Most energy transfers to and from ATP involve adding or removing that third phosphate.

When ATP breaks down, it loses one phosphate group as *inorganic phosphate* (P_i). The remainder of the molecule becomes adenosine diphosphate (ADP). Most of the energy given off is lost as heat, a major contributor to body temperature. A small but life-sustaining fraction of the energy, however, is used to do many kinds of work in the body.

$$\text{ATP} + \text{H}_2\text{O} \longrightarrow \text{ADP} + \text{P}_i + \text{Energy} \begin{array}{l} \nearrow \text{Heat} \\ \searrow \text{Work} \end{array}$$

ATP is a short-lived molecule, usually consumed within 60 seconds of its formation. The entire amount of ATP in the body would support life for less than 1 minute if it were not continually replenished. At a moderate rate of physical activity, a full day's supply of ATP would weigh twice as much as you do. Even if you never got out of bed, you would need about 45 kg (99 lb) of ATP to stay alive for a day. The reason cyanide is so lethal is that it halts ATP synthesis.

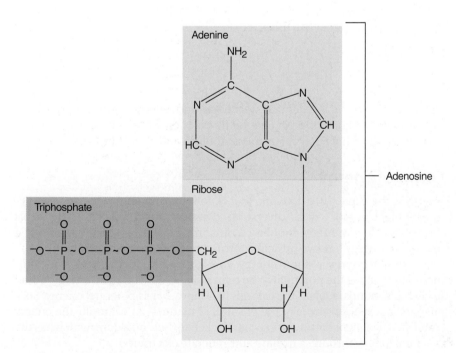

Figure 2.20 Adenosine Triphosphate (ATP). ATP transfers energy from one chemical reaction to another. The transferable energy is carried in the last two P~O covalent bonds, indicated by wavy lines.

Before You Go On

Answer these questions from memory. Reread the preceding section if there are too many you don't know.

9. Which reaction—dehydration synthesis or hydrolysis—converts a polymer to its monomers? Which one converts monomers to a polymer? Explain your answer.

10. What is the chemical name of blood sugar? What carbohydrate is polymerized to form starch and glycogen?

11. What is the main chemical similarity between carbohydrates and lipids? What are the main differences between them?

12. Explain the statement, All proteins are polypeptides but not all polypeptides are proteins.

13. Use the lock and key analogy to explain why excessively acidic body fluids (acidosis) could destroy enzyme function.

14. How does ATP change structure in the process of releasing energy?

2.4 Energy and Chemical Reactions

Expected Learning Outcomes

When you have completed this section, you should be able to:

a. understand how chemical reactions are symbolized by chemical equations;

b. list and define the fundamental types of chemical reactions;

c. define *metabolism* and its two subdivisions; and

d. define *oxidation* and *reduction* and relate these to changes in the energy content of a molecule.

The essence of life is chemical reactions and energy transfers—in short, **metabolism**. We conclude this chapter by examining the basic types of chemical reactions and how they relate to metabolism and energy transfer.

Forms of Energy

Energy is the capacity to do work, that is, to move matter or change its structure. Biological work is as diverse as digesting starch, moving chemicals through a cell membrane, building a protein, circulating blood, or running a marathon. Energy has two principle forms—potential and kinetic.

Potential energy is stored energy, energy that is not doing work at the moment but that has the potential to be released and do work. The energy stored in chemical bonds, such as the calories in a candy bar, is potential energy. So is energy that exists by virtue of the position of matter—for example, the energy stored in a flashlight battery or in water backed up behind a dam, which has the potential to flow through a turbine and generate electricity.

Kinetic energy is the energy of motion or change—for example, the electrical current produced when you turn on a flashlight, the energy of sodium ions flowing through the membrane of a nerve cell, the energy of muscle contraction, and the heat generated by exercising muscles.

Energy is routinely converted from one form to another, often through chemical reactions. In muscle contraction, for example, the chemical (potential) energy of glucose is released by chemical reactions that produce muscular motion and heat. Some energy is lost as heat in every chemical reaction, and then is no longer available to do useful work.

Classes of Chemical Reactions

A **chemical reaction** is a process in which a covalent or ionic bond is formed or broken. The course of a chemical reaction is symbolized by a **chemical equation,** which typically shows the **reactants** on the left, the **products** on the right, and an arrow pointing from the reactants to the products. For example, consider the reaction that turns wine sour when it is exposed to air for several days:

$$CH_3CH_2OH \; + \; O_2 \; \longrightarrow \; CH_3COOH \; + \; H_2O$$

$$\text{Ethanol} \qquad \text{Oxygen} \qquad \text{Acetic acid} \qquad \text{Water}$$

Ethanol and oxygen are the reactants, and acetic acid and water are the products. Not all reactions are shown with the arrow pointing from left to right. In complex biochemical equations, reaction chains are often written vertically or even in circles.

Chemical reactions can be classified as *decomposition, synthesis,* or *exchange reactions.* In **decomposition reactions**, a large molecule breaks down into two or more smaller ones (fig. 2.21a); symbolically, AB → A + B. When you eat a potato, for example, digestive enzymes decompose its starch into thousands of glucose molecules, and most cells further decompose glucose to water and carbon dioxide. One molecule of starch ultimately yields about 36,000 molecules of H_2O and CO_2.

Synthesis reactions are the opposite—two or more small molecules combine to form a larger one; symbolically, A + B → AB (fig. 2.21b). When the body synthesizes proteins, for example, it typically combines several hundred amino acids into one protein molecule.

Reversible reactions can go in either direction under different circumstances and are represented with paired or double-headed arrows. For example, carbon dioxide combines with water to produce carbonic acid, which in turn decomposes into bicarbonate ions and hydrogen ions:

$$CO_2 \; + \; H_2O \; \rightleftharpoons \; H_2CO_3 \; \rightleftharpoons \; HCO_3^- \; + \; H^+$$

| Carbon dioxide | Water | Carbonic acid | Bicarbonate ion | Hydrogen ion |

This reaction appears in this book more often than any other, especially where we discuss respiratory, urinary, and digestive physiology.

In the absence of upsetting influences, reversible reactions exist in a state of **equilibrium**, in which the ratio of products to reactants is stable. The carbonic acid reaction, for example, normally maintains a 20:1 ratio of bicarbonate ions to carbonic acid molecules. This equilibrium can be upset, however, by a surplus of hydrogen ions, which drive the reaction to the left, or adding carbon dioxide and driving it to the right.

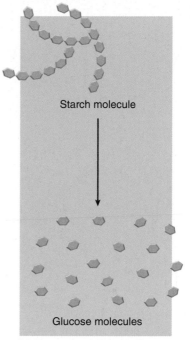

Starch molecule

Glucose molecules

(a) Decomposition reaction

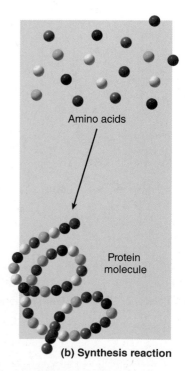

Amino acids

Protein molecule

(b) Synthesis reaction

Figure 2.21 Decomposition and Synthesis Reactions. (a) In a decomposition reaction, large molecules are broken down into simpler ones. (b) In a synthesis reaction, smaller molecules are joined to form larger ones.

Metabolism, Oxidation, and Reduction

Metabolism has two divisions—catabolism and anabolism. **Catabolism**[14] (ca-TAB-oh-lizm) consists of energy-releasing decomposition reactions. Such reactions break covalent bonds, produce smaller molecules from larger ones, and release energy that can be used for other physiological work. Energy-releasing reactions are called *exergonic*[15] reactions. If you break down energy-storage molecules (fats and carbohydrates) to run a race, you get hot. The heat signifies that exergonic reactions are occurring. The earlier-described decomposition of ATP to ADP + P_i is an example of catabolism.

Anabolism[16] (ah-NAB-oh-lizm) consists of energy-storing synthesis reactions, such as the production of protein or fat. Reactions that require an energy input, such as these, are called *endergonic*[17] reactions. Anabolism is driven by the energy that catabolism releases, so endergonic and exergonic processes, anabolism and catabolism, are inseparably linked.

Oxidation is any chemical reaction in which a molecule gives up electrons and releases energy. A molecule is *oxidized* by this process, and whatever molecule takes the electrons from it is an **oxidizing agent (electron acceptor)**. The term *oxidation* stems from the fact that the electron acceptor is often oxygen. The rusting of iron, for example, is a slow oxidation process in which oxygen is added to iron to form iron oxide (Fe_2O_3). Many oxidation reactions, however, do not involve oxygen at all. For example, when yeast ferments glucose to alcohol, no oxygen is required; indeed, the alcohol *contains less oxygen* than the sugar originally did, but it is *more oxidized* and contains less energy than the sugar:

$$C_6H_{12}O_6 \longrightarrow 2\,CH_3CH_2OH + 2\,CO_2$$

$$\text{Glucose} \qquad \text{Ethanol} \qquad \text{Carbon dioxide}$$

Reduction is a chemical reaction in which a molecule gains electrons and energy. When a molecule accepts electrons, it is said to be *reduced;* a molecule that donates electrons to another is therefore called a **reducing agent (electron donor)**. The oxidation of one molecule is always accompanied by the reduction of another.

Table 2.7 summarizes the variety of energy-transfer chemical reactions in the body.

---

[14] *cata* = down, to break down

[15] *ex, exo* = out; *erg* = work

[16] *ana* = up, to build up

[17] *end* = in ; *erg* = work

Table 2.7 Energy-Transfer Reactions in the Human Body

Exergonic reactions	Reactions in which there is a net release of energy. The products have less total energy than the reactants did.
Oxidation	An exergonic reaction in which electrons are removed from a reactant. The product is then said to be oxidized.
Decomposition	A reaction such as digestion or ATP hydrolysis, in which larger molecules are broken down into smaller ones.
Catabolism	The sum of all decomposition reactions in the body.
Endergonic reactions	Reactions in which there is a net input of energy. The products have more total energy than the reactants did.
Reduction	An endergonic reaction in which electrons are donated to a reactant. The product is then said to be reduced.
Synthesis	A reaction such as protein and glycogen synthesis, in which two or more smaller molecules are combined into a larger one.
Anabolism	The sum of all synthesis reactions in the body.

CAREER SPOTLIGHT

Medical Technologist

A medical technologist, or "med tech" (MT), is one who performs diagnostic tests on body fluids and tissue specimens to provide patient information to physicians and other clinicians. MTs work in hospital laboratories, doctors' offices, independent diagnostic laboratories, and the biotechnology industry. They work with such specimens as blood, urine, cerebrospinal fluid, pericardial fluid, and bone marrow. Using microscopic examination, analytical instruments, and other methods, they perform blood counts, liver function tests, urinalysis, comprehensive metabolic panels, blood

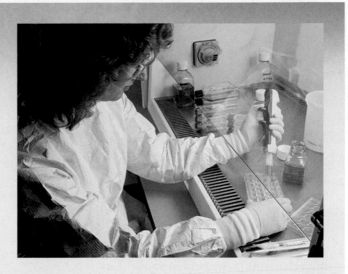

and tissue typing, coagulation panels, lipid profiles, bacteriological cultures, and other assays. Most MTs are generalists, but in large institutions they may specialize in such areas as immunology, hematology, or cytogenetics.

Medical technology education is a 4-year undergraduate program usually leading to a bachelor's degree in medical technology. Typically, it is a "3+1" program with 3 years of classroom instruction and 1 year of clinical rotation; the latter consists of hands-on training in various laboratory departments. Such programs are taught in many universities and hospitals. States vary on their certification requirements, but there are multiple national certifying organizations (the American Society for Clinical Pathology and others), and several states also require a state license. Graduate education is available for career advancement.

For links to further information on medical technology programs and careers, see appendix B.

Apply What You Know

When sodium chloride is formed, an electron transfers from a sodium atom to a chlorine atom (see fig. 2.2). Therefore, which element—sodium or chlorine—is oxidized? Which one is reduced? Explain your answers.

Before You Go On

Answer these questions from memory. Reread the preceding section if there are too many you don't know.

15. Other than the examples already given in this chapter, give an example of potential energy and how it might be transformed to kinetic energy.

16. Distinguish between a decomposition reaction and a synthesis reaction, and state one role played by each type in the human body.

17. Define *metabolism, catabolism, and* anabolism.

18. What does oxidation *mean? What does* reduction *mean? Which of them is endergonic and which is exergonic?*

19. Suppose you eat a hamburger and your body digests the starch of the bun to glucose and digests the protein and fat to amino acids and fatty acids. Identify which of the following terms pertain to these digestive processes—catabolism, reduction, oxidation, anabolism, endergonic, exergonic.

Study Guide

Assess Your Learning Outcomes

To test your knowledge, discuss the following topics with a study partner or in writing, ideally from memory.

2.1 Atoms, Ions, and Molecules (p. 32)

1. The definition of a chemical *element*, and the most common elements in the human body
2. The significance of minerals in human structure and function
3. How protons, neutrons, and electrons are arranged in the structure of an atom, and the special significance of valence electrons
4. How chemical isotopes differ from each other, the relationship of radioactivity to certain isotopes, and the medical uses and risks of ionizing radiation
5. The differences between an atom and an ion and between a cation and an anion
6. The definition of *electrolytes*, their physiological roles, and the most significant electrolytes in human physiology
7. The definition of *free radicals*, their relevance to human physiology, and antioxidants as protection from free radical damage
8. Distinctions between atoms, molecules, and compounds
9. How isomers resemble and differ from each other
10. Three types of chemical bonds and the differences between them
11. Why polar covalent bonds are necessary for the existence of hydrogen bonds, and the role of hydrogen bonds in molecular structure

2.2 Water, Mixtures, and pH (p. 38)

1. The properties of water that make it so important to life, and how these arise from the polarity of its covalent bonds
2. The concepts of hydrophilic and hydrophobic substances
3. How *calorie* is defined in relation to water
4. The difference between a mixture and a compound
5. How solutions, colloids, and suspensions differ from each other, and examples of each type in the human body
6. The meaning of *acid* and *base*
7. The pH scale and how acidity and alkalinity are related to their respective numerical ranges on the scale

2.3 Organic Compounds (p. 43)

1. The definitions of *organic chemistry* and *biochemistry*
2. The meaning of *monomers* and *polymers,* how they relate to each other, and the roles of dehydration synthesis and hydrolysis in converting one to the other
3. The defining characteristics and functions of carbohydrates
4. The most relevant monosaccharides, disaccharides, and polysaccharides, and their dietary sources and functions
5. The defining characteristics and functions of lipids
6. The distinction between saturated and unsaturated fatty acids
7. The structure of triglycerides and phospholipids; the similarity and difference between the two; and their respective functions
8. The structure and functions of steroids
9. How amino acids differ from each other and the structural properties that they all have in common
10. The relationships between amino acids, peptides, polypeptides, and proteins
11. The various three-dimensional shapes that proteins assume after their amino acid sequence is assembled
12. How the shape of a protein is related to its function and to denaturation
13. What enzymes are and what universal role they play in cellular function
14. The lock-and-key analogy for enzyme–substrate interaction
15. The reason enzymes are so sensitive to variations in temperature and pH
16. The characteristics that nucleotides have in common and how nucleotides differ from each other
17. The double-helix structure and complementary base pairing of DNA
18. How RNA differs structurally and functionally from DNA
19. Three major components of adenosine triphosphate (ATP) and what role ATP plays in cellular metabolism

2.4 Energy and Chemical Reactions (p. 56)

1. The definition of *energy* and the distinction between potential and kinetic energy
2. How chemical reactions are represented; the meaning of *reactants* and *products*
3. Reversible reactions and chemical equilibrium
4. The difference between catabolism and anabolism, and an example of each
5. The essence of oxidation and reduction, how they relate to catabolism and anabolism, and how they affect the energy content of a molecule

Testing Your Recall

1. A substance that ___ is considered to be a chemical compound.
   a. contains at least two different elements
   b. contains at least two atoms
   c. has covalent bonds
   d. has any type of chemical bond
   e. has a stable valence shell

2. An ionic bond is formed when
   a. two anions meet.
   b. two cations meet.
   c. an anion interacts with a cation.
   d. electrons are unequally shared between nuclei.
   e. electrons transfer completely from one atom to another.

3. The ionization of a sodium atom to produce Na^+ is an example of
   a. oxidation.
   b. reduction.
   c. catabolism.
   d. anabolism.
   e. decomposition.

4. The weakest chemical bonds, easily disrupted by temperature and pH changes, are
   a. polar covalent bonds.
   b. nonpolar covalent bonds.
   c. hydrogen bonds.
   d. ionic bonds.
   e. disulfide bonds.

5. A substance capable of dissolving freely in water is
   a. hydrophilic.
   b. hydrophobic.
   c. hydrolyzed.
   d. hydrated.
   e. amphipathic.

6. A carboxyl group is symbolized
   a. —OH.
   b. —NH_2.
   c. —CH_3.
   d. —CH_2OH.
   e. —COOH.

7. The only polysaccharide synthesized in the human body is
   a. cellulose.
   b. glycogen.
   c. cholesterol.
   d. starch.
   e. triglyceride.

8. Disulfide bonds are sometimes important in stabilizing the three-dimensional shape of
   a. the DNA double helix.
   b. polysaccharides.
   c. proteins.
   d. steroids.
   e. phospholipids.

9. Which of the following functions is more characteristic of carbohydrates than of proteins?
   a. contraction
   b. energy storage
   c. catalyzing reactions
   d. immune defense
   e. intercellular communication

10. The feature that most distinguishes a lipid from a carbohydrate is that a lipid has
   a. more phosphate.
   b. more sulfur.
   c. a lower ratio of carbon to oxygen.
   d. a lower ratio of oxygen to hydrogen.
   e. a greater molecular weight.

11. When an atom gives up an electron and acquires a positive charge, it becomes a/an _____.

12. Dietary antioxidants are important because they neutralize _____.

13. Any substance that increases the rate of a reaction without being consumed by it is a/an _____. In the human body, _____ serve this function.

14. All the synthesis reactions in the body form a division of metabolism called _____.

15. A chemical reaction that produces water as a by-product is called _____.

16. The suffix _____ denotes a sugar, while the suffix _____ denotes an enzyme.

17. The amphipathic lipids of cell membranes are called _____.

18. _____ is a nucleotide highly important in transferring energy from one chemical reaction pathway to another.

19. Metabolic reactions that break down large molecules into smaller ones and release energy are called _____.

20. A substance acted upon and changed by an enzyme is called the enzyme's _____.

Answers in appendix A

True or False

Determine which five of the following statements are false, and briefly explain why.

1. The monomers of a polysaccharide are called amino acids.

2. An emulsion is a mixture of two liquids that separate from each other on standing.

3. ATP provides better long-term storage of energy than triglycerides do.

4. If a pair of shared electrons is more attracted to one nucleus than to the other, the electrons form a polar covalent bond.

5. Amino acids are joined by a unique type of bond called a peptide bond.

6. A saturated fat is defined as a fat to which no more carbon can be added.

7. Catabolism is an energy-releasing oxidation process.

8. Cellulose is an important source of dietary calories.

9. Two isomers have identical chemical properties, because their chemical behavior depends on the number and types of atoms present, not on how the atoms are arranged.

10. A solution of pH 8 has one-tenth the hydrogen ion concentration of a solution with pH 7.

Answers in appendix A

Testing Your Comprehension

1. How would the important life-sustaining properties of water change if it had nonpolar covalent bonds instead of polar covalent? Explain.

2. In one form of radioactive decay, a neutron breaks down into a proton and electron and emits a gamma ray. Is this an endergonic or exergonic reaction, or neither? Is it an anabolic or catabolic reaction, or neither? Explain both answers.

3. Some metabolic conditions such as diabetes mellitus cause disturbances in the acid–base balance of the body, which gives the body fluids an abnormally low pH. Explain how this could affect the ability of enzymes to control biochemical reactions in the body.

The Cellular Level of Organization

Dividing cancer cells from an adenocarcinoma. Adenocarcinoma is a tumor arising from glands in the mucous membrane of an organ such as the lung.

BASE CAMP

Before ascending to the next level, be sure you're properly equipped with a knowledge of these concepts from earlier chapters:

- The levels of human structure (p. 14)
- Phospholipids (p. 48)
- Protein structure and function (p. 49)
- DNA and RNA (p. 53)
- Adenosine triphosphate (ATP) (p. 55)

Anatomy & Physiology **REVEALED**®
aprevealed.com

Module 2: Cells and Chemistry

The most important revolution in the history of medicine was the realization that all bodily functions result from cellular activity. By extension, nearly every disorder of the body is now recognized as stemming from a dysfunction at the cellular level. New technology and methods of study have generated empowering changes in our understanding of the inner workings of cells. This has paved the way for a deeper understanding of the structure and function of the human body and mechanisms of disease, and thus for more informed and effective strategies of therapy. Our study of cell structure and function in this chapter will lay a foundation for understanding the rest of this book.

3.1 The General Structure of Cells

Expected Learning Outcomes
When you have completed this section, you should be able to:

a. outline and define the major structural components of a cell;
b. name the fluids that occur inside and outside a cell; and
c. state the size range of human cells and explain why cell size is limited.

Cytology,[1] the study of cellular structure and function, got its start in the seventeenth century when inventors Robert Hooke (1635–1703) and Antony van Leeuwenhoek (1632–1723) crafted microscopes adequate for seeing individual cells. Cytology made little further progress, however, until improved optics and tissue staining techniques were developed in the nineteenth century. Even then, the material between the nucleus and cell surface was thought to be little more than a gelatinous mixture of chemicals and vaguely defined particles. When the first biologically useful electron microscopes were developed in the mid-twentieth century, their vastly superior magnification and resolution showed cells to be crowded with a maze of passages, compartments, and fibers. Figure 3.1 depicts this in a very generalized form, but should not blind you to the diversity of cellular form and function. There are about 200 kinds of cells in the human body, with a variety of shapes, sizes, and functions.

Cell Components

The major components of a cell are as follows:

Plasma membrane
Cytoplasm
  Cytoskeleton
  Organelles
  Inclusions
  Cytosol

---

[1] *cyto* = cell; *logy* = study of

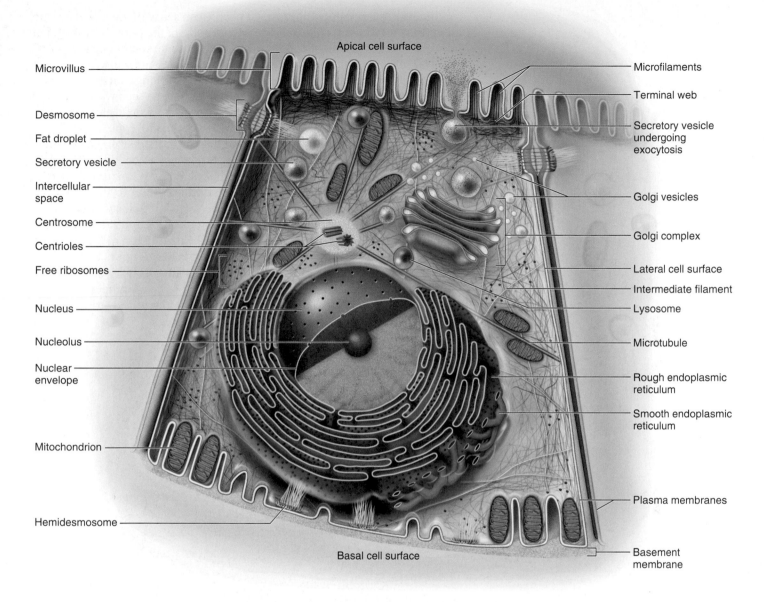

Microvillus

Desmosome

Fat droplet

Secretory vesicle

Intercellular space

Centrosome

Centrioles

Free ribosomes

Nucleus

Nucleolus

Nuclear envelope

Mitochondrion

Hemidesmosome

Apical cell surface

Basal cell surface

Microfilaments

Terminal web

Secretory vesicle undergoing exocytosis

Golgi vesicles

Golgi complex

Lateral cell surface

Intermediate filament

Lysosome

Microtubule

Rough endoplasmic reticulum

Smooth endoplasmic reticulum

Plasma membranes

Basement membrane

Figure 3.1 Structure of a Generalized Cell.
The cytoplasm is usually more crowded with organelles than is shown here. The organelles are not all drawn to the same scale. Components of the cytoskeleton are shown in greater detail in figure 3.9.
AP|R

The **plasma membrane (cell membrane)** forms a cell's surface boundary. The cell's largest organelle is typically the nucleus. The material between the plasma membrane and the nucleus is the **cytoplasm.**[2] It contains the *cytoskeleton,* a supportive framework of protein filaments and tubules; an abundance of *organelles,* diverse structures that perform various metabolic tasks for the cell; and *inclusions,* which are accumulated cell products such as lipids and pigments or internalized foreign matter such as dust and bacteria. The cytoskeleton, organelles, and inclusions are embedded in a clear gel called the **cytosol.**

The cytosol is also called **intracellular fluid (ICF).** All body fluids not contained in the cells are collectively called the **extracellular fluid (ECF).** The ECF located between the cells is also called **tissue fluid.** Extracellular fluids also include blood plasma, lymph, and cerebrospinal fluid.

Cell Sizes

The most useful unit of measurement for designating cell sizes is the **micrometer (μm)**—one-millionth (10^{-6}) of a meter, one-thousandth (10^{-3}) of

---

[2] *cyto* = cell; *plasm* = formed, molded

a millimeter. The smallest objects most people can see with the naked eye are about 100 µm, which is about one-quarter the size of the period at the end of this sentence. A few human cells fall within this range, such as egg cells and some fat cells, but most human cells are about 10 to 15 µm wide. The longest human cells are nerve cells (sometimes over a meter long) and muscle cells (up to 30 cm long), but these are usually too slender to be seen with the naked eye. With a good light microscope, one can see objects as small as 0.5 µm. Objects smaller than that generally require an electron microscope.

For a variety of reasons, cells cannot attain unlimited size. If a cell grew excessively large, it would rupture like an overfilled water balloon. Also, molecules could not diffuse from place to place fast enough to support its metabolism. The time required for diffusion is proportional to the square of distance, so if cell diameter doubled, the travel time for molecules within the cell would increase fourfold. For example, if it took 10 seconds for a molecule to diffuse from the surface to the center of a cell with a 10 µm radius, then we increased this cell to a radius of 1 mm, it would take 278 hours to reach the center—far too slow to support the cell's life activities. Having organs composed of many small cells instead of fewer large ones has another advantage: the death of one or a few cells is of less consequence to the structure and function of the whole organ.

Before You Go On

Answer these questions from memory. Reread the preceding section if there are too many you don't know.

1. Which term refers to all the cell contents between the plasma membrane and nucleus: cytosol, cytoplasm, nucleoplasm, or extracellular fluid?
2. About how big would a cell have to be for you to see it without a microscope? Are any cells actually this big? If so, what ones?
3. Explain why cells cannot grow to an indefinitely large size.

3.2 The Cell Surface

Expected Learning Outcomes

When you have completed this section, you should be able to:

a. identify the components of the plasma membrane that encloses each cell;
b. state the functions of each molecular component of the plasma membrane;
c. describe the composition, appearance, and functions of the glycocalyx that coats each cell;
d. describe the structure and functions of microvilli, cilia, and flagella;
e. name the different types of junctions that connect cells to each other, and describe their functions; and
f. explain the processes for moving material through the plasma membrane.

A great deal of physiology takes place at the cell surface—for example, the binding of signal molecules such as hormones, the stimulation of cellular activity, the attachment of cells to each other, and the transport of materials into and out of cells. This, then, is where we begin our study of cellular structure and function. Like an explorer discovering a new island, we will examine the interior of the cell only after we have investigated its boundary.

The Plasma Membrane

The plasma membrane defines the boundary of a cell and governs its interactions with other cells. It controls the passage of materials into and out of the cell and maintains differences in chemical composition between the ECF and ICF. Several organelles are enclosed in one or two similar membranes, but the term *plasma membrane* refers exclusively to the membrane at the cell surface.

Membrane Lipids

The plasma membrane is an oily, two-layered lipid film with proteins embedded in it (fig. 3.2). About 75% of the lipid molecules are **phospholipids.** In chapter 2, we saw that phospholipids are *amphipathic*—they have a hydrophilic phosphate head and two hydrophobic fatty acid tails. The phosphate heads face the water on both the inside and outside of the cell, thus forming a sandwichlike *phospholipid bilayer.* The fatty acid tails form the middle of the sandwich, as far away from the surrounding water as possible. The phospholipids are not stationary but highly fluid—drifting laterally from place to place, spinning on their axes, and flexing their tails.

About 20% of the lipid molecules are **cholesterol,** which affects membrane fluidity. If there is too much cholesterol, it inhibits the action of enzymes and other proteins in the membrane; too little, and plasma membranes

Figure 3.2 The Plasma Membrane. Review the relationship between the yellow phospholipid symbols here and phospholipid structure in figure 2.13 (p. 48). AP|R

- *Why do all the phospholipid heads face the ECF and ICF and not the middle of the membrane?*

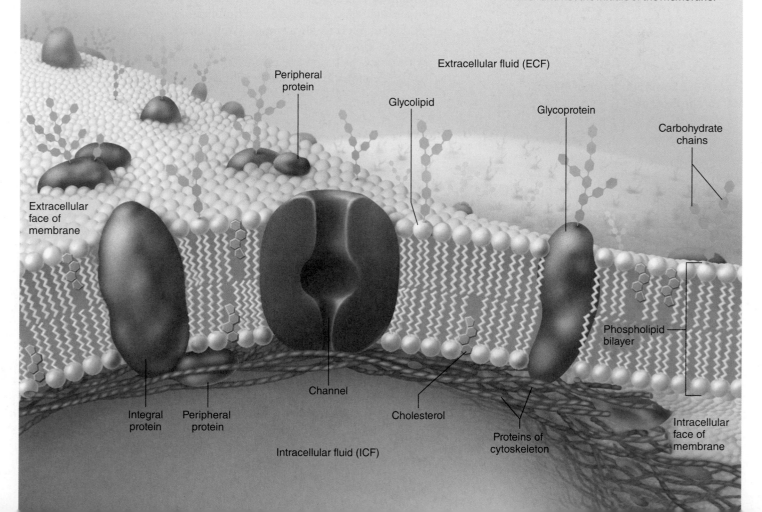

Extracellular fluid (ECF)

Peripheral protein

Glycolipid

Glycoprotein

Carbohydrate chains

Extracellular face of membrane

Phospholipid bilayer

Integral protein

Peripheral protein

Channel

Cholesterol

Proteins of cytoskeleton

Intracellular face of membrane

Intracellular fluid (ICF)

become excessively fragile. This is one of several reasons why cholesterol, in spite of its undeservedly bad reputation in health science, is indispensable to human survival.

The remaining 5% of the lipids are **glycolipids**—phospholipids with short carbohydrate chains bound to the extracellular surface. Essentially, every body cell is "sugar-coated."

Membrane Proteins

The types of proteins associated with the plasma membrane vary considerably from cell to cell, in contrast to the lipid portion, which has the same basic composition regardless of cell type. Proteins give membranes specific functional abilities and contribute greatly to the functional differences between cell types.

Proteins that penetrate into the membrane, often all the way through to the other side, are called **integral proteins.** Most of these are **glycoproteins,** which, like glycolipids, have carbohydrate chains linked to them. **Peripheral proteins** are those that do not protrude into the phospholipid but adhere to either face of the membrane, usually the intracellular face.

The functions of membrane proteins are highly diverse and are among the most interesting aspects of cell physiology. They include the following.

- **Receptors** (fig. 3.3a). Receptor proteins receive and bind chemical signals from other cells, such as hormones, neurotransmitters, and growth factors.
- **Enzymes** (fig. 3.3b). Enzymes carry out chemical reactions at the membrane surface, such as degrading signal molecules after the message is received and breaking dietary nutrients down into forms that the intestine can absorb.
- **Channel proteins** (fig. 3.3c). Channel proteins have tunnels through them that allow water and hydrophilic solutes to enter or leave a cell. Some channels are always open, while others, called **gates** (fig. 3.3d), can open or close and thus allow material to enter or leave the cell at specific times. Such gates are responsible for firing of the heart's pacemaker, muscle contraction, and the transmission of nerve signals.

Figure 3.3 Some Functions of Plasma Membrane Proteins.

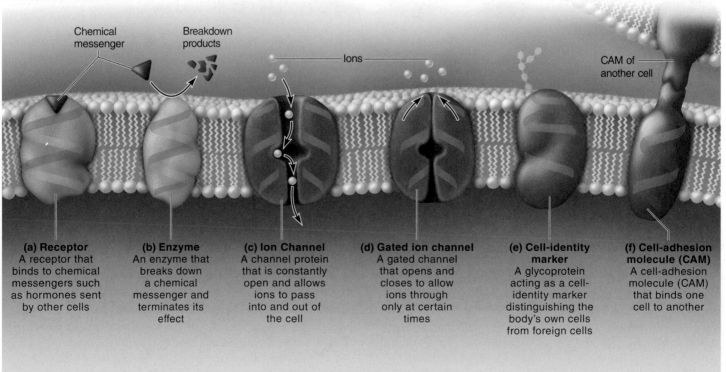

(a) Receptor
A receptor that binds to chemical messengers such as hormones sent by other cells

(b) Enzyme
An enzyme that breaks down a chemical messenger and terminates its effect

(c) Ion Channel
A channel protein that is constantly open and allows ions to pass into and out of the cell

(d) Gated ion channel
A gated channel that opens and closes to allow ions through only at certain times

(e) Cell-identity marker
A glycoprotein acting as a cell-identity marker distinguishing the body's own cells from foreign cells

(f) Cell-adhesion molecule (CAM)
A cell-adhesion molecule (CAM) that binds one cell to another

- **Carriers** (see fig. 3.7b, c). Carriers actively bind a substance on one side of the membrane and release it on the other side. They transport glucose, amino acids, sodium, potassium, calcium, and many other substances into and out of cells.
- **Cell-identity markers** (fig. 3.3e). Glycoproteins and glycolipids are a cell's identification tags, genetically unique to an individual (or to identical twins). They enable the immune system to distinguish what belongs to one's own body from what does not, so it can selectively attack such things as bacteria and parasites.
- **Cell-adhesion molecules (CAMs)** (fig. 3.3f). CAMs are membrane proteins that link cells to each other and to extracellular material; they bind a tissue together like a coupling between two railroad cars. Special events such as sperm–egg binding and the binding of an immune cell to a cancer cell also require CAMs.

The Glycocalyx

All cells are covered with a fuzzy carbohydrate coat called the **glycocalyx**[3] (GLY-co-CAIL-icks) (fig. 3.4). It consists of short chains of sugars belonging to the glycolipids and glycoproteins. The glycocalyx has multiple functions. It includes the cell-adhesion molecules previously mentioned. It also cushions the plasma membrane and protects it from physical and chemical injury, somewhat like the styrofoam "peanuts" in a shipping carton. It functions in cell identity and thus in the body's ability to distinguish its own healthy cells from diseased cells, invading organisms, and transplanted tissues. Human blood types and transfusion compatibility are determined by the glycocalyx.

Microvilli, Cilia, and Flagella

Most cells have surface extensions of one or more types called *microvilli, cilia,* and *flagella.* These aid in absorption, movement, and sensory processes.

Microvilli

Microvilli[4] (MY-cro-VIL-eye; singular, *microvillus*) are extensions of the plasma membrane that serve primarily to increase its surface area. On many cells, they are little more than tiny bumps on the surface. They are better developed in cells specialized for absorption, such as the epithelial cells of the small intestine and kidney tubules. Here they can be very dense and appear as a surface fringe called the **brush border** (fig. 3.4), named for its resemblance to a hairbrush. A brush border gives a cell up to 40 times more surface area. On cells of the taste buds and inner ear, microvilli are well developed but less numerous, and instead of absorptive functions, they serve sensory roles such as detecting food chemicals or sound.

Individual microvilli can't be distinguished very well with the light microscope because they are only 1 to 2 μm long. With the electron microscope, however, they typically look like finger-shaped projections of the cell surface. They often show little internal structure, but sometimes have a bundle of stiff supportive filaments of protein.

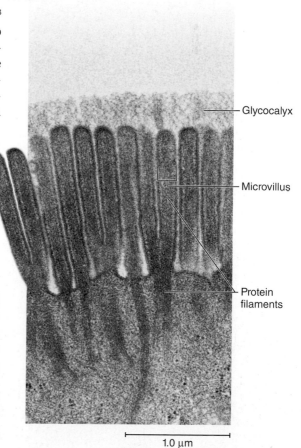

Glycocalyx

Microvillus

Protein filaments

1.0 μm

Figure 3.4 The Glycocalyx and Brush Border. The microvilli are anchored by protein microfilaments, which occupy the core of each microvillus and project into the cytoplasm. The glycocalyx is composed of short carbohydrate chains (oligosaccharides) bound to the membrane phospholipids and proteins.

- *Name an organ of the body from which this cell might have come. Explain your answer.*

---

[3] *glyco* = sugar; *calyx* = cup, vessel
[4] *micro* = small; *villi* = hairs

Cilia

Cilia (SIL-ee-uh; singular, *cilium*[5]) are hairlike processes about 7 to 10 μm long. Nearly every cell has a solitary, nonmotile *primary cilium.* Its function in some cases is still a mystery, but apparently many of them are sensory, serving as the cell's "antenna" for monitoring nearby conditions. The light-absorbing parts of the retinal cells in the eye are modified primary cilia; in the inner ear, they play a role in the senses of motion and balance; and in kidney tubules, they are thought to monitor fluid flow. Odor molecules bind to nonmotile cilia on the sensory cells of the nose.

Motile (moving) cilia are less widespread, but very abundant in the tissues where they do occur—notably the mucous membranes of the respiratory tract and uterine (fallopian) tubes. Cilia here typically number 50 to 200 per cell (fig. 3.5a). They beat in synchronized waves that sweep across the surface of an epithelium, always in the same direction. In the respiratory tract, they move mucus from the lungs and trachea up to the throat, where it is swallowed. In the uterine tubes, they move an egg or embryo toward the uterus, like people in a stadium passing a beach ball overhead from hand to hand.

Cilia exhibit a distinctive central core called the **axoneme**[6] (ACK-so-neem), composed of a pinwheel-like array of thin protein cylinders called *microtubules.* There are two central microtubules surrounded by a ring of nine microtubule pairs (fig. 3.5b, c). The central microtubules stop at the cell surface, but the others continue a short distance into the cell as part of a **basal body** that anchors the cilium. Each pair of peripheral microtubules is equipped with little motor proteins that produce the beating motion of the cilium.

Flagella

A **flagellum**[7] (fla-JEL-um) resembles a long solitary cilium. It has an identical axoneme and basal body, but it is stiffened by a sheath of microfilaments in a space between the axoneme and plasma membrane. The only functional flagellum in humans is the whiplike tail of a sperm cell, measuring about 50 μm long (see fig. 19.13, p. 651). Sperm use the tail to crawl like a snake up the mucous membrane of the uterus and uterine tubes in their quest to find an egg.

> ### Apply What You Know
>
> Kartagener syndrome is a hereditary disease in which cilia and flagella lack the motor protein and therefore cannot move. How do you think Kartagener syndrome will affect a man's ability to father a child? How might it affect his respiratory health? Would it affect a woman's fertility? Explain your answers.

Cell Junctions

In complex multicellular organisms, no cell is independent of the others. Proteins at the cell surface form **cell junctions** that link cells together and attach them to the extracellular material. Such attachments enable cells to grow and divide normally, resist stress, and communicate with each other. Without them, cardiac muscle cells would pull apart when they contracted, and every swallow of food would scrape away the lining of the esophagus. The three principal

---

[5] *cilium* = eyelash

[6] *axo* = axis; *neme* = thread

[7] *flagellum* = whip

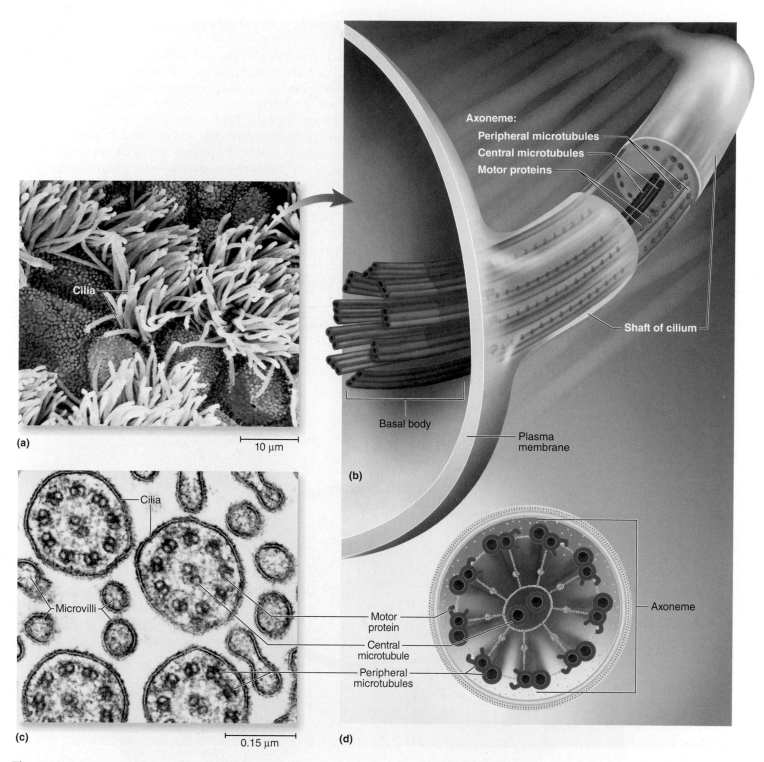

Figure 3.5 The Structure of Cilia. (a) Inner surface of the trachea (SEM). Several nonciliated, mucus-secreting goblet cells (orange) are visible among the ciliated cells. The goblet cells have short microvilli. (b) Three-dimensional structure of a cilium. (c) Cross section of a few cilia and microvilli (TEM). (d) Cross-sectional structure of a cilium. Note the relative sizes of cilia and microvilli in parts (a) and (c).

types of junctions between cells are tight junctions, desmosomes, and gap junctions (fig. 3.6).

A **tight junction** completely encircles an epithelial cell near its upper end and joins it tightly to adjacent cells, somewhat like the plastic harness on a six-pack of soda cans. It is formed by the fusion of the plasma membranes of adjacent cells, sealing off the intercellular space and making it difficult for substances to leak between the cells. Tight junctions ensure that absorbed materials must pass *through* the epithelial cells and not *between* them, thus enabling the epithelial cells to chemically process them. In the stomach and intestines, they also prevent digestive juices from seeping between epithelial cells and digesting the underlying tissue, and they help prevent bacteria from invading the tissues.

A **desmosome**[8] (DEZ-mo-some) is a protein patch that holds cells together at a specific point. We can compare a tight junction and a desmosome, respectively, to a zipper and a snap on a pair of jeans. Desmosomes are not continuous and therefore cannot prevent substances from passing between cells. They serve to keep cells from pulling apart and thus enable a tissue to

---

[8] *desmo* = band, bond, ligament; *som* = body

Figure 3.6 Cell Junctions.

- *Which type of junction allows materials to pass directly from cell to cell?*

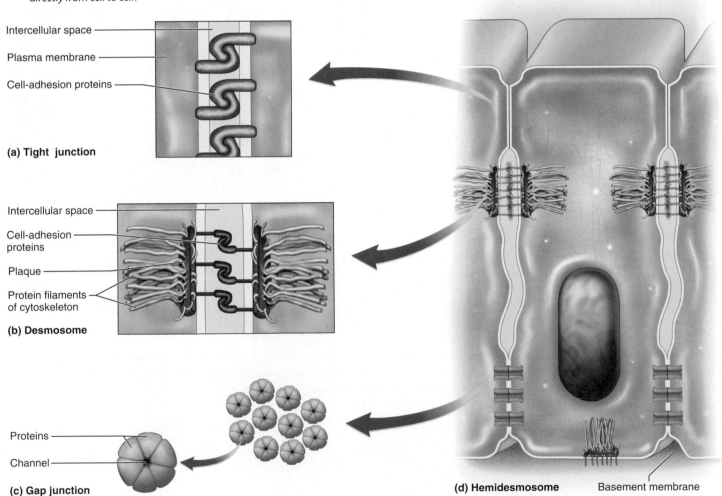

Intercellular space

Plasma membrane

Cell-adhesion proteins

(a) Tight junction

Intercellular space

Cell-adhesion proteins

Plaque

Protein filaments of cytoskeleton

(b) Desmosome

Proteins

Channel

(c) Gap junction

(d) Hemidesmosome   Basement membrane

Clinical Application 3.1

WHEN DESMOSOMES FAIL

We often get our best insights into the importance of a structure from the dysfunctions that occur when it breaks down. Desmosomes are destroyed in a disease called *pemphigus vulgaris*[9] (PEM-fih-gus vul-GAIR-iss), in which the immune system launches a misguided attack on the desmosome proteins, especially in the skin. The resulting breakdown of desmosomes between the cells leads to widespread blistering of the skin and oral mucosa, loss of tissue fluid, and sometimes death. The condition can be controlled with drugs that suppress the immune system, but such drugs compromise the body's ability to fight off infections.

resist mechanical stress. Desmosomes are common in such tissues as the epidermis and cardiac muscle. A desmosome consists of a thick protein plaque on the inner surface of the plasma membranes of two cells, directly across from each other, with protein filaments spanning the space from the plaque of one cell to the plaque of the other. Half-desmosomes called *hemidesmosomes* anchor an epithelial cell to the basement membrane below it.

Apply What You Know

Why wouldn't desmosomes be suitable as the sole intercellular junctions between epithelial cells of the stomach?

A **gap junction** is formed by a ring of six proteins arranged somewhat like the segments of an orange, surrounding a water-filled channel. Ions, glucose, amino acids, and other small solutes can diffuse through the channel directly from the cytoplasm of one cell into the next. In the human embryo, nutrients pass from cell to cell through gap junctions until the circulatory system forms and takes over the role of nutrient distribution. In cardiac muscle, gap junctions allow electrical excitation to pass directly from cell to cell so that the cells contract in near unison.

Transport Through the Plasma Membrane

To maintain the internal conditions necessary for life, a cell must regulate what materials are allowed to pass in and out through the plasma membrane. The membrane is said to be **selectively permeable** because it allows some substances through but holds back others, especially those too large to pass through its protein channels. Here we will survey eight methods of transport through a cell's surface. Although our emphasis here is on the plasma membrane, most of these processes also occur at some of the internal membranes of a cell, such as those that enclose the mitochondria and endoplasmic reticulum.

---

[9] *pemphigus* = blistering; *vulgaris* = common

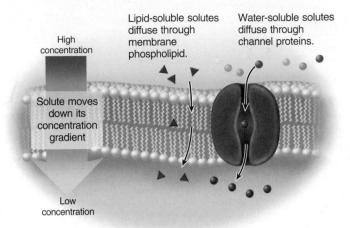

(a) Simple diffusion

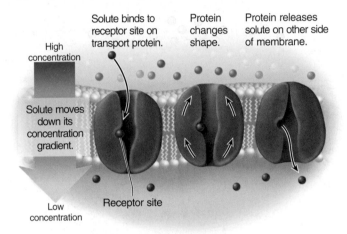

(b) Facilitated diffusion

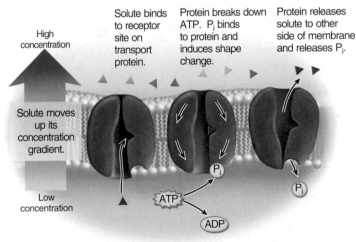

(c) Active transport

Figure 3.7 Some Modes of Membrane Transport. AP|R

- *Which of these processes can also occur in synthetic membranes such as cellophane or dialysis tubing?*

Simple Diffusion

Simple diffusion (fig. 3.7a) is the net movement of particles from a place of high concentration to a place of low concentration—in other words, *down a concentration gradient*. This is how oxygen and steroid hormones enter cells and potassium ions leave, for example. The cell doesn't expend energy to achieve this; all molecules are in spontaneous random motion, and this alone provides the energy for their diffusion through space. Molecules can diffuse through both living membranes (the plasma membrane) and nonliving ones (such as dialysis tubing and cellophane) if the membrane has large enough gaps or pores. Nonpolar solutes such as oxygen and carbon dioxide, and hydrophobic substances such as steroids, diffuse through the lipid regions of the plasma membrane; hydrophilic solutes such as salts, however, can diffuse only through the water-filled protein channels of the membrane.

Osmosis

Osmosis[10] (oz-MO-sis) is the net movement of water through a selectively permeable membrane from the side where there is a relatively low concentration of solutes to the side where there is a higher solute concentration. The reason it occurs is that when water molecules move to the high-solute side, they tend to loosely associate with the solute molecules. Becoming less mobile, they are less able to break free and cross back to the low-solute side; there are more water molecules on the low-solute side free to pass through the membrane than there are on the high-solute side. Thus, if the fluids on two sides of a cell membrane differ in the concentration of dissolved matter (and these solutes cannot penetrate the membrane), water tends to pass by osmosis from the more dilute to the less dilute side. Osmosis plays a key role in many aspects of homeostasis. For example, cell volume is maintained when there is an appropriate balance in water content on the two sides of the membrane. Blood capillaries absorb fluid from the tissues by osmosis, thereby removing metabolic wastes from the tissues and preventing the tissues from swelling with excess fluid.

> **Apply What You Know**
>
> In dehydration, the relative solute concentration of the extracellular fluid (ECF) increases. How would this affect the volume (sizes) of the cells in that ECF? Explain.

Facilitated Diffusion

The next two processes, facilitated diffusion and active transport, are called *carrier-mediated transport* because they employ carrier proteins in the plasma membrane. **Facilitated**[11] **diffusion** (fig. 3.7b) is the movement of a solute through a cell membrane, down its concentration gradient, with the aid of a carrier. It does not involve any ATP expenditure by the cell. The carrier binds a

---

[10] *osm* = push, thrust; *osis* = condition, process

[11] *facil* = easy

Clinical Application 3.2

CALCIUM CHANNEL BLOCKERS

Membrane channels may seem only an abstract concept until we see how they relate to disease and drug design. For example, drugs called *calcium channel blockers* are often used to treat high blood pressure (hypertension). How do they work? The walls of the arteries contain smooth muscle that constricts to narrow the vessels and raise blood pressure, or relaxes to let them widen and reduce blood pressure. Excessive, widespread vasoconstriction (vessel narrowing) can cause hypertension, so one approach to the prevention of hypertension is to inhibit this. In order to constrict, smooth muscle cells open calcium channels in the plasma membrane. The inflow of calcium activates the proteins of muscle contraction. Calcium channel blockers act, as their name says, by preventing calcium channels from opening and thereby preventing arterial constriction.

particle on the side of a membrane where the solute is more concentrated, then releases it on the side where it is less concentrated. The carrier transports solutes that otherwise couldn't pass through the membrane or would pass less efficiently. One use of facilitated diffusion is to absorb the sugars and amino acids from digested food.

Active Transport

Active transport (fig. 3.7c) is a process that employs a carrier protein and uses energy from ATP to move a solute through the membrane *up its concentration gradient*—that is, from the side where it is less concentrated to the side where it is already more concentrated. Active transport requires ATP because moving particles up a gradient requires an energy input, like getting a wagon to roll uphill. If a cell stops producing ATP, owing to cell death or poisoning, active transport ceases immediately.

An especially important active-transport process is the **sodium–potassium (Na^+–K^+) pump.** Sodium is normally much more concentrated in the ECF than in the ICF, and potassium is more so in the ICF. Yet cells continually pump more Na^+ out and more K^+ into the cell. The Na^+–K^+ pump binds three sodium ions from the ICF and ejects them from the cell, then binds two potassium ions from the ECF and releases these into the cell. It repeats the process over and over, using one ATP molecule for each cycle. The Na^+–K^+ pump plays roles in controlling cell volume (water follows Na^+ by osmosis); generating body heat (ATP consumption releases heat); maintaining the electrical excitability of your nerves, muscles, and heart; and providing energy for other transport pumps to draw upon in moving such solutes as glucose through the plasma membrane. About half of the calories that you "burn" every day are used just to operate your Na^+–K^+ pumps.

Vesicular Transport

All of the processes discussed up to this point move molecules or ions individually through the plasma membrane. In **vesicular transport,** however, cells move larger particles or droplets of fluid through the membrane in bubblelike *vesicles*. All vesicular processes that bring matter into a cell are called **endocytosis**[12] (EN-doe-sy-TOE-sis) and those that export material from a cell

---

[12] *endo* = into; *cyt* = cell; *osis* = process

are called **exocytosis**[13] (EC-so-sy-TOE-sis). Like active transport, all forms of vesicular transport require ATP. Figure 3.8 illustrates the following modes of vesicular transport.

There are three forms of endocytosis: *phagocytosis, pinocytosis,* and *receptor-mediated endocytosis*. In **phagocytosis**[14] (FAG-oh-sy-TOE-sis), or "cell eating," a cell reaches out with footlike **pseudopods** and surrounds a particle such as a bacterium or a bit of cell debris. It engulfs the particle and enzymatically degrades it (fig. 3.8a). Phagocytosis is carried out especially by white blood cells and *macrophages,* which are described in chapter 12. Some macrophages consume as much as 25% of their

---

[13] *exo* = out of; *cyt* = cell; *osis* = process
[14] *phago* = eating; *cyt* = cell; *osis* = process

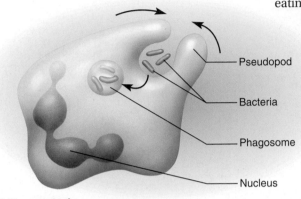

Pseudopod

Bacteria

Phagosome

Nucleus

(a) Phagocytosis

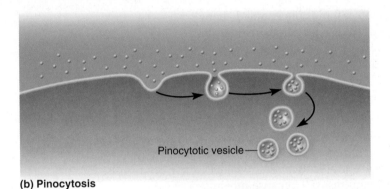

Pinocytotic vesicle

(b) Pinocytosis

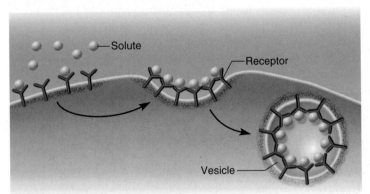

Solute

Receptor

Vesicle

(c) Receptor-mediated endocytosis

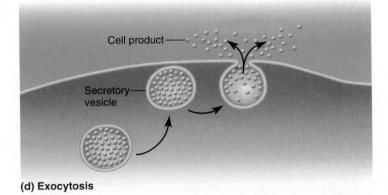

Cell product

Secretory vesicle

(d) Exocytosis

Figure 3.8 Modes of Vesicular Transport. (a) Phagocytosis. A white blood cell engulfing bacteria with its pseudopods. (b) Pinocytosis. A cell imbibing droplets of extracellular fluid. (c) Receptor-mediated endocytosis. The Ys in the plasma membrane represent membrane receptors. The receptors bind a solute in the extracellular fluid, then cluster together. The membrane caves in at that point until a vesicle pinches off into the cytoplasm bearing the receptors and bound solute. (d) Exocytosis. A cell releasing a secretion or waste product. **AP|R**

own volume in material per hour, thus living up to their name[15] and playing a vital role in cleaning up the tissues.

Pinocytosis[16] (PIN-oh-sy-TOE-sis), or "cell drinking," occurs in all human cells. In this process, dimples form in the plasma membrane and progressively cave in until they pinch off as vesicles containing droplets of extracellular fluid (ECF) (fig. 3.8b). One use of pinocytosis is seen in kidney tubule cells; they use this method to reclaim the small amount of protein that filters out of the blood, thus preventing protein from being lost in the urine.

Receptor-mediated endocytosis (fig. 3.8c) is more selective. It enables a cell to take in specific molecules from the ECF with a minimum of unnecessary fluid. Solute molecules in the ECF bind to specific receptor proteins on the plasma membrane. The receptors then cluster together and the membrane sinks in at this point, creating a pit. The pit soon pinches off to form a vesicle in the cytoplasm. One use of receptor-mediated endocytosis is the absorption of insulin from the blood. Hepatitis, polio, and AIDS viruses "trick" cells into admitting them by receptor-mediated endocytosis.

Exocytosis (fig. 3.8d) is the process of discharging material from a cell by vesicular transport. It is used, for example, by digestive glands to secrete enzymes, by breast cells to secrete milk, and by sperm cells to release enzymes for penetrating an egg. It resembles endocytosis in reverse. A secretory vesicle in the cell migrates to the surface and fuses with the plasma membrane. A pore opens up that releases the product from the cell, and the empty vesicle usually becomes part of the plasma membrane. In addition to releasing cell products, exocytosis is the cell's way of replacing the bits of membrane removed by endocytosis.

Before You Go On

Answer these questions from memory. Reread the preceding section if there are too many you don't know.

4. *What property of phospholipid molecules causes them to organize themselves into a bilayer?*

5. *Which of these is important in determining blood transfusion compatibility: cell-adhesion molecules, membrane carriers, membrane cholesterol, the glycocalyx, or microvilli? Explain.*

6. *Compare and contrast microvilli and cilia in terms of their structure, function, and general location.*

7. *Which type of cell junction best serves to keep food from scraping away the lining of your oral cavity? Which type enables a cardiac muscle cell to directly stimulate the next one? Which type best serves to keep your digestive enzymes from eroding the tissues beneath your intestinal lining?*

8. *What membrane transport processes get all the necessary energy from the spontaneous movement of molecules? What ones require ATP as a source of energy? Which processes are carrier-mediated? Which ones are not?*

---

[15] *macro* = big; *phage* = eater

[16] *pino* = drinking; *cyt* = cell; *osis* = process

3.3 The Cell Interior

Expected Learning Outcomes

When you have completed this section, you should be able to:

a. describe the cytoskeleton, its three components, and its functions;

b. give some examples of cell inclusions and explain how inclusions differ from organelles;

c. list the main organelles of a cell and explain their functions; and

d. describe how cells synthesize, process, package, and secrete proteins.

We now probe more deeply into the cell to study the structures in the cytoplasm. These are classified into three groups—cytoskeleton, inclusions, and organelles—all embedded in the clear, gelatinous cytosol. If we think of a cell as being like an office building, the cytoskeleton would be the steel beams and girders that hold it up and define its shape and size; the plasma membrane and its channels would be the building's exterior walls and doors; and organelles would be the interior rooms that divide the building into compartments with different functions.

The Cytoskeleton

The **cytoskeleton** is a network of protein filaments and tubules. It often forms a very dense supportive web in the cytoplasm (fig. 3.9). It structurally supports a cell, determines its shape, organizes its contents, and—going beyond our office building analogy—transports substances within the cell and contributes to movements of the cell as a whole. It is connected to proteins of the plasma membrane and they in turn are connected to proteins external to the cell, so there is a strong structural continuity from the cytoplasm to the extracellular material.

The cytoskeleton is composed of *microfilaments, intermediate filaments,* and *microtubules.* **Microfilaments** are about 6 nm thick and are made of the protein *actin.* They form a dense fibrous mesh called the **terminal web** on the internal side of the plasma membrane. The oily plasma membrane is spread out over the terminal web like butter on a slice of bread. It is thought that the membrane would break up into little droplets without this support. Actin is also the supportive protein in the cores of the microvilli (see fig. 3.4) and plays a role in muscle contraction and other cell movements.

Intermediate filaments (8–10 nm in diameter) are thicker and stiffer than microfilaments. They contribute to the strength of the desmosomes and include the tough protein *keratin* that fills the cells of the epidermis and gives strength to the skin.

Microtubules (25 nm in diameter) are hollow cylinders of protein. They hold organelles in place, form bundles that maintain cell shape and rigidity, and act somewhat like monorails to guide organelles and molecules to specific destinations in a cell. They form the axonemes of cilia and flagella as well as the *centrioles* and *mitotic spindle* involved in cell division (described later).

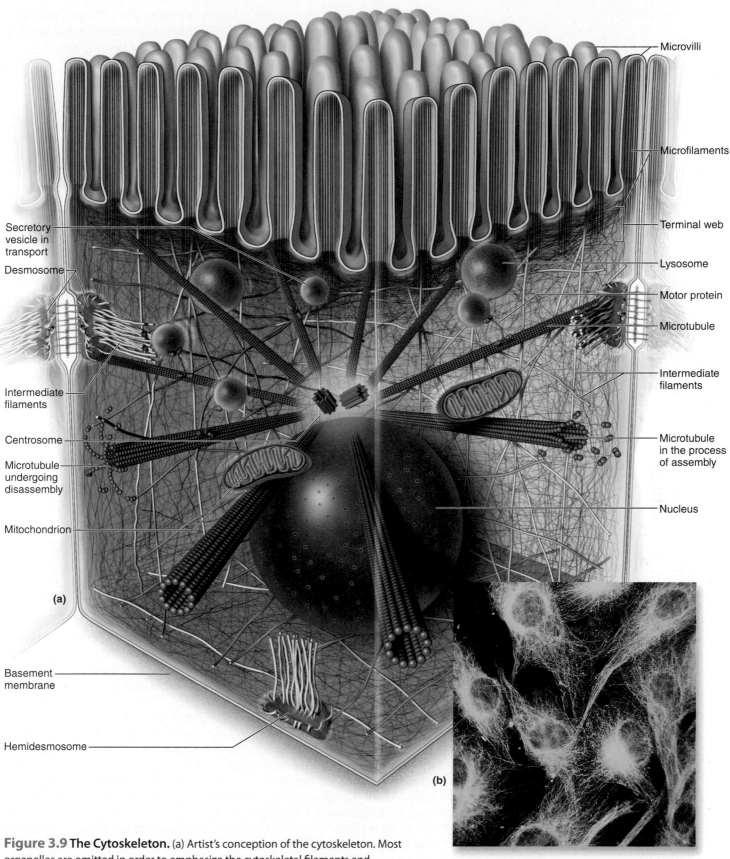

Figure 3.9 The Cytoskeleton. (a) Artist's conception of the cytoskeleton. Most organelles are omitted in order to emphasize the cytoskeletal filaments and microtubules. (b) Cells with their cytoskeleton selectively labeled with a fluorescent dye. These show that the cytoskeleton is actually far denser than showing in part (a).

If you think of intermediate filaments as being like stiff uncooked spaghetti, you could, by comparison, think of microfilaments as being like fine angelhair pasta and microtubules as being like tubular penne pasta.

Inclusions

Inclusions are of two kinds: stored cellular products such as pigments, fat globules, and glycogen granules; and foreign bodies such as viruses, bacteria, and dust particles or other debris phagocytized by a cell. Inclusions are never enclosed in a membrane, and unlike the organelles and cytoskeleton, they are not essential to cell survival.

Organelles

Organelles (literally "little organs") are to the cell what organs are to the body—metabolically active structures that play individual roles in the survival of the whole (see fig. 3.1). A cell may have 10 billion protein molecules, some of which are potent enzymes with the potential to destroy the cell if they're not contained and isolated from other cellular components. You can imagine the enormous problem of keeping track of all this material, directing molecules to the correct destinations, and maintaining order against the incessant tendency toward disorder. Cells maintain order partly by compartmentalizing their contents in organelles. Figure 3.10 shows the most important ones.

The Nucleus

The **nucleus** (fig. 3.10a) is the largest organelle and usually the only one visible with the light microscope. Most cells have only one nucleus, but there are exceptions. Mature red blood cells have none; they are *anuclear.*[17] A few cell types are *multinuclear*—having 2 to 50 nuclei—including skeletal muscle cells, some liver cells, and certain bone-dissolving cells.

The nucleus is surrounded by a **nuclear envelope** consisting of two parallel membranes. The envelope is perforated with **nuclear pores** formed by a ring-shaped complex of proteins. These proteins regulate molecular traffic into and out of the nucleus and bind the two membranes together.

The nucleus contains the **chromosomes**[18]—threadlike bodies of DNA and protein—and is therefore the genetic control center of cellular activity. Most of our cells have 46 chromosomes. In nondividing cells, they are in the form of very fine filaments broadly dispersed throughout the nucleus, visible only with the electron microscope. Collectively, this material is called **chromatin** (CRO-muh-tin). The nuclei also usually exhibit one or more dense masses called **nucleoli** (singular, *nucleolus*), where subunits of the *ribosomes* (described shortly) are made before they are transported out to the cytoplasm.

Endoplasmic Reticulum

The term *endoplasmic reticulum (ER)* literally means "little network within the cytoplasm." The ER is a system of interconnected channels called **cisternae**[19] (sis-TUR-nee) enclosed by a membrane (fig. 3.10b). In areas called **rough endoplasmic reticulum**, the cisternae are flat, parallel, and covered with ribosomes, which give it its rough or granular appearance. The rough ER is continuous with the outer membrane of the nuclear envelope, and adjacent cisternae are connected by transverse bridges. In areas called

---

[17] *a* = without; *nucle* = nucleus

[18] *chromo* = color; *some* = body

[19] *cistern* = reservoir

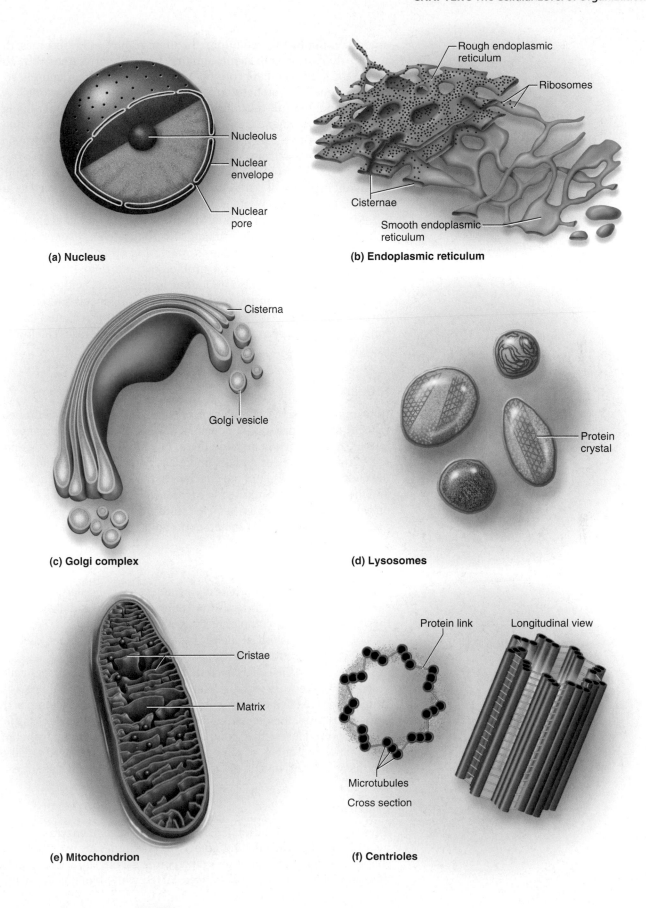

(a) **Nucleus**

Nucleolus

Nuclear envelope

Nuclear pore

(b) **Endoplasmic reticulum**

Rough endoplasmic reticulum

Ribosomes

Cisternae

Smooth endoplasmic reticulum

(c) **Golgi complex**

Cisterna

Golgi vesicle

(d) **Lysosomes**

Protein crystal

(e) **Mitochondrion**

Cristae

Matrix

(f) **Centrioles**

Protein link

Longitudinal view

Microtubules

Cross section

Figure 3.10 Major Organelles. AP|R

- *Which of these six organelles directly participate in protein synthesis and secretion?*

smooth endoplasmic reticulum, the membrane lacks ribosomes, the cisternae are more tubular in shape, and they branch more extensively. The cisternae of the smooth ER are continuous with those of the rough ER, so the two are functionally different parts of the same cytoplasmic network.

The endoplasmic reticulum synthesizes steroids and other lipids, detoxifies alcohol and other drugs, and manufactures the cell's internal and surface membranes. The rough ER produces the phospholipids and proteins of the plasma membrane. It also synthesizes proteins that are destined to be secreted from the cell or packaged in organelles called *lysosomes*. Rough ER is most abundant in cells that synthesize large amounts of protein, such as antibody-producing cells and cells of the digestive glands.

Most cells have only scanty smooth ER, but it is relatively abundant in cells that engage extensively in detoxification, such as liver and kidney cells. Long-term abuse of alcohol, barbiturates, and other drugs leads to tolerance partly because the smooth ER proliferates and detoxifies the drugs more quickly. Smooth ER is also abundant in cells that synthesize steroid hormones, for example in the testes and ovaries. Skeletal and cardiac muscle contain extensive networks of smooth ER, which store calcium ions when the muscle is at rest. Upon stimulation, the smooth ER releases calcium to trigger muscle contraction.

Ribosomes

Ribosomes are small granules of protein and **ribosomal ribonucleic acid (rRNA);** they are produced in the nucleus, but most of them are then exported into the cytoplasm and function there. Ribosomes "read" coded genetic messages from the nucleus and assemble amino acids into proteins specified by the code. Many ribosomes are attached to the surface of the nuclear envelope and the rough ER, and make proteins that will either be packaged in lysosomes or, as in digestive enzymes, secreted from the cell. Others lie free in the cytoplasm and make enzymes and other proteins for internal use by the cell.

Golgi Complex

The **Golgi**[20] (GOAL-jee) **complex** is a small cluster of cisternae that synthesize carbohydrates and put the finishing touches on protein and glycoprotein synthesis (fig. 3.10c). The complex resembles a stack of pita bread. Typically, it consists of about six cisternae, slightly separated from each other, each of them a flattened, slightly curved sac with swollen edges.

The edges of the Golgi cisternae pinch off membranous sacs called **Golgi vesicles,** like the waxy globules that break away from the main mass in a lava lamp. These are filled with the complex's secretory product. Some of them become lysosomes, the organelles discussed next. Some migrate to the plasma membrane and fuse with it, contributing fresh protein and phospholipid to the membrane. Still others become **secretory vesicles** that store a cell product, such as breast milk, mucus, or digestive enzymes, for later release by exocytosis.

Lysosomes

A **lysosome**[21] (LY-so-some) (fig. 3.10d) is a package of enzymes enclosed in a membrane. Although often round or oval, lysosomes are extremely variable in shape. Lysosomal enzymes break down proteins, nucleic acids, carbohydrates, phospholipids, and other substances. When white blood cells called *neutro-*

---

[20] Camillo Golgi (1843–1926), Italian histologist
[21] *lyso* = loosen, dissolve; *some* = body

phils phagocytize bacteria, a lysosome releases enzymes to digest and destroy the microbes. Lysosomes also digest worn-out mitochondria and other organelles. They sometimes carry out a sort of "cell suicide" called *apoptosis* (AP-oh-TOE-sis) *(programmed cell death),* in which cells that are no longer needed undergo a prearranged death. The uterus, for example, weighs about 900 g at full-term pregnancy, and shrinks by apoptosis to 60 g within 5 or 6 weeks after birth because surplus cells are digested by their own lysosomes.

Peroxisomes

Peroxisomes resemble lysosomes (and are not illustrated) but contain different enzymes and are not produced by the Golgi complex. They are especially abundant in liver and kidney cells. Their primary function is to break down fatty acids into two-carbon molecules that can be used as an energy source for ATP synthesis. They also neutralize free radicals, detoxify alcohol and other drugs, and kill bacteria. They are named for the hydrogen peroxide (H_2O_2) they produce in the course of these reactions.

Mitochondria

Mitochondria[22] (MY-toe-CON-dree-uh) (singular, *mitochondrion*) are organelles specialized for ATP synthesis. They have a variety of shapes: spheroid, rod-shaped, bean-shaped, or threadlike (fig. 3.10e). Like the nucleus, a mitochondrion is surrounded by a double membrane. The inner membrane usually has folds called **cristae**[23] (CRIS-tee), which project like shelves across the organelle. Cristae bear the enzymes that produce most of the ATP. The space between the cristae is called the **mitochondrial matrix.** It contains enzymes, ribosomes, and a small DNA molecule called **mitochondrial DNA.**

Mitochondria have an interesting history. There is conclusive evidence that they once were free-living bacteria that were internalized by another primitive cell, evolved to become permanent residents, and eventually became indispensable to the life of the host cell. The inner mitochondrial membrane apparently was the original bacterial plasma membrane, whereas the outer one was the membrane that the host cell produced in the course of phagocytosis. Like the DNA of modern bacteria, mitochondrial DNA (mtDNA) is a circular loop rather than an open-ended linear molecule. It is genetically different from the DNA in the cell's nucleus and is replicated independently of the host (nuclear) DNA. We hear a lot these days about mtDNA in the news and even in television crime dramas, as it is often used to link a suspect to crime-scene evidence. Mutations in mtDNA are responsible for some muscle, heart, and eye diseases. mtDNA is also important in evolutionary research for establishing relationships among species and a timetable of some evolutionary events; for example, as a "molecular clock," mtDNA has established that *Homo sapiens* originated in Africa about 200,000 years ago.

Centrioles

A **centriole** (SEN-tree-ole) is a short cylindrical assembly of microtubules arranged in nine groups of three microtubules each (fig. 3.10f). Near the nucleus, most cells have a small, clear patch of cytoplasm called the **centrosome**[24] containing a pair of centrioles (see fig. 3.1). These centrioles play a role in cell division described later. The basal bodies of cilia and flagella are often considered to be solitary centrioles.

---

[22] *mito* = thread; *chondr* = grain

[23] *crista* = crest

[24] *centro* = central; *some* = body

Protein Synthesis

One of the most significant activities of a cell is to produce proteins, and we are now in a position to understand how several of the foregoing organelles interact to do this. The proteins of a cell can be for internal structural use, such as proteins of the cytoskeleton and plasma membrane; they include enzymes that control the cell's internal metabolism; and in many cases they are secreted from the cell to serve elsewhere, as in the case of digestive enzymes, antibodies, and some hormones such as insulin and growth hormone.

The essence of protein synthesis (fig. 3.11) is that DNA in the nucleus contains codes (genes) for how to construct the needed proteins. Synthesis begins with making a small copy of a gene, called *messenger RNA (mRNA).* This usually migrates into the cytoplasm, where ribosomes read the code and assemble amino acids in the right order to make that protein. The step from DNA to mRNA is called *transcription,* and the step from mRNA to protein is called *translation.* The next two sections give further details of these two steps.

Transcription

Protein synthesis begins with **transcription** in the cell's nucleus (fig. 3.11, panel ①). At the site of a particular gene, an enzyme unzips the double helix of DNA and exposes the gene's nitrogenous bases. Another enzyme *(RNA polymerase)* reads these bases (panel ②) and creates a parallel molecule of **messenger RNA (mRNA),** which is more or less a mirror image of the gene. Where the enzyme finds a thymine (T) on the DNA, it adds the complementary base adenine (A) to the mRNA. Where it finds a guanine (G) on the DNA, it adds cytosine (C) to the mRNA. Where it finds adenine, however, it adds uracil (U) to the mRNA; remember from chapter 2 that RNA does not contain any thymine, but substitutes uracil for this base. Therefore, if the enzyme had read a DNA base sequence of TACCGTCCA, it would produce a complementary mRNA sequence of AUGGCAGGU. A typical mRNA is several hundred bases long and some are up to 10,000.

The mRNA is "edited" in the nucleus to remove some noncoding ("nonsense") segments. The coding segments are spliced together to make a mature, coding mRNA molecule, which is then usually exported from the nucleus, through a nuclear pore, into the cytoplasm. (Some remains in the nucleus and is translated there.)

Translation

The next step of protein synthesis is **translation,** in which the genetic code of mRNA is translated into a new language, the amino acid sequence of a protein (fig. 3.11, panel ③). This is where ribosomes enter the story. As mRNA emerges from a nuclear pore, two ribosomal subunits—large and small—assemble on it and begin reading the coded message of the mRNA.

The genetic code is in the form of **codons**—three-base segments of mRNA. There are 64 of these, most of which stand for a particular amino acid to be inserted into a protein. For example, the sequence AUG represents methionine, GCA represents alanine, and GGU represents glycine. With 64 codons in the genetic code and only 20 amino acids composing proteins, there is some redundancy in the code. Codons GGU, GGA, and GGG all represent glycine, for example. AUG represents not only methionine, but like the capital letter at the beginning of a sentence, it also signifies that the ribosome is to begin reading the message and assembling the protein at that point. Three codons (UAG, UGA, and UAA) represent no amino acids but serve like the period at the end of

Figure 3.11 Protein Synthesis. AP|R

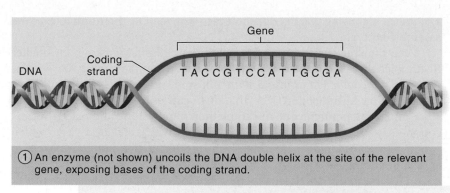

① An enzyme (not shown) uncoils the DNA double helix at the site of the relevant gene, exposing bases of the coding strand.

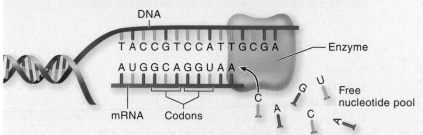

② **Transcription.** A second enzyme reads bases of the coding strand, draws RNA nucleotides from a free nucleotide pool, and assembles these in order specified by the gene to make mRNA.

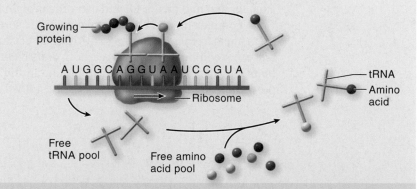

③ **Translation.** A two-part ribosome binds to the mRNA and moves along it, reading its codons. tRNA binds free amino acids and delivers them to the ribosome. The ribosome assembles the amino acids in order specified by the genetic code in the codons. After delivering its amino acid, the free tRNA leaves the ribosome to repeat the process.

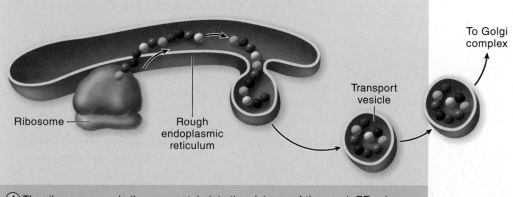

④ The ribosome spools the new protein into the cisterna of the rough ER, where it may be further processed and packaged in transport vesicles. Transport vesicles take it to the Golgi complex for further processing.

a sentence, signifying that the ribosome is to stop reading there and release the assembled protein.

Each time the ribosome reads a particular codon, it must bind a smaller RNA molecule called **transfer RNA (tRNA)** with a complementary base series called an *anticodon*. For example, if the ribosome reads GGU on the mRNA, it binds a tRNA with anticodon CCA. That tRNA carries a glycine, which the ribosome adds to the protein. The ribosome then moves along and reads the next codon—say, GCA. It would then bind a tRNA with anticodon CGU, which would deliver an alanine to be added to the protein. In this manner, one amino acid after another is added to the protein until the ribosome reaches a stop codon. At that point, it releases the protein and the ribosome itself detaches from the mRNA and breaks up into its two subunits.

Protein Processing and Secretion

Several organelles are involved in producing, packaging, and secreting proteins (fig. 3.11, panel ④). If a protein is destined to be packaged in a lysosome or to be secreted from the cell, the ribosome–mRNA complex "docks" on the surface of the rough endoplasmic reticulum and the protein reels into the ER cisterna as it is assembled. In the rough ER, enzymes cut certain amino acid segments from the protein, splice segments together, and perform other alterations.

The altered protein is then shuffled into **transport vesicles**—bubblelike organelles that bud off the ER and carry the protein to the nearest cisterna of the Golgi complex (fig. 3.12). The Golgi complex sorts these proteins and passes them along from one cisterna to the next. It may further cut and splice them or add other components such as carbohydrate chains. The Golgi complex finally buds off Golgi vesicles, which can remain in the cell as lysosomes or release products from the cell by exocytosis, as described earlier.

Before You Go On

Answer these questions from memory. Reread the preceding section if there are too many you don't know.

9. Which cytoskeletal component gives support to the plasma membrane? Which component can be compared to a monorail on which intracellular materials can travel from place to place in the cytoplasm? Which type lends toughness to your skin surface?

10. Which two organelles are surrounded by a double membrane? Which of these has shelflike infoldings of the inner membrane? What is the purpose of that organelle?

11. A red blood cell (RBC) escapes from the circulation into the tissues, where a macrophage finds it and phagocytizes it. What is that RBC to the macrophage at this point—an organelle or an inclusion? What is the basis for your decision?

12. Name three organelles involved in protein synthesis and describe their respective roles. Explain the difference between transcription and translation and state where in the cell each of these occurs.

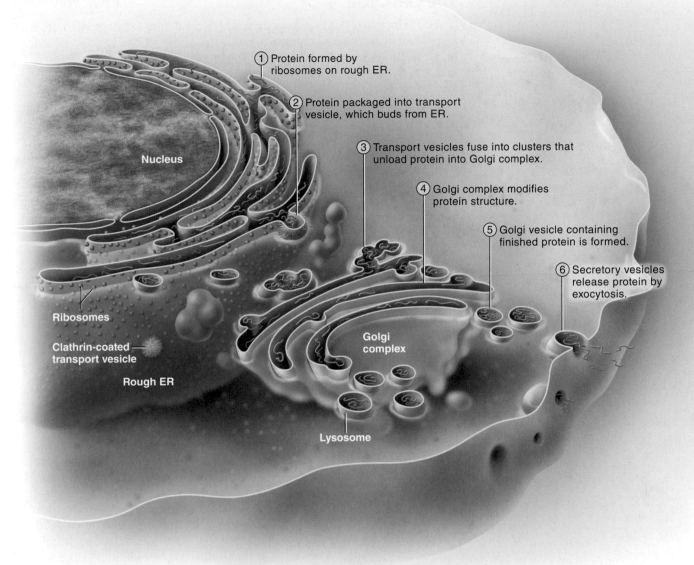

① Protein formed by ribosomes on rough ER.

② Protein packaged into transport vesicle, which buds from ER.

③ Transport vesicles fuse into clusters that unload protein into Golgi complex.

④ Golgi complex modifies protein structure.

⑤ Golgi vesicle containing finished protein is formed.

⑥ Secretory vesicles release protein by exocytosis.

Nucleus

Ribosomes

Clathrin-coated transport vesicle

Rough ER

Golgi complex

Lysosome

Figure 3.12 Organelles of Protein Synthesis and Secretion. The steps in protein synthesis and secretion are numbered ① through ⑥.

3.4 The Life Cycle of Cells

Expected Learning Outcomes

When you have completed this section, you should be able to:

a. describe the stages of a cell's life cycle and list the events that define each stage; and

b. name the stages of mitosis and describe what occurs in each.

The Cell Cycle

A basic principle of cell biology is that all cells arise from existing cells, both in an evolutionary sense, reaching across billions of years to the earliest cell, and in our lifetimes, when we grow from a single-celled fertilized egg to a fetus, infant, and adult. Most cells have a finite life span, during which they must accurately duplicate their DNA and separate the copies into two daughter cells. The life cycle of a cell extends from one division to the next. This **cell cycle** is divided into four main phases: G_1, S, G_2, and M (fig. 3.13).

The **first gap (G_1) phase** is an interval between cell division (the "birth" of two new cells from a parent cell) and DNA replication. During this time, a new cell synthesizes proteins, grows, and carries out its preordained tasks for the body, such as secreting enzymes if it's a digestive gland cell. Cells in G_1 also begin to replicate their centrioles in preparation for the next division, and they accumulate the materials needed in the next phase to replicate their DNA.

The **synthesis (S) phase** is a period in which a cell carries out DNA replication, doubling its DNA content in preparation for the upcoming cell division. Remember the law of complementary base pairing in DNA (see p. 53). If you knew that one strand of the DNA double helix had a base sequence ATCGCA, you could predict from this law that the one across from it must read TAGCGT. This predictability also enables a cell to produce duplicate copies of its DNA. In the S phase, each DNA molecule in the nucleus begins to "unzip" at several places along its length, separating the double helix into two strands (fig. 3.14, panel ①). At each of these sites, an enzyme called *DNA polymerase* (green in fig. 3.14, panel ②) reads the base sequence on one strand and assembles nucleotides in the right order to make a complementary strand. At each replication site, two of these enzymes work simultaneously, moving along their DNA strands in opposite directions. The end result of this process going on in numerous places at once on the 46 chromosomes is that, by the end of the S phase, the cell has two complete sets of identical DNA molecules. Each DNA molecule is composed of one strand from the old DNA and one that has been newly made by the polymerase (fig. 3.14, panel ③). Each chromosome, at that point, has two identical DNA molecules, which are available to be divided up between daughter cells at the next cell division.

The **second gap (G_2) phase** is a relatively brief interval between DNA replication and cell division. In G_2, a cell finishes replicating its centrioles and synthesizes enzymes that control cell division.

The **mitotic (M) phase** is the period in which a cell undergoes mitosis—it replicates its nucleus, divides its DNA into two identical sets (one per nucleus), and pinches in half to form two genetically identical daughter cells. The details of this phase are considered in the next section. Phases G_1, S, and G_2 are collectively called **interphase**—the time between cell divisions.

The length of the cell cycle varies greatly from one cell type to another. Cultured connective tissue cells called *fibroblasts* divide about once a day and spend about 18 to 24 hours in interphase and 1 to 2 hours in mitosis. Stomach and skin cells divide rapidly, bone and cartilage cells slowly,

Figure 3.13 The Cell Cycle.

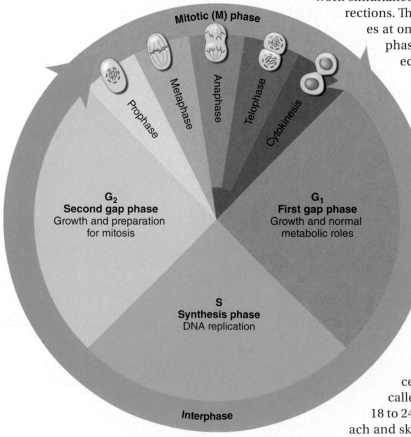

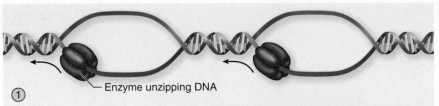

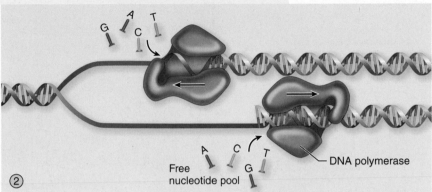

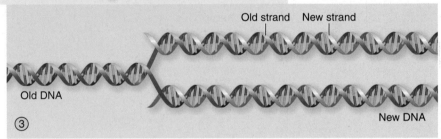

Figure 3.14 DNA Replication. ① Enzyme unzips DNA. ② DNA polymerase assembles nucleotides to make a complementary strand. ③ The process results in two identical DNA molecules. **AP|R**

and skeletal muscle and nerve cells not at all. Some cells leave the cell cycle and stand by without dividing for days, years, or the rest of one's life. The balance between cells that are actively cycling and those on standby is an important factor in determining the number of cells in the body. An inability to stop cycling and go on standby is characteristic of cancer cells (see Perspectives on Health, p. 92).

Mitosis

Cells divide by two mechanisms called mitosis and meiosis. Meiosis is restricted to one purpose, the production of eggs and sperm, and is therefore treated in chapter 19 on reproduction. **Mitosis** serves all the other functions of cell division: the development of a fertilized egg (one cell) into an individual composed of some 40 trillion cells; continued growth of all the organs after birth; the replacement of cells that die; and the repair of damaged tissues. Four phases of mitosis are recognizable—*prophase, metaphase, anaphase,* and *telophase* (fig. 3.15).

In **prophase,**[25] the chromosomes coil into short, dense rods that are easier to distribute to daughter cells than the long, delicate chromatin of interphase. At this stage, a chromosome consists of two genetically identical bodies called **sister chromatids,** joined together at a pinched spot called the **centromere**

---

[25] *pro* = first

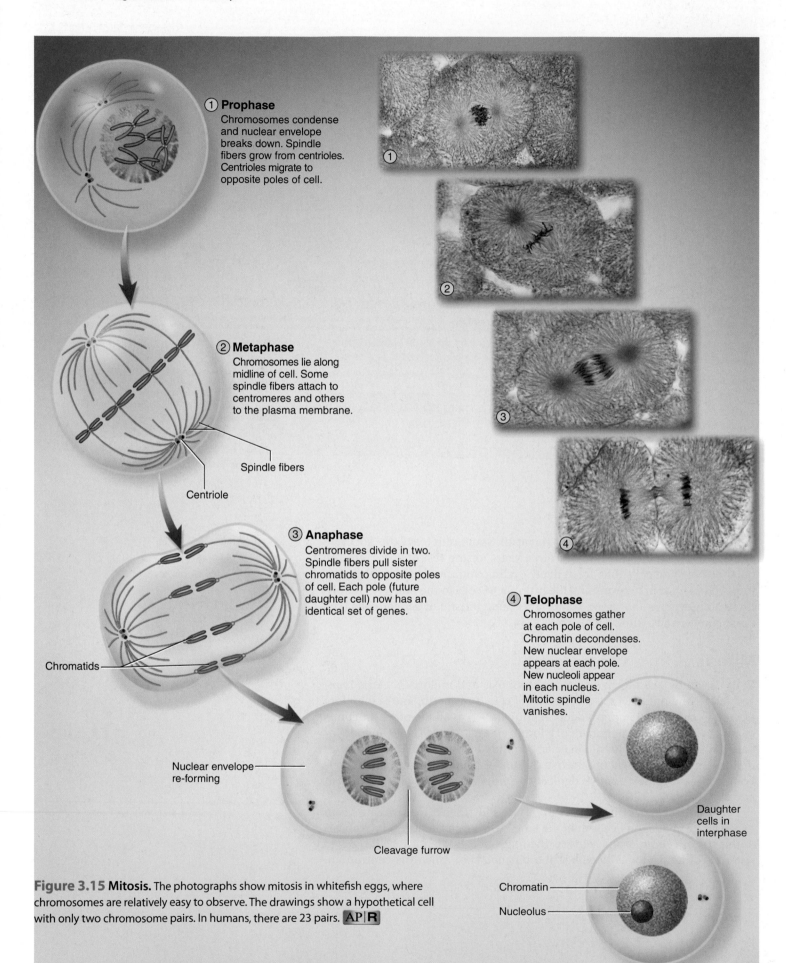

① **Prophase**
Chromosomes condense and nuclear envelope breaks down. Spindle fibers grow from centrioles. Centrioles migrate to opposite poles of cell.

② **Metaphase**
Chromosomes lie along midline of cell. Some spindle fibers attach to centromeres and others to the plasma membrane.

Spindle fibers

Centriole

③ **Anaphase**
Centromeres divide in two. Spindle fibers pull sister chromatids to opposite poles of cell. Each pole (future daughter cell) now has an identical set of genes.

④ **Telophase**
Chromosomes gather at each pole of cell. Chromatin decondenses. New nuclear envelope appears at each pole. New nucleoli appear in each nucleus. Mitotic spindle vanishes.

Chromatids

Nuclear envelope re-forming

Cleavage furrow

Daughter cells in interphase

Chromatin

Nucleolus

Figure 3.15 Mitosis. The photographs show mitosis in whitefish eggs, where chromosomes are relatively easy to observe. The drawings show a hypothetical cell with only two chromosome pairs. In humans, there are 23 pairs. AP|R

(fig. 3.16). There are 46 chromosomes, two chromatids per chromosome, and one molecule of DNA in each chromatid—or 92 DNA molecules in all. The nuclear envelope disintegrates during prophase and releases the chromosomes into the cytosol. The centrioles begin to sprout elongated microtubules called **spindle fibers,** which push the centrioles apart as they grow. Eventually, one pair of centrioles comes to lie at each pole of the cell. Microtubules grow toward the chromosomes and attach to the centromeres. The spindle fibers then tug the chromosomes back and forth until they line up along the midline of the cell.

In **metaphase,**[26] the chromosomes are aligned on the cell equator, oscillating slightly and awaiting a signal that stimulates each chromosome to split in two at the centromere. The spindle fibers now form a lemon-shaped array called the **mitotic spindle,** with long microtubules reaching out from each centriole to the chromosomes, and shorter microtubules anchoring the assembly to the inside of the plasma membrane at each end of the cell.

Anaphase[27] begins with activation of an enzyme that cleaves each centromere in two, separating the sister chromatids from each other. Each chromatid is now regarded as a separate, single-stranded *daughter chromosome.* One daughter chromosome migrates to each pole of the cell, with its centromere leading the way and the arms trailing behind. Since sister chromatids are genetically identical, and since each daughter cell receives one chromatid from each chromosome, the daughter cells of mitosis are genetically identical. Almost every cell of your body is genetically identical (with a few exceptions such as eggs, sperm, and certain immune cells).

In **telophase,**[28] the chromatids cluster on each side of the cell. The rough ER produces a new nuclear envelope around each cluster, and the chromatids begin to uncoil and return to the thinly dispersed chromatin form. The mitotic spindle breaks up and vanishes.

Telophase is the end of nuclear division but overlaps with cytokinesis[29] (SY-toe-kih-NEE-sis), division of the cytoplasm. Cytokinesis is achieved by a motor protein pulling on the membrane skeleton. This creates a crease called the *cleavage furrow* around the equator of the cell, and the cell eventually pinches in two. Interphase has now begun for these new cells; the life cycle begins anew.

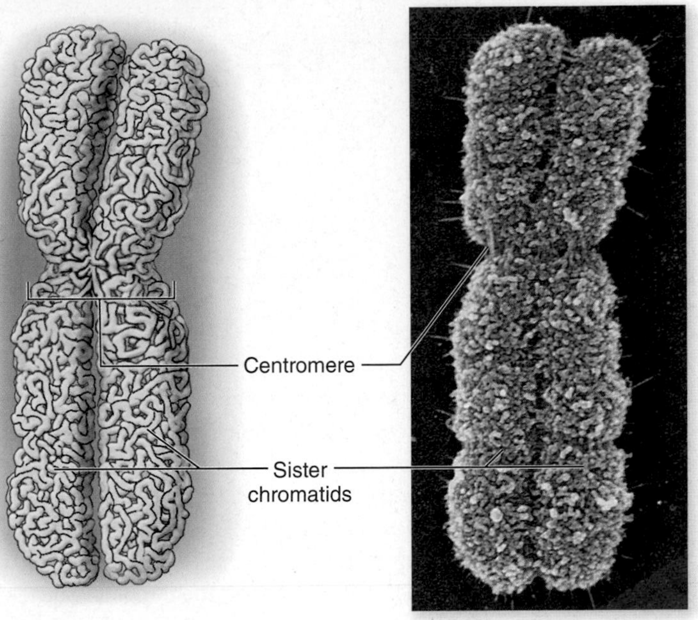

(a)   (b)

Figure 3.16 Chromosomes. (a) Drawing of a metaphase chromosome. (b) Scanning electron micrograph.

- *What are the chromatids composed of besides DNA?*

Before You Go On

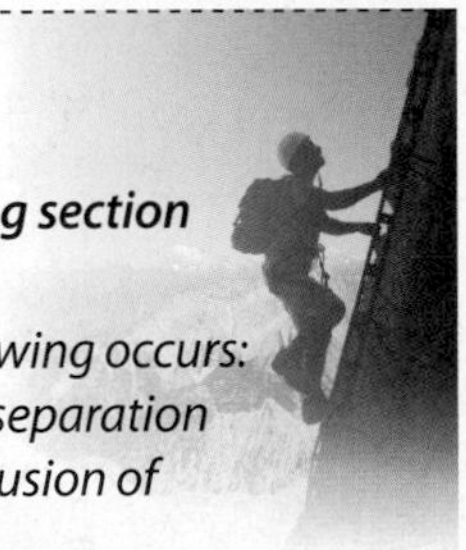

Answer these questions from memory. Reread the preceding section if there are too many you don't know.

13. Identify the stage of the cell cycle where each of the following occurs: (a) DNA replication, (b) onset of centriole replication, (c) separation of the two sets of DNA into separate nuclei, and (e) conclusion of centriole replication.

14. Identify the stage of mitosis in which each of the following occurs: (a) double-stranded chromosomes align on the cell equator awaiting a signal to split in two, (b) chromosomes coil and thicken and their chromatids become visible, (c) the centromeres split and the chromatids separate, (d) new nuclear envelopes form around each chromosome set.

---

[26] *meta* = next in a series

[27] *ana* = apart

[28] *telo* = end, final

[29] *cyto* = cell; *kinesis* = action, motion

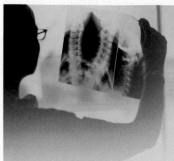

Cancer

Of all things that can go wrong with cellular function, cancer is probably the most dreaded. It's typically the first thing that comes to mind when one is found to have a tumor, or *neoplasm*.[30] The next question to arise in a worried patient's mind is often, Is it benign or malignant? If a tumor is *benign*[31] (be-NINE), this means the cellular mass is surrounded by a fibrous capsule, the cells can't freely break loose and spread to other sites in the body, and the tumor is relatively slow-growing. Even some benign tumors can be deadly, but most can be surgically removed, producing full recovery.

A *malignant*[32] tumor is fast-growing, lacks a fibrous capsule, and has cells capable of breaking free and spreading to other organs (*metastasizing*[33]). The word *cancer* refers only to malignant tumors. Cancer was named by Hippocrates, who compared the distended veins in some breast tumors to the outstretched legs of a crab.[34]

Most cancer is caused by mutations (changes in DNA or chromosome structure), which can be induced by chemicals, viruses, or radiation, or simply occur through errors in DNA replication during the cell cycle. Agents that cause mutation are called *mutagens*,[35] and those that induce cancer are also called *carcinogens*.[36] Cancer stems from mutations in two gene families, the oncogenes and tumor-suppressor genes. *Oncogenes*[37] are mutated genes that promote the secretion of excessive growth factors (chemicals that stimulate cell division) or excessive sensitivity of target cells to growth factors. *Tumor-suppressor (TS) genes* inhibit the development of cancer by opposing oncogenes, promoting DNA repair, and other means. Cancer occurs when TS genes are unable to perform this protective function. Oncogenes are like an accelerator to the cell cycle, while TS genes are like a brake. With either a "stuck accelerator" or "faulty brakes," cells divide at a rate exceeding cell death, thus amassing to form a tumor.

Untreated cancer is almost always fatal. Tumors destroy healthy tissue; they can grow to block major blood vessels or respiratory airways; they can damage blood vessels and cause hemorrhaging; they can compress and kill brain tissue; and they tend to drain the body of nutrients and energy as they hungrily consume a disproportionate share of the body's oxygen and nutrients. Treatments for cancer include surgery, chemotherapy, and radiation therapy.

Genomic Medicine

Genomic medicine is the application of our knowledge of the genome—one individual's complete set of genes—to the prediction, diagnosis, and treatment of disease. It is relevant to disorders as diverse as cancer, Alzheimer disease, schizophrenia, obesity, and even a person's susceptibility to nonhereditary diseases such as AIDS and tuberculosis.

As the technology of gene sequencing continues to be improved, geneticists expect that we may soon be able to sequence any person's entire genome for less than $1,000. Why would we want to? Because knowing one's genome could dramatically change clinical care. It may allow clinicians to forecast a person's risk of disease and to predict its course; mutations in a single gene can affect the severity of such diseases as hemophilia, muscular dystrophy, cancer, and cystic fibrosis. Genomics should also allow for earlier detection of diseases and for earlier, more effective clinical intervention. Drugs that are safe for most people can have serious side effects in others, owing to genetic variations in drug metabolism. Genomics may therefore provide a basis for choosing the safest or most effective drug and for adjusting dosages for different patients on the basis of their genetic makeup. Genomics is also playing a new role in drug design; *pharmacogenetics* is the science of examining the effects of genetic variation on drug action.

Knowing the sites of disease-producing mutations expands the potential for *gene-substitution therapy*. This is a procedure in which cells are removed from a patient with a genetic disorder, supplied with a normal gene in place of the defective one, and reintroduced to the body. The hope is that these genetically modified cells will proliferate and provide the patient with a gene product that he or she was lacking—perhaps insulin for a patient with diabetes or a blood-clotting factor for a patient with hemophilia. Attempts at gene therapy have been marred by some tragic setbacks, however, and still face great technical difficulties.

Although genomics is in its infancy, it has generated high hopes that the twenty-first century may be an era of *personalized medicine,* with treatments tailored to the genetic constitution of the individual. Yet genomic medicine is introducing new problems in medical ethics and law. Should your genome be a private matter between you and your physician? Or should an insurance company be entitled to know your genome before issuing health or life insurance to you, so they can know your risk of contracting a catastrophic illness, adjust the cost of your coverage, or even deny coverage? Should a prospective employer have the right to know your genome before offering employment? These are areas in which biology, politics, and law converge to shape public policy.

---

[30] *neo* = new; *plasm* = growth, formation
[31] *benign* = mild, gentle
[32] *mal* = bad, evil
[33] *meta* = beyond; *stas* = being stationary
[34] *cancer* = crab
[35] *muta* = change; *gen* = to produce
[36] *carcino* = cancer; *gen* = to produce
[37] *onco* = tumor

CAREER SPOTLIGHT

Cytotechnologist

A cytotechnologist is a person who examines tissue specimens for signs of cancer and other diseases. Cytotechnologists work under the supervision of pathologists in hospitals, clinics, commercial diagnostic laboratories, universities, and public health agencies. Using automated tissue preparation, microscopy, computer-assisted screening, and other techniques, a cytotechnologist scans a slide; marks areas of cellular abnormality such as changes in cell size, shape, color, or nuclear volume; and refers abnormal specimens to the pathologist for diagnosis. The tissue specimens studied by cytotechnologists include Pap smears and biopsies of breast, lymph node, liver, thyroid, lung, and many other tissues. To become a cytotechnologist, one must first earn a baccalaureate degree with coursework in biology, chemistry, and mathematics or statistics; attend a 1- to 2-year program in cytotechnology training at a university or hospital; and pass a board certification exam. The advanced training includes tissue-preparation techniques, microscopy, cytochemistry, anatomy and physiology, histology, immunology, and other subjects. Sophisticated molecular diagnostic methods also are a fast-growing area of cytotechnology education and practice. See appendix B for further career information.

Study Guide

Assess Your Learning Outcomes

To test your knowledge, discuss the following topics with a study partner or in writing, ideally from memory.

3.1 The General Structure of Cells (p. 64)

1. Fundamental components of a cell
2. Intracellular and extracellular fluids
3. The typical size range of human cells and what factors limit cell size

3.2 The Cell Surface (p. 66)

1. Molecular components and organization of the plasma membrane
2. Varieties and functions of the plasma membrane proteins
3. The composition, location, and functions of a cell's glycocalyx
4. Structural and functional distinctions between microvilli, cilia, and flagella
5. Structural distinctions and respective advantages of three types of cell junctions—tight junctions, desmosomes, and gap junctions
6. The eight modes of transport through a plasma membrane and how they differ with respect to the use of carrier proteins, direction of movement of the transported substances, and demand for ATP

3.3 The Cell Interior (p. 78)

1. Components and functions of the cytoskeleton
2. Types of cell inclusions and how inclusions differ from organelles
3. What organelles have in common and how they differ, as a class, from other cellular components
4. Structure of the nucleus, particularly of its nuclear envelope, chromatin, and nucleoli
5. Two forms of endoplasmic reticulum, their spatial relationship, their structural similarities and differences, and their functional differences
6. The composition, appearance, locations, and function of ribosomes
7. Structure of the Golgi complex and its role in the synthesis, packaging, and secretion of cell products

8. Similarities and differences between lysosomes and peroxisomes in structure, contents, and functions
9. Structure, function, and evolutionary origin of mitochondria, and the significance of mitochondrial DNA
10. Structure, locations, and functions of centrioles
11. The processes of genetic transcription and translation, including the roles of mRNA, rRNA, and tRNA
12. How the amino acid sequence of a protein is represented by the codons of mRNA
13. How proteins are processed and secreted after their assembly on a ribosome

3.4 The Life Cycle of Cells (p. 87)

1. Four phases of the cell cycle and the main events in each phase
2. How DNA is replicated in preparation for mitosis
3. Functions of mitosis
4. Four stages of mitosis; changes in chromosome structure and distribution that occur in each stage; and the role of centrioles and the mitotic spindle
5. The mechanism and result of cytokinesis

Testing Your Recall

1. The clear, structureless gel in a cell is its
   a. nucleoplasm.
   b. endoplasm.
   c. cytoplasm.
   d. neoplasm.
   e. cytosol.

2. New nuclei form and a cell begins to pinch in two during
   a. prophase.
   b. metaphase.
   c. interphase.
   d. telophase.
   e. anaphase.

3. The amount of ___ in a plasma membrane affects its fluidity.
   a. phospholipid
   b. cholesterol
   c. glycolipid
   d. glycoprotein
   e. integral protein

4. Cells specialized for absorption of matter from the extracellular fluid are likely to show an abundance of
   a. lysosomes.
   b. microvilli.

c. mitochondria.
d. secretory vesicles.
e. ribosomes.

5. A ___ serves as a mechanical linkage between adjacent cells but does not obstruct the movement of materials through the space between cells.
   a. glycocalyx
   b. phospholipid bilayer
   c. tight junction
   d. gap junction
   e. desmosome

6. The word root *phago-* means
   a. eating.
   b. drinking.
   c. emitting fluid.
   d. intracellular.
   e. extracellular.

7. The amount of DNA in a cell doubles during
   a. prophase.
   b. metaphase.
   c. anaphase.
   d. the S phase.
   e. the G_2 phase.

8. Fusion of a secretory vesicle with the plasma membrane and release of the vesicle's contents is
   a. exocytosis.
   b. receptor-mediated endocytosis.
   c. active transport.
   d. pinocytosis.
   e. phagocytosis.

9. Most cellular membranes are made by
   a. the nucleus.
   b. the cytoskeleton.
   c. enzymes in the peroxisomes.
   d. the endoplasmic reticulum.
   e. replication of existing membranes.

10. Which of the following is/are not involved in protein synthesis?
    a. ribosomes
    b. centrioles
    c. mRNA
    d. rough endoplasmic reticulum
    e. codons

11. Most human cells are 10 to 15 _____ wide.

12. When a hormone cannot enter a cell, it binds to a _____ at the cell surface.

13. _____ are channels in the plasma membrane that open or close in response to various stimuli.

14. Cells are somewhat protected from mechanical trauma by a spongy carbohydrate surface coat called the _____.

15. The separation of sister chromatids from each other marks the _____ stage of mitosis.

16. The majority of molecules that compose the plasma membrane are _____.

17. Two human organelles that are surrounded by a double membrane are the _____ and _____.

18. Liver cells can detoxify alcohol with two organelles, the _____ and _____.

19. Cells adhere to each other and to extracellular material by means of membrane proteins called _____.

20. A macrophage would use the process of _____ to engulf a dying tissue cell.

Answers in appendix A

True or False

Determine which five of the following statements are false, and briefly explain why.

1. The shape of a cell is determined more by its cytoskeleton than by its plasma membrane.

2. DNA replication occurs during mitosis.

3. A cell can release matter by exocytosis but not by phagocytosis or pinocytosis.

4. The hydrophilic heads of membrane phospholipids are in contact with both the ECF and ICF.

5. Water-soluble substances usually must pass through channel proteins to enter a cell.

6. Cells must use ATP to move substances down a concentration gradient.

7. Osmosis is a type of active transport involving water.

8. Cilia and flagella have an axoneme but microvilli do not.

9. Desmosomes enable substances to pass from cell to cell.

10. A nucleolus is an organelle within the nucleoplasm.

Answers in appendix A

Testing Your Comprehension

1. Breast milk contains both sugar (lactose) and proteins (albumin and casein). Identify which organelles of the mammary gland cells are involved in synthesizing and secreting these components, and describe the structural pathway from synthesis to release from the cell.

2. A person with lactose intolerance cannot digest lactose, so instead of being absorbed by the small intestine, this sugar passes undigested into the large intestine. Here, it causes diarrhea among other signs. Which of the membrane transport processes do you think is most directly involved in the diarrhea? On that basis, explain why the diarrhea occurs.

3. Consider a cardiac muscle cell, an enzyme-producing pancreatic cell, a phagocytic white blood cell, and a hormone-secreting cell of the ovary. Which of these would you expect to show the greatest number of lysosomes? Mitochondria? Rough endoplasmic reticulum? Smooth endoplasmic reticulum? Explain each answer.

Chapter 4

Histology—
The Tissue Level
of Organization

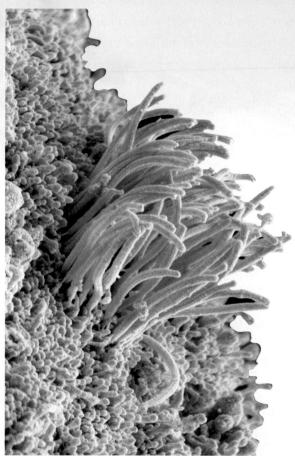

Cilia on a cell of the uterine (fallopian) tube. The synchronized beating of such cilia serves to move an egg or embryo toward the uterus.

BASE CAMP

Before ascending to the next level, be sure you're properly equipped with a knowledge of these concepts from earlier chapters:

- Body cavities and membranes (p. 18)
- Cilia and microvilli (p. 69)
- Secretory vesicles and exocytosis (p. 77)

Anatomy & Physiology **REVEALED**
aprevealed.com

Module 3: Tissues

With its 50 trillion cells and thousands of organs, the human body may seem to be a structure of forbidding complexity. Fortunately for our health, longevity, and self-understanding, the biologists of past generations were not discouraged by this, but discovered patterns that make it more understandable. One of these is the fact that these trillions of cells belong to only 200 types or so, and are organized into tissues that fall into just four broad categories—epithelial, connective, nervous, and muscular.

An organ is a structure with discrete boundaries that is composed of two or more of these tissue types (usually all four). Organs derive their function not from their cells alone but from how the cells are organized into tissues. This chapter describes the structural and functional characteristics of the major human tissues, and later chapters describe the histological organization of the respective organ systems.

4.1 The Study of Tissues

Expected Learning Outcomes

When you have completed this section, you should be able to:

a. name the four primary classes of adult tissues;

b. visualize the three-dimensional shape of a structure from a two-dimensional tissue section; and

c. interpret descriptive terms for various shapes of cells.

The Primary Tissue Classes

Histology[1] is the study of tissues and how they are arranged into organs. A tissue is a group of similar cells and cell products that arise from the same region of an embryo and work together to perform a specific structural or physiological role in an organ. The four *primary tissues*—epithelial, connective, nervous, and muscular (table 4.1)—differ from each other in the types and functions of their cells, the characteristics of the **matrix (extracellular material)** that surrounds the cells, and the relative amount of space occupied by cells versus matrix.

The matrix is nonliving matter secreted by the tissue cells. It is composed of fibrous proteins and, usually, a clear gel variously known as **ground substance** or **extracellular fluid (ECF).** In summary, a tissue is composed of cells and matrix, and the matrix is composed of fibers and ground substance.

---

[1] *histo* = tissue; *logy* = study of

Table 4.1 The Four Primary Tissue Classes

Type	Definition	Representative Locations
Epithelial	Tissue composed of layers of closely spaced cells; covers organ surfaces, forms glands, and serves for protection, secretion, and absorption	Epidermis; lining of digestive tract; liver and other glands
Connective	Tissue with usually more matrix than cell volume; often specialized to support, bind, and protect organs	Tendons and ligaments, cartilage, fat, bone, blood
Nervous	Tissue containing excitable cells specialized for rapid transmission of information to other cells	Brain, spinal cord, nerves
Muscular	Tissue composed of elongated, excitable cells specialized for contraction and movement	Skeletal muscles; heart; walls of uterus, bladder, intestines, and other internal organs

Interpreting Tissue Sections

Students of anatomy and physiology often study **histological sections**—tissues that have been cut into very thin slices, generally one or two cells thick, then mounted on microscope slides and dyed with various **stains.** These stains lend contrast to such structures as cell nuclei, cytoplasm, and extracellular fibers, but one must remember that these colors are entirely artificial; they do not represent the natural colors of the organs.

A tissue section shows a three-dimensional object as a two-dimensional image. You must keep this in mind and try to translate the microscopic image into a mental image of the whole structure, like trying to imagine what a whole loaf of bread looks like if you had seen only a few slices. Like the boiled egg and elbow macaroni in figure 4.1, an object may look quite different when it is cut at various levels, or *planes of section*. A coiled tube, such as a gland of

Figure 4.1 Three-Dimensional Interpretation of Two-Dimensional Images. (a) A boiled egg. Note that grazing sections (upper left and right) would miss the yolk, just as a tissue section may miss the nucleus of a cell and create an illusion that the cell did not have one. (b) Elbow macaroni, which resembles many curved ducts and tubules. A section far from the bend would give the impression of two separate tubules; a section near the bend would show two interconnected lumina (interior spaces); and a section still farther down could miss the lumen completely. (c) A coiled gland in three dimensions and as it would look in a vertical tissue section.

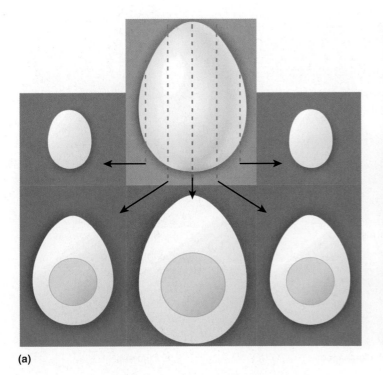

(a)

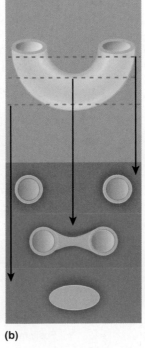

(b)

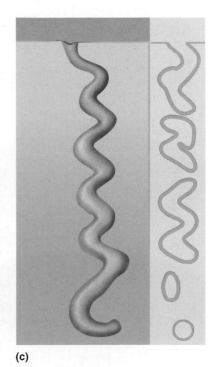

(c)

the uterus (fig. 4.1c), is often broken up into multiple portions since it meanders in and out of the plane of section. With experience, however, one recognizes that the separated pieces are parts of a single tube winding its way to the organ surface.

> **Apply What You Know**
>
> How would a cylindrical blood vessel look in two dimensions if cut lengthwise (a longitudinal section); if cut perpendicular to this (a cross section); and if cut on a slant between these two (an oblique section)?

Cell Shapes

Descriptions of organ and tissue structure often refer to the shapes of the cells by the following terms (fig. 4.2):

- **Squamous**[2] (SQUAY-mus)—a thin, flat, scaly shape. Squamous cells line the esophagus and form the surface layer (epidermis) of the skin.
- **Cuboidal**—squarish and about equal in height and width. Good examples are found in the kidney tubules and liver.
- **Columnar**—distinctly taller than wide, such as the inner lining cells of the stomach and intestines.
- **Polygonal**[3]—having irregular, angular shapes with four, five, or more sides. Squamous, cuboidal, and columnar cells often look polygonal when viewed from above rather than from the side.

---

[2] *squam* = scale; *ous* = characterized by
[3] *poly* = many; *gon* = angles

Figure 4.2 Terminology of Cell Shapes.

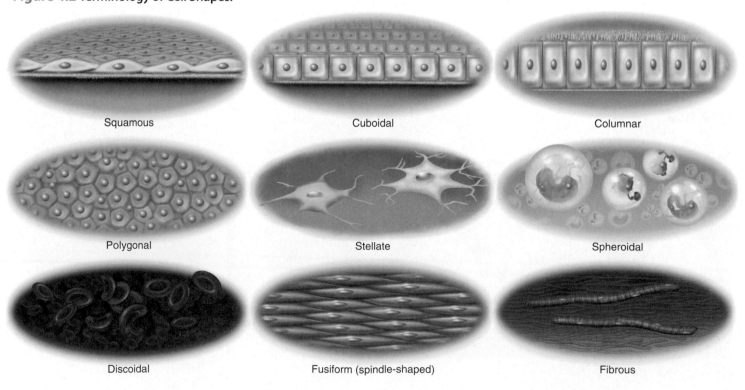

Squamous

Cuboidal

Columnar

Polygonal

Stellate

Spheroidal

Discoidal

Fusiform (spindle-shaped)

Fibrous

- **Stellate**[4]—having multiple pointed processes projecting from the body of a cell, giving it a somewhat starlike shape. Most nerve cells are stellate.
- **Spheroidal** to **ovoid**—round to oval, as in egg cells and white blood cells.
- **Discoidal**—disc-shaped, as in red blood cells.
- **Fusiform**[5] (FEW-zih-form)—spindle-shaped; elongated, with a thick middle and tapered ends, as in smooth muscle cells.
- **Fibrous**—long, slender, and threadlike, as in skeletal muscle cells.

Before You Go On

Answer these questions from memory. Reread the preceding section if there are too many you don't know.

1. *Classify each of the following into one of the four primary tissue classes: the epidermis, fat, the spinal cord, most heart tissue, bones, tendons, blood, and the inner lining of the stomach.*
2. *What are tissues composed of in addition to cells?*
3. *What is the term for a thin, stained slice of tissue mounted on a microscope slide?*

4.2 Epithelial Tissue

Expected Learning Outcomes

When you have completed this section, you should be able to:

a. describe the properties that distinguish epithelium from other tissue classes;

b. list and classify eight types of epithelium, distinguish them from each other, and state where each type can be found in the body;

c. discuss how the structure of each type of epithelium relates to its function; and

d. recognize epithelial types from specimens or photographs.

Epithelial[6] **tissue** consists of a flat sheet of closely spaced cells, one or more cells thick, like bricks in a wall. The upper surface is usually exposed to the environment; to an internal space of a body cavity; or to the **lumen,** or internal space, of a hollow organ. Epithelium covers the body surface, lines body cavities, forms the external and internal linings of many organs, and constitutes most gland tissue. Epithelia contain no blood vessels, but they almost always lie on a layer of loose connective tissue and depend on its blood vessels for nourishment and waste removal.

---

[4] *stell* = star; *ate* = characterized by

[5] *fusi* = spindle; *form* = shape

[6] *epi* = upon; *theli* = nipple, female

Between an epithelium and the underlying connective tissue is a layer called the **basement membrane,** composed mainly of protein. It anchors the epithelium to the connective tissue, regulates the exchange of materials between the epithelium and the underlying tissues, and binds growth factors from below that regulate epithelial development.

The cell surface attached to the basement membrane is called the **basal surface;** the upper surface, opposite from this, is the **apical surface;** and the **lateral surfaces** between these two form the sides of the cell. You could compare these to the floor, roof, and walls of a house, respectively. These surfaces often differ in their functions (such as absorption, secretion, attachment, or intercellular communication) and therefore differ in the components of their plasma membranes. Epithelial cells often have cilia, microvilli, or both on their apical surface.

Epithelia are classified into two broad categories—*simple* and *stratified*—with four types in each category. In a simple epithelium, every cell touches the basement membrane, whereas in a stratified epithelium, some cells rest on top of other cells (like a multistory apartment building) and do not extend to the basement membrane (fig. 4.3).

Simple Epithelia

A simple epithelium has only one layer of cells. Three types of simple epithelia are named for the shapes of their cells: **simple squamous** (thin scaly cells), **simple cuboidal** (squarish or rounded cells), and **simple columnar** (tall narrow cells). In the fourth type, **pseudostratified columnar,** not all cells reach the free surface; the shorter cells are covered over by the taller ones, but all of them reach the basement membrane—much like trees in a forest that reach different heights but are all anchored in the soil below. Simple columnar and pseudostratified epithelia often produce protective mucous coatings. The mucus is secreted by wineglass-shaped **goblet cells.** Table 4.2 (figs. 4.4 to 4.7) illustrates and summarizes the structural and functional differences between the four

Figure 4.3 Epithelial Types and Cell Shapes. Pseudostratified columnar epithelium is a special type of simple epithelium that gives a false impression of multiple cell layers. AP|R

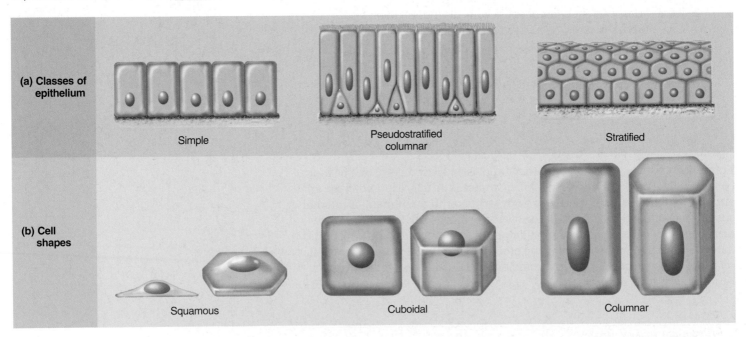

(a) Classes of epithelium

Simple

Pseudostratified columnar

Stratified

(b) Cell shapes

Squamous

Cuboidal

Columnar

Table 4.2 Simple Epithelia

Simple Squamous Epithelium	Simple Cuboidal Epithelium
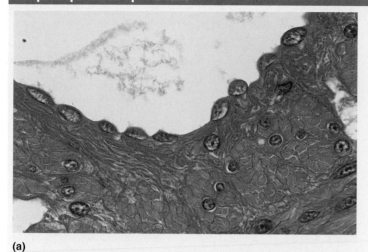 (a)	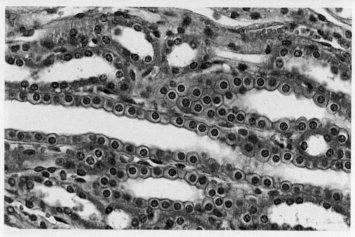 (a)
Squamous epithelial cells Nuclei of smooth muscle Basement membrane 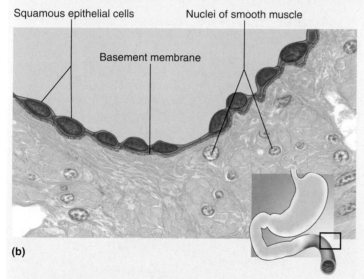 (b)	Lumen of kidney tubule Cuboidal epithelial cells Basement membrane 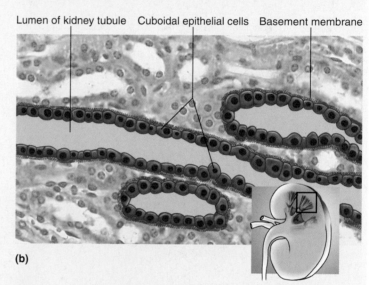 (b)

Figure 4.4 Simple Squamous Epithelium on the External Surface of the Small Intestine (x400). AP|R

Microscopic appearance: Single layer of thin cells, shaped like fried eggs with a bulge where the nucleus is located; nucleus somewhat flattened in the plane of the cell, like an egg yolk; cytoplasm may be so thin it is hard to see in tissue sections; in surface view, cells have angular contours and nuclei appear round

Representative locations: Air sacs (alveoli) of lungs; glomerular capsules of kidneys; some kidney tubules; inner lining (endothelium) of heart and blood vessels; serous membranes of stomach, intestines, and some other viscera; surface layer (mesothelium) of pleurae, pericardium, peritoneum, and mesenteries

Functions: Allows rapid diffusion or transport of substances through membranes; secretes lubricating serous fluid

Figure 4.5 Simple Cuboidal Epithelium in the Kidney Tubules (x400). AP|R

Microscopic appearance: Single layer of squarish or rounded cells; in glands, cells often pyramidal and arranged like segments of an orange around a central space; spherical, centrally placed nuclei; often with a brush border of microvilli in some kidney tubules; ciliated in bronchioles of lung

Representative locations: Liver, thyroid, mammary, salivary, and other glands; many gland ducts; most kidney tubules; bronchioles

Functions: Absorption and secretion; production of protective mucous coat; movement of respiratory mucus

(continued on next page)

Table 4.2 Simple Epithelia *(continued)*

Simple Columnar Epithelium	Pseudostratified Columnar Epithelium

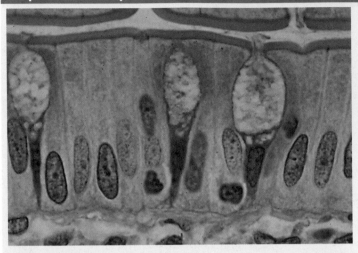

(a)

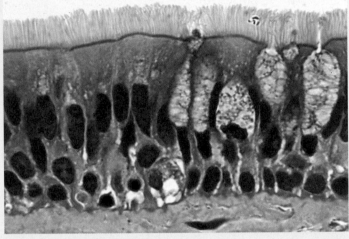

(a)

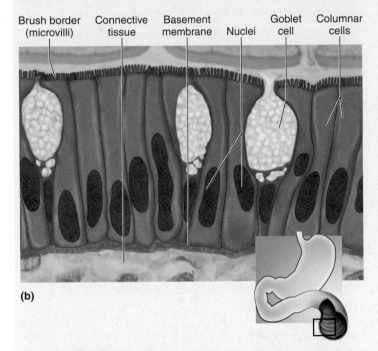

Brush border (microvilli)   Connective tissue   Basement membrane   Nuclei   Goblet cell   Columnar cells

(b)

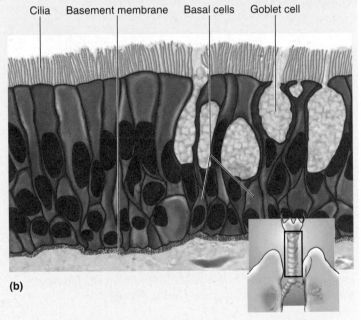

Cilia   Basement membrane   Basal cells   Goblet cell

(b)

Figure 4.6 Simple Columnar Epithelium on the Internal Surface (Mucosa) of the Small Intestine (x400). AP|R

Microscopic appearance: Single layer of tall, narrow cells; oval or sausage-shaped nuclei, vertically oriented, usually in basal half of cell; apical portion of cell often shows secretory vesicles visible with TEM; often shows a brush border of microvilli; ciliated in some organs; may possess goblet cells

Representative locations: Inner lining of stomach, intestines, gallbladder, uterus, and uterine tubes; some kidney tubules

Functions: Absorption and secretion; movement of egg and embryo in uterine tube; secretion of mucus

Figure 4.7 Ciliated Pseudostratified Columnar Epithelium in the Mucosa of the Trachea (x400). AP|R

Microscopic appearance: Looks multilayered; some cells do not reach free surface but all cells reach basement membrane; nuclei at several levels in deeper half of epithelium; often with goblet cells; often ciliated

Representative locations: Respiratory tract from nasal cavity to bronchi; portions of male reproductive tract

Functions: Secretes and propels respiratory mucus

types of simple epithelium. The magnifications given in these tables, such as ×400, are the magnifications at which the tissue would show this level of detail if you were at the microscope; they are not the magnifications attained by enlarging the images to the size presented on the page.

Stratified Epithelia

Stratified epithelia range from 2 cell layers to 20 or more. Three of the stratified epithelia are named for the shapes of their surface cells: **stratified squamous, stratified cuboidal,** and **stratified columnar epithelia.** The deeper cells may be of a different shape than the surface cells. The fourth type, **transitional epithelium,** was named when it was thought to represent a transitional stage between stratified squamous and stratified columnar epithelium. This is now known to be untrue, but the name has persisted. Stratified columnar epithelium is rare (occurring in short transitional zones where one epithelium type grades into another) and we will not consider it any further. The other three types are illustrated and summarized in table 4.3.

The most widespread epithelium in the body is stratified squamous epithelium, which deserves further discussion. Its deepest cells are cuboidal to columnar and undergo continual mitosis. Their daughter cells push toward the surface and become flatter (more *squamous*, or scaly) as they migrate farther upward, until they finally die and flake off (fig. 4.8). Their separation from the surface is called **exfoliation;**[7] the study of exfoliated cells is called *exfoliate cytology*—for example, in a Pap smear (see fig. 19.8, p. 643).

There are two kinds of stratified squamous epithelia—keratinized and nonkeratinized. A **keratinized**[8] epithelium, found in the epidermis, is covered with a layer of compact, dead squamous cells. These cells are packed with the durable protein *keratin* and coated with a water repellent. The skin surface is therefore relatively dry, it retards water loss from the body, and it resists penetration by disease organisms. The tongue, oral mucosa, esophagus, vagina, and a few other internal surfaces are covered with the **nonkeratinized** type, which lacks the surface layer of dead cells. This type provides a surface that is, again, abrasion-resistant, but also moist and slippery. These characteristics are well suited to resist stress produced by the chewing and swallowing of food and by sexual intercourse and childbirth.

Table 4.3 (figs. 4.9 to 4.12) illustrates and summarizes the characteristics of the most common types of stratified epithelium.

---

[7] *ex* = out of, from; *foli* = leaf; *ation* = process (to strip away leaves)

[8] *kerat* = horn (named for animal horns)

Figure 4.8 Exfoliation of Squamous Cells from the Mucosa of the Vagina. [© Dr. Richard Kessel & Dr. Randy Kardon/Tissues & Organs/Visuals Unlimited, Inc.]

• *What are some other places in the body where you would expect to find similar exfoliation?*

Table 4.3 Stratified Epithelia

Stratified Squamous Epithelium—Keratinized	Stratified Squamous Epithelium—Nonkeratinized

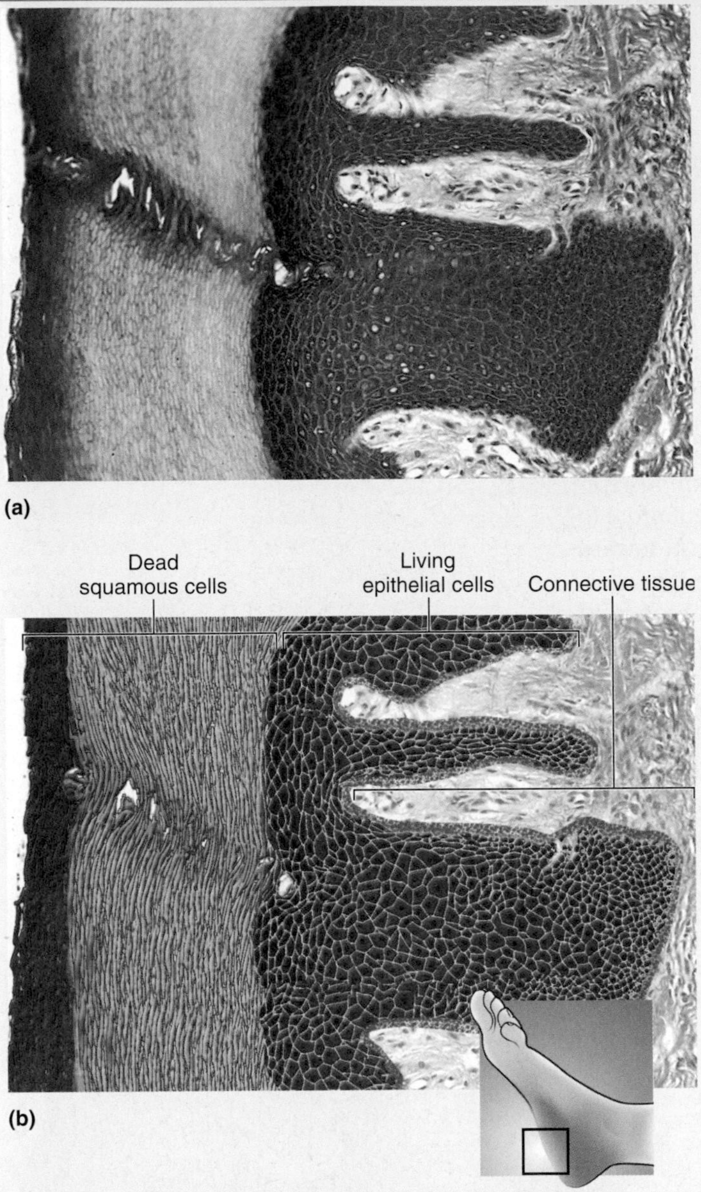

(a)

Dead squamous cells · Living epithelial cells · Connective tissue

(b)

Figure 4.9 Keratinized Stratified Squamous Epithelium on the Sole of the Foot (x400). AP|R

Microscopic appearance: Multiple cell layers with cells becoming increasingly flat and scaly toward surface; surface covered with a layer of compact dead cells without nuclei; basal cells may be cuboidal to columnar

Representative locations: Epidermis; palms and soles are especially heavily keratinized

Functions: Resists abrasion; retards water loss through skin; resists penetration by pathogenic organisms

(a)

Living epithelial cells · Connective tissue

(b)

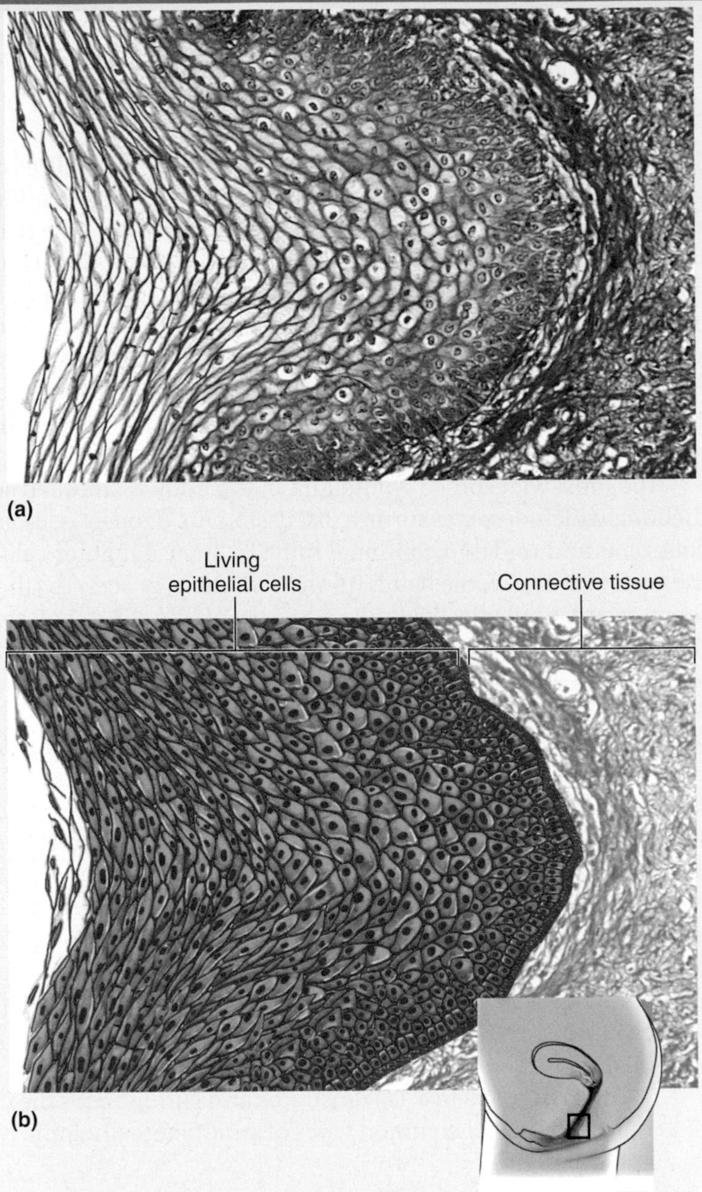

Figure 4.10 Nonkeratinized Stratified Squamous Epithelium in the Mucosa of the Vagina (x400). AP|R

Microscopic appearance: Same as keratinized epithelium but without the surface layer of dead cells

Representative locations: Tongue, oral mucosa, esophagus, anal canal, vagina

Functions: Resists abrasion and penetration by pathogenic organisms while providing a moist slippery surface

Table 4.3 Stratified Epithelia *(continued)*

Stratified Cuboidal Epithelium	Transitional Epithelium

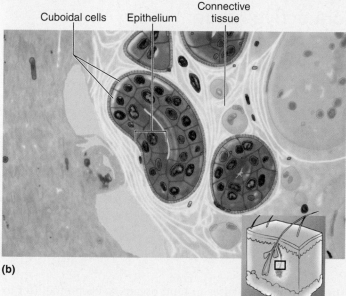

(a)

(a)

(b)

Cuboidal cells   Epithelium   Connective tissue

Basement membrane   Connective tissue   Binucleate epithelial cell

(b)

Figure 4.11 Stratified Cuboidal Epithelium in the Duct of a Sweat Gland (x400). AP|R

Microscopic appearance: Two or more layers of cells; surface cells squarish or rounded

Representative locations: Sweat gland ducts; egg-producing vesicles (follicles) of ovaries; sperm-producing ducts (seminiferous tubules) of testes

Functions: Contributes to sweat secretion; secretes ovarian hormones; produces sperm

Figure 4.12 Transitional Epithelium in the Kidney (x400). AP|R

Microscopic appearance: Somewhat resembles stratified squamous epithelium, but surface cells are rounded, not flattened, and often bulge above surface; typically five or six cells thick when relaxed, two or three cells thick when stretched; cells may be flatter and thinner when epithelium is stretched (as in a distended bladder); some cells have two nuclei

Representative locations: Limited to urinary tract after birth—part of kidney, ureter, bladder, part of urethra; found in umbilical cord of fetus

Functions: Stretches to allow filling of urinary tract; protects deeper cells and tissues from osmotic damage by concentrated urine

4.3 Connective Tissue

Expected Learning Outcomes

When you have completed this section, you should be able to:

a. describe the properties that most connective tissues have in common;

b. discuss the types of cells found in connective tissue;

c. explain what the matrix of a connective tissue is and describe its components;

d. list 10 types of connective tissue, describe their cellular components and matrix, and explain what distinguishes them from each other; and

e. visually recognize each type from specimens or photographs.

Overview

Connective tissue serves in most cases to bind organs to each other (for example, the way a tendon connects a muscle to a bone) or to support and protect organs. The volume of the extracellular matrix is greater than the volume occupied by its cells. Most cells are not in direct contact with each other, but are separated by extracellular material. Most connective tissue is richly supplied with blood vessels. This is the most abundant, widely distributed, and histologically variable of the four primary tissues.

The functions of connective tissue include the following:

- **Binding of organs.** Tendons bind muscle to bone, ligaments bind one bone to another, and fat holds the kidneys and eyes in place.
- **Support.** Bones support the body, and cartilage supports the ears, nose, trachea, and bronchi.
- **Physical protection.** Bones protect delicate organs such as the brain, lungs, and heart; fat and fibrous capsules around the kidneys and eyes cushion these organs.
- **Immune protection.** Connective tissue cells attack foreign invaders.
- **Movement.** Bones provide the lever system for body movement, cartilages are involved in movement of the vocal cords, and cartilages on bone surfaces ease joint movements.

- **Storage.** Fat is the body's major energy reserve; bone is a reservoir of calcium and phosphorus that can be drawn upon when needed.
- **Heat production.** Brown fat generates heat in infants and children.
- **Transport.** Blood transports gases, nutrients, wastes, hormones, and blood cells.

Fibrous Connective Tissue

The most diverse connective tissues are in a class called *fibrous connective tissue* (or *connective tissue proper*). Nearly all connective tissues contain fibers, but they are especially conspicuous in this class. Its most common fiber type is composed of **collagen,**[9] the body's most abundant protein. Collagenous (col-LADJ-eh-nus) fibers are tough, flexible, and resist stretching. They compose such animal products as gelatin, leather, and glue. Tendons, ligaments, and the deep layer of the skin (dermis) are made mainly of collagen, but less visible collagen fibers pervade the matrix of cartilage and bone. In fresh tissue, collagenous fibers have a glistening white appearance, as seen in tendons and some cuts of meat (fig. 4.13).

Thin, glycoprotein-coated collagen fibers called **reticular**[10] **fibers** form the matrix of a fibrous connective tissue called **reticular tissue,** which provides a spongelike framework for such organs as the spleen and lymph nodes. **Elastic fibers** are composed of a stretchy protein called *elastin.* These fibers are thinner than collagenous fibers, and they branch and rejoin. Elastic fibers account for the ability of the skin, lungs, and arteries to spring back after they are stretched.

The dominant cells of fibrous connective tissue are **fibroblasts**[11]—large cells that often taper at the ends and show slender, wispy branches. They produce the fibers and ground substance that form the matrix. Also common in fibrous connective tissues are large phagocytic cells called **macrophages,**[12] which engulf and destroy bacteria, other foreign particles, and dead or dying cells of our own body. **Leukocytes,**[13] or **white blood cells (WBCs),** spend most of their time in the fibrous connective tissues, crawling about and providing various forms of defense against bacteria, toxins, and other foreign agents. Fat cells, or **adipocytes,** also appear in isolation or in small clusters in fibrous connective tissue, although when they dominate an area, the tissue is called *adipose tissue.*

The cells and fibers of connective tissue are embedded in ground substance. This material has little to no microscopic structure of its own, but is a fairly uniform material that ranges from a fluid or gel in some connective tissues to the rubbery texture of cartilage and stony texture of bone. Its texture results primarily from large protein–carbohydrate complexes and the water they absorb and retain, although in bone, calcium phosphate and other minerals harden the matrix. The ground substance of a fibrous connective tissue absorbs compressive forces and, like the styrofoam packing in a shipping carton, protects the more delicate cells from mechanical injury.

Types of Fibrous Connective Tissue

Fibrous connective tissue is divided into two broad categories according to the relative abundance of fiber: *loose* and *dense.* In **loose connective tissue,** the fibers are widely spaced, running in apparently random directions; the cells tend to be widely separated; and there is an abundance of ground substance, which looks like empty space in routine tissue specimens. These characteristics are especially conspicuous in *areolar*[14] *tissue* (AIR-ee-OH-lur) (fig. 4.14). In **dense connective tissue,** most of the space is occupied by closely packed fibers. Tables 4.4 and 4.5 (figs. 4.14 to 4.17) summarize the types of loose and dense fibrous connective tissues.

[9] *colla* = glue; *gen* = producing
[10] *ret* = network; *icul* = little
[11] *fibro* = fiber; *blast* = producing
[12] *macro* = big; *phage* = eater
[13] *leuko* = white; *cyte* = cell
[14] *areola* = little space

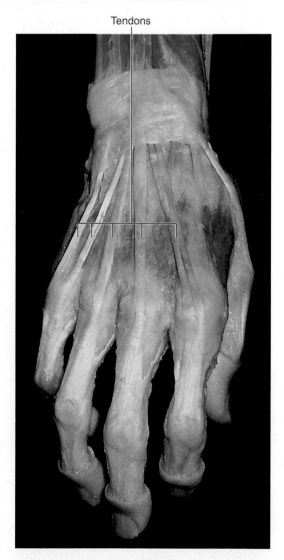

Figure 4.13 Tendons of the Hand. The white glistening appearance results from the collagen of which tendons are composed. The braceletlike band across the wrist is also composed of collagen.

Tendons

Table 4.4 Loose Connective Tissues

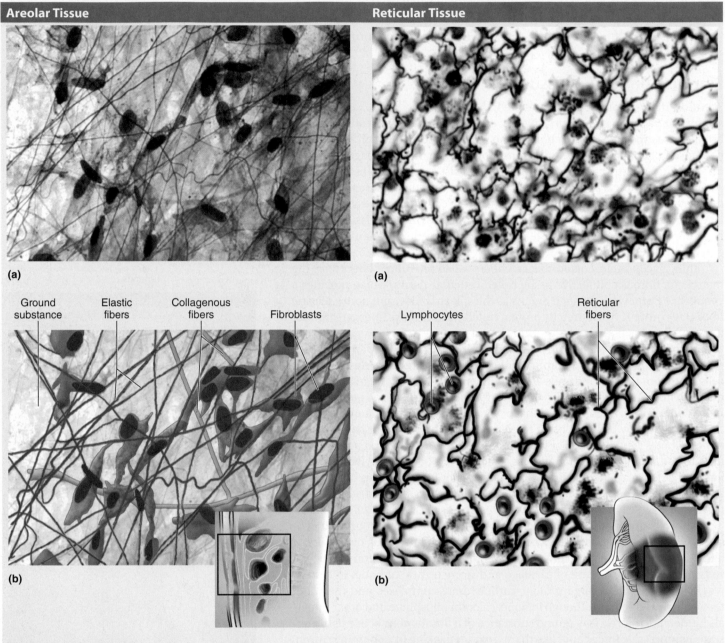

| Areolar Tissue | Reticular Tissue |

(a)

Ground substance   Elastic fibers   Collagenous fibers   Fibroblasts

Lymphocytes   Reticular fibers

(b)

Figure 4.14 Areolar Tissue in a Spread of the Mesentery (x400). AP|R

Figure 4.15 Reticular Tissue of the Spleen (x400). AP|R

Microscopic appearance: Loose, random-looking arrangement of predominantly collagenous and elastic fibers; scattered cells of various types; abundant ground substance; numerous blood vessels

Representative locations: Underlying nearly all epithelia; surrounding blood vessels, esophagus, and trachea; fascia between muscles; mesenteries; visceral layers of pericardium and pleura

Functions: Loosely binds epithelia to deeper tissues; allows passage of nerves and blood vessels through other tissues; provides an arena for immune defense; and its blood vessels provide nutrients and waste removal for overlying epithelia

Microscopic appearance: Loose network of reticular fibers, reticular cells, and fibroblasts, infiltrated with numerous lymphocytes and other blood cells

Representative locations: Lymph nodes, spleen, thymus, bone marrow

Functions: Supportive stroma (framework) for lymphatic organs

Table 4.5 Dense Connective Tissues

Dense Regular Connective Tissue	Dense Irregular Connective Tissue

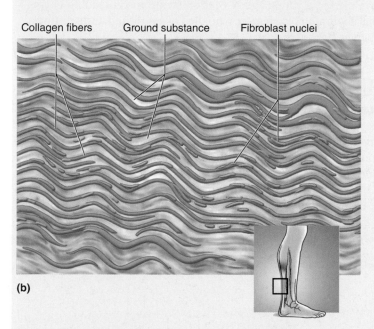

(a)

Collagen fibers   Ground substance   Fibroblast nuclei

(b)

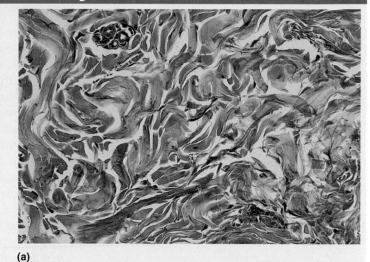

(a)

Bundles of collagen fibers   Gland ducts   Fibroblast nuclei   Ground substance

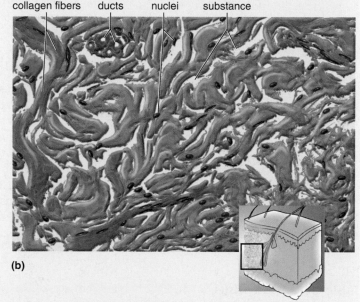

(b)

Figure 4.16 Dense Regular Connective Tissue of a Tendon (x400). AP|R

Figure 4.17 Dense Irregular Connective Tissue in the Dermis of the Skin (x400). AP|R

Microscopic appearance: Usually has the form of densely packed, parallel, often wavy collagen fibers; slender fibroblast nuclei compressed between collagen bundles; few cells other than fibroblasts; scanty ground substance; few blood vessels; has the form of parallel wavy sheets of elastic tissue in arteries

Representative locations: Tendons, ligaments, vocal cords, arteries

Functions: Ligaments tightly bind bones together, stabilize joints, determine their range of motion, and resist stress; tendons attach muscle to bone and transfer muscular tension to bones; parallel arrangement of fibers is an adaptation to forces acting in a single consistent direction, as when a tendon pulls a bone; stretches and recoils in arteries to accommodate surges in blood pressure and relieve pressure on smaller downstream vessels

Microscopic appearance: Densely packed collagen fibers running in seemingly random directions; scanty open space (ground substance); few visible cells; long fibers in the tissue appear as short, chopped-up pieces in thin histological sections, as in photo

Representative locations: Deeper portion of dermis of skin; capsules around viscera such as liver, kidney, spleen; fibrous sheaths around cartilages, bones, and nerves

Functions: Durable, hard to tear; variable orientation of fibers withstands stresses applied in unpredictable directions

Clinical Application 4.1

MARFAN SYNDROME—A CONNECTIVE TISSUE DISEASE

Serious anatomical and functional abnormalities can result from hereditary errors in the structure of connective tissue proteins. *Marfan*[15] *syndrome,* for example, results from the mutation of a gene on chromosome 15 that codes for a glycoprotein called *fibrillin,* the structural scaffold for elastic fibers. Clinical signs of Marfan syndrome include unusually tall stature, long limbs and spidery fingers, abnormal spinal curvature, and a protruding "pigeon breast." Some other signs include hyperextensible joints, hernias of the groin, and visual problems resulting from abnormally long eyeballs and deformed lenses. More seriously, victims exhibit a weakening of the heart valves and arterial walls. The aorta, where blood pressure is the highest, is sometimes enormously dilated close to the heart, and may suddenly rupture. Marfan syndrome is present in about 1 out of 20,000 live births and kills most victims by their mid-30s. Abraham Lincoln's tall, gangly physique and spindly fingers led some authorities to claim that he had Marfan syndrome; however, the evidence is inconclusive. A number of star athletes have died at a young age of Marfan syndrome, including Olympic volleyball champion Flo Hyman, who died of a ruptured aorta during a game in Japan in 1986, at the age of 31.

Adipose Tissue

Adipose tissue, or **fat,** is connective tissue in which adipocytes are the dominant cell type (table 4.6, fig. 4.18). The cells are tightly packed together and the narrow spaces between them are occupied by areolar tissue, reticular tissue, and blood capillaries. Adipose tissue serves primarily as an energy reservoir. It undergoes a constant turnover of stored triglyceride, with an equilibrium between synthesis and hydrolysis, energy storage and energy use. Adipose tissue also provides thermal insulation, anchors and cushions such organs as the eyeballs and kidneys, and contributes to body contours such as the female breasts and hips.

[15] Antoine Bernard-Jean Marfan (1858–1942), French physician

Table 4.6 Adipose Tissue

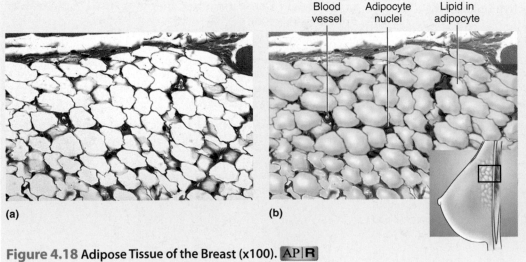

Microscopic appearance: Dominated by adipocytes—large, empty-looking cells with thin margins; tissue sections often very pale because of scarcity of stained cytoplasm; adipocytes shrunken; nucleus pressed against plasma membrane; blood vessels often present

Representative locations: Subcutaneous fat beneath skin; breast; heart surface; mesenteries; surrounding organs such as kidneys and eyes

Functions: Energy storage; thermal insulation; protective cushion for some organs; filling space, shaping body; heat production by brown fat

Figure 4.18 Adipose Tissue of the Breast (x100). AP|R

Most adipose tissue is a type called *white fat.* Until recently, only fetuses, infants, and children were thought to also have a heat-generating tissue called *brown fat,* which accounts for up to 6% of an infant's weight. Discovery of brown fat in adults has generated excitement because it burns calories and may contribute to weight loss. Brown fat gets its color from an unusual abundance of mitochondria, and stores lipid in the form of multiple globules rather than one large one. When brown fat breaks down its triglycerides, it releases energy only as heat, and does not produce ATP.

Supportive Connective Tissue

Cartilage and bone are less flexible than other connective tissues and thus provide physical support for various organs and the body as a whole.

Cartilage

Cartilage is a supportive connective tissue with a rubbery matrix. In meat, we know it as gristle. Cartilage gives shape to the external ear, the tip of the nose, and the larynx (voice box)—the most easily palpated cartilages in the body. The cells of cartilage are called called **chondrocytes**[16] (CON-dro-sites). They secrete the matrix and surround themselves with it until they become trapped in little cavities called **lacunae**[17] (la-CUE-nee). Chondrocytes often occur in little clusters called *cell nests,* descended from the same mother cell. Cartilage is free of blood capillaries, so its nutrition and waste removal depend on diffusion through the stiff matrix. Because this is such a slow process, chondrocytes have low rates of metabolism and cell division, and injured cartilage heals slowly. The matrix contains collagen fibers that range in thickness from invisibly fine to conspicuously coarse. Differences in the fibers provide a basis for classifying cartilage into three types: *hyaline*[18] *cartilage* (HY-uh-lin), *elastic cartilage,* and *fibrocartilage* (table 4.7, figs. 4.19 to 4.21). The first two types are usually covered with a sheath of dense irregular connective tissue called the **perichondrium**[19] (PERR-ee-CON-dree-um).

Bone

Bone, or **osseous tissue,** is a connective tissue with a hard, calcified matrix. Most tissue sections presented for study are of a type called **compact (dense) bone,** an opaque white tissue of the skeletal surface (table 4.8, fig. 4.22). This tissue is arranged in microscopic cylinders that surround **central canals,** which run longitudinally through the shafts of long bones such as the femur. Blood vessels and nerves travel through the central canals. The bone matrix is deposited in concentric layers called **lamellae** around each central canal; in cross sections, these look like the layers of an onion slice. Bone cells, called **osteocytes,**[20] occupy little cavities called **lacunae** between these layers. The structure of compact bone will be examined in further detail in chapter 6.

At the organ level, a bone such as the humerus or femur consists of an outer shell of compact bone enclosing a more porous type of osseous tissue called **spongy bone.** Bone marrow occupies the interior spaces of the spongy bone and the hollow shafts of the long bones. The bone as a whole is covered with a tough fibrous **periosteum** (PERR-ee-OSS-tee-um) similar to the perichondrium of cartilage.

---

[16] *chondro* = cartilage, gristle; *cyte* = cell

[17] *lacuna* = lake, cavity

[18] *hyal* = glass

[19] *peri* = around; *chondri* = cartilage, gristle

[20] *osteo* = bone; *cyte* = cell

Table 4.7 Cartilage

Hyaline Cartilage	Elastic Cartilage	Fibrocartilage

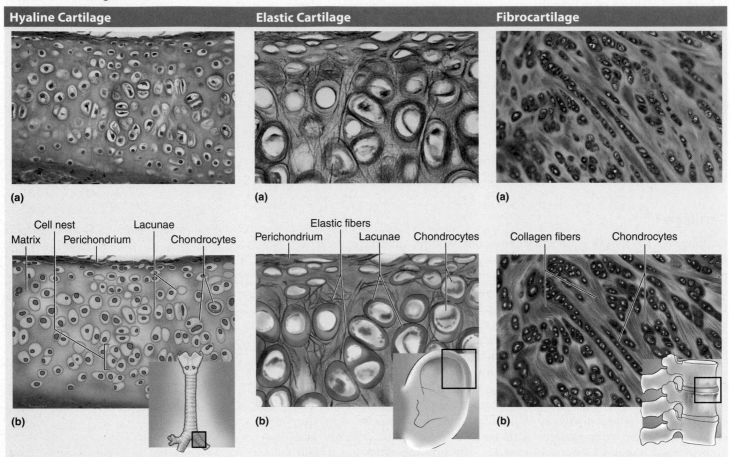

Figure 4.19 Hyaline Cartilage of the Fetal Skeleton (x400). AP|R

Microscopic appearance: Clear, glassy matrix, often stained light blue or pink in tissue sections; fine, dispersed collagen fibers, not usually visible; chondrocytes and lacunae often grouped in cell nests; usually covered by perichondrium

Representative locations: Forms a thin articular cartilage, lacking perichondrium, over the ends of bones at movable joints; costal cartilages attach ends of the ribs to the breastbone; forms supportive rings and plates around trachea and bronchi; forms a boxlike enclosure around the larynx; forms much of the fetal skeleton

Functions: Eases joint movements; holds airway open during respiration; moves vocal cords during speech; a precursor of bone in the fetal skeleton and growth zones of the long bones of children

Figure 4.20 Elastic Cartilage of the External Ear (x1,000). AP|R

Microscopic appearance: Elastic fibers form weblike mesh amid lacunae; always covered by perichondrium

Representative locations: External ear; epiglottis

Functions: Provides flexible, elastic support; gives shape to the ear

Figure 4.21 Fibrocartilage of an Intervertebral Disc (x400). AP|R

Microscopic appearance: Coarse, parallel bundles of collagen fibers similar to those of tendon; rows of chondrocytes in lacunae between collagen bundles; never has a perichondrium

Representative locations: Pubic symphysis (anterior joint between two halves of pelvic girdle); intervertebral discs that separate bones of vertebral column; menisci, or pads of shock-absorbing cartilage, in knee joint; at points where tendons insert on bones near articular hyaline cartilage

Functions: Resists compression and absorbs shock in some joints; often a transitional tissue between dense connective tissue and hyaline cartilage (for example, at some tendon–bone junctions)

Table 4.8 Compact Bone

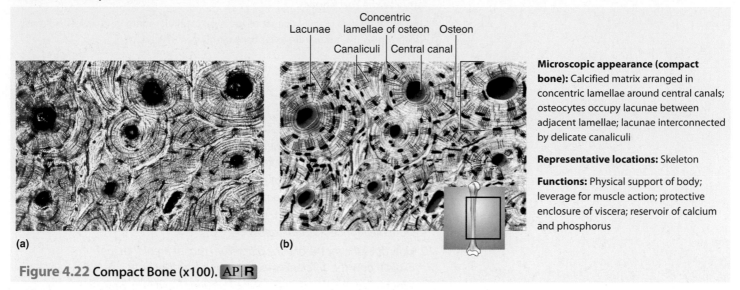

Lacunae
Concentric lamellae of osteon   Osteon
Canaliculi   Central canal

Microscopic appearance (compact bone): Calcified matrix arranged in concentric lamellae around central canals; osteocytes occupy lacunae between adjacent lamellae; lacunae interconnected by delicate canaliculi

Representative locations: Skeleton

Functions: Physical support of body; leverage for muscle action; protective enclosure of viscera; reservoir of calcium and phosphorus

(a)   (b)

Figure 4.22 Compact Bone (x100). AP|R

Blood

Blood (table 4.9, fig. 4.23) is a liquid connective tissue that travels through tubular vessels. Its primary function is to transport cells and dissolved matter such as oxygen and nutrients from place to place. Blood consists of a ground substance called **plasma,** and cells and cell fragments collectively called **formed elements.** The formed elements are of three kinds: (1) **Red blood cells (RBCs),** or **erythrocytes**[21] (eh-RITH-ro-sites), transport oxygen. (2) **White blood cells (WBCs),** or **leukocytes,** combat infections and other foreign or unwanted agents. (3) **Platelets** function in blood clotting and blood vessel repair and maintenance. Leukocytes are of five kinds distinguished by function, nuclear shape, and cytoplasmic appearance: *neutrophils, eosinophils, basophils, lymphocytes,* and *monocytes.* All of these are described in more detail in chapter 12.

[21] *erythro* = red; *cyte* = cell

Table 4.9 Blood

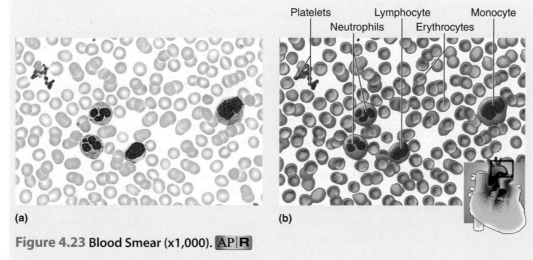

Platelets   Lymphocyte   Monocyte
Neutrophils   Erythrocytes

Microscopic appearance: Erythrocytes (RBCs) appear as pale pink discs with light centers and no nuclei; leukocytes (WBCs) are slightly larger, are much fewer, and have variously shaped nuclei, usually stained violet; platelets are cell fragments with no nuclei, about one-quarter the diameter of erythrocytes

Representative locations: Contained in heart and blood vessels; WBCs also occupy loose connective tissues

Functions: Transports gases, nutrients, wastes, chemical signals, and heat throughout body; provides defensive leukocytes; contains clotting agents to minimize bleeding; platelets secrete growth factors that promote tissue maintenance and repair

(a)   (b)

Figure 4.23 Blood Smear (x1,000). AP|R

Apply What You Know

Connective tissues usually consist of more extracellular material than cell volume. Which connective tissues are exceptions to this generalization?

Before You Go On

Answer these questions from memory. Reread the preceding section if there are too many you don't know.

8. What features do most or all connective tissues have in common to set this class apart from others?

9. List the cell and fiber types found in fibrous connective tissues and state their functional differences.

10. How does the matrix of bone differ from that of areolar tissue? Explain how the differences relate to the tissue functions.

11. What is areolar tissue? How can it be distinguished from any other kind of connective tissue?

12. Describe some similarities, differences, and functional relationships between hyaline cartilage and bone.

13. What are the three basic classes of formed elements in blood, and what are their respective functions?

4.4 Nervous and Muscular Tissue— The Excitable Tissues

Expected Learning Outcomes

When you have completed this section, you should be able to:

a. explain what distinguishes excitable tissues from other tissues;

b. name the cell types that compose nervous tissue;

c. identify the major parts of a nerve cell;

d. visually recognize nervous tissue from specimens or photographs;

e. name the three kinds of muscular tissue and describe the differences between them; and

f. identify types of muscular tissue from specimens or photographs.

Excitability is a characteristic of all living cells, but it is developed to its highest degree in nervous and muscular tissue, which are therefore described as excitable tissues. The basis for their excitation is an electrical charge on the plasma membrane. Nervous and muscular tissues exhibit quick electrical responses when stimulated. Nerve cells respond to stimulation by rapidly transmitting signals to other cells. Muscle cells respond by developing tension, often shortening and pulling on other tissues.

Nervous Tissue

Nervous tissue (table 4.10, fig. 4.24) consists of **neurons** (NOOR-ons), or nerve cells, and a much greater number of supporting cells called **glial** (GLEE-ul) **cells,** which protect and assist the neurons. Neurons are specialized to detect stimuli, respond quickly, and transmit information rapidly to other cells. A typical neuron has a prominent **cell body** that houses the nucleus; several short filamentous processes called **dendrites** that receive signals and transmit messages to the cell body; and one long process, the **axon** or **nerve fiber,** that sends outgoing signals to other cells. Variations in the structure of neurons and nervous tissue are described in chapters 8 and 9.

Muscular Tissue

Muscular tissue consists of elongated cells that are specialized to contract and exert a force on other tissues and organs. Not only do movements of the body as a whole depend on muscle, but so do such processes as digestion, waste elimination, breathing, speech, facial expression, and blood circulation. There are three types of muscle—*skeletal, cardiac,* and *smooth* (table 4.11, figs. 4.25 to 4.27). Skeletal and cardiac muscle exhibit fine, transverse, light and dark bands called **striations**[22] (stry-AY-shuns), resulting from the internal arrangement of the proteins involved in their contraction; thus, they are called **striated muscle.** The contractile proteins are less uniformly arranged in smooth muscle, which therefore lacks striations. Skeletal muscle is usually under voluntary control, and is thus called **voluntary muscle.** Cardiac and smooth muscle are called **involuntary muscle** because we usually cannot consciously control their contractions.

Most smooth muscle is found in the walls of hollow organs such as the stomach, intestines, uterus, and urinary bladder, where it is also called **visceral muscle.** The cells of skeletal muscle are commonly called **muscle fibers** because of their extraordinarily long, slender shape, whereas the shorter cells of cardiac and smooth muscle are often called **myocytes.**[23] Skeletal muscle fibers are further discussed in chapter 7 and cardiac myocytes in chapter 13.

---

[22] *striat* = striped

[23] *myo* = muscle; *cyte* = cell

Table 4.10 Nervous Tissue AP|R

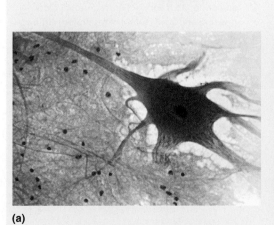
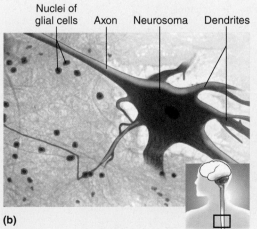

Nuclei of glial cells   Axon   Neurosoma   Dendrites

Microscopic appearance: Most nervous tissue sections or smears show a few large neurons, usually with rounded or stellate cell bodies and fibrous processes (axon and dendrites) extending from the somas; neurons are surrounded by a greater number of much smaller glial cells

Representative locations: Brain, spinal cord, nerves, ganglia

Function: Internal communication

(a)   (b)

Figure 4.24 A Neuron and Glial Cells of the Spinal Cord (x400). AP|R

Table 4.11 Muscular Tissue AP|R

Skeletal Muscle	Cardiac Muscle	Smooth Muscle

(a)

Nuclei   Striations   Muscle fiber

(b)

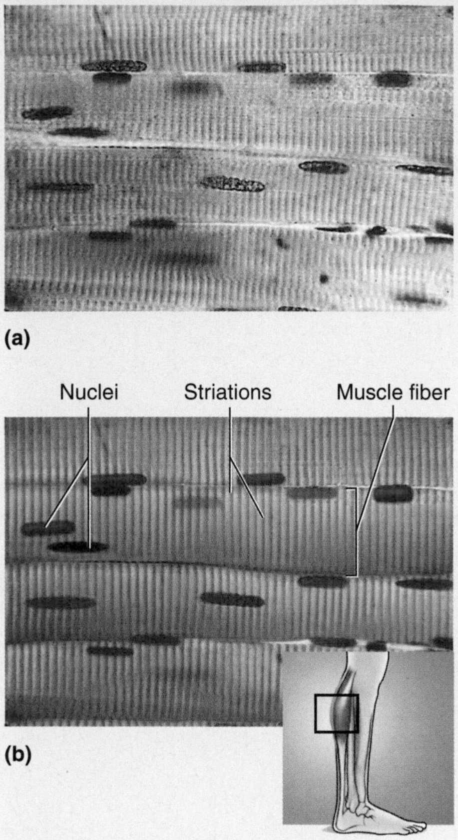

Figure 4.25 Skeletal Muscle (x400).
AP|R

Microscopic appearance: Long, threadlike, unbranched muscle fibers, with relatively parallel appearance in longitudinal tissue sections; striations; each cell with multiple nuclei lying near the plasma membrane

Representative locations: Skeletal muscles, mostly attached to bones but also in the tongue, esophagus, and voluntary *sphincters*[24] (circular muscles) of the eyelids, urethra, and anus

Functions: Body movements, facial expression, posture, breathing, speech, swallowing, control of urination and defecation, and assistance in childbirth; under voluntary control

(a)

Intercalated discs   Striations   Glycogen

(b)

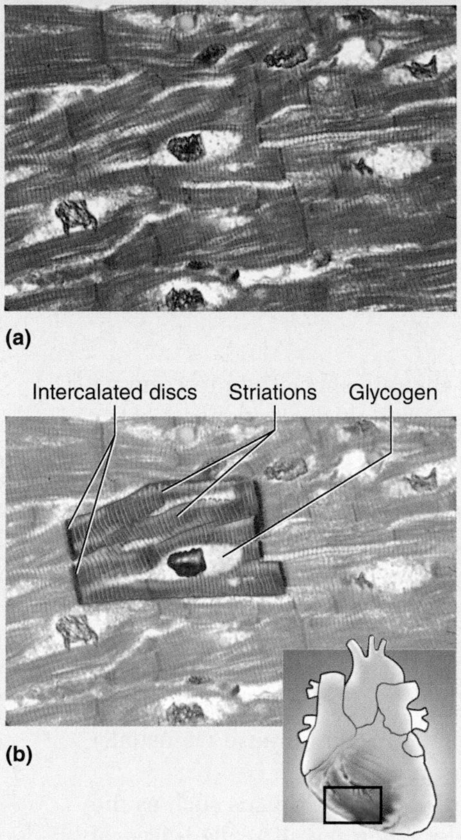

Figure 4.26 Cardiac Muscle (x400).
AP|R

Microscopic appearance: Short branched cells with a less parallel appearance in tissue sections; striations; thicker dark bands called *intercalated*[25] discs, containing mechanical and electrical junctions, where cells meet end to end; one nucleus per cell, centrally located and often surrounded by a light zone of glycogen

Representative locations: Heart only

Functions: Pumping of blood; under involuntary control

(a)

Nuclei   Muscle cells

(b)

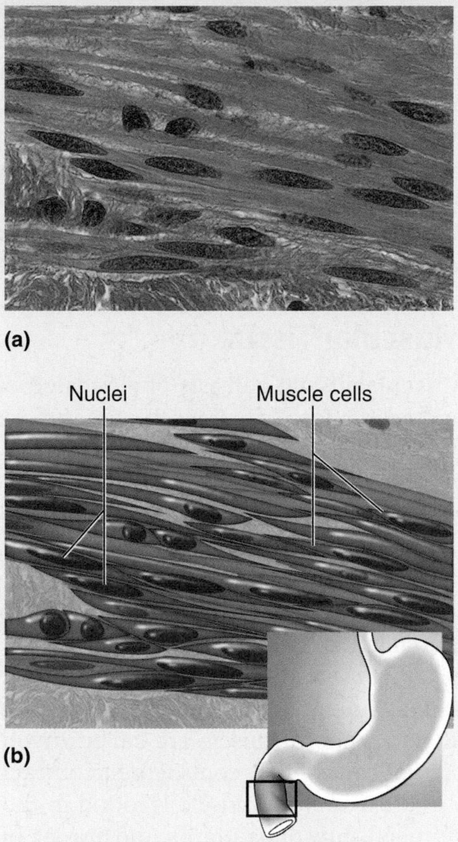

Figure 4.27 Smooth Muscle of the Intestinal Wall (x1,000). AP|R

Microscopic appearance: Short fusiform myocytes overlapping each other; nonstriated; one nucleus per cell, centrally located

Representative locations: Usually found as sheets of tissue in walls of viscera; also in iris and associated with hair follicles; involuntary sphincters of urethra and anus

Functions: Swallowing; contractions of stomach and intestines; retention and expulsion of feces and urine; labor contractions; control of blood pressure and flow; control of respiratory airflow; control of pupillary diameter; erection of hairs; under involuntary control

---

[24] *sphinc* = squeeze, bind tightly

[25] *inter* = between; *calated* = inserted

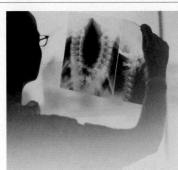

PERSPECTIVES ON HEALTH

Tissue Engineering

Tissue repair is not only a natural process but also a lively area of research in biotechnology. *Tissue engineering* is the artificial production of tissues and organs in the laboratory for implantation into the human body. The process commonly begins with building a scaffold (supportive framework) of collagen or biodegradable polyester, sometimes in the shape of a desired organ such as a blood vessel or ear. The scaffold is seeded with human cells and put in a "bioreactor" to grow. The bioreactor supplies nutrients, oxygen, and growth factors. It may be an artificial chamber or the body of a human patient or laboratory animal. When a lab-grown tissue reaches a certain point, it is implanted into the patient.

Tissue-engineered skin grafts have long been on the market. Scientists are not yet close to anything as complex as a lab-grown heart, but some are working on components such as valves, coronary arteries, patches of cardiac tissue, and whole heart chambers. Others have grown liver, bone, ureter, tendon, intestinal, and breast tissue in the laboratory. Scientists at the University of Massachusetts and the Massachusetts Institute of Technology have grown a "human" outer ear on the back of a mouse. They seeded a polymer scaffold with human cartilage cells and grew it on an immunodeficient mouse unable to reject the human tissue. They see potential in growing ears and noses for cosmetic treatment of children with birth defects or who have suffered disfiguring injuries from playground fights, accidents, or animal bites. In 2006, scientists at Children's Hospital in Boston reported that seven patients from ages 4 to 19 were living with tissue-engineered urinary bladders. These had been engineered almost entirely, other than the basal region where the ureters join the bladder, from cells harvested elsewhere in the body to replace bladders that had been defective.

The Stem Cell Controversy

One of the most controversial scientific and religious issues in the last several years revolves around histology—stem cell research. Stem cells are immature cells with the ability to differentiate into one or more types of mature, specialized tissues. *Adult stem (AS) cells* exist in most of the body's tissues, where they multiply and replace older cells that are lost to damage or normal cellular turnover. *Embryonic stem (ES) cells* compose human embryos up to about 150 cells. There is hope that stem cells can be manipulated to replace a broad range of tissues, such as cardiac muscle damaged by a heart attack, injured spinal cords, brain cells lost to Parkinson and Alzheimer diseases, or the insulin-secreting cells needed by people with diabetes mellitus.

Embryonic stem cells are less limited than adult stem cells in developmental potential; they are called *pluripotent* because they can develop into any type of fetal or adult cell. ES cells are often obtained from the excess embryos created in fertility clinics when a couple attempts to conceive a child by in vitro fertilization (IVF). AS cells, in contrast, are difficult to harvest from mature organs and maintain in culture. Furthermore, it remains uncertain whether AS cells can be manipulated into producing all cell types needed to treat such a broad range of diseases; they have already started down a path that limits their developmental versatility. For such reasons, ES cells have generated the greatest hope that they could be made to grow into new tissues of almost any kind, using cells that would otherwise have been discarded and wasted by IVF clinics.

However, stem cell technology has been embroiled in political, religious, and ethical debate. Some would argue that since the excess embryos of IVF clinics are destined to be destroyed, it would seem sensible to use them to save lives and restore health. Others argue, however, that potential medical benefits cannot justify the destruction of a human embryo, even one that consists of scarcely more than 100 cells. Further scientific advances, however, are beginning to defuse the controversy to some degree. Cell biologists have developed methods to make adult stem cells reverse course on their developmental paths, going back to a pluripotent state that would enable them to go down new roads. In principle, an AS cell that was originally destined to become smooth muscle, for example, might be made to revert to a pluripotent state, then be chemically guided down a path leading to nervous tissue for Alzheimer patients or persons paralyzed by spinal cord injury. Such modified adult cells, now called *induced pluripotent stem (iPS) cells,* have begun to show great promise for both medical benefit and reduced controversy.

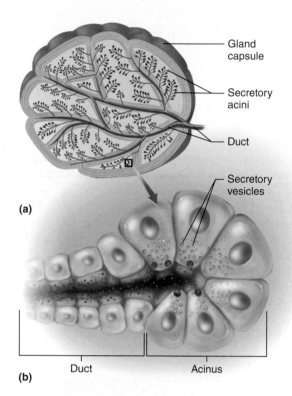

(a)

(b)

Duct   Acinus

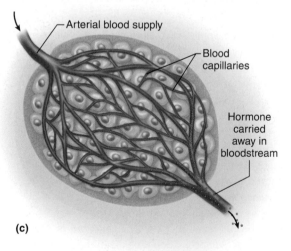

(c)

Figure 4.28 Exocrine and Endocrine Glands.
(a) General structure of an exocrine gland. The duct branches repeatedly until its finest branches terminate in sacs of secretory cells called acini. (b) Detail of an exocrine acinus and beginning of a duct. (c) General structure of an endocrine gland. The secretions of an endocrine gland (hormones) enter blood capillaries in the gland and are carried away in the bloodstream.

Before You Go On

Answer these questions from memory. Reread the preceding section if there are too many you don't know.

14. What do nervous and muscular tissue have in common? What is the primary function of each?

15. What two principal cell types compose nervous tissue, and how can they be distinguished from each other?

16. Name the three kinds of muscular tissue, describe how to distinguish them from each other in microscopic appearance, and state a location and function for each one.

4.5 Glands and Membranes

Expected Learning Outcomes

When you have completed this section, you should be able to:

a. describe or define various kinds of glands;

b. compare modes of glandular secretion; and

c. describe the structure of mucous and serous membranes.

Glands

A **gland** is a cell or organ that releases substances for use elsewhere in the body or for elimination from the body. Its product is called a **secretion** if it is useful to the body (such as a digestive enzyme or hormone) and an **excretion** if it is a waste product (such as urine or sweat).

Gland Types

Glands are classified as exocrine or endocrine. **Exocrine**[26] (EC-so-crin) **glands** release a secretion onto an epithelial surface and usually into the lumen (internal cavity) of another organ. Typically, the secretion is produced by a group of cells that form a microscopic secretory sac called an **acinus** (ASS-ih-nus) (fig. 4.28a, b). One side of the acinus leads into an epithelial **duct.** These ducts converge until one or a few main ducts exit the gland. Enclosing the gland, there is usually a fibrous **capsule** that sets it off from surrounding tissues. Extensions of the capsule often subdivide the gland into lobes and smaller compartments, and branches of the duct typically travel through these capsular extensions. Some examples of exocrine glands are sweat, mammary, and salivary glands, and the kidneys and liver.

 Endocrine[27] (EN-doe-crin) **glands** have no ducts, but instead, release their products into the bloodstream (fig. 4.28c). Their secretions are called **hormones,** and serve as chemical signals to other organs. Some examples of

[26] *exo* = out; *crin* = to separate, secrete

[27] *endo* = in, into; *crin* = to separate, secrete

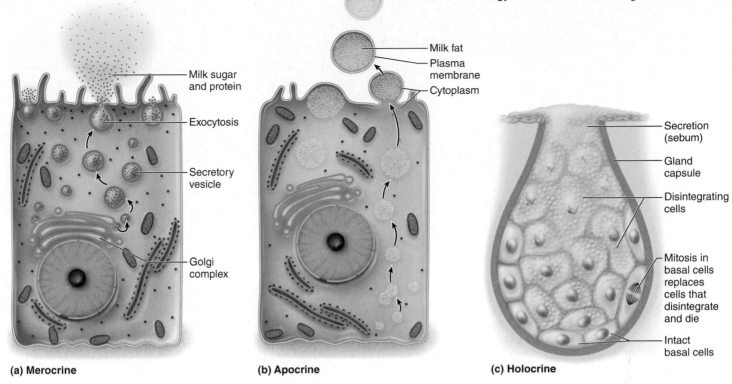

(a) Merocrine

Milk sugar and protein

Exocytosis

Secretory vesicle

Golgi complex

(b) Apocrine

Milk fat

Plasma membrane

Cytoplasm

(c) Holocrine

Secretion (sebum)

Gland capsule

Disintegrating cells

Mitosis in basal cells replaces cells that disintegrate and die

Intact basal cells

endocrine glands are the pituitary, thyroid, and adrenal glands. Some glands, such as the pancreas, ovaries, and liver, have both exocrine and endocrine functions. Endocrine glands are discussed in further detail in chapter 11; the rest of this discussion concerns exocrine glands.

Exocrine glands are classified by the nature of their secretions. **Serous** (SEER-us) **glands** produce thin, watery fluids such as perspiration, milk, tears, and digestive juices. **Mucous glands,** found in the digestive and respiratory tracts among other places, secrete a glycoprotein called *mucin* (MEW-sin). After it is secreted, mucin absorbs water, swells, and forms sticky *mucus.* (Note that *mucus,* the secretion, is spelled differently from *mucous,* the adjective form of the word.) **Mixed glands,** such as the two pairs of salivary glands in the floor of the mouth, contain both serous and mucous cells and produce a mixture of secretions.

Exocrine glands are further classified according to how they release their secretions. Most of them are **merocrine**[28] (MERR-oh-crin) glands, which release their products by means of exocytosis (fig. 4.29a). These include the tear glands, pancreas, and most others. Mammary glands secrete the sugar and protein of milk by this method, but secrete the milk fat by another method called **apocrine**[29] secretion (fig. 4.29b). Lipids coalesce from the cytosol into a single droplet that buds from the cell surface, covered by a layer of plasma membrane and a very thin film of cell cytoplasm. Sweat glands of the axillary (armpit) region were once thought to use the apocrine method as well. They are commonly called *apocrine sweat glands* even though it is now known that they secrete by the ordinary merocrine method.

Holocrine[30] **glands,** in which the gland cells break down entirely and *become* the secretion, are relatively rare (fig. 4.29c). The principal example of these is the oil-producing *sebaceous* (seh-BAY-shus) *glands* of the skin (see chapter 5). Others are found in the margins of the eyelids.

Figure 4.29 Modes of Exocrine Secretion.
(a) Merocrine secretion in a cell of the mammary gland, secreting milk sugar (lactose) and proteins by exocytosis. (b) Apocrine secretion of fat by a cell of the mammary gland. Fat droplets coalesce in the cytosol and then bud off from the cell surface with a thin coating of cytoplasm and plasma membrane. (c) Holocrine secretion by a sebaceous (oil) gland of the scalp. In this method, entire gland cells break down and become the secretion (sebum).

- *Which of these glands would require the highest rate of mitosis in its secretory cells? Why?*

---

[28] *mero* = part; *crin* = to separate, secrete

[29] *apo* = from, off, away; *crin* = to separate, secrete

[30] *holo* = whole, entire; *crin* = to separate, secrete

Membranes

Mucous membranes (fig. 4.30a) line tracts of the body that open to the exterior environment: the digestive, respiratory, urinary, and reproductive tracts. A mucous membrane, also called a **mucosa,** consists of an epithelium (usually nonkeratinized stratified squamous, simple columnar, or pseudostratified columnar) overlying a layer of areolar tissue and often a thin layer of smooth muscle. Mucous membranes have absorptive, secretory, and protective functions. They are often covered with mucus secreted by goblet cells, multicellular mucous glands, or both. The mucus traps bacteria and foreign particles, which keeps them from invading the tissues and aids in their removal from the body. This is an important defense for body passages open to the germ- and debris-laden air. The epithelium of a mucous membrane may also include absorptive, ciliated, and other types of cells.

Serous (SEER-us) **membranes** (fig. 4.30b; see also fig. 4.4) line the thoracic and abdominal cavities and cover the external surfaces of organs such as the stomach and intestines. They include the pleura around each lung, the pericardium around the heart, and the peritoneum of the abdominal cavity. They are composed of a simple squamous epithelium resting on a thin layer of areolar connective tissue. Serous membranes produce watery **serous fluid,** which arises from the blood and derives its name from the fact that it is similar to blood serum in composition.

Figure 4.30 Mucous and Serous Membranes. (a) A mucous membrane (mucosa) such as the inner lining of the trachea. (b) A serous membrane (serosa) such as the external lining of the small intestine.

- *Where are some other places one could find each type of membrane?*

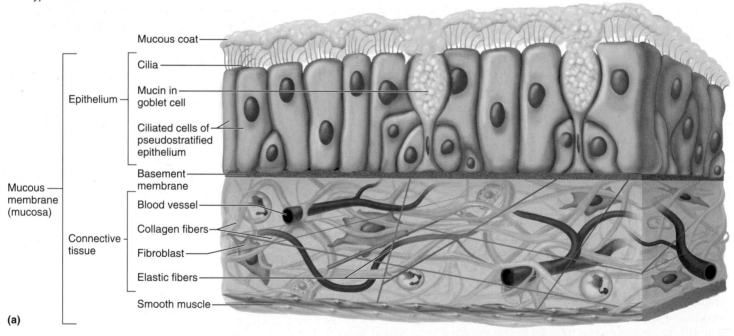

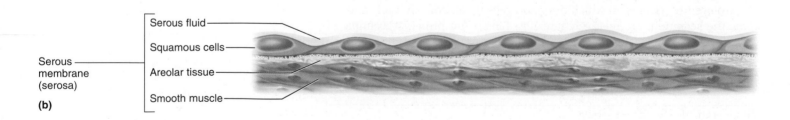

(a)

(b)

Before You Go On

Answer these questions from memory. Reread the preceding section if there are too many you don't know.

17. Distinguish between endocrine and exocrine glands, and give an example of each.

18. Distinguish between serous and mucous secretions.

19. Contrast the merocrine and holocrine methods of secretion, and name a gland product produced by each method.

20. Distinguish between serous and mucous membranes, and state where each may be found.

4.6 Tissue Growth, Development, Repair, and Death

Expected Learning Outcomes

When you have completed this section, you should be able to:

a. name and describe the modes of tissue growth;

b. distinguish between the ways the body repairs damaged tissues; and

c. name and describe some modes and causes of tissue shrinkage and death.

Tissue Growth

Tissues grow either because their cells increase in number or because the existing cells grow larger. Most embryonic and childhood growth involves **hyperplasia**[31] (HY-pur-PLAY-zhuh), tissue growth through cell multiplication. Skeletal muscles and adipose tissue, however, grow through **hypertrophy**[32] (hy-PUR-truh-fee), the enlargement of preexisting cells. However "bulked up" and muscular a person becomes, or however obese, it is through a growth in the size of muscle or fat cells, not an increase in their number. **Neoplasia**[33] (NEE-oh-PLAY-zhuh) is the development of a tumor (neoplasm), whether benign or malignant, composed of abnormal, nonfunctional tissue.

---

[31] *hyper* = above, beyond; *plas* = growth

[32] *hyper* = above, beyond; *trophy* = nourishment

[33] *neo* = new; *plas* = growth

Clinical Application 4.2

KELOIDS

In some people, healing skin wounds exhibit excessive fibrosis and produce raised, shiny scars called *keloids* (fig. 4.31). Keloids extend beyond the boundaries of the original wound, and if they are surgically removed, they tend to return or even worsen. Keloids may result from the excessive secretion of a fibroblast-stimulating growth factor by macrophages and platelets. They occur most often on the upper trunk and earlobes and in dark-skinned adults. Some tribespeople practice *scarification*—scratching or cutting the skin in patterns to induce keloid formation as a way of decorating the body.

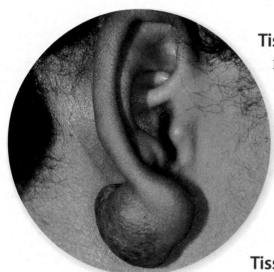

Figure 4.31 A Keloid.

Tissue Repair

Damaged tissues can be repaired in two ways: *regeneration* or *fibrosis*. **Regeneration** is the replacement of dead or damaged cells by the same type of cells as before, thus restoring normal function to the organ. Most skin injuries (cuts, scrapes, and minor burns) heal by regeneration. The liver also regenerates remarkably well. **Fibrosis** is the replacement of damaged tissue with scar tissue, composed mainly of collagen produced by fibroblasts. Scar tissue helps to hold an organ together, but it does not restore normal function. Examples include the healing of severe cuts and burns, the healing of muscle injuries, scars left in the cardiac muscle by earlier heart attacks, and scarring of the lungs in tuberculosis.

Tissue Shrinkage and Death

The shrinkage of a tissue through a loss in cell size or number is called **atrophy**[34] (AT-ro-fee). It can result from lack of use of an organ, as when muscles shrink from lack of exercise, or from simple aging.

Necrosis[35] (neh-CRO-sis) is the pathological death of tissue due to trauma, toxins, infection, and so forth. One form of necrosis is **infarction,** the sudden death of tissue that occurs when its blood supply is cut off—for example, the *myocardial infarction (MI)* of a heart attack and the *cerebral infarction* of a stroke. Another form of necrosis is **gangrene,** a more slowly developing necrosis usually resulting from an insufficient blood supply and often involving infection of the tissue. A **bed sore (decubitus ulcer)** is a form of gangrene that occurs when immobilized persons, such as those confined to a hospital bed or wheelchair, are unable to move and continual pressure on the skin cuts off blood flow to an area. Bed sores occur most often in places where a bone comes close to the body surface, such as the hips, sacral region, and ankles.

Apoptosis[36] (AP-oh-TOE-sis), or **programmed cell death,** is the normal death of cells that have completed their function and best serve the body by dying and getting out of the way. Apoptosis disposes of unneeded tissue in embryonic development. For example, the foot and hand of a fetus start as paddlelike structures. Separation of the fingers and toes occurs by apoptosis of cells that form the initial webs between them. Apoptosis also shrinks the uterus after a woman gives birth.

---

[34] *a* = without; *troph* = nourishment

[35] *necr* = death; *osis* = process

[36] *apo* = away; *ptosis* = falling

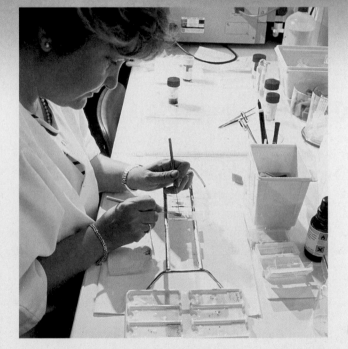

Before You Go On

Answer these questions from memory. Reread the preceding section if there are too many you don't know.

21. *Tissues can grow through an increase in cell size or cell number. What are the respective terms for these two kinds of growth?*

22. *Explain the difference between infarction and gangrene.*

23. *Distinguish between atrophy, necrosis, and apoptosis, and describe a circumstance under which each of these forms of tissue loss may occur.*

24. *Distinguish between regeneration and fibrosis. Which process restores normal cellular function? What good is the other process if it does not restore function?*

CAREER SPOTLIGHT

Histotechnician

Histotechnicians prepare tissues by sectioning, staining, and other methods, to produce specimens like those in the photographs in this chapter. Most histotechnicians work in hospitals, clinics, and diagnostic laboratories, preparing specimens to be examined by a pathologist or cytotechnologist for evidence of disease. A career in this field requires a knowledge of biology and chemistry. Good manual dexterity and hand–eye coordination are needed for work with delicate and tiny specimens and sophisticated laboratory equipment. A histotechnician often must work quickly (but carefully), because a surgical team may be waiting in an operating room for biopsy results that depend on the skills of the histotechnician and pathologist team. Histotechnicians are also highly valued team members in industrial research, veterinary medicine, and forensics laboratories.

   To become a histotechnician requires at least an associate degree and training in a hospital or histotechnician program. With a bachelor's degree and further training, one can become a histotechnologist, with responsibility for more advanced techniques and more potential for administrative authority. See appendix B for further information.

Study Guide

Assess Your Learning Outcomes

To test your knowledge, discuss the following topics with a study partner or in writing, ideally from memory.

4.1 The Study of Tissues (p. 98)

1. The scope of histology
2. The definition of *tissue* and the four primary types of tissues
3. Three basic components of a tissue
4. How tissues are prepared for microscopic study, and how the sectioning of tissues relates to one's three-dimensional interpretation of their microscopic appearance
5. The terminology of common cell shapes

4.2 Epithelial Tissue (p. 101)

1. Characteristics of epithelium as a class
2. Where epithelial tissue is found
3. What separates an epithelium from adjacent tissues
4. Terminology of the three surfaces of epithelial cells
5. The distinction between simple and stratified epithelia
6. Four kinds of simple epithelium, defining characteristics of each, where each type can be found, and functional advantages of the various types
7. Four kinds of stratified epithelium, defining characteristics of each, where each type can be found, and functional advantages of the various types
8. Structural and functional differences between keratinized and nonkeratinized stratified squamous epithelium, and where each can be found

4.3 Connective Tissue (p. 108)

1. Characteristics of connective tissue as a class
2. Functions of connective tissues
3. Three principal components of fibrous connective tissue
4. Cell and fiber types commonly seen in fibrous connective tissues, and the function of each
5. Variations in the ground substance of fibrous connective tissue

6. Types of loose connective tissue, the differences between them, and examples of their locations
7. Types of dense connective tissue, the differences between them, and examples of their locations
8. Structure and functions of adipose tissue, and the functional distinction between the two types of adipose tissue
9. Structure of cartilage, its three types, and their differences in composition and location
10. Structure of compact bone and its relationship with spongy bone, bone marrow, and periosteum
11. Components and functions of the blood

4.4 Nervous and Muscular Tissue— The Excitable Tissues (p. 116)

1. Why nervous and muscular tissues are considered to be "excitable" tissues
2. Two types of cells that compose nervous tissue, and the difference between them
3. The basic structure of a neuron
4. Three types of muscular tissue, where they are found, and how they differ in structure and function

4.5 Glands and Membranes (p. 120)

1. The difference between a secretion and an excretion
2. Differences between an endocrine and an exocrine gland; examples of each
3. Differences between serous, mucous, and mixed glands; examples of each
4. Differences between merocrine, holocrine, and apocrine modes of secretion; examples of each
5. Differences between mucous and serous membranes; examples of each

4.6 Tissue Growth, Development, Repair, and Death (p. 123)

1. Differences between hyperplasia, hypertrophy, and neoplasia, and which of them is pathological
2. Two modes of tissue repair and how they differ
3. Tissue atrophy and its causes
4. Types of tissue necrosis, and how necrosis differs from apoptosis
5. Some normal developmental functions of apoptosis

Testing Your Recall

1. Transitional epithelium is found in
   a. the urinary system.
   b. the respiratory system.
   c. the digestive system.
   d. the reproductive system.
   e. all of the above.

2. The external surface of the stomach is covered by
   a. a mucous membrane.
   b. a serous membrane.
   c. the pleura.
   d. a smooth muscle.
   e. a basement membrane.

3. Which of these is the most widespread type of epithelium?
   a. simple cuboidal
   b. simple columnar
   c. pseudostratified columnar
   d. stratified squamous
   e. stratified cuboidal

4. A seminiferous tubule of the testis is lined with _____ epithelium.
   a. simple cuboidal
   b. pseudostratified columnar
   c. stratified squamous
   d. transitional
   e. stratified cuboidal

5. Which of these cells is specialized to engulf and destroy foreign matter?
   a. an erythrocyte
   b. a fibroblast
   c. a macrophage
   d. an adipocyte
   e. a myocyte

6. Tendons are composed of
   a. areolar tissue.
   b. skeletal muscle.
   c. reticular tissue.
   d. dense irregular connective tissue.
   e. dense regular connective tissue.

7. The collagen of areolar tissue is produced by
   a. macrophages.
   b. fibroblasts.
   c. lymphocytes.
   d. leukocytes.
   e. chondrocytes.

8. Thermal insulation is one of the functions of
   a. adipose tissue.
   b. areolar tissue.
   c. muscular tissue.
   d. stratified squamous epithelium.
   e. fibrocartilage.

9. The shape of the external ear is due to
   a. skeletal muscle.
   b. elastic cartilage.
   c. fibrocartilage.
   d. articular cartilage.
   e. hyaline cartilage.

10. Any gland that releases its product by exocytosis is called
    a. a mucous gland.
    b. a holocrine gland.
    c. a merocrine gland.
    d. a serous gland.
    e. an apocrine gland.

11. Any form of pathological tissue death is called _____.

12. The peritoneum has a _____ type of epithelium.

13. Osteocytes and chondrocytes occupy little cavities called _____.

14. Bones are covered with a fibrous sheath called the _____.

15. Tendons and ligaments are made mainly of the protein _____.

16. Most of the cells in nervous tissue are called _____.

17. An epithelium rests on a layer called the _____ between its deepest cells and the underlying connective tissue.

18. Fibers and ground substance make up the _____ of a connective tissue.

19. A _____ gland is one in which the secretion forms by complete breakdown of the gland cells.

20. Any epithelium in which every cell touches the basement membrane is called a _____ epithelium.

Answers in appendix A

True or False

Determine which five of the following statements are false, and briefly explain why.

1. The esophagus is protected from abrasion by a keratinized stratified squamous epithelium.

2. All cells of a pseudostratified columnar epithelium contact the basement membrane.

3. Not all skeletal muscle is attached to bones.

4. Fibers are part of the matrix but not part of the ground substance of a connective tissue.

5. In all connective tissues, the matrix occupies more space than the cells do.

6. Adipocytes are limited to adipose tissue.

7. A single exocrine gland can be serous, mucous, or both.

8. Apoptosis is a normal, healthy tissue change but infarction is not.

9. Nerve and muscle cells are the body's only electrically excitable cells.

10. Cartilage is always covered by a fibrous perichondrium.

Answers in appendix A

Testing Your Comprehension

1. A woman in labor is often told to push. In doing so, is she consciously contracting her uterus to expel the baby? Justify your answer based on the muscular composition of the uterus.

2. One major tenet of biology is the cell theory, which states in part that all bodily structure and function are based on cells. The structural properties of bone, cartilage, and tendons, however, are due more to their extracellular material than to their cells. Is this an exception to the cell theory? Why or why not?

3. The epithelium of the respiratory tract is mostly of the pseudostratified columnar ciliated type, but in the alveoli—the tiny air sacs where oxygen and carbon dioxide are exchanged between the blood and inhaled air—the epithelium is simple squamous. Explain the functional significance of this histological difference. That is, why don't the alveoli have the same kind of epithelium as the rest of the respiratory tract?

Chapter

5

The Integumentary System

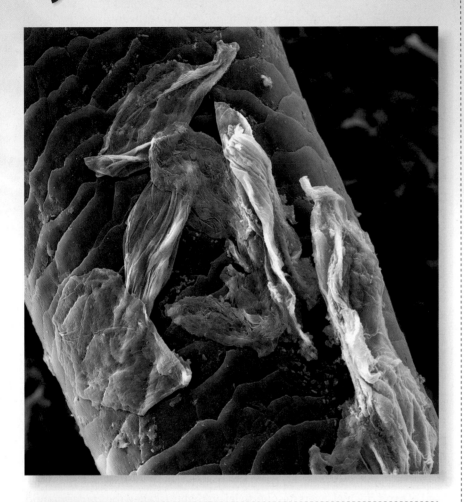

A human hair with a few loose squamous epidermal cells clinging to it. The overlapping scales of the hair surface help prevent hairs from being pulled easily from the scalp.

BASE CAMP

Before ascending to the next level, be sure you're properly equipped with a knowledge of these concepts from earlier chapters:

- Desmosomes (p. 72)
- Stratified squamous epithelium (p. 106)
- Collagen (p. 109)
- Areolar and dense irregular connective tissue (p. 110–111)
- Merocrine, apocrine, and holocrine glands (p. 120)

Anatomy & Physiology | REVEALED®
aprevealed.com

Module 4: Integumentary

The **integumentary**[1] **system** consists of the skin *(integument),* hair, nails, and cutaneous glands. We pay more attention to this organ system than to any other. It is, after all, the most visible one, and its appearance strongly affects our social interactions. Few people venture out of the house without first looking in a mirror to see if their skin and hair are presentable. Aside from aesthetics, inspection of the skin, hair, and nails is a significant part of a physical examination. The integumentary system can provide clues not only to its own health but also to deeper disorders such as liver cancer, anemia, lung disease, and heart failure. Also, care of the integumentary system is an important part of a total plan of in-patient care, since it strongly affects a person's self-image and sense of well-being.

5.1   The Skin and Subcutaneous Tissue

Expected Learning Outcomes

When you have completed this section, you should be able to:

a. describe the functions of the skin;
b. name the cell types of the epidermis and identify their functions;
c. name and describe the layers of the epidermis and dermis;
d. explain what accounts for various normal and pathological skin colors; and
e. name the various physical markings of the skin.

The skin is a membrane that covers the external surface of the body. It is composed of two principal layers: a superficial epithelium called the *epidermis* and a deeper connective tissue layer called the *dermis* (fig. 5.1). Between the skin and muscles is a looser connective tissue layer called the *hypodermis.* This is not part of the skin, but is customarily studied in conjunction with it.

Functions

The foremost role of the skin is protection, but its functions are much more diverse than commonly supposed.

- **Resistance to trauma and infection.** The skin bears the brunt of most physical injuries to the body, but it resists and recovers from trauma better than other organs do. The epidermal cells are packed with a tough

---

[1] *integument* = covering

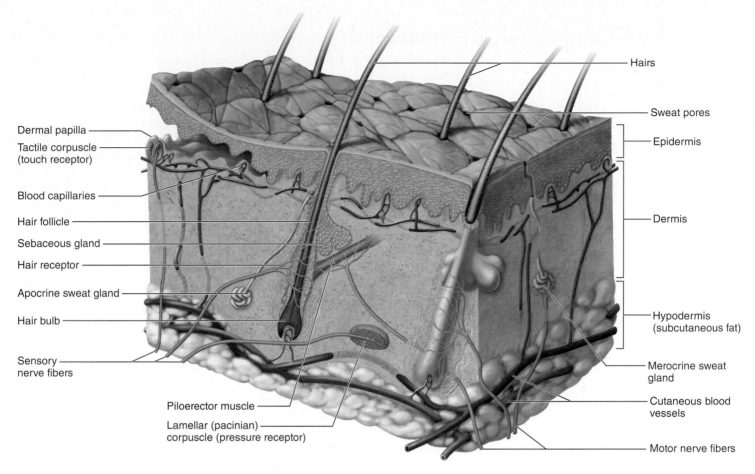

Dermal papilla

Tactile corpuscle
(touch receptor)

Blood capillaries

Hair follicle

Sebaceous gland

Hair receptor

Apocrine sweat gland

Hair bulb

Sensory
nerve fibers

Piloerector muscle

Lamellar (pacinian)
corpuscle (pressure receptor)

Hairs

Sweat pores

Epidermis

Dermis

Hypodermis
(subcutaneous fat)

Merocrine sweat
gland

Cutaneous blood
vessels

Motor nerve fibers

**Figure 5.1 Structure of the Skin and
Accessory Organs.** AP|R

protein called **keratin** and linked by strong desmosomes. Few infectious
organisms can penetrate the intact skin. Bacteria and fungi live on the
surface in moderate numbers, but their populations are kept in check by
the relative dryness of the skin and by a protective acidic film called the
acid mantle.

- **Water retention.** The skin prevents the body from losing excess water, as
  well as from absorbing too much of it when you are swimming or bathing.
  In spite of this, however, we lose about 400 mL of water per day through
  the skin even in the absence of sweating, and potentially much more
  when we sweat.

- **Vitamin D synthesis.** Epidermal cells carry out the first step in vitamin
  D synthesis, a process completed by the kidneys and liver. Vitamin D is
  needed for bone development and maintenance, as detailed in chapter 6.

- **Sensation.** The skin is the body's largest sense organ. It is equipped with a
  variety of nerve endings that react to heat, cold, touch, temperature, texture, pressure, vibration, and injury. The skin senses and nerve endings
  are more fully described in chapter 10.

- **Thermoregulation.** The skin plays a key role in stabilizing body temperature by regulating heat exchange with the environment. It does this by increasing or reducing blood flow close to the body surface and by sweating
  in response to overheating, as detailed in chapter 18.

- **Nonverbal communication.** The skin is an important means of communication. We possess numerous small facial muscles that act on collagen
  fibers of the dermis to produce subtle and varied expressions.

Structure

The skin (fig. 5.2) is the body's largest organ. In adults, it covers up to 2 m² of body surface and accounts for about 15% of the body weight. Most of the skin is 1 to 2 mm thick—comparable to the cover of a hardcover book—but it ranges from 0.5 mm on the eyelids to 6 mm between the shoulder blades. The difference is due mainly to variation in the thickness of the dermis. However, the skin is classified as thick or thin skin based on the thickness of the epidermis alone. **Thick skin** covers the palms, soles, and corresponding surfaces of the fingers and toes—areas subject to the greatest mechanical stress. Here the epidermis is about 0.5 mm thick, owing to a thick surface layer of dead cells called the *stratum corneum*. Thick skin has sweat glands but no hair follicles or sebaceous (oil) glands. The rest of the body is covered with **thin skin,** which has an epidermis about 0.1 mm thick, with a thin stratum corneum. It possesses hair follicles, sebaceous glands, and sweat glands. It may seem contradictory that the thickest skin is on the upper back, yet that is classified as thin skin. This is because the term *thick skin* refers to epidermis only, whereas the thickness of the back skin is due to the dermis.

The Epidermis

The **epidermis**[2] is a keratinized stratified squamous epithelium; that is, its surface consists of dead cells packed with keratin. Most cells of the epidermis are **keratinocytes** (keh-RAT-ih-no-sites), which are packed with the tough protein *keratin* (fig. 5.2). The basal layer of the epidermis also has pigment-producing cells called **melanocytes; tactile cells,** which are specialized for the sense of touch; and **stem cells,** which divide and replace other epidermal cells that die. Higher in the epidermis, the skin also has **dendritic cells,** which detect foreign matter and microbes that invade the skin and alert the immune system to ward off infection.

Figure 5.2 Layers and Cell Types of the Epidermis. Compare this to the photograph in figure 4.9, p. 106. **AP|R**

---

[2] *epi* = above, upon; *derm* = skin

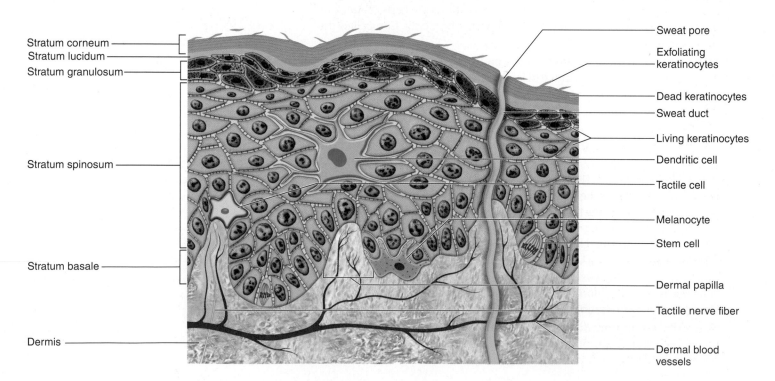

Stratum corneum

Stratum lucidum

Stratum granulosum

Stratum spinosum

Stratum basale

Dermis

Sweat pore

Exfoliating keratinocytes

Dead keratinocytes

Sweat duct

Living keratinocytes

Dendritic cell

Tactile cell

Melanocyte

Stem cell

Dermal papilla

Tactile nerve fiber

Dermal blood vessels

Clinical Application 5.1

TRANSDERMAL ABSORPTION

The ability of the skin to absorb chemicals makes it possible to administer several medicines as ointments or lotions, or by means of adhesive patches that release the medicine steadily through a membrane. For example, inflammation can be treated with hydrocortisone ointment, nitroglycerine patches are used to relieve heart pain, nicotine patches are used to help overcome tobacco addiction, and other medicated patches are used to control high blood pressure and motion sickness.

Unfortunately, the skin can also be a route for absorption of poisons. These include toxins from poison ivy and other plants; metals such as mercury, arsenic, and lead; and solvents such as carbon tetrachloride, acetone (nail polish remover), paint thinner, and pesticides. Some of these can cause brain damage, liver failure, or kidney failure, which is good reason for using protective gloves when handling such substances.

The epidermis is usually composed of four layers, but five in thick skin. From the deepest layer to the surface, these are as follows:

1. **Stratum basale** (bah-SAIL-ee). This is a deep, single layer of stem cells, keratinocytes, melanocytes, and tactile cells. Stem cells here divide and maintain their own population while producing new keratinocytes. All dead keratinocytes that flake off the skin surface (exfoliate) are replaced by mitosis in this layer.

2. **Stratum spinosum.** This is the thickest layer of epidermis, consisting of many layers of keratinocytes and dendritic cells. As keratinocytes are pushed upward by dividing cells below, they cease dividing and synthesize keratin. Keratin accumulation causes the cells to flatten and assume a squamous shape.

3. **Stratum granulosum.** This consists of three to five layers of flat keratinocytes with coarse, dark-staining granules. These cells aggregate their keratin filaments into thick, tough bundles and produce a water barrier that enables the skin to resist water loss from the body. The cells die here as the water barrier cuts them off from nutrients from below.

4. **Stratum lucidum.** This is present only in the thick skin of the palms and soles. It is a thin, clear (lucid) layer of dead cells with no nuclei or other visible internal structure.

5. **Stratum corneum.** This consists of up to 30 layers of dead, keratin-packed keratinocytes that give the skin much of its toughness. The dead cells **exfoliate** (flake off) from the surface as *dander* at a rate compensated by the production of new cells in the stratum basale. A keratinocyte usually lives 30 to 40 days from its "birth" in the stratum basale to its exfoliation from the stratum corneum.

The Dermis

The **dermis** ranges from 0.2 mm thick in the eyelids to 4 mm thick in the palms and soles and 5 or 6 mm between the shoulders. It consists mainly of collagen but also contains elastic and reticular fibers and the usual cells of fibrous connective tissue (see p. 109). It is well supplied with blood vessels, sweat glands, sebaceous glands, and nerve endings. The hair and nails are rooted in the dermis. In the face, skeletal muscles attach to dermal collagen fibers and produce such expressions as a smile or a wrinkle of the forehead. By moving the lips, they also aid in speech.

The boundary between the epidermis and dermis is histologically conspicuous and usually wavy. The upward waves are fingerlike extensions of dermis called **dermal papillae.**[3] They interlock with the epidermis like corrugated cardboard, an arrangement that resists slippage of the epidermis across the dermis and gives the skin more resistance to stress. If you look closely at the back of your hand, you will see delicate furrows that divide the skin into tiny rectangular to rhomboidal areas. The dermal papillae produce the raised areas between the furrows. These papillae contain an abundance of blood capillaries, the sole source of nourishment to the bloodless epidermis.

> ### Apply What You Know
> Would you expect to find more prominent dermal papillae in the palms or the forehead? Explain your answer.

There are two zones of dermis called the papillary and reticular layers. The **papillary** (PAP-ih-lerr-ee) **layer** is a thin zone of areolar tissue in and near the dermal papillae. The loosely organized tissue of this layer allows leukocytes to move around easily and attack bacteria or other invaders introduced through breaks in the skin. The deeper **reticular**[4] **layer** constitutes about four-fifths of the dermis. It consists of dense irregular connective tissue (see fig. 4.17, p. 111), composed mainly of thick bundles of collagen but also containing elastic fibers, fibroblasts, and small clusters of adipocytes. Sweat glands, nail roots, and hair follicles are embedded in this layer. Leather is made from this layer of an animal hide, which attests to its toughness.

The Hypodermis

The dermis blends into the underlying **hypodermis**[5] **(subcutaneous tissue),** which typically exhibits looser connective tissue and more adipose tissue **(subcutaneous fat).** The hypodermis binds the skin to the muscles or other underlying tissues. Subcutaneous fat pads the body, serves as an energy reservoir, and provides thermal insulation. It averages about 8% thicker in women than in men and differs in distribution between the sexes.

Skin Color

The most significant factor in skin color is the brown to black **melanin.**[6] Melanocytes produce it, but transfer it to the keratinocytes, where it accumulates. People of different skin colors have essentially the same number of melanocytes, but in dark skin, the melanocytes produce greater quantities of melanin, the melanin in the keratinocytes is relatively spread out, and it breaks down more slowly. Thus, melanized cells may be seen throughout the epidermis (fig. 5.3). In light skin, the melanin is less abundant and it breaks down more rapidly. Melanin screens ultraviolet radiation, which otherwise has the potential to cause DNA damage (mutations) and skin cancer. Melanocytes accelerate their production of melanin in response to ultraviolet rays—hence the

---

[3] *papilla* = little nipple

[4] *reticul* = little network

[5] *hypo* = below; *derm* = skin

[6] *melano* = black

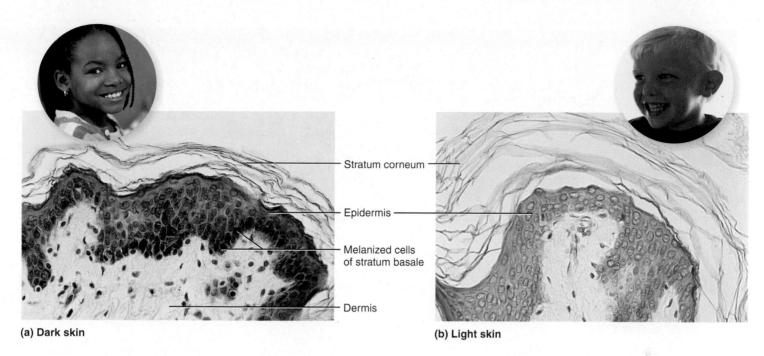

(a) Dark skin　　　　　　　　　　　　　　　　　**(b) Light skin**

Labels: Stratum corneum, Epidermis, Melanized cells of stratum basale, Dermis

suntanning effect—and keratinocytes concentrate it on the "sunny side" of the nucleus, like a protective parasol over the DNA. A suntan fades as melanin is degraded in older keratinocytes and as the keratinocytes migrate to the surface and exfoliate. Local concentrations of melanin produce freckles and moles.

Figure 5.3 Variations in Skin Pigmentation.
(a) The stratum basale shows heavy deposits of melanin in dark skin. (b) Light skin shows little to no visible melanin.

- *Which of the five types of epidermal cells are the melanized cells in part (a)?*

> ***Apply What You Know***
>
> Skin cancer is relatively rare in people with dark skin. Other than possible differences in behavior, such as intentional suntanning, what could be the reason for this?

Other factors in skin color are hemoglobin and carotene. **Hemoglobin,** the red pigment of blood, imparts reddish to pink hues to the skin as blood vessels show through the white of the dermal collagen. **Carotene**[7] is a yellow pigment acquired from egg yolks and yellow and orange vegetables. It can become concentrated to various degrees in the stratum corneum and subcutaneous fat. It is often most conspicuous in skin with the thickest stratum corneum, such as on the heel and in "corns" or calluses of the feet.

The skin may also exhibit abnormal colors of diagnostic value. **Cyanosis**[8] is blueness of the skin resulting from oxygen deficiency in the blood, as in cases of drowning, emphysema and some other lung diseases, and reduced dermal blood flow in cold weather. **Erythema**[9] (ERR-ih-THEE-muh) is abnormal redness of the skin, as in sunburn, exercise, hot weather, and embarrassment. A **hematoma**[10] is a bruise, a mass of clotted blood showing through the skin. Its causes range from accidental trauma to physical abuse, platelet deficiencies, anticoagulant drugs, and hemophilia. **Pallor** is a temporary pale or ashen color that occurs when there is so little cutaneous blood flow that skin color is dominated by the white dermal collagen. It can result from emotional stress,

---

[7] *carot* = carrot

[8] *cyan* = blue; *osis* = condition

[9] *eryth* = red; *em* = blood

[10] *hemat* = blood; *oma* = mass

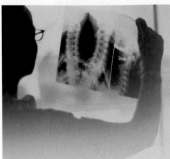

PERSPECTIVES ON HEALTH

The integument is our most visible and vulnerable organ system, exposed to trauma, infection, radiation, and injurious chemicals. Consequently, the integumentary system needs and receives more medical attention than any other organ system. The study and clinical treatment of this system is called *dermatology*.[11]

Skin Cancer

The stage for skin cancer is set when UV radiation causes mutations in the DNA of epidermal cells. There are three forms of skin cancer, differing in the cells in which they originate. *Basal cell carcinoma*[12] begins in the stratum basale and eventually invades the dermis. It is the most common but least dangerous form of skin cancer, because it seldom metastasizes. *Squamous cell carcinoma* arises in the stratum spinosum. The chance of recovery is good with early detection and surgical removal, but if neglected, it can metastasize to lymph nodes and can be lethal. Basal and squamous cell carcinomas produce similar lesions with raised edges and a concave center (fig. 5.4a). *Melanoma* (fig. 5.4b) is the rarest form of skin cancer (5% of cases) but also the most deadly, because it metastasizes quickly. It usually arises from melanocytes of a pre-existing mole. A useful guideline for recognizing melanoma is the American Cancer Society's *ABCD rule: A* for

[11] *dermato* = skin; *logy* = study of
[12] *carcin* = cancer; *oma* = tumor

asymmetry (one side of the lesion looks different from the other), *B* for border (the contour is not uniform but wavy or scalloped), *C* for color (often a mixture or brown, black, tan, and sometimes red and blue), and *D* for diameter (greater than 6 mm, about the diameter of a standard pencil eraser).

Burns

Burns are the leading cause of accidental death. They can be caused by fires, kitchen spills, excessively hot bath water, electrical shock, strong acids and bases, sunlight, and other forms of radiation. **First-degree burns,** such as most sunburns, damage only the epidermis. They are marked by redness, slight edema (swelling), and pain, but soon heal and seldom leave scars. **Second-degree burns** involve the entire epidermis and part of the dermis. They may be red, tan, or white, and are blistered and very painful. They may take up to several months to heal, and sometimes leave scars. Some sunburns and many scalds are second degree. First- and second-degree burns are also called **partial-thickness burns** because they do not penetrate all the way through the skin. **Third-degree (full-thickness) burns** destroy the epidermis and all of the dermis. Since nothing is left of the dermis in the burned area, the skin can regenerate only from the unburned edges. These burns often require skin grafts and typically result in scarring or severe disfigurement. The most urgent concerns in treating burn patients are fluid replacement and infection control, since a severe burn destroys the barrier to fluid loss and invading microbes.

(a) Basal cell carcinoma

(b) Melanoma

Figure 5.4 Skin Cancer. (a) Basal cell carcinoma. The lesions of squamous cell carcinoma look similar to this. (b) Melanoma, showing the typical border irregularity, asymmetry, and discoloration.

- *Which of these two cancers is the more common? Which is the more dangerous?*

circulatory shock, cold temperatures, or severe anemia. **Albinism**[13] is a hereditary lack of pigmentation in the skin, hair, and eyes owing to an inability to synthesize melanin. It is characterized by milky white hair and skin and blue-gray eyes. **Jaundice**[14] is yellowing of the skin and whites of the eyes. It occurs in various liver diseases, such as cancer, hepatitis, and cirrhosis, and in some blood diseases with a rapid rate of erythrocyte breakdown.

Before You Go On

Answer these questions from memory. Reread the preceding section if there are too many you don't know.

1. *To what besides pathogenic organisms does the skin present a barrier?*
2. *Describe two ways in which the skin helps to regulate body temperature.*
3. *Name the five kinds of epidermal cells and state their functions.*
4. *List the five layers of epidermis from deep to superficial. What are the distinctive features of each?*
5. *What are the two layers of the dermis? What type of tissue composes each layer?*
6. *Name the pigments responsible for normal skin colors, and explain how certain conditions can produce pathological discoloration of the skin.*

5.2 Accessory Organs

Expected Learning Outcomes

When you have completed this section, you should be able to:

a. describe the histology of a hair and its follicle;
b. explain how a hair grows;
c. discuss the purposes served by various kinds of hair;
d. describe the structure and function of nails;
e. name two types of sweat glands, and describe the structure and function of each; and
f. describe the location, structure, and function of other glands of the skin.

The hair, nails, and cutaneous glands are **accessory organs (appendages)** of the skin. All of them originate in the epidermis of the embryo, but penetrate deeply into the dermis or even the hypodermis of mature skin.

----

[13] *alb* = white; *ism* = state, condition
[14] *jaun* = yellow

Hair

A hair *(pilus)* is a slender filament of keratinized cells that grows from an oblique tube called a **hair follicle.** The keratinocytes of the epidermis contain a pliable *soft keratin,* but hair and nails are made of a stronger *hard keratin.* Hair occurs almost everywhere except the lips, nipples, parts of the genitals, palms and soles, and lateral surfaces of the fingers and toes. The density of hairs in a given area does not differ much from one person to another or even between the sexes. Differences in apparent hairiness are due mainly to differences in hair texture and pigmentation.

Structure of the Hair and Follicle

A hair follicle is a diagonal tube of epithelium and connective tissue that originates deep in the dermis or hypodermis. The **shaft** of a hair is the portion above the skin surface. All that below the surface, within the follicle, is called the **root.** At the lower end of the root is a swelling called the **bulb** (fig. 5.5a, b). Except near the bulb, all the hair tissue is dead. A bit of vascular connective tissue called the **dermal papilla** grows into the bulb and provides the hair with its sole source of nutrition. Just above this is the **hair matrix,** the exclusive site of cellular mitosis and hair growth.

The surface layer of a hair, called the **cuticle** (fig. 5.5c), is composed of thin scaly cells that overlap each other like roof shingles and resist pulling of the hair out of its follicle, as when you brush your hair. Beneath this is the **cortex,** which makes up most of the bulk of the hair. It is composed of several layers of elongated keratinized cells. The thickest hairs, such as those of the eyebrows and lashes, have a central core called the **medulla,** composed of loosely arranged cells and air spaces. The medulla is lacking from thin scalp and body hairs. In cross section, straight hair is relatively round, whereas curly or kinky hair is flatter, from ovoid to ribbonlike in shape.

The hair follicle has a **root sheath** composed of a layer of epithelium (an extension of the epidermis) alongside the hair and a layer of condensed dermal connective tissue around the epithelial layer. Nerve fibers called **hair receptors** coil around the follicle and respond to hair movements, as when an ant crawls across one's arm bending one hair after another. Also associated with each hair follicle is a smooth muscle called a **piloerector**[15] **(pilomotor) muscle.** It contracts in response to stimuli such as cold, fear, and touch, pulling the follicle into a more vertical position and making the hair stand upright.

Hair color derives from different types and proportions of melanin. Black and brown hair are colored by a form of melanin called *eumelanin.* Another, sulfur-rich melanin called *pheomelanin* causes various shades of blond in moderate amounts, and orange to red hair if more abundant. Gray and white hair get their color from a combination of pigment scarcity in the cortex and air in the medulla.

Hair Growth

Hair grows by mitosis of cells in the hair matrix. Hair and follicle cells together are pushed toward the skin surface as the cells below them multiply. The hair cells become progressively keratinized and die as they are pushed upward, away from the blood supply in the dermal papilla. A typical scalp hair in a young adult grows for 6 to 8 years at a rate of about 1 mm per 3 days (10–18 cm/year). It then stops growing and the follicle atrophies over a period of 2 or

---

[15] *pilo* = hair

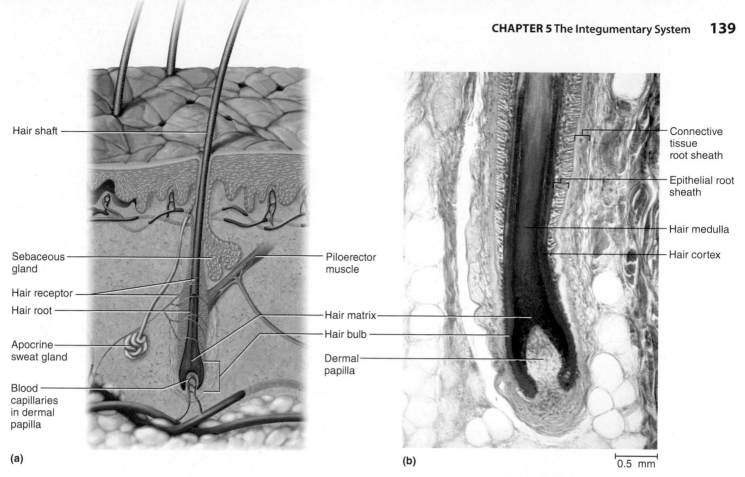

Hair shaft

Sebaceous gland

Hair receptor

Hair root

Apocrine sweat gland

Blood capillaries in dermal papilla

Piloerector muscle

Hair matrix

Hair bulb

Dermal papilla

(a)

Connective tissue root sheath

Epithelial root sheath

Hair medulla

Hair cortex

(b)

0.5 mm

Figure 5.5 Structure of a Hair and its Follicle. (a) Anatomy of the follicle and associated structures. (b) Light micrograph of the base of a hair follicle. (c) Electron micrograph of a hair emerging from its follicle. Note the exfoliating epidermal cells encircling the follicle like rose petals. **AP|R**

- *In light of your knowledge of hair, discuss the validity of an advertising claim that a shampoo will "nourish your hair." Where and how does a hair get its sole nourishment?*

3 weeks. Finally, the hair enters a resting phase 1 to 3 months long, during which it may fall out on its own or be pushed out by a new hair growing beneath it in the same follicle. We typically lose 50 to 100 scalp hairs per day.

Functions of Hair

In comparison to other mammals, the relative hairlessness of humans is so unusual that it raises the question, Why do we have any hair at all? Except on the scalp, it is too sparse to have the heat-retaining function that it serves in other mammals. Scalp hair does, however, help to retain heat in an area where there is an abundant flow of warm blood just under the cranium, but little to no insulating fat. It also protects the scalp from sunburn and skin cancer. Beard, pubic, and axillary hair begin growing at puberty and visually advertise sexual maturity. They also absorb aromatic secretions of specialized scent glands in these areas, discussed shortly. Hair on the trunk and limbs is largely vestigial (an evolutionary remnant of hair that was more functional in our prehuman ancestors), but still serves a sensory role: stimulation of the hair receptors alerts one to parasites such as ticks and fleas crawling on the skin. The eyebrows accentuate facial expressions produced by muscles that move the skin between and above the

(c)

Clinical Application 5.2

THE FINGERNAILS IN CLINICAL DIAGNOSIS

During a physical examination, a physician may inspect the fingernails for signs of disorders including, but not limited to, the integumentary system. Integumentary diseases such as psoriasis, tinea (fungal infection of the skin), and onychomycosis (nail fungus) produce changes in the nails such as thickening, pitting, loss of luster, and yellow discoloration. Cyanosis is especially visible in the nail beds, and could indicate anything from cardiovascular disease to a cold examining room. Cardiovascular disease and lung cancer can cause *clubbing* of the fingers—bulbous distal segments, unusually convex nail plates, and spongy nail folds. Cirrhosis of the liver, diabetes mellitus, and congestive heart failure can cause *Terry's nails*, in which the nails are whitened but with a band of reddish brown just behind the free edge. Iron deficiency anemia produces *spoon nails*, in which the lateral edges of the nail are raised and the center is depressed. Several acute, severe illnesses can produce transverse depressions called *Beau's lines*. From the location of Beau's lines and the growth rate of the nails, a physician can estimate how long ago an illness occurred.

eyes. They may also help to reduce the glare of sunlight and keep sweat and debris out of the eyes. Finally, stout protective *guard hairs (vibrissae)* block foreign particles from the nostrils, auditory canals, and eyes. The eyelashes and blink reflex shield the eyes from debris.

Nails

Fingernails and toenails are clear, hard derivatives of the stratum corneum. They are composed of very thin dead cells, densely packed together and filled with parallel fibers of hard keratin. Flat nails allow for more fleshy and sensitive fingertips than other mammals have, while they also serve as strong keratinized "tools" that can be used for digging, grooming, picking apart food, and other manipulations.

The anatomical features of a nail are shown in figure 5.6. The **nail plate** is the visible portion covering the fingertip; the **nail bed** is the epidermis underlying most of the plate; the **nail fold** is a thickening of the skin around the margins of the plate; and the **free edge** is the part of the plate that extends beyond the nail bed at the tip of the finger. The **nail matrix** is a growth zone concealed beneath the skin at the proximal edge of the nail. The nail grows by mitosis in the nail matrix, adding new cells to the plate. Fingernails grow at a rate of about 1 mm per week and toenails somewhat more slowly.

Cutaneous Glands

The skin has five types of glands: *merocrine sweat glands, apocrine sweat glands, sebaceous glands, ceruminous glands,* and *mammary glands* (fig. 5.7). The two types of sweat glands are also collectively called *sudoriferous* (SOO-dor-IF-er-us) *glands.* Mammary glands are discussed in chapter 19.

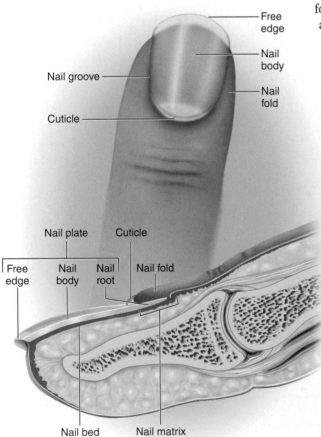

Figure 5.6 Anatomy of a Fingernail. AP|R

Lumen   Secretory cells

Figure 5.7 Cutaneous Glands. (a) Apocrine sweat glands have a large lumen and a duct that conveys their aromatic secretion into a hair follicle. (b) Merocrine sweat glands have a relatively narrow lumen and a duct that opens by way of a pore on the skin surface. (c) Sebaceous glands have cells that break down in entirety to form an oily secretion that is released into the hair follicle. **AP|R**

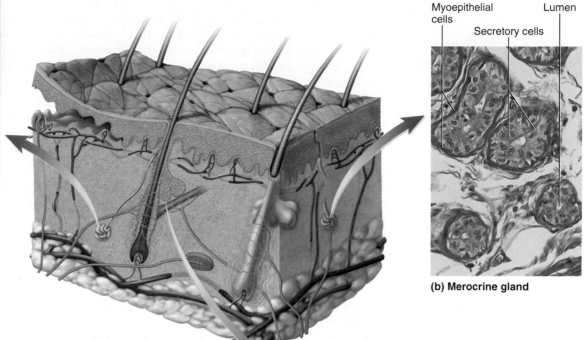

(a) Apocrine gland

Myoepithelial cells   Lumen

Secretory cells

(b) Merocrine gland

Gland   Hair follicle

(c) Sebaceous gland

Sweat Glands

Most sweat glands are a type called **merocrine**[16] **(eccrine) sweat glands.** These are widespread in the skin and produce watery perspiration that serves to cool the body. Each begins with a twisted coil in the dermis or hypodermis and an undulating duct that ascends to a sweat pore on the skin surface. Sweat begins as a filtrate of the blood plasma produced deep in the gland, and is modified by cells of the gland duct. Sweat is about 99% water; it contains potassium, sodium, chloride, ammonia, urea, and lactic acid, and it has a pH ranging from 4 to 6. Each day, these glands secrete about 500 mL of *insensible perspiration,* which does not produce noticeable wetness of the skin. Under conditions of exercise or heat, however, one may lose as much as a liter of perspiration per hour. Sweating with visible wetness of the skin is called **diaphoresis**[17] (DY-uh-for-EE-sis).

Apocrine[18] **sweat glands** occur in the groin, anal region, axilla, areola, and male beard area, although they are sparse or absent from some Asian ethnic groups. Their ducts lead into hair follicles rather than directly to the skin surface. The *apocrine* name comes from an earlier misunderstanding about their method of secretion (see p. 120); it is now known that they produce sweat in the same way that merocrine glands do—by exocytosis. The secretory part of an apocrine gland, however, stores up secretion and has a much larger lumen than that of a merocrine gland, so these glands have continued to be referred to as apocrine glands to distinguish them from the merocrine type. Apocrine sweat contains more fatty acids than merocrine sweat, making it thicker and more cloudy.

[16] *mero* = part; *crin* = to separate

[17] *dia* = through; *phoresis* = carrying

[18] *apo* = away from; *crin* = to separate

Apocrine sweat glands are scent glands that respond especially to stress and sexual stimulation. They do not develop until puberty, and they apparently correspond to the scent glands that develop in other mammals on attainment of sexual maturity. In women, they enlarge and shrink in phase with the menstrual cycle. Fresh apocrine sweat is considered attractive or arousing in some cultures, where it is as much a part of courtship as artificial perfume is to other people. Like other mammalian scent glands, human apocrine glands produce *pheromones*—chemicals that influence the physiology or behavior of other members of the species. The pubic and axillary hair probably serve to retain the secretion and prevent it from evaporating too quickly, so it can accumulate and exercise a stronger pheromonal effect.

Sebaceous Glands

Sebaceous[19] (see-BAY-shus) **glands** produce an oily secretion called **sebum** (SEE-bum). They are flask-shaped, with short ducts that usually open into a hair follicle. These are holocrine glands whose secretion is formed by the breakdown of entire gland cells (see fig. 4.29c, p. 120). Sebum keeps the skin and hair from becoming dry, brittle, and cracked. The sheen of well-brushed hair is due to sebum distributed by the hairbrush.

Ceruminous Glands

Ceruminous (seh-ROO-mih-nus) **glands** are found only in the auditory (ear) canal, where their secretion combines with sebum and dead epidermal cells to form earwax, or **cerumen.**[20] They are simple, coiled, tubular glands with ducts leading to the skin surface. Cerumen keeps the eardrum pliable, waterproofs the auditory canal, and has a bactericidal effect.

Before You Go On

Answer these questions from memory. Reread the preceding section if there are too many you don't know.

7. *Name and define the three regions of a hair from its base to its tip, and the three layers of a thick hair seen in cross section.*

8. *State the functions of the hair papilla, hair receptors, and piloerector.*

9. *State a reasonable theory for the functions of hair of the eyebrows, eyelashes, scalp, nostrils, and axilla.*

10. *Define or describe the nail plate, nail fold, and nail matrix.*

11. *How do merocrine and apocrine sweat glands differ in structure and function?*

12. *What other type of gland is associated with hair follicles? How does its mode of secretion differ from that of sweat glands?*

---

[19] *seb* = fat, tallow; *aceous* = possessing
[20] *cer* = wax

Aging of the Integumentary System

Aging is especially noticeable in the integumentary system, and nearly all people have concerns or complaints about it as they age. The hair turns grayer and thinner. The skin and hair become drier because of atrophy of the sebaceous glands. The loss of collagen and elastin from the dermis makes the skin thinner, translucent, and looser. Aged skin also has fewer blood vessels than younger skin, and those that remain are more fragile. Consequently, aged skin bruises more easily; the skin can become reddened as broken vessels leak into the connective tissue; and poorer circulation results in slower healing of skin injuries. The density of cutaneous nerve endings declines by two-thirds from the age of 20 to the age of 80. The skin is consequently less sensitive to touch, pressure, and injurious stimuli. Dendritic cells of the skin also decline in number, leaving the skin more susceptible to recurring infections.

The atrophy of cutaneous blood vessels, sweat glands, and subcutaneous fat makes it more difficult to regulate body temperature in old age. Older people are more vulnerable to hypothermia in cold weather and heatstroke in hot weather. Heat waves and cold spells take an especially heavy toll among elderly poor people, who suffer from a combination of reduced homeostasis and inadequate housing.

In addition to all these normal aspects of growing older, the skin also suffers *photo-aging*—degenerative changes from a lifetime of exposure to UV radiation. This accounts for more than 90% of the integumentary changes that people find medically troubling or cosmetically disagreeable: age spots, wrinkling, yellowing and mottling of the skin, and skin cancer. All of these affect the face, hands, and arms more than areas of the body that receive less exposure to the sun.

CAREER SPOTLIGHT

Dermatology Nurse

Dermatology nursing is one of many specialties in the nursing profession. Dermatology nurses specialize in skin and wound care; assist dermatologists in the medical, surgical, and cosmetic treatment of the skin; and assist in treating conditions as diverse as psoriasis, burns, skin cancer, skin grafts, and gunshot wounds. Entry into the field of dermatology nursing requires a bachelor's degree and licensure as a registered nurse (RN). With 2 years of experience in dermatology nursing as an RN, one may sit for a board exam to become a *dermatology nurse certified (DNC)* or a *dermatology certified nurse practitioner (DCNP)*. See appendix B for further details on careers in dermatology nursing.

CONNECTIVE ISSUES

Ways in Which the Integumentary System Affects Other Organ Systems

All Systems

The integumentary system serves all other systems by providing a physical barrier to environmental hazards and minimizing water loss from the body.

Skeletal System

The role of the skin in vitamin D synthesis promotes calcium absorption needed for bone growth and maintenance.

Muscular System

Vitamin D synthesis promotes absorption of calcium needed for muscle contraction; skin dissipates heat generated by muscles.

Nervous System

Sensory impulses from the skin are transmitted to the nervous system.

Endocrine System

Vitamin D acts as a hormone.

Circulatory System

Dermal vasoconstriction diverts blood from the skin to other organs; the skin prevents loss of fluid from the cardiovascular system.

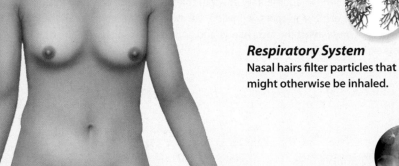

Lymphatic and Immune Systems

Dendritic cells detect foreign substances and alert the immune system.

Respiratory System

Nasal hairs filter particles that might otherwise be inhaled.

Urinary System

The sweat glands complement the urinary system by excreting salts and some nitrogenous wastes.

Digestive System

Vitamin D promotes calcium absorption by the small intestine.

Reproductive System

Cutaneous receptors respond to sexual stimuli; mammary glands produce milk to nourish infants; apocrine glands produce scents with subtle sexual functions.

Study Guide

Assess Your Learning Outcomes

To test your knowledge, discuss the following topics with a study partner or in writing, ideally from memory.

5.1 The Skin and Subcutaneous Tissue (p. 130)

1. The organs that constitute the integumentary system
2. Six functions of the skin
3. The two layers of skin proper and the layer of tissue immediately beneath the skin
4. The locations of *thick* and *thin skin,* their relative thickness, and other structural differences between them
5. Five types of cells in the epidermis, their locations, and the function of each
6. Five layers of the epidermis, characteristics of each one, and which one is missing from thin skin
7. The life history of a keratinocyte from its birth at the base of the epidermis to its death and exfoliation from the epidermal surface; the relationship between these events, the toughness of the epidermis, and the epidermal role as a water barrier
8. Composition and layers of the dermis; dermal papillae and their significance
9. The difference between the hypodermis and dermis; functions of the hypodermis; and an alternative name for the hypodermis in regions where it is mostly adipose tissue
10. The pigments that account for the normal variation in skin colors; names of six pathological changes in skin color and what accounts for each

5.2 Accessory Organs (p. 137)

1. Structures that are considered to be the *accessory organs* of the integumentary system
2. Definitions of *hair* and *hair follicle*
3. How the keratin of hair differs from that of the epidermis
4. The distribution of body hair and hairless regions of the skin
5. Three regions of a hair from base to tip; the sites of hair nutrition, cell division, and growth
6. Three zones of a hair seen in a cross section, and how cross-sectional shape relates to the texture of hair
7. The chemical basis for brown, black, blond, red, gray, and white hair
8. Structure of a hair follicle and its associated nerve ending and muscle
9. The process and duration of hair growth
10. Functional interpretations of scalp, beard, pubic, axillary, eyebrow, trunk, eyelash, and nasal hairs
11. Structure of a nail and nail bed
12. Some functions of the nails and the advantage of the flat nails of humans over the claws of other mammals
13. Similarities and differences in structure, function, and locations of merocrine and apocrine sweat glands
14. How sweat is produced, what it contains, and how diaphoresis differs from insensible perspiration
15. How sebaceous glands differ from sweat glands in location, structure, mode of secretion, and function
16. The location and function of ceruminous glands and composition of cerumen

Testing Your Recall

1. Cells of the ___ are keratinized and dead.
   a. papillary layer
   b. stratum spinosum
   c. stratum basale
   d. stratum corneum
   e. stratum granulosum

2. The outermost layer of a hair is called
   a. the medulla.
   b. the cuticle.
   c. the root sheath.
   d. the papilla.
   e. the cortex.

3. Which of the following skin conditions or appearances would most likely result from liver failure?
   a. pallor
   b. erythema
   c. cyanosis
   d. jaundice
   e. melanization

4. Thick skin can be found on
   a. the palms.
   b. the region between the shoulder blades.
   c. the elbows and knees.
   d. the chest.
   e. the cheeks.

5. The integumentary system includes all of the following except
   a. the epidermis.
   b. the nails.
   c. the sweat glands.
   d. the dermis.
   e. the subcutaneous tissue.

6. The thickest *living* layer of the epidermis is
   a. the stratum basale.
   b. the stratum spinosum.
   c. the stratum granulosum.
   d. the stratum lucidum.
   e. the stratum corneum.

7. Which of the following is a scent gland?
   a. a holocrine gland
   b. a sebaceous gland
   c. an apocrine gland
   d. a ceruminous gland
   e. a merocrine gland

8. ___ are skin cells with a sensory role.
   a. Tactile cells
   b. Dendritic cells
   c. Stem cells
   d. Melanocytes
   e. Keratinocytes

9. The function of epidermal dendritic cells is
   a. immune defense.
   b. protection from UV radiation.
   c. water retention.
   d. the sense of touch.
   e. mitotic replacement of dead keratinocytes.

10. Red hair gets its color primarily from
   a. eumelanin.
   b. keratin.
   c. hemoglobin.
   d. collagen.
   e. pheomelanin.

11. _____ is sweating without noticeable wetness of the skin.

12. A muscle that causes a hair to stand erect is called a/an _____.

13. Two common word roots that refer to the skin in medical terminology are _____ and _____.

14. Blueness of the skin due to low oxygen concentration in the blood is called _____.

15. Upward projections of the dermis along the dermal–epidermal boundary are called _____.

16. Cerumen is more commonly known as _____.

17. The holocrine glands that secrete into a hair follicle are called _____.

18. The most abundant protein of the epidermis is ___, while the most abundant protein of the dermis is _____.

19. The epidermis, liver, and kidneys work together to synthesize _____, which promotes absorption of dietary calcium.

20. Epidermal melanin is produced by melanocytes but accumulates in the _____.

Answers in appendix A

True or False

Determine which five of the following statements are false, and briefly explain why.

1. Dander consists of dead keratinocytes.

2. The term *integument* means only the skin, but *integumentary system* refers also to the hair, nails, and cutaneous glands.

3. The dermis is composed mainly of keratin.

4. Vitamin D is synthesized by certain cutaneous glands.

5. Cells of the stratum granulosum cannot undergo mitosis.

6. Dermal papillae are better developed in skin that is subject to a lot of mechanical stress than in skin that is subject to less stress.

7. The three layers of the skin are the epidermis, dermis, and hypodermis.

8. People of African descent have a much higher density of epidermal melanocytes than do people of northern European descent.

9. Pallor indicates a genetic lack of melanin.

10. Apocrine scent glands develop at the same time in life as the pubic and axillary hair.

Answers in appendix A

Testing Your Comprehension

1. Many organs of the body contain numerous smaller organs, perhaps even thousands. Describe an example of this in the integumentary system.

2. Explain how the complementarity of form and function is reflected in the fact that the dermis has two histological layers and not just one.

3. Why is it important for the epidermis to be effective, but not too effective, in screening out UV radiation?

The Skeletal System

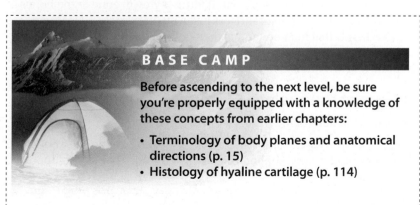

Colorized X-ray of a human knee prosthesis. The durable bonding of such artificial joints with natural bone has been a major achievement of biomedical engineering.

BASE CAMP

Before ascending to the next level, be sure you're properly equipped with a knowledge of these concepts from earlier chapters:

- Terminology of body planes and anatomical directions (p. 15)
- Histology of hyaline cartilage (p. 114)

Anatomy & Physiology REVEALED®
aprevealed.com

Module 5: Skeletal System

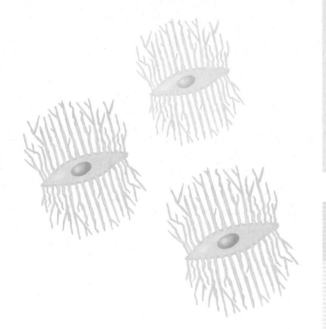

In art and history, nothing has symbolized death more than a skull or skeleton.[1] Bones and teeth are the most durable remains of a once-living body and the most vivid reminder of the impermanence of life. The clean dry bones presented for laboratory study indeed suggest that the skeleton is nonliving—an inert scaffold for the body, like the steel girders of a building. Seeing it in such a sanitized form makes it easy to forget that the living skeleton is made of dynamic tissues, full of cells—that it continually remodels itself and interacts physiologically with all of the other organ systems of the body. The skeleton is permeated with nerves and blood vessels, which attests to its sensitivity and metabolic activity. Indeed, if the skeleton were nonliving, a fractured bone would never heal.

This chapter concerns a family of biomedical sciences related to the skeleton—**osteology**,[2] the study of bones; **arthrology**,[3] the study of joints; and **kinesiology**,[4] the study of musculoskeletal movement.

6.1 Skeletal Structure and Function

Expected Learning Outcomes

When you have completed this section, you should be able to:

    a. identify the organs and tissues that compose the skeletal system;

    b. describe several functions of the skeletal system;

    c. define terms for the gross anatomy of bones;

    d. distinguish the two types of bone marrow; and

    e. identify features of the microscopic anatomy of bone, including the two types of bone tissue, four types of bone cells, and components of the noncellular matrix.

The **skeletal system** is composed of bones, cartilages, and ligaments joined tightly to form a strong, flexible framework for the body. Cartilage is the forerunner of most bones in embryonic development; it forms a growth zone in the bones of children; and it covers many joint surfaces in the mature skeleton. **Ligaments** are collagenous bands that hold bones together at the joints. **Tendons** are structurally similar to ligaments but attach muscle to bone. The skeletal system also includes **bone marrow,** the soft bloody or fatty material enclosed in the bones.

Functions of the Skeletal System

The principal functions of the skeletal system are as follows:

- **Support.** Bones of the limbs and vertebral column support the body; the mandible and maxilla support the teeth; and some viscera are supported by nearby bones.

---

[1] *skelet* = dried up

[2] *osteo* = bone; *logy* = study of

[3] *arthro* = joint; *logy* = study of

[4] *kinesio* = movement; *logy* = study of

- **Protection.** Bones enclose and protect the brain, spinal cord, lungs, heart, and pelvic viscera.
- **Movement.** Bones provide attachment and leverage for the muscular system, allowing for such actions as limb movement and ventilation of the lungs.
- **Blood formation.** Red bone marrow is the major producer of blood cells.
- **Storage.** The skeleton is the body's main reservoir of calcium and phosphorus, among other minerals. It stores these when they are available in ample quantity and releases them when they are needed for other functions. The fatty bone marrow present in many bones also serves as one of the body's reserves of stored fuel.

Osseous (Bone) Tissue

The hard, calcified tissue of a bone is called **osseous**[5] **tissue.** It exists in two forms: *compact (dense) bone,* which is solidly filled with opaque matrix, and *spongy bone,* which is a porous lattice honeycombed with spaces. Spongy bone is always found in the interior of a bone, covered with compact bone of the surface (fig. 6.1).

Bone Cells

Osseous tissue consists of cells and matrix, like any other connective tissue, and the matrix consists of fibers and ground substance. There are four kinds of bone cells (fig. 6.2):

1. **Osteogenic**[6] **cells** occur on the bone surface, beneath the fibrous connective tissue membranes that cover a bone. They are stem cells that give

---

[5] *os, osse, oste* = bone
[6] *osteo* = bone; *genic* = producing

Figure 6.1 Two Types of Osseous Tissue. In this frontal section of the hip joint, we can see the dense white shell of compact bone on the surfaces of the femur and hipbone, enclosing more porous spongy bone.

- *In this dried specimen, the spongy bone appears full of air spaces. What would occupy that space in a living bone?*

Figure 6.2 Bone Cells and Their Development. (a) Osteogenic cells give rise to osteoblasts, which deposit matrix around themselves and transform into osteocytes. (b) Osteoclasts form by the fusion of bone marrow stem cells.

- *Why do osteoblasts have more rough endoplasmic reticulum than osteogenic cells do?*

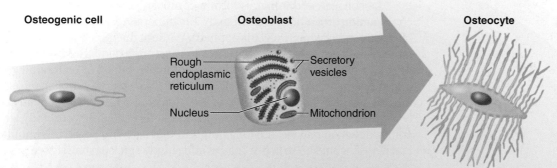

Osteogenic cell — Osteoblast — Osteocyte
Rough endoplasmic reticulum — Secretory vesicles
Nucleus — Mitochondrion

(a) Osteocyte development

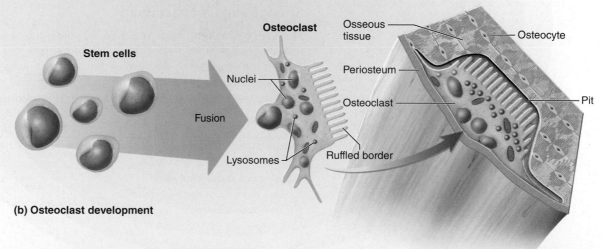

Osteoclast — Osseous tissue — Osteocyte
Stem cells
Nuclei — Periosteum
Fusion — Osteoclast — Pit
Lysosomes — Ruffled border

(b) Osteoclast development

rise to osteoblasts. They are the only bone cells capable of dividing and making more bone cells.

2. **Osteoblasts**[7] develop from osteogenic cells and lie in a single layer on the bone surface, somewhat resembling a cuboidal epithelium. Their role is to synthesize the organic matter of bone and deposit matrix. In keeping with this, they exhibit an abundance of mitochondria and rough endoplasmic reticulum.

3. **Osteocytes** are former osteoblasts that have become trapped in the matrix they deposited. They reside in cavities called **lacunae**[8] (la-COO-nee), which are connected to each other by slender channels called **canaliculi**[9] (CAN-uh-LIC-you-lye). Each osteocyte has delicate cytoplasmic processes that reach into the canaliculi to meet the processes of neighboring osteocytes. Osteocytes pass nutrients, wastes, and chemical signals to each other through gap junctions at the tips of their cytoplasmic processes.

4. **Osteoclasts**[10] are bone-dissolving cells that develop from a separate line of bone marrow stem cells. They are exceptionally large (up to 150 μm in diameter), have multiple nuclei, and lie on the bone surface like osteoblasts. On the side facing the bone, an osteoclast has an unusual comb-like row of infoldings of the plasma membrane, called the *ruffled border*. Along this surface, the cell secretes hydrochloric acid and enzymes to dissolve osseous tissue. Osteoclasts often lie in pits that they have eroded into the bone surface.

Bone Matrix

The **matrix** of osseous tissue is the stony matter that surrounds the osteocytes and lacunae. It is about one-third organic and two-thirds inorganic matter by weight. The organic matter includes collagen and large protein–carbohydrate complexes. The inorganic matter is about 85% calcium phosphate, with lesser amounts of calcium carbonate, magnesium, potassium, and fluoride.

The collagen and minerals form a composite material that gives bones a combination of flexibility and strength. The minerals resist compression (crumbling or sagging when weight is applied). Without the mineral component, the bones would be rubbery and could not support one's body. This is seen to some degree in the mineral-deficient, easily deformed bones of childhood *rickets* and the similar adult disease, *osteomalacia* ("soft bones"). Collagen, on the other hand, gives bones the ability to resist tension so the bone can bend slightly without snapping. When you run, for example, each time your body weight comes down on one leg, the long bones in that leg bend slightly. Without collagen, they would be so brittle they would simply crack or shatter (see p. 54).

Spongy and Compact Bone

Spongy (cancellous) bone (fig. 6.3a, b) consists of a porous lattice of slender rods and plates called **trabeculae**[11] (tra-BECK-you-lee). Although calcified and hard, it is permeated by tiny spaces that give it a spongelike appearance. These spaces are filled with bone marrow and small blood vessels. Spongy bone is well designed to impart strength to a bone while adding a minimum of weight.

---

[7] *osteo* = bone; *blast* = form, produce

[8] *lac* = lake, hollow place; *una* = little

[9] *canal* = canal, channel; *icul* = little

[10] *osteo* = bone; *clast* = destroy, break down

[11] *trabe* = plate; *cul* = little

The trabeculae are not randomly arranged as it might seem at a glance, but aligned along planes that give the bone the best strain resistance.

Compact (dense) bone (fig. 6.3a, c) forms the outer shell that surrounds the spongy bone. It prevents the bone marrow from seeping out and provides solid attachment surfaces for muscles, tendons, and ligaments. At the surfaces of a bone, it is organized in parallel layers like plywood, laid down by the surface osteoblasts. Deeper in the bone, most of it is organized in cylindrical units of structure called **osteons (haversian[12] systems).** In cross sections, they look like onion slices—layers called **lamellae** arranged concentrically around a **central (osteonic** or **haversian) canal.** The canal contains small blood vessels and nerves. The osteocytes occupy lacunae between the lamellae.

---

[12] Clopton Havers (1650–1702), English anatomist

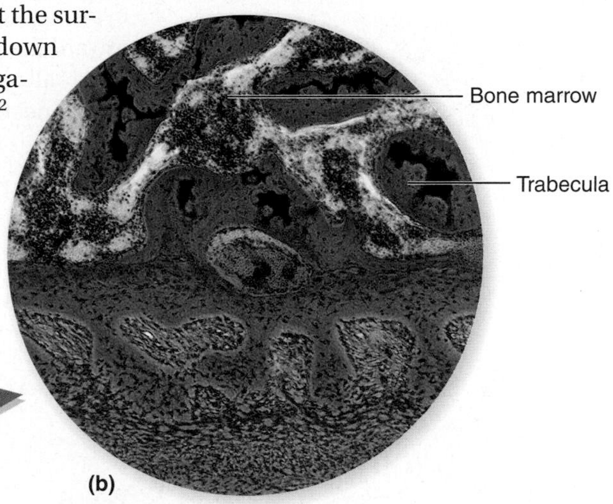

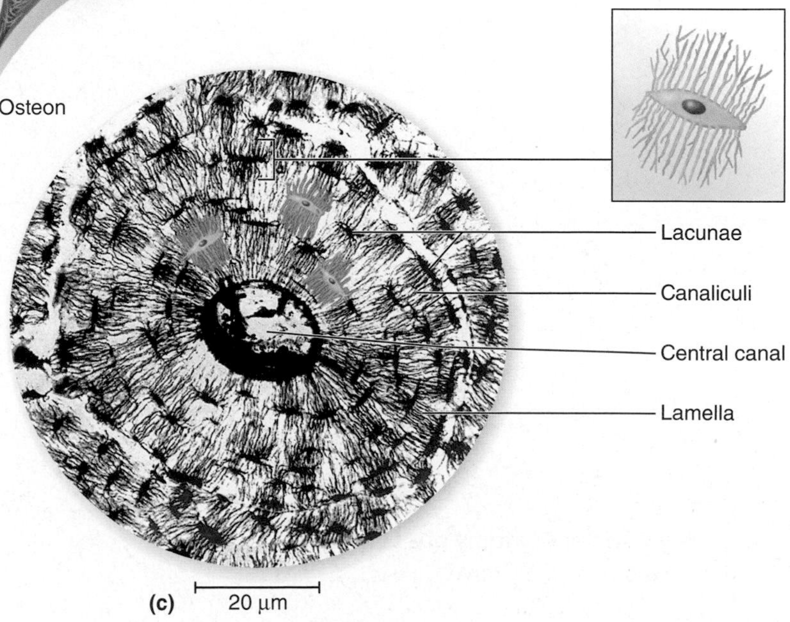

Bone marrow

Trabecula

(b)

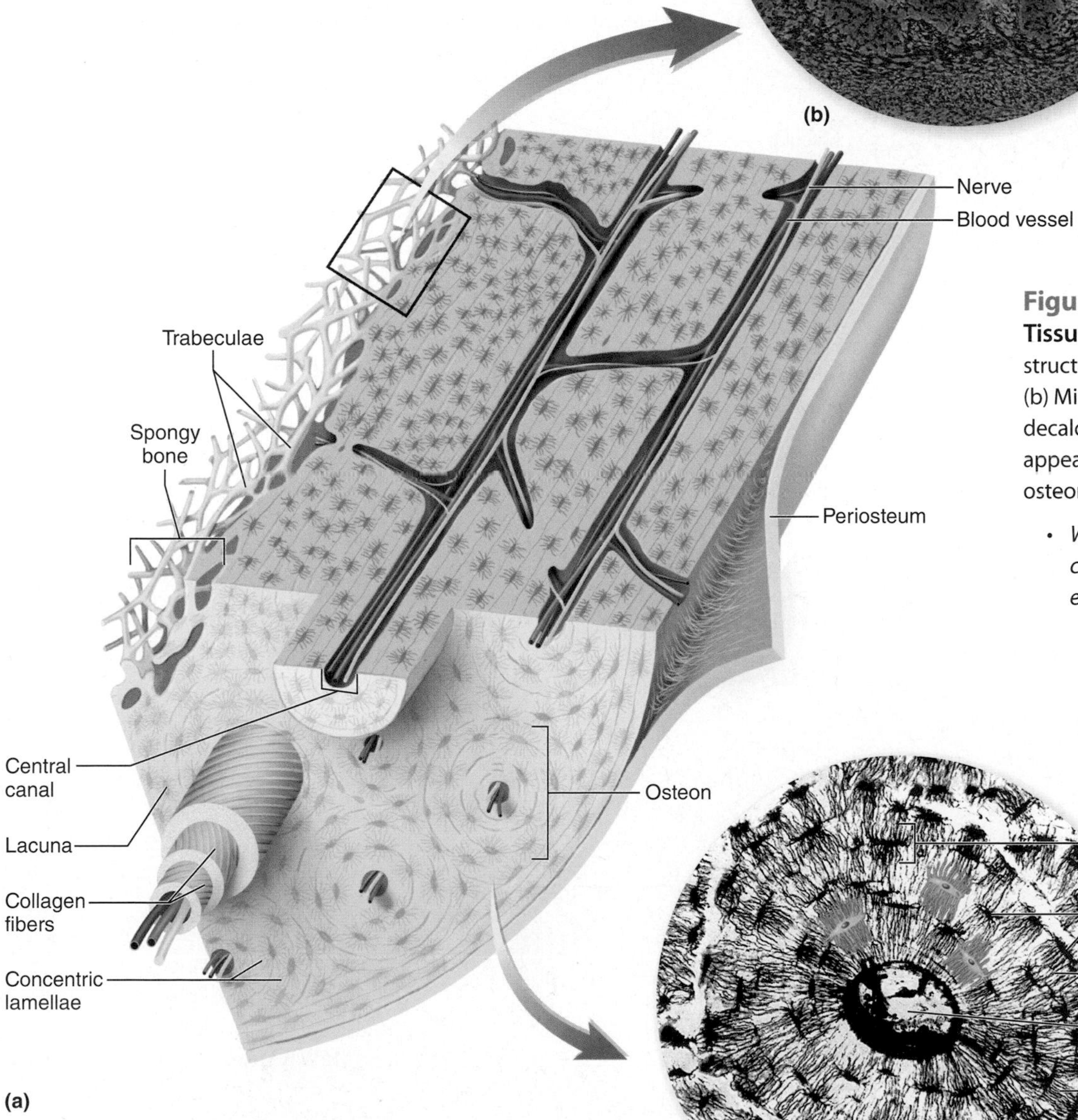

Nerve

Blood vessel

Trabeculae

Spongy bone

Periosteum

Central canal

Lacuna

Collagen fibers

Concentric lamellae

(a)

Osteon

Figure 6.3 Histology of Osseous Tissue. (a) The three-dimensional structure of compact bone. (b) Microscopic appearance of decalcified spongy bone. (c) Microscopic appearance of a cross section of an osteon of dried compact bone. **AP|R**

- *Which type of bone, spongy or compact, has more surface area exposed to osteoclast action?*

Lacunae

Canaliculi

Central canal

Lamella

(c)   20 µm

Blood vessels enter a bone through minute pores in the surface and lead to the central canals. The wall of a central canal looks as if it were pierced with innumerable pinholes. The innermost osteocytes receive nutrients from the bloodstream by way of these pores and pass them along to more remote osteocytes through the network of canaliculi. Wastes travel in the other direction, toward the central canal, to be removed by the bloodstream. In spongy bone, nearly all osteocytes are close to the blood supply in the marrow cavity, so there is little need for central canals or osteons.

Gross Anatomy of Bones

Most bones of the limbs are *long bones* specialized for leverage and movement. Figure 6.4 shows the anatomy of one example, the femur. The elongated midsection is called the **shaft** or **diaphysis**[13] (dy-AF-ih-sis), and each

---

[13] *dia* = across; *physis* = growth; originally named for a ridge on the shaft of the tibia

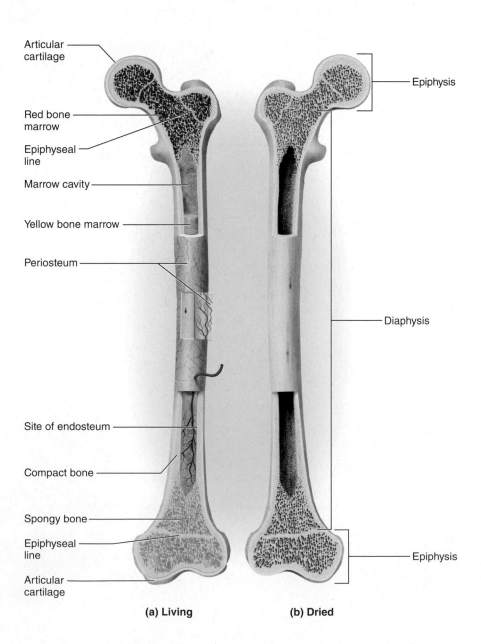

Figure 6.4 General Anatomy of a Long Bone.
The femur is shown as an example.

(a) Living   (b) Dried

expanded end is called a **head** or **epiphysis**[14] (eh-PIF-ih-sis). The shaft provides leverage whereas the heads are enlarged to strengthen a joint and provide added surface area for the attachment of tendons and ligaments. The ends where two bones meet are covered with a thin layer of hyaline cartilage called the **articular cartilage,** which eases joint movements. Each head and nearby region of the shaft are filled with spongy bone. **Bone marrow** occupies the **medullary** (MED-you-lerr-ee) **cavity** of the shaft and the spaces in the spongy bone of the head. **Epiphyseal lines** of an adult bone mark the former growth zones of the child's bone.

Flat bones (fig. 6.5) act as protective armorlike plates covering delicate organs beneath, such as the sternum anterior to the heart and the cranial bones enclosing the brain. Some furnish broad attachment surfaces for muscles, as in the case of the scapula (shoulder blade) and hip bone. In flat bones, two layers of compact bone enclose a middle layer of spongy bone like a sandwich.

Some bones fit the description of neither long nor flat bones and are sometimes called *short* and *irregular* bones (such as wrist bones and vertebrae, respectively).

Externally, a bone is covered with a fibrous sheath called the **periosteum**[15] (see fig. 6.3a). Collagen fibers of the periosteum are continuous with the tendons that bind muscle to bone and penetrate as well into the bone itself. The periosteum thus provides strong attachment and continuity from muscle to tendon to bone. There is no periosteum over the articular cartilage. The internal surface of a bone is lined with **endosteum,**[16] a thin layer of reticular connective tissue. Periosteum covers both sides of a flat bone; in the skull, for example, it covers both the surface facing the brain and the surface facing the scalp, whereas endosteum covers the spongy bone trabeculae in the middle layer.

There are two types of bone marrow—red and yellow. **Red bone marrow,** which serves to produce blood cells and platelets, fills nearly every bone of a child's skeleton, but has a more limited distribution in adults (fig. 6.6). It is an organ in itself, consisting of a soft but very structured network of delicate blood vessels surrounded by reticular tissue and islands of blood-forming cells. As we mature, much of the red bone marrow is gradually replaced by fatty **yellow bone marrow,** like the fat seen at the center of a ham bone. Yellow marrow dominates the long limb bones of the adult.

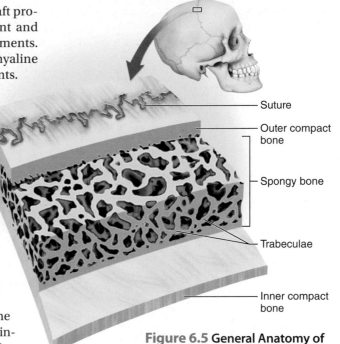

Suture
Outer compact bone
Spongy bone
Trabeculae
Inner compact bone

Figure 6.5 General Anatomy of a Flat Bone. The example here is the parietal bone from the top of the skull. **AP|R**

> ### Apply What You Know
>
> Predict at least two functional problems that would result if a mutation caused someone's bones to fail to develop the outer shell of compact bone.

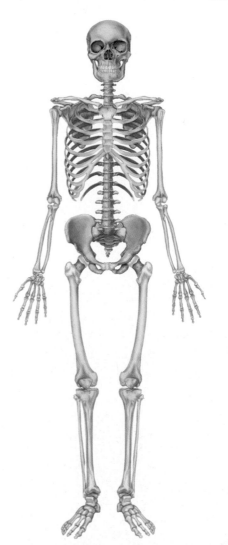

---

[14] *epi* = upon, above; *physis* = growth

[15] *peri* = around; *oste* = bone

[16] *endo* = within; *oste* = bone

Figure 6.6 Distribution of Red and Yellow Bone Marrow. In an adult, red bone marrow occupies the marrow cavities of the axial skeleton and proximal heads of the humerus and femur. Yellow bone marrow occurs in the long bones of the limbs.

6.2 Bone Development and Metabolism

Expected Learning Outcomes

When you have completed this section, you should be able to:

a. describe the two processes of bone formation from embryonic tissues;

b. describe how bones are remodeled throughout life; and

c. identify hormones and other factors that govern bone deposition, bone resorption, and calcium balance.

Ossification

The formation of bone is called **ossification** (OSS-ih-fih-CAY-shun). There are two methods of ossification—*intramembranous* and *endochondral*. Both begin with a soft embryonic connective tissue called *mesenchyme* (MEZ-en-kime), which is a forerunner of our adult bone, muscle, blood, and many other tissues between the skin and the gut.

Intramembranous Ossification

Intramembranous[17] (IN-tra-MEM-bruh-nus) **ossification** produces the flat bones of the skull and most of the clavicle (collarbone). Follow its stages in figure 6.7 as you read the correspondingly numbered descriptions here.

---

[17] *intra* = within; *membran* = membrane

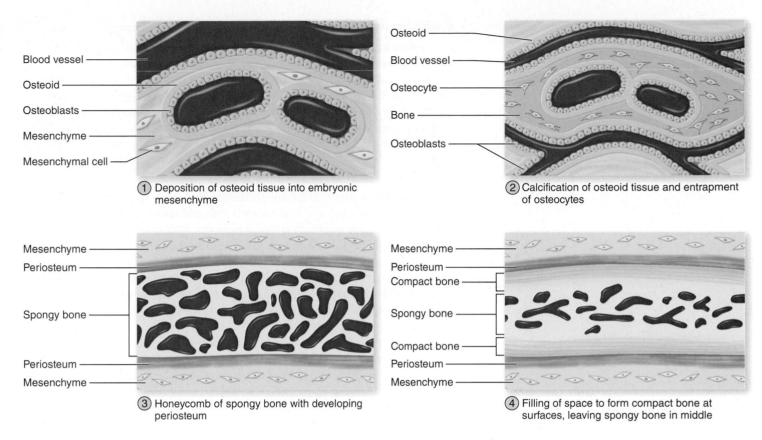

Blood vessel —
Osteoid —
Osteoblasts —
Mesenchyme —
Mesenchymal cell —

① Deposition of osteoid tissue into embryonic mesenchyme

Osteoid —
Blood vessel —
Osteocyte —
Bone —
Osteoblasts —

② Calcification of osteoid tissue and entrapment of osteocytes

Mesenchyme —
Periosteum —

Spongy bone —

Periosteum —
Mesenchyme —

③ Honeycomb of spongy bone with developing periosteum

Mesenchyme —
Periosteum —
Compact bone —

Spongy bone —

Compact bone —
Periosteum —
Mesenchyme —

④ Filling of space to form compact bone at surfaces, leaving spongy bone in middle

Figure 6.7 **Intramembranous Ossification.** The figures are drawn to different scales, with the highest magnification and detail at the beginning and backing off for a broader overview at the end of the process.

① Mesenchyme first condenses into a soft sheet of tissue permeated with blood vessels—the *membrane* to which *intramembranous* refers. Mesenchymal cells line up along the blood vessels and secrete a soft collagenous tissue called *osteoid (prebone)* in the direction away from the vessel. Osteoid resembles bone but is not yet hardened by minerals.

② Calcium phosphate and other minerals crystallize on the collagen fibers of the osteoid and harden the matrix into a network of spongy bone trabeculae. Continued osteoid deposition and mineralization squeezes the blood vessels and future bone marrow into narrower and narrower spaces. As osteoblasts become trapped in their own hardening matrix, they become osteocytes.

③ While the foregoing processes are going on, more of the mesenchyme adjacent to the developing bone condenses and forms a fibrous periosteum on each surface. The spongy bone becomes a honeycomb of slender calcified trabeculae.

④ At the surfaces, osteoblasts beneath the periosteum deposit layers of bone, fill in the spaces between trabeculae, and create a zone of compact bone on each side, as well as thickening the bone overall. This process gives rise to the sandwichlike structure typical of a flat cranial bone—a layer of spongy bone between two surface layers of compact bone.

Intramembranous ossification also plays an important role in the lifelong thickening, strengthening, and remodeling of the long bones discussed next. Throughout the skeleton, it is the method of depositing new tissue on the bone surface even past the age where our bones can no longer grow in length.

Endochondral Ossification

Endochondral[18] (EN-doe-CON-drul) **ossification** produces most other bones, including the vertebrae, ribs, scapulae, pelvic bones, bones of the limbs, and some parts of the skull. It is a method in which mesenchyme first transforms into a hyaline cartilage "model" in the approximate shape of the bone to come. The cartilage is then broken down and replaced by osseous tissue. Many bones have complex shapes with several centers of endochondral ossification, so we illustrate this in figure 6.8 with one of the anatomically simplest bones—a *metacarpal bone* from the palm of the fetal hand. Correlate the following steps with the figure as you read the story.

① The forerunner of the future bone is a body of hyaline cartilage that approximates its shape. It is covered with a fibrous perichondrium.

② In a *primary ossification center* near the middle of this cartilage, chondrocytes begin to inflate and die, while the thin walls between them calcify. Meanwhile, cells of the perichondrium become osteoblasts and deposit a thin layer of bone around the cartilage model, forming a collar that reinforces the shaft of the future bone. The fibrous sheath around this collar is then considered to be periosteum. It thickens and grows toward the ends of the bone.

③ Blood vessels grow inward from the periosteum and invade the ossification center. Osteoclasts arrive in this blood flow and digest calcified tissue in the shaft, hollowing it out and creating the *primary marrow cavity*. Osteoblasts also arrive and deposit layers of bone lining the cavity, thickening the shaft. About this time, a *secondary ossification center* develops

---

[18] *endo* = within; *chondr* = cartilage

Figure 6.8 Endochondral Ossification. The metacarpal bones of the hand exhibit a relatively uncomplicated case of endochondral ossification. Many other bones have two or more epiphyseal plates.

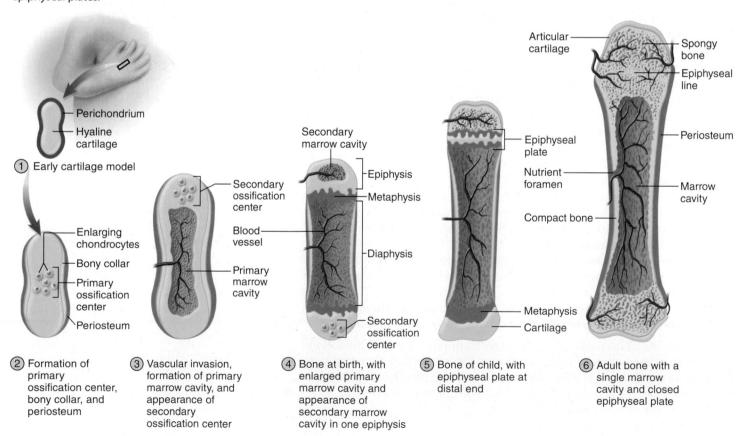

Perichondrium
Hyaline cartilage
① **Early cartilage model**

Enlarging chondrocytes
Bony collar
Primary ossification center
Periosteum

② **Formation of primary ossification center, bony collar, and periosteum**

Secondary ossification center
Blood vessel
Primary marrow cavity
Periosteum

③ **Vascular invasion, formation of primary marrow cavity, and appearance of secondary ossification center**

Secondary marrow cavity
Epiphysis
Metaphysis
Diaphysis
Secondary ossification center

④ **Bone at birth, with enlarged primary marrow cavity and appearance of secondary marrow cavity in one epiphysis**

Epiphyseal plate
Nutrient foramen
Compact bone
Metaphysis
Cartilage

⑤ **Bone of child, with epiphyseal plate at distal end**

Articular cartilage
Spongy bone
Epiphyseal line
Periosteum
Marrow cavity

⑥ **Adult bone with a single marrow cavity and closed epiphyseal plate**

at one end of the bone, resembling the primary ossification center in the center.

(4) By the time of a child's birth, the secondary ossification center has hollowed out a secondary marrow cavity at that end, and another ossification center develops at the other end of the bone. At each end of the primary marrow cavity is a transitional zone called a **metaphysis** (meh-TAFF-ih-sis), shown in violet, where cartilage is undergoing replacement by bone.

(5) Throughout childhood and adolescence, the primary and secondary marrow cavities are separated by a wall called the **epiphyseal plate.** It consists of a middle layer of hyaline cartilage with a metaphysis on each side. The plate functions as a growth zone enabling the individual to grow in limb length and height.

(6) By the late teens to early twenties, the reserve cartilage in the epiphyseal plate is depleted and the primary and secondary marrow cavities are no longer separated. The bones can grow no longer, and one attains his or her maximum adult height. The only remnant of the original cartilage model is the articular cartilage that covers the joint surfaces of the bone.

Growth and Remodeling

Bones continue to grow and remodel themselves throughout life, changing size and shape to accommodate the changing forces applied to the skeleton. For example, in children the limbs become longer; the curvature of the cranium changes to accommodate a growing brain; and many bones develop surface bumps, spines, and ridges as a child begins to walk and the muscles exert tension on the bones. Tension on the skeleton stimulates an increase in bone mass. On average, therefore, the bones of athletes and people engaged in heavy manual labor have greater mass than the bones of sedentary people. In tennis players, the bones of the racket arm and the clavicle on that side are often more robust than those of the other arm.

The bones and cartilages of the skeleton grow by two methods. One is **interstitial**[19] **growth,** in which chondrocytes multiply, enlarge, and secrete new matrix between them. Interstitial growth in the epiphyseal plate adds to the length of a bone until the plate is depleted in early adulthood. In the metaphysis on each side of the epiphyseal plate, facing the two marrow cavities, chondrocytes die, their lacunae break open and merge into longitudinal channels, and osseous tissue is deposited in these channels to form new osteons of compact bone. In children, the epiphyseal plate appears on X-rays as a transparent line across the end of a bone, creating the appearance of a gap separating the epiphysis from the diaphysis. In the late teens to late twenties—at different ages in different bones—the cartilage is depleted and the gap is filled in with bone. The plates are said to "close" because there is no longer a gap between the epiphysis and diaphysis. The bones can then grow no longer and a person can grow no taller. The epiphyseal plate leaves behind an internal mark, the *epiphyseal line* seen in figure 6.4.

Appositional[20] **growth** is the other means of cartilage and bone growth, in which new matrix is deposited on the surface of the tissue. Mature bone is limited to this type of growth, because osteocytes have no room to spare for the deposition of more matrix between them. The only way an adult bone can grow, therefore, is by adding more surface tissue. Appositional bone growth occurs by intramembranous ossification. Osteoblasts in the periosteum deposit

---

[19] *inter* = between; *stit* = to place, stand

[20] *ap* = *ad* = to, near; *posit* = to place

osteoid on the bone surface, calcify it, and become trapped in it as osteocytes. This osseous tissue is laid down in layers parallel to the surface, not in cylindrical osteons like the tissue deeper in the bone. While deposition occurs at the outer surface of a bone, osteoclasts dissolve bone on the inner surface and thus enlarge the medullary cavity.

Therefore, we see that flat bones develop by intramembranous ossification alone, whereas long bones develop by a combination of endochondral and intramembranous ossification. AP|R

Mineral Homeostasis

Even after a bone is fully formed, it remains a metabolically active organ with many roles to play. Not only is it involved in its own maintenance, growth, and remodeling; but it also exerts a profound influence on the rest of the body by exchanging minerals with the extracellular fluid. The skeleton is the body's primary reservoir of calcium and phosphate. These minerals are used for much more than bone structure, and can be withdrawn from bone when needed elsewhere. Phosphate is a component of DNA, RNA, ATP, phospholipids, and other compounds. Calcium is involved in muscle contraction, blood clotting, exocytosis, nervous communication, and cellular responses to hormones, among other processes.

Mineral **deposition** is a crystallization process in which osteoblasts extract calcium, phosphate, and other ions from the blood and deposit them in the osseous tissue. **Resorption** is a process in which the osteoclasts dissolve bone, releasing minerals into the blood and making them available for other uses. There is a critical balance between bone deposition and resorption. If one process outpaces the other, or if both of them occur too rapidly, various bone deformities and other developmental abnormalities occur, some of them quite grotesque and crippling.

The concentration of calcium in the blood and tissue fluid must be regulated within narrow limits. A calcium deficiency, or **hypocalcemia**[21] (HY-po-cal-SEE-me-uh), causes dysfunctions ranging from muscle tremors to tetanus—inability of the muscle to relax. One can suffocate from tetanus in the larynx *(laryngospasm)*. A calcium excess, or **hypercalcemia,**[22] depresses nervous, muscular, and cardiac function.

Calcium homeostasis is regulated mainly by two hormones: calcitriol and parathyroid hormone, both of which raise blood calcium levels. **Calcitriol** is the most active form of vitamin D. It raises blood calcium by promoting calcium absorption from digested food, reducing urinary loss of calcium, and stimulating osteoclasts to release calcium from the bones to the blood. Even though it promotes bone resorption, it is also necessary for bone deposition because without its effects on the kidneys and small intestine, blood calcium levels are too low for normal deposition.

Parathyroid hormone (PTH) is secreted by the parathyroid glands, which adhere to the posterior surface of the thyroid. The parathyroids respond to a drop in blood calcium level by secreting PTH. PTH then stimulates bone resorption by osteoclasts, promotes calcium reabsorption by the kidneys, and promotes calcitriol synthesis. All of these effects raise blood calcium levels.

---

[21] *hypo* = below normal; *calc* = calcium; *emia* = blood condition
[22] *hyper* = above normal; *calc* = calcium; *emia* = blood condition

Clinical Application 6.1

BONE TUMORS

The skeletal system is subject to a variety of benign and malignant tumors. *Osteochondroma* is a benign tumor of bone and cartilage, often forming outgrowths called *bone spurs* at the ends of the long bones. *Osteoma* is a benign tumor occurring especially in the flat bones of the skull, sometimes growing into the orbits or sinuses. *Osteosarcoma* is far more serious. This is the most common and deadly form of bone cancer. It occurs most often in boys and young men, and arises especially in the tibia, femur, and humerus. In about 10% of cases, it metastasizes to the lungs and other organs. Untreated metastatic osteosarcoma is usually fatal within a year. *Chondrosarcoma* is a slow-growing malignant tumor of hyaline cartilage. Because of the lack of blood flow to cartilage, it is unresponsive to chemotherapy and must be removed surgically.

Calcitonin is secreted by the thyroid gland in response to elevated blood calcium levels. It promotes deposition of the excess calcium into the skeleton and thereby lowers the blood calcium level. It has little influence on adult bone, but plays a more significant role in regulating blood calcium in children and may help to reduce bone loss in pregnant women.

Many other hormones, growth factors, and vitamins affect osseous tissue in complex ways that are still not well understood. Growth hormone, estrogen, and testosterone promote ossification and bone growth, especially in adolescence. Bone deposition is also promoted by thyroid hormone, insulin, and at least 12 growth factors that are produced in the bone itself and stimulate nearby bone cells. Not only vitamin D but also vitamins A and C are needed for bone deposition and repair.

Before You Go On

Answer these questions from memory. Reread the preceding section if there are too many you don't know.

8. Explain how a soft sheet of mesenchyme turns into a flat bone such as a cranial bone, and how this process accounts for the sandwichlike, three-layered structure of such a bone.

9. Explain how a cartilage model is replaced by osseous tissue to form a long bone such as the humerus. Explain how this process accounts for the medullary cavity, articular cartilage, and epiphyseal plate of a bone.

10. Explain why fully developed bones must remain subject to remodeling, and how such remodeling is achieved.

11. Describe the effects of calcitriol, parathyroid hormone, and other hormones and vitamins on bone and calcium metabolism.

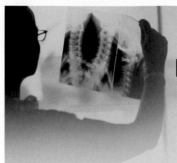

PERSPECTIVES ON HEALTH

Among the most familiar disorders of the skeletal system are fractures, osteoporosis, and arthritis. Others have been briefly described in Clinical Applications 6.1 and 6.2.

Fractures

Many of us have broken a bone at one time or another. Perhaps surprisingly, the most commonly broken bone is the clavicle, because it is so close to the surface and because people often reach out with their arms to break a fall or brace themselves against an impact. When a fracture results from an unusual stress on a bone, such as a fall or auto accident, it is called a *stress fracture*. When a bone has been weakened by some other condition, such as bone cancer or osteoporosis, and it fractures under a stress that a healthy bone would withstand, it is called a *pathologic fracture*. Any fracture that breaks through the skin is called an *open fracture;* if it does not break the skin, it is a *closed fracture.*

When a bone fractures, it bleeds and the blood clot *(fracture hematoma)* soon becomes infiltrated with fibroblasts, macrophages, osteoclasts, and osteogenic cells, which clean up the tissue fragments and begin the repair process. Deposition of collagen and fibrocartilage turns the hematoma into a *soft callus.* In 4 to 6 weeks, osteoblasts deposit calcified tissue, a *hard callus,* around this to splint the broken bone pieces together. It is especially important during this time that the bone be immobilized by traction or casting to prevent reinjury. The hard callus persists for 3 to 4 months. During this time, osteoclasts dissolve and remove remaining fragments of broken bone while osteoblasts build spongy bone to bridge the gap between the broken ends. This spongy bone then fills in by intramembranous ossification to solidify the repair.

Clinically, most fractures are set by *closed reduction,* requiring no surgery but often using a cast to immobilize the healing bones. More serious fractures may require *open reduction,* in which bone pieces are surgically exposed and joined by plates, screws, or pins.

Osteoporosis

Osteoporosis[23] is the most common of all bone diseases. It is a loss of bone tissue (especially spongy bone) to the point that the bones become unusually porous and brittle. It is usually associated with aging, with postmenopausal white women being at greatest risk; however, it can occur in either sex and at any age. In early space missions, osteoporosis was a significant problem among astronauts, who spent several days under microgravity conditions with too little bone-sustaining stress on the skeleton. Osteoporosis compromises a person's physical activity and presents a high risk of disabling bone fractures, often leading to death among the elderly. Fractures of the wrist and hip are common, as are crushing compression fractures of the vertebrae. Vertebral fractures commonly result in such spinal deformities as kyphosis (see p. 171). The fracture of osteoporotic bones can result from stresses as slight as sitting down too quickly or supporting the body on one's wrists when rising from an armchair.

After menopause, women are especially at risk because the ovaries cease to secrete estrogen, which normally stimulates bone deposition. Other risk factors include smoking, insufficient exercise, inadequate calcium and protein intake, vitamin C deficiency, and diabetes mellitus.

Arthritis

Gradually declining bone density can be a silent, insidious menace to one's health, but a more noticeable effect of aging is the degenerative joint disease called *osteoarthritis (OA).* Nicknamed "wear-and-tear arthritis," OA sets in as the synovial joints contain less lubricating synovial fluid and the articular cartilages soften and degenerate. As the cartilages become roughened by wear, joint movements may be accompanied by crunching or crackling sounds called *crepitus.* As OA progresses, exposed bone surfaces tend to develop spurs that grow into the joint cavity, restrict movement, and cause pain—especially in the fingers, spine, hips, and knees. About 85% of people over age 70 experience some degree of OA. It is seldom crippling, but in extreme cases it can immobilize the hip.

Rheumatoid arthritis (RA) is less common but more often crippling. It results from an autoimmune attack against the joint tissues. It begins when a misguided antibody called *rheumatoid factor* attacks the synovial membranes. Inflammatory cells accumulate in the joint and produce enzymes that degrade the articular cartilage. As the cartilage degenerates, the joint begins to ossify. The bones sometimes become

---

[23] *osteo* = bone; *porosis* = a state of being porous

solidly fused and immobilized, a condition called *ankylosis* (fig. 6.9). Rheumatoid arthritis is named for the fact that symptoms tend to flare up and subside periodically.[24] It affects more women than men, and because it often begins between the ages of 30 and 40, it can cause decades of pain and disability. There is no cure; it is possible only to slow its progression with anti-inflammatory drugs and physical therapy.

A hereditary arthritis more common in men is *gouty arthritis*. *Gout* is the accumulation of uric acid crystals in the joints, especially of the great toe. Uric acid irritates the articular cartilage and synovial membrane, causing swelling, tissue generation, and sometimes joint fusion.

Rheumatism is a broad term for any pain in the supportive and locomotor organs, including muscles, tendons, ligaments, and bones. It includes arthritis, bursitis, gout, and many other conditions. Physicians who specialize in the rheumatic diseases are called *rheumatologists*.

---

[24] *rheumat* = tending to change

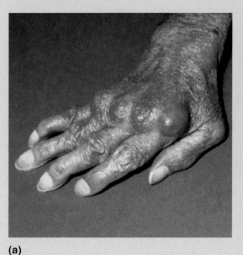

(a)

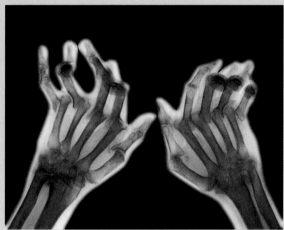

(b)

Figure 6.9 Rheumatoid Arthritis (RA). (a) A severe case with immobilizing ankylosis of the joints. (b) X-ray of ankylosis of the hands.

- *Name some specific bones in part (b) that no longer properly articulate with each other.*

6.3 The Axial Skeleton

Expected Learning Outcomes

When you have completed this section, you should be able to:

a. define terms for the surface features and markings of bones;

b. name and define the two basic subdivisions of the skeleton;

c. name and describe the cavities within the skull;

d. identify each of the skull bones, name their features, and name some of the main sutures that join them;

e. describe the general features of the vertebral column, general structure of a vertebra and intervertebral disc, and regional differences between the vertebrae; and

f. describe the anatomy of the ribs and sternum, regional differences between the ribs, and the articulation of the ribs with the vertebral column and sternum.

Table 6.1 Bone Markings

Term	Description and Example
Canal	A tubular passage or tunnel in a bone (auditory canal of the ear)
Condyle	A rounded knob (occipital condyles of the skull)
Crest	A narrow ridge (iliac crest of the pelvis)
Epicondyle	A flare superior to a condyle (medial epicondyle of the femur)
Facet	A smooth joint surface that is flat or only slightly concave or convex (articular facets of the vertebrae)
Fissure	A slit through a bone (orbital fissures behind the eye)
Foramen	A hole through a bone, usually round (foramen magnum of the skull)
Fossa	A shallow, broad, or elongated basin (infraspinous fossa of the scapula)
Process	Any bony prominence (mastoid process of the skull)
Sinus	A cavity within a bone (frontal sinus of the skull)
Spine	A sharp, slender, or narrow process (spine of the scapula)
Tubercle	A small, rounded process (greater tubercle of the humerus)
Tuberosity	A rough surface (tibial tuberosity)

We turn our attention now to the 206 named bones that make up the adult skeleton. Descriptions of these bones include a variety of ridges, spines, bumps, depressions, holes, and joint surfaces, collectively called *bone markings*. It is important to know the names of these features because later descriptions of joints, muscle attachments, and the routes traveled by nerves and blood vessels are based on this terminology. The most common of these features are defined in table 6.1, and several are illustrated in figure 6.10. Other terms specific to one or a few bones are introduced when those bones are described.

Figure 6.10 Bone Markings. Most of these features also occur on many other bones of the body.

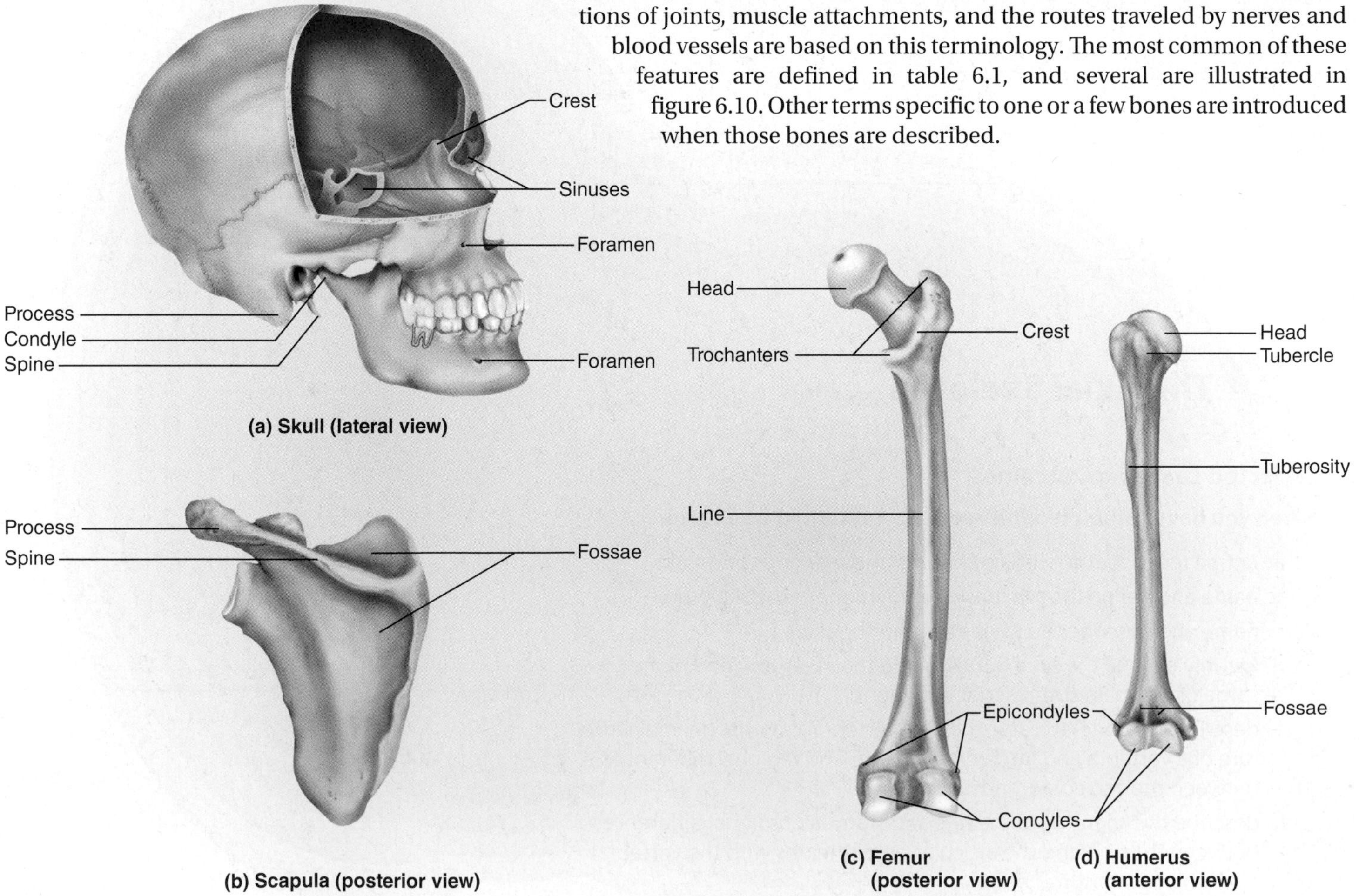

(a) Skull (lateral view)

(b) Scapula (posterior view)

(c) Femur (posterior view)

(d) Humerus (anterior view)

The skeleton is divided into two regions, axial and appendicular (fig. 6.11). The **axial skeleton** forms the central supporting axis of the body and includes the skull, vertebral column, ribs, and sternum. The **appendicular skeleton** includes the bones of the upper limb and pectoral girdle, and bones of the lower limb and pelvic girdle.

Figure 6.11 The Adult Skeleton. The appendicular skeleton is colored green, and the rest is axial skeleton. AP|R

- *The appendicular skeleton is not entirely within the appendicular region of the body. Explain.*

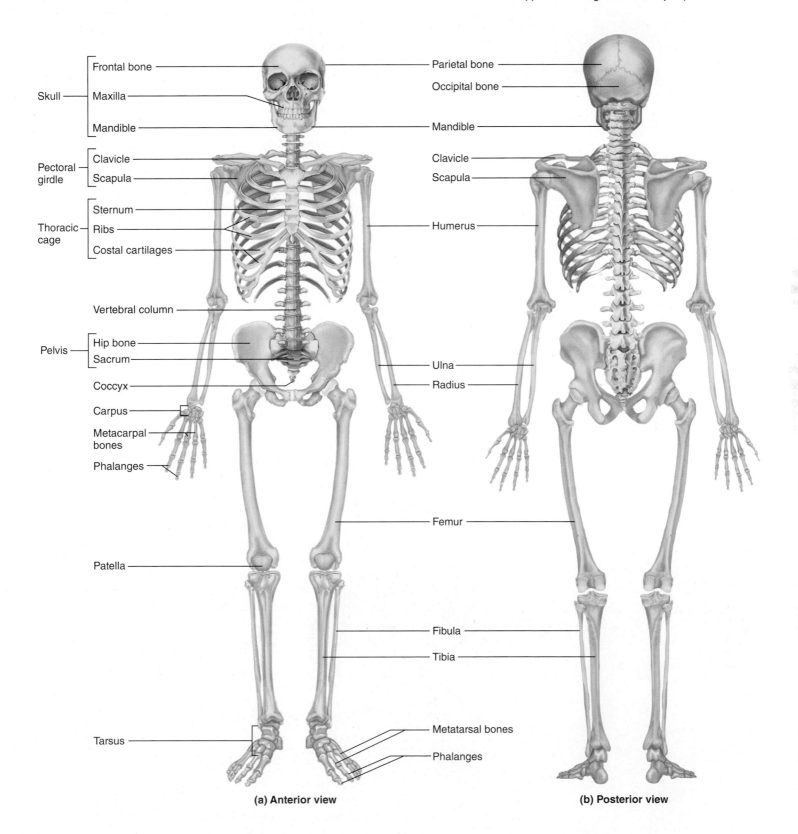

(a) Anterior view   (b) Posterior view

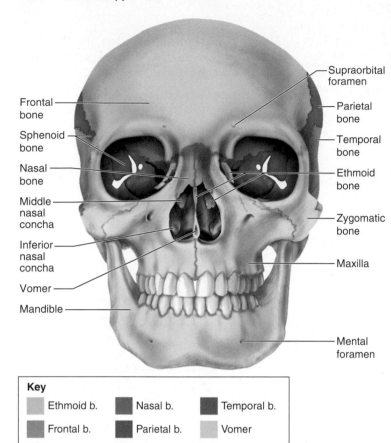

Frontal bone
Sphenoid bone
Nasal bone
Middle nasal concha
Inferior nasal concha
Vomer
Mandible

Supraorbital foramen
Parietal bone
Temporal bone
Ethmoid bone
Zygomatic bone
Maxilla
Mental foramen

Key

▢ Ethmoid b.	▢ Nasal b.	▢ Temporal b.
▢ Frontal b.	▢ Parietal b.	▢ Vomer
▢ Mandible	▢ Sphenoid b.	▢ Zygomatic b.
▢ Maxilla		

Figure 6.12 The Skull, Anterior View. AP|R

Figure 6.13 The Paranasal Sinuses.

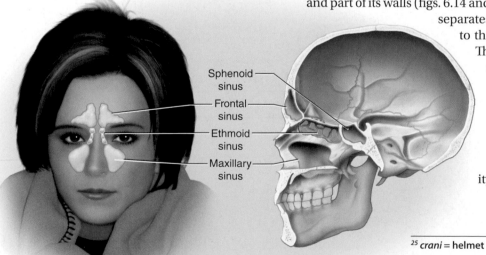

Sphenoid sinus
Frontal sinus
Ethmoid sinus
Maxillary sinus

The Skull

The skull (figs. 6.12 to 6.17) is the most complex part of the skeleton. It consists of one freely movable bone, the mandible, and 21 relatively immobile bones connected by **sutures** (SOO-chers)—joints that appear as seams on the skull surface. Most bones of the skull are perforated by **foramina** (singular, *foramen*)—holes that allow passage for nerves and blood vessels to go to and from the brain.

The skull contains several prominent cavities—the **cranial cavity,** which encloses the brain; the **orbits** (eye sockets); the **nasal cavity;** the **oral (buccal) cavity;** the **middle-** and **inner-ear cavities;** and **paranasal sinuses,** air-filled spaces connected to the nasal cavity and named for the bones in which they occur—the **frontal, sphenoid, ethmoid,** and **maxillary sinuses** (fig. 6.13).

Eight bones of the skull are called *cranial bones* and the other 14 are called *facial bones.* The **cranial bones** form the **cranium.**[25] They are distinguished by being in direct contact with the fibrous membranes, the *meninges,* that enclose the brain. They are as follows:

1 frontal bone	1 occipital bone
2 parietal bones	1 sphenoid bone
2 temporal bones	1 ethmoid bone

The Frontal Bone

The **frontal bone** (figs. 6.12, 6.14, and 6.15b) is the broad bone that forms the forehead, the roof of the orbit, and the anterior one-third of the roof of the cranial cavity. In the region of the eyebrows it has a thickened ridge with a small hole, the **supraorbital foramen,** for the passage of a nerve, artery, and vein. The frontal bone also contains the frontal sinus, although this is absent from some people. An infant has separate right and left frontal bones, which normally fuse into a single bone. The two bones remain separate, however, in some adults.

The Parietal Bones

The right and left **parietal** (pa-RYE-eh-tul) **bones** form most of the cranial roof and part of its walls (figs. 6.14 and 6.15). They begin at the **coronal**[26] **suture** that separates them from the frontal bone and they extend to the **lambdoid**[27] (LAM-doyd) **suture** at the rear. The two parietal bones are separated by a longitudinal **sagittal suture.**

The Temporal Bones

The **temporal bones** (fig. 6.14) form the region around the ears (the *temples*) and part of the lower wall and floor of the cranial cavity. They derive their name from the fact that

---

[25] *crani* = helmet

[26] *corona* = crown

[27] Shaped like the Greek letter lambda (λ)

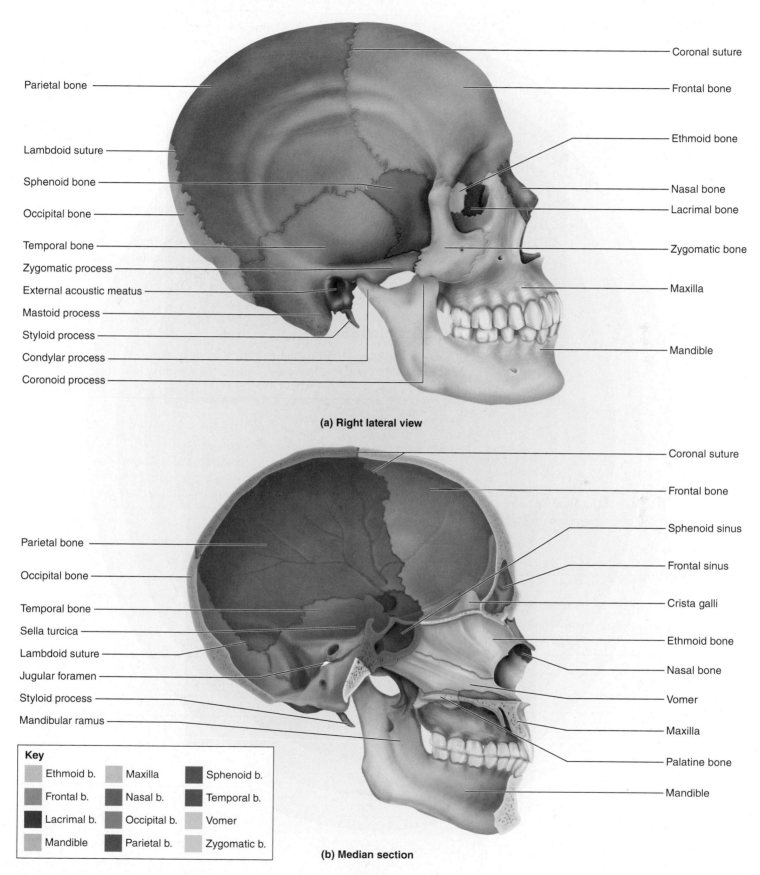

Coronal suture

Parietal bone

Frontal bone

Ethmoid bone

Lambdoid suture

Sphenoid bone

Nasal bone

Occipital bone

Lacrimal bone

Temporal bone

Zygomatic bone

Zygomatic process

External acoustic meatus

Maxilla

Mastoid process

Styloid process

Condylar process

Mandible

Coronoid process

(a) Right lateral view

Coronal suture

Parietal bone

Frontal bone

Sphenoid sinus

Occipital bone

Frontal sinus

Temporal bone

Crista galli

Sella turcica

Lambdoid suture

Ethmoid bone

Jugular foramen

Nasal bone

Styloid process

Vomer

Mandibular ramus

Maxilla

Key

Ethmoid b.	Maxilla	Sphenoid b.
Frontal b.	Nasal b.	Temporal b.
Lacrimal b.	Occipital b.	Vomer
Mandible	Parietal b.	Zygomatic b.

Palatine bone

Mandible

(b) Median section

Figure 6.14 The Skull, Lateral Surface and Median Section. AP|R

• *List all bones that articulate with the temporal bone.*

people often develop their first gray hairs on the temples with the passage of time.[28] Their principal features follow:

- The **zygomatic process,** a spine that sweeps outward from the temple to form part of the **zygomatic arch.** The zygomatic bone in the middle and the zygomatic process of the maxilla anteriorly complete the arch.
- The **external acoustic meatus** (me-AY-tus), or **auditory** (ear) **canal.**
- The **mastoid process,** a blunt downward growth that you can palpate as a prominent lump behind your earlobe.
- The **styloid process,** a deeper, more slender and pointed process that provides attachment for a muscle of the throat.
- The **mandibular fossa,** a depression on the inferior surface where the mandible articulates with the cranium (fig. 6.15a).

Also visible on the inferior surface are two foramina, the **carotid canal** and **jugular foramen.** These are passages for the internal carotid artery and internal jugular vein, respectively. On its medial surface, within the cranial cavity, the temporal bone forms a prominent diagonal ridge that contains the middle- and inner-ear cavities, which house the organs of hearing and balance (fig. 6.15b).

The Occipital Bone

The **occipital** (oc-SIP-ih-tul) **bone** forms the rear of the skull and much of its base (figs. 6.14 and 6.15). Its most conspicuous feature is a large opening, the

---

[28] *tempor* = time

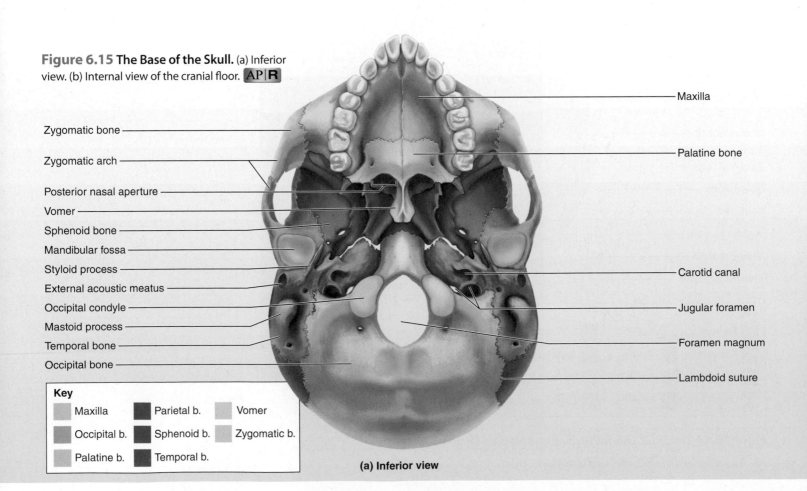

Figure 6.15 The Base of the Skull. (a) Inferior view. (b) Internal view of the cranial floor. AP|R

Zygomatic bone

Zygomatic arch

Posterior nasal aperture

Vomer

Sphenoid bone

Mandibular fossa

Styloid process

External acoustic meatus

Occipital condyle

Mastoid process

Temporal bone

Occipital bone

Maxilla

Palatine bone

Carotid canal

Jugular foramen

Foramen magnum

Lambdoid suture

Key

Maxilla	Parietal b.	Vomer
Occipital b.	Sphenoid b.	Zygomatic b.
Palatine b.	Temporal b.	

(a) Inferior view

foramen magnum,[29] which admits the spinal cord to the cranial cavity. On each side of the foramen magnum is a smooth knob, the **occipital condyle** (CON-dile), where the skull rests on the vertebral column.

The Sphenoid Bone

The **sphenoid**[30] (SFEE-noyd) **bone** (figs. 6.14 and 6.15) has a complex shape with a thick median **body** and outstretched **wings,** which give the bone as a whole a ragged mothlike shape. The body contains a pair of sphenoid sinuses and has a saddlelike structure named the **sella turcica**[31] (SEL-la TUR-sih-ca), which houses the pituitary gland. The wings form part of the lateral surface of the cranium anterior to the temporal bone and part of the wall of the orbit. Each wing exhibits an **optic canal,** which permits passage of the optic nerve from the eye, and a gash, the **superior orbital fissure** (fig. 6.17), which allows passage of nerves that supply the muscles of eye movement. On its inferior side, the sphenoid exhibits the paired openings of the nasal cavity, the **posterior nasal apertures.**

The Ethmoid Bone AP|R

The **ethmoid**[32] (ETH-moyd) **bone** is located between the orbits and forms the roof of the nasal cavity. It is honeycombed with *air cells* that collectively constitute the ethmoid sinus. A vertical **perpendicular plate** of the ethmoid

[29] *foramen* = hole; *magnum* = large

[30] *sphen* = wedge; *oid* = resembling

[31] *sella* = saddle; *turcica* = Turkish

[32] *ethmo* = sieve, strainer; *oid* = resembling

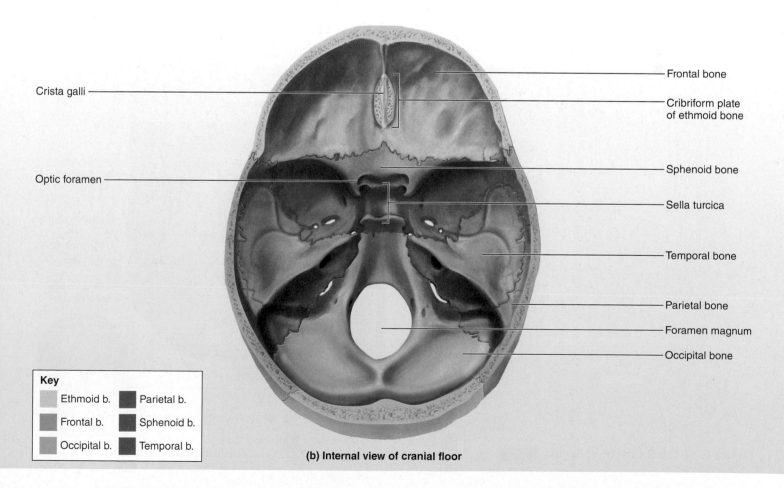

Crista galli

Optic foramen

Frontal bone

Cribriform plate of ethmoid bone

Sphenoid bone

Sella turcica

Temporal bone

Parietal bone

Foramen magnum

Occipital bone

Key

Ethmoid b.

Frontal b.

Occipital b.

Parietal b.

Sphenoid b.

Temporal b.

(b) Internal view of cranial floor

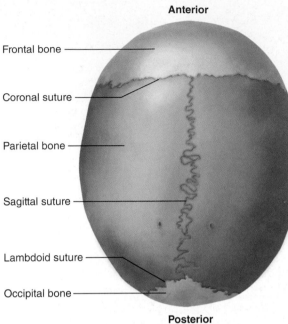

Anterior

Frontal bone

Coronal suture

Parietal bone

Sagittal suture

Lambdoid suture

Occipital bone

Posterior

Figure 6.16 The Calvaria (Skullcap), Superior View. AP|R

forms the superior part of the **nasal septum,** which divides the nasal cavity into right and left spaces called the **nasal fossae** (FOSS-ee) (fig. 6.14b). Three curled, scroll-like **nasal conchae**[33] (CON-kee) project into each fossa from the lateral wall (see fig. 15.2, p. 506). The superior and middle conchae are extensions of the ethmoid bone. The inferior concha is a separate bone discussed later. Within the cranial cavity, one can see only a small superior part of the ethmoid bone (fig. 6.15b). It exhibits a pair of horizontal **cribriform**[34] (CRIB-rih-form) **plates** marked by numerous perforations for passage for the nerve fibers of smell. The cribriform plates are separated by a median ridge called the **crista galli**[35] (GAL-eye).

The 14 **facial bones** shape the face and support internal structures of the oral and nasal cavities, but have no contact with the meninges or brain. They are as follows:

2 maxillae	2 nasal bones
2 palatine bones	2 inferior nasal conchae
2 zygomatic bones	1 vomer
2 lacrimal bones	1 mandible

The Maxillae

The two **maxillae** (mac-SILL-ee) form the upper jaw and support its teeth (see figs. 6.12 and 6.14). Even though the teeth are preserved with the skull, they are not bones. They are discussed in chapter 17. Each maxilla extends upward to form the floor and medial wall of the orbit. Within the orbit, it exhibits a gash, the **inferior orbital fissure** (fig. 6.17), which is a passage for blood vessels and a nerve to the eye muscles. Behind the teeth, the maxillae turn inward as a shelf that forms about four-fifths of the **hard palate,** the anterior part of the roof of the mouth (fig. 6.15a).

---

[33] *conchae* = conchs (large marine snails)

[34] *cribri* = sieve; *form* = in the shape of

[35] *crista* = crest; *galli* = of a rooster

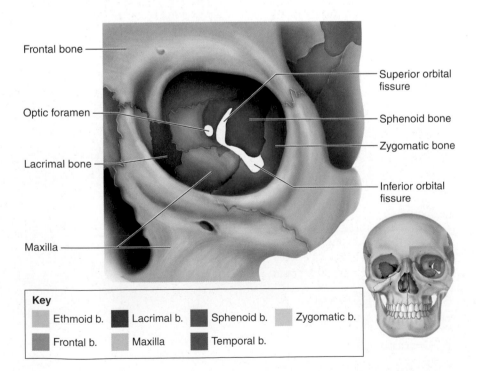

Frontal bone

Optic foramen

Lacrimal bone

Maxilla

Superior orbital fissure

Sphenoid bone

Zygomatic bone

Inferior orbital fissure

Key

Ethmoid b.	Lacrimal b.	Sphenoid b.	Zygomatic b.
Frontal b.	Maxilla	Temporal b.	

Figure 6.17 The Orbital (Eye) Region. AP|R

The Palatine Bones

Two small L-shaped **palatine bones** form the posterior one-fifth of the hard palate, part of the wall of the nasal cavity, and part of the floor of the orbit (see fig. 6.15a).

The Zygomatic Bones

The **zygomatic**[36] **bones** form the angles of the cheeks inferolateral to the eyes, and part of the lateral wall of each orbit (see fig. 6.14a). Each zygomatic bone has an inverted T shape. The prominent *zygomatic arch* ("cheekbone") that flares from each side of the skull is formed by the union of the zygomatic bone, temporal bone, and maxilla (see fig. 6.15a).

The Lacrimal Bones

The tiny **lacrimal**[37] (LACK-rih-mul) **bones** form part of the medial wall of each orbit (fig. 6.17). They house membranous sacs that collect tears from the eyes and drain them into the nasal cavity.

The Nasal Bones

Two small rectangular **nasal bones** form the bridge of the nose and support cartilages that shape the lower portion of the nose (see figs. 6.12, 6.14). The nasal bones are often fractured by blows to the nose.

The Inferior Nasal Conchae

The three conchae of the nasal cavity were mentioned earlier. The superior and middle conchae are parts of the ethmoid bone. The **inferior nasal concha,** the largest of the three, is a separate bone (fig. 6.12; see also fig. 15.2, p. 506).

The Vomer

The **vomer** forms the lower part of the nasal septum and joins the perpendicular plate of the ethmoid bone, which forms the upper part. The name *vomer* literally means "plowshare," which refers to the resemblance of this bone to the blade of a plow (see fig. 6.14b).

The Mandible

The **mandible** (see figs. 6.12 and 6.14) supports the lower teeth and provides attachment for muscles of mastication. The horizontal portion, carrying the teeth, is the **body;** the posterior part that rises to meet the floor of the cranium is the **ramus** (RAY-mus); and the body and ramus meet at a corner called the **angle.** The ramus is divided by a U-shaped notch at its superior end into an anterior **coronoid process** and posterior **condylar** (CON-dih-lur) **process.** The condylar process is capped by a knob, the **mandibular condyle,** which meets the temporal bone to form the hinge of the jaw, called the **temporomandibular joint (TMJ).** The coronoid process and angle are important attachment points for chewing muscles discussed in chapter 7. At the anterolateral surface of the body is a hole, the **mental foramen,** for the passage of nerves and blood vessels of the chin.

---

[36] *zygo* = to join, unite
[37] *lacrim* = tear, to cry

Bones Associated with the Skull

Seven bones are closely associated with the skull but not considered part of it. These are the three **auditory ossicles**[38] in each middle ear (see chapter 10) and the **hyoid**[39] **bone** beneath the chin. The hyoid serves for the attachment of muscles that control the mandible, tongue, and larynx; the larynx is suspended from the hyoid by a ligament.

The Vertebral Column

The **vertebral column** physically supports the skull and trunk; protects the spinal cord; absorbs stresses produced by walking, running, and lifting; and provides attachment for the limbs, thoracic cage, and postural muscles. Although commonly called the backbone, it is not a single bone but a flexible chain of 33 **vertebrae** and 23 cartilaginous **intervertebral discs** (fig. 6.18). The adult vertebral column averages about 71 cm (28 in.) long, with the discs accounting for about one-quarter of this.

The vertebrae are divided into five groups: 7 *cervical vertebrae* in the neck, 12 *thoracic vertebrae* in the chest, 5 *lumbar vertebrae* in the lower back,

---

[38] *os* = bone; *icle* = little
[39] *hy* = the letter U; *oid* = resembling

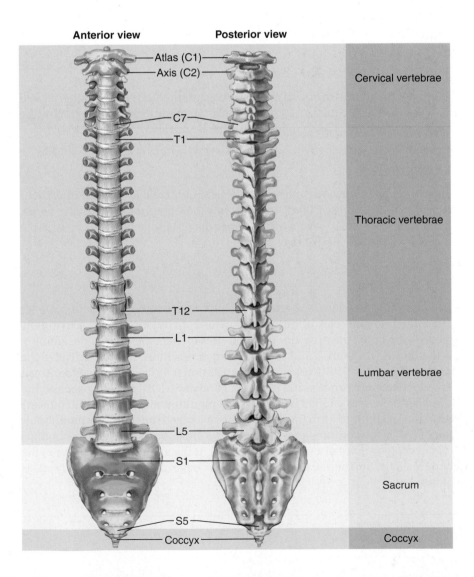

Figure 6.18 The Vertebral Column. AP|R

5 *sacral vertebrae* at the base of the spine, and 4 tiny *coccygeal vertebrae.* About 1 person in 20, however, has 4 lumbar and 6 sacral vertebrae, or 6 lumbar and 4 sacral. In adults, the sacral and coccygeal vertebrae are fused together into single bones, the *sacrum* and *coccyx;* the other three groups of vertebrae remain separate bones. To help remember that the first three groups are 7, 12, and 5 in number, think of a typical work day: go to work at 7, get off for lunch at 12, and go home at 5.

The vertebral column is C-shaped at birth. Beyond the age of 3 years, it is slightly S-shaped, with four bends called the **cervical, thoracic, lumbar,** and **pelvic curvatures** (fig. 6.19a). This shape makes sustained bipedal walking possible because the trunk of the body does not lean forward, as it does in primates such as a chimpanzee. The head is balanced over the body's center of gravity, and the eyes are directed straight forward. Three common abnormalities of the spinal curvatures (fig. 6.19b) are a lateral curvature called *scoliosis;* a "hunchbacked" exaggeration of the thoracic curvature, clinically known as *kyphosis;* and a "swaybacked" exaggeration of the lumbar curvature, clinically called *lordosis.*

General Structure of a Vertebra

A representative vertebra and intervertebral disc are shown in figure 6.20. The most obvious feature of a vertebra is the **body (centrum).** This is the weight-bearing portion. Its rough superior and inferior surfaces provide firm attachment to the intervertebral discs. Posterior to the body is an opening called the **vertebral foramen,** enclosed in a **vertebral arch.** The arch consists of a pair of flat plates called *laminae* supported on a pair of pillars called *pedicles.* Collectively, these foramina form the **vertebral canal,** a passage for the spinal cord. Extending from the apex of the arch, a projection called the **spinous process** is directed toward the rear and downward. You can see and palpate the spinous processes as a row of bumps along the spine of a living person. A pair of **transverse processes** extend laterally from the arch. The spinous and transverse processes provide attachment for ligaments of the spine and muscles of the back.

Figure 6.19 Adult Spinal Curvatures. (a) Normal curvatures. (b) Common abnormalities. Scoliosis often results from a developmental abnormality of the spine. Kyphosis most often results from osteoporosis. Lordosis commonly occurs in pregnancy and obesity.

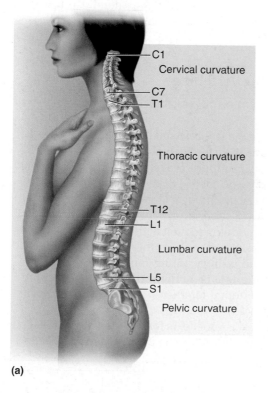

C1
Cervical curvature
C7
T1
Thoracic curvature
T12
L1
Lumbar curvature
L5
S1
Pelvic curvature

(a)

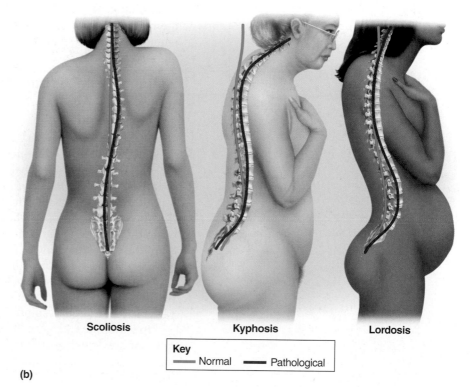

Scoliosis　　　Kyphosis　　　Lordosis

Key
—— Normal　　—— Pathological

(b)

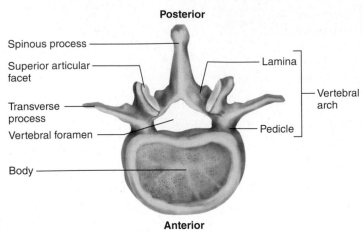

Posterior

Spinous process

Superior articular facet

Transverse process

Vertebral foramen

Body

Lamina

Vertebral arch

Pedicle

Anterior

(a) 2nd lumbar vertebra (L2)

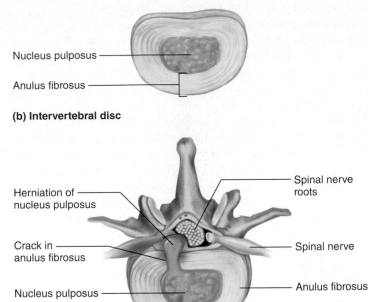

Nucleus pulposus

Anulus fibrosus

(b) Intervertebral disc

Herniation of nucleus pulposus

Crack in anulus fibrosus

Nucleus pulposus

Spinal nerve roots

Spinal nerve

Anulus fibrosus

(c) Herniated disc

Figure 6.20 A Representative Vertebra and Intervertebral Disc, Superior Views. (a) A typical vertebra. (b) An intervertebral disc, oriented the same way as the vertebral body in part (a) for comparison. (c) A herniated disc, showing compression of the spinal nerve roots by the nucleus pulposus oozing from the disc. See p. 175 for a description of the intervertebral disc.

> **Apply What You Know**
>
> The lower we look on the vertebral column, the larger the vertebral bodies and inter-vertebral discs are. What is the functional significance of this?

A pair of **superior articular processes** project upward from one vertebra and meet a similar pair of **inferior articular processes** that project downward from the one above (see fig. 6.22b, c). Each process has a flat surface (facet) facing that of the adjacent vertebra. These processes restrict twisting of the vertebral column. Where two vertebrae meet, they exhibit a lateral gap between them called the **intervertebral foramen.** This allows passage for spinal nerves emerging from the spinal cord.

Regional Characteristics of the Vertebrae

The vertebrae differ from one region of the vertebral column to another, as follows.

Cervical vertebrae (C1–C7) are the smallest and lightest. The first two (C1 and C2) have unique structures that allow for head movements (fig. 6.21). Vertebra C1 is called the **atlas** because it supports the head in a manner reminiscent of the giant of Greek mythology who carried the heavens on his shoulders. It is little more than a delicate ring surrounding a large vertebral foramen. On the upper surface is a pair of concave **superior articular facets** that meet the occipital condyles of the skull. In a nodding motion of the skull, as in gesturing "yes," the occipital condyles rock back and forth on these facets. Vertebra C2, the **axis,** allows rotation of the head as in gesturing "no." Its most distinctive feature is a prominent knob called the **dens** (pronounced "denz"), or **odontoid**[40] **process,** which projects into the vertebral foramen of C1. The axis is the first vertebra that exhibits a spinous process.

In vertebrae C2 to C6, the spinous process is forked—a feature not seen in C7 or any lower vertebrae (fig. 6.22a). All seven cervical vertebrae have a

[40] *dens* = *odont* = tooth; *oid* = resembling

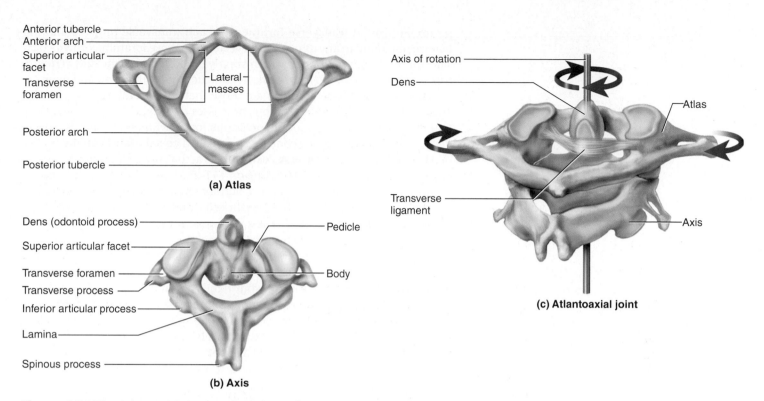

Anterior tubercle
Anterior arch
Superior articular facet
Transverse foramen
Lateral masses
Posterior arch
Posterior tubercle

(a) Atlas

Dens (odontoid process)
Superior articular facet
Transverse foramen
Transverse process
Inferior articular process
Lamina
Spinous process
Pedicle
Body

(b) Axis

Axis of rotation
Dens
Atlas
Transverse ligament
Axis

(c) Atlantoaxial joint

Figure 6.21 The Atlas and Axis, Vertebrae C1 and C2. (a) The atlas, C1, which directly supports the skull. (b) The axis, C2, which inserts into the vertebral foramen of the atlas. (c) Articulation of the two vertebrae and rotation of the atlas. This movement turns the head from side to side, as in gesturing "no." Note the ligament holding the dens of the axis in place. **AP|R**

- *What serious consequence could result from the rupture of the transverse ligament in an automobile accident?*

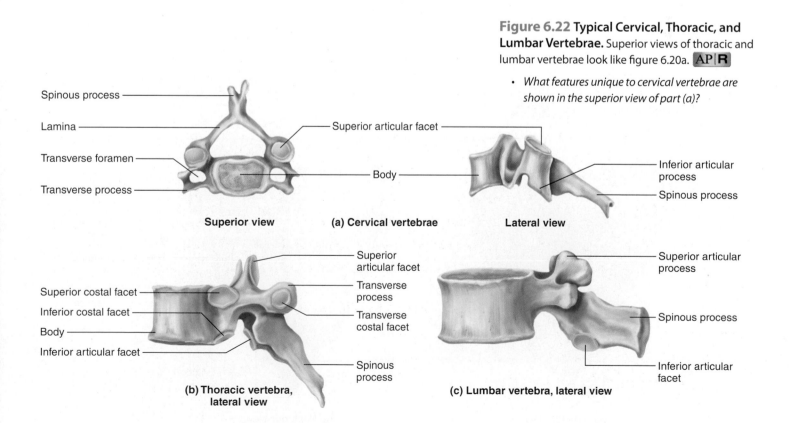

Spinous process
Lamina
Transverse foramen
Transverse process
Superior articular facet
Body

Superior view   **(a) Cervical vertebrae**   **Lateral view**

Inferior articular process
Spinous process

Figure 6.22 Typical Cervical, Thoracic, and Lumbar Vertebrae. Superior views of thoracic and lumbar vertebrae look like figure 6.20a. **AP|R**

- *What features unique to cervical vertebrae are shown in the superior view of part (a)?*

Superior articular facet
Transverse process
Transverse costal facet
Superior costal facet
Inferior costal facet
Body
Inferior articular facet
Spinous process

(b) Thoracic vertebra, lateral view

Superior articular process
Spinous process
Inferior articular facet

(c) Lumbar vertebra, lateral view

prominent round **transverse foramen** in each transverse process, another feature not found in any other class of vertebrae. These foramina provide passage for the *vertebral arteries,* which supply blood to the brain, and *vertebral veins,* which drain blood from superficial structures of the head and neck.

Thoracic vertebrae (T1–T12) are those that support the ribs (fig. 6.22b). They have relatively pointed spinous processes that angle sharply downward, and they exhibit small smooth depressions for attachment of the ribs. These depressions include slightly concave spots called **costal**[41] **facets** on the bodies of the vertebrae, and **transverse costal facets** at the ends of the transverse processes of vertebrae T1 to T10 (lacking on T11–T12). In most cases, a rib inserts at the gap between adjacent thoracic vertebrae, so each vertebra contributes one-half of the articular surface. Vertebrae T1 and T10 to T12, however, have complete costal facets on the bodies; the corresponding ribs articulate on the vertebral bodies instead of between vertebrae.

Lumbar vertebrae (L1–L5) have a thick, stout body and a blunt, squarish spinous process (fig. 6.22c). For the most part, their superior articular processes face medially (like the palms of your hands about to clap), and the inferior processes face laterally, toward the superior processes of the next vertebra. This arrangement enables the lumbar region of the spine to resist twisting. Exceptions to this orientation of the articular facets are seen at the joints between vertebrae T12 and L1, and between L5 and the sacrum.

Sacral vertebrae (S1–S5) begin to fuse around age 16, and by age 26 they are fully fused into a bony plate, the **sacrum** (SAY-krum) (fig. 6.23). The sacrum forms the posterior wall of the pelvic cavity. It was named *sacrum*[42] for its prominence as the largest and most durable bone of the vertebral column. It has four pairs of large **sacral foramina** on the anterior and posterior surfaces, for emergence of spinal nerves to the pelvic organs and lower limbs. The anterior surface of the sacrum is relatively smooth and concave and has four *transverse lines* that indicate where the five vertebrae have fused. The posterior surface is very rough. The spinous processes of the vertebrae fuse into a posterior ridge called the **median sacral crest,** and the transverse processes into a less prominent **lateral sacral crest** on each side. At the point

---

[41] *costa* = rib; *al* = pertaining to

[42] *sacr* = great, prominent

Figure 6.23 The Sacrum and Coccyx. AP|R

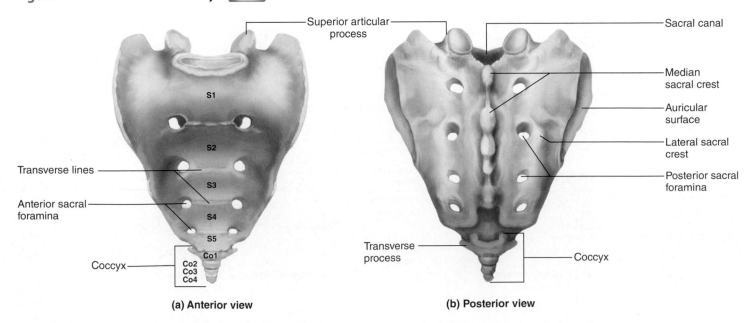

(a) Anterior view

Superior articular process
S1
S2
Transverse lines
S3
Anterior sacral foramina
S4
S5
Co1
Coccyx — Co2 Co3 Co4

(b) Posterior view

Sacral canal
Median sacral crest
Auricular surface
Lateral sacral crest
Posterior sacral foramina
Transverse process
Coccyx

where the sacrum articulates on each side with the hip bone—the **sacroiliac (SI) joint**—each bone has an ear-shaped **auricular**[43] **surface**. A **sacral canal** runs through the sacrum and contains spinal nerve roots.

Coccygeal (coc-SIDJ-ee-ul) **vertebrae** (Co1–Co4) fuse by the age of 20 to 30 into a single small triangular bone, the **coccyx**[44] (COC-six) (fig. 6.23). The Roman anatomist Claudius Galen (c. 130–c. 200) named it this because he thought it resembled the beak of a cuckoo. The coccyx provides attachment for muscles of the pelvic floor. It is a vestige of the bones which, in other mammals, continue as the vertebrae of the tail; hence it is colloquially known as our "tailbone."

Intervertebral Discs

An **intervertebral disc** is a pad consisting of an inner gelatinous **nucleus pulposus** surrounded by a ring of fibrocartilage, the **anulus fibrosus** (see fig. 6.20b). The discs bind adjacent vertebrae together, enhance spinal flexibility, support the weight of the body, and absorb shock. Under stress—for example, when you lift a heavy weight—the discs bulge laterally. Excessive stress can cause a *herniated disc*—a condition in which the anulus fibrosus cracks and the nucleus pulposus oozes out. The extruded gel may put painful pressure on the spinal cord or a spinal nerve (fig. 6.20c).

The Thoracic Cage

The **thoracic cage** (fig. 6.24) consists of the thoracic vertebrae, sternum, and ribs. It encloses the heart and lungs and provides attachment for the pectoral girdle and upper limbs. The ribs protect not only the thoracic organs but also the spleen, most of the liver, and to some extent the kidneys. The thoracic cage

---

[43] *aur* = ear; *icul* = little

[44] *coccyx* = cuckoo

Figure 6.24 The Thoracic Cage and Pectoral Girdle, Anterior View. AP|R

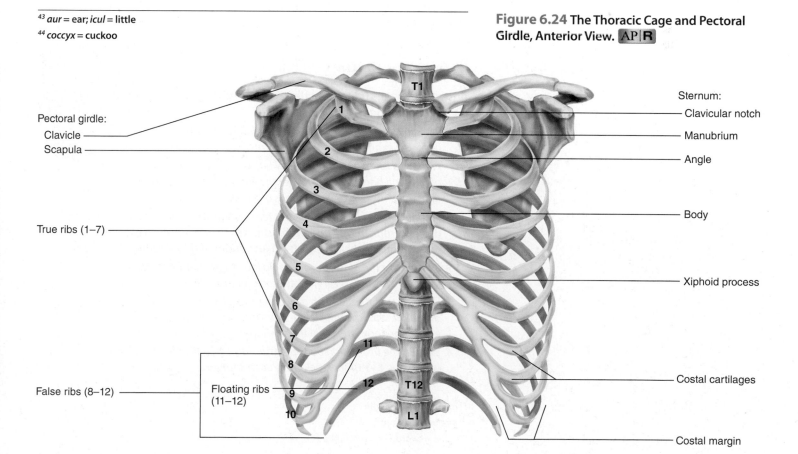

Pectoral girdle:
Clavicle
Scapula

True ribs (1–7)

False ribs (8–12)

Floating ribs (11–12)

Sternum:
Clavicular notch
Manubrium
Angle

Body

Xiphoid process

Costal cartilages

Costal margin

is rhythmically expanded by the respiratory muscles to create a vacuum that draws air into the lungs.

Sternum

The **sternum** (breastbone) is a bony plate anterior to the heart. It consists of three parts: a short superior **manubrium**[45] (ma-NOO-bree-um) with a shape resembling the knot in a necktie; an elongated swordlike **body (gladiolus**[46]**);** and at the inferior end, a small, daggerlike **xiphoid**[47] (ZIF-oyd or ZY-foyd) **process.** The first pair of ribs articulates with the manubrium; the second pair articulates at the junction of the manubrium and body; and the next eight pairs articulate with the body only. The clavicles (collarbones) articulate with **clavicular notches** on the superolateral corners of the manubrium.

Ribs

There are 12 pairs of **ribs.** Each attaches posteriorly to the vertebral column, and all but the last two arch around the flank of the chest and attach by way of a cartilaginous strip, the **costal cartilage,** to the sternum. Most of the ribs are curved, flattened blades, squared off at the distal ends where the costal cartilages begin, and with the broad surfaces of the blades oriented vertically. Rib 1, however, is a horizontal plate, and ribs 11 and 12 are pointed at the distal ends.

Ribs 1 to 7 are called **true ribs** because each has its own costal cartilage. Ribs 8 to 10 attach to the costal cartilage of rib 7, and ribs 11 and 12 have no costal cartilages at all. Ribs 8 to 12 are therefore called **false ribs,** and ribs 11 and 12 are also called **floating ribs** for lack of any connection to the sternum. The first 10 ribs articulate with the vertebrae at two points: the vertebral bodies and the transverse costal facets. Ribs 11 and 12 articulate with the vertebral bodies only. Thoracic vertebrae T11 and T12 therefore have no transverse costal facets.

Before You Go On

Answer these questions from memory. Reread the preceding section if there are too many you don't know.

12. *Explain the distinction between cranial bones and facial bones; name six examples of each and locate them on the skull.*

13. *Prepare a list of all boldfaced key terms in the foregoing descriptions of the skull bones and try to identify each of those features on unlabeled models or illustrations of the skull.*

14. *Prepare a list of all boldfaced key terms in the foregoing descriptions of the sternum, ribs, and vertebrae and try to identify each of those features on unlabeled models or illustrations of the skeleton.*

15. *Summarize the features you would look for to identify a single isolated vertebra as cervical, thoracic, or lumbar.*

16. *What part of the sternum articulates with the clavicles? What parts articulate with the ribs? What part articulates with no other bones except for another part of the sternum?*

17. *Explain the distinction between true, false, and floating ribs. State which ribs, by number, belong in each category.*

---

[45] *manubrium* = handle

[46] *gladiolus* = sword

[47] *xipho* = sword; *oid* = resembling

6.4 The Appendicular Skeleton

Expected Learning Outcomes

When you have completed this section, you should be able to:

a. describe the bones of the pectoral girdle and upper limb, and the major features of the individual bones; and

b. do the same for the bones of the pelvic girdle and lower limb.

The *appendicular skeleton* consists of bones of the upper and lower limbs and the pectoral and pelvic girdles that attach the limbs to the axial skeleton. Injuries to the appendicular skeleton are especially common in athletics, recreation, and the workplace, and they are quite disabling because we depend so much on the limbs for mobility and the ability to manipulate objects.

The Pectoral Girdle and Upper Limb

The **pectoral girdle** (shoulder girdle) supports the arm. It consists of a *clavicle* and a *scapula* on each side of the body.

The Clavicle

The **clavicle,**[48] or collar bone (fig. 6.25), is a slightly S-shaped bone easily seen and palpated on the upper thorax. Its superior surface is relatively rounded and smooth and its inferior surface flatter and slightly rough. The medial **sternal end** has a rounded, hammerlike head that articulates with the manubrium of the sternum. The lateral **acromial end** is markedly flattened and articulates with the acromion of the scapula. The clavicles brace the shoulders; without them, the pectoralis major muscles would pull the shoulders forward and medially.

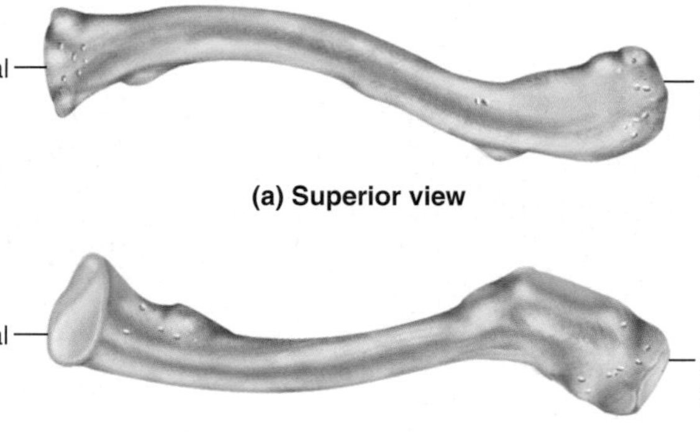

The Scapula

The **scapula,** or shoulder blade (fig. 6.26), is a triangular plate that overlies ribs 2 to 7 on the upper back. The triangle is bounded by *superior, medial,* and

---

[48] *clav* = hammer, club; *icle* = little

(a) Superior view

Sternal end — Acromial end

(b) Inferior view

Sternal end — Acromial end

Figure 6.25 The Right Clavicle (Collarbone). AP|R

Figure 6.26 The Right Scapula. AP|R

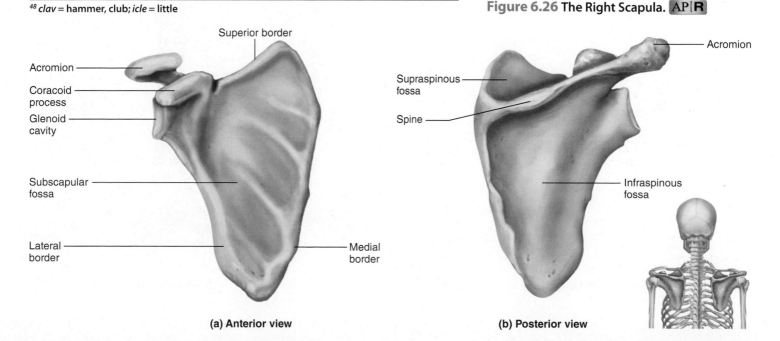

Superior border

Acromion

Coracoid process

Glenoid cavity

Subscapular fossa

Lateral border

Medial border

Supraspinous fossa

Spine

Acromion

Infraspinous fossa

(a) Anterior view

(b) Posterior view

lateral borders; its broad anterior surface, called the **subscapular fossa,** is slightly concave and relatively featureless. The posterior surface has a prominent transverse ridge called the **spine,** a deep indentation superior to the spine called the **supraspinous fossa,** and a broad surface inferior to it called the **infraspinous fossa.**[49] The fossae are occupied mainly by muscles of the *rotator cuff,* which are concerned with movements of the arm (see p. 238).

The most complex part of the scapula is the region of the shoulder, where it has three main features: (1) The **acromion**[50] (ah-CRO-me-on), a platelike extension of the scapular spine, forms the apex of the shoulder. It articulates with the clavicle and is the sole point through which the arm and scapula attach to the axial skeleton. (2) The **coracoid**[51] (COR-uh-coyd) **process,** shaped like a bent finger, provides attachment for the biceps and some other muscles of the arm. (3) The **glenoid**[52] (GLEN-oyd) **cavity** is a shallow socket that articulates with the rounded head of the humerus.

The Humerus

There are 30 bones in each of the upper limbs: the *humerus* in the arm proper, the *radius* and *ulna* in the forearm, 8 *carpal bones* in the wrist, and 5 *metacarpal bones* and 14 *phalanges* in the hand.

[49] *supra* = above; *infra* = below

[50] *acr* = extremity, point; *omi* = shoulder

[51] *corac* = crow; *oid* = resembling

[52] *glen* = pit, socket

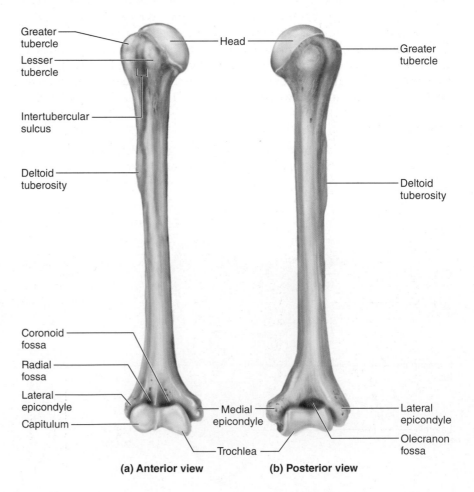

Greater tubercle

Lesser tubercle

Intertubercular sulcus

Deltoid tuberosity

Coronoid fossa

Radial fossa

Lateral epicondyle

Capitulum

Head

Medial epicondyle

Trochlea

Greater tubercle

Deltoid tuberosity

Lateral epicondyle

Olecranon fossa

(a) Anterior view

(b) Posterior view

Figure 6.27 The Right Humerus. AP|R

The **humerus** is the only bone of the *brachial region,* from shoulder to elbow, and is the largest bone of the upper limb. It has a hemispherical **head** that inserts into the glenoid cavity of the scapula (fig. 6.27). Lateral to the head are two muscle attachments called the **greater** and **lesser tubercles,** with an *intertubercular groove* between them that accommodates a tendon of the biceps muscle. The shaft has a rough area called the **deltoid tuberosity** on its lateral surface. This is the insertion for the deltoid muscle of the shoulder.

The distal end of the humerus has two smooth condyles—the **capitulum**[53] (ca-PIT-you-lum) on the lateral side and the pulleylike **trochlea**[54] (TROCK-lee-uh) on the medial side. Immediately proximal to these condyles, the humerus flares to form two bony processes, the **lateral** and **medial epicondyles,** which are easily palpated at the widest point of your elbow. The distal end also shows three deep pits or fossae. The **olecranon** (oh-LEC-ruh-non) **fossa** on the posterior side accommodates the ulna when the elbow is extended. On the anterior surface, a medial pit called the **coronoid fossa** and a lateral pit called the **radial fossa** accommodate the heads of the ulna and radius, respectively, when the elbow is flexed.

The Radius

The **radius** extends from elbow to wrist on the lateral side of the forearm, ending just proximal to the base of the thumb (fig. 6.28). Its proximal head is a distinctive disc that rotates freely on the capitulum of the humerus when the palm is

[53] *capit* = head; *ulum* = little
[54] *troch* = wheel, pulley

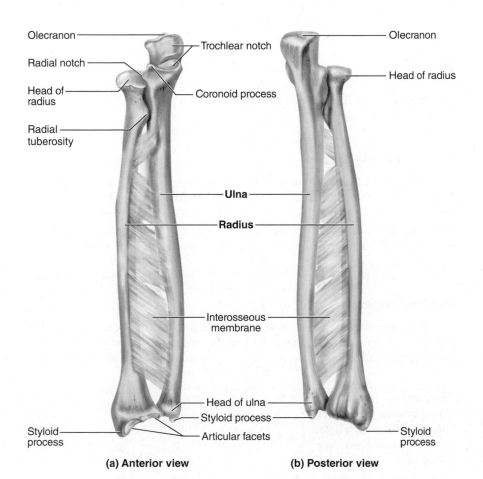

(a) Anterior view (b) Posterior view

Figure 6.28 The Right Radius and Ulna. AP|R

turned forward and back. Just distal to the head is a process called the **radial tuberosity,** an attachment for the tendon of the biceps muscle. The flared distal end of the radius articulates with three of the carpal bones, and has a small point called the **styloid process** that can be palpated on the lateral side of the wrist.

The Ulna

The **ulna**[55] is the medial bone of the forearm (fig. 6.28). Its proximal end has a wrenchlike shape, with a deep C-shaped **trochlear notch** that wraps around the trochlea of the humerus. The posterior wall of the notch is formed by a prominent bony point called the **olecranon.** If you rest with your chin in your hands and your elbows on a table, it is the olecranon that contacts the table. The anterior wall of the notch is formed by a less prominent **coronoid process.** Medially, the head of the ulna has a less conspicuous **radial notch,** which accommodates the edge of the head of the radius. At the distal end of the ulna is a **styloid process** similar to that of the radius**.** It can be palpated on the medial side of the wrist.

A fibrous sheet called the **interosseous**[56] (IN-tur-OSS-ee-us) **membrane** loosely joins the radius and ulna along the length of their shafts. If you stand and lean forward and support your weight with your hands on a table, about 80% of the force is borne by the radius. The interosseous membrane pulls the ulna upward and transfers some of this force through the ulna to the humerus, thus distributing your weight more evenly across the elbow joint. Such weight distribution is also important in four-legged animals.

The Carpal Bones

The **carpal bones** in the base of the hand are arranged in two rows of four bones each (fig. 6.29). They allow movements of the hand from side to side and anterior to posterior. The carpal bones of the proximal row, starting at the lateral (thumb) side, are the **scaphoid, lunate, triquetral** (tri-QUEE-trul), and **pisiform** (PY-sih-form). Translating from the Latin, these words mean boat-, moon-, triangle-, and pea-shaped, respectively. The bones of the distal row, again starting on the lateral side, are the **trapezium,**[57] **trapezoid, capitate,**[58] and **hamate.**[59] The hamate can be recognized by a prominent hook, or *hamulus,* on the palmar side.

The Metacarpal Bones

The **metacarpal bones**[60] occupy the palmar region. They are numbered I through V with metacarpal I at the base of the thumb and metacarpal V at the base of the little finger (fig. 6.29). On a skeleton, the metacarpals look like extensions of the fingers, so that the fingers seem much longer than they really are. Each is divided into a *base, body,* and *head;* the heads form the knuckles of a clenched fist.

The Phalanges

The **phalanges** (fah-LAN-jeez) are the bones of the fingers; the singular is *phalanx* (FAY-lanks). They are identified by roman numerals corresponding to the numbering of the metacarpals, preceded by *proximal, middle,* and *distal.* Digits

---

[55] *ulna* = elbow

[56] *inter* = between; *osse* = bones

[57] *trapez* = table, grinding surface

[58] *capit* = head; *ate* = possessing

[59] *ham* = hook; *ate* = *possessing*

[60] *meta* = beyond; *carp* = wrist

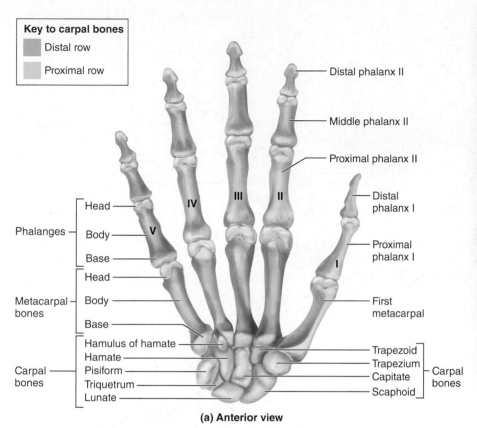

Key to carpal bones
- Distal row
- Proximal row

Distal phalanx II

Middle phalanx II

Proximal phalanx II

Distal phalanx I

Proximal phalanx I

Phalanges
- Head
- Body
- Base

IV   III   II

V

I

Metacarpal bones
- Head
- Body
- Base

First metacarpal

Carpal bones
- Hamulus of hamate
- Hamate
- Pisiform
- Triquetrum
- Lunate

Trapezoid
Trapezium
Capitate
Scaphoid

Carpal bones

(a) Anterior view

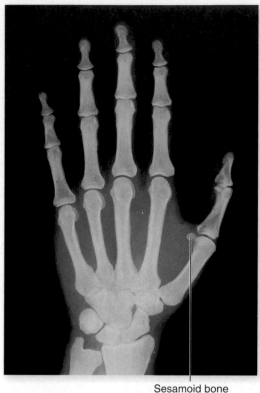

Sesamoid bone

(b) X-ray of adult hand

Figure 6.29 **The Right Hand.** (a) Anterior view with carpal bones color-coded to distinguish the proximal and distal rows. (b) X-ray of an adult hand. Sesamoid bones are small "extra" bones that often develop within tendons after birth. The pisiform bone and patella (kneecap) are also sesamoid bones. AP|R

II through IV have three bones, called the proximal, middle, and distal phalanges. Most people with wedding rings, for example, wear them encircling the left proximal phalanx IV. The **pollex** (thumb) has only two phalanges, proximal and distal (fig. 6.29). Like the metacarpals, each phalanx has a *base, body, and head.*

The Pelvic Girdle and Lower Limb

The adult **pelvic**[61] **girdle** is composed of three bones: the right and left *hip bones* and the *sacrum* (fig. 6.30a). The sacrum has already been described, since it is part of the vertebral column. Posteriorly, the hip bones are joined to the sacrum at the sacroiliac joints; anteriorly, they are joined to each other at the *pubic symphysis,* just superior to the genitalia. The symphysis consists of a median pad of fibrocartilage called the *interpubic disc* and the adjacent bone on each side.

The pelvic girdle plus the associated muscles and ligaments constitute the bowl-like **pelvis,** which has two regions. The broad superior region between the flare of the hips is the **greater (false) pelvis;** it contains and supports mainly the lower intestines. A somewhat round opening called the **pelvic inlet** leads into a narrower inferior space called the **lesser (true) pelvis,** which contains mainly the rectum; the urinary bladder; and in women, the uterus. The edge of the pelvic inlet is called the **pelvic brim.** The lower opening of the lesser pelvis is called the **pelvic outlet.** The inlet and outlet refer to the passage of an infant during birth, as it descends into and through the lesser pelvis.

The pelvis is the most *sexually dimorphic* part of the skeleton—that is, the one whose anatomy most differs between the sexes. The average male pelvis is more robust (heavier and thicker) than the female's owing to the forces exerted on the bone by stronger muscles. The female pelvis is adapted to the needs of

[61] *pelv* = basin, bowl

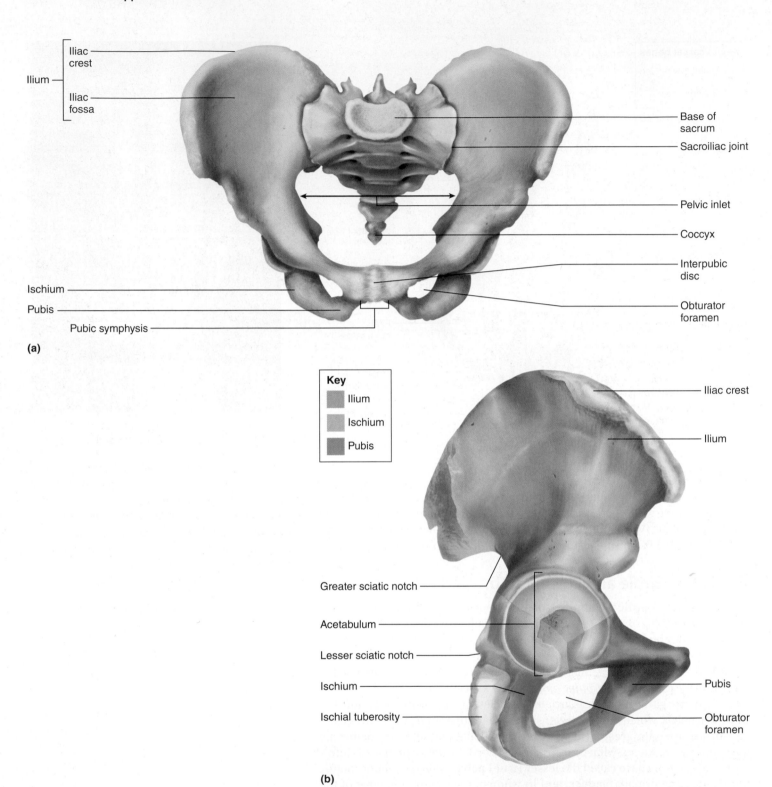

Figure 6.30 The Pelvic Girdle. (a) Anterosuperior view. The pelvic girdle consists of the two hip bones and the sacrum. (b) The right hip bone, lateral view. The three childhood bones that fuse to form the adult hip bone are identified by color. **AP|R**

pregnancy and childbirth. It is wider and shallower and has a larger, rounder pelvic inlet and outlet for passage of the infant. These and several other features serve as aids in the sexual identification of skeletal remains.

The Hip Bone

The **hip bone** is also known as the *os coxae*[62] (oss COC-see) or *coxal bone.* It forms by the fusion of three childhood bones called the *ilium* (ILL-ee-um), *is-*

---

[62] *os* = bone; *coxae* = of the hip

chium (ISS-kee-um), and *pubis* (PEW-biss), identified by color in figure 6.30b. The largest of these is the **ilium**[63] (ILL-ee-um). If you rest your hand on your hip, you are supporting it on the upper margin of the ilium, called the **iliac crest.** Laterally, the most prominent feature of the hip bone is the **acetabulum**[64] (ASS-eh-TAB-you-lum), a deep socket for the head of the femur. The ilium is that part of the hip bone from the iliac crest to the superior part of the acetabulum. A prominent feature of the posterior margin of the ilium is the deep **greater sciatic** (sy-AT-ic) **notch,** named for the sciatic nerve that passes through here on its way down the posterior side of the thigh. The posterolateral surface of the ilium is relatively rough because it serves for attachment of several muscles of the buttocks and thighs. The anteromedial surface, by contrast, is the smooth, slightly concave **iliac fossa,** covered by the broad *iliacus muscle.*

The **ischium** forms the inferoposterior portion of the hip bone. It is a roughly C-shaped bone that contributes most of the posterior wall of the acetabulum. Its lowermost part is the thick, rough-surfaced **ischial tuberosity,** which you are probably sitting on as you read this. It can be palpated by sitting on your fingers.

The **pubis** (pubic bone) is the most anterior portion of the hip bone. It contributes the anterior part of the acetabulum. The **pubic symphysis**[65] is the point where the anterior bodies of the two pubic bones are joined by a cartilaginous **interpubic disc.** The ischium and pubis encircle the **obturator**[66] **foramen,** a large round-to-triangular hole below the acetabulum, closed in life by a ligamentous sheet.

The number and arrangement of bones in the lower limb are similar to those of the upper limb, and again number 30 bones per limb: the femur in the thigh, the patella (kneecap), the tibia and fibula in the leg proper, 7 tarsal bones in the ankle, and 5 metatarsal bones and 14 phalanges in the foot. The bones of the lower limb, unlike those of the upper, are adapted for weight bearing and locomotion, and are therefore shaped and articulated differently. Most notably, the tarsal bones are thoroughly integrated into the arch of the foot, unlike the carpal bones, which are limited to the base the hand.

The Femur

The **femur** (FEE-mur) is the body's longest and strongest bone (fig. 6.31). It has a hemispherical head that inserts into the acetabulum. Distal to the head is a constricted **neck** and two massive, rough processes of the upper femur called the **greater** and **lesser trochanters** (tro-CAN-turs); these are insertions for the powerful muscles of the hip. The anterior surface of the shaft is smooth and rounded, but the posterior surface is marked by a strongly developed ridge, the **linea aspera,** an attachment for strong *adductor muscles* of the thigh. The distal end of the femur flares into **medial** and **lateral epicondyles,**

---

[63] *ilium* = flank, loin

[64] *acetabulum* = vinegar cup

[65] *sym* = together; *physis* = growth

[66] *obtur* = to close, stop up; *ator* = that which

Figure 6.31 The Right Femur and Patella. AP|R

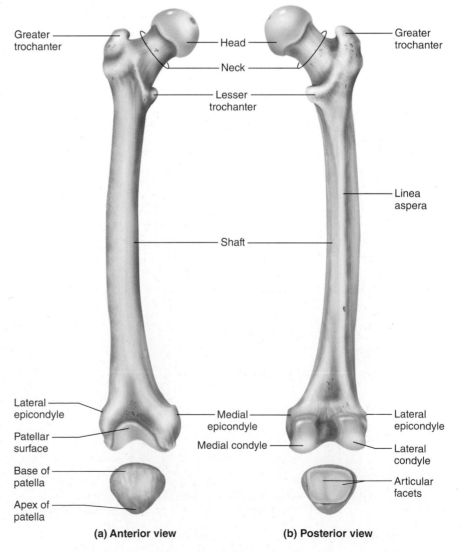

(a) **Anterior view**   (b) **Posterior view**

which serve as sites of muscle and ligament attachment and which can be palpated at the widest point of the knee. Distal to these are two smooth wheel-like surfaces of the knee joint, the **medial** and **lateral condyles.** On the anterior side of the femur, a smooth medial depression called the **patellar surface** accommodates the patella.

The Patella

The **patella,**[67] or kneecap (fig. 6.31), is a roughly triangular bone that develops within the quadriceps tendon of the knee. It is cartilaginous at birth and ossifies between 3 and 6 years of age. It has a broad superior **base,** a pointed inferior **apex,** and a pair of shallow **articular facets** on its posterior surface where it articulates with the femur. The patella glides up and down on the patellar surface of the femur when the knee is flexed and extended.

The Tibia

The **tibia** is the thick, strong bone on the medial side of the leg proper (fig. 6.32). (Remember that in anatomical terminology, *leg* refers only to the *crural region* between the knee and ankle.) The tibia is the only weight-bearing bone of this region. Its broad superior head has two fairly flat surfaces, the **medial** and **lateral condyles.** The condyles of the femur rock on these surfaces when the knee is flexed. The rough anterior surface of the tibia, the **tibial tuberosity,** can be palpated just below the patella. This is where the patellar ligament inserts and the quadriceps muscle of the thigh exerts its pull when it extends the knee, as in kicking a football. Distal to this, the shaft has a sharply angular **anterior border,** which can be palpated in the shin. At the ankle, just above the brim of a standard dress shoe, one can palpate a prominent bony knob on each side. These knobs are the **medial malleolus**[68] (MAL-ee-OH-lus) of the tibia and the **lateral malleolus** of the fibula.

The Fibula

The **fibula** (fig. 6.32) is a slender lateral strut that helps to stabilize the ankle. It does not bear any weight. The fibula is somewhat thicker and broader at its proximal end, the **head,** than at the distal end. The distal expansion is the lateral malleolus just mentioned. The shafts of the tibia and fibula are connected through a flexible interosseous membrane similar to that of the forearm.

The Ankle and Foot

The **tarsal bones** of the ankle are arranged in proximal and distal groups somewhat like the carpal bones of the wrist (fig. 6.33). Because of the load-bearing role of the ankle, however, they are arranged quite differently from the carpal bones and are thoroughly integrated into the arches of the foot. The proximal group includes

Figure 6.32 The Right Tibia and Fibula. AP|R

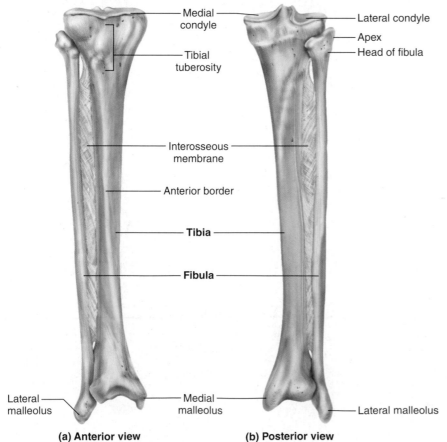

Medial condyle
Tibial tuberosity
Interosseous membrane
Anterior border
Tibia
Fibula
Lateral malleolus

Lateral condyle
Apex
Head of fibula
Medial malleolus
Lateral malleolus

(a) Anterior view   (b) Posterior view

---

[67] *pat* = pan; *ella* = little

[68] *malle* = hammer; *olus* = little

the **calcaneus**[69] (cal-CAY-nee-us) of the heel; the **talus,** a superior bone that articulates with the tibia; and the short wide **navicular** bone anterior to the talus. The distal group forms a row of four bones. Proceeding from medial to lateral, these are the **first, second,** and **third cuneiforms**[70] (cue-NEE-ih-forms) and the **cuboid.** The cuboid is the largest of the four.

The remaining bones of the foot are similar in arrangement and name to those of the hand. The **metatarsals**[71] are similar to the metacarpals. They are numbered I to V from medial to lateral, metatarsal I being proximal to the great toe. (Roman numeral I refers to the largest digit in both the hand and foot.) Bones of the toes, like those of the fingers, are called **phalanges.** The great toe is

---

[69] *calc* = stone, chalk

[70] *cunei* = wedge; *form* = in the shape of

[71] *meta* = beyond; *tars* = ankle

Figure 6.33 The Right Foot. (a) Superior (dorsal) view. (b) Inferior (plantar) view. (c) Medial view. (d) X-ray of lateral view. The white arrow represents the lateral longitudinal arch. AP|R

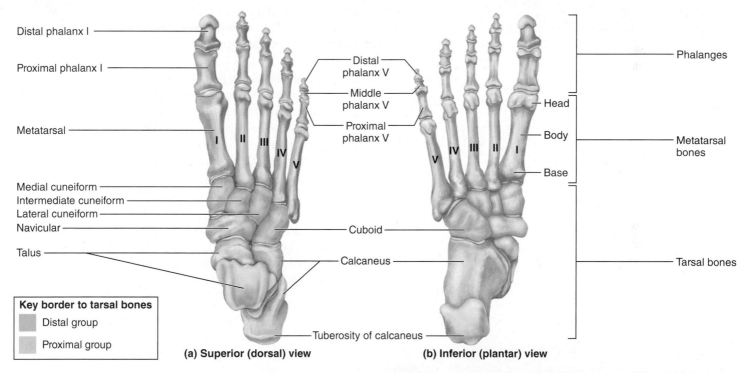

(a) Superior (dorsal) view

(b) Inferior (plantar) view

Key border to tarsal bones
Distal group
Proximal group

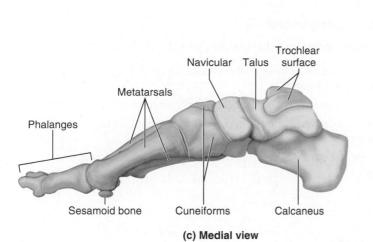

(c) Medial view

(d) X-ray of right foot, lateral view

the **hallux** and contains only two bones, the proximal and distal phalanx I. The other toes each contain a proximal, middle, and distal phalanx.

Three strong, springy foot arches absorb shock as the body jostles up and down during walking and running. The **medial longitudinal arch** extends from the heel to the great toe; the **lateral longitudinal arch** extends from the heel to the little toe (fig. 6.33d); and the **transverse arch** extends from side to side at the level where the metatarsal bones articulate with the distal row of tarsal bones (the cuboid and cuneiforms). These arches are held together by short, tough ligaments. Excessive weight or repetitive stress can weaken and stretch these ligaments, resulting in the condition commonly called *fallen arches* or *flat feet (pes planus*[72]*)*. This condition makes it harder to tolerate prolonged standing and walking.

Before You Go On

Answer these questions from memory. Reread the preceding section if there are too many you don't know.

18. *Prepare a list of all boldfaced key terms in the foregoing descriptions of the upper and lower limbs and the two limb girdles; try to identify each of those features on unlabeled models or illustrations of the skeleton.*

19. *What structure is found in the intertubercular groove of the humerus? What structure attaches to its deltoid tuberosity? What bones articulate with its head, capitulum, and trochlea, respectively?*

20. *Describe the features you can use to distinguish a radius from an ulna, and how you can tell the proximal from the distal end of each bone.*

21. *How many carpal bones are there? How many tarsal bones? If there are 30 bones in both the upper and the lower limb, what makes up for the unequal number of carpal and tarsal bones?*

22. *Give the anatomical (osteological) name for each of the following bony structures that are easily palpated on the body surface: the apex of the shoulder; the point of the elbow; the bumps on each side of the wrist; the bumps on each side of the ankle; the heel; and the "ball" of the foot at the base of the great toe.*

6.5 Joints

Expected Learning Outcomes

When you have completed this section, you should be able to:

a. define and describe three major classes of joints, the subclasses of each, and examples;

b. identify the anatomical components seen in most synovial joints;

c. define and demonstrate the types of movement that occur at synovial joints; and

d. describe the basic anatomy of the shoulder, elbow, hip, and knee joints.

[72] *pes* = foot; *planus* = flat

A **joint,** or **articulation,** is any point at which two bones meet, regardless of whether they are movable at that point. Your shoulder, for example, is a very movable joint, whereas the skull sutures are immovable joints.

Classification

Joints are classified according to the manner in which the adjacent bones are connected, with corresponding differences in how freely the bones can move. There are three major structural categories of joints: fibrous, cartilaginous, and synovial.

Fibrous Joints

A **fibrous joint,** or **synarthrosis**[73] (SIN-ar-THRO-sis), is a point at which adjacent bones are bound by collagen fibers that emerge from the matrix of one bone, cross the space between them, and penetrate into the matrix of the other. There are three kinds of fibrous joints: sutures, syndesmoses, and gomphoses. A **suture** is a fibrous joint between two bones of the skull, marked by a surface line such as the coronal, lambdoid, and sagittal sutures described earlier in this chapter (see fig. 6.16). A **syndesmosis**[74] (SIN-dez-MO-sis) is a joint at which two bones are bound by longer collagenous fibers than in a suture, such as the interosseous membranes binding the radius and ulna in the forearm and the tibia and fibula in the leg (see figs. 6.28 and 6.32). The greater mobility of these membranes allows for such actions as pronation and supination of the forearm (see fig. 6.41). **Gomphoses** are the joints that bind the teeth to the jaw bones.

Cartilaginous Joints

A **cartilaginous joint,** or **amphiarthrosis**[75] (AM-fee-ar-THRO-sis), is a point at which two bones are linked by cartilage. If the bones are joined by hyaline cartilage, the joint is called a **synchondrosis**[76] (SIN-con-DRO-sis). An example is the attachment of rib 1 to the sternum by a hyaline costal cartilage (see fig. 6.24). (Ribs 2 through 10, however, are joined to the sternum by synovial joints.) If two bones are joined by fibrocartilage, the joint is called a **symphysis**[77] (SIM-fih-sis). One example is the pubic symphysis, in which the right and left pubic bones are joined by the cartilaginous interpubic disc (see fig. 6.30a). Another is the intervertebral discs joining adjacent vertebrae to each other.

Synovial Joints

A **synovial** (sih-NO-vee-ul) **joint,** or **diarthrosis**[78] (DY-ar-THRO-sis), is what usually comes to mind when one thinks of a skeletal joint. Synovial joints include the most familiar and movable joints of the body, such as the shoulder, elbow, knuckles, hip, and knee joints, as well as some less obvious examples among the carpal and tarsal bones. Synovial joints are the most structurally complex type of joint, and are the most likely to develop uncomfortable and crippling dysfunctions. They are therefore the most important joints for professionals such as physical and occupational therapists, nurses, fitness trainers, and athletic and dance coaches to understand well.

The relative mobility of synovial joints stems partly from the fact that the two bones are separated by a narrow space, the **joint (articular) cavity;** the space contains a slippery lubricant called **synovial**[79] **fluid** (fig. 6.34), similar to raw egg white

[73] *syn* = together; *arthr* = joined; *osis* = condition

[74] *syn* = together; *desm* = band; *osis* = condition

[75] *amphi* = on all sides; *arthr* = joined; *osis* = condition

[76] *syn* = together; *chondr* = cartilage; *osis* = condition

[77] *sym* = together; *physis* = growth

[78] *dia* = separate, apart; *arthr* = joint; *osis* = condition

[79] *ovi* = egg

Figure 6.34 Structure of a Simple Synovial Joint. AP|R

- *Why is a meniscus unnecessary in an interphalangeal joint?*

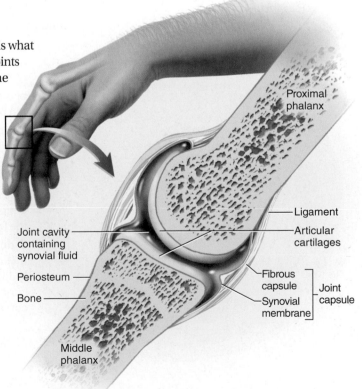

Proximal phalanx

Ligament

Articular cartilages

Joint cavity containing synovial fluid

Periosteum

Bone

Fibrous capsule

Synovial membrane

Joint capsule

Middle phalanx

in consistency. The facing bone surfaces are covered with articular cartilages about 2 mm thick. A connective tissue **joint (articular) capsule** encloses the cavity; its inner layer, the *synovial membrane,* secretes the fluid and retains it in the joint cavity.

In some synovial joints, especially those subjected to a lot of weight or pressure, fibrocartilage grows inward from the joint capsule and forms a pad between the bones. In the jaw, between the distal ends of the radius and ulna, and at both ends of the clavicle, the pad crosses the entire joint capsule and is called an **articular disc.** In the knee, two cartilages extend inward from the left and right but do not entirely cross the joint. Each is called a **meniscus**[80] because of its crescent shape. Such cartilages absorb shock and pressure and improve the fit between the bones, guide the bones across each other, and stabilize the joint, reducing the chance of dislocation.

Accessory structures associated with a synovial joint include tendons, ligaments, and bursae. A **tendon** is a cord or sheet of tough, collagenous connective tissue that attaches a muscle to a bone. Tendons are often the most important structures in stabilizing a joint. A **ligament** is a similar tissue that attaches one bone to another. A **bursa**[81] is a fibrous sac filled with synovial fluid, located between adjacent muscles, between bone and skin, or where a tendon passes over a bone (see fig. 6.43). Bursae cushion muscles, help tendons slide more easily over the joints, and sometimes enhance the mechanical effect of a muscle by modifying the direction in which its tendon pulls. They are especially numerous in the hand, knee, and foot. Inflammation of a bursa is called *bursitis.* It is usually caused by overuse or repeated trauma to a joint, and it occurs most commonly in the shoulder. A *bunion* is bursitis of the great toe.

There are six classes of synovial joints, distinguished by patterns of motion determined by the shapes of the articulating bone surfaces (fig. 6.35). A bone's movement at a joint can be described with reference to three mutually perpendicular planes in space (*x, y,* and *z*). If the bone can move in only one plane, the joint is said to be *monaxial;* if it can move in two planes, the joint is *biaxial;* and if three, it is *multiaxial.* The six synovial joints are described here in descending order of mobility: one multiaxial type (ball-and-socket), three biaxial types (condylar, saddle, and gliding), and two monaxial types (hinge and pivot).

1. **Ball-and-socket joints** occur at the shoulder and hip, where one bone has a smooth hemispherical head that fits within a cuplike depression (socket) on the other. The head of the humerus fits into the glenoid cavity of the scapula, and the head of the femur fits into the acetabulum of the hip bone. These are the only multiaxial joints of the skeleton.

2. **Condylar (ellipsoid) joints** have an oval convex surface on one bone that fits into a complementary depression on the next—for example, the joints between the phalanges and metacarpal bones at the base of the fingers. As biaxial joints, they can move in two planes. To demonstrate this, hold your hand with your palm facing you. Make a fist, and these joints flex in the sagittal plane. Fan your fingers apart, and they move in the frontal plane.

3. **Saddle joints** occur at the base of the thumb between the trapezium and metacarpal I, and between the clavicle and sternum. Each bone has a saddle-shaped surface—concave in one direction (like the front-to-rear curvature of a horse saddle) and convex in the other (like the left-to-right curvature of a saddle). These joints are biaxial. The thumb, for example,

---

[80] *men* = moon, crescent; *iscus* = little

[81] *burs* = purse

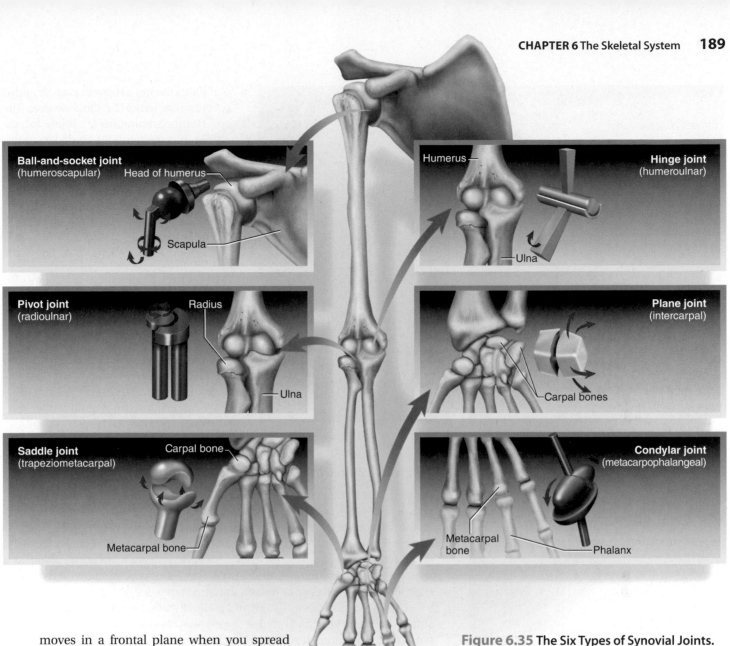

Ball-and-socket joint
(humeroscapular)   Head of humerus

Scapula

Hinge joint
(humeroulnar)

Humerus

Ulna

Pivot joint
(radioulnar)   Radius

Ulna

Plane joint
(intercarpal)

Carpal bones

Saddle joint
(trapeziometacarpal)   Carpal bone

Metacarpal bone

Condylar joint
(metacarpophalangeal)

Metacarpal bone   Phalanx

Figure 6.35 The Six Types of Synovial Joints.
All six have representatives in the upper limb.
Mechanical models show the types of motion
possible at each joint.

moves in a frontal plane when you spread your fingers and in a sagittal plane when you move it as if to grasp the handle of a hammer. The saddle joint is what makes the human thumb opposable (able to encircle and grasp objects).

4. At a **plane (gliding) joint**, the articular surfaces are flat or only slightly concave and convex. Plane joints occur between the wrist and ankle bones. The adjacent bones slide over each other and have rather limited movement. Their movements are complex but usually biaxial.

5. At a **hinge joint,** one bone has a convex surface that fits into a concave depression of the other one. Hinge joints are essentially monaxial—like a door hinge, they can move freely in one plane, but they have very little movement in any other. Examples include the elbow, knee, and interphalangeal (finger and toe) joints.

6. **Pivot joints** are monaxial joints in which a bone spins on its longitudinal axis. The principal examples are the joints between the first two vertebrae (see fig. 6.21) and between the radius and ulna at the elbow. The skull and atlas pivot on the axis when you shake your head "no," and the radius spins on its axis when you rotate your forearm.

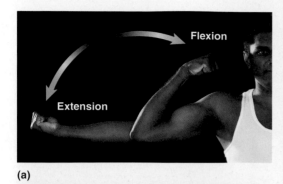

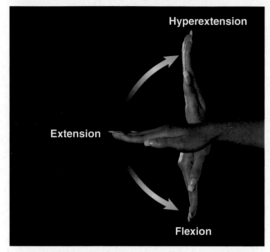

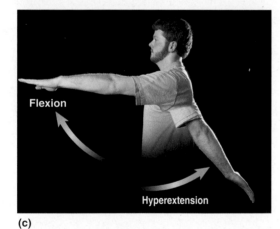

(a)

(b)

(c)

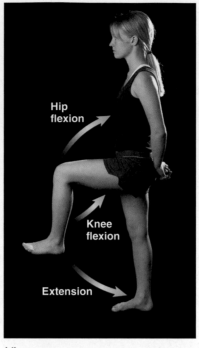

(d)

Figure 6.36 Flexion and Extension. (a) Flexion and extension of the elbow. (b) Flexion, extension, and hyperextension of the wrist. (c) Flexion and hyperextension of the shoulder. (d) Hip and knee flexion and extension.

Joints are not as easy to classify as this scheme may make it seem, however. The jaw (temporomandibular) joint, for example, has elements of condylar, hinge, and plane joints. Some other joints also show aspects of two or more of these six types.

Movement

Kinesiology, physical therapy, and related fields have a specific vocabulary for movements of the synovial joints. You will need a command of these terms to understand the muscle actions in chapter 7.

When one is standing in anatomical position, each joint is said to be in its **zero position.** Joint movements can be described as deviating from the zero position or returning to it.

Flexion, Extension, and Hyperextension

Most acts of flexion, extension, and hyperextension are movements that increase or decrease a joint angle in the sagittal plane (fig. 6.36). **Flexion** decreases the joint angle, as in bending the elbow or knee. The meaning of *flexion* is less obvious in the ball-and-socket joints of the shoulder and hip. Flexion of the shoulder consists of raising the arm from zero position in a sagittal plane, as if to point in front of you or toward the ceiling. Flexion of the hip entails raising the thigh, as in placing your foot on the next step when ascending stairs.

Extension straightens a joint and generally returns a body part to zero position—for example, straightening the elbow or knee to move the arm or thigh back to zero position. In stair climbing, both the hip and knee extend when lifting the body to the next step. **Hyperextension** extends a joint beyond zero position. For example, raising the back of your hand as if admiring a new ring hyperextends the wrist. If you move your arm to a position posterior to the shoulder, such as reaching for something in your back pocket, you hyperextend your shoulder.

> **Apply What You Know**
>
> Some synovial joints have articular surfaces or ligaments that prevent them from being hyperextended. Try hyperextending some of your synovial joints and list a few for which this is impossible.

Abduction and Adduction

Abduction[82] (ab-DUC-shun) (fig. 6.37) is movement of a body part in the frontal plane away from the midline of the body—for example, raising the arm away from one side of the body or standing spread-legged. To abduct the fingers is to

[82] *ab* = away; *duc* = to carry, lead

spread them apart. **Adduction**[83] (ah-DUC-shun) is movement toward the median plane, returning the body part to zero position.

Elevation and Depression

Elevation is movement that raises a body part vertically. The mandible is elevated when biting off a piece of food, and the clavicle and scapula are elevated when lifting a suitcase. The opposite of elevation is **depression,** such as lowering the mandible to open the mouth or lowering the shoulders.

Protraction and Retraction

Protraction[84] is anterior movement of a body part on a horizontal plane, and

(a) Abduction

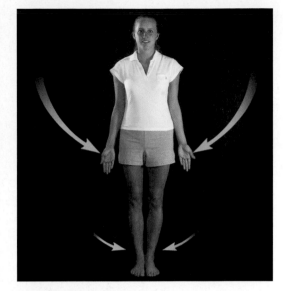

(b) Adduction

Figure 6.37 Abduction and Adduction.

retraction[85] is posterior movement (fig. 6.38). You protract your shoulders when you reach out to hug someone or push open a door. They are retracted when one stands at military attention. Such exercises as bench presses and push-ups involve repeated protraction and retraction of the shoulders. Most of us have some degree of overbite, so we protract the mandible to make the edges of the incisors meet when biting off a piece of food, then retract it to make the molars meet and grind food between them.

Circumduction

Circumduction[86] (fig. 6.39) is movement in which one end of an appendage remains relatively stationary while the other end makes a circular motion. The appendage as a whole thus describes a conical space. For example, if an artist standing at an easel reaches out and draws a circle on the canvas, the shoulder

[83] *ad* = toward; *duc* = to carry, lead
[84] *pro* = forward; *trac* = pull, draw
[85] *re* = back; *trac* = pull, draw
[86] *circum* = around; *duc* = to carry, lead

(a) Protraction

(b) Retraction

Figure 6.38 Protraction and Retraction of the Shoulder.

Figure 6.39 Circumduction of the Upper Limb.

Figure 6.40 Rotation of the Humerus.
(a) Medial (internal) rotation. (b) Lateral (external) rotation.

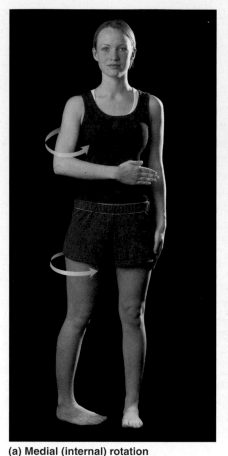

(a) Medial (internal) rotation   **(b) Lateral (external) rotation**

remains relatively stationary while the hand makes a circle. The extremity as a whole thus exhibits circumduction. A baseball player winding up for the pitch circumducts the arm in a more extreme "windmill" fashion.

Rotation

Rotation is a movement in which a bone spins on its longitudinal axis. Figure 6.40 shows limb movements that occur in **lateral (external)** and **medial (internal) rotation** of the femur and humerus. Turning the head from side to side is called **right** and **left rotation.** Powerful right and left rotation at the waist is important in such actions as baseball pitching and golf. Good examples of the lateral and medial rotation of the humerus are its movements in the forehand and backhand strokes of tennis.

Supination and Pronation

Supination[87] (SOO-pih-NAY-shun) (fig. 6.41a) is rotation of the forearm so that the palm faces forward or upward; in anatomical position, the forearm is supinated. Supination is the movement you would usually make with the right hand to turn a doorknob clockwise or drive a screw into a piece of wood. **Pronation**[88] (fig. 6.41b) is rotation of the forearm so that the palm faces toward

[87] *supin* = to lay back

[88] *pron* = to bend forward

the rear or downward. As an aid to memory, think of it this way: you are *prone* to stand in the most comfortable position, which is with the palm *pronated*. If you were holding a bowl of *soup* in the palm of your hand, your forearm would have to be *supinated* to avoid spilling it.

Opposition and Reposition

Opposition[89] is movement of the thumb to approach or touch the fingertips (hence we say humans have an *opposable thumb*), and **reposition**[90] is its movement back to zero position, parallel to the index finger. Opposition enables the hand to grasp objects and is the single most important hand function.

Dorsiflexion and Plantar Flexion

Special names are given to vertical movements at the ankle (fig. 6.42a). **Dorsiflexion** (DOR-sih-FLEC-shun) is a movement in which the toes are raised (as one might do to apply toenail polish). Dorsiflexion occurs in each step you take as your foot comes forward. It prevents your toes from scraping on the ground and results in a *heel strike* when the foot touches down in front of you. **Plantar flexion** is a movement that points the toes downward, as in standing on tiptoe or pressing the gas pedal of a car. This motion also produces the *toe-off* in each step you take, as the heel of the foot behind you lifts off the ground.

Inversion and Eversion

Inversion[91] is a foot movement that tilts the soles medially toward each other; **eversion**[92] tilts the soles away from each other (fig. 6.42b, c). These movements are common in fast sports such as tennis and football, and sometimes result in ankle sprains. They are also important in walking across uneven surfaces, for example in mountain hiking on a rocky trail. These terms also refer to congenital deformities of the feet, which are often corrected by orthopedic shoes or braces.

---

[89] *op* = against; *posit* = to place
[90] *re* = back; *posit* = to place
[91] *in* = inward; *version* = turning
[92] *e* = outward; *version* = turning

(a) Supination   (b) Pronation

Figure 6.41 Supination and Pronation of the Forearm.

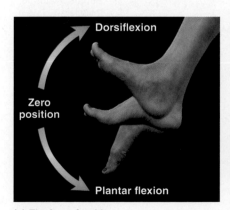  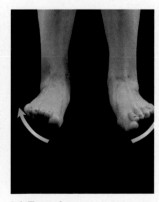

(a) Flexion of ankle   (b) Inversion   (c) Eversion

Dorsiflexion

Zero position

Plantar flexion

Figure 6.42 Movements of the Foot. (a) Dorsiflexion and plantar flexion. (b) Inversion. (c) Eversion.

Anatomy of Selected Synovial Joints

We now examine the gross anatomy of a few representative synovial joints at the shoulder, elbow, hip, and knee. These joints most often require medical attention, and they have a strong bearing on athletic performance.

The Shoulder

At the shoulder, the head of the humerus inserts into the shallow glenoid cavity of the scapula, thus forming the **glenohumeral joint** (fig. 6.43). The glenoid cavity is somewhat deeper than it appears on a dried skeleton because of a ring of fibrocartilage around its margin called the **glenoid labrum.**[93] This joint is stabilized mainly by a tendon of the biceps muscle, which extends like a strap through the groove between the tubercles of the humerus, over the shoulder, and to the scapula. Four shoulder muscles at this joint contribute another stabilizing complex, the **rotator cuff,** described in chapter 7 (p. 238).

This is the most mobile joint of the body, and serves with the elbow to position the hand for a given task. Its mobility, however, comes at considerable cost to stability. The shallowness of the glenoid cavity and looseness of the joint capsule make the shoulder highly subject to dislocation and other injuries. The shoulder is often dislocated when the arm is outstretched and receives a blow from above, such as from a heavy object falling off a shelf. Children may suffer shoulder dislocations as a result of an adult jerking or lifting them by one arm.

The Elbow

The elbow is a hinge with three joints (fig. 6.44): (1) the **humeroulnar joint** where the trochlear notch of the ulna encircles the trochlea of the humerus; (2) the **humeroradial joint** where the disc-shaped head of the radius meets the capitulum of the humerus; and (3) the **proximal radioulnar joint** where the edge of the radial head meets the ulna. Only the first two joints contribute to the elbow hinge, but the last two act in forearm supination and pronation. In these actions, the upper discoidal surface of the radial head swivels on the

---

[93] *labrum* = lip

Figure 6.43 The Shoulder Joint. Bursae are shown in green. Gray bands not otherwise labeled are ligaments. The supraspinatus and subscapularis tendons are two tendons of the rotator cuff, a common site of athletic and workplace injuries.
AP|R

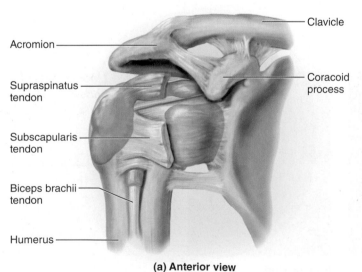

Acromion

Supraspinatus tendon

Subscapularis tendon

Biceps brachii tendon

Humerus

Clavicle

Coracoid process

(a) Anterior view

Acromion

Subdeltoid bursa

Deltoid muscle

Humerus

Clavicle

Supraspinatus muscle

Capsular ligament

Glenoid labrum

Synovial membrane

Glenoid cavity of scapula

Glenoid labrum

(b) Frontal section

capitulum of the humerus, and the edge of the radial head spins like a tire in a notch of the ulna. A loop of fiber called the **anular ligament** attaches to the ulna at both its ends and loops around the neck of the radius, holding the radius against the radial notch but allowing it to spin freely.

On the posterior side of the elbow, there is a prominent **olecranon bursa** that eases the movement of tendons over the elbow. This is a common site of painful inflammation (*bursitis of the elbow,* or *olecranon bursitis*) resulting from impact injuries or repetitive motions on one's job or in sports such as tennis and golf.

The Hip

The **hip (coxal) joint** is the point where the head of the femur inserts into the acetabulum of the hip bone (fig. 6.45). Because the coxal joints bear much of the body's weight, they have deep sockets and are much more stable than the shoulder joint. The depth of the socket is somewhat greater than we see on dried bones because of a horseshoe-shaped ring of fibrocartilage, the **acetabular labrum,** attached to its rim. The joint is enclosed in three strong *pubofemoral* and *iliofemoral* ligaments (fig. 6.45a) that twist when one stands, pulling the head of the femur tightly into the acetabulum and stabilizing the joint. This is important to human bipedalism because it allows one to stand with minimal continual muscle exertion at the joint. Because of the depth of the acetabulum and tautness of these ligaments, hip dislocations are rare. Hip fractures are common in the elderly, however, usually at the femoral neck just below the head.

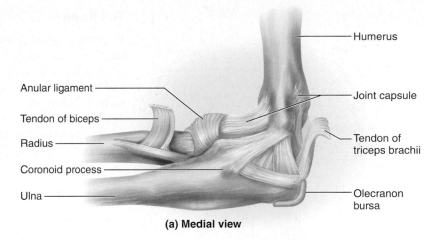

(a) Medial view

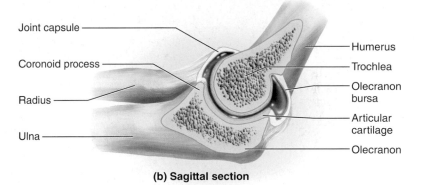

(b) Sagittal section

Figure 6.44 The Elbow Joint. AP|R

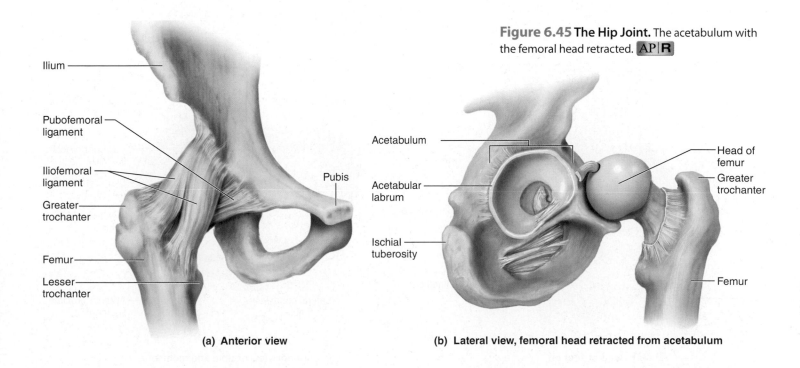

(a) Anterior view

Figure 6.45 The Hip Joint. The acetabulum with the femoral head retracted. AP|R

(b) Lateral view, femoral head retracted from acetabulum

The Knee

The **tibiofemoral (knee) joint** is the largest and most complex synovial joint (fig. 6.46). Anteriorly, it is covered by the *patellar ligament* and *quadriceps femoris tendon* of the quadriceps muscle of the thigh (see fig. 7.20, p. 243). The lateral and posterior aspects are enclosed by the joint capsule. The knee is surrounded by about 13 bursae. It is stabilized mainly by the quadriceps tendon in front and the tendon of one of the hamstring muscles, the *semimembranosus,* on the rear of the thigh. Developing strength in these muscles therefore reduces the risk of knee injury.

Figure 6.46 The Knee Joint. Bursae are shown in green, articular cartilages in light blue, and tendons and ligaments in gray. AP|R

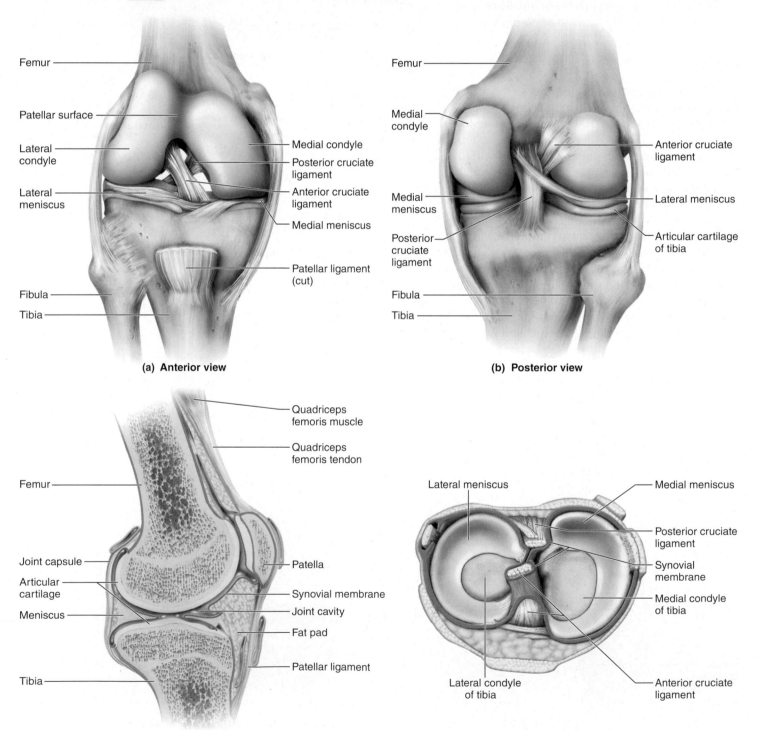

(a) **Anterior view**

(b) **Posterior view**

(c) **Sagittal section**

(d) **Superior view of tibia and menisci**

Clinical Application 6.2

KNEE INJURIES AND ARTHROSCOPIC SURGERY

The knee is highly vulnerable to rotational and horizontal stress, especially when it is flexed (as in skiing or running) and receives a blow from behind or from the side. The most common injuries are to a meniscus or the anterior cruciate ligament (ACL). Knee injuries heal slowly because ligaments and tendons have a very scanty blood supply and cartilage usually has no blood vessels at all. A damaged ACL is often replaced with a tissue graft from the patellar ligament or from a hamstring tendon.

The diagnosis and surgical treatment of knee injuries have been greatly improved by *arthroscopy,* a procedure in which a slender viewing instrument *(arthroscope)* and surgical instruments are inserted through small incisions in the knee. Shreds of torn menisci can be removed in this way. Arthroscopic surgery produces much less tissue damage than conventional surgery and enables patients to recover more quickly.

The joint cavity contains two crescent-shaped cartilages called the **lateral** and **medial menisci** (singular, *meniscus*) (fig. 6.46d). They absorb pressure from the weight of the body and stabilize the knee as described earlier. Deep within the joint cavity is a pair of ligaments called the **anterior cruciate**[94] (CROO-she-ate) **ligament (ACL)** and **posterior cruciate ligament (PCL).** These are called *cruciate* for the fact that they cross each other in the form of an X, and are called *anterior* and *posterior* for their attachments on the respective surfaces of the tibia. When the knee is extended, the ACL is pulled tight and prevents hyperextension. The PCL prevents the femur from sliding off the front of the tibia and prevents the tibia from being displaced backward. The most common knee injuries, often requiring surgical treatment, are to the ACL and menisci (see Clinical Application 6.2).

An important aspect of human bipedalism is the ability to "lock" the knees and stand erect without tiring the extensor muscles of the leg. When the knee is extended to the fullest degree allowed by the ACL, the femur rotates medially on the tibia. This action locks the knee, and in this state all the major knee ligaments are twisted and taut. To unlock the knee, a small posterior *popliteus muscle* rotates the femur laterally and untwists the ligaments.

Before You Go On

Answer these questions from memory. Reread the preceding section if there are too many you don't know.

23. Explain the distinction between fibrous and cartilaginous joints and give an example of each.

24. Explain what unique structural features of a synovial joint make some of these the most freely movable joints of the body.

25. Give one example of each of the following synovial joint types: ball-and-socket, condyloid, gliding, hinge, pivot, and saddle.

26. Explain why both the shoulder and hip sockets are deeper in life than they appear on a dried skeleton.

27. What function is served by the cruciate ligaments of the knee? What function is served by the menisci?

---

[94] *cruci* = cross; *ate* = characterized by

Aging of the Skeletal System

Nearly everyone experiences some degree of bone loss, called *osteopenia,*[95] with age. It begins as early as the 30s, as osteoblasts become less active than osteoclasts and bone deposition fails to keep pace with the rate of resorption. The aging of other organ systems also contributes to aging of the skeleton. As the skin ages, its role in vitamin D synthesis declines by as much as 75%, and declining levels of vitamin D mean we absorb less calcium from our food. As muscle mass and strength decline, the bones are subjected to less stress and are therefore less stimulated to deposit osseous tissue. One skeletal condition that becomes less common in old age, however, is herniated intervertebral discs. The discs become more fibrous in old age and have less nucleus pulposus, so they are less likely to rupture. Their degeneration, however, contributes to back pain and stiffness as the vertebral column becomes less supple. The thoracic cage also stiffens with age, because the costal cartilages tend to ossify. As a consequence, breathing becomes shallower and more laborious.

---

[95] *osteo = bone; penia = deficiency*

CAREER SPOTLIGHT

Orthopedic Nurse

An orthopedic nurse is a specialist in the prevention and treatment of musculoskeletal disorders. He or she has achieved RN or LPN certification and then completed advanced training in orthopedic nursing. Many achieve certification as an orthopedic nurse certified (ONC), orthopedic nurse practitioner (ONP-C), or orthopedic clinical nurse specialist (OCNS-C); some go as far as master's and doctoral degrees. Orthopedic nurses must be skilled in such practices as traction, casting, external therapy, continuous passive motion therapy, and monitoring of neurological and circulatory status. Their work ranges from neonatal to geriatric care, in settings as diverse as hospitals, nursing homes, physician outpatient practices, industry, academics, and home health agencies. Another growing career in skeletal health is bone densitometrist, a radiologic technologist who specializes in assessment of bone density, especially in the evaluation and treatment of osteoporosis. See appendix B for links to further information on careers in orthopedic nursing and bone densitometry.

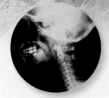

CONNECTIVE ISSUES

Ways in Which the Skeletal System Affects Other Organ Systems

Integumentary System
Bones close to the body surface, such as skull bones, support and shape the skin.

Muscular System
Bones provide the attachment sites for most muscles and furnish leverage, by means of which muscles exert most of their actions. Muscle contraction depends on calcium, for which the skeleton is the body's primary reservoir.

Nervous System
The cranium and vertebral column protect the brain and spinal cord. Neural function also is strongly affected by the skeleton's storage and release of calcium.

Endocrine System
Bones protect the endocrine glands of the head, chest, and pelvic cavity. Some hormones employ calcium in stimulating their target cells.

Circulatory System
The red bone marrow is the ultimate source of all blood cells. Calcium is needed for contraction of the heart. Osteoblasts and osteoclasts maintain blood calcium homeostasis.

Lymphatic and Immune Systems
The red bone marrow is the origin of all immune cells.

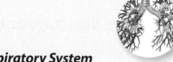

Respiratory System
Bones support the respiratory passages through the nasal cavity. The ribs protect the lungs, and rib movements are crucial to pulmonary ventilation.

Urinary System
The skeleton supports and protects organs of the urinary system.

Digestive System
The skeleton provides protective enclosure for some of the digestive organs.

Reproductive System
The skeleton provides protective enclosure for some reproductive organs. Pelvic anatomy contributes to the difficulty and complications of childbirth.

Study Guide

Assess Your Learning Outcomes

To test your knowledge, discuss the following topics with a study partner or in writing, ideally from memory.

6.1 Skeletal Structure and Function (p. 148)

1. Bones and other components of the skeletal system
2. Five functions of the skeletal system
3. Differences in location and structure between compact and spongy bone
4. Four kinds of bone cells, their respective functions, and the relationship of osteocytes to the bone matrix and to each other
5. The components of the bone matrix and their functional significance
6. The histology of spongy and compact bone
7. Features of the gross anatomy of long and flat bones
8. Differences between the two types of bone marrow

6.2 Bone Development and Metabolism (p. 154)

1. The basic difference between intramembranous and endochondral ossification, and examples of bones that develop by each method
2. Stages of intramembranous and endochondral ossification; how each process accounts for the mature anatomy of a flat cranial bone and a long limb bone, respectively
3. The difference between interstitial and appositional growth of cartilage and bone; why osseous tissue is limited to the latter method; and how the former method contributes to child and adolescent growth
4. Why epiphyseal plates are detectable in an X-ray of a child but not in an X-ray of a middle-aged adult
5. Terms for the addition of minerals to the skeleton and their removal from it; the cells that carry out these two processes, and how they do it
6. Terms for a deficiency and an excess of calcium in the blood, and the pathological consequences of such imbalances
7. Effects and relative importance of calcitriol, parathyroid hormone, and calcitonin on the skeleton
8. Other hormones and nutrients that affect skeletal metabolism

6.3 The Axial Skeleton (p. 161)

1. The appearance to be expected when you see any of the terms for bone surface markings in table 6.1
2. The usual number of named bones in the adult skeleton, and which groups of bones belong to the axial and appendicular skeleton
3. The cavities of the skull
4. The two categories of skull bones, how each category is defined, and the category to which any named bone of the skull belongs
5. Major features of any of the following cranial bones, including the ability to recognize and name these bones and their features in illustrations and laboratory materials: the frontal, parietal, temporal, occipital, sphenoid, and ethmoid bones
6. Major features of any of the following facial bones, including the ability to recognize and name these bones and their features in illustrations and laboratory materials: the maxillae, mandible, vomer, and inferior nasal conchae; and the palatine, zygomatic, lacrimal, and nasal bones
7. The locations and functions of the auditory ossicles and the hyoid bone
8. The usual number of vertebrae and intervertebral discs, the five groups of vertebrae, and the number of vertebrae in each group
9. The four curvatures of the adult spine, their functional significance, and the three main pathological deviations from normal
10. The general structural features seen in most cervical through lumbar vertebrae, and how their structure relates to the anatomy of the spinal cord and spinal nerves
11. Structural features that distinguish cervical, thoracic, and lumbar vertebrae from each other, and how one can identify a single isolated vertebra on the basis of these features
12. Unique features and names of the first two cervical vertebrae, and how their anatomy relates to movements of the head
13. Anatomy of the sacrum and coccyx, and how certain features of these can be seen as derived from features of the vertebrae higher on the spine
14. The structure of an intervertebral disc and how it relates to the condition known as a herniated disc
15. The three components of the thoracic cage
16. The three parts of the sternum, all bones that articulate with it, and the locations of these articulations
17. The distinction between true ribs, false ribs, and floating ribs, and which ribs belong to each category
18. How the anatomy of ribs 1, 11, and 12 differs from the more typical anatomy of ribs 2 through 10
19. How the ribs articulate with the vertebrae, and what features on the vertebrae mark these points of articulation

6.4 The Appendicular Skeleton (p. 177)

1. The names of the bones that form the pectoral and pelvic girdles
2. The shape and function of the clavicle and how one can distinguish the lateral from the medial end and superior from inferior surface of an isolated clavicle
3. Names and locations of the three fossae of the scapula; what landmark separates the two posterior fossae; and in general terms, what occupies these three fossae in a living person
4. Names and structures of the processes and joint cavity at the lateral end of the scapula
5. Names and locations of the 30 bones of the upper limb
6. Major anatomical features of the humerus, radius, and ulna, especially at the proximal and distal ends of each bone
7. The fibrous membrane between the radius and ulna, and its functional significance
8. Names of the four carpal bones of the proximal row, in order from lateral to medial, and the same for the four bones of the distal row
9. Structure of the metacarpal bones and phalanges, and the system for giving each of these 19 bones a unique name and number
10. Components of the pelvic girdle; the distinction between *pelvic girdle* and *pelvis;* and differences between the male and female pelvis
11. Anatomy of the hip bone; the childhood bones that fuse to form the adult hip bone; and articulations of the hip bone with the sacrum and femur
12. Anatomical features of the femur, patella, tibia, and fibula, especially at the proximal and distal ends of the long bones
13. Names and locations of the tarsal bones, and similarities and differences between the tarsal and carpal bones
14. The system for naming and numbering the metatarsal bones and phalanges of the foot
15. The three foot arches and their relationship to human bipedalism and locomotion

6.5 Joints (p. 186)

1. The differences between fibrous, cartilaginous, and synovial joints; synonyms for each
2. Two types of fibrous joints, how they differ, and where they are found
3. Two types of cartilaginous joints, how they differ, and where they are found
4. The general features of synovial joints and the modifications seen in synovial joints that bear unusual weight or pressure
5. Comparisons and contrasts between tendons and ligaments

6. The structure, function, and location of bursae
7. The meaning of monaxial, biaxial, and multiaxial synovial joints, and examples of each
8. Six kinds of synovial joints, how they differ, and where each can be found
9. Definition, examples, and ability to demonstrate joint flexion, extension, and hyperextension
10. The same for abduction and adduction
11. The same for elevation and depression
12. The same for protraction and retraction
13. The same for circumduction
14. The same for lateral and medial rotation, and right and left rotation
15. The same for supination and pronation
16. The same for opposition and reposition of the thumb
17. The same for dorsiflexion and plantar flexion of the ankle
18. The same for inversion and eversion of the foot
19. The anatomical name of the shoulder joint; the features of its articular bone surfaces; its relationship to the rotator cuff; and the advantage and disadvantage of its great mobility
20. The three synovial joints found at the elbow; the features of its articular bone surfaces and how they relate to elbow and forearm movements; and the location and clinical significance of the olecranon bursa
21. Similarities and differences between the hip joint and shoulder joint; the role of cartilages in stabilizing these two joints; and the effects of standing on the hip joint
22. The anatomical name of the principle knee joint; the features of its articular bone surfaces; the muscles most responsible for stabilizing the knee joint; the structure, importance, and clinical significance of the menisci and cruciate ligaments; and the mechanisms of locking and unlocking the knees when one is standing

Testing Your Recall

1. Which cells secrete hydrochloric acid, resorb osseous tissue, and raise the blood calcium concentration?
   a. chondrocytes
   b. osteocytes
   c. osteogenic cells
   d. osteoblasts
   e. osteoclasts

2. The medullary cavity of an adult bone may contain
   a. adipose tissue.
   b. hyaline cartilage.
   c. periosteum.
   d. osteocytes.
   e. articular cartilages.

3. Which of the following movements are unique to the foot?
   a. dorsiflexion and inversion
   b. elevation and depression
   c. circumduction and rotation
   d. abduction and adduction
   e. opposition and reposition

4. Which of the following joints has anterior and posterior cruciate ligaments?
   a. the shoulder
   b. the elbow
   c. the hip
   d. the knee
   e. the ankle

5. Which of these is the bone of the heel?
   a. cuboid
   b. calcaneus
   c. navicular
   d. trochlear
   e. talus

6. ___ is secreted in response to a falling blood calcium level, and raises the level by promoting osteoclast activity.
   a. Parathyroid hormone
   b. Calcitonin
   c. Growth hormone
   d. Estrogen
   e. Calcitriol

7. A child jumps to the ground from the top of a playground "jungle gym." His leg bones do not shatter mainly because they have
   a. more osteons than older bone.
   b. young, resilient osteocytes.
   c. an abundance of calcium phosphate.
   d. collagen fibers.
   e. a strong periosteum.

8. One long bone meets another at its
   a. diaphysis.
   b. epiphyseal plate.
   c. periosteum.
   d. metaphysis.
   e. epiphysis.

9. The spinal cord passes through the ___ of the vertebrae.
   a. vertebral foramina
   b. intervertebral foramina
   c. transverse foramina
   d. obturator foramina
   e. bodies

10. The principal facial bone between the orbit and upper teeth is the
    a. frontal bone.
    b. zygomatic bone.
    c. maxilla.
    d. mandible.
    e. parietal bone.

11. The mastoid process and external acoustic meatus are parts of the _____ bone.

12. Osteocytes contact each other through channels called _____ in the bone matrix.

13. A bone increases in diameter only by _____ growth, the addition of new surface lamellae.

14. The lubricant in the shoulder and hip joints is called _____.

15. A calcium deficiency called _____ can cause death by suffocation.

16. _____ are bone cells that secrete collagen and stimulate calcium phosphate deposition.

17. _____ is the science of body movement.

18. The femur is prevented from slipping sideways off the tibia in part by a pair of crescent-shaped cartilages called the lateral and medial _____.

19. Bones of the skull are joined along lines called _____.

20. A herniated disc occurs when a ring called the _____ cracks.

Answers in appendix A

True or False

Determine which five of the following statements are false, and briefly explain why.

1. There are more carpal bones than tarsal bones.

2. The growth zone of a child's long bones is the articular cartilage.

3. Most ligaments connect a muscle to a bone.

4. On a living person, it would be possible to palpate the muscles in the infraspinous fossa but not those of the subscapular fossa.

5. Reaching behind you to take something out of your hip pocket involves hyperextension of the elbow.

6. Most bones develop from hyaline cartilage.

7. Synovial fluid is secreted by the bursae.

8. In strict anatomical terminology, the words *arm* and *leg* both refer to regions with only one bone.

9. There is no meniscus in the elbow joint.

10. Canaliculi in the bone matrix allow osteocytes to communicate with each other.

Answers in appendix A

Testing Your Comprehension

1. Most osteocytes of an osteon are far removed from blood vessels, but are still able to respond to hormones in the blood. Explain how it is possible for hormones to reach and stimulate these cells.

2. How does the regulation of blood calcium concentration exemplify negative feedback and homeostasis?

3. Name the action that would occur at each of the following joints in the indicated situation. (For example, the shoulder in picking up a suitcase. Answer: elevation.) (a) The arm when you raise it to rest your hand on the back of a sofa on which you're sitting. (b) Your neck when you look up at a plane in the sky. (c) Your tibia when you turn the toes of one foot to touch the heel of the other foot. (d) Your humerus when you reach up to scratch the back of your head. (e) A bowler's backswing. (f) A basketball player's foot as she makes a jump shot. (g) Your shoulder when you pull back on the oars of a rowboat. (h) Your elbow when lifting a barbell. (j) A soccer player's knee when kicking the ball. (k) Your index finger when dialing an old rotary telephone. (l) Your thumb when you pick up a tiny bead between your thumb and index finger.

The Muscular System

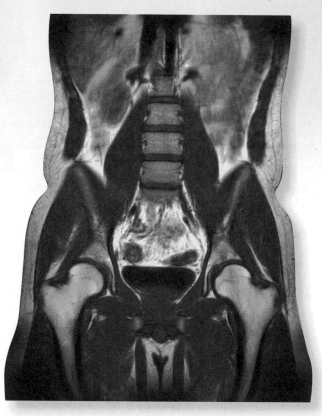

A colorized frontal MRI scan, showing muscles of the lumbar, pelvic, and upper femoral regions. Return and identify as many muscles as you can after reading this chapter.

BASE CAMP

Before ascending to the next level, be sure you're properly equipped with a knowledge of these concepts from earlier chapters:

- The functions of membrane proteins, especially receptors and ion gates (p. 68)
- Desmosomes and gap junctions (pp. 72–73)
- Gross anatomy of the skeleton (pp. 159–184)
- Terminology of joint movements (pp. 188–191)

Anatomy & Physiology | REVEALED®
aprevealed.com

Module 6: Muscular System

Movement is a fundamental characteristic of all living things, but reaches its highest development in animals because of their muscular tissue. Muscular tissue is composed of elongated cells that contract when stimulated. A muscle cell is essentially a device for converting the chemical energy of ATP into the mechanical energy of contraction.

Understanding structure and function of the muscular system is of central importance in several fields of health care and fitness. Physical and occupational therapists must be well acquainted with the muscular system to design and carry out rehabilitation programs. Nurses and other health-care providers often move patients who are physically incapacitated, and to do this safely and effectively requires an understanding of joints and muscles. Athletes, coaches, exercise physiologists, dancers, therapists, sports medicine professionals, and amateur fitness enthusiasts use their knowledge of muscle anatomy and physiology to maximize performance and minimize injury.

In this chapter, we will examine the mechanism of muscle contraction at the cellular and molecular levels, which explains such aspects of muscle performance as warm-up, strength, endurance, and fatigue. We will then study the behavior of muscles taken as a whole, and finally the anatomy and functions of several of the major muscle groups of the human body. Although cardiac and smooth muscle are not part of the muscular system, they will also be discussed in this chapter for comparison to skeletal muscle. Cardiac muscle is discussed at greater length in chapter 13.

7.1 Muscular Tissue and Cells

Expected Learning Outcomes
When you have completed this section, you should be able to:

a. identify diverse functions of muscular tissue;

b. describe the structure of a skeletal muscle fiber and relate this to its function; and

c. describe the nerve–muscle relationship in skeletal muscle.

There are three types of muscular tissue: skeletal, cardiac, and smooth, compared in table 4.11 (p. 118). The term **muscular system** refers only to the skeletal muscles, which are the primary focus of this chapter. However, we will begin with some characteristics that the three types have in common, and look more closely at their differences later in the chapter.

The Functions of Muscles

The functions of muscular tissue go well beyond the obvious:

- **Movement.** This includes externally visible movements of the head, trunk, and limbs and less conspicuous internal actions such as the movements of breathing, propulsion of contents of the digestive tract, expulsion of urine from the bladder and a fetus from the uterus, the pumping of blood, and the dilation and constriction of blood vessels to regulate blood pressure and flow. Muscular movements also play important roles in communication: speech, writing, gestures, and facial expressions.

- **Stability.** It is also important that the muscles prevent unwanted movement, as in maintaining posture, holding certain bones such as the scapula and humerus in place, and holding one bone still while another one moves.

- **Control of body openings and passages.** Ring-shaped **sphincter muscles** around the eyelids and pupils control admission of light to the eye; others regulate waste elimination; still others control the movement of food, bile, and other materials through the body.

- **Heat generation.** Body heat is necessary for enzymes to function and regulate one's metabolism. The skeletal muscles produce 20% to 30% of our body heat at rest and up to 40 times as much during exercise.

- **Glycemic control.** This means the regulation of blood glucose within normal limits. The skeletal muscles play a significant role in stabilizing blood sugar levels by absorbing a large share of one's glucose. In old age, in obesity, and when muscles become deconditioned and weakened, people suffer an increased risk of type 2 diabetes mellitus because of the decline in this glucose-buffering function.

Skeletal Muscle Fibers

Skeletal muscle may be defined as voluntary striated muscle that is usually attached to one or more bones. It is called **voluntary** because it is usually subject to conscious control, and **striated** because it has alternating light and dark transverse bands, or **striations** (fig. 7.1), that reflect the overlapping arrangement of the internal proteins that enable it to contract. Skeletal muscle cells are perhaps the most internally complex, tightly organized of all human cells. In order to understand muscle function, one must know how the organelles and contractile proteins of a muscle fiber are arranged.

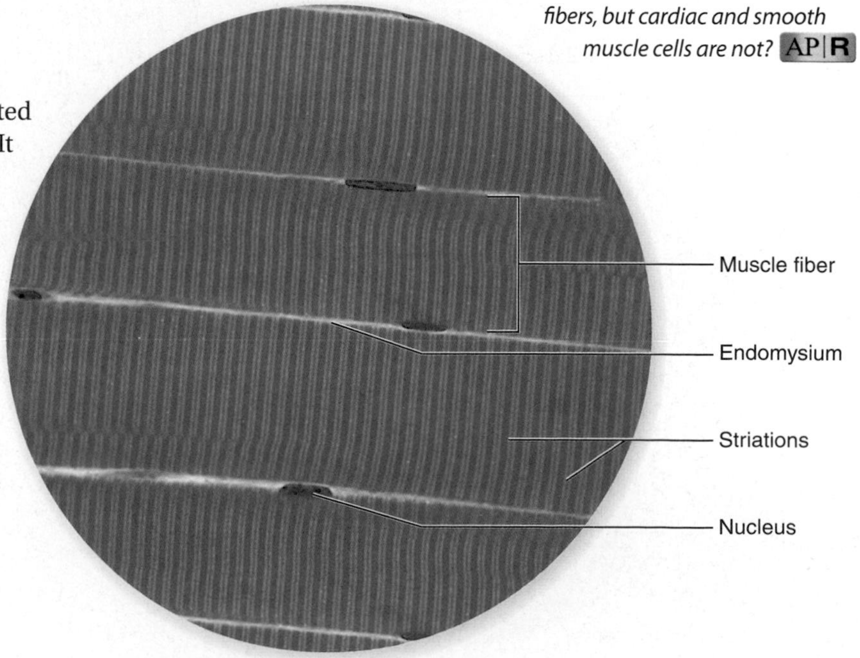

Figure 7.1 Skeletal Muscle Fibers (×1,000).

- *Why are skeletal muscle cells called fibers, but cardiac and smooth muscle cells are not?* AP|R

Muscle fiber

Endomysium

Striations

Nucleus

Structure of the Muscle Fiber

Skeletal muscle cells are called **muscle fibers** because of their exceptionally long slender shape: typically about 100 μm in diameter and 3 cm (30,000 μm) long, but sometimes up to 500 μm thick and 30 cm long. Each muscle fiber has multiple nuclei pressed against the inside of the plasma membrane, reserving the deeper part of the cell mainly for thick bundles of contractile protein (fig. 7.2).

These bundles, called **myofibrils**, number from several dozen to a thousand or more. Packed between them are numerous mitochondria, a network of smooth endoplasmic reticulum, deposits of the high-energy carbohydrate **glycogen,** and a red oxygen-storing pigment, **myoglobin**. Stored glycogen and oxygen are employed to produce ATP during periods of exercise.

The plasma membrane, called the **sarcolemma,**[1] has tunnel-like infoldings called **transverse (T) tubules** that penetrate through the fiber and emerge on the other side. The function of a T tubule is to carry an electrical current from the surface of the cell to the interior when the cell is stimulated.

The smooth endoplasmic reticulum of a muscle fiber is called **sarcoplasmic reticulum (SR)**. It forms a web around each myofibril, and alongside the

---

[1] *sarco* = flesh, muscle; *lemma* = husk

Figure 7.2 Structure of a Skeletal Muscle Fiber. This is a single cell containing 11 myofibrils (9 shown at the left end and 2 cut off at midfiber). A few myofilaments are shown projecting from the myofibril at the left. Their finer structure is shown in figure 7.3.

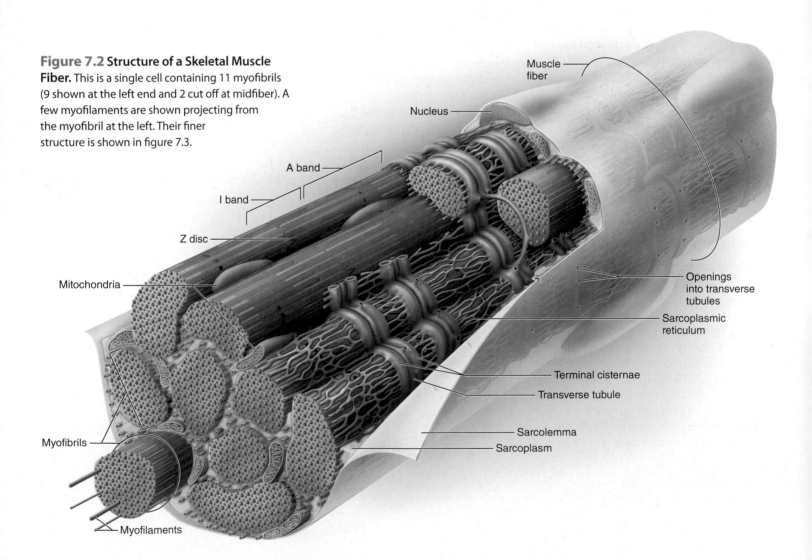

T tubules it exhibits dilated sacs called **terminal cisternae**. The SR is a reservoir for calcium ions; when the muscle fiber is stimulated, the SR releases a flood of calcium into the cytosol to activate the contraction process.

Myofilaments and Striations

Understanding muscle contraction requires a closer look at the myofibrils that pack the muscle fiber. Each of these is a bundle of parallel protein microfilaments called **myofilaments** (fig. 7.3). There are two main kinds of myofilaments.

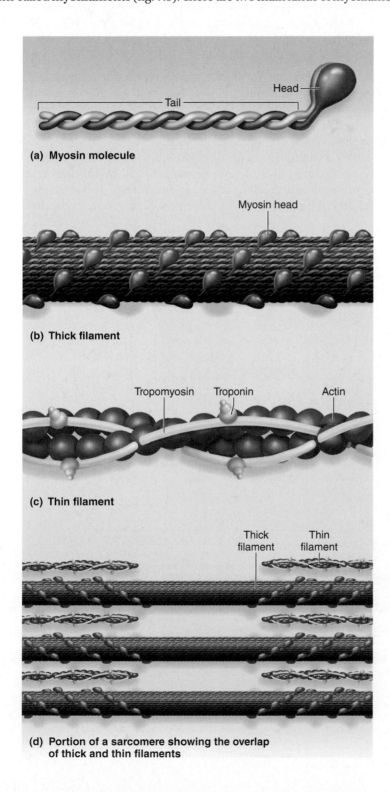

(a) **Myosin molecule**

(b) **Thick filament**

(c) **Thin filament**

(d) **Portion of a sarcomere showing the overlap of thick and thin filaments**

Figure 7.3 Molecular Structure of Thick and Thin Filaments. (a) A single myosin molecule consists of two intertwined polypeptides forming a filamentous tail and a double globular head. (b) A thick filament consists of 200 to 500 myosin molecules bundled together with the heads projecting outward in a helical array. (c) A thin filament consists of two intertwined chains of globular actin molecules, smaller filamentous tropomyosin molecules, and a calcium-binding protein called troponin associated with the tropomyosin. (d) A region of overlap between the thick and thin myofilaments.

1. **Thick filaments** are made of several hundred molecules of a protein called myosin. A **myosin** molecule is shaped like a golf club, with two polypeptides intertwined to form a shaftlike *tail,* and a double globular *head* projecting from it at an angle. A thick filament may be likened to a bundle of 200 to 500 such "golf clubs," with their heads directed outward in a helical array around the bundle.

2. **Thin filaments**, about half as wide as the thick filaments, are composed mainly of two intertwined strands of a protein called **actin**. Each actin is like a bead necklace—a string of globular subunits. A thin filament also has about 50 shorter molecules of a regulatory protein called **tropomyosin**, and each tropomyosin has a small calcium-binding protein called **troponin** attached to it.

Myosin and actin are not unique to muscle; they occur in nearly all cells, where they function in motility, mitosis, and transport of intracellular materials. In skeletal and cardiac muscle they are especially abundant, however, and are organized into a precise array that accounts for the striations of these two muscle types.

The thick and thin filaments are held in line with each other by springy *elastic filaments* composed of a protein called *titin.* Their highly regular pattern of overlap gives striated muscle an appearance of dark **A bands** alternating with lighter **I bands** (fig. 7.4). (These names refer to the way these bands

Figure 7.4 Muscle Striations and Myofilament Overlap. (a) Five myofibrils of a single muscle fiber, showing the striations in the relaxed state (electron micrograph). (b) The overlapping pattern of thick and thin myofilaments that accounts for the striations seen in part (a).

AP|R

• *Why are A bands darker than I bands in the photo?*

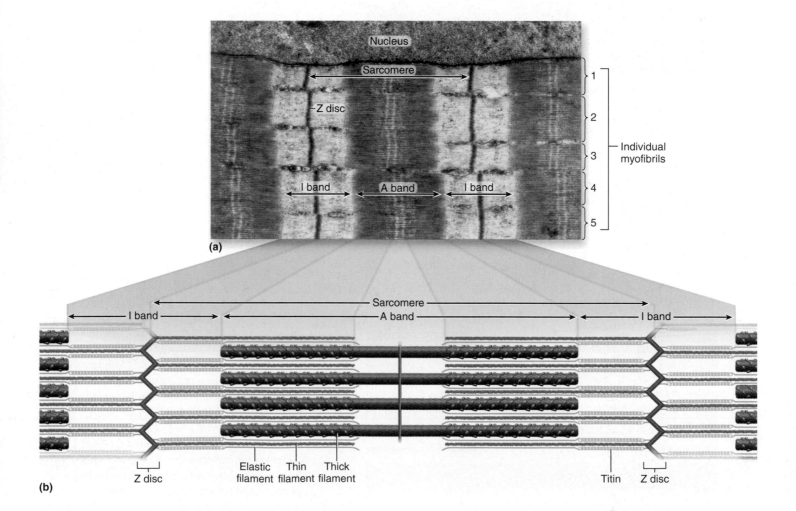

(a)

(b)

affect polarized light, but to help remember which band is which, think "d**A**rk" and "l**I**ght.") Each I band consists only of thin filaments. The I band is bisected by a thin dark line called the **Z disc**,[2] a plaque of protein that provides anchorage for the thin filaments. The darker A bands are regions in which thick and thin filaments overlap, accounting for the denser apearance of these bands. The middle of an A band is a little lighter than the rest, however; thin filaments do not reach this far, so the middle of the A band is composed of myosin only.

Each segment of a myofibril from one Z disc to the next is called a **sarcomere**[3] (SAR-co-meer), the functional unit of the muscle fiber. A typical myofibril 3 cm long has about 10,000 sarcomeres, end to end. A muscle shortens because its sarcomeres shorten and pull the Z discs closer to each other. A huge protein called *dystrophin* links the thin filaments to the inner surface of the sarcolemma. When the sarcomeres shorten, dystrophin pulls on the sarcolemma and the whole cell shortens. Genetic defects in dystrophin underlie the disease *muscular dystrophy*.

The Nerve–Muscle Relationship

Skeletal muscle cannot contract unless it is stimulated by a nerve (or artificially with electrodes). If its nerve connections are severed or poisoned, a muscle is paralyzed. Thus, muscle contraction cannot be understood without first understanding the relationship between nerve and muscle cells.

The nerve cells that stimulate skeletal muscles, called **motor neurons**, are located in the brainstem and spinal cord. Their axons, called **motor nerve fibers**, lead to the muscles. At its distal end, each axon branches to multiple muscle fibers, but each muscle fiber receives only one nerve fiber (fig. 7.5). Each motor neuron stimulates all the muscle fibers of its group to contract at once, so one motor neuron and all muscle fibers supplied by it are called a **motor unit**. Some motor units are very large, with one nerve fiber branching to as many as 1,000 muscle fibers. These motor units, found for example in muscles of the thigh, produce powerful contractions but not fine motor control. Others are very small, with one nerve fiber supplying as few as 3 to 6 muscle fibers. These motor units are not strong, but produce finely controlled muscular actions for such purposes as eye and hand movements.

One advantage of having multiple motor units in a muscle is that the nervous system can generate variable muscle contractions by activating a variable number of motor units—more motor units for a stronger contraction, for example. Another advantage is that the motor units can "work in shifts." Muscle fibers fatigue when subjected to continual stimulation. If all of the fibers in your postural muscles fatigued at once, for example, you might collapse. To prevent this, other motor units take over while the fatigued ones rest, and the muscle as a whole can sustain long-term contraction.

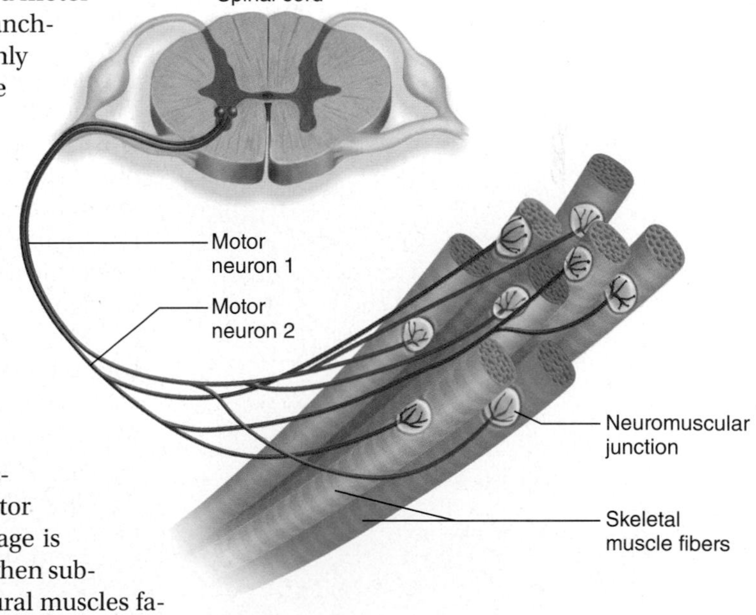

Figure 7.5 **Motor Units.** A motor unit consists of one motor neuron and all skeletal muscle fibers that it supplies. Two motor units are represented here by the red and blue nerve and muscle fibers. Note that the muscle fibers of a motor unit are not clustered together but distributed through the muscle and commingled with the fibers of other motor units.

---

[2] *Z = Zwichenscheibe* (German) = "between disc"

[3] *sarco* = muscle; *mere* = part, segment

The point where the end of a nerve fiber meets another cell is called a **synapse** (SIN-aps). When the second cell is a skeletal muscle fiber, the synapse is also called a **neuromuscular junction (NMJ)**, or **motor end plate** (fig. 7.6). Each branch of the nerve fiber ends in a bulbous swelling called a **synaptic knob**, which is nestled in a depression on the muscle fiber. The two cells do not actually touch each other but are separated by a tiny gap, the **synaptic cleft**.

The synaptic knob contains membrane-bounded sacs called **synaptic vesicles**, which contain a signaling chemical called **acetylcholine (ACh)** (ASS-eh-till-CO-leen). ACh is one of many signaling chemicals called neurotransmitters employed by the nervous system; others are introduced in chapter 8. When the nerve fiber releases ACh, the ACh diffuses quickly across the synaptic cleft and binds to proteins called **ACh receptors** on the surface of the muscle fiber. This binding stimulates the muscle fiber to contract. Also found on the muscle fiber and in the synaptic cleft is an enzyme called **acetylcholinesterase (AChE)** (ASS-eh-till-CO-lin-ESS-ter-ase). It quickly breaks down ACh to stop stimulation of the muscle and allow it to relax. Interference with ACh action can weaken or paralyze the muscles (see Clinical Application 7.1).

Clinical Application 7.1

NEUROMUSCULAR TOXINS AND PARALYSIS

Some forms of muscle paralysis are caused by toxins that interfere with functions at the neuromuscular junction. Some pesticides, for example, contain *cholinesterase inhibitors,* which bind to AChE and prevent it from degrading ACh. As a result, ACh lingers in the synapse and overstimulates the muscle, causing *spastic paralysis*—a state of continual contraction. This can suffocate a person if the laryngeal muscles contract spasmodically and close the airway, or if the respiratory muscles cannot relax and allow free breathing. *Tetanus* (lockjaw) is a form of spastic paralysis caused by a bacterial toxin. The toxin acts in the spinal cord to cause overstimulation of the motor neurons, causing them to set off unwanted muscle contractions.

Flaccid paralysis is a state in which the muscles are limp and cannot contract. One agent that causes this is *curare* (cue-RAH-ree), a poison extracted from certain plants and used by some South American natives to poison blowgun darts. Curare binds to the ACh receptors but does not stimulate the muscle. Thus, it prevents ACh from acting. Curare has been used to treat muscle spasms in some neurological disorders.

Figure 7.6 **The Neuromuscular Junction.** (a) Neuromuscular junctions, with muscle fibers slightly teased apart. (b) Structure of a single neuromuscular junction. AP|R

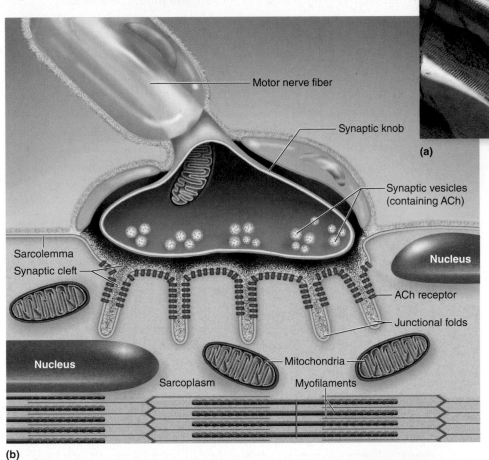

(a)

(b)

Before You Go On

Answer these questions from memory. Reread the preceding section if there are too many you don't know.

1. *Describe some functions of the muscular system other than external movements of the body.*

2. *Explain, in terms of thick and thin myofilaments, why skeletal muscle appears striated under the microscope. Identify the bands of a sarcomere and which contractile proteins are in each band.*

3. *What are the functions of synaptic vesicles, acetylcholine, and acetylcholinesterase?*

7.2 Physiology of Skeletal Muscle

Expected Learning Outcomes

When you have completed this section, you should be able to:

a. describe how a nerve fiber stimulates a muscle fiber and initiates contraction;

b. explain the mechanisms of contraction and relaxation;

c. describe and explain twitch, summation, and other aspects of muscle behavior;

d. contrast isometric and isotonic contraction;

e. describe two ways in which muscle meets the energy demands of exercise;

f. discuss the factors that cause muscle fatigue and limit endurance;

g. distinguish between fast and slow types of muscle fibers; and

h. identify some variables that determine muscular strength.

Muscle Excitation, Contraction, and Relaxation

Movement results from a series of events that begin at the neuromuscular junction and continue in the thick and thin filaments within the muscle fiber. We can view the cycle of muscle contraction and relaxation as occurring in four stages: excitation, excitation–contraction coupling, contraction, and relaxation.

Excitation

Excitation is the process of converting an electrical nerve signal to an electrical signal in the muscle fiber. At rest, a muscle fiber has a voltage, or electrical charge difference, across its sarcolemma. In principle, this is much like the 12 volts of a car battery or 1.5 volts of a flashlight battery, although in cells it is much smaller—about 90 millivolts, with the inside of the sarcolemma being negative to the outside (therefore represented –90 mV). Excitation of the muscle fiber occurs in the following steps, here numbered to match figure 7.7.

① A nerve signal arrives at the synapse and stimulates synaptic vesicles to release acetylcholine (ACh) into the synaptic cleft.

② ACh diffuses across the cleft and binds to ACh receptors in the sarcolemma. Each receptor is a gated channel that opens in response to ACh, letting sodium ions (Na^+) quickly diffuse into the muscle fiber and potassium ions (K^+) diffuse out.

③ Na^+ and K^+ diffusion electrically excites the sarcolemma and initiates a wave of voltage changes called *action potentials* (discussed in more depth in chapter 8). Action potentials spread in all directions away from the neuromuscular junction—like ripples spreading out in a pond when you drop a stone into it—and pass down into the T tubules. The muscle fiber is now excited.

Excitation–Contraction Coupling

The next two steps are links between the surface excitation of the muscle fiber and the onset of contraction, so they are called **excitation–contraction coupling.**

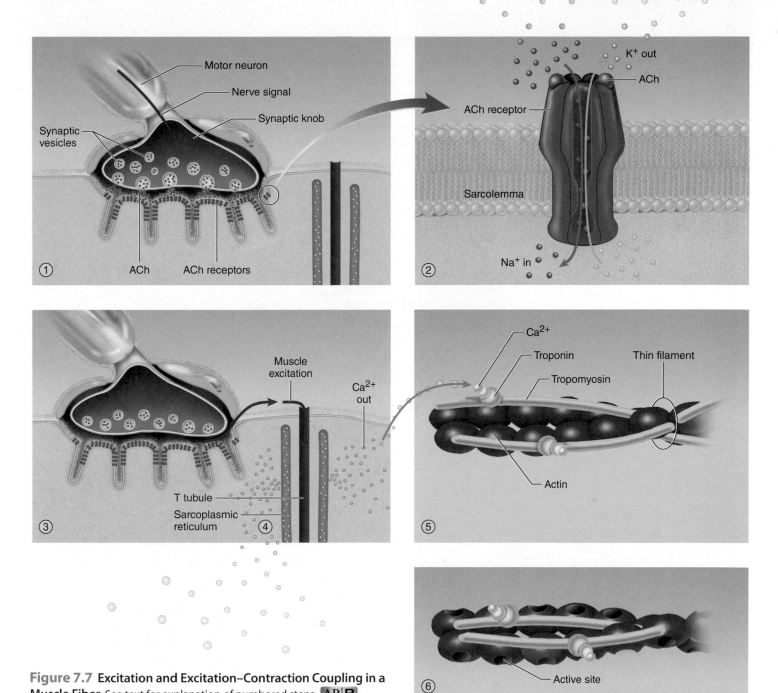

Figure 7.7 Excitation and Excitation–Contraction Coupling in a Muscle Fiber. See text for explanation of numbered steps. **AP|R**

④ Excitation of a T tubule membrane opens calcium (Ca²⁺) channels in the adjacent terminal cisternae of the sarcoplasmic reticulum. Calcium escapes the sarcoplasmic reticulum and floods the cytosol of the muscle fiber.

⑤ Ca²⁺ binds to troponin molecules (calcium receptors) on the thin filaments of the sarcomeres.

⑥ Calcium–troponin binding induces the associated tropomyosin molecule to change shape and move out of the way of active sites on the actin of the thin filaments. With these active sites now exposed to myosin, muscle contraction can begin.

Contraction

Contraction is the step in which the muscle fiber develops tension and may shorten. (Muscles often "contract," or develop tension, without shortening, as we see later.) It was once thought that a muscle cell contracted because its thick or thin filaments became shorter. Now it is understood that these filaments do not shorten; they slide across each other (fig. 7.8). The mechanism of contraction is therefore called the *sliding filament theory*.

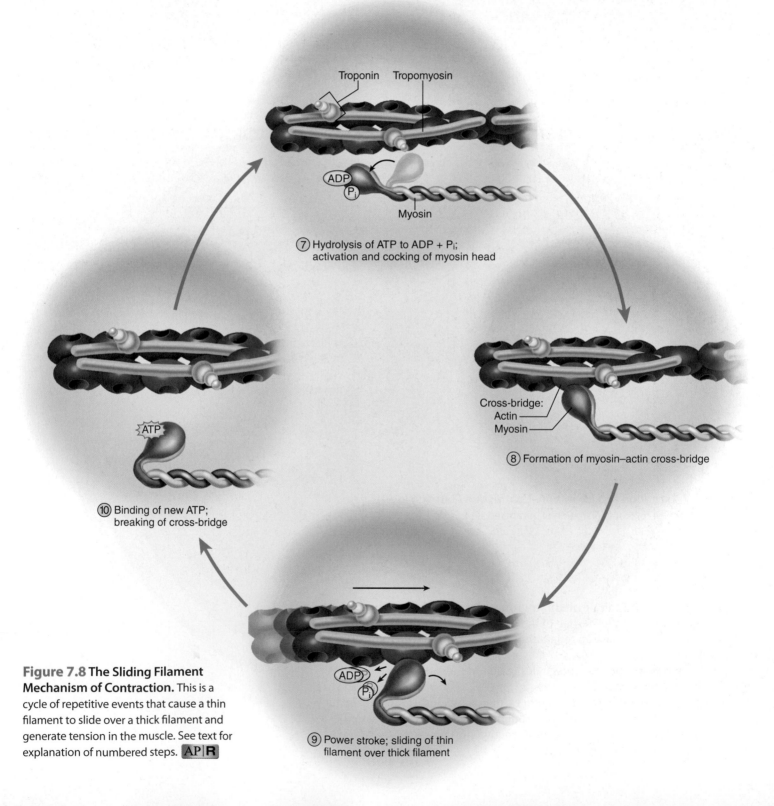

Troponin   Tropomyosin

ADP
Pi

Myosin

⑦ Hydrolysis of ATP to ADP + Pi;
activation and cocking of myosin head

Cross-bridge:
Actin
Myosin

⑧ Formation of myosin–actin cross-bridge

ATP

⑩ Binding of new ATP;
breaking of cross-bridge

ADP
Pi

⑨ Power stroke; sliding of thin
filament over thick filament

Figure 7.8 The Sliding Filament Mechanism of Contraction. This is a cycle of repetitive events that cause a thin filament to slide over a thick filament and generate tension in the muscle. See text for explanation of numbered steps. **AP|R**

⑦ Each myosin head binds an ATP molecule and splits it into ADP and a phosphate (P_i) group. This causes the head to cock from a flexed position (like a bent elbow) to an extended, high-energy position.

⑧ The cocked myosin head binds to an active site on the thin filament, like straightening your elbow to reach out and grasp a rope in your hand. The link between a myosin head and actin filament is called a **cross-bridge**.

⑨ Myosin then releases the ADP and P_i and flexes into its original bent, low-energy position, tugging the thin filament along with it (like bending your elbow and tugging on the rope). This is called the **power stroke**. The head remains bound to actin until it binds to a new ATP.

⑩ Upon binding more ATP, the myosin head releases the actin. (Many other myosin heads retain their hold on the actin while some release it, so the actin does not slip back to the starting point.) The head is now prepared to repeat the whole process—it will split the ATP, recock (the **recovery stroke**), attach to a new site farther down the thin filament, and produce another power stroke.

This cycle repeats at a rate of about five strokes per second, with each stroke consuming one ATP molecule. Hundreds of myosin heads "crawl" along each thin filament at once, each head taking tiny jerky "steps" but with the net effect being a smooth motion, much like the smooth crawling of a millipede resulting from the tiny steps taken by hundreds of legs in succession. As the thin filament is pulled along the thick filament, it pulls the Z disc along behind it. The cumulative effect is that all the sarcomeres shorten. Through their linkage to the sarcolemma (via dystrophin), they shorten the entire cell.

Figure 7.9 compares the appearance of sarcomeres in relaxed and contracted states. Note that in the contracted state, the sarcomere is shorter, but the thick and thin filaments are the same length as they are at rest. The sarcomere is shorter not because the filaments are shorter, but because the filaments overlap more extensively. (The overstretched state in the figure is discussed later.)

Relaxation

When its work is done and the nerve stops stimulating it, a muscle fiber relaxes and returns to its resting length. This is achieved by the following steps (not illustrated):

⑪ When nerve signals cease, the synaptic knob stops releasing ACh.

⑫ Acetylcholinesterase breaks down the ACh that is already in the synapse, so stimulation of the muscle fiber ceases.

⑬ The sarcoplasmic reticulum reabsorbs Ca^{2+}, taking it back into storage. This is done by active-transport pumps, which require ATP; therefore, ATP is required not only for muscle contraction but also for its relaxation.

⑭ Deprived of Ca^{2+}, the troponin–tropomyosin complex shifts back into the resting position shown in figure 7.7, where it blocks myosin from binding to actin. The muscle fiber thus ceases to produce or maintain tension. The muscle relaxes and may return to its resting length.

Apply What You Know

Chapter 2 noted that one of the most important properties of proteins is their ability to change shape (conformation) when other chemicals bind to them. Explain the relevance of this property to at least two proteins involved in muscle contraction.

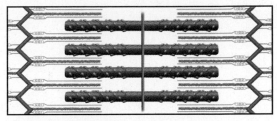

(a) Relaxed sarcomere

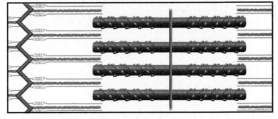

(b) Overstretched relaxed sarcomere

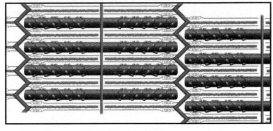

(c) Fully contracted sarcomere

Figure 7.9 Relaxed and Contracted Sarcomeres. (a) A relaxed sarcomere, with partially overlapping thick and thin filaments and wide I bands. This is the state typical of resting muscle tone. (b) Overstretched resting sarcomere. Here there is very little thick–thin filament overlap and the muscle would respond very weakly if stimulated. Resting muscle tone serves to avoid this state. (c) Fully contracted sarcomeres. Thick filaments abut Z discs and the sarcomeres would not be capable of contracting any further even if stimulated.

- *If you looked at an electron micrograph of muscle fibers in the state depicted in part (c), which band of the striations would be absent? Explain.*

Clinical Application 7.2

RIGOR MORTIS

Rigor mortis[4] is the hardening of the muscles and stiffening of the body that begins 3 to 4 hours after death. It occurs partly because the sarcoplasmic reticulum deteriorates upon death, releasing calcium into the cytosol. Also, calcium leaks into the cell from the extracellular fluid through the deteriorating sarcolemma. The calcium ions activate myosin–actin cross-bridging. Once bound to actin, myosin cannot release it without first binding an ATP molecule—and of course ATP is unavailable in a dead body. Thus, the thick and thin filaments remain rigidly cross-linked until the myofilaments begin to decay. Rigor mortis peaks about 12 hours after death and then diminishes over the next 48 to 60 hours.

Whole-Muscle Contraction

Now you know how an individual muscle cell shortens. Our next objective is to consider how this relates to the action of the muscle as a whole.

Muscle Twitch and Tetanus

The minimum contraction exhibited by a muscle is called a **muscle twitch**, a single cycle of contraction and relaxation (fig. 7.10a). A twitch is very brief, lasting as little as 7 milliseconds (ms) in the fastest muscles and no more than 100 ms (0.1 s) even in the slowest ones. It is too brief and too weak to do useful muscular work such as to move a joint. Useful work depends on **summation**, or the addition of multiple twitches, which occurs when multiple nervous stimuli arrive in rapid succession.

When stimuli arrive close together—say, 10 or 20 per second—each twitch is stronger than the one before (fig. 7.10b); this phenomenon is called *treppe*[5] (TREP-eh). One possible reason for this is that the sarcoplasmic reticulum does not have time to reabsorb all the calcium it released on the previous twitch, so calcium levels build to higher and higher levels with each new stimulus, enabling more and more cross-bridges to form. At a still higher stimulus frequency, such as 20 to 40/s, each new stimulus arrives before the previous twitch is over, so the muscle relaxes only partially between stimuli. Each new twitch "rides piggyback" on the previous one and generates higher tension (fig. 7.10c). This phenomenon is called **wave summation** because it results from one wave of contraction added to another. It produces a state of sustained fluttering contraction called **incomplete tetanus**. The reason for the smoothness of overall muscle contraction despite the "flutter" in an individual motor unit is that motor units work in shifts; when one motor unit relaxes, another contracts and takes over so that the muscle does not lose tension. (*Complete tetanus,* a state of constant spasmodic tension, is achieved only by artificial stimulation with electrodes and does not occur naturally in the body.)

We can see, then, that variation in muscle tension can result from differences in the frequency of stimulation by a nerve fiber. Another factor in muscle tension is the number of motor units activated at once. If a gentle contraction is needed, as in lifting a teacup, the brain and spinal cord activate relatively few motor units in the muscle. For a stronger contraction, as in lifting a barbell, the nervous system activates a greater number of motor units, and activates larger ones (with more muscle fibers per motor neuron).

---

[4] *rigor* = rigidity; *mortis* = of death

[5] *treppe* = staircase

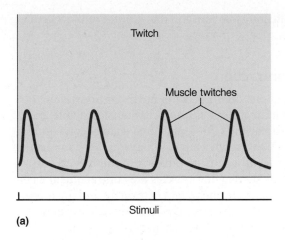

(a)

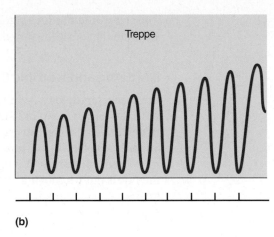

(b)

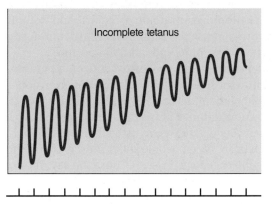

(c)

Figure 7.10 Stimulus Frequency and Muscle Tension. (a) An individual muscle twitch produced by a single stimulus. (b) At a moderate frequency of repetitive stimulation, the muscle relaxes fully between contractions, but successive twitches are stronger. (c) Wave summation and incomplete tetanus. At a somewhat higher stimulus frequency, the muscle does not have time to relax completely between twitches and the force of each twitch builds on the previous one.

Muscle Tone

Muscles normally do not relax to their maximum lengths. A low level of stimulation from the nervous system keeps even relaxed muscles in a state of partial contraction called **muscle tone (tonus)**. By keeping the muscles firm even at rest, muscle tone helps to stabilize the joints. It also ensures that the thick and thin filaments always overlap enough so that when the muscle is called into action, a sufficient number of myosin heads can bind to actin and initiate contraction (fig. 7.9a). If a muscle is excessively stretched, there is little myosin–actin overlap. Then when the muscle is stimulated, very few myosin heads can find anything to attach to, and the contraction is weak; little tension is generated (fig. 7.9b).

Contraction is also weak if the muscle is overly shortened prior to stimulation (fig. 7.9c). In this case, myosin cannot slide very far along the actin, if at all, before it runs up against the Z discs and can go no farther; the stimulated muscle is capable of little or no additional contraction. Therefore, to produce the strongest possible contraction when needed, it is important that a resting muscle be stretched to an optimal length, but not too far. The relationship between resting length and the ability to generate tension is called the **length–tension relationship**.

We take advantage of this principle in recreational applications such as bicycle design and a runner's crouch. A bicycle designed to make the rider lean forward over the handle bars does more than cut wind resistance. It stretches the gluteal muscles to their optimal length to produce a more powerful downstroke of the pedal. Similarly, a sprinter crouching in readiness for the starter's gun flexes the waist at an angle that produces optimal stretch in the gluteal muscles and a more powerful push-off at the start of the race. The gluteal muscles would produce significantly less power if the sprinter or bicycle rider was standing or sitting upright.

Isometric and Isotonic Contraction

In muscle physiology, "contraction" doesn't always mean the shortening of a muscle—it may mean only that the muscle is producing tension, while an external resistance may cause it to stay the same length or even to become longer. Suppose you lift a heavy box of books from a table. When you first contract the muscles of your arms, you can feel the tension building in them even though the box is not yet moving. At this point, your muscles are contracting at a cellular level, but their tension is resisted by the weight of the load and the muscle produces no external movement. This phase is called **isometric**[6] **contraction**—contraction without a change in length. Once muscle tension rises enough to move the box of books, the muscle begins to shorten while maintaining constant tension. This phase is called **isotonic**[7] **contraction**. Isometric and isotonic contraction are both phases of normal muscular action. Isometric contraction is also important in muscles that stabilize joints and maintain posture.

Isotonic contraction is further divided into two forms: concentric and eccentric. In **concentric contraction**, a muscle shortens as it maintains tension—for example, when the biceps brachii contracts and flexes the elbow such as during biceps curls. In **eccentric contraction**, a muscle maintains tension as it lengthens, such as the biceps when you lower a weight to the floor but maintain enough control not to simply let it drop. When people suffer muscle injuries in weight lifting, it is usually during the eccentric phase, because the sarcomeres and connective tissues of the muscle are pulling in one direction while the weight is pulling the muscle in the opposite direction.

Muscle Metabolism

All muscle contraction depends on ATP; this is the only energy source that myosin can use, so no other molecule can serve in its place. The question of providing energy to muscles during exercise therefore comes down to alternative ways of providing ATP.

There are two mechanisms for generating ATP: *anaerobic fermentation* and *aerobic respiration*. Anaerobic fermentation is a pathway in which glucose is ultimately converted to lactic acid. For each glucose consumed, this pathway produces a net yield of 2 ATP. Anaerobic fermentation has the advantage that it does not use oxygen. Therefore it is a way for a muscle to produce ATP during times when demand is high (in short bursts of exercise) but oxygen cannot be delivered fast enough to meet the needs of aerobic respiration. Its disadvantage is that the ATP yield is low. The lactic acid generated by anaerobic fermentation was long thought to be a significant factor in muscle fatigue, but recent research has cast doubt on this, showing fatigue to be independent of lactic acid level in the muscle and finding other, more complicated explanations for it.

In the alternative pathway, aerobic respiration, an intermediate product of glucose metabolism called *pyruvic acid* is oxidized in the mitochondria to carbon dioxide and water. In skeletal muscle, this method yields another 30 ATP molecules for each original glucose; thus it is far more efficient than anaerobic fermentation. Its end products are carbon dioxide and water. Its disadvantage is that it requires oxygen, so aerobic respiration cannot function adequately

---

[6] *iso* = same, uniform; *metr* = length

[7] *iso* = same, uniform; *ton* = tension

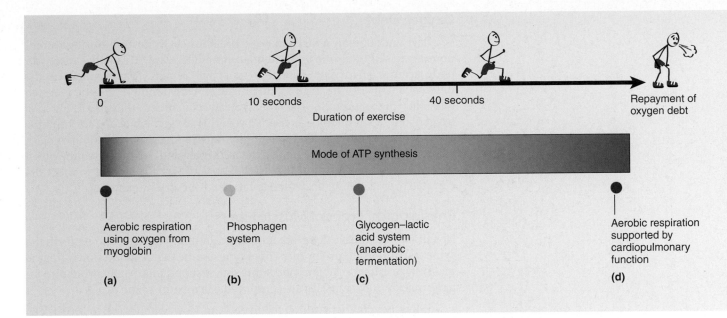

during periods when exercise is so intense that the respiratory and cardiovascular system cannot keep pace with the needs of the muscular system.

In brief, intense exercise such as a 100-meter dash or a basketball play, muscle uses the oxygen stored in its myoglobin to make ATP aerobically (fig. 7.11a). As this ATP is consumed, it breaks down to ADP. The muscle fiber converts some of this ADP back to ATP by borrowing a phosphate group, either from a phosphate-storage molecule called *creatine phosphate,* or from other ADP molecules. This mechanism, called the *phosphagen system,* can supply enough ATP for about 6 seconds of sprinting or fast swimming, or about 1 minute of brisk walking (fig. 7.11b).

When these sources of ATP are exhausted, the muscle gradually shifts to anaerobic fermentation (the *glycogen–lactic acid system*), which provides enough ATP for about 30 to 40 seconds of maximum activity, as in running completely around a baseball diamond (fig. 7.11c). In time, a faster heartbeat and faster, deeper respiration catch up with the demands of the muscle, providing enough oxygen for it to shift back to aerobic respiration. Aerobic respiration usually takes over after about 40 seconds (fig. 7.11d). In exercises lasting more than 10 minutes, such as bicycling or marathon running, more than 90% of the ATP is produced aerobically.

Figure 7.11 Modes of ATP Production During Exercise.

Fatigue and Endurance

Muscle fatigue is the progressive weakness that results from prolonged use of the muscles. This is what sets a limit, for example, on how many chin-ups or bench presses a person can perform. Writer's cramp (not so much a cramp as tiring of the muscles) is another example. There are numerous causes of fatigue, including depletion of ATP and acetylcholine, increasingly uncontrolled leakage of calcium from the sarcoplasmic reticulum, and accumulation of potassium in the extracellular fluid, which reduces the excitability of the muscle fiber.

Endurance, or tolerance of prolonged exercise, depends on several factors: a muscle's supply of myoglobin and glycogen; its density of blood capillaries; the supply of organic nutrients such as fatty acids and amino acids; the number of mitochondria; one's maximal rate of oxygen uptake (which depends on physical conditioning); and simple will power and other psychological factors.

Oxygen Debt

You have undoubtedly noticed that you breathe heavily not only during a strenuous exercise but also for an extended time afterward. This is because your body accrues an oxygen debt that must be "repaid." **Oxygen debt** is the difference between the resting rate of oxygen consumption and the elevated rate following an exercise. The body typically consumes an extra 11 L or so of oxygen after a strenuous exercise is over. This oxygen is needed for multiple purposes: to rebuild oxygen levels in the muscle myoglobin, blood hemoglobin, and blood plasma; to replenish the resting levels of ATP and creatine phosphate in the muscles; to oxidize and dispose of lactic acid; and to meet the elevated overall metabolic demand of the body heated by exercise.

Physiological Types of Muscle Fibers

Not all skeletal muscle fibers are metabolically alike or adapted to perform the same task. There are two primary fiber types: some that respond slowly but are relatively resistant to fatigue, and some that respond more quickly but also fatigue quickly. Most muscles are composed of a mixture of the two.

- **Slow-twitch fibers** exhibit relatively long, slow twitches (up to 100 ms), so they are not well adapted for quick responses. On the other hand, they are well adapted for aerobic respiration and do not fatigue easily. Their aerobic adaptations include abundant mitochondria, myoglobin, and blood capillaries. Myoglobin and blood capillaries give the tissue a deep red color, so muscles made predominantly of slow-twitch fibers are sometimes called *red muscles*. Examples include postural muscles of the back and a deep calf muscle called the *soleus*.

- **Fast-twitch fibers** are well adapted for quick responses (twitches as fast as 7.5 ms), but not for fatigue resistance. They are especially important in sports such as basketball that require stop-and-go activity and frequent changes of pace. They are rich in enzymes for anaerobic fermentation and for regenerating ATP from creatine phosphate. Their sarcoplasmic reticulum releases and reabsorbs Ca^{2+} quickly, which partially accounts for their quick, forceful contractions. They have more glycogen but less myoglobin and fewer mitochondria and blood capillaries than slow-twitch fibers, so they are relatively pale. Muscles composed predominantly of fast-twitch fibers are sometimes called *white muscles*. Examples include the *biceps brachii* of the arm, muscles of eye movement, and a superficial calf muscle called the *gastrocnemius*.

People with different types and levels of physical activity differ in the proportion of one fiber type to another even in the same muscle. For example, one study of male athletes found the quadriceps muscle of the thigh in marathon runners to be 82% slow-twitch and 18% fast-twitch fibers, but quite the opposite in sprinters and jumpers: 37% slow- and 63% fast-twitch. The ratio of fiber types seems to be genetically determined; one person might be a "born sprinter" and another a "born marathoner." One who is born with a genetic predisposition for predominantly slow-twitch fibers in the powerful muscles of the thighs might become a world-class marathoner, but be unable to cross over to a competitive level of sprinting, and vice versa.

Sometimes two or more muscles act across the same joint and superficially seem to have the same function. For example, the gastrocnemius and soleus muscles both insert on the heel through the same *calcaneal (Achilles) tendon*, so they have the same action—to lift the heel and raise the body, as in walking, running, and jumping. But they are not redundant. The gastrocnemius is a white muscle adapted for quick, powerful movements such as jumping, whereas the soleus is a red muscle that does most of the work in endurance exercises such as running, skiing, and climbing.

Muscular Strength and Conditioning

Muscular strength depends on a variety of anatomical and physiological factors. We have just seen that sustained or repetitive contraction can cause muscle fatigue, a gradual loss of strength; muscle strength therefore depends in part on fatigue resistance. Strength is also proportional to the diameter of a muscle at its thickest point (such as the bulge at the middle of the biceps brachii), which is why body building increases strength. There are two types of exercise that improve muscle performance.

Resistance exercise, such as weight lifting, is the contraction of muscles against a load that resists movement. A few minutes of resistance exercise at a time, a few times each week, is enough to stimulate muscle growth. Muscle fibers are incapable of mitosis, so this growth does not result from an increase in the number of muscle cells; rather, it results from an increase in the size of preexisting cells.

Endurance (aerobic) exercise, such as running and swimming, improves fatigue resistance. In slow-twitch fibers, especially, it results in increased stores of glycogen, greater numbers of mitochondria, and a greater density of blood capillaries—all of which promote faster production of ATP and therefore greater fatigue resistance. Endurance exercise also increases the oxygen-transport capacity of the blood and enhances the efficiency of the cardiovascular, respiratory, and nervous systems.

Resistance exercise does not significantly improve fatigue resistance, and endurance exercise does not significantly increase muscular strength. Optimal performance comes from *cross-training*, which incorporates elements of both types.

Apply What You Know

Review hyperplasia and hypertrophy in chapter 4. To which of these processes would you attribute the growth of a weight lifter's muscles? Explain.

Before You Go On

Answer these questions from memory. Reread the preceding section if there are too many you don't know.

4. What is the role of calcium in muscle contraction? Where does it come from and what triggers its release from that source?

5. What process makes the thin myofilaments "slide" over the thick ones during muscle contraction? What is the role of ATP in that process?

6. Explain why high-frequency stimulation makes a muscle contract with more tension than a single twitch produces.

7. Describe what roles isometric and isotonic contraction play when you lift a heavy object.

8. What is the difference between anaerobic fermentation and aerobic respiration? Identify a physical activity that would depend mainly on each type of ATP production.

9. Describe some tasks for which slow-twitch fibers are more effective than fast-twitch fibers, and vice versa.

10. Identify the benefits and shortcomings of both resistance and endurance exercise.

7.3 Cardiac and Smooth Muscle

Expected Learning Outcomes

When you have completed this section, you should be able to:

a. describe the special functional roles of cardiac and smooth muscle; and

b. explain how the structure of these two forms of muscle accounts for their functional properties.

Until now we have been concerned with skeletal muscle. The other two types of muscular tissue are cardiac and smooth.

Cardiac Muscle

Cardiac muscle (fig. 7.12) is limited to the heart, where its function is to pump blood. Knowing that, we can predict the properties that it must have: (1) It must contract with a regular rhythm; (2) the muscle cells of a given heart chamber must be well synchronized so the chamber can effectively expel blood; (3) each contraction must last long enough to expel blood from the chamber; (4) it must function in sleep and wakefulness, without fail and without need of conscious

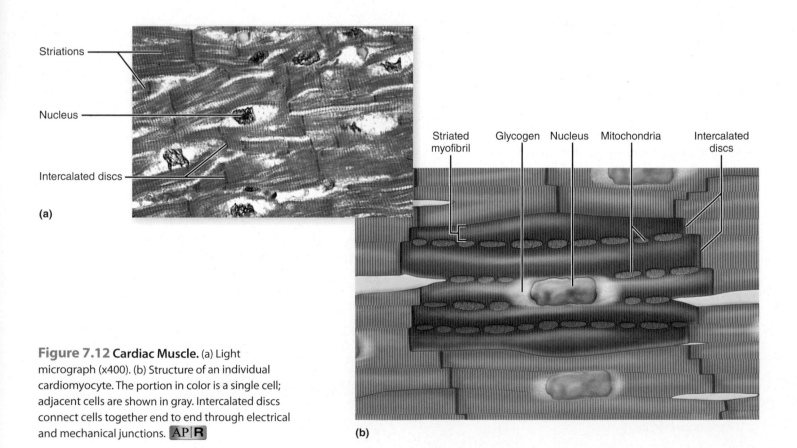

Striations

Nucleus

Intercalated discs

(a)

Striated myofibril   Glycogen   Nucleus   Mitochondria   Intercalated discs

Figure 7.12 Cardiac Muscle. (a) Light micrograph (x400). (b) Structure of an individual cardiomyocyte. The portion in color is a single cell; adjacent cells are shown in gray. Intercalated discs connect cells together end to end through electrical and mechanical junctions. AP|R

(b)

attention; and (5) it must be highly resistant to fatigue. These functional properties are keys to understanding how cardiac muscle differs structurally and physiologically from skeletal muscle.

Cardiac muscle is striated like skeletal muscle, but it is **involuntary** (it contracts automatically, without requiring one's conscious control and attention) and the heart is **autorhythmic**[8] (it has a self-maintained rhythm of contraction). The rhythm is set by a pacemaker in the wall of the upper heart. Although the heart receives nerves from the brainstem and spinal cord, they only modify the rate and force of the heartbeat; they do not initiate each beat.

The cells of cardiac muscle, called **cardiomyocytes**, are not long fibers like those of skeletal muscle, but are shorter and appear roughly rectangular in tissue sections. They are slightly forked or notched at the ends, and each fork links one cardiomyocyte to the next. The linkages, called **intercalated** (in-TUR-kuh-LAY-ted) **discs**, appear in stained tissue sections as dark lines thicker than the striations. An intercalated disc has electrical *gap junctions* that allow each myocyte to directly stimulate its neighbors, and has *desmosomes* and other mechanical junctions that keep the myocytes from pulling apart when the heart contracts.

Cardiac muscle does not exhibit quick twitches like skeletal muscle; rather, its contractions are prolonged for as much as a quarter of a second (250 ms). This means that each heart chamber sustains its contraction long enough to effectively expel blood.

Cardiac muscle uses aerobic respiration almost exclusively. It is very rich in myoglobin and glycogen, and it has especially large mitochondria that fill about 25% of the cell, compared to smaller mitochondria occupying about 2% of a skeletal muscle fiber. Because it makes little use of anaerobic fermentation, cardiac muscle is very resistant to fatigue. The skeletal muscles in your hand would quickly tire if you squeezed every 0.8 s, but the heart contracts at this rate, and often faster, for a lifetime without fatigue.

Cardiac muscle is explored to greater depth in chapter 13.

Smooth Muscle

Smooth muscle (fig. 7.13) is involuntary, like cardiac, but it lacks striations; the latter quality is the reason it is called smooth. It occurs in the walls of the blood vessels and many body-cavity organs (viscera)—such as the, respiratory,

---

[8] *auto* = self

Figure 7.13 Smooth Muscle. (a) Light micrograph (x400). (b) Layered arrangement of smooth muscle in a cross section of the esophagus. Many hollow organs have alternating layers of smooth muscle like this. **AP|R**

- *For what purpose does the esophagus require such a muscular wall?*

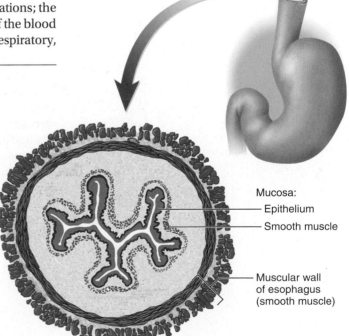

Mucosa:
- Epithelium
- Smooth muscle

Muscular wall of esophagus (smooth muscle)

Nucleus   Myocyte

(a)   (b)

urinary, digestive, and reproductive organs—where it is also called **visceral muscle**. A slightly different form of smooth muscle occurs in the iris of the eye and is associated with the hair follicles of the skin.

Smooth muscle cells, or **myocytes**, are much shorter than skeletal muscle fibers; have only one nucleus located near the middle of the cell; and are fusiform in shape—thick in the middle and tapered at the ends.

Some functions of smooth muscle, and its special properties, can be inferred by reflecting on organs such as the stomach, urinary bladder, and uterus, whose walls are composed primarily of this tissue. Such organs must be able to relax and stretch a great deal as they fill with food, urine, or a growing fetus. Recall that when skeletal muscle is overly stretched, its contractions become weak, owing to the length–tension relationship. But smooth muscle must be able to contract forcefully even when it is greatly stretched; it is not subject to the length–tension relationship. Furthermore, when such organs are empty, they do not collapse into flaccid bags. Their smooth muscle maintains a state of contraction called **smooth muscle tone**, which keeps the walls of the organs firm.

Special structural characteristics of smooth muscle cells underlie their distinctive functional properties. Thick and thin myofilaments are present, but they do not overlap each other with the regularity that produces the striations of skeletal and cardiac muscle. Sarcomeres and Z discs are absent. The thin filaments slide over thick filaments during contraction, but instead of pulling on Z discs and shortening a sarcomere, they pull on plaques of protein called *dense bodies* scattered throughout the cytoplasm and on the inner surface of the sarcolemma. In doing so, they cause the myocyte to contract with a twisting motion, like wringing out a wet towel.

Skeletal muscle must be within ±30% of its optimum length in order to contract strongly when stimulated. Smooth muscle, by contrast, can be anywhere from half to twice its resting length and still contract powerfully. One reason for this is that there are no Z discs to halt the sliding of thin filaments over the thick filaments. Another is that the irregular arrangement of the thick and thin filaments ensures that even in stretched muscle, there is enough filament overlap that myosin can bind to actin and initiate contraction.

Smooth muscle contracts in response to nervous stimulation as well as such stimuli as stretch (as in a full bladder), hormones (for example, during labor contractions), and blood levels of CO_2, O_2, and pH. As in the other muscle types, contraction is triggered by calcium ions (Ca^{2+}), energized by ATP, and achieved by the sliding of thin filaments over the thick filaments. There are no T tubules and there is very little sarcoplasmic reticulum. Smooth muscle obtains its calcium from the extracellular fluid by way of calcium channels in the sarcolemma. During relaxation, calcium is pumped back out of the cell by active transport.

Smooth muscle contracts slowly, with tension peaking about 0.5 second after stimulation and then declining over a period of 1 to 2 seconds. Its myosin does not always detach from actin immediately; it can remain attached for prolonged periods without consuming more ATP. Smooth muscle therefore maintains tension while consuming only 1/10 to 1/300 as much ATP as skeletal muscle does; this makes it very resistant to fatigue.

Unlike skeletal and cardiac muscle, smooth muscle is fully capable of mitosis. Thus, an organ such as the pregnant uterus can grow by adding more myocytes, and injured smooth muscle regenerates much better than the other two muscle types.

Before You Go On

Answer these questions from memory. Reread the preceding section if there are too many you don't know.

11. Explain why cardiac muscle resists fatigue better than skeletal muscle does. Do the same for smooth muscle.

12. What are the functions of intercalated discs?

13. Two things necessary to the action of skeletal muscle are T tubules and Z discs. Smooth muscle lacks both of these, yet it still contracts. Explain how smooth muscle compensates for the absence of these components.

7.4 Anatomy of the Muscular System

Expected Learning Outcomes

When you have completed this section, you should be able to:

a. describe the relationship of muscle fibers to connective tissues in a muscle;

b. define the *origin, insertion,* and *belly* of a muscle;

c. explain how muscles act in groups to govern the movements of a joint;

d. interpret some Latin terms commonly used in the names of muscles; and

e. identify or describe the locations and functions of several major skeletal muscles of the body.

General Aspects of Muscle Anatomy

The rest of this chapter is concerned with skeletal muscles. Before we examine the anatomy of individual muscles, we will survey general features of gross anatomy that muscles have in common, the functional relationships of different muscles to each other, and principles of naming skeletal muscles.

Figure 7.14 Muscle–Connective Tissue Relationships. (a) The muscle–bone attachment. (b) Muscle fascicles in the tongue. Vertical fascicles passing between the superior and inferior surfaces of the tongue are seen alternating with cross-sectioned horizontal fascicles that pass from the tip to the rear of the tongue. A fibrous perimysium can be seen between the fascicles, and endomysium can be seen between the muscle fibers within each fascicle (c.s. = cross section; l.s. = longitudinal section). (c) Muscle compartments in the leg. Fibrous fasciae separate muscle groups into tight-fitting compartments along with their blood vessels and nerves.

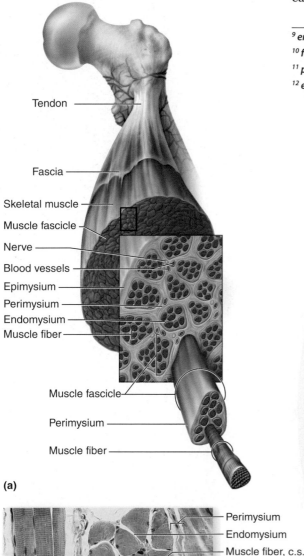

(a)

(b)

Connective Tissues and Muscle Attachments

A skeletal muscle is composed of both muscular tissue and fibrous connective tissue (fig. 7.14). Each muscle fiber is enclosed in a thin sleeve of loose connective tissue called the **endomysium**[9] (EN-doe-MIZ-ee-um). This tissue slightly separates muscle fibers and allows room for blood capillaries and nerve fibers to reach every muscle fiber. As an insulator, it also prevents electrical activity in one muscle fiber from directly stimulating adjacent fibers.

Muscle fibers are arranged in bundles called **fascicles**[10] (FASS-ih-culs) (the grain visible in a cut of meat such as roast beef). Each fascicle is surrounded and set off from neighboring ones by a connective tissue sheath called the **perimysium**,[11] thicker than the endomysium. The muscle as a whole is surrounded by still another connective tissue layer, the **epimysium**.[12] Adjacent muscles are separated from each other and from the skin by fibrous sheets called **fasciae** (FASH-ee-ee) (singular, *fascia*).

---

[9] *endo* = within; *mys* = muscle
[10] *fasc* = bundle; *icle* = little
[11] *peri* = around; *mys* = muscle
[12] *epi* = upon, above; *mys* = muscle

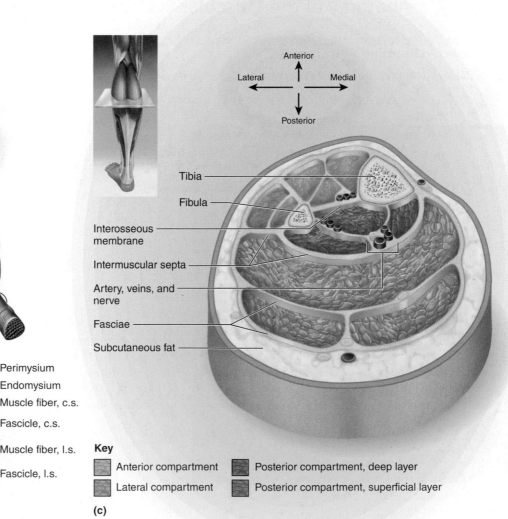

Key

Anterior compartment	Posterior compartment, deep layer
Lateral compartment	Posterior compartment, superficial layer

(c)

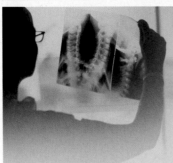

PERSPECTIVES ON HEALTH

The muscular system suffers fewer disorders than any other organ system, but muscle soreness, cramps, and injuries are common, and some people suffer muscular dystrophy and other serious muscle diseases.

Muscle Pain

The most common form of muscle pain is *delayed onset muscle soreness (DOMS)*—the pain and stiffness commonly felt from hours to a day after strenuous muscular work. DOMS is associated with microtrauma to the muscles, including disrupted plasma membranes and sarcomere structure. *Cramps* are painful muscle spasms triggered by heavy exercise, extreme cold, dehydration, electrolyte loss, low blood glucose, or inadequate blood flow to the muscles.

Muscle Injuries

Every year, thousands of athletes from the high school to professional levels sustain some type of injury to their muscles, as do the increasing numbers of people who have taken up running, weight lifting, and other forms of physical conditioning. Long, hard running and repetitive kicking (as in soccer and football), for example, sometimes result in *pulled hamstrings*—strained hamstring muscles or tears in the hamstring tendons, often with a hematoma (blood clot) in the fascia around the muscles. *Shinsplints* are various injuries producing pain in the crural region, such as inflammation of the tibialis posterior tendon (tendinitis) or periosteum of the tibia. Shinsplints commonly result from unaccustomed jogging or other vigorous activity of the legs after a period of relative inactivity.

Rotator cuff injuries are tears in the tendons of any of the rotator cuff muscles (see p. 238), most often the tendon of the supraspinatus.

Such injuries are usually caused by strenuous circumduction of the shoulder, shoulder dislocation, or persistent use of the arm in a position above horizontal. They are common among baseball pitchers, bowlers, swimmers, weight lifters, and those who play racquet sports. Recurrent inflammation of a rotator cuff tendon can cause the tendon to degenerate and then rupture in response to moderate stress. The injury is not only painful but also makes the shoulder joint unstable and subject to dislocation.

Compartment syndrome is a disorder in which overuse, muscle strain, or trauma damages the blood vessels in a muscle compartment. Since the fasciae cannot stretch, a blood clot *(contusion)* or tissue fluid in the compartment compresses the muscles, nerves, and vessels. The reduced blood flow can lead to nerve destruction within 2 to 4 hours and death of muscular tissue if the condition remains untreated for 6 hours or more. While nerves can regenerate if blood flow is restored, muscle death is irreversible. Depending on its severity, compartment syndrome may be treated with immobilization of the limb, rest, or an incision to drain fluid and relieve compartment pressure.

Muscular Dystrophy

Duchenne muscular dystrophy (DMD), unlike the foregoing conditions, is hereditary. It involves progressive replacement of muscular tissue with adipose and fibrous tissue. Most cases occur in males and are diagnosed between the ages of 2 and 10 years. When a child with DMD begins to walk, he may fall frequently and have difficulty standing up again. DMD tends to affect the hips first, then the lower limbs, and then progresses to the abdominal and spinal muscles. Patients are usually confined to a wheelchair by early adolescence and rarely live past the age of 20. Genetic screening can determine whether a prospective parent is a DMD carrier, and can serve as a basis for determining the risk of having a child with DMD.

Some fasciae separate groups of functionally related muscles into **muscle compartments**. A muscle compartment also contains the nerves and blood vessels that supply the muscle group. The tight binding of muscles by these fasciae contributes to a clinical problem called *compartment syndrome*, a common athletic injury (see Perspectives on Health on this page).

Most skeletal muscles attach to the bones by way of fibrous cords or sheets called **tendons**. Collagen fibers of the tendon continue into the periosteum and bone matrix. You can easily palpate tendons at the heel and wrist to feel their texture. In some cases, the tendon is a broad sheet called an **aponeurosis**[13] (AP-oh-new-RO-sis). This term originally referred to the tendon located beneath the scalp, but now it also refers to similar tendons associated with certain abdominal, lumbar, hand, and foot muscles (see fig. 7.19a).

Most muscles attach to a different bone at each end, so either the muscle or its tendon spans at least one joint. When the muscle contracts, it moves one bone relative to the other. The attachment at the relatively stationary end is called the **origin**, or **head**, of the muscle; its attachment at the more mobile end is called the **insertion**. Many muscles, such as the biceps brachii, are narrow at the origin and insertion and have a thicker middle region called the **belly** (fig. 7.15).

Origin and *insertion* are admittedly flawed terms, because the end that moves may differ from one joint action to another. For example, the quadriceps femoris arises mainly on the femur and connects at its distal end to the tibia just below the knee. If you kick a soccer ball, this muscle moves the tibia more than the femur, so the femur would be considered its origin and the tibia its insertion. But when you stand up from a chair, the tibia remains stationary and the femur moves, so by the foregoing principle, we would now have to regard the tibia as the origin and the femur as the insertion. This chapter presents the conventionally accepted origins and insertions of the skeletal muscles, but some anatomists are beginning to abandon origin–insertion terminology and, in some cases, are speaking instead of proximal and distal attachments.

Coordinated Action of Muscle Groups

The movement produced by a muscle is called its **action**. Skeletal muscles seldom act alone; instead, they function in groups whose combined actions produce the coordinated motion of a joint (fig. 7.15). Muscles can be classified into four categories according to their actions, but it must be stressed that a particular muscle can act in a certain way during one joint action and in a different way during other actions of the same joint.

1. The **prime mover (agonist)** is the muscle that produces most of the force during a particular joint action. In flexing the elbow, for example,

---

[13] *apo* = upon, above; *neuro* = nerve

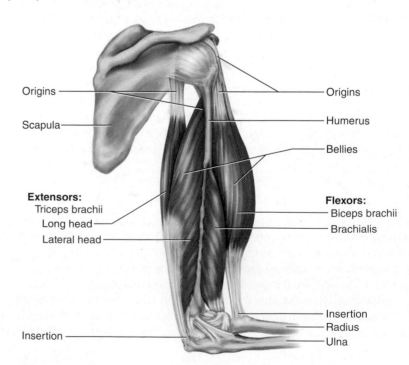

Figure 7.15 Synergistic and Antagonistic Muscle Pairs. The biceps brachii and brachialis muscles are synergists in elbow flexion. The triceps brachii is an antagonist of those two muscles and is the prime mover in elbow extension.

- *Suppose you threw a dart at a dart board. Which of these muscles would be the prime mover in that action, and which would act as a brake or antagonist?*

the prime mover is the *brachialis,* which lies on the anterior side of the humerus deep to the biceps.

2. A **synergist**[14] (SIN-ur-jist) is a muscle that aids the prime mover. The biceps brachii, for example, works with the brachialis as a synergist to flex the elbow. The actions of a prime mover and its synergist are not necessarily identical and redundant. A synergist may stabilize a joint and restrict unwanted movements, or modify the direction of a movement so that the action of the prime mover is more coordinated and specific.

3. An **antagonist**[15] is a muscle that opposes the prime mover. The brachialis and triceps brachii, for example, represent an **antagonistic pair** of muscles that act on opposite sides of the elbow. The brachialis flexes the elbow and the triceps extends it. Moreover, when either of these muscles contracts, the other one exerts a braking action that moderates the joint movement and makes it smoother and better coordinated. For example, if you extend your arm to reach out and pick up a cup of tea, your triceps acts as the prime mover of elbow extension and your brachialis and biceps act as antagonists to slow the extension and stop it at the appropriate point. Which member of the group acts as the agonist depends on the motion under consideration. In flexion of the elbow, the brachialis is the prime mover and the triceps is the antagonist; when the elbow is extended, their roles are reversed.

4. A **fixator** is a muscle that restricts a bone from moving. To *fix* a bone means to hold it steady, allowing another muscle attached to it to pull on something else. For example, if no fixators were involved, contraction of the biceps brachii would tend not only to flex the elbow but also to pull the scapula forward. There are fixators (the *rhomboids*), however, that attach the scapula to the vertebral column and hold it firmly in place so that the force of the biceps acts almost entirely on the elbow joint.

How Muscles Are Named

The remainder of this chapter is mostly a descriptive inventory of major skeletal muscles. Long muscle names such as *flexor carpi radialis brevis* may seem intimidating at first, but they are really more of a help than an obstacle to understanding if you gain a little familiarity with the words most commonly used in naming muscles. This name means "a short *(brevis)* flexor of the wrist *(flexor carpi)* that lies alongside the radius *(radialis)*." Muscles are typically named for such characteristics as their action, location, shape, size, orientation, and number of heads. Footnotes throughout the following pages will help give you insight into the muscle names.

With the growing popularity of weight training, many people know some of the skeletal muscles by slang terms—the "glutes," "abs," and so forth. Table 7.1 relates these common terms to the formal anatomical names of the muscles.

The muscular system consists of approximately 600 muscles (fig. 7.16). We begin our survey with those of the axial region—the head, neck, abdomen, and back. The general location and everyday actions of these muscles will be described first. Table 7.2 then summarizes the muscles with more specific information on origins and insertions and more formal terminology for the muscle actions. Throughout this series of illustrations, muscles discussed in the text are labeled in boldface; a few others not discussed are labeled for context and for reference in comparing to other books.

Table 7.1 Slang Terms for Major Skeletal Muscles

Slang Term	Anatomical Term
Abs	Abdominal obliques, rectus abdominis
Calf muscles	Gastrocnemius, soleus
Delts	Deltoid
Glutes	Gluteus maximus, medius, and minimus
Hamstrings	Biceps femoris, semimembranosus, semitendinosus
Lats	Latissimus dorsi
Pecs	Pectoralis major
Quads	Quadriceps femoris
Six-pack	Rectus abdominis

---

[14] *syn* = together; *erg* = work

[15] *ant* = against; *agonist* = competitor

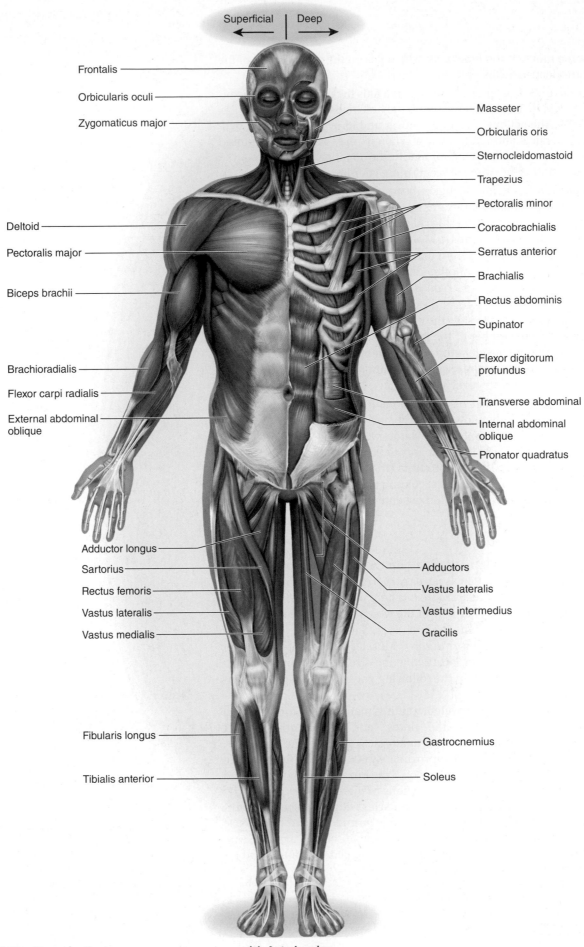

Superficial | Deep

Frontalis

Orbicularis oculi

Zygomaticus major

Masseter

Orbicularis oris

Sternocleidomastoid

Trapezius

Pectoralis minor

Deltoid

Pectoralis major

Coracobrachialis

Serratus anterior

Brachialis

Biceps brachii

Rectus abdominis

Supinator

Flexor digitorum
profundus

Brachioradialis

Flexor carpi radialis

External abdominal
oblique

Transverse abdominal

Internal abdominal
oblique

Pronator quadratus

Adductor longus

Sartorius

Rectus femoris

Vastus lateralis

Vastus medialis

Adductors

Vastus lateralis

Vastus intermedius

Gracilis

Fibularis longus

Gastrocnemius

Tibialis anterior

Soleus

Figure 7.16 The Muscular System.

(a) Anterior view

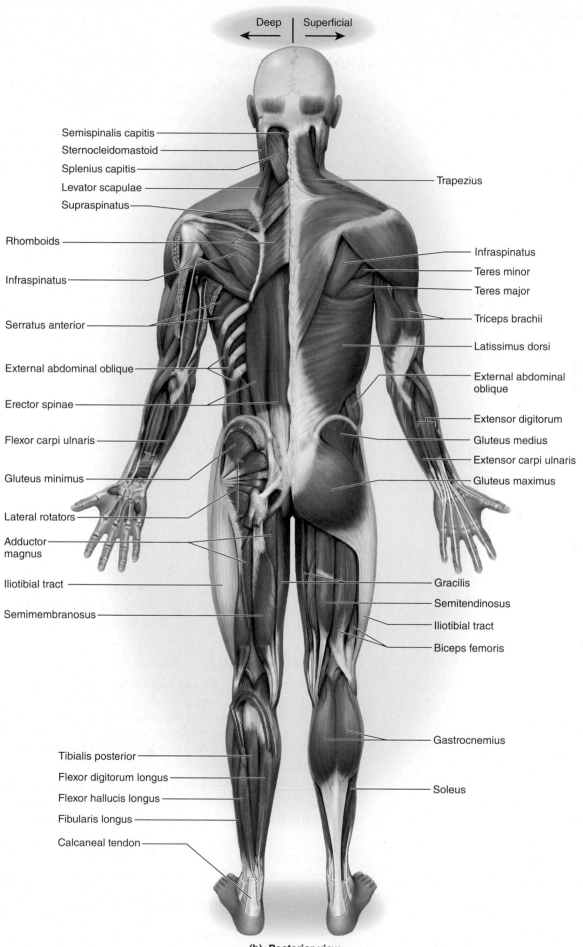

Deep | Superficial

Semispinalis capitis

Sternocleidomastoid

Splenius capitis

Levator scapulae

Supraspinatus

Rhomboids

Infraspinatus

Serratus anterior

External abdominal oblique

Erector spinae

Flexor carpi ulnaris

Gluteus minimus

Lateral rotators

Adductor magnus

Iliotibial tract

Semimembranosus

Tibialis posterior

Flexor digitorum longus

Flexor hallucis longus

Fibularis longus

Calcaneal tendon

Trapezius

Infraspinatus

Teres minor

Teres major

Triceps brachii

Latissimus dorsi

External abdominal oblique

Extensor digitorum

Gluteus medius

Extensor carpi ulnaris

Gluteus maximus

Gracilis

Semitendinosus

Iliotibial tract

Biceps femoris

Gastrocnemius

Soleus

(b) Posterior view

Muscles of the Face

One of the most striking contrasts between a human face and that of a rat, horse, or dog, for example, is the variety and subtlety of human facial expression. This is made possible by a complex array of small muscles that insert in the dermis and tense the skin when they contract. These muscles produce effects as diverse as a smile, a scowl, a frown, and a wink. They add subtle shades of meaning to our spoken words and are enormously important in nonverbal communication.

Frontalis (frun-TAY-lis) (fig. 7.17a). The frontalis is a muscle of the forehead that inserts in the skin beneath the eyebrows. It expressively elevates the eyebrows and wrinkles the skin of the forehead.

Orbicularis oculi[16] (or-BIC-you-LERR-is OC-you-lye) (fig. 7.17a). This is a circular muscle (sphincter) located in and beyond the eyelid. It serves to blink and close the eyes and promotes the flow of tears and their drainage into the nasal cavity.

Orbicularis oris[17] (or-BIC-you-LERR-is OR-is) (fig. 7.17a). This is an array of small interlacing muscles located in the lips, encircling the mouth. It closes the lips (but not the jaw) and, with harder contraction, protrudes them as in kissing. It is important in speech.

Zygomaticus[18] (ZY-go-MAT-ih-cus) (fig. 7.17a). Several muscles converge on the lips and make the mouth the most expressive part of the face. The *zygomaticus major* and *minor* originate on the zygomatic bone and insert on the orbicularis oris at the corner of the mouth. They draw the corner of the mouth laterally and upward as in smiling and laughing. Other muscles that elevate the upper lip include the *levator anguli oris* at the corners of the mouth and the *levator labii superioris* along the middle of the lip.

Depressor anguli oris[19] (de-PRESS-ur ANG-you-lye OR-is) (fig. 7.17a). This muscle arises from the mandible and inserts on the corner of the mouth. It draws the corner of the mouth laterally and downward in opening the mouth and in sad or frowning expressions.

Buccinator[20] (BUCK-sin-AY-tur) (fig. 7.17a). This is the muscle of the cheek. It has multiple functions in airflow, drinking, and chewing. It is used in drinking through a straw, nursing by infants, blowing out air or spitting out liquid, playing a wind instrument, helping to control airflow in speech, and pushing and retaining food between the teeth during chewing. When one closes the mouth, the buccinator retracts the cheek to protect it from being bitten.

Muscles of Chewing

There are four paired muscles of mastication: the temporalis, masseter, and medial and lateral pterygoids. Collectively, they have four actions on the mandible: *elevation,* to bite; *protraction,* a forward movement of the mandible to align the incisors in biting; *retraction,* to draw the mandible posteriorly and align the molars in chewing; and *lateral* and *medial excursion,* left–right movements that grind food between the molars.

Temporalis[21] (TEM-po-RAY-liss) (fig. 7.17a). This is a broad fan-shaped muscle on the side of the cranium. It originates on faint ridges along the side of the frontal, parietal, and occipital bones. It converges toward the zygomatic

---

[16] *orb* = circle; *ocul* = eye

[17] *orb* = circle; *oris* = of the mouth

[18] refers to the zygomatic arch

[19] *depressor* = that which lowers; *angul* = corner, angle

[20] *bucc* = cheek

[21] *orb* = circle; refers to the temporal bone

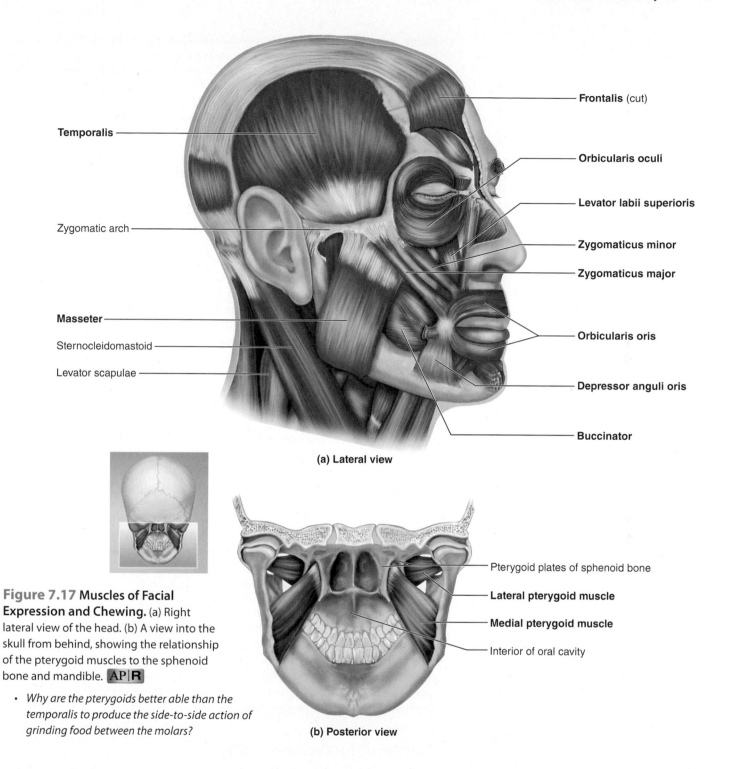

Frontalis (cut)

Temporalis

Orbicularis oculi

Levator labii superioris

Zygomatic arch

Zygomaticus minor

Zygomaticus major

Masseter

Sternocleidomastoid

Orbicularis oris

Levator scapulae

Depressor anguli oris

Buccinator

(a) Lateral view

Pterygoid plates of sphenoid bone

Lateral pterygoid muscle

Medial pterygoid muscle

Interior of oral cavity

(b) Posterior view

Figure 7.17 Muscles of Facial Expression and Chewing. (a) Right lateral view of the head. (b) A view into the skull from behind, showing the relationship of the pterygoid muscles to the sphenoid bone and mandible. **AP|R**

- *Why are the pterygoids better able than the temporalis to produce the side-to-side action of grinding food between the molars?*

arch, passes behind (medial) to it, and inserts on the coronoid process of the mandible. Its main action is to elevate and retract the mandible.

Masseter[22] (ma-SEE-tur) (fig. 7.17a). The masseter is shorter and superficial to the temporalis, arising from the zygomatic arch and inserting laterally on the angle of the mandible. It is a thick muscle easily palpated at the rear of your

[22] *masset* = chew

cheek. It is a synergist of the temporalis in elevating the mandible, but it also produces some degree of protraction and excursion.

Pterygoids[23] (TERR-ih-goyds) (fig. 7.17b). The *lateral* and *medial pterygoid* muscles arise from the *pterygoid plates* of the sphenoid bone on the inferior surface of the cranium. They pass obliquely and insert on the inner (medial) surface of the mandible. They elevate and protract the mandible and move it from side to side for grinding food.

Muscles of Head and Neck Movement

Muscles that move the head originate on the vertebral column, thoracic cage, and pectoral girdle and insert on the cranial bones. The principal flexor of the neck is the *sternocleidomastoid,* which courses up the neck from an anterior to a lateral position. The principal extensors are the *trapezius, splenius capitis,* and *semispinalis capitis,* located on the rear of the neck.

Sternocleidomastoid[24] (STIR-no-CLY-do-MAST-oyd) (figs. 7.16a, 7.17a). This is the prime mover of neck flexion. It originates on the sternum and medial half of the clavicle, passes obliquely up the neck, and inserts behind the ear on the mastoid process. It is a thick, cordlike muscle often visible on the side of the neck, especially when the head is turned to one side and slightly extended. It rotates the head to the left and right or tilts it upward and toward the opposite side, as in looking over one's shoulder, depending on what other muscles do at the same time. When both sternocleidomastoids contract at once, the neck flexes forward—for example, when you look down to eat or read.

Trapezius (tra-PEE-zee-us) (fig. 7.16b). This is a vast triangular muscle of the neck and upper back; together, the right and left trapezius muscles form a trapezoid. The trapezius has a broad origin extending from the *occipital protuberance*—a median bump at the rear of the skull—to vertebra T4. The muscle converges toward the shoulder, where it inserts on the spine and acromion of the scapula and the lateral end of the clavicle. Either trapezius working alone flexes the neck to one side. When they contract together, the trapezius muscles extend the neck, as in looking straight upward. The trapezius also has actions on the scapula and arm described later, under limb muscles.

Muscles of the Trunk AP|R

Here we consider muscles of the abdominal wall as well as back muscles that act on the vertebral column. The trunk muscles also include muscles of the pelvic floor, and muscles of respiration—the *intercostal muscles* between the ribs and the *diaphragm* between the thoracic and abdominal cavities, which are described in chapter 15. Some prominent muscles seen on the trunk, such as the *pectoralis major* of the mammary region, act on the arm and are therefore considered later as appendicular muscles.

Muscles of the Abdomen AP|R

Four pairs of sheetlike muscles reinforce the anterior and lateral abdominal walls. They support the intestines and other viscera against the pull of gravity;

---

[23] *pteryg* = wing; *oid* = shaped
[24] *sterno* = sternum; *cleido* = clavicle; *mastoid* = mastoid process of skull

stabilize the vertebral column during heavy lifting; and as abdominal compressors, aid in expelling organ contents during respiration, urination, defecation, vomiting, and childbirth. They are the *rectus abdominis, external* and *internal abdominal obliques,* and *transverse abdominal.*

Unlike the thoracic cavity, the abdominal cavity lacks a protective bony enclosure. However, the fascicles of these three muscle layers run in different directions (horizontally, vertically, and obliquely), lending strength to the abdominal wall much like a sheet of plywood, in which wood fibers run in different directions from layer to layer. The tendons of the abdominal muscles are aponeuroses. At its inferior margin, the aponeurosis of the external oblique forms a strong, cordlike *inguinal ligament.*

Rectus[25] **abdominis** (REC-tus ab-DOM-ih-nis) (fig. 7.16a). This is a median muscular strap extending vertically from the pubis to the sternum and lower ribs. It flexes the waist, as in bending forward or doing sit-ups; stabilizes the pelvic region during walking; and compresses the abdominal viscera. It is separated into four segments by fibrous *tendinous intersections* that give the abdomen a segmented appearance in well-muscled individuals. The rectus abdominis is enclosed in a fibrous sleeve called the *rectus sheath,* and the right and left rectus muscles are separated by a vertical fibrous strip called the *linea alba.*[26]

Abdominal obliques (fig. 7.16a). The **external abdominal oblique** is the most superficial muscle of the lateral abdominal wall. Its fascicles run anteriorly and downward. Deep to it is the **internal abdominal oblique**, whose fascicles run anteriorly and upward. These muscles maintain posture; flex and rotate the vertebral column, as in twisting at the waist; support the viscera; aid in deep breathing and loud vocalizations such as singing and public speaking; and compress the abdominal viscera.

Transverse abdominal (fig. 7.16a). This is the deepest of the abdominal group. It crosses the abdomen horizontally, like the cummerbund of a tuxedo, and acts primarily as an abdominal compressor.

Muscles of the Back `AP|R`

The most superficial muscles of the back are the trapezius of the upper back and latissimus dorsi of the lower back (see fig. 7.16b). These act on the arm and are therefore discussed later under the appendicular musculature. The strongest of the deeper back muscles is the following.

Erector spinae (eh-RECK-tur SPY-nee) (fig. 7.16b). This is a complex, multipart muscle group with cervical, thoracic, and lumbar portions and three parallel, vertical columns on each side of the spine. It maintains posture, straightens the spine after one bends at the waist, and is employed in arching the back; thus, as the name implies, it is important in standing and sitting erect. Unilateral contraction flexes the waist laterally, and its superior (cervical) portion contributes to rotation of the head.

Table 7.2 summarizes the axial muscles and provides more detail on their origins and insertions.

---

[25] *rect* = straight

[26] *linea* = line; *alb* = white

Table 7.2　Axial Muscles

Name	Origin	Insertion	Principal Actions
Facial Muscles			
Buccinator (fig. 7.17a)	Lateral aspect of maxilla and mandible	Orbicularis oris muscle; submucosa of cheek and lips	Compresses cheek; pushes food between teeth; creates suction; expels air or liquid from mouth; retracts cheek when mouth is closing so it is not bitten
Depressor anguli oris (fig. 7.17a)	Mandible	Corner of mouth	Draws corner of mouth laterally and downward
Frontalis (fig. 7.17a)	Subcutaneous tissue of scalp	Subcutaneous tissue of eyebrows	Raises eyebrows; wrinkles forehead
Orbicularis oculi (fig. 7.17a)	Lacrimal bone; adjacent regions of frontal bone and maxilla; median angle of eyelid	Upper and lower eyelids; skin adjacent to orbit	Closes eye; promotes flow of tears
Orbicularis oris (fig. 7.17a)	Connective tissue near angle of lips	Mucosa of lips	Closes lips; protrudes lips as in kissing; aids in speech
Zygomaticus major and minor (fig. 7.17a)	Zygomatic bone	Superolateral corner of mouth	Draw corner of mouth laterally and upward, as in smiling
Muscles of Chewing			
Masseter (fig. 7.17a)	Zygomatic arch	Lateral surface of mandibular ramus and angle	Elevates and protracts mandible in biting; aids in mandibular excursion (grinding of food)
Pterygoids (fig. 7.17b)	Pterygoid plates of sphenoid bone; palatine bone; maxilla	Medial surface of mandible near angle and mandibular condyle	Protract mandible in biting and produce mandibular excursion in chewing
Temporalis (fig. 7.17a)	Lateral surface of occipital, temporal, and frontal bones	Mandible at coronoid process and anterior border of ramus	Elevates and retracts mandible; aids in excursion
Muscles of Head–Neck Movement			
Sternocleidomastoid (figs. 7.16, 7.17a)	Clavicle and superior margin of sternum	Mastoid process of skull	Unilateral contraction rotates head left or right and directs gaze over opposite shoulder; bilateral contraction tilts head downward
Trapezius (fig. 7.16b)	External occipital protuberance of skull; ligament of nuchal region; spinous processes of vertebrae C7–T4	Clavicle; acromion and spine of scapula	Flexes neck to one side; extends neck (as in looking upward). See other functions in table 7.3.
Muscles of the Abdomen			
Abdominal obliques (fig. 7.16a)	Ribs 5–12; inguinal ligament, iliac crest, and thoracolumbar fascia	Xiphoid process, linea alba, pubis, ribs 10–12	Flexes abdomen in sitting up; rotates trunk at waist; compresses abdomen
Rectus abdominis (fig. 7.16a)	Pubis	Xiphoid process of sternum; costal cartilages of lower ribs	Flexes waist in sitting up; stabilizes pelvis in walking; compresses abdomen
Transverse abdominal (fig. 7.16a)	Inguinal ligament, iliac crest, thoracolumbar fascia, costal cartilages	Xiphoid process, linea alba, pubis	Compresses abdomen
Muscles of the Back			
Erector spinae (fig. 7.16b)	Extensive origin on ribs, vertebrae, sacrum, iliac crest	Vertebrae and ribs	Extends vertebral column as in arching back; laterally flexes vertebral column

Muscles of the Pectoral Girdle and Upper Limb

The rest of this chapter concerns muscles that act on the limbs and limb girdles—that is, the appendicular skeleton. As with the axial muscles, we begin with a general description of muscle locations and everyday actions, and table 7.3 will summarize these muscles and provide more detail as to origins and insertions. We begin with muscles that originate on the axial skeleton and act on the scapula.

Pectoralis minor (PECK-toe-RAY-lis) (fig. 7.16a). This is a deep muscle of the mammary region. It originates on ribs 3 through 5 and inserts on the coracoid process of the scapula. It draws the scapula laterally and forward, protracting the shoulder as in reaching out for a door handle. Acting with other muscles, it depresses the shoulder, as in reaching down to pick up a suitcase.

Serratus anterior (serr-AY-tus) (fig. 7.16a). This muscle originates on all or nearly all of the ribs, wraps around the chest between the rib cage and scapula, and inserts on the medial border of the scapula. It holds the scapula against the rib cage, protracts it, and is the prime mover in forward thrusting, throwing, and pushing; it is nicknamed the "boxer's muscle" for these actions. It also elevates the shoulder, as in lifting a suitcase.

Trapezius (tra-PEE-zee-us) (fig. 7.16b). The trapezius was discussed earlier for its actions on the head, but it also has important actions on the scapula. Its action depends on whether its superior, middle, or inferior portions contract and whether it acts alone or with other muscles. It can either elevate or depress the shoulder; it retracts the scapula when one draws back the shoulders, as in pulling on a rope; and it can stabilize the scapula and shoulder during arm movements. Normal depression of the scapula occurs mainly by gravitational pull, but the trapezius and serratus anterior can cause faster, more forcible depression, as in swimming, hammering, and rowing a boat.

Levator scapulae (leh-VAY-tur SCAP-you-lee) (fig. 7.16b, 7.18a). The levator scapulae works with the superior portion of the trapezius to elevate the scapula, for example when one carries a heavy weight on the shoulder or lifts a suitcase. Depending on the action of other muscles, it can also flex the neck laterally, retract the scapula and brace the shoulder, or rotate the scapula to lower the apex of the shoulder.

Rhomboids (ROM-boyds) (see fig. 7.16b). The **rhomboids** originate on the lower cervical and upper thoracic vertebrae and insert on the medial border of the scapula. They retract and elevate the scapula by drawing it toward the spine and fix the scapula when the biceps brachii contracts so that the biceps flexes the elbow. Otherwise the biceps would tend to pull the scapula forward.

Muscles Acting on the Humerus

Nine muscles cross the glenohumeral (shoulder) joint and insert on the humerus. The first two of these, the *pectoralis major* and *latissimus dorsi,* are considered *axial muscles* because they originate primarily on the axial skeleton. They bear the primary responsibility for attaching the arm to the trunk. The axilla (armpit) is the depression between these two muscles.

Pectoralis major (PECK-toe-RAY-lis) (fig. 7.18a). This is the thick, superficial, fleshy muscle of the mammary region. It adducts the humerus and rotates it medially; flexes the shoulder as in reaching out to hug someone; and aids in climbing, pushing, and throwing.

Latissimus dorsi[27] (la-TISS-ih-mus DOR-sye) (fig. 7.18b). This is a broad muscle of the back that extends from the waist to the axilla. It extends the shoulder as in reaching behind you to clasp your hands behind your back; thus, it is an antagonist of the pectoralis major. It pulls the body upward in climbing, and produces strong downward strokes of the arm as in hammering or swimming; it is sometimes nicknamed the "swimmer's muscle."

The rest of these are called *scapular muscles* because they originate on the scapula. The first four provide the tendons that form the **rotator cuff** of the shoulder (fig. 7.18d). They are nicknamed the "SITS muscles" after the first letters of their formal names—the *supraspinatus, infraspinatus, teres minor,* and *subscapularis.* The subscapularis fills most of the subscapular fossa on the anterior surface of the scapula. The other three originate on the posterior surface. The tendons of these muscles merge with the joint capsule of the shoulder as they pass en route to the humerus. They insert on the proximal end of the humerus, forming a partial sleeve around it. The rotator cuff reinforces the joint capsule and holds the head of the humerus in the glenoid cavity. The rotator cuff, especially the tendon of the supraspinatus, is easily damaged by strenuous circumduction, as in bowling and baseball pitching (see Perspectives on Health, p. 227).

Supraspinatus (SOO-pra-spy-NAY-tus) (fig. 7.18b). This muscle occupies the *supraspinous fossa* above the scapular spine on the posterior side of the scapula. It abducts the humerus and resists downward slippage of the humerus when one is carrying weight.

Infraspinatus (IN-fra-spy-NAY-tus) (fig. 7.18b). This muscle occupies most of the *infraspinous fossa* below the scapular spine. It extends and laterally ro-

---

[27] *latissimus* = broadest; *dorsi* = of the back

Figure 7.18 Muscles of the Pectoral, Scapular, and Brachial Regions. (a) Anterior view of superficial pectoral and brachial muscles. (b) Posterior view of muscles that act on the shoulder and elbow joints. (c) Anterior view of scapular and brachial muscles, with rib cage removed to see facing surface of scapula behind it. (d) Left lateral view of rotator cuff muscles in relation to scapula. AP|R

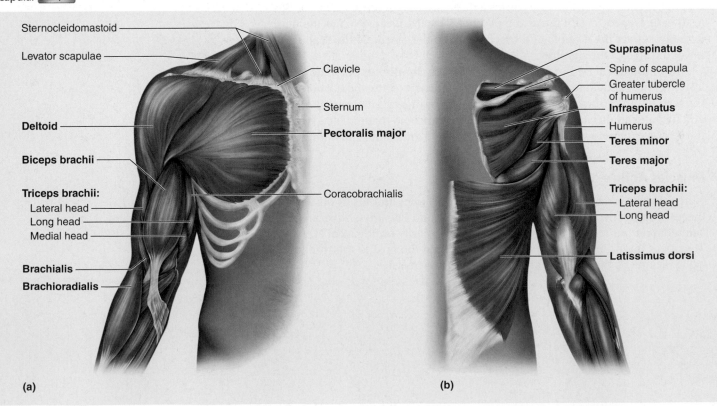

(a)

(b)

tates the humerus and prevents the humeral head from slipping upward out of the glenoid cavity.

Teres minor (TARE-eez) (fig. 7.18b). This muscle is found on the posterior side of the scapula near its inferior border, between the infraspinatus and teres major. It adducts and laterally rotates the humerus, and prevents it from slipping out of the glenoid cavity when the arm is abducted.

Subscapularis (SUB-SCAP-you-LARE-is) (fig. 7.18c). This muscle is found on the anterior side of the scapula, filling the subscapular fossa. It medially rotates the humerus and otherwise acts similarly to the teres minor.

The following two scapular muscles do not form part of the rotator cuff but are also important in movements of the arm at the shoulder.

Deltoid (DEL-toyd) (fig. 7.18a). This is the thick muscle that caps the shoulder; it is a commonly used site for intramuscular drug injections. The deltoid acts like three different muscles, contributing greatly to the shoulder mobility that characterizes humans and other primates in contrast to other mammals. Its anterior portion flexes the shoulder, its posterior portion extends it, and its lateral portion abducts it. Abduction by the deltoid is antagonized by the combined action of the pectoralis major and latissimus dorsi.

Teres major (TARE-eez) (fig. 7.18b). The teres major is located at the inferior border of the scapula on its posterior side. It extends from the scapula to the shaft of the humerus. It medially rotates the humerus and draws it backward, as in the arm-swinging movement that accompanies walking.

Muscles Acting on the Forearm

The elbow and forearm are capable of four motions: flexion, extension, pronation, and supination. Flexion is achieved by the *biceps brachii, brachialis,* and *brachioradialis;* extension by the *triceps brachii;* pronation by the *pronator teres* and *pronator quadratus;* and supination by the *supinator.*

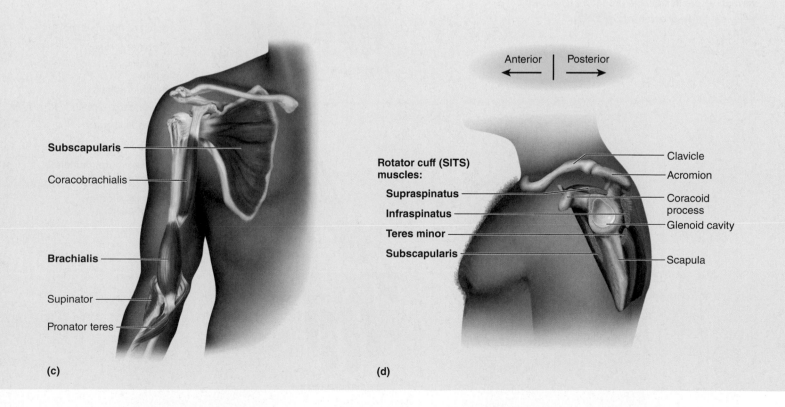

Subscapularis

Coracobrachialis

Brachialis

Supinator

Pronator teres

(c)

Anterior | Posterior

Rotator cuff (SITS) muscles:

Supraspinatus

Infraspinatus

Teres minor

Subscapularis

Clavicle

Acromion

Coracoid process

Glenoid cavity

Scapula

(d)

Biceps brachii[28] (BY-seps BRAY-kee-eye) (figs. 7.15, 7.18a). This muscle is the bulge on the anterior side of the arm, superficial to the brachialis. It has two heads that arise from tendons on the lateral angle of the scapula. Distally, the muscle converges near the elbow to a single tendon that inserts on the radius and fascia of the forearm. Despite its conspicuousness and interest to body builders, it is less powerful than the brachialis and acts as a synergist in flexing the elbow. It also contributes to supination of the forearm.

Brachialis (BRAY-kee-AL-is) (figs. 7.15, 7.18c). This is a deep muscle lying along the anterior surface of the lower half of the humerus. It extends from humerus to ulna and is the most powerful flexor of the elbow.

Brachioradialis (BRAY-kee-oh-RAY-dee-AL-is) (fig. 7.18a). This muscle flexes the elbow although it is located in the forearm; it is the thick muscular mass on the radial (lateral) side of the forearm, near the elbow. It is much less powerful than the biceps and brachialis.

Triceps brachii (TRY-seps BRAY-kee-eye) (figs. 7.15, 7.18b). This is a three-headed muscle on the posterior side of the humerus. It extends from the scapula and humerus to the olecranon of the ulna, and is the prime mover of elbow extension.

Pronators. Pronation of the forearm is achieved by the **pronator teres** (PRO-nay-tur TARE-eez) near the elbow and **pronator quadratus** (kwa-DRAY-tus) near the wrist (figs. 7.18c, 7.19b).

Supinator (SOO-pih-NAY-tur) (figs. 7.16a, 7.18c). Supination of the forearm is achieved by the biceps brachii, as noted earlier, and the *supinator* muscle slightly distal to the elbow.

---

[28] *bi* = two; *ceps* = head; *brachi* = arm. Note that biceps is singular; there is no such word as bicep. The plural form is bicipites (by-SIP-ih-teez).

Figure 7.19 Muscles of the Forearm.
(a) Anterior view of superficial flexors. (b) Anterior view of deeper flexors and pronators. (c) Posterior view of superficial extensors. **AP|R**

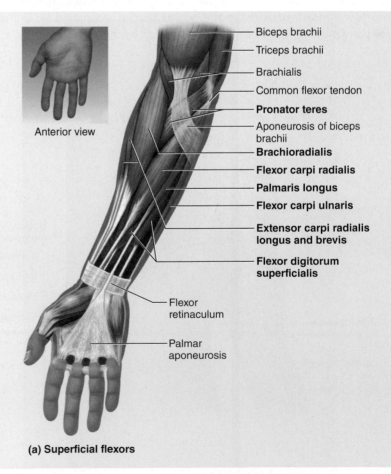

Anterior view

Biceps brachii
Triceps brachii
Brachialis
Common flexor tendon
Pronator teres
Aponeurosis of biceps brachii
Brachioradialis
Flexor carpi radialis
Palmaris longus
Flexor carpi ulnaris
Extensor carpi radialis longus and brevis
Flexor digitorum superficialis
Flexor retinaculum
Palmar aponeurosis

(a) Superficial flexors

Muscles Acting on the Wrist and Hand

The hand carries out more subtle and sophisticated movements than any other part of the body; the wrist and hand are therefore acted upon by at least 15 *extrinsic muscles* in the forearm and 18 *intrinsic muscles* in the hand itself. Relatively few of them can be considered here.

The bellies of the extrinsic muscles form the fleshy roundness of the proximal forearm. Their actions are mainly flexion and extension. Flexors are on the anterior side. Their tendons pass through the wrist and insert mainly on carpal and metacarpal bones and the phalanges. Most of these tendons pass under a transverse ligament that crosses the wrist like a bracelet, the **flexor retinaculum**[29] (fig. 7.19a). One exception is the *palmaris longus* muscle, which is relatively weak and even absent from some people; its tendon passes over the retinaculum. The extensor muscles are on the posterior side of the forearm, and their tendons pass under a similar **extensor retinaculum** on that side of the wrist (fig. 7.16b). Although the forearm muscles are numerous and complex, most of their names suggest their actions, and from their actions, their approximate locations in the forearm can generally be deduced.

Following are three important flexors of the wrist and fingers. These are especially important to the strength of one's grip.

Flexor carpi radialis (FLEX-ur CAR-pye RAY-dee-AL-is) (fig. 7.19a). This is a superficial muscle on the radial (lateral) side of the forearm, bordered laterally by the brachioradialis and medially by (in most people) the palmaris longus. It inserts on metacarpals II and III and flexes the wrist anteriorly and laterally.

---

[29] *retinac* = retainer, bracelet; *cul* = little

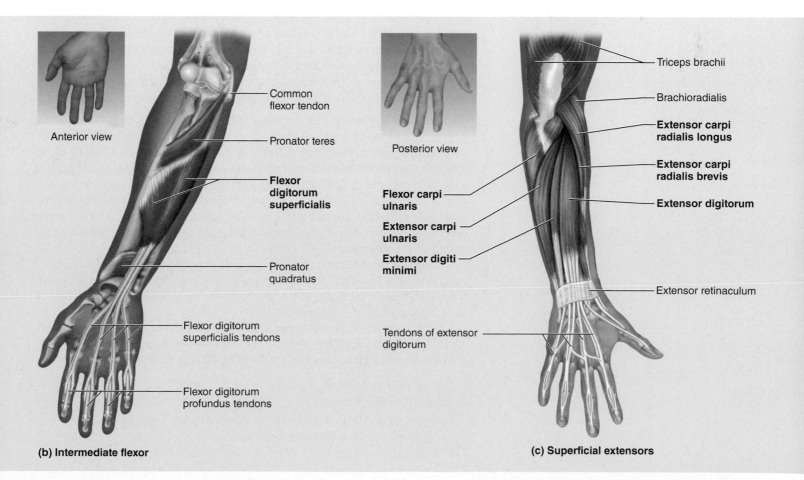

Anterior view

Common flexor tendon

Pronator teres

Flexor digitorum superficialis

Pronator quadratus

Flexor digitorum superficialis tendons

Flexor digitorum profundus tendons

(b) Intermediate flexor

Posterior view

Flexor carpi ulnaris

Extensor carpi ulnaris

Extensor digiti minimi

Tendons of extensor digitorum

Triceps brachii

Brachioradialis

Extensor carpi radialis longus

Extensor carpi radialis brevis

Extensor digitorum

Extensor retinaculum

(c) Superficial extensors

Clinical Application 7.3

CARPAL TUNNEL SYNDROME

Prolonged, repetitive motion of the wrist and fingers can cause tissues in the carpal tunnel to become inflamed, swollen, or fibrotic. Since the carpal tunnel cannot expand, swelling compresses the *median nerve* of the wrist, which passes through the tunnel with the flexor tendons. This pressure causes tingling and muscular weakness in the palm and lateral side of the hand and pain that may spread ("radiate") to the arm and shoulder. This condition, called **carpal tunnel syndrome**, is common among pianists, meat cutters, and others who spend long hours making repetitive wrist motions. It is treated with aspirin and other anti-inflammatory drugs, immobilization of the wrist, and sometimes surgical removal of part or all of the flexor retinaculum to relieve pressure on the nerve.

Flexor carpi ulnaris (FLEX-ur CAR-pye ul-NAIR-is) (fig. 7.19a). This is a superficial muscle on the ulnar (medial) side of the forearm. It inserts on some of the wrist bones and metacarpal V, and flexes the wrist anteriorly and medially.

Flexor digitorum superficialis (FLEX-ur DIDJ-ih-TORE-um SOO-per-FISH-ee-AY-lis) (fig. 7.19b). This muscle lies deep to the previous two. It has four tendons leading to the middle phalanges of digits II through V. Depending on the action of other muscles, it flexes the fingers as in making a fist, or flexes just the wrist. Beneath it, the *flexor digitorum profundus* (fig. 7.16a) has similar tendinous insertions and performs a similar action, but it is the only muscle that can independently flex the distal segments of those four fingers.

A frequent site of discomfort and disability is the **carpal tunnel**, a tight space between the carpal bones and flexor retinaculum. The flexor tendons passing through the tunnel are enclosed in tubular bursae called *tendon sheaths* that enable them to slide back and forth quite easily, although this region is very subject to injury (see Clinical Application 7.3).

The extensors on the posterior side of the forearm share a single tendon arising from the humerus. We will consider only four of them:

Extensor digitorum (ex-TEN-sur DIDJ-ih-TORE-um) (fig. 7.19c). This broad, superficial extensor on the midline of the forearm has four distal tendons leading to the phalanges of digits II through V. Its tendons are easily seen and palpated on the back of the hand when the fingers are extended and the wrist is hyperextended. It extends the wrist and extends and spreads the fingers. The little finger and the index finger also have their own extensors, respectively called the *extensor digiti minimi* and *extensor indicis,* and there are several muscles for thumb movement alone. Loss of the thumb is the most disabling of all hand injuries.

Extensor carpi ulnaris (ex-TEN-sur CAR-pye ul-NARE-is) (fig. 7.19c). This muscle lies medial to the extensor digitorum (on the ulnar side). It extends the wrist and flexes it medially, and it can hold the wrist straight when the fist is clenched or the hand grips an object.

Extensor carpi radialis longus and brevis (ex-TEN-sur CAR-pye RAY-dee-AY-lis LAWN-gus and BREV-is) (fig. 7.19c). These muscles lie side by side, lateral to the extensor digitorum. Both of them extend the wrist and flex it laterally.

Apply What You Know

Andre is a renowned symphony pianist. Explain why his career would result in more than average strength in his extensor digitorum and flexor digitorum profundus muscles.

Muscles of the Pelvic Girdle and Lower Limb

The largest and strongest muscles in the body are found in the lower limb. Unlike those of the upper limb, they are adapted less for precision than for the strength needed to stand, maintain balance, walk, and run. Several of them cross and act upon two or more joints, such as the hip and knee.

Muscles Acting on the Hip and Femur

There are four principal groups of muscles that act on the femur, producing movement at the hip joint: anterior muscles that flex the hip; gluteal muscles of the buttocks, which extend the hip and abduct the thigh; additional abductors on the lateral side of the femur; and adductors on the medial side. The two principal anterior hip muscles are the *iliacus* and *psoas major*. They converge on a single tendon that inserts on the femur, so they have the same action and are often referred to collectively as the *iliopsoas*.

  Iliopsoas (IL-ee-oh-SO-us) (fig. 7.20). The **iliacus** (ih-LY-uh-cus) is a broad muscle that fills the iliac fossa of the pelvis, and the **psoas** (SO-us) **major** originates mainly on the lumbar vertebrae. Together, these muscles flex the hip as in stair climbing and in the forward swing of the leg when walking; they flex the trunk if the femur is fixed, as in leaning forward in a chair or sitting up in bed; and they balance the trunk during sitting.

  Gluteal muscles (figs. 7.21). The gluteus maximus, medius, and minimus are the muscles of the buttocks. The **gluteus maximus** is the largest muscle of this group. It is a hip extensor that produces the backswing of the leg in walking; it helps you stand straight after bending forward at the waist; and it provides most of the lift when you climb stairs and most of the thrust in pushing on a bicycle pedal. The **gluteus medius** and **minimus** abduct the femur, for example when moving the feet apart to stand in a spread-legged position. They are also important in walking; when one leg is lifted from the ground, these muscles on the other side of the body flex the waist to shift the body weight toward the leg still on the ground, and thus prevent one from falling over.

  Thigh adductors (fig. 7.20). The medial side of the thigh has several adductor muscles—the *adductor longus, adductor brevis, adductor magnus, pectineus,* and *gracilis.* Collectively, these muscles originate on the pubis and ischium of the hip bone and insert along the posterior to medial surfaces of the shaft of the femur, except that the gracilis inserts medially on the tibia just below the knee. In addition to adducting the thigh, most of these muscles aid in flexing the hip, and the gracilis aids in flexing the knee. Imagine the importance of the adductors in riding a horse.

  Thigh rotators and abductors (fig. 7.21). The lateral side of the hip has a group of short muscles collectively called the **lateral rotators.** They originate on the hip bone and insert, along with the gluteus medius and minimus, on or near the greater trochanter of the femur. They rotate the femur laterally, as in standing with the toes pointed outward, and like those two gluteal muscles, they aid in shifting the body weight and maintaining balance when one is walking. Some of them aid the gluteus medius and minimus in abducting the thigh.

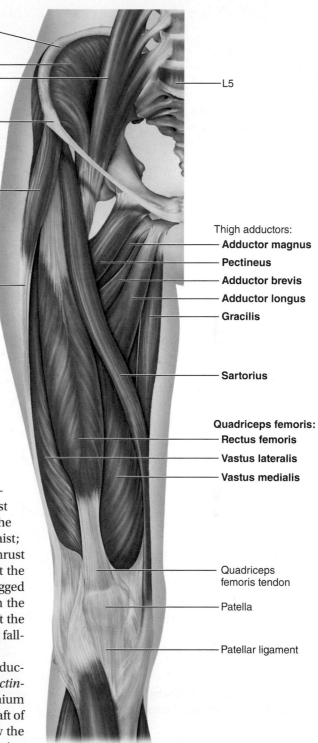

Figure 7.20 Anterior Muscles of the Hip and Thigh. AP|R

- *What two joints are acted upon by the rectus femoris?*

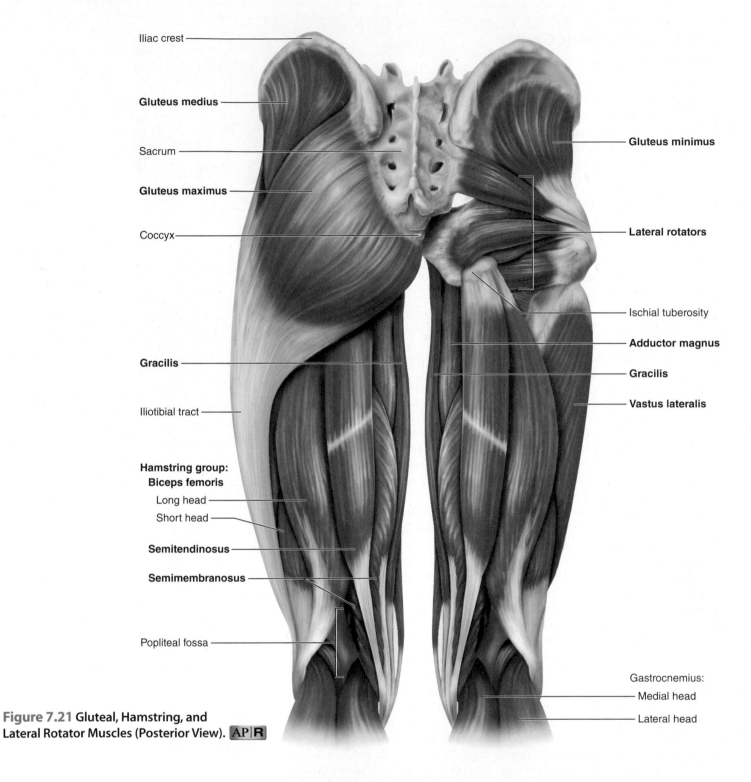

Iliac crest

Gluteus medius

Sacrum

Gluteus maximus

Coccyx

Gracilis

Iliotibial tract

Hamstring group:
Biceps femoris

Long head

Short head

Semitendinosus

Semimembranosus

Popliteal fossa

Gluteus minimus

Lateral rotators

Ischial tuberosity

Adductor magnus

Gracilis

Vastus lateralis

Gastrocnemius:

Medial head

Lateral head

Figure 7.21 Gluteal, Hamstring, and Lateral Rotator Muscles (Posterior View). AP|R

Muscles Acting on the Knee

The fleshiness of the thigh is formed mostly by the four-part **quadriceps femoris** on the anterior side and three muscles, the **hamstring** group, on the posterior side. Also crossing the thigh anteriorly is the **sartorius**.

Quadriceps femoris (QUAD-rih-seps FEM-oh-ris) (fig. 7.20). This large muscle consists of four heads that arise on the ilium and the shaft of the femur and converge on a single tendon at the knee. The heads are the **rectus femoris** in the middle; flanked by the **vastus lateralis** and **vastus medialis** on the lateral and medial sides; and the **vastus intermedius**, which is located between

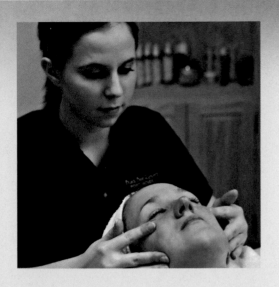

CAREER SPOTLIGHT

Massage Therapist

A massage therapist manipulates muscles and other soft tissues by kneading, gliding, and compression in order to promote relaxation; relieve pain; stimulate blood and lymph flow; and address issues of stress, anxiety, and tension. Massage therapists work with clients as diverse as athletes, people with soft-tissue injuries, pregnant women, and children, but most often, simply with persons seeking relaxation and a sense of wellness. There are more than 80 types of massage therapy, such as Swedish massage, deep pressure, and acupressure, and spas often combine massage with hydrotherapy, aromatherapy, and other methods. Most massage therapists are employed by health spas—a rapidly growing industry—but other work settings include fitness centers, sports clinics, retirement homes, rehabilitation centers, colleges, chiropractic offices, hotels, malls, cruise ships, and clients' homes and workplaces. Most states require one to be a Certified Massage Therapist (CMT) or equivalent before granting a license to practice. Candidates must complete a training program of usually 500 to 1,000 hours of instruction, entailing both hands-on practice and classroom instruction in anatomy and physiology, kinesiology, first aid, CPR, business practices, ethics, and legal issues. One must also pass a licensing examination with both written and demonstration components. See appendix B for further information.

these and deep to the rectus femoris (therefore not illustrated). The **quadriceps (patellar) tendon** extends from the distal end of this muscle group to arch over the knee and enclose the patella; it continues beyond the patella as the **patellar ligament,** ending on the anterior surface of the tibia just below the knee. This change in terminology reflects the definition of a tendon as connecting muscle to bone (quadriceps to patella) and a ligament as connecting bone to bone (patella to tibia). The quadriceps extends the knee, as in kicking a football or standing up from a chair, and the rectus femoris helps to flex the hip.

Sartorius[30] (sar-TORE-ee-us) (fig. 7.20). This straplike muscle is the longest in the body. It originates on the lateral side of the hip, crosses diagonally over the quadriceps femoris, and inserts medially on the tibia. Nicknamed the "tailor's muscle," it is employed in crossing the legs and aids in flexing the hip and knee.

Hamstrings (fig. 7.21). *Hamstrings* is a colloquial name for the three major muscles on the posterior side of the thigh—the **biceps femoris** (BY-seps FEM-oh-ris), **semimembranosus** (SEM-ee-MEM-bra-NO-sus), and **semitendinosus** (SEM-ee-TEN-dih-NO-sus). The pit at the rear of the knee *(popliteal fossa)* is bordered laterally by the biceps tendon and medially by the tendons of the other two. The hamstrings flex the knee, and aided by the gluteus maximus, they extend the hip during walking and running. Hamstring injuries are common among sprinters, soccer players, and other athletes who rely on quick acceleration (see Perspectives on Health, p. 227). The hamstrings are important in supporting the body weight. Chronic low-back pain can be relieved by exercises to stretch those tendons and strengthen the hamstring and abdominal muscles.

Muscles Acting on the Foot

The fleshy mass of the leg proper (below the knee) is formed by a group of *crural muscles,* which act on the ankle and foot.

[30] *sartor* = tailor

Anterior group (fig. 7.22a). Two principal muscles on the anterior side of the leg are the **extensor digitorum longus** (ex-TEN-sur DIDJ-ih-TORE-um LAWN-gus) and **tibialis** (TIB-ee-AY-lis) **anterior**. Their tendons pass under two **extensor retinacula** on the anterior side of the ankle. These muscles dorsiflex the foot, and the first of them extends all of the toes except the great toe, which has its own extensor. Dorsiflexion prevents the toes from scuffing on the ground during walking.

Posterior group (fig. 7.22b, c). The major posterior (calf) muscles are the superficial **gastrocnemius**[31] (GAS-trock-NEE-me-us), whose two heads form the prominent bulge of the upper calf, and the deeper **soleus**.[32] Both of these muscles converge on the same tendon, the **calcaneal (Achilles) tendon**, which inserts on the calcaneus (heel bone). Together these muscles are sometimes called the **triceps surae**[33] (TRY-seps SOOR-ee). They plantar flex the foot, as in standing on tiptoe or jumping. Plantar flexion also provides lift and forward thrust in walking. Also aiding in plantar flexion are the **tibialis posterior**, the **flexor digitorum longus**, and two **fibularis** (FIB-you-LARE-is) muscles (*fibularis brevis* and *fibularis longus*) (fig. 7.16b).

Like the hand, the foot has 19 small *intrinsic muscles* that support the foot arches and flex, extend, abduct, and adduct the toes.

---

[31] *gastro* = belly; *cnem* = leg
[32] named for its resemblance to a flatfish, the sole
[33] *tri* = three; *ceps* = head; *sura* = calf of leg

Figure 7.22 Muscles of the Leg. (a) Superficial anterior group. (b) Superficial posterior muscles. (c) Gastrocnemius removed to show the deeper posterior soleus muscle. AP|R

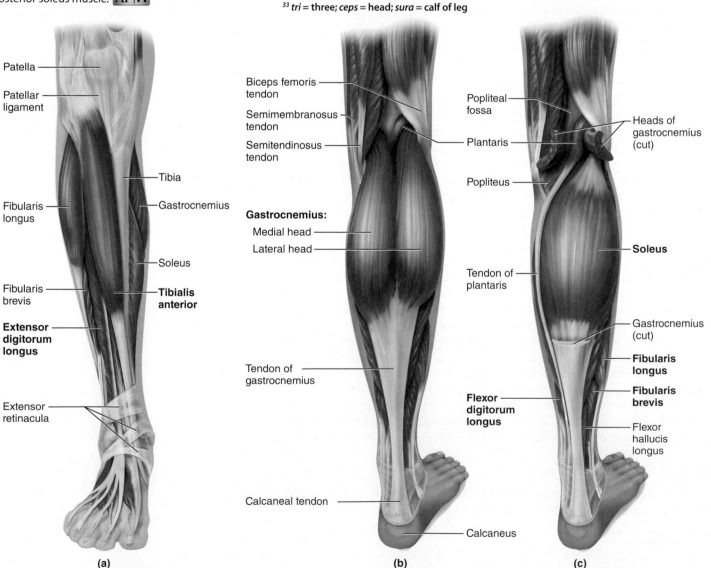

Patella
Patellar ligament
Tibia
Fibularis longus
Gastrocnemius
Soleus
Fibularis brevis
Tibialis anterior
Extensor digitorum longus
Extensor retinacula

(a)

Biceps femoris tendon
Semimembranosus tendon
Semitendinosus tendon
Gastrocnemius:
Medial head
Lateral head
Tendon of gastrocnemius
Calcaneal tendon
Calcaneus

(b)

Popliteal fossa
Plantaris
Popliteus
Heads of gastrocnemius (cut)
Soleus
Tendon of plantaris
Gastrocnemius (cut)
Fibularis longus
Fibularis brevis
Flexor digitorum longus
Flexor hallucis longus

(c)

Table 7.3 summarizes the appendicular muscles and provides more detailed information on their origins and insertions.

Table 7.3 Appendicular Muscles

Name	Origin	Insertion	Action
Muscles Acting on the Scapula			
Levator scapulae (fig. 7.16b, 7.18a)	Vertebrae C1–C4	Medial border and superior angle of scapula	Elevates scapula; tilts apex of shoulder downward; braces shoulder; flexes neck laterally
Pectoralis minor (fig. 7.16a)	Ribs 3–5	Coracoid process of scapula	Draws scapula laterally and forward around chest; lowers apex of shoulder
Rhomboids (fig. 7.16b)	Vertebrae C7–T5	Medial border of scapula	Retract and elevate scapula; rotate apex of shoulder downward; fix scapula during arm movements; brace shoulder
Serratus anterior (fig. 7.16a)	Most or all ribs	Medial border of scapula	Draws scapula laterally and forward around chest; raises apex of shoulder; prime mover in forward thrusting, throwing, and pushing
Trapezius (fig. 7.16b)	External occipital protuberance of skull; ligament of nuchal region; spinous processes of vertebrae C7–T4	Clavicle; acromion and spine of scapula	Different parts of trapezius raise or lower apex of shoulder, retract scapula, and stabilize scapula during arm movements. See other functions in table 7.2.
Rotator Cuff Muscles Acting on the Humerus			
Infraspinatus (fig. 7.18b)	Infraspinous fossa of scapula	Greater tubercle of humerus	Extends and laterally rotates humerus and prevents it from sliding upward
Subscapularis (fig. 7.18c)	Subscapular fossa of scapula	Lesser tubercle of humerus, anterior surface of joint capsule	Medially rotates humerus and prevents it from sliding upward
Supraspinatus (fig. 7.18b)	Supraspinous fossa of scapula	Greater tubercle of humerus	Abducts humerus and resists its downward displacement when carrying heavy weight
Teres minor (fig. 7.18b)	Lateral border of scapula	Greater tubercle of humerus, posterior surface of joint capsule	Adducts and laterally rotates humerus
Other Muscles Acting on the Humerus			
Deltoid (fig. 7.18a)	Clavicle; spine and acromion of scapula	Deltoid tuberosity of humerus	Different parts abduct, flex, and medially rotate humerus, and extend and laterally rotate it
Latissimus dorsi (fig. 7.18b)	Vertebrae T7–L5; ribs 9 or 10 through 12; thoracolumbar fascia; iliac crest	Floor of intertubercular groove of humerus	Extends shoulder; pulls body upward in climbing; produces strong downward thrusts as in hammering and swimming; adducts and medially rotates humerus
Pectoralis major (fig. 7.18a)	Clavicle, sternum, costal cartilages 1–6	Lateral margin of intertubercular groove of humerus	Adducts and medially rotates humerus; flexes shoulder as in reaching forward; aids in climbing , pushing, and throwing
Teres major (fig. 7.18b)	Inferior angle of scapula	Medial margin of intertubercular groove of humerus	Adducts and medially rotates humerus; extends shoulder
Muscles Acting on the Forearm			
Biceps brachii (fig. 7.15, 7.18a)	Supraglenoid tubercle and coracoid process of scapula	Radial tuberosity	Strongly supinates forearm; synergizes elbow flexion; secures head of humerus in glenoid cavity
Brachialis (fig. 7.15, 7.18c)	Anterior aspect of lower humerus	Coronoid process and tuberosity of ulna	Flexes elbow
Brachioradialis (fig. 7.18a)	Lateral supracondylar ridge of humerus	Distal radius near styloid process	Flexes elbow

(continued on next page)

Table 7.3 Appendicular Muscles *(continued)*

Name	Origin	Insertion	Action
Pronator quadratus (fig. 7.19b)	Anterior aspect of lower ulna	Anterior aspect of lower radius	Pronates forearm
Pronator teres (fig. 7.18c, 7.19b)	Near medial epicondyle of humerus; coronoid process of ulna	Lateral midshaft of radius	Pronates forearm
Supinator (fig. 7.18c, 7.19b)	Lateral epicondyle of humerus; shaft of upper ulna	Shaft of upper radius	Supinates forearm
Triceps brachii (fig. 7.15, 7.18b)	Inferior margin of glenoid cavity; posterior aspect of humerus	Olecranon of ulna	Extends elbow; long head adducts humerus
Flexors Acting on the Wrist and Hand			
Flexor carpi radialis (fig. 7.19a)	Medial epicondyle of humerus	Base of metacarpals II and III	Flexes wrist anteriorly and laterally (radial flexion)
Flexor carpi ulnaris (fig. 7.19a)	Medial epicondyle of humerus	Pisiform, hamate, and metacarpal V	Flexes wrist anteriorly and medially (ulnar flexion)
Flexor digitorum superficialis (fig. 7.19b)	Medial epicondyle of humerus; radius; coronoid process of ulna	Middle phalanges II–V	Flexes wrist and knuckles II–V
Extensors Acting on the Wrist and Hand			
Extensor carpi radialis longus and brevis (fig. 7.19c)	Lateral supracondylar ridge and lateral epicondyle of humerus, respectively	Base of metacarpals II and III, respectively	Extend wrist and aid in radial flexion
Extensor carpi ulnaris (fig. 7.19c)	Lateral epicondyle of humerus; posterior side of ulnar shaft	Base of metacarpal V	Extends wrist and aids in ulnar flexion
Extensor digitorum (fig. 7.19c)	Lateral epicondyle of humerus	Posterior (dorsal) side of phalanges II–V	Extends wrist and fingers II–V
Muscles Acting on the Hip and Femur			
Adductor longus and brevis (fig. 7.20)	Pubis	Posterior shaft of femur	Adduct and medially rotate femur; flex hip
Adductor magnus (fig. 7.20)	Ischium and pubis	Posterior shaft of femur	Adducts and medially rotates femur; posterior part extends hip
Gluteus maximus (fig. 7.21)	Posterior surface of ilium, sacrum, and coccyx	Gluteal tuberosity of femur	Extends hip joint, important in backswing of stride, abducts and laterally rotates femur
Gluteus medius and minimis (fig. 7.21)	Posterior surface of ilium	Greater trochanter of femur	Abduct and medially rotate femur, maintain balance by shifting body weight during walking
Gracilis (fig. 7.20)	Pubis and ischium	Medial surface of upper tibia	Adducts femur; flexes knee
Iliopsoas (fig. 7.20)	Iliac crest and fossa; sacrum; vertebral bodies T12–L5	Lesser trochanter and adjacent shaft of femur	Flexes hip as in bending at waist or raising thigh
Pectineus (fig. 7.20)	Pubis	Posterior surface of upper femur	Adducts femur; flexes hip
Muscles Acting on the Knee			
Biceps femoris (fig. 7.21)	Ischial tuberosity and posterior surface of shaft of femur	Head of fibula	Flexes knee; extends hip; laterally rotates leg
Quadriceps femoris (fig. 7.20)	Ilium; posterior and anterior aspects of shaft of femur	Tibial tuberosity	Extends knee; flexes hip
Sartorius (fig. 7.20)	Anterior superior spine of ilium	Medial aspect of tibial tuberosity	Flexes hip and knee; used in crossing legs
Semimembranosus (fig. 7.21)	Ischial tuberosity	Medial condyle of tibia; lateral condyle of femur	Flexes knee; extends hip
Semitendinosus (fig. 7.21)	Ischial tuberosity	Medial surface of upper tibia	Same as semimembranosus
Muscles Acting on the Foot			
Extensor digitorum longus (fig. 7.22a)	Lateral condyle of tibia, shaft of fibula, interosseous membrane	Middle and distal phalanges II–V	Extends toes II–V; dorsiflexes foot
Fibularis brevis and fibularis longus (fig. 7.22c)	Shaft of fibula; lateral condyle of tibia	Base of metatarsal V; metatarsal I; medial cuneiform	Plantar flex and evert foot; maintain plantar concavity
Flexor digitorum longus (figs. 7.16b, 7.22c)	Posterior surface of tibial shaft	Distal phalanges II–V	Flexes toes II–V; plantar flexes foot
Gastrocnemius (fig. 7.22b)	Condyles and popliteal surface of femur; capsule of knee joint	Calcaneus	Flexes knee; plantar flexes foot

Table 7.3 Appendicular Muscles *(continued)*

Name	Origin	Insertion	Action
Soleus (fig. 7.22c)	Posterior surface of proximal one-fourth of fibula and middle one-third of tibia	Calcaneus	Plantar flexes foot
Tibialis anterior (fig. 7.22a)	Lateral margin of tibia; interosseous membrane	Medial cuneiform; metatarsal I	Dorsiflexes and inverts foot
Tibialis posterior (fig. 7.16b)	Posterior surface of proximal half of tibia, fibula, and interosseous membrane	Navicular; medial cuneiform; metatarsals II–IV	Plantar flexes and inverts foot

---

Before You Go On

Answer these questions from memory. Reread the preceding section if there are too many you don't know.

14. *Consider a contracting muscle fiber pulling on its individual endomysium. Describe the connective tissue elements through which this force must be transmitted before it moves a bone.*

15. *How does a synergist modify the action of a prime mover? How does an antagonist modify it? What role does a fixator play?*

16. *Name a muscle employed in each of the following: a wink, a kiss, a smile, and a frown.*

17. *Name the three muscle layers of the lateral abdominal region from superficial to deep, and describe the actions of these muscles.*

18 *Name the muscles of the rotator cuff and describe their locations.*

19. *Identify three antagonists of the triceps brachii and describe what actions the triceps and these antagonists perform.*

20. *Describe the functions of the gluteal muscles, hamstrings, and quadriceps femoris.*

21. *List two muscles and their actions involved in each of the following activities: sit-ups, dribbling a basketball, kicking a soccer ball, and rising from a chair.*

Aging of the Muscular System

One of the most common changes in old age is the replacement of lean body mass (muscle) with fat, accompanied by loss of muscular strength. Muscular strength and mass peak in the 20s, and by the age of 80, most people have only half as much strength and endurance. Many people over age 75 cannot lift a 4.5 kg (10 lb) weight, making such simple tasks as carrying a bag of groceries very difficult. Tasks such as buttoning the clothes also take more time and effort. The loss of strength is a major contributor to falls, fractures, and dependence on others for living assistance. Fast-twitch muscle fibers show the earliest and greatest atrophy, thus increasing reaction time, slowing the reflexes, and reducing coordination.

There are multiple reasons for the loss of strength. Aged muscle has fewer myofibrils; more disorganized sarcomeres; smaller mitochondria; and reduced amounts of ATP, myoglobin, glycogen, and creatine phosphate. Increased adipose and fibrous tissue in the muscles limits their movement and blood circulation. In addition, there are fewer motor neurons in the spinal cord, so some muscle atrophy may result from reduced innervation. Even the neurons that do remain produce less acetylcholine and stimulate the muscles less effectively.

Even though people typically lose muscle mass and function as they age, these effects are noticeably less in people who continue to exercise throughout life. For example, studies show that even moderate exercise can help elderly people maintain muscle mass and improve balance; it even seems to improve mental agility.

CONNECTIVE ISSUES

Ways in Which the Muscular System and Other Muscular Tissues Affect Other Organ Systems

Integumentary System
Superficial muscles support and shape the skin; facial muscles act on the skin to produce facial expressions.

Skeletal System
Muscles move the bones, stabilize the joints, and produce stress that stimulates bone ossification and remodeling; the shapes of mature bones are determined partly by the muscular stresses applied to them.

Nervous System
Muscles give expression to thoughts, emotions, and motor commands that arise in the central nervous system.

Endocrine System
Exercise stimulates the secretion of stress hormones; skeletal muscles protect some endocrine organs; muscle mass affects the body's insulin sensitivity.

Circulatory System
Skeletal muscle contractions help to move blood through veins; exercise stimulates the growth of new blood vessels; cardiac muscle pumps blood and smooth muscle governs blood vessel diameter and blood and flow.

Lymphatic and Immune Systems
Muscle contractions promote lymph flow; exercise elevates levels of immune cells and antibodies; excessive exercise can inhibit immune responses.

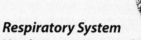

Respiratory System
Muscle contractions ventilate the lungs; skeletal muscles of the larynx and pharynx and smooth muscle of the trachea and bronchial tree regulate airflow; CO_2 generated by exercising muscles stimulates heavier breathing.

Urinary System
Skeletal and smooth muscles control urination; muscles of the pelvic floor support the bladder.

Digestive System
Skeletal muscles enable chewing and swallowing and control defecation; smooth muscle propels material through the digestive tract and also controls defecation; abdominal and lumbar muscles protect the lower digestive organs.

Reproductive System
Muscles contribute to erection and orgasm; abdominal and pelvic muscles aid in childbirth.

Study Guide

Assess Your Learning Outcomes

To test your knowledge, discuss the following topics with a study partner or in writing, ideally from memory.

7.1 Muscular Tissue and Cells (p. 204)

1. Five functions of muscular tissue
2. Why skeletal muscle is described as striated and voluntary
3. General appearance and typical dimensions of muscle fibers
4. Description of the sarcolemma, transverse tubules, sarcoplasmic reticulum, terminal cisternae, myofibrils, and myofilaments and their locations in a muscle fiber
5. The roles of glycogen and myoglobin in muscle fibers
6. The composition, molecular arrangement, and functions of thick and thin myofilaments
7. The role of the proteins myosin, actin, tropomyosin, and troponin in muscle fibers
8. Structural relationships of the A bands, I bands, Z discs, and sarcomeres; how the arrangement of the myofilaments accounts for the A and I bands
9. Why a sarcomere is considered the functional unit of a muscle fiber; the relationship between thin filaments, dystrophin, and the shortening of a sarcomere
10. Motor neurons and their relationship with skeletal muscle fibers
11. What a motor unit is, why it is called that, and the functional advantages of a muscle being composed of many motor units
12. Structural components of a neuromuscular junction and the function of each
13. The locations and roles of acetylcholine and acetylcholinesterase

7.2 Physiology of Skeletal Muscle (p. 212)

1. The four stages of muscle contraction and relaxation
2. How an arriving nerve signal excites a muscle fiber
3. Events that link muscle excitation to the onset of contraction
4. How a muscle fiber shortens; the roles of myosin, actin, tropomyosin, calcium, troponin, and ATP in this process
5. How a muscle relaxes; the roles of acetylcholinesterase, the sacroplasmic reticulum, ATP, and tropomyosin in this process

6. Why a single muscle twitch cannot perform useful muscular work; the effect of summation of twitches on muscle performance
7. How muscle tone is maintained and why this is important
8. Why skeletal muscle is subject to the length–tension relationship, and what bearing this has on muscle performance
9. The difference between isometric and isotonic muscle contraction, and how these relate to ordinary muscle actions; the roles of concentric and eccentric contraction in weight lifting
10. The two metabolic pathways for producing the ATP needed for muscle contraction, and the advantage and disadvantage of each
11. How a muscle shifts from one mode of ATP production to another over the course of a prolonged exercise
12. Causes of muscle fatigue and variables that improve endurance
13. Why a person continues to breathe heavily after the end of a strenuous exercise
14. How slow- and fast-twitch muscle fibers differ metabolically; the advantage and limitation of each; and examples of muscles in which each type is abundant
15. The cellular effects and respective benefits of resistance and endurance exercise

7.3 Cardiac and Smooth Muscle (p. 222)

1. The properties of cardiac muscle that enable it to reliably pump blood
2. The structure of cardiomyocytes and how they differ from skeletal muscle fibers
3. Significance of the prolonged contraction of cardiomyocytes
4. The significance of aerobic respiration in cardiac muscle, and how this is reflected in the internal structure of the muscle cell
5. Locations of smooth muscle
6. Reasons why smooth muscle is not limited by the length–tension relationship, and why this is important to its function
7. Similarities and differences between smooth muscle myocytes, cardiomyocytes, and skeletal muscle fibers
8. Why smooth muscle tone is important in hollow organs
9. How smooth muscle tissue differs from skeletal and cardiac muscle in its capacity for repair

7.4 Anatomy of the Muscular System (p. 225)

1. The connective tissue components of a skeletal muscle and their spatial relationship to the muscle fibers

2. Attachments between skeletal muscle and bone; the difference between an aponeurosis and other tendons

3. How the origin and insertion of a muscle are defined, and why the distinction between these is an imperfect one

4. The four functional categories of muscles that may be found at a joint, and how these four types modify each other's actions

5. Why the Latin names of muscles can be helpful in understanding their locations and functions once one becomes accustomed to them

6. The proper anatomical names of several muscles that have popular nicknames among body builders

7. The locations, origins, insertions, and actions of the following facial muscles—the frontalis, orbicularis oculi, orbicularis oris, zygomaticus, depressor anguli oris, and buccinator—and some ordinary activities in which these muscles are involved

8. The same for the following muscles of mastication: temporalis, masseter, and pterygoids

9. The same for the following muscles of head and neck movement: sternocleidomastoid and trapezius

10. The same for the following abdominal muscles: rectus abdominis, external and internal abdominal obliques, and transverse abdominal

11. The same for the erector spinae

12. The same for the following muscles acting on the scapula: pectoralis minor, serratus anterior, trapezius, levator scapulae, and rhomboids

13. The same for additional muscles that act on the humerus: pectoralis major, latissimus dorsi, deltoid, and teres major

14. The same for the four rotator cuff muscles: the supraspinatus, infraspinatus, teres minor, and subscapularis; why their tendons are called the rotator cuff, and some actions that often cause rotator cuff injuries

15. The same for the following muscles acting on the elbow or forearm: biceps brachii, brachialis, brachioradialis, triceps brachii, pronator teres, pronator quadratus, and supinator

16. The difference between intrinsic and extrinsic muscles of the hand, and why there are so many hand muscles

17. The locations and functional significance of the flexor retinaculum, extensor retinaculum, and carpal tunnel

18. The locations, origins, and insertions, and actions of the following wrist and finger flexors—flexor carpi radialis, flexor carpi ulnaris, flexor digitorum superficialis, and flexor digitorum profundus—and some ordinary activities in which these muscles are involved

19. The same for the following wrist and finger extensors: extensor digitorum, extensor carpi ulnaris, extensor carpi radialis longus, and extensor carpi radialis brevis

20. The same for the two components of the iliopsoas muscle complex: the iliacus and psoas major

21. The same for the three gluteal muscles: gluteus maximus, medius, and minimus

22. The same for the adductor and abductor groups of thigh muscles, considering each of the two groups collectively

23. The same for the four heads of the quadriceps femoris; the sartorius; and the three hamstring muscles

24. The same for the anterior muscles of the leg: extensor digitorum longus and tibialis anterior

25. The same for the posterior muscles of the leg: gastrocnemius, soleus, tibialis posterior, flexor digitorum longus, fibularis brevis, and fibularis longus

26. Definition of the intrinsic muscles of the foot, and their major functions

Testing Your Recall

1. The relatively light middle portion of the A band of skeletal muscle
   a. has myosin but no actin.
   b. is the site of the Z disc.
   c. is light because of the glycogen stored there.
   d. has thin myofilaments but no thick myofilaments.
   e. marks the boundary between one sarcomere and the next.

2. Before a muscle fiber can contract, ATP must bind to
   a. a Z disc.
   b. the myosin head.
   c. tropomyosin.
   d. troponin.
   e. actin.

3. ACh receptors are found in
   a. the synaptic vesicles.
   b. the terminal cisternae.
   c. the thick filaments.
   d. the thin filaments.
   e. the sarcolemma.

4. In the activation of muscle contraction, acetylcholine is released from
   a. the terminal cisternae.
   b. the sarcolemma.
   c. synaptic vesicles.
   d. the synaptic cleft.
   e. intercalated discs.

5. Slow-twitch fibers have all of the following *except*
   a. an abundance of myoglobin.
   b. an abundance of glycogen.
   c. high fatigue resistance.
   d. a deep red color.
   e. a high capacity to synthesize ATP aerobically.

6. Which of these muscles is an extensor of the neck?
   a. external oblique
   b. sternocleidomastoid
   c. trapezius
   d. temporalis
   e. latissimus dorsi

7. Which of the following muscles does *not* extend the hip joint?
   a. quadriceps femoris
   b. gluteus maximus
   c. biceps femoris
   d. semitendinosus
   e. semimembranosus

8. Both the gastrocnemius and _____ muscles insert on the heel by way of the calcaneal tendon.
   a. semimembranosus
   b. flexor digitorum longus
   c. tibialis anterior
   d. soleus
   e. quadriceps femoris

9. Which of the following muscles contributes the most to smiling?
   a. levator palpebrae superioris
   b. orbicularis oris
   c. zygomaticus minor
   d. masseter
   e. mentalis

10. Four scapular muscles contribute tendons to the rotator cuff, including all of these *except*
    a. the supraspinatus.
    b. the teres minor.
    c. the subscapularis.
    d. the levator scapulae.
    e. the infraspinatus.

11. Thick myofilaments consist mainly of the protein _____.

12. The neurotransmitter that stimulates skeletal muscle is _____.

13. A state of continual partial muscle contraction is called _____.

14. A bundle of muscle fibers surrounded by perimysium and seen as the grain in a cut of meat is called a/an _____.

15. Skeletal muscle contraction requires the release of _____ ions from the sarcoplasmic reticulum.

16. Cardiac and smooth muscle are called _____ because they can contract without nervous stimulation and are usually not subject to conscious control.

17. A circular muscle that regulates a body opening or passage is called a/an _____.

18. The knee is extended by a four-part muscle called the _____ on the anterior side of the thigh.

19. The largest muscle of the back, extending from the midthoracic to lumbar region and converging on the axilla, nicknamed the swimmer's muscle, is the _____.

20. The large bulge of the calf, just below the popliteal region, is a two-headed muscle called the _____.

Answers in appendix A

True or False

Determine which five of the following statements are false, and briefly explain why.

1. Each motor neuron supplies just one muscle fiber.

2. To initiate muscle contraction, calcium ions must bind to the myosin heads.

3. Slow-twitch fibers are relatively resistant to fatigue.

4. Thin filaments get shorter when a muscle contracts.

5. The heart is autorhythmic, but skeletal muscles are not.

6. The origin of the sternocleidomastoid muscle is the mastoid process of the skull.

7. To push someone away from you, you would use the serratus anterior more than the trapezius.

8. Excitation–contraction coupling refers to binding of the myosin head to actin.

9. The zygomaticus major can be considered an antagonist of the depressor anguli oris.

10. In lifting barbells by flexion of the elbow, the brachialis muscles provide more power than the biceps brachii muscles do.

Answers in appendix A

Testing Your Comprehension

1. Without ATP, relaxed muscle cannot contract and a contracted muscle cannot relax. Explain why.

2. Why would skeletal muscle be unsuitable for the wall of the urinary bladder? Explain how this illustrates the complementarity of form and function at a cellular and molecular level.

3. Radical mastectomy, once a common treatment for breast cancer, involved removal of the pectoralis major along with the breast. What functional impairments would result from this? What synergists could a physical therapist train a patient to use to recover some lost function?

The Nervous System I

NERVE CELLS, THE SPINAL CORD, AND REFLEXES

Nerve cells growing in laboratory culture. These were produced in an experiment to culture nervous tissue and study neural regeneration for the treatment of spinal paralysis.

Chapter Outline

BASE CAMP

Before ascending to the next level, be sure you're properly equipped with a knowledge of these concepts from earlier chapters:

- Plasma membrane structure (p. 67)
- Gated channels (p. 68)
- Acetylcholine (ACh) and ACh receptors (p. 210)

Anatomy & Physiology | **REVEALED**®
aprevealed.com

Module 7: Nervous System

The human body is thought to have at least 50 trillion cells. If all of them acted independently without regard to what others were doing, the result would be physiological chaos and death. Therefore, like any other multicellular organism, we must have a means of coordinating their activities. Two systems of internal communication serve this role: (1) the endocrine system, which uses hormones as intercellular signals and is considered in chapter 11, and (2) the nervous system, which uses a combination of electrical and chemical signals and is the subject of the next three chapters. In this chapter, we will study nerve cells, the spinal cord and its nerves, and spinal reflexes. Chapter 9 will survey the brain, its associated nerves, and the autonomic nervous system; and chapter 10 concerns the sense organs.

8.1   Cells and Tissues of the Nervous System

Expected Learning Outcomes

When you have completed this section, you should be able to:

a. state three fundamental functions of the nervous system;
b. define the two major anatomical subdivisions of the nervous system;
c. define three functional categories of neurons;
d. describe the structure of a generalized neuron and the common variations on this structure;
e. discuss the supporting cells of nervous tissue and their functions;
f. discuss the myelin sheath that envelops many nerve fibers, including its function and how it is produced; and
g. distinguish between the gray and white matter of the central nervous system.

Functions and Divisions of the Nervous System

The functions of the nervous system can be summarized in three steps: sensory, integrative, and motor. The **sensory** function is the ability to respond to stimuli within and around the body, and to generate signals that carry information about those stimuli to the spinal cord or brain. The **integrative** function is the ability to receive and process that information, store and retrieve it, and make decisions as to whether or how to respond to it. The **motor** function is the ability to issue outgoing signals to muscle and gland cells to produce a response.

The nervous system has two main anatomical subdivisions (fig. 8.1).

1. The **central nervous system (CNS)** is composed of the brain and spinal cord and is enclosed and protected by bone—the cranium and vertebral column. It carries out the integrative functions of the nervous system.

2. The **peripheral nervous system (PNS)** is composed of nerves leading to and from the CNS. The PNS provides the CNS with pathways of signal input and output, connecting it to the body's sense organs, muscles, and glands. Thus, it carries out both the sensory and motor functions.

Both the CNS and PNS have further anatomical and functional subdivisions, which we will encounter as we progress through these chapters.

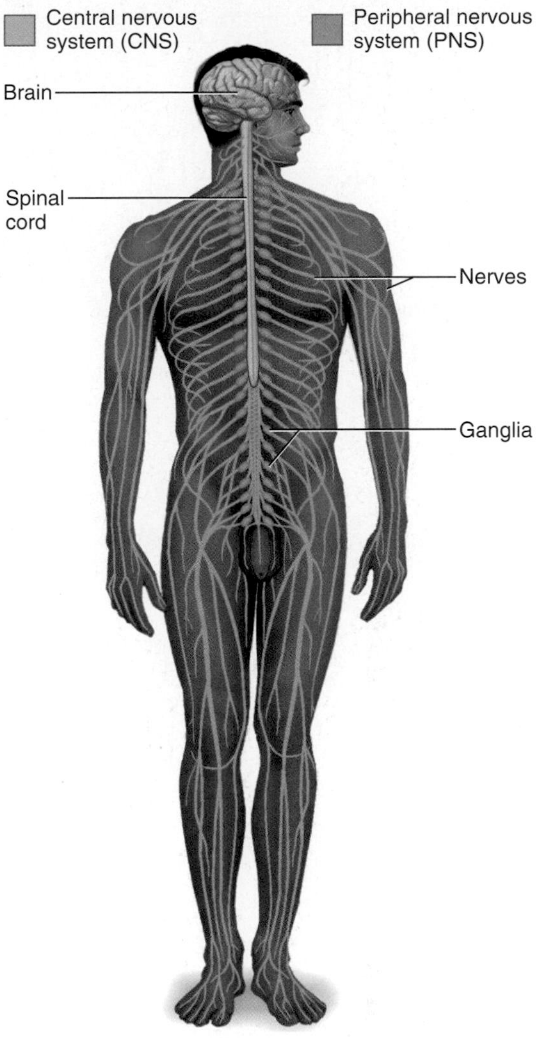

Neurons (Nerve Cells)

The agents of communication in the nervous system are **neurons (nerve cells).** A neuron usually consists of a more or less globular or stellate cell body where its nucleus is located, and two or more long fibrous processes that reach out to other cells.

Basic Structure of Neurons

Neurons are highly variable in shape, but a good starting point for understanding their structure is a *spinal motor neuron,* a cell that originates in the spinal cord and leads to a skeletal muscle—the type of neuron that enables us to move the body (fig. 8.2). The control center of the neuron is its **neurosoma**[1] (**soma** or **cell body**). It contains the nucleus and is therefore the cell's site of genetic control. It also has most of the usual organelles found in other cells: mitochondria, lysosomes, a Golgi complex, and an extensive rough endoplasmic reticulum. There are no centrioles, however, and for this and other reasons, neurons cannot undergo mitosis. Neurons have a dense supportive cytoskeleton composed of protein bundles called *neurofibrils.*

Arising from the soma, there are several thick arms that divide repeatedly into fine branches. They are named **dendrites**[2] after their striking similarity to the branches and twigs of a leafless tree in winter. They are the "receiving end" of a neuron, the primary route by which it receives signals from other neurons. Some neurons have only one dendrite, and some have thousands.

On one side of the soma is a conical mound called the *axon hillock,* which gives rise to a long process called the **axon (nerve fiber).** This is the neuron's output pathway for signals that it sends to other cells. An axon is roughly cylindrical and relatively unbranched except at its distal end. A neuron never has more than one axon, and some neurons in the brain and retina have none. At the distal end, an axon usually exhibits a profusion of fine branches, with each twig ending with a bulb called a *synaptic knob.* This will be the focus of our attention when we consider how a neuron transmits a signal to the next cell across a junction called the *synapse.*

Simplified textbook drawings cannot do justice to the structure of a neuron, but an analogy may create a useful impression of its dimensions. A typical neuron has a soma from 5 to 135 μm in diameter, whereas the axon is about 1 to 20 μm in diameter and is sometimes more than 1.5 m (5 ft) long, yet too slender to see without a microscope. If we scale the soma up to the size of a tennis ball, such an axon would be up to 1.7 km (1 mile) long and a little narrower than a garden hose, and the dendrites could densely fill a 30-seat classroom from wall

Figure 8.1 The Nervous System.

---

[1] *soma* = body

[2] *dendr* = tree, branch; *ite* = little

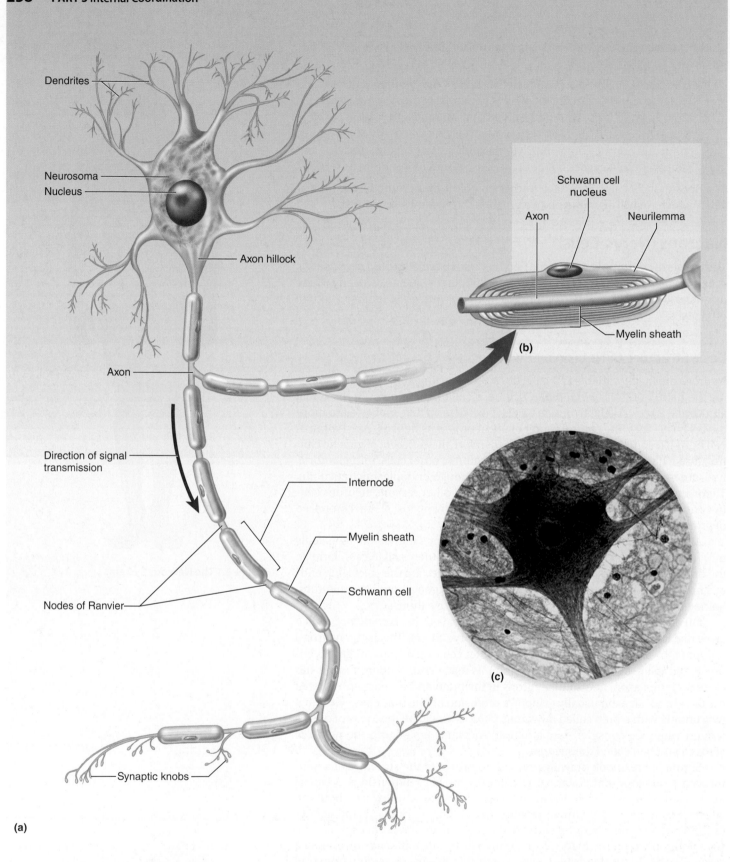

Figure 8.2 A Representative Neuron. (a) A multipolar neuron such as a spinal motor neuron. (b) Detail of the myelin sheath. The Schwann cells and myelin are explained later in this chapter. (c) Photograph of this neuron type. **AP|R**

to wall and ceiling to floor, like an impenetrable briar thicket. Such proportions are all the more impressive when we consider that the neuron must assemble molecules and organelles in that "tennis ball" soma and deliver them through its "mile-long hose" to the end of the axon. Neurons use mobile *motor proteins* to "walk" up and down the axon carrying substances to and from the soma—a process called **axonal transport.**

Apply What You Know

The axon of a neuron has a dense cytoskeleton. Considering the functions of a cytoskeleton you studied in chapter 3, suggest at least two reasons that this cytoskeleton is so necessary to an axon.

Structural and Functional Classes of Neurons

Not all neurons fit the preceding basic description. Neurons are classified structurally according to the number of processes that arise from the soma (fig. 8.3):

1. **Multipolar** neurons, like the one just described, have one axon and multiple dendrites. This is the most common type and includes most neurons of the brain and spinal cord.
2. **Bipolar** neurons have one axon and one dendrite. These include sensory neurons of hearing, smell, and vision.
3. **Unipolar** neurons have only one process leading away from the soma. It branches like a T a short distance away, with one branch bringing signals from sources such as the skin and joints and the other branch leading to the spinal cord. The dendrites are the branching sensory tips at the end of the former branch, and the rest of the fiber is considered to be the axon.

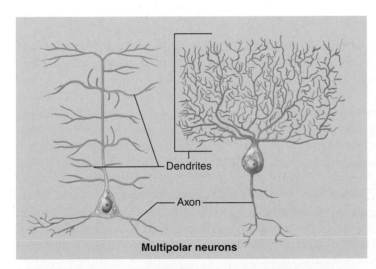

Multipolar neurons

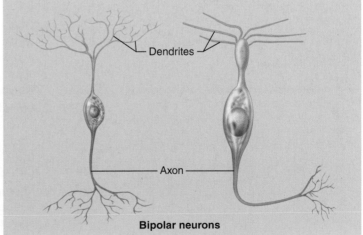

Bipolar neurons

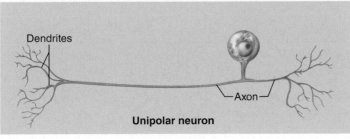

Unipolar neuron

Figure 8.3 Variation in Neuron Structure. Upper left frame, two multipolar neurons of the brain—a pyramidal cell (left) and a Purkinje cell. Upper right frame, two bipolar neurons—a bipolar cell of the retina (left) and an olfactory neuron of the nose. Lower left frame: a unipolar neuron of the type involved in the senses of touch and pain.

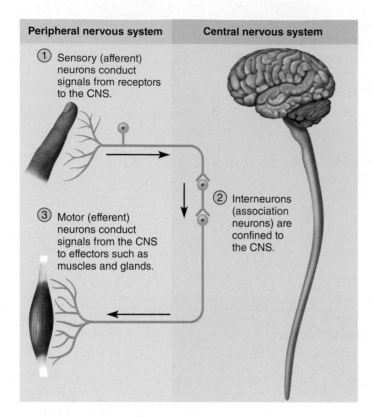

Peripheral nervous system	Central nervous system

① Sensory (afferent) neurons conduct signals from receptors to the CNS.

② Interneurons (association neurons) are confined to the CNS.

③ Motor (efferent) neurons conduct signals from the CNS to effectors such as muscles and glands.

Figure 8.4 Functional Classes of Neurons. All neurons can be classified as sensory, motor, or interneurons based on their location and the direction of signal conduction.

Functionally, neurons fit into three basic classes corresponding to the sensory, integrative, and motor functions of the nervous system (fig. 8.4):

1. **Sensory (afferent[3]) neurons** are specialized to detect stimuli and transmit information about them to the CNS. They lead from the eyes, ears, skin, joints, internal organs, and other sources to the brain or spinal cord. The word *afferent* refers to the fact that signals in these cells travel from relatively remote localities in the body toward the CNS.

2. **Interneurons (association neurons)** perform the integrative functions of the nervous system—processing, storing, and retrieving information and "making decisions" about how the body should respond to a given stimulus or situation. They are contained entirely within the CNS and are the most abundant of all neurons.

3. **Motor (efferent[4]) neurons** are specialized to carry outgoing signals from the CNS to the cells and organs that carry out its commands. They begin in the CNS and extend to muscle and gland cells elsewhere. The word *efferent* refers to the fact that these cells carry nerve signals away from the CNS.

Most sensory neurons are unipolar or bipolar; interneurons and motor neurons are generally multipolar.

Neuroglia (Supporting Cells)

Although neurons carry out the communicative function of the nervous system, they are outnumbered at least 10 to 1 by supporting cells called **neuroglia[5]** (noor-OG-lee-uh), or **glial** (GLEE-ul) cells. These cells perform various protective and "housekeeping" functions for the nervous system and aid the neurons in their functions. There are four kinds of glial cells in the central nervous system (fig. 8.5).

1. **Oligodendrocytes[6]** (OL-ih-go-DEN-dro-sites) are large bulbous cells with as many as 15 armlike processes, giving them an appearance somewhat like an octopus. Each arm reaches out to a nearby nerve fiber and spirals around it to form a layer of insulation called the *myelin sheath,* which we'll examine later.

2. **Ependymal[7] cells** are cuboidal cells that line the internal, fluid-filled cavities of the brain and spinal cord. They produce a liquid called *cerebrospinal fluid* (CSF) that fills these cavities and bathes the CNS surface. They have cilia on their surfaces that help to circulate the CSF. Functions of the CSF are discussed in chapter 9.

3. **Microglia** are small phagocytic cells that wander through the CNS tissue searching out and destroying tissue debris, infectious microorganisms, and other foreign matter. They become concentrated in areas damaged by infection, trauma, or stroke.

---

[3] *af,* from *ad* = toward; *fer* = to carry

[4] *ef,* from *ex* = out; *fer* = to carry

[5] *glia* = glue

[6] *oligo* = a few; *dendro* = branches; *cyte* = cell

[7] *ep* = above, upper; *endym* = garment, wrap

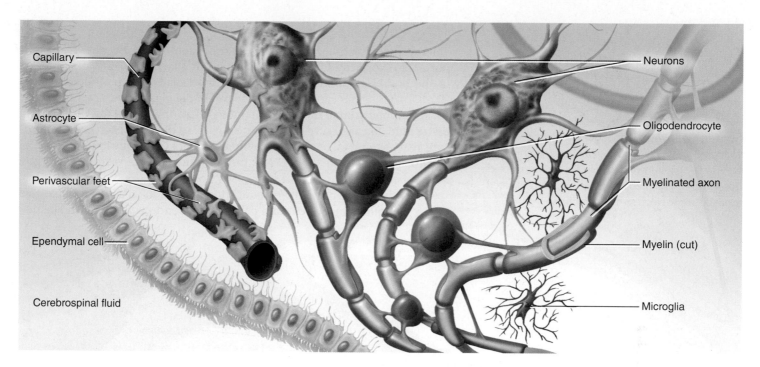

Figure 8.5 Neuroglia of the Central Nervous System.

- *What cells in the peripheral nervous system perform the same function as the oligodendrocytes seen in this figure?*

4. **Astrocytes,**[8] the most abundant glial cells in the CNS, constitute over 90% of the brain tissue in some areas. They cover the brain surface and most surfaces of the neurons except where one neuron meets another at a synapse. They have the most diverse roles of all glial cells. They form a supportive framework for CNS tissue; they issue *perivascular feet* that envelop blood vessels and help to form a *blood–brain barrier,* described in chapter 9; they secrete *nerve growth factors* to promote neuron growth and synapse formation; they stabilize the chemical environment of the CNS; they convert blood glucose to lactate, which then nourishes the neurons; and they form scar tissue in damaged regions of the CNS.

Two more glial cell types occur in the peripheral nervous system.

1. **Satellite cells** surround the cell bodies of peripheral neurons. They insulate them electrically and regulate their chemical environment.

2. **Schwann**[9] **cells** (pronounced "shwon") wrap around nerve fibers of the PNS, enclosing each in a sleeve called the **neurilemma**[10] (NOOR-ih-LEM-ah). In most cases, a Schwann cell also spirals repeatedly around the nerve fiber, depositing layer after layer of its own membrane between the neurilemma and axon. This wrapping constitutes the myelin sheath of a peripheral nerve fiber. That is, Schwann cells perform for the PNS what oligodendrocytes do for the CNS.

Schwann cells are also necessary for the regeneration and healing of damaged nerve fibers. Since they are absent from the CNS and oligodendrocytes cannot perform this regenerative role, damaged fibers in the CNS are lost forever. Being enclosed in bone, however, nervous tissue of the CNS is injured far less frequently than nerves of the PNS.

---

[8] *astro* = starlike; *cyte* = cell

[9] Theodor Schwann (1810–82), German histologist

[10] *neuri* = nerve; *lemma* = sheath, husk

The Myelin Sheath

The **myelin** (MY-eh-lin) **sheath** bears closer examination because it is so important in signal conduction by nerve fibers. It is formed by oligodendrocytes in the brain and spinal cord and Schwann cells in the peripheral nerves (fig. 8.6). It is composed of layer upon layer of the plasma membrane of those cells, wrapped around the nerve fiber and insulating it from the surrounding tissue fluid, like electrical tape around a wire. In the CNS, each oligodendrocyte reaches out to several nearby nerve fibers and contributes a myelin sheath to each of them. In the PNS, one Schwann cell wraps around only one nerve fiber.

In both cases, a nerve fiber is much longer than the reach of a single glial cell, so it requires many Schwann cells or oligodendrocytes to myelinate one

Figure 8.6 Myelination. (a) A Schwann cell of the PNS, wrapping repeatedly around an axon to form the multilayered myelin sheath. The myelin spirals outward away from the axon as it is laid down. The outermost coil of the Schwann cell constitutes the neurilemma. (b) An oligodendrocyte of the CNS wrapping around the axons of multiple neurons. Here, the myelin spirals inward toward the axon as it is laid down. (c) A myelinated axon (top) and unmyelinated axon (bottom) (electron micrograph). **AP|R**

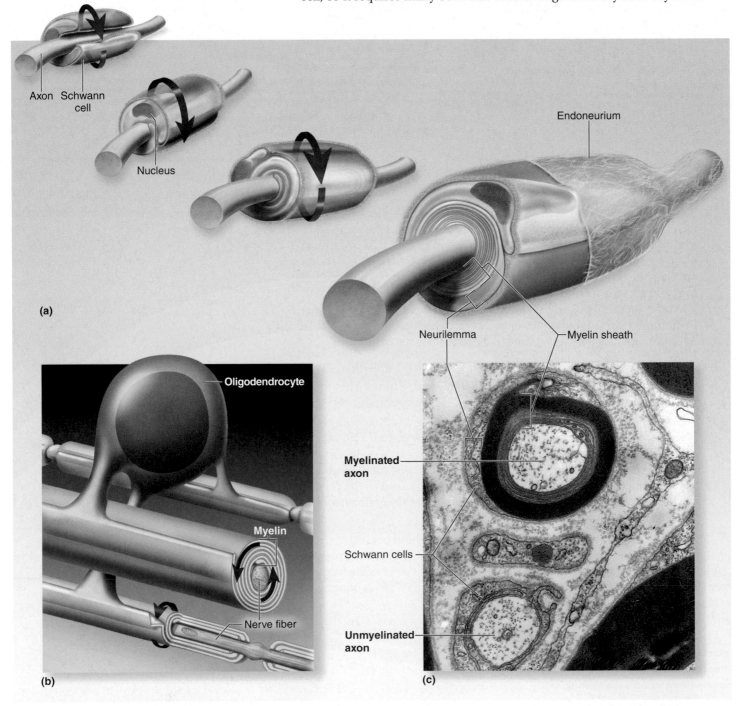

Clinical Application 8.1

MULTIPLE SCLEROSIS

Multiple sclerosis[11] (MS) is a degenerative disorder of the myelin sheath. Its cause remains uncertain, but it is widely thought to be an autoimmune reaction triggered by a virus in genetically susceptible individuals. It is characterized by deterioration of the oligodendrocytes and myelin sheaths of the CNS, especially in individuals between the ages of 20 and 40. MS is named for the hardened scar tissue that replaces the myelin. This degeneration and scarring disrupts nerve conduction, with effects that depend on what part of the CNS is involved—double vision, blindness, speech defects, neurosis, tremors, or numbness, for example. Patients experience variable cycles of milder and worse symptoms until they eventually become bedridden. As yet there is no cure, but there is conflicting evidence of how much it shortens a person's life expectancy, if at all. Some MS victims die within 1 year of diagnosis, but many live with it for 25 or 30 years.

nerve fiber. Consequently, the myelin sheath is segmented. The gaps between segments are called **nodes of Ranvier**[12] (RON-vee-AY), and the myelin-covered segments are called **internodes** (see fig. 8.2). The internodes are about 0.2 to 1.0 mm long. The relevance of nodes and internodes to nerve signal conduction will become apparent later in this chapter.

Forms of Nervous Tissue

In the central nervous system, we find two fundamental types of nervous tissue called white and gray matter. **White matter** consists of bundles of nerve fibers called **tracts** that travel up and down the spinal cord, between one region of the brain and another, or between brain and cord. Many fibers of these tracts are myelinated; myelin gives the white matter a glistening, pearly white color. All fibers in one tract have a similar origin, destination, and function. We'll take a closer look at tracts of the spinal cord in this chapter and tracts of the brain in chapter 9. There are no nerves in the brain or spinal cord. The body's bundles of nerve fibers are called *nerves* in the PNS and *tracts* in the CNS.

Gray matter is where the neurosomas, dendrites, and synapses are located. There is relatively little myelin here, so this tissue has a duller color in fresh nervous tissue. It is the information-processing part of the CNS, whereas the white matter is more like a telephone cable—a bundle of "wires" (nerve fibers) carrying signals from place to place but not processing signals along the way.

In the spinal cord, white matter forms the surface tissue and the gray matter forms the inner core. In the brain, by contrast, white matter forms most of the deep tissue, whereas gray matter forms a surface layer as well as a few deep masses embedded in the white matter. In the CNS, these deep gray matter masses surrounded by white matter are called **nuclei** (not to be confused with cell nuclei). Some authorities call them *ganglia,* but this book will follow the practice of reserving that term for the PNS, where we will visit the subject later in this chapter. A given nucleus of CNS gray matter typically has a specific function such as regulating your respiratory rhythm or controlling your appetite.

[11] *scler* = hard, tough; *osis* = condition
[12] L. A. Ranvier (1835–1922), French histologist and pathologist

Before You Go On

Answer these questions from memory. Reread the preceding section if there are too many you don't know.

1. Which nervous system function pertains to information input to the CNS? Which function pertains to its output to muscles and glands? What are the terms for the input and output neurons that correspond to these two functions?
2. Distinguish between the central and peripheral nervous systems and identify the components of each.
3. Make a sketch of a generalized multipolar neuron and label its dendrites, soma, axon, and synaptic knobs.
4. What types of neuroglia occur in the central nervous system? What are their functions? Name the types in the peripheral nervous system and state their functions.
5. Describe the structure of the myelin sheath found around many nerve fibers.
6. What is the difference between the gray and white matter of the central nervous system? Which of these is the seat of the CNS's integrative function?

8.2 The Physiology of Neurons

Expected Learning Outcomes

When you have completed this section, you should be able to:

a. explain how a neuron maintains a resting potential across its plasma membrane;
b. describe the mechanism behind the voltage changes that occur when a neuron is stimulated;
c. explain how a neuron conducts a signal from the soma to the end of its axon;
d. explain how a neuron transmits information to the next cell; and
e. describe some simple networks in which neurons work in groups.

Producing a Resting Membrane Potential

Nerve signals are essentially electrical events created by the movement of sodium and potassium ions. Every living cell is like a little biological battery. Just as a battery has a positive and a negative pole, a cell has a positive charge on the outer surface of the plasma membrane and a negative charge on the inner surface. In neurons, this charge difference, called the **resting membrane potential (RMP),** is a readiness for action, a starting point for generating a nerve signal. Thus, to understand how nerve cells communicate, the first thing we must address is how neurons produce the RMP; how does the battery get "charged up"?

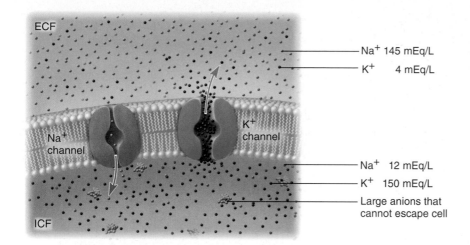

Figure 8.7 Ionic Basis of the Resting Membrane Potential. Note that sodium ions are much more concentrated in the extracellular fluid (ECF) than in the intracellular fluid (ICF), while potassium ions are more concentrated in the ICF. Large anions unable to penetrate the plasma membrane give the cytoplasm a negative charge relative to the ECF. Milliequivalents per liter (mEq/L) is a unit of concentration that reflects the number of particles per given volume of fluid and the electrical charge on each particle (see appendix C).

In a battery, the charge difference results from a difference in the concentration of electrons at the two poles. In a cell, it results from a difference in the concentration of sodium, potassium, and other ions on the two sides of the membrane. Sodium ions (Na^+) are about 12 times as concentrated on the outside of the membrane as on the inside, and potassium ions (K^+) are almost 38 times as concentrated on the inside (fig. 8.7). In addition, the cell contains larger negative ions (anions) such as phosphates, proteins, and nucleic acids that cannot pass out through the membrane. All of these inequalities of ion distribution collectively make the interior of the cell negatively charged relative to the exterior.

When a battery has such an unequal charge distribution, we say it is charged; when a cell membrane does so, we say it is **polarized.** Electrical charges are measured in volts (V) or fractions of a volt (millivolts, or mV). For example, an automobile battery typically has a charge of 12 V and a flashlight battery, 1.5 V. The charges across a cell membrane are much lower—typically an RMP of about –70 mV in neurons. The minus sign indicates the negative internal side of the membrane.

The plasma membrane includes integral proteins that act as channels for Na^+ and K^+, and the slow constant leakage of these ions through the membrane always threatens to abolish the RMP. However, the sodium–potassium (Na^+-K^+) pump in the membrane continually ejects 3 Na^+ and pumps in 2 K^+ for each cycle of the pump, thereby maintaining the negative internal charge (see p. 75). The Na^+-K^+ pump requires ATP to do this, and accounts for about 70% of the energy requirement of the nervous system.

Generating an Action Potential

The first step in nerve signaling is excitation of a neuron, usually at a dendrite. Often, the stimulus is a chemical (ligand) that binds to channel proteins. In response, the channels open and allow sodium ions to flow into the cell like water through a faucet that has just been opened (fig. 8.8). Since they are positive ions, the inflowing Na^+ cancels some of the internal negative charge and the membrane potential drifts from the –70 mV resting voltage toward zero. Such a shift to a less negative voltage is called **depolarization.** Any such voltage change occurring near the point of stimulation is called a **local potential.**

Figure 8.8 Excitation of a Neuron by a Chemical Stimulus. When the chemical (ligand) binds to a receptor on the neuron, the receptor acts as a ligand-gated channel that opens and allows Na^+ to diffuse into the cell. This depolarizes the plasma membrane.

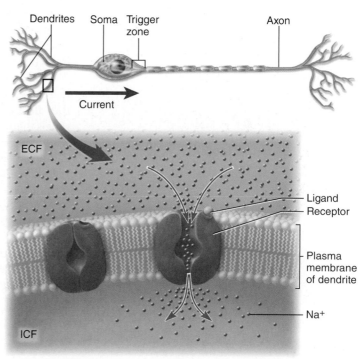

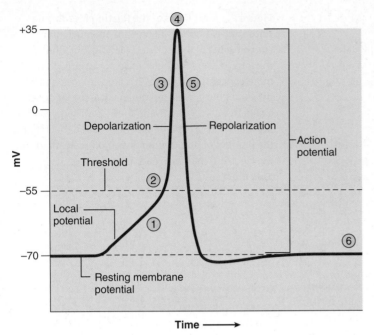

① Stimulation of the neuron creates a gradually rising local potential.

② If a threshold voltage is reached at the axon hillock, voltage-gated Na^+ channels open and initiate an action potential.

③ Na^+ inflow causes a rapidly rising voltage. Na^+ channels begin to close at 0 mV, but the voltage typically reaches about +35 mV by the time all of them close.

④ Membrane voltage typically reaches about +35 mV by the time all Na^+ channels close. By the time, K^+ channels have opened.

⑤ K^+ outflow causes the voltage to drop back into the negative range, slightly overshooting the original resting membrane potential (RMP).

⑥ After the action potential has passed, ion diffusion restores the original RMP.

Figure 8.9 An Action Potential. AP|R

When Na^+ enters a neuron, it diffuses under the plasma membrane to adjacent regions of the cell, changing the voltage there as well. A wave of excitation thus travels from the point of stimulation toward the cell's axon. The axon hillock and nearby region of the axon have a high density of **voltage-gated channels** for sodium and potassium. If the local potential spreads this far and attains a minimum voltage called **threshold,** it opens these channels and produces a more dramatic shift in membrane voltage called an **action potential** (fig. 8.9). A neuron is said to "fire" when it produces an action potential.

As Na^+ rushes into the cell, it neutralizes the negative charges near the inside of the membrane, so the membrane voltage shoots rapidly upward. The more it rises, the more voltage-gated Na^+ channels open; the more gates that open, the more Na^+ enters the cell. Thus, there occurs a rapidly accelerating positive feedback loop that causes the membrane voltage to rise very quickly. As the voltage passes 0 mV, the Na^+ channels start to close, but by the time all of them do so, the voltage generally has risen to about +35 mV.

Voltage-gated potassium channels in the hillock also open, but more slowly. Potassium ions rapidly leave the cell when these gates open, both because they flow down their concentration gradient and because the incoming Na^+ repels them, like giving an extra push. This effect lasts beyond the +35 mV peak where the Na^+ channels have closed. The outflow of K^+ makes the interior of the cell turn negative again; the voltage drops to and slightly beyond the original resting membrane potential. The shift of the voltage back into the negative numbers is called **repolarization.** For a short time afer the action potential, the membrane of the neuron is in a **refractory period** in which it cannot be stimulated to fire again.

In sum, stimulation of a neuron leads to a local potential, the depolarization of a short section of the membrane. If the stimulus is strong enough, the signal will flow all the way to the axon hillock, where it may stimulate the opening of voltage-gated channels and produce an action potential.

Conducting a Nerve Signal to Its Destination

We have seen how a nerve signal is initiated; now we examine how it travels to its final destination. The action potential is a voltage spike over a limited area of plasma membrane. However, it triggers another action potential in the membrane immediately ahead of it, and that action potential triggers another, and so forth. Thus, we get a chain reaction of one action potential after another along the length of a nerve fiber. This chain reaction constitutes the **nerve signal (nerve impulse).** An illuminating analogy to this is standing up a long row of dominoes and pushing the first one over. When that domino falls, it pushes over the second, and so forth—and the chain reaction produces a wave of energy traveling to the end of the line. No one domino moves to the other end of the line; a falling domino is a local event. Similarly, an action potential is a local event, but it triggers the next one and, like the row of falling dominoes, we get a wave of energy traveling from one end of the axon to the other. That traveling wave is the nerve signal (fig. 8.10). Action potentials do not travel; nerve signals do.

An unmyelinated nerve fiber has a high density of voltage-gated Na$^+$ and K$^+$ channels along its entire length, so everywhere an action potential occurs, it can excite another one immediately ahead of it until the signal arrives at the end of the fiber. Any region of membrane that has just fired is in its refractory period and cannot immediately fire again. This prevents a nerve signal from backing up and returning to the soma; it can only go forward to the distal end of the axon.

Things are different in a myelinated fiber. There are very few voltage-gated ion channels in the myelin-covered internodes—not enough to generate an action potential. No action potentials could occur in the internodes anyway, since myelin insulates the nerve fiber from the surrounding extracellular fluid (ECF) and no significant amount of Na$^+$ can get to the nerve fiber. At each node of Ranvier, however, the nerve fiber is exposed to the ECF, and here is where we find the greatest concentration of ion channels. This is the only place at which a myelinated fiber can generate action potentials.

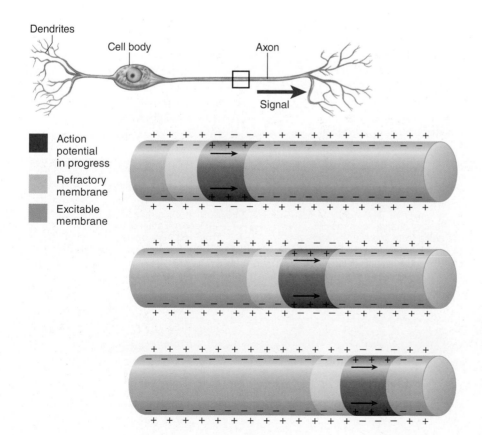

Figure 8.10 Signal Conduction in an Unmyelinated Nerve Fiber. Action potentials are occurring in areas colored red; note the reversal of membrane polarity in those areas. Yellow areas are in a refractory period of reduced excitability, and green areas of membrane are fully excitable, ready to respond. AP|R

- *What prevents a nerve signal from backing up and returning to the cell body?*

From one node to the next, the signal travels by a very rapid process that is beyond the scope of this book. It becomes weaker with distance, however, and by the time it reaches the next node of Ranvier, it is just strong enough to open the voltage-gated sodium channels there. A new inflow of Na⁺ generates action potentials at this node and boosts the signal back to its original strength. Since action potentials occur only at the nodes, it appears as if the nerve signal were jumping from node to node; this mode of conduction has been called **saltatory**[13] **conduction** (fig. 8.11).

Nerve signals travel more rapidly in myelinated fibers than in unmyelinated ones of comparable diameter. This is because the transfer along the internodes is faster than the generation of action potentials at the nodes, and for most of the distance that it travels, a nerve signal passes through internodes. It slows down slightly at each node of Ranvier, but the advantage of the nodes is to reamplify the signal back to its original strength. Internodes have the advantage of speed; nodes of Ranvier have the advantage of renewing signal strength. Even if a nerve signal travels a meter or more from neurosoma to the end of the axon, it is just as strong at the end as it was at the beginning—much like the last domino in a line falling just as forcefully as the first one did.

[13] from *saltare* = to leap or dance

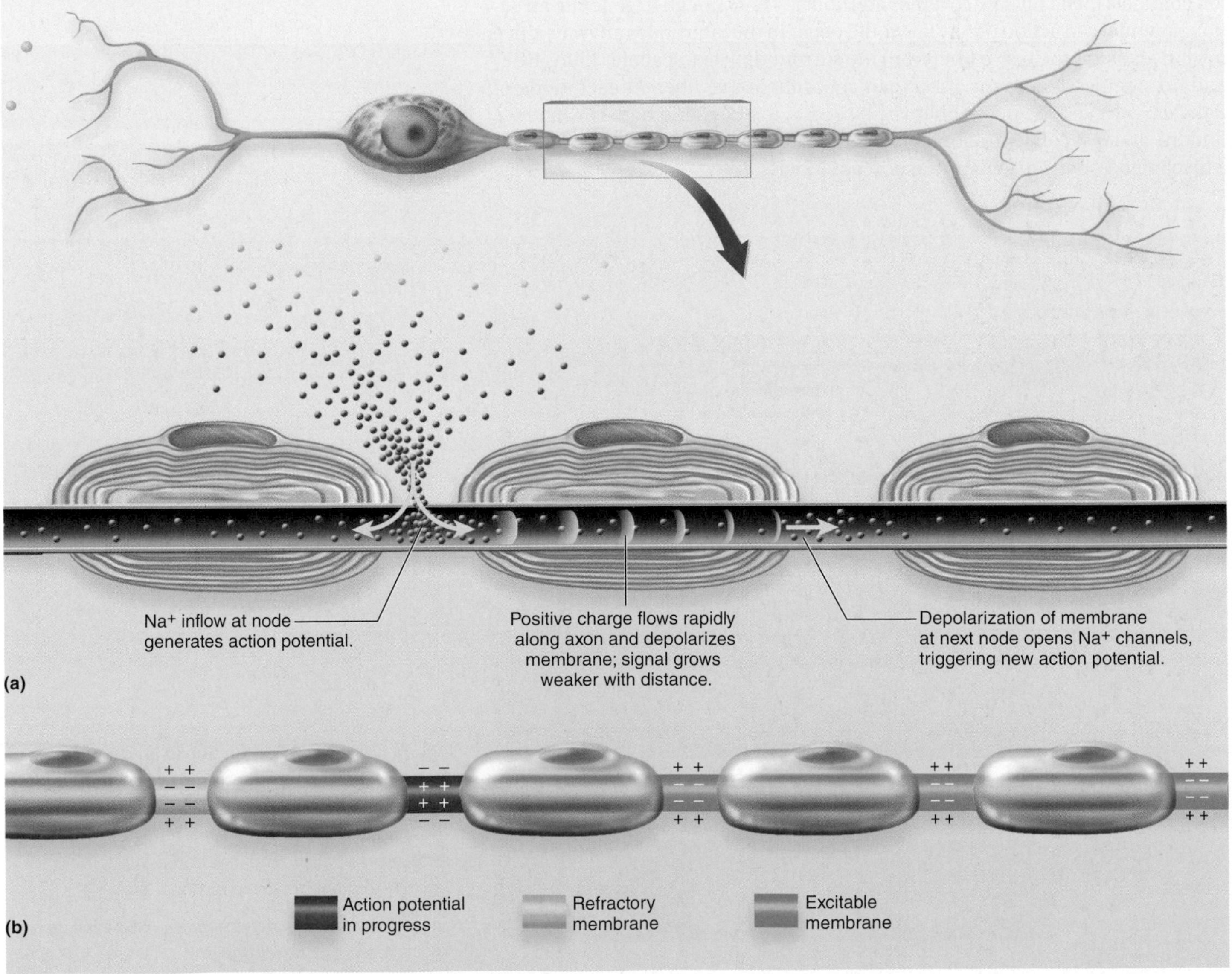

Figure 8.11 Signal Conduction in a Myelinated Nerve Fiber. (a) Sodium inflow at a node of Ranvier initiates a force that is quickly transferred from ion to ion along the internode to the next node, getting weaker with distance. At the next node, Na⁺ ion (charge) density is just sufficient to open new voltage-gated channels and repeat the process. (b) Action potentials can occur only at nodes of Ranvier, so the nerve signal appears to jump from node to node. **AP|R**

Na⁺ inflow at node generates action potential.

Positive charge flows rapidly along axon and depolarizes membrane; signal grows weaker with distance.

Depolarization of membrane at next node opens Na⁺ channels, triggering new action potential.

(a)

(b)

Action potential in progress | Refractory membrane | Excitable membrane

How fast does a nerve signal travel? This depends on the diameter of the nerve fiber and the presence or absence of myelin. The slowest fibers are small unmyelinated ones (2–4 μm in diameter), which carry signals at a speed of 0.5 to 2.0 m/s. Even in a person 6 ft (or 2 m) tall, however, this is fast enough to deliver a signal almost anywhere in the body within 1 or 2 seconds—quite adequate for such processes as digestive secretion or focusing the eye. A fiber of the same diameter but with myelin conducts signals 3 to 15 m/s, and large myelinated fibers (up to 20 μm in diameter) conduct signals as fast as 120 m/s. Large myelinated fibers are therefore important in cases that require quick responses, such as the reflexes described later in this chapter.

Transmitting a Nerve Signal to the Next Cell

All good things must come to an end, and so it is with the wave of action potentials; they reach the end of the nerve fiber, where there is a narrow gap between it and the next cell. How is the neuron to communicate with the cell across the gap? In the early twentieth century, many biologists refused to believe such gaps existed. They thought it was like a cut in a wire and would stop the signal at that point.

But this is where the chemical part of nervous communication comes in. The point where the two cells meet is called a **synapse** (fig. 8.12). The nerve signal arrives at the synapse by way of the **presynaptic neuron,** and the desirable effect is to stimulate the next cell, the **postsynaptic neuron.** The gap between the two is called the **synaptic cleft.**

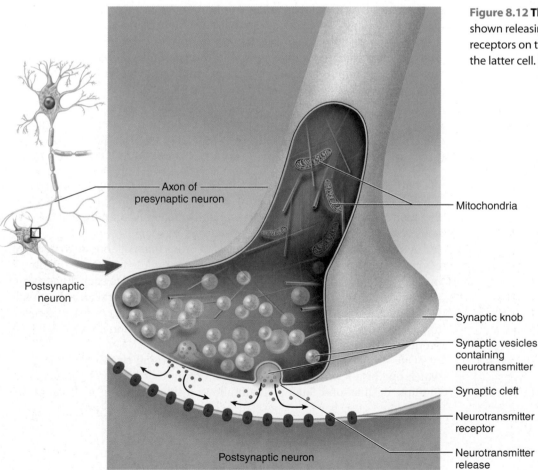

Figure 8.12 The Synapse. The presynaptic neuron is shown releasing a neurotransmitter, which binds to receptors on the postsynaptic neuron and may excite the latter cell. AP|R

Axon of presynaptic neuron

Postsynaptic neuron

Mitochondria

Synaptic knob

Synaptic vesicles containing neurotransmitter

Synaptic cleft

Neurotransmitter receptor

Neurotransmitter release

Postsynaptic neuron

The presynaptic neuron ends in a dilated tip called a **synaptic knob.** In this knob are numerous spheroidal organelles called **synaptic vesicles** loaded with chemicals called **neurotransmitters.** The arrival of a nerve signal at the end of the presynaptic fiber triggers several synaptic vesicles to undergo exocytosis, releasing their neurotransmitter into the synaptic cleft. Neurotransmitter molecules diffuse across the cleft and bind to receptors on the membrane of the postsynaptic neuron.

One such neurotransmitter is **acetylcholine (ACh).** In chapter 7, we studied its binding to receptors on the surface of a muscle fiber. The receptors are membrane channels that open in response to ACh and allow Na^+ and K^+ to flow through, producing a local potential in the postsynaptic cell. If local potentials in the postsynaptic cell are great enough to depolarize it to a critical voltage called its **threshold,** that cell fires and forwards the signal to another cell farther down the line.

Each synapse slows down a nerve signal, creating a **synaptic delay** of typically 0.5 ms. Synapses, however, are the points at which the nervous system makes all of its decisions. Without them, we would have no memories, no ability to determine when another neuron fires or when a gland or muscle cell responds—indeed, little or no ability to control the body at all. The decision-making role of synapses more than compensates for the small price we pay in synaptic delay.

Stimulation of the postsynaptic cell is very brief. The neurotransmitter in the cleft may be chemically broken down by enzymes in the cleft or reabsorbed by the presynaptic neuron, or it may simply diffuse away from the cleft into the surrounding tissue fluid. A failure to remove neurotransmitter and stop synaptic transmission can result in consequences as severe as muscle spasms, paralysis, and suffocation (see Clinical Application 7.1, p. 210).

Acetylcholine is but one neurotransmitter. More than a hundred others are known, such as norepinephrine, serotonin, and dopamine, which play key roles in mood, alertness, muscular control, glandular secretion, and other functions. The importance of dopamine in motor control and Parkinson disease is discussed in Perspectives on Health on page 272.

Neural Pools and Circuits—Neurons Working in Groups

We have so far considered neurons in isolation or in a simple linear series joined by synapses. However, neurons actually function in tangled webs of interaction called *neural pools* and *circuits.* The functional complexity of the nervous system arises from the way these neurons interact.

A **neural pool** is a localized cluster of neurons that collaborate to perform a specific physiological function, such as generating the breathing rhythm or monitoring body temperature. Neural pools are connected to each other, as are neurons within a pool, through varied pathways called **neural circuits,** much like the wiring or circuit boards of electronic devices connect transistors, resistors, capacitors, and other simple components into functionally sophisticated products.

One simple type of neural circuit is the **diverging circuit** (fig. 8.13a), in which one or a few neurons produce an output that ultimately branches to multiple destinations. Such a circuit allows a small group of neurons in the

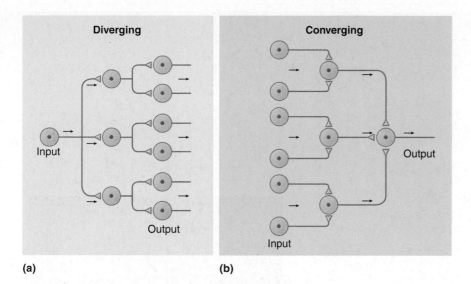

Figure 8.13 Neural Circuits. Arrows indicate the direction of the nerve signal.

brain, for example, to issue commands that ultimately reach thousands of muscle fibers to produce coordinated body movements.

Conversely, many neurons work in **converging circuits** (fig. 8.13b), in which multiple *input neurons* converge on fewer and fewer neurons, and ultimately one or a few *output neurons*. For example, your sense of balance—knowing whether your body is tipping and whether you need to take any corrective action—depends on input from your eyes, your inner ears, and stretch receptors in your neck that monitor the head–neck angle (so you know whether your whole body is falling or you are just cocking your head to one side). The neural pool that receives all this input can then integrate it and determine whether the brain should issue an output to certain muscles to catch your balance.

Before You Go On

Answer these questions from memory. Reread the preceding section if there are too many you don't know.

7. What is the difference between a resting membrane potential and an action potential? Explain the role of sodium and potassium ions in each of these.

8. Explain how an unmyelinated nerve fiber conducts a signal from the beginning to the end of an axon. How does the process differ in a myelinated nerve fiber? Which one is faster?

9. Describe the structure of a synapse.

10. What is the role of a neurotransmitter in communication from one neuron to the next? Name some neurotransmitters.

11. What is the difference between a converging circuit and a diverging circuit? Describe at least one bodily process in which each of these may be involved.

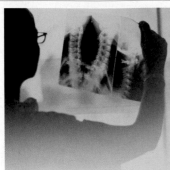

PERSPECTIVES ON HEALTH

Alzheimer and Parkinson diseases are the two most common degenerative nervous system disorders. Both are associated with neurotransmitter deficiencies.

Alzheimer Disease

Alzheimer[14] disease (AD) may begin before the age of 50 with signs so slight and ambiguous that early diagnosis is difficult. A common early sign is memory loss, especially for recent events. A person with AD may ask the same questions repeatedly, show a reduced attention span, and become disoriented and lost in previously familiar places. Family members often feel helpless as they watch their loved one's personality gradually deteriorate beyond recognition. The AD patient may become moody, confused, or even paranoid, combative, and hallucinatory. Some lose the ability to read, write, talk, walk, and eat. Death ensues from complications of confinement and immobility, such as pneumonia.

AD affects about 11% of the U.S. population over the age of 65; the incidence rises to 47% by age 85. It accounts for nearly half of all nursing home admissions and is a leading cause of death among the elderly. AD claims about 100,000 lives per year in the United States.

Diagnosis can be confirmed by autopsy. There is atrophy of some of the gyri (folds) of the cerebral cortex and the hippocampus, an important center of memory. Nerve cells exhibit *neurofibrillary tangles*—dense masses of broken and twisted cytoskeleton (fig. 8.14). The number of tangles is proportional to the severity of the disease. In the intercellular spaces, there are *senile plaques* consisting of aggregations of cells, altered nerve fibers, and a core of *β-amyloid protein*—the breakdown product of a glycoprotein of plasma membranes. This protein is widely believed to be the crucial factor that triggers all the other aspects of AD pathology.

Three genes on chromosomes 1, 14, and 21 have been implicated in various forms of early- and late-onset AD. Interestingly, persons with Down syndrome (trisomy 21), who have three copies of chromosome 21 instead of the usual two, tend to show early-onset Alzheimer disease. Nongenetic (environmental) factors also seem to be involved.

As for treatment, considerable attention now focuses on trying to halt β-amyloid formation or stimulate the immune system to clear

[14] **Alois Alzheimer (1864–1915), German neurologist**

(a)

Shrunken gyri
Wide sulci

(b)

Neurons with neurofibrillary tangles
Senile plaque

Figure 8.14 Alzheimer Disease (AD). (a) Brain of a person who died of AD. Note the shrunken folds of cerebral tissue (gyri) and wide gaps (sulci) between them, indicating atrophy of the brain. (b) Cerebral tissue from a person with AD. Neurofibrillary tangles are present within the neurons, and a senile plaque is evident in the extracellular matrix.

β-amyloid from the brain tissue, but clinical trials in both of these approaches have been suspended until certain serious side effects can be resolved. AD patients show deficiencies of acetylcholine (ACh) and nerve growth factor (NGF). Some patients show improvement when treated with NGF or cholinesterase inhibitors, but results so far have been modest.

Parkinson Disease

*P*arkinson[15] *disease (PD),* or *parkinsonism,* is a progressive loss of motor function beginning in a person's 50s or 60s. It is due to degeneration of dopamine-releasing neurons in a brainstem tissue called the *substantia nigra.* A gene has been identified for a hereditary form of PD, but most cases are nonhereditary and of little-known cause.

Dopamine is an inhibitory neurotransmitter that normally prevents excessive activity in motor centers of the brain called the *basal nuclei.* Degeneration of dopamine-releasing neurons leads to an excessive ratio of ACh to dopamine, causing hyperactivity of the basal nuclei. As a result, a person with PD suffers involuntary muscle contractions. These take such forms as shaking of the hands (tremor) and compulsive "pill-rolling" motions of the thumb and fingers. In addition, the facial muscles may become rigid and produce a staring, expressionless face with a slightly open mouth. The patient's range of motion diminishes. He or she takes smaller steps and develops a slow, shuffling gait with a forward-bent posture and a tendency to fall forward. Speech becomes slurred and handwriting becomes cramped and eventually illegible. Tasks such as buttoning clothes and preparing food become increasingly laborious.

Patients cannot be expected to recover from PD, but its effects can be alleviated with drugs and physical therapy. Treatment with dopamine is ineffective, but drugs that enhance the action of the brain's self-produced dopamine are of some benefit. The dopamine precursor, levodopa (L-dopa), affords some relief from symptoms, but it does not slow progression of the disease and it has undesirable side effects on the liver and heart. It is effective for only 5 to 10 years of treatment. A newer drug, deprenyl, retards neural degeneration and delays the development of symptoms. In severe cases that are unresponsive to medication, certain surgical treatments can help to quell the muscle tremors.

---

[15] **James Parkinson (1755–1824), British physician**

8.3   The Spinal Cord, Spinal Nerves, and Reflexes

Expected Learning Outcomes

When you have completed this section, you should be able to:

a. describe the functional anatomy of the spinal cord;

b. name and describe the three protective membranes that envelop the spinal cord;

c. describe the basic anatomy of nerves and ganglia, and anatomy of the spinal nerves in particular;

d. state the universal characteristics of reflexes and distinguish between somatic and visceral reflexes;

e. identify the basic components of a somatic reflex arc; and

f. explain the mechanisms of stretch and withdrawal reflexes.

The Spinal Cord—Functional Anatomy

The **spinal cord** (fig. 8.15) is a cylinder of nervous tissue enclosed in the vertebral canal, extending for about 45 cm (18 in.) from the foramen magnum of the skull to the level of the first or second lumbar vertebra, just below the ribs (farther in infants and children). It is divided into **cervical, thoracic, lumbar,** and **sacral regions.** The last of these is named for the fact that its nerves emerge from the vertebral canal in the sacral region, not because the spinal cord itself extends that far. The cord averages about 1.8 cm wide (comparable to the little finger), but it widens slightly in the lower cervical and lumbar regions—the **cervical** and **lumbar enlargements**—because of the greater mass of neurons here to control the complex movements of the upper and lower limbs.

At regular intervals along its length, the spinal cord gives rise to a pair of **spinal nerves,** which pass through the intervertebral foramina between adjacent vertebrae. The nerves then branch out to the skin, muscles, bones, and viscera of their respective levels of the body. Near the cord, each spinal nerve divides into two **roots**—anterior and posterior—which arch around to the respective side of the cord. Each root breaks up into half a dozen **rootlets** that enter or leave the adjacent cord. A length of spinal cord served by one pair of spinal nerves and their rootlets is called a **segment** of the cord. There are 31 pairs of spinal nerves and therefore 31 segments of the cord. At its lower end, the spinal cord tapers to a blunt point called the **medullary cone.** The lumbar enlargement and medullary cone give rise to a bundle of nerves that occupy the vertebral canal from there through the sacrum. The nerve bundle somewhat resembles a horse's tail and is therefore named the **cauda equina**[16] (CAW-duh ee-KWY-nuh).

---

[16] *cauda* = tail; *equin* = horse

Figure 8.15 The Spinal Cord, Posterior View.
(a) Overview of spinal cord structure. (b) Detail of the spinal cord and associated structures. **AP|R**

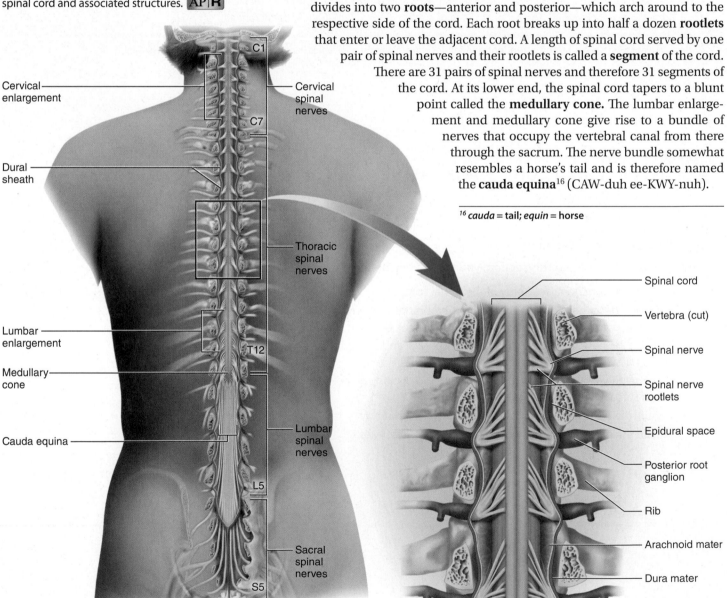

Cervical enlargement
Dural sheath
Lumbar enlargement
Medullary cone
Cauda equina

C1
Cervical spinal nerves
C7
Thoracic spinal nerves
T12
Lumbar spinal nerves
L5
Sacral spinal nerves
S5
Co1

Spinal cord
Vertebra (cut)
Spinal nerve
Spinal nerve rootlets
Epidural space
Posterior root ganglion
Rib
Arachnoid mater
Dura mater

(a)

(b)

Apply What You Know

Spinal cord injuries commonly result from fractures of vertebrae C5 to C6 (a "broken neck"), but never from fractures of L3 to L5. Explain these contrasting observations.

In cross section, the cord is elliptical in shape and exhibits a central butterfly- or H-shaped core of gray matter surrounded by white matter (fig. 8.16). A pair of median grooves almost divides the cord in half, but they stop short of a gray matter bridge (**gray commissure**) that connects the right and left halves of

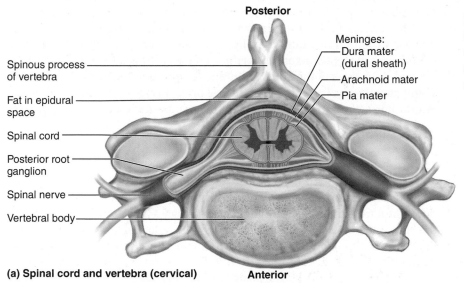

Posterior

Spinous process of vertebra

Fat in epidural space

Spinal cord

Posterior root ganglion

Spinal nerve

Vertebral body

Meninges:
Dura mater (dural sheath)
Arachnoid mater
Pia mater

(a) Spinal cord and vertebra (cervical)   **Anterior**

Figure 8.16 Cross-Sectional Anatomy of the Spinal Cord. (a) Relationship to the vertebra, meninges, and spinal nerve. (b) Detail of the spinal cord, meninges, and spinal nerves. (c) Cross section of the lumbar spinal cord with spinal nerves. AP|R

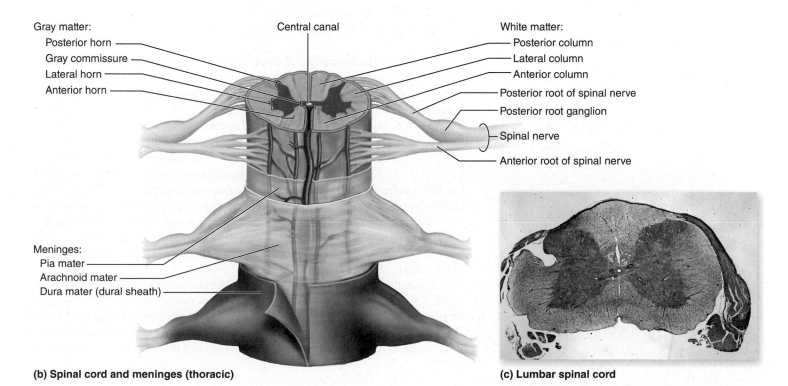

Gray matter:
Posterior horn
Gray commissure
Lateral horn
Anterior horn

Central canal

White matter:
Posterior column
Lateral column
Anterior column
Posterior root of spinal nerve
Posterior root ganglion
Spinal nerve
Anterior root of spinal nerve

Meninges:
Pia mater
Arachnoid mater
Dura mater (dural sheath)

(b) Spinal cord and meninges (thoracic)

(c) Lumbar spinal cord

the cord. In the middle of the gray commissure, there is often a **central canal** filled with cerebrospinal fluid, but in most regions of the adult spinal cord, the canal is collapsed and closed.

The Spinal Gray Matter

In the gray matter, the two posterior "wings of the butterfly" are called the **posterior (dorsal) horns.** They extend to the surface of the cord and serve as receiving points for incoming sensory information. The two anterior wings are called the **anterior (ventral) horns.** They are wider than the posterior horns and extend only partially toward the spinal cord surface. They are the locations of the somas of the motor neurons depicted in figure 8.2. The axons of these neurons extend to the cord surface and emerge there to become part of a spinal nerve. Between the incoming nerve fibers of the posterior horn and the somas of the motor neurons in the anterior horn, there are usually many interneurons forming complex networks in the gray matter of the cord, including some whose axons cross through the commissure and connect the left and right sides of the cord. From the second thoracic to the first lumbar segment of the cord, there is also a lateral projection of gray matter called the **lateral horn,** the origin of neurons of the sympathetic nervous system described in chapter 9.

The Spinal White Matter

The white matter of the spinal cord consists of nerve fibers traveling up and down the cord, serving as communication pathways between different levels of the body and between the spinal cord and brain. Some of these are axons originating in brain neurons, and others originate from the neurons in the gray matter of the cord. These bundles of nerve fibers are arranged in three **columns** on each side of the cord: (1) the **posterior (dorsal) column** between the posterior horn of gray matter and the posterior median groove, (2) the **lateral column** between the posterior and anterior horns, and (3) the **anterior (ventral) column** arching around the anterior horn and extending as far as the anterior median groove.

A functional knowledge of these columns is crucial in understanding spinal cord injuries, because often a patient loses specific sensory or motor abilities depending on which column, and which of its subdivisions, has been injured. Each column is divided into functionally distinct tracts of white matter. Some of them are **ascending tracts** that carry sensory information to higher levels of the cord and the brain. Most of these have names with *spino-* as the first word root, followed by a root that denotes their destinations in the brain—for example, the *spinothalamic tract,* ending in a part of the brain called the thalamus. Others are **descending tracts** that carry motor signals from the brain down the cord to meet the motor neurons that control muscles and glands. These are named with a word root that denotes the point of origin in the brain, followed by the root *-spinal;* an example is the *corticospinal tract,* carrying nerve fibers that originate in the cerebral cortex of the brain.

A peculiar and clinically important point about these spinal tracts is that most of them cross over from the left side of the cord or brainstem to the right, or vice versa, along the way. Therefore, sensory signals originating on the left side of the body are transmitted to the right side of the brain, and motor signals originating in the right side of the brain command muscles on the left side of

Clinical Application 8.2

SPINA BIFIDA

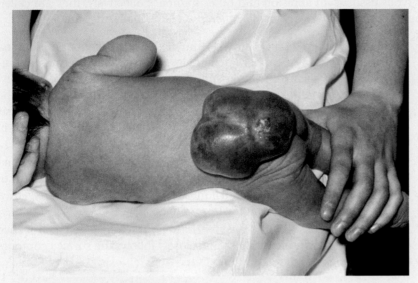

Figure 8.17 **Spina Bifida Cystica.**

Spina bifida (SPY-nuh BIF-ih-duh) occurs in about 1 out of 1,000 live births. It is a congenital defect in which one or more vertebrae fail to form a complete vertebral arch for enclosure of the spinal cord. This is especially common in the lumbosacral region. In its most serious form, *spina bifida cystica*,[17] a sac protrudes from the spine and may contain meninges, cerebrospinal fluid, and parts of the spinal cord and nerve roots (fig. 8.17). In extreme cases, inferior spinal cord function is absent, causing lack of bowel control and paralysis of the lower limbs and urinary bladder. The last of these conditions can lead to chronic urinary infections and renal failure. Spina bifida cystica usually requires surgical closure within 72 hours of birth. The prognosis ranges from a nearly normal, productive life, to a lifetime of treatment for multisystem complications, or to infant death in extreme cases.

A woman can reduce the risk of having a child with spina bifida by taking ample folic acid—obtainable from green leafy vegetables, lentils, black beans, and enriched bread and pasta, or in tablet form. However, this must be done before pregnancy; spina bifida arises in the first 4 weeks of development, so by the time a woman knows she is pregnant, it is too late for folic acid supplements to prevent the disorder. This is the reason that the Food and Drug Administration decided in 1996 to add folic acid to flour and other grain products, a policy that has dramatically reduced the incidence of neural tube defects.

the body, and vice versa. Such crossing over is called **decussation**.[18] It explains why a person who has had a stroke that kills brain tissue on one side of the brain experiences paralysis or sensory loss on the opposite side of the body.

---

[17] *spina* = spine; *bifida* = forked, divided in two; *cystica* = in the form of a sac

[18] *decuss* = to cross, form an X

The Spinal Meninges—Protective Membranes AP|R

The brain and spinal cord are covered by three fibrous membranes that lie between the nervous tissue and bone (fig. 8.16b). They help to protect the delicate nervous tissue from abrasion and other trauma. Collectively, they are called **meninges** (men-IN-jeez) (singular, *meninx*). Individually, they are

1. The **dura mater**[19] (DOOR-uh MAH-tur), a tough collagenous membrane about as thick as a rubber kitchen glove. This is the outermost membrane. It lies against the inner surface of the bone in the cranial cavity, but in the vertebral canal it forms a loose-fitting *dural sac* around the spinal cord and is separated from the bone by a fat-filled *epidural space.* Anesthetics are often introduced into this space to deaden the pain of childbirth or surgery *(epidural anesthesia).*
2. The **arachnoid**[20] **mater** (ah-RACK-noyd), a delicate middle layer named for a loose, webby appearance suggestive of a spider's web. It consists of a simple squamous epithelium adhering to the inside of the dura, and a loose mesh of fibers extending inward to the pia.
3. The **pia**[21] **mater** (PEE-uh), the innermost layer, a transparent connective tissue of microscopic thinness that closely follows the surface contours of the brain and spinal cord.

Spinal Nerves—Communicating with the Rest of the Body

In the peripheral nervous system, we find two kinds of structures: nerves and ganglia. A **nerve** is a bundle of nerve fibers and connective tissue wrappings, with internal blood vessels. It resembles a string that frays into progressively finer branches the farther away from the CNS we look. Nerves have a pearly white color due mainly to their fibrous connective tissues.

Each nerve fiber in a nerve is wrapped in a thin layer of loose connective tissue called the **endoneurium**[22] (fig. 8.18). In most nerves, the fibers are gathered in bundles called **fascicles,**[23] each wrapped in a sheath of flat, epithelium-like cells called the **perineurium.**[24] Several fascicles are then bundled together and wrapped in an outer, fibrous sleeve called the **epineurium**[25] to compose the nerve as a whole. The tough epineurium protects the nerve from stretching and injury. Nerves have a high metabolic rate and need a plentiful blood supply. These three layers of connective tissues ensure that blood vessels can reach every nerve fiber.

If we compare a nerve to a string, then a **ganglion**[26] is like a knot in the string—a swelling, usually near one end of the nerve, that contains the cell bodies of the peripheral neurons. In some ganglia, the neurons form synaptic contacts with each other, and the ganglion is therefore an information-processing center in the PNS.

---

[19] *dura* = tough; *mater* = mother, womb
[20] *arachn* = spider, spider web; *oid* = like
[21] *pia* = through mistranslation, now construed as tender, thin, or soft
[22] *endo* = within; *neuri* = nerve
[23] *fasc* = bundle; *icle* = little
[24] *peri* = around
[25] *epi* = upon
[26] *gangli* = knot

Figure 8.18 Anatomy of a Nerve. (a) A spinal nerve and its association with the spinal cord. (b) Cross section of a nerve (electron micrograph). Myelinated nerve fibers appear in the photograph as white rings and unmyelinated fibers as solid gray. [Part (b) © Dr. Richard Kessel & Dr. Kardon/Tissues and Organs/Visuals Unlimited, Inc.] **AP|R**

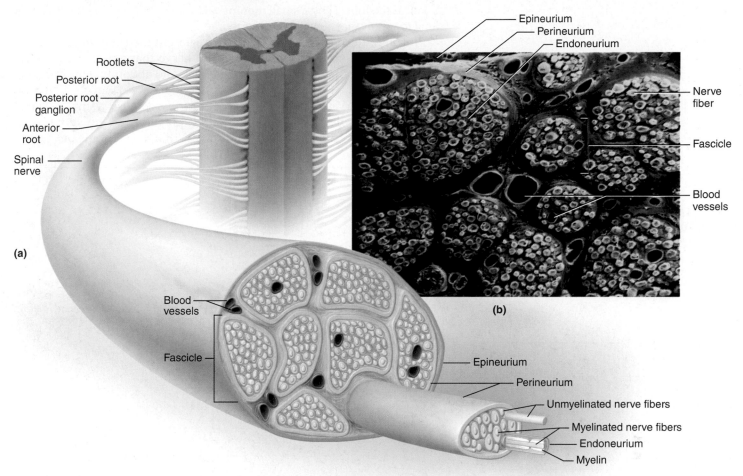

The spinal nerves were mentioned in the foregoing discussion of the gross anatomy of the spinal cord. They provide a means for the cord, and therefore the whole CNS, to communicate with the rest of the body. In this section, we take a closer look at these nerves and their relationship to spinal cord anatomy.

Each spinal nerve originates from the cord as six to eight *nerve rootlets* emerging from the posterior horn of the cord and a similar number arising from the anterior horn (see figs. 8.15 and 8.16). The posterior rootlets converge to form a single **posterior (dorsal) root** of the nerve, which immediately expands into a swelling called the **posterior (dorsal) root ganglion;** it narrows again as it approaches the intervertebral foramen. The anterior rootlets also converge into a single **anterior (ventral) root,** but with no ganglion. The posterior and anterior roots unite into a single spinal nerve and pass through the intervertebral foramen.

Nerve roots are shortest in the cervical region and pass almost horizontally to the nearest intervertebral foramen. From infancy to adulthood, however, the spine grows faster and farther than the spinal cord does, and the nerve roots must grow and sweep downward to stay with their corresponding intervertebral foramen. Therefore, the roots become progressively longer the lower on the spinal cord we look. The roots from spinal segments L2 to Co1 form the cauda equina and emerge from the vertebral canal below the lowest part of the spinal cord.

After emerging from the intervertebral foramen, each spinal nerve branches into a **posterior ramus**[27] that supplies the skin, muscles, and joints of the back; an **anterior ramus** that supplies the anterior and lateral skin and muscles as well as the limbs; and a small **meningeal branch** that reenters the vertebral canal to supply the meninges and vertebrae.

The anterior ramus differs from one region of the trunk to another. In the thoracic region, it forms an **intercostal nerve** that travels between the ribs, innervating the intercostal muscles, skin, and abdominal muscles. Everywhere else, anterior rami from multiple spinal nerves join each other to form a web of nerves called a *plexus*. There are five of these: a small **cervical plexus** deep in the neck; a large **brachial plexus** near the shoulder; a **lumbar plexus** in the lower back; and below that, a **sacral plexus** and fi-

---

[27] *ramus* = branch

Figure 8.19 The Spinal Nerve Roots and Plexuses, Posterior View.

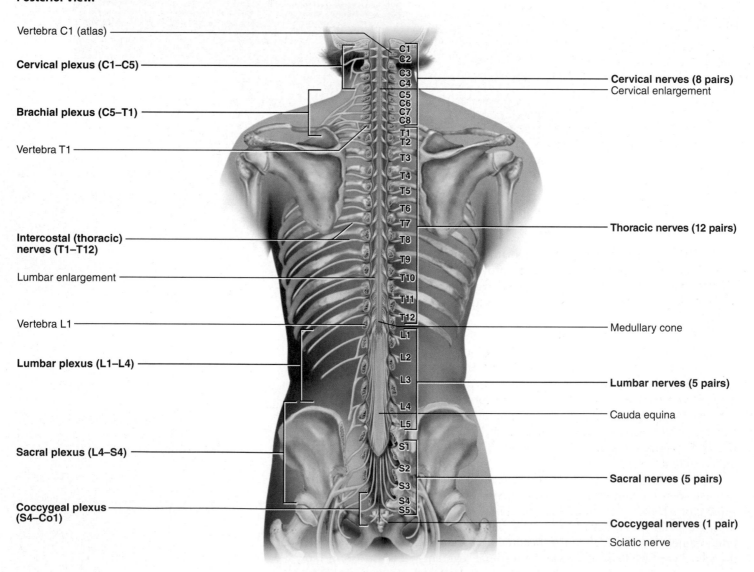

Vertebra C1 (atlas)

Cervical plexus (C1–C5)

Brachial plexus (C5–T1)

Vertebra T1

Intercostal (thoracic) nerves (T1–T12)

Lumbar enlargement

Vertebra L1

Lumbar plexus (L1–L4)

Sacral plexus (L4–S4)

Coccygeal plexus (S4–Co1)

C1
C2
C3
C4
C5
C6
C7
C8
T1
T2
T3
T4
T5
T6
T7
T8
T9
T10
T11
T12
L1
L2
L3
L4
L5
S1
S2
S3
S4
S5

Cervical nerves (8 pairs)
Cervical enlargement

Thoracic nerves (12 pairs)

Medullary cone

Lumbar nerves (5 pairs)

Cauda equina

Sacral nerves (5 pairs)

Coccygeal nerves (1 pair)

Sciatic nerve

Table 8.1 Structures Innervated by the Spinal Nerve Plexuses AP|R

Plexus	Structures Innervated
Cervical plexus	*Sensory:* Skin of external ear and from surrounding region of head and neck, underside of chin, and shoulder; anterior chest, diaphragm, pleurae, and pericardium
	Motor: Diaphragm and a few muscles associated with larynx and hyoid bone
Brachial plexus	*Sensory:* Skin, muscles, and joints of shoulder and upper limb
	Motor: Muscles of upper limb
Lumbar plexus	*Sensory:* Skin of lower abdominal, gluteal, and genital regions; skin, muscles, and joints of lower limb
	Motor: Muscles of abdominal and pelvic regions and thigh
Sacral and coccygeal plexuses	*Sensory:* Skin of gluteal and perineal regions; genitals; skin, muscles, and joints of gluteal region and lower limb
	Motor: Muscles of gluteal and perineal regions and lower limb

nally a tiny **coccygeal plexus.** The nerves that emerge from these plexuses **innervate** (supply) the skin, muscles, and other structures at their respective regions of the trunk and issue nerves to the limbs. Figure 8.19 shows a general view of these plexuses, and table 8.1 summarizes the structures innervated by each one.

Somatic Reflexes

A **reflex** may be defined as a quick, involuntary, stereotyped reaction of a gland or muscle to stimulation. This definition sums up four important properties: (1) Reflexes are *quick,* usually involving simple neural pathways with relatively few interneurons and little synaptic delay. (2) They are *involuntary,* meaning they occur without our intent, often without our awareness, and they are difficult or impossible to suppress. (3) They are *stereotyped,* meaning they occur in essentially the same way every time, unlike our more variable voluntary and learned behaviors. (4) They require *stimulation;* they are not spontaneous actions like muscle tics but responses to sensory input.

Some reflexes involve cardiac muscle, smooth muscle, or glands, such as acceleration of the heartbeat in fear, contraction of the esophagus in swallowing, or the secretion of tears in response to irritation of the cornea. These are called **visceral reflexes.** They are controlled by a division of the nervous system called the *autonomic nervous system* and will be discussed in chapter 9. Other reflexes involve skeletal muscles, such as pulling your hand back from a hot stove or the knee-jerk reflex so often evoked in a medical examination. These are called **somatic reflexes** and are the type we'll discuss here. They are sometimes called *spinal reflexes,* but this is a misleading expression because (1) many visceral reflexes also involve the spinal cord, and (2) many somatic reflexes are mediated more by the brain than by the spinal cord.

The Reflex Arc

Somatic reflexes usually involve a relatively simple neural pathway called a **reflex arc.** Figure 8.20 illustrates a reflex arc passing through the spinal cord and exemplifies the essential components of an arc:

- A *somatic receptor,* a sensory nerve ending or simple sense organ in the skin, a muscle, or a tendon.
- An *afferent nerve fiber,* the axon of a unipolar neuron that carries signals from the receptor to the spinal cord or brainstem.
- An *integrating center,* a point of synaptic contact between neurons in the gray matter of the spinal cord or brainstem. In most reflex arcs, this involves one or more interneurons, but in the simplest arcs, the afferent fiber may synapse directly with the efferent fiber. These simple arcs allow for quicker but less complex or refined responses.
- An *efferent nerve fiber,* which carries motor signals to a skeletal muscle.
- An *effector,* which in this case is the skeletal muscle. In reflexes in general, *effector* means any organ or cell carries out the final response.

The Stretch Reflex and Reciprocal Inhibition

The workings of a reflex arc can be exemplified most simply by considering the test so commonly performed in routine medical examinations, the *patellar tendon ("knee-jerk") reflex* (fig. 8.20). The doctor taps on the patellar ligament just below the kneecap with a rubber reflex hammer. This stretches the ligament and its continuation, the quadriceps tendon leading to the quadriceps femoris muscle of the thigh. At the junction between the muscle and tendon, and less abundantly elsewhere in the muscle, are specialized stretch receptors called **muscle spindles.** The sudden stretch produced by the hammer excites the spindle's nerve fibers, which conduct signals to the spinal cord.

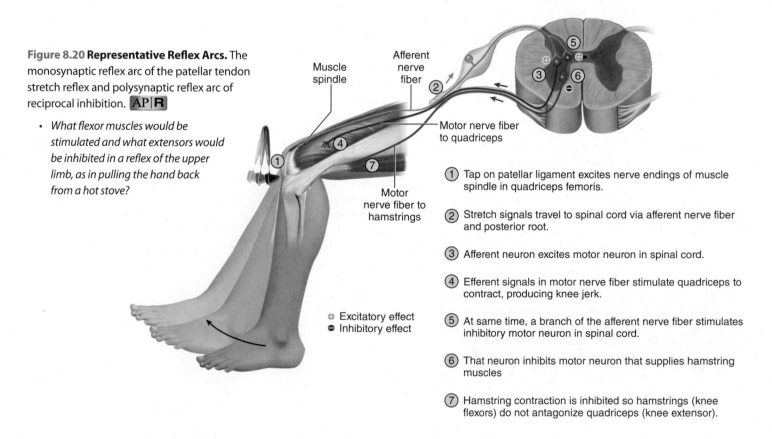

Figure 8.20 Representative Reflex Arcs. The monosynaptic reflex arc of the patellar tendon stretch reflex and polysynaptic reflex arc of reciprocal inhibition. AP|R

- *What flexor muscles would be stimulated and what extensors would be inhibited in a reflex of the upper limb, as in pulling the hand back from a hot stove?*

Muscle spindle

Afferent nerve fiber

Motor nerve fiber to quadriceps

Motor nerve fiber to hamstrings

⊕ Excitatory effect
⊖ Inhibitory effect

1 Tap on patellar ligament excites nerve endings of muscle spindle in quadriceps femoris.

2 Stretch signals travel to spinal cord via afferent nerve fiber and posterior root.

3 Afferent neuron excites motor neuron in spinal cord.

4 Efferent signals in motor nerve fiber stimulate quadriceps to contract, producing knee jerk.

5 At same time, a branch of the afferent nerve fiber stimulates inhibitory motor neuron in spinal cord.

6 That neuron inhibits motor neuron that supplies hamstring muscles

7 Hamstring contraction is inhibited so hamstrings (knee flexors) do not antagonize quadriceps (knee extensor).

In the cord, these nerve fibers synapse directly with motor neurons. Thus, such an arc is called a **monosynaptic reflex arc** because there is only one synapse between the sensory and motor neurons. Those neurons transmit signals immediately back to the muscle fibers of the quadriceps, which contract abruptly, pulling on the tibia and producing the twitch of the knee. By observing the strength of the response, a physician can get clues to the health of the nervous system as well as to disorders such as hormone and electrolyte imbalances, diabetes, and alcoholism.

In order for the quadriceps to produce the knee jerk, the hamstrings must be simultaneously inhibited from contracting and fighting the quadriceps. This is achieved through a **polysynaptic reflex arc** in which input from the muscle spindle also inhibits the spinal motor neurons that would otherwise stimulate the hamstrings. It is called polysynaptic because there are at least one interneuron and multiple synapses between the afferent and efferent neurons. This particular example—enabling an extensor muscle (quadriceps) to act by forcing its antagonists (hamstrings) to relax—is called a **reciprocal inhibition reflex.**

Some natural functions of stretch reflexes are to respond to joint movements and produce corrections in muscle tension that make our actions better coordinated, to maintain posture, and to prevent your head from tipping forward or your jaw from hanging open (unless you are nodding off in class). The vigorous and precisely coordinated movements of dance and athletics would be impossible without stretch reflexes.

The Flexor Reflex

Another well-known reflex is the **flexor (withdrawal) reflex,** which occurs for example when you pull back your hand upon burning your fingers, or your leg when you've stepped on something sharp (fig. 8.21). The protective function of this reflex requires more than the quick jerk and return to normal that we see in the stretch reflex; it requires the coordinated action of multiple muscles and a more sustained contraction so you don't put your hand or foot right back in harm's way. The flexor reflex also involves a polysynaptic reflex arc. Some signals follow pathways with relatively few synapses and thus head out to the muscles relatively quickly. Others follow pathways with more synapses and delays, and thus follow a little behind the first. Consequently, the muscles receive a more prolonged output from the spinal cord and produce a more sustained contraction. In the case of a leg withdrawal as in the figure, note that the knee flexors (hamstrings) are excited, whereas the extensor (quadriceps) is inhibited.

There are other kinds of somatic reflexes that we cannot explore in detail. For example: (1) A *crossed-extension reflex* causes you to stiffen one leg when you lift the other from the ground, so you don't fall over when withdrawing a limb from pain. (2) An *intersegmental reflex arc* travels up and down the spinal cord to produce reactions at a level different from the one where the stimulus is received, such as tilting your torso to the left when you raise the right leg, in order to keep your balance. (3) A *tendon reflex* inhibits excessive muscle contractions to prevent the rupturing of tendons.

Figure 8.21 The Flexor Reflex. A pain stimulus triggers a flexor reflex, which results in contraction of flexor muscles of the injured limb and withdrawal of the limb from danger. The polysynaptic reflex arc produces a more sustained, protective response than the quick jerk of the stretch reflex in figure 8.20.

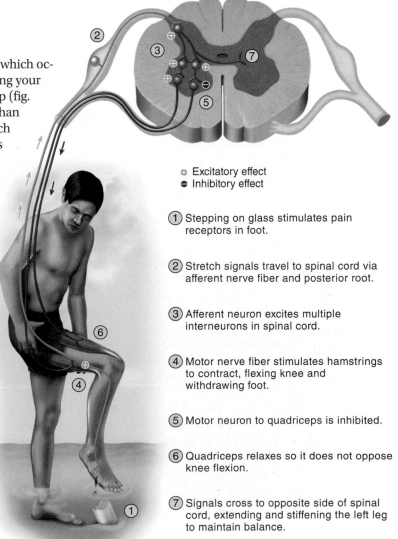

⊕ Excitatory effect
⊖ Inhibitory effect

① Stepping on glass stimulates pain receptors in foot.

② Stretch signals travel to spinal cord via afferent nerve fiber and posterior root.

③ Afferent neuron excites multiple interneurons in spinal cord.

④ Motor nerve fiber stimulates hamstrings to contract, flexing knee and withdrawing foot.

⑤ Motor neuron to quadriceps is inhibited.

⑥ Quadriceps relaxes so it does not oppose knee flexion.

⑦ Signals cross to opposite side of spinal cord, extending and stiffening the left leg to maintain balance.

Before You Go On

Answer these questions from memory. Reread the preceding section if there are too many you don't know.

12. What are the four regions of the spinal cord? Which two of them have enlargements, and why?

13. At what level of the vertebral column does the adult spinal cord end? What exists in the vertebral canal below that point?

14. Describe the butterfly-shaped components of the spinal gray matter and name the columns of white matter that surround the gray. How do the spinal gray matter and white matter differ in function?

15. From superficial to deep, name and describe the three spinal meninges.

16. What are the fascicles of a nerve? How do they relate to the endoneurium, perineurium, and epineurium?

17. Describe the roots and rami of a spinal nerve and how these relate to the vertebrae.

18. What are the five basic components of a reflex arc?

19. How do a stretch reflex and a withdrawal reflex differ in their neural circuits and functions?

CAREER SPOTLIGHT

Occupational Therapist

An occupational therapist (OT) is a health-care professional who serves people who are impaired in various ways in order to restore or sustain the skills needed to lead independent, productive, and satisfying lives. OTs work with a highly diverse range of clients—persons with birth defects, spinal cord injuries, hip replacements or artificial limbs, stroke, visual impairments, alcohol or drug dependence, mental retardation, emotional disturbances, and other conditions. The objectives of occupational therapy range from the most basic skills—dressing, grooming, bathing, cooking, eating—to logical reasoning, social poise, computer use, driving a car, strength conditioning, use of adaptive equipment such as wheelchairs, caring for a home, advancing one's education, or holding a job. Many occupational therapists specialize in certain types of disability or in certain age groups, ranging from premature infants to the elderly. OTs work in hospitals, outpatient clinics, rehabilitation centers, fitness centers, hospices, schools, and clients' homes.

To qualify for a career in occupational therapy, one must complete a graduate degree and a period of supervised fieldwork, and pass a national certification exam. Many colleges and universities offer master's and doctoral programs in OT. Prerequisites for admission vary from school to school, but may require biology, anatomy and physiology, kinesiology, physics, psychology, sociology, and other courses, as well as volunteer or paid work experience with disabled persons. See appendix B for further career information.

Study Guide

Assess Your Learning Outcomes

To test your knowledge, discuss the following topics with a study partner or in writing, ideally from memory.

8.1 Cells and Tissues of the Nervous System (p. 256)

1. The three fundamental functions of the nervous system
2. Distinction between the central and peripheral nervous system, and the components of each
3. The structure of a generalized neuron, and the most common variations on that basic structure
4. Three functional categories of neurons—sensory neurons, interneurons, and motor neurons
5. Six kinds of neuroglia, whether each kind is located in the CNS or PNS, and the functions performed by each
6. Structure and function of the myelin sheath, and how it is produced in the CNS and PNS
7. The distinction between the gray and white matter of the CNS

8.2 The Physiology of Neurons (p. 264)

1. What is meant by saying a cell has a resting membrane potential and that its plasma membrane is electrically polarized
2. Distinctions between a resting membrane potential, a local potential, and an action potential
3. How the resting membrane potential is produced by the distribution of cations and anions on the two sides of the plasma membrane, and how the Na$^+$–K$^+$ pump maintains this potential
4. The mechanism by which stimulation of a neuron produces a local potential
5. The mechanism by which a local potential can lead to an action potential; the role of voltage-gated ion channels in producing action potentials
6. The phases of an action potential and the ion movements that account for each
7. The difference between an action potential and a nerve signal
8. How nerve signals are conducted in unmyelinated nerve fibers, and how the process differs in a myelinated nerve fiber
9. Why myelinated nerve fibers conduct signals faster than unmyelinated ones do

10. The structure of a synapse where neurons meet
11. How a neuron transmits information across a synapse to the next cell
12. The definition of *neurotransmitter* and some examples
13. The significance of neural pools and circuits in information processing by the nervous system
14. The distinction between a converging and a diverging circuit and some bodily processes that involve each type

8.3 The Spinal Cord, Spinal Nerves, and Reflexes (p. 273)

1. Gross anatomy of the spinal cord
2. The relationship between the spinal cord and vertebral canal, and the contents of the adult vertebral canal below the termination of the spinal cord
3. The anatomical relationship of a spinal nerve to the spinal cord and how this relates to the concept of a spinal cord segment
4. The shape of the spinal gray matter and the terminology and functional significance of its horns
5. Three zones of white matter on each side of the spinal cord; how it is organized into columns and tracts; and the two fundamental types of spinal tracts
6. Differences between ascending and descending spinal cord tracts, and an example of each
7. The meaning of *decussation* and its significance in the relationship between CNS function and the sensory and motor functions of the body outside of the CNS
8. Names, composition, and structural relationships of the three meninges of the spinal cord
9. Names, composition, and structural relationships of the three connective tissues of a nerve, and how these relate to nerve fibers and fascicles
10. Where ganglia occur, what composes them, and how they relate structurally and functionally to nerves
11. Anatomy of a spinal nerve, including its two roots, its rootlets, the posterior root ganglion, and its three rami, and its spatial relationship to the spinal cord and vertebrae
12. The five plexuses of spinal nerves, where they are located, and the general destinations or origins of nerve fibers that lead to or exit from each plexus
13. General properties of a reflex and the distinctions between visceral and somatic reflexes
14. Fundamental components of a somatic reflex arc

15. The mechanism of a stretch reflex, the role of muscle spindles, and why it is advantageous for stretch reflexes to be monosynaptic

16. The mechanism of a flexor reflex and why it is advantageous for flexor reflexes to be polysynaptic

17. Some general purposes of crossed-extension, intersegmental, and tendon reflexes, and the importance of reciprocal inhibition in certain reflexes

Testing Your Recall

1. Myelinated nerve fibers produce action potentials only at their
   a. nodes of Ranvier.
   b. internodes.
   c. synapses.
   d. dendrites.
   e. terminal arborizations.

2. The posterior root ganglion next to the spinal cord contains the somas of
   a. efferent neurons.
   b. interneurons.
   c. unipolar neurons.
   d. multipolar neurons.
   e. motor neurons.

3. An oligodendrocyte serves CNS nerve fibers in the same way that a/an ___ serves PNS nerve fibers.
   a. Schwann cell
   b. microglia cell
   c. endoneurium
   d. ependymal cell
   e. satellite cell

4. An action potential begins with
   a. a flow of K⁺ out of a neuron.
   b. a flow of Ca²⁺ into a neuron.
   c. the release of a neurotransmitter.
   d. a flow of Na⁺ out of a neuron.
   e. a flow of Na⁺ into a neuron.

5. A synaptic vesicle is a storage site for
   a. calcium ions.
   b. sodium ions.
   c. neurotransmitter receptors.
   d. neurotransmitter.
   e. acetylcholinesterase.

6. The spinal cord is divided into all of the following regions *except*
   a. sacral.
   b. coccygeal.
   c. lumbar.
   d. cervical.
   e. thoracic.

7. Neurons that carry sensory information to the spinal cord are called
   a. efferent neurons.
   b. effluent neurons.
   c. interneurons.
   d. affluent neurons.
   e. afferent neurons.

8. The nerve fibers that stimulate skeletal muscles below the neck arise from neurosomas in
   a. the posterior horn.
   b. the anterior horn.
   c. the lateral horn.
   d. the posterior root ganglion.
   e. the brain.

9. A bundle of nerve fibers within a nerve is called
   a. the gray matter of the nerve.
   b. an endoneurium.
   c. a fascicle.
   d. the white matter of the nerve.
   e. a perineurium.

10. Jerking your hand back when you are burned is an example of
    a. a flexor reflex.
    b. a stretch reflex.
    c. decussation.
    d. a visceral reflex.
    e. a monosynaptic reflex.

11. The outermost and toughest of the meninges is the _____.

12. The brain and spinal cord compose the _____ nervous system.

13. Sensory neurons are also called _____ neurons because they carry signals toward the CNS.

14. Muscles of the forearm are controlled by nerves arising from the _____ plexus.

15. In a myelinated peripheral nerve fiber, the gap between one Schwann cell and the next is called a/an _____.

16. The sensory neurons for smell have the _____ shape, meaning they possess only one dendrite and one axon.

17. Neuroglia called _____ cells line the internal cavities of the brain and spinal cord and secrete cerebrospinal fluid.

18. A nerve signal is a chain reaction of one _____ after another, analogous to a row of falling dominoes.

19. Neurons communicate with each other by means of chemical secretions called _____.

20. The posterior and anterior horns of the spinal cord are composed of a type of nervous tissue called _____.

Answers in appendix A

True or False

Determine which five of the following statements are false, and briefly explain why.

1. As suggested by its name, the spinothalamic tract is an ascending tract of the spinal cord.

2. The arachnoid mater lies between the pia mater and the nervous tissue of the spinal cord.

3. The outermost layer of a nerve is a fibrous covering called the epineurium.

4. Spinal nerves of the thoracic region form a weblike nerve plexus adjacent to the vertebral column.

5. All somatic reflex arcs pass through the spinal cord.

6. Muscle spindles are stretch receptors that activate stretch reflexes.

7. There are no interneurons in the peripheral nervous system.

8. *Neuron* is another name for *nerve fiber*.

9. The myelin sheath of brain and spinal cord neurons is produced by Schwann cells.

10. A neuron can have many dendrites, but never more than one axon.

Answers in appendix A

Testing Your Comprehension

1. The local anesthetics tetracaine and procaine (Novocain) prevent voltage-gated Na⁺ channels from opening. Explain why this would block the conduction of pain signals in a sensory nerve.

2. Mr. Richards has had a stroke that damaged some of the tissue on the left side of his brain and paralyzed his right arm. He doesn't understand why his *right* arm is paralyzed when the neurologist told him the stroke was on the *left* side of his brain, and he feels sure his neurologist is wrong. If you were his nurse or therapist, how would you explain this to him in simple, lay terms?

3. In a spinal tap, a needle is inserted between two of the lower lumbar vertebrae into the vertebral canal in order to withdraw a sample of cerebrospinal fluid. From what you know of spinal cord anatomy, deduce why it is safer to perform the procedure in this region rather than higher on the vertebral column.

Chapter

9

The Nervous System II

THE BRAIN, CRANIAL NERVES, AND AUTONOMIC NERVOUS SYSTEM

Module 7: Nervous System

An MRI scan of the left cerebral hemisphere of the human brain. This is color enhanced to accentuate the folds (gyri) and grooves (sulci) of the cerebral tissue.

BASE CAMP

Before ascending to the next level, be sure you're properly equipped with a knowledge of these concepts from earlier chapters:

- Anatomy of the cranium (p. 162–166)
- Glial cells and their functions (p. 260)
- Spinal meninges (p. 278)
- Structure of nerves and ganglia (p. 278)

The mystique of the brain continues to intrigue modern biologists and psychologists even as it did the philosophers of antiquity. Aristotle thought it was merely a radiator for cooling the blood, but Hippocrates correctly surmised that it was the seat of "our pleasures, joy, laughter and jests, as well as our sorrows, pains, griefs and tears. Through it, in particular, we think, see, hear, and distinguish the ugly from the beautiful, the bad from the good, the pleasant from the unpleasant."

Brain function is strongly associated with what it means to be alive and human. With its hundreds of neural pools and trillions of synapses, the brain performs sophisticated tasks beyond the present understanding of science. Here we can only briefly explore its major anatomical features and where in the brain certain key functions reside.

We also will explore the connections of the central nervous system to the rest of the body through 12 pairs of cranial nerves, which arise from the floor of the brain and emerge from the skull; and we'll examine the sympathetic and parasympathetic nervous systems, which regulate many of our unconscious bodily processes such as heart rate and digestion.

9.1 Overview of the Brain

Expected Learning Outcomes

When you have completed this section, you should be able to:

a. describe the major subdivisions and anatomical landmarks of the brain;

b. describe the locations of the gray and white matter of the brain;

c. name and describe the three meninges that enclose the brain;

d. describe the system of chambers and channels in the brain and the flow of cerebrospinal fluid through this system; and

e. describe the brain's blood supply and the selective barrier between the blood and brain tissue.

Figure 9.1 Surface Anatomy of the Brain.
(a) Superior view of the cerebral hemispheres. (b) Left lateral view. (c) The partially dissected brain of a cadaver. Part of the left hemisphere is cut away to expose the insula. The arachnoid mater is removed from the anterior half of the brain to expose the gyri and sulci. The arachnoid with its blood vessels is seen in the posterior half. Blood vessels of the brainstem are left in place. **AP|R**

Major Landmarks

A broad overview of the brain's gross anatomy will provide important points of reference as we progress through a more detailed study. The largest, most conspicuous part of the brain is the **cerebrum** (seh-REE-brum or SER-eh-brum), composed of a pair of half-globes called the **cerebral hemispheres** (fig. 9.1). The cerebral surface is marked by thick folds called **gyri**[1] (JY-rye; singular, *gyrus*)

---

[1] *gyr* = turn, twist

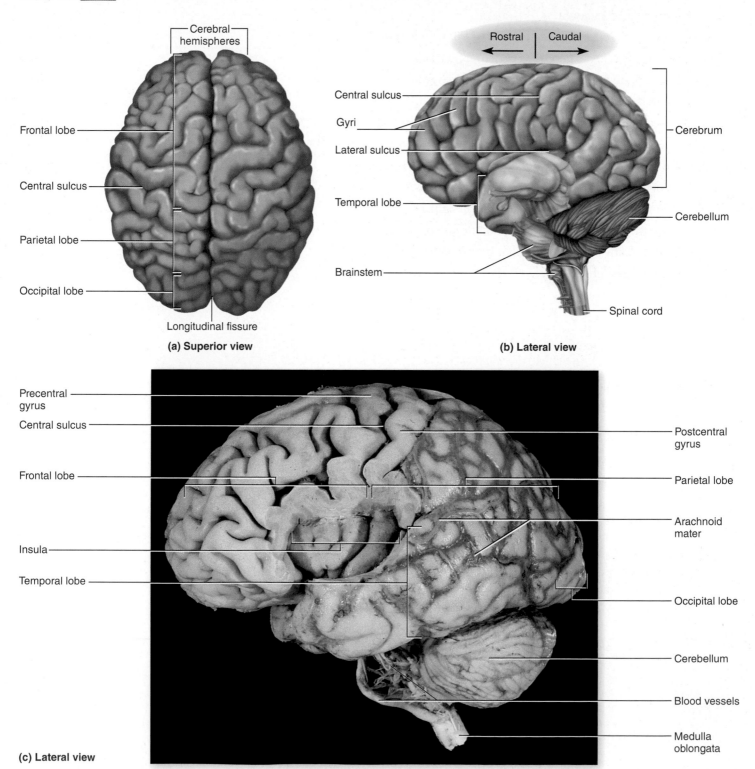

(a) Superior view

(b) Lateral view

(c) Lateral view

separated by shallow grooves called **sulci**[2] (SUL-sye; singular, *sulcus*). The left and right hemispheres are separated by a very deep groove, the **longitudinal fissure.** At the bottom of this fissure, the hemispheres are connected by a prominent bundle of nerve fibers called the **corpus callosum**[3]—a conspicuous landmark in the bisected brain (fig. 9.2).

---

[2] *sulc* = furrow, groove

[3] *corpus* = body; *call* = thick

Figure 9.2 Medial Aspect of the Brain. (a) Major anatomical landmarks of the medial surface. (b) Median section of the cadaver brain. **AP|R**

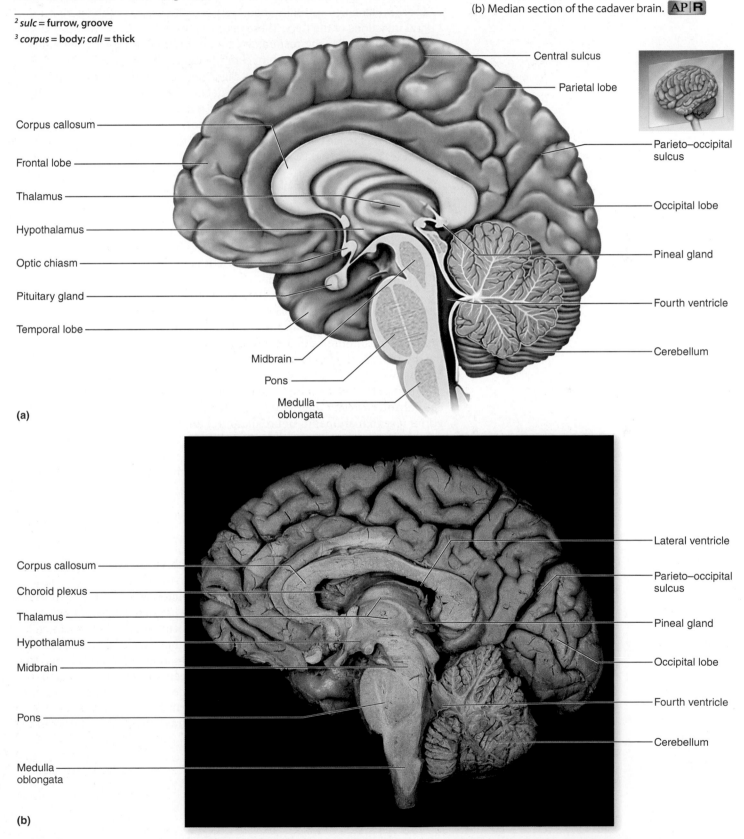

(a)

(b)

Clinical Application 9.1

MENINGITIS

Meningitis—inflammation of the meninges—is one of the most serious diseases of infancy and childhood. It is caused by a variety of bacterial and viral infections of the CNS. These usually invade by way of the nose and throat, often following respiratory, throat, or ear infections. Meningitis can cause swelling of the brain, cerebral hemorrhaging, and sometimes death within mere hours of onset. Signs and symptoms include high fever, stiff neck, drowsiness, intense headache, and vomiting. Death can occur so suddenly that infants and children with a high fever should receive immediate medical attention. Freshman college students show a slightly elevated incidence of meningitis, especially those living in crowded dormitories rather than off campus.

The **cerebellum**[4] (SER-eh-BEL-um) is the second-largest part of the brain. It lies at the rear of the head beneath the cerebrum. Like the cerebrum, it is composed of right and left hemispheres. It is marked by slender folds called **folia**[5] separated by shallow sulci.

The third part of the brain is the **brainstem,** the smallest of all and yet least dispensable to survival. Strokes and injuries here are more likely to be fatal than comparable injuries to the cerebrum or cerebellum. We can liken the brain to a mushroom, with the cerebrum being the cap and the brainstem being the stalk. At the foramen magnum of the skull, the brainstem connects to the spinal cord.

In describing the brain, anatomists often use the directional terms **caudal** and **rostral.** *Caudal* literally means "toward the tail" (as in rats, a favorite research animal in neuroanatomy) and refers to anything that is toward the rear of the head (or down the spinal cord). *Rostral* literally means "toward the nose" and refers to anything closer to the human forehead. You will see these terms used in some of the art in this chapter (such as fig. 9.1b) to help orient you to the brain.

Meninges

Like the spinal cord, the brain is enveloped in three connective tissue membranes, the **meninges,** which lie between the nervous tissue and bone (fig. 9.3). The thickest and toughest of these is the **dura mater,** which lines the inside of the cranium and forms a fibrous sac around the brain. It consists of two layers of tissue that are separated in some places to form blood-filled spaces called **dural sinuses.** These are veins that collect blood that has circulated through the brain tissue. The dura mater folds inward to form tough fibrous walls between the two cerebral hemispheres, another between the two cerebellar hemispheres, and a horizontal shelf between the cerebrum and cerebellum. In chapter 8, we saw that the dura mater of the vertebral canal is separated from the bone by a fat-filled epidural space. By contrast, this does not exist in the

---

[4] *cereb*= brain; *ellum*= little

[5] *foli* = leaf

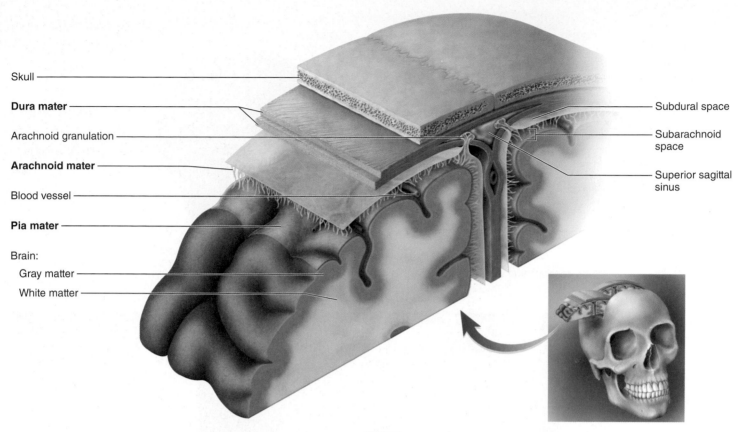

Skull

Dura mater

Arachnoid granulation

Arachnoid mater

Blood vessel

Pia mater

Brain:
　Gray matter
　White matter

Subdural space

Subarachnoid space

Superior sagittal sinus

Figure 9.3 The Meninges of the Brain. Frontal section of the head. AP|R

- *In what ways does the dura mater here differ from the dura mater enclosing the spinal cord?*

cranium; the dura is pressed against the cranial bones and attached to them at a few places, such as encircling the foramen magnum.

　Beneath the dura is an arachnoid mater similar to that of the spinal cord; the two membranes are separated in some places by a *subdural space*. Beneath the arachnoid mater, a *subarachnoid space* separates it from the innermost membrane, the pia mater. The pia mater closely follows all the contours of the brain surface, and is so thin and transparent that in most places it is visible only with a microscope. Delicate filaments extend from the arachnoid mater to the pia mater and suspend the brain in its bath of cerebrospinal fluid.

Gray and White Matter

Probing now into the tissues of the brain itself, we find that they are divided into gray and white matter like those of the spinal cord. In the brain, however, most of the gray matter is on the surface, where it forms a layer called the **cortex** of the cerebrum and cerebellum (fig. 9.4c). Smaller masses of gray matter called **nuclei** occur more deeply, surrounded by white matter. The gray matter contains the cell bodies, dendrites, and synapses of the neurons, so it is the site of all information processing, memory, thought, and decision making by the brain. The white matter is composed of **tracts** (bundles) of myelinated nerve fibers connecting one region of the CNS to another. The aforementioned corpus callosum is the largest tract. Tracts get their glistening white color from the myelin.

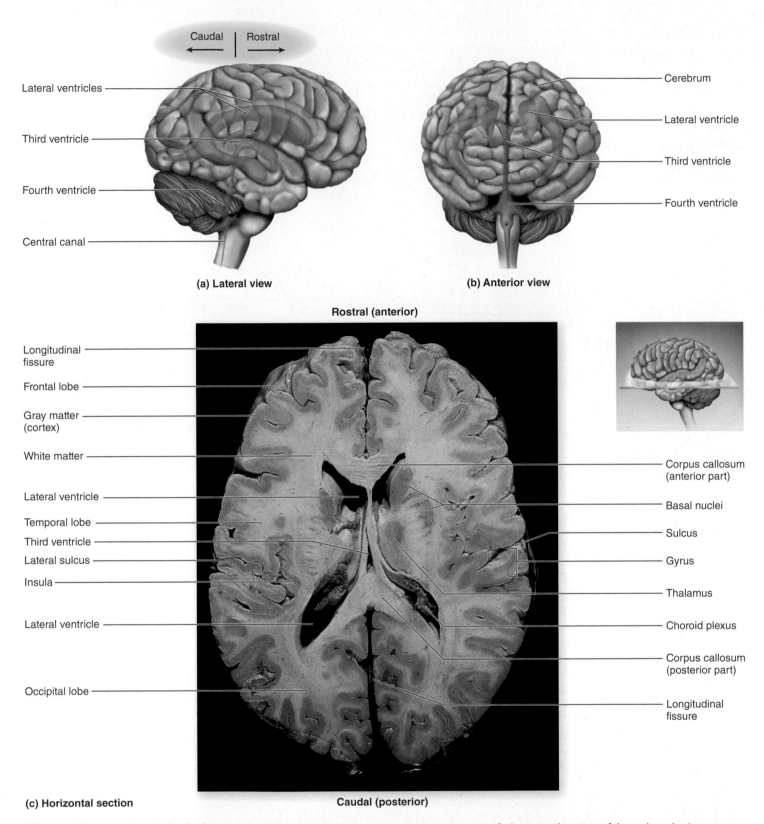

Caudal | Rostral

(a) Lateral view

Lateral ventricles

Third ventricle

Fourth ventricle

Central canal

Cerebrum

Lateral ventricle

Third ventricle

Fourth ventricle

(b) Anterior view

Rostral (anterior)

Longitudinal fissure

Frontal lobe

Gray matter (cortex)

White matter

Lateral ventricle

Temporal lobe

Third ventricle

Lateral sulcus

Insula

Lateral ventricle

Occipital lobe

Corpus callosum (anterior part)

Basal nuclei

Sulcus

Gyrus

Thalamus

Choroid plexus

Corpus callosum (posterior part)

Longitudinal fissure

(c) Horizontal section

Caudal (posterior)

Figure 9.4 Ventricles of the Brain. (a) Right lateral view. (b) Anterior view. (c) Superior view of a horizontal section of the cadaver brain, showing the lateral ventricles and some other features of the cerebrum. Note the thin layer of gray matter (cerebral cortex) overlying the cerebral white matter. AP|R

Ventricles and Cerebrospinal Fluid AP|R

Sections through the brain reveal four internal chambers called **ventricles** (fig. 9.4c)—a *lateral ventricle* within each cerebral hemisphere, a single *third ventricle* in a narrow space between the the hemispheres, and a *fourth ventricle* between the cerebellum and brainstem. A small pore connects each lateral ventricle to the third, and a canal through the brainstem connects the third ventricle to the fourth. The system then continues into the spinal cord as the central canal. These chambers and canals are filled with **cerebrospinal fluid (CSF),** which also fills the subarachnoid space and bathes the external surface of the brain and spinal cord.

CSF is produced by the filtration of blood plasma through the brain surface into the subarachnoid space; by the layer of *ependymal cells* that lines the ventricles; and by a spongy mass of blood capillaries, the **choroid** (CORE-oyd) **plexus,** in each of the four ventricles. The CSF produced in the ventricles circulates from the lateral ventricles to the third, and down the brainstem canal to the fourth ventricle and spinal cord. The fourth ventricle has three pores through which CSF exits the internal spaces and enters the subarachnoid space. Here, it bathes the outer surfaces of the brain and spinal cord. It is eventually reabsorbed into the bloodstream through outgrowths of the arachnoid mater called **arachnoid granulations.**

Although CSF is a filtrate of the blood plasma, it is lower in potassium and has hardly any protein. Its pH and the concentration of ions such as potassium, sodium, and chloride are maintained within strict ranges because they strongly affect neuron function. The CSF also provides buoyancy for the brain so it does not rest heavily in the floor of the cranium, cushions the brain to protect it from minor jolts or blows to the head, and continually rinses metabolic wastes from the brain and spinal cord.

Blood Supply and the Blood–Brain Barrier

The brain is only 2% of the body weight, yet it has such a high metabolic rate that it demands 15% of the blood supply. A mere 4-minute cessation of cerebral blood flow can cause irreversible brain damage. It receives its supply through two **internal carotid arteries** that begin near the angle of the mandible and enter the cranial cavity through the *carotid foramina* in the base of the skull, and through two **vertebral arteries** that pass up the back of the neck through the transverse foramina of the cervical vertebrae and enter the cranial cavity through the foramen magnum. These four arteries converge on a loop of blood vessels called the **cerebral arterial circle (circle of Willis[6])** on the floor of the brain, surrounding the pituitary gland (see fig. 13.20, p. 453). Three pairs of **cerebral arteries** arise from this circle and distribute blood throughout the cerebral tissue.

Despite its critical importance to the energy-hungry brain, the blood is also a source of antibodies, toxins, and other agents that are potentially harmful to the irreplaceable brain tissue. Consequently, there is a **blood–brain barrier (BBB)** that strictly regulates what substances can get from the bloodstream into the tissue fluid of the brain. The BBB consists mainly of tight junctions between the epithelial cells of the cerebral blood capillaries. The BBB is freely permeable to such necessities as oxygen and glucose (and such drugs as alcohol and nicotine), but it prevents many other substances from passing through. While the BBB provides valuable protection for the brain tissue, it also has an unfortunate aspect: It makes it very difficult to deliver drugs for the treatment of brain diseases. For example, the BBB is an obstacle to treating brain cancer with chemotherapy.

---

[6] **Thomas Willis (1621–75), English anatomist**

Before You Go On

Answer these questions from memory. Reread the preceding section if there are too many you don't know.

1. What are the three principal subdivisions of the brain? Which ones have gyri and sulci, and what do those terms mean?
2. Where are the gray matter and white matter of the brain located?
3. Describe the source of CSF, the route that it takes through and around the CNS, and its functions.
4. What is the blood–brain barrier, and why is it important?

9.2 Principal Divisions of the Brain

Expected Learning Outcomes

When you have completed this section, you should be able to:

a. list the parts of the brainstem and describe their structure and function;
b. describe the structure and function of the cerebellum;
c. describe the locations of the gray matter and tracts of white matter in the cerebrum;
d. list the five lobes of the cerebrum, identify their anatomical boundaries, and state their functions; and
e. describe the limbic system and identify its major functions.

In our regional survey of the brain, we will start at the rear and work forward, from brainstem to cerebrum. This has the advantage of beginning with regions that are relatively simple, anatomically and functionally, and working our way up to the most complex and mysterious part of the brain, the cerebrum—the seat of such complex functions as thought, memory, and emotion. AP|R

The Brainstem

The brainstem (fig. 9.5) is an elongated pillar of nervous tissue extending vertically from the foramen magnum to the cerebrum. From inferior to superior, it consists of the medulla oblongata, pons, midbrain, and diencephalon (although not all authorities include the diencephalon). The brainstem serves in part as the "information highway" between the cerebrum and the lower body; it contains prominent tracts of nerve fibers carrying signals coming up from the spinal cord and down from the cerebrum.

Embedded among these tracts are at least 100 pairs of nuclei and neural pools—small masses of gray matter that control many fundamental physiological functions of the body, especially such basic processes as breathing, swallowing, and the heartbeat. Many of these nuclei form a loosely organized core

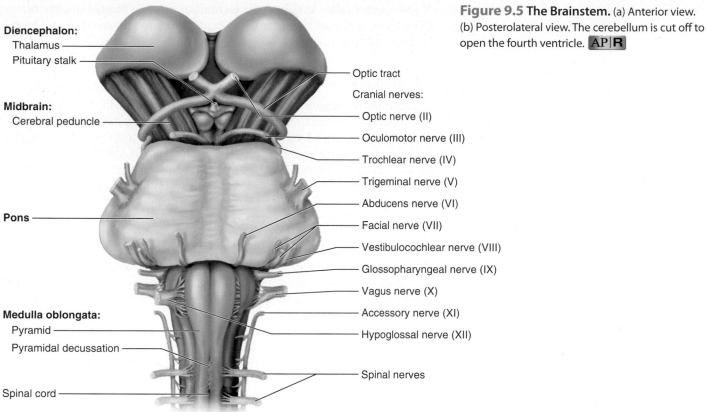

Diencephalon:
Thalamus
Pituitary stalk

Optic tract

Cranial nerves:

Optic nerve (II)

Midbrain:
Cerebral peduncle

Oculomotor nerve (III)

Trochlear nerve (IV)

Trigeminal nerve (V)

Abducens nerve (VI)

Pons

Facial nerve (VII)

Vestibulocochlear nerve (VIII)

Glossopharyngeal nerve (IX)

Vagus nerve (X)

Medulla oblongata:
Pyramid

Accessory nerve (XI)

Pyramidal decussation

Hypoglossal nerve (XII)

Spinal cord

Spinal nerves

(a) Anterior view

Figure 9.5 The Brainstem. (a) Anterior view.
(b) Posterolateral view. The cerebellum is cut off to
open the fourth ventricle. **AP|R**

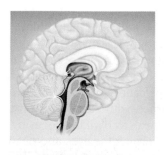

Diencephalon:
Thalamus
Pineal gland

Optic tract

Midbrain:
Superior colliculus
Inferior colliculus
Cerebral peduncle

Pons

Fourth ventricle

Cerebellar
peduncles

**Medulla
oblongata**

Spinal cord

(b) Posterolateral view

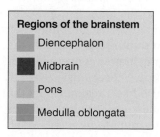

Regions of the brainstem

Diencephalon

Midbrain

Pons

Medulla oblongata

of gray matter called the **reticular formation** running throughout the brainstem. We'll survey the brainstem from bottom to top, with an emphasis on the functions of these nuclei.

The Medulla Oblongata

The **medulla oblongata**[7] looks superficially like a slightly wider continuation of the spinal cord. It contains fiber tracts that carry sensory information from the spinal cord to higher levels of the brain, and motor commands from higher brain centers to the spinal cord. The motor signals travel predominantly through a pair of anterior ridges called **pyramids,** shaped somewhat like two parallel baseball bats. When you consciously will your muscles to contract, the motor signals pass through here. Ridges on the posterior surface of the medulla carry sensory signals destined for the cerebrum.

Amid these tracts of white matter are gray matter nuclei concerned with controlling the heart rate and blood pressure; the respiratory rhythm, coughing, and sneezing; the digestive tract functions of swallowing, salivation, gastrointestinal secretion, gagging, and vomiting; and roles in speech, sweating, and tongue and head movements.

The Pons

The **pons**[8] (pronounced "ponz") is a distinctly wider segment of the brainstem superior to the medulla. It is the most significant relay center for signals going to the cerebellum. It also contains nuclei concerned with sleep, hearing, equilibrium, taste, eye movements, facial expressions, facial sensation, respiration, swallowing, bladder control, and posture.

The Midbrain

The **midbrain** is a relatively short bridge between the pons and higher brainstem centers. It has a pair of stalks called **cerebral peduncles**[9] (peh-DUN-culs) that anchor the cerebral hemispheres to the brainstem, and a pair of *cerebellar peduncles* that receive the most output from the cerebellum. The posterior surface of the midbrain has four conspicuous humps called *colliculi.* The two **superior colliculi**[10] (col-LICK-you-lye) are concerned with visual attention, tracking moving objects with the eyes, and such reflexes as blinking, focusing, dilation and constriction of the pupil, and quickly turning the eyes and head toward something that catches your attention in your peripheral vision. The two **inferior colliculi** receive signals of hearing from the inner ear and relay them to other parts of the brain, and mediate such reflexes as jumping when startled by a sudden noise and turning the head in response to a sound. The midbrain also has a prominent region of **central gray matter** concerned with filtering pain stimuli. It determines whether we become consciously aware of pain and how intensely we feel it. The midbrain also has a region called the **substantia nigra**[11] (sub-STAN-she-uh NY-gruh), a dark gray to black nucleus pigmented with melanin. It produces the neurotransmitter *dopamine,* which acts on the *basal nuclei* of the cerebrum (described later) to inhibit unwanted muscular movements. Parkinson disease (see p. 273) results from degeneration of the substantia nigra and the resulting deficiency of dopamine.

---

[7] *medulla* = pith, inner core; *oblongata* = elongated

[8] *pons* = bridge

[9] *ped* = foot; *uncle* = little

[10] *colli* = hill; *cul* = little

[11] *substantia* = substance; *nigra* = black

The Diencephalon

The main components of the diencephalon are the thalamus and the hypothalamus, which will be discussed here, and the pineal gland, an endocrine gland that we will consider in chapter11. Authorities vary on whether they consider the diencephalon to be part of the brainstem.

Each side of the brain has a **thalamus,**[12] an ovoid mass perched at the superior end of the brainstem beneath the cerebral hemisphere (fig. 9.5). The thalami constitute about four-fifths of the diencephalon. In about 70% of people, they are joined medially by a narrow bridge of tissue. Each thalamus contains over 20 nuclei with various roles in memory, emotion, motor control, and sensation. Its main function is to process incoming sensory information and relay it to the appropriate places in the cerebrum. For example, it receives visual signals from the optic nerve, processes them, and relays information to the posterior region of the cerebral cortex, where we first become aware of what we see. Similarly, it processes and relays signals for taste, smell, hearing, equilibrium, vision, touch, pain, pressure, heat, and cold to appropriate processing centers in the cerebrum. Thus, the thalamus is often nicknamed "the gateway to the cerebral cortex."

The **hypothalamus,**[13] as its name implies, lies beneath the thalamus. It is shaped somewhat like a flattened funnel, forming the floor and the lateral walls of the third ventricle (see fig. 9.2). The pituitary gland is attached by a stalk to its floor. The hypothalamus is a major control center for many instinctive and automatic functions of the body including water balance, blood pressure, hunger and thirst, digestive functions, metabolism, stress responses, shivering and sweating, sleep, memory, growth, emotion, reproductive cycles, sexual behavior, childbirth, and lactation.

It may seem confusing that some of these functions were already attributed to centers in the medulla and pons, but it is not contradictory. Signals for a function such as modulating the heart rate can originate in the hypothalamus, travel through nerve fibers to the *cardiac center* of the medulla, and then continue from the medulla down the spinal cord and out *cardiac nerves* that lead to the heart. Thus, the hypothalamus and medulla are both involved in the path of cardiac control.

The Cerebellum

The **cerebellum** is the second-largest part of the brain (fig. 9.6). It is attached to the brainstem by three pairs of stalks called the **cerebellar peduncles.** Even though it is only 10% of the brain mass, it contains over 50% of its neurons. As noted earlier, the cerebellum consists of right and left hemispheres connected by a narrow bridge. Each hemisphere has slender parallel folia separated by shallow sulci. The cerebellum has a surface cortex of gray matter and a deeper layer of white matter. In a sagittal section, the white matter exhibits a branching, fernlike pattern called the

Figure 9.6 The Cerebellum. (a) Median section, showing relationship to the brainstem. (b) Superior view. **AP|R**

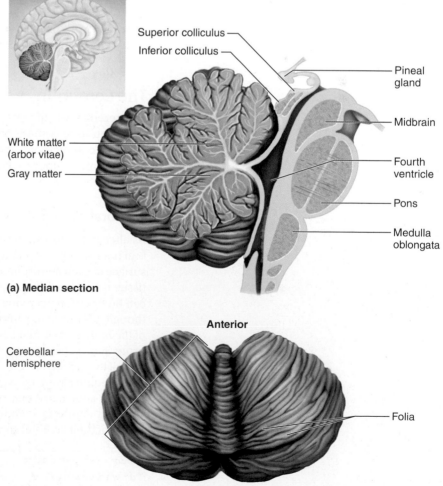

Superior colliculus
Inferior colliculus
Pineal gland
Midbrain
Fourth ventricle
Pons
Medulla oblongata
White matter (arbor vitae)
Gray matter

(a) Median section

Anterior
Cerebellar hemisphere
Folia

(b) Superior view   Posterior

---

[12] *thalamus* = chamber, inner room

[13] *hypo* = below

arbor vitae (AR-bur VEE-tee).[14] Each hemisphere also has four **deep nuclei,** masses of gray matter embedded in the white matter. All input to the cerebellum goes to the cortex and nearly all of its output comes from the deep nuclei. It sends and receives messages to and from the pons, midbrain, and spinal cord through the peduncles.

Until recently, the function of the cerebellum was thought to be largely limited to motor coordination. People with injuries to the cerebellum often exhibit a movement disorder called **ataxia,**[15] in which they have an uncoordinated, stumbling gait and may be unable to perform such tasks as climbing stairs. But with modern brain-imaging techniques such as PET scans and functional MRI, it has become clear that motor coordination is only one part of a much broader cerebellar role. Its more general function seems to be the evaluation of various forms of sensory input. It is highly engaged in feeling textures with the fingertips, in strategizing movements (such as knowing just when and where to reach to catch a flying baseball), in judging the three-dimensional shape of an object from a two-dimensional image (spatial perception), in judging the passage of time, and in predicting how much the eyes must move to compensate for head movements and remain visually fixed on an object. Cerebellar lesions impair a person's ability to distinguish differences in pitch between two sounds and to distinguish between similar-sounding words such as *rabbit* and *rapid.* People with cerebellar lesions also have difficulty planning and scheduling tasks. They tend to overreact emotionally and have difficulty with impulse control. Many children with attention-deficit/hyperactivity disorder (ADHD) have abnormally small cerebellums.

> **Apply What You Know**
>
> Discuss the multifaceted importance of the cerebellum to a tennis player, a concert pianist, and a student driving to class in city traffic.

The Cerebrum

The **cerebrum** is the largest and most conspicuous part of the brain. It enables you to turn these pages, read and comprehend the words, remember ideas, talk about them, and take an examination. It is the seat of your sensory perception, memory, thought, judgment, and voluntary motor control, among other functions.

Cerebral Tissues

As already described, the cerebrum consists of two large half-globes, the cerebral hemispheres, separated by the deep *longitudinal fissure* (see fig. 9.1a). The surface of each hemisphere is marked by conspicuous gyri and sulci. Its surface tissue is a layer of gray matter called the **cerebral cortex,**[16] which contains the cell bodies of the cerebral neurons, their dendrites, and their synapses. Even though it is only 2 to 3 mm thick, the cortex constitutes about 40% of the mass of the brain and contains 14 to 16 billion neurons. Elaborate folding of the cerebrum allows the cranium to contain three times as much cortex as we would have if the cerebrum had a smooth surface. If spread out flat, the area of cortex would equal about 4½ pages of this book, or 2,500 cm². Animals with less behavioral sophistication, variability, and adaptability have relatively smooth-surfaced cerebrums. Only whales, dolphins, and apes rival humans in the extent of their cerebral convolutions.

---

[14] *arbor* = tree; *vitae* = of life

[15] *a* = without; *taxi* = order

[16] *cortex* = bark, rind

Yet most of the volume of the cerebrum is white matter (see fig. 9.4c), which has no information-processing capability. White matter is composed of neuroglia and myelinated nerve fibers (axons) leading to and from the cell bodies of the cortex. These fibers are bundled into tracts that connect different gyri or larger regions within each hemisphere, pass between higher and lower brain centers and the spinal cord, and connect the right and left hemispheres through the thick C-shaped corpus callosum. It is chiefly through the corpus callosum that the left and right hemispheres coordinate their activities with each other.

Lobes of the Cerebrum

Some cerebral gyri are highly variable from one person's brain to another, and between the right and left hemispheres of the same person. Others, however, are quite consistent and serve as landmarks that universally define the functional *lobes* of any human cerebrum. There are five lobes in each hemisphere, mostly named for the overlying cranial bones (fig. 9.7):

1. The **frontal lobe** begins immediately behind the frontal bone, superior to the eyes. Its posterior boundary is a wavy vertical groove, the **central sulcus,** at the top of the head. A wide region of the frontal lobe just anterior to this sulcus is concerned with voluntary motor functions. Further anterior to this, between the motor region and the forehead, is the **prefrontal cortex.** This is little developed in other mammals but very prominent in humans, and is concerned with memory, mood, emotion, motivation, and abstract thought such as foresight, planning, problem solving, and social judgment.

2. The **parietal lobe** forms the uppermost part of the brain and underlies the parietal bone. Its anterior boundary is the central sulcus and its posterior boundary is the **parieto–occipital sulcus,** visible on the medial surface of each hemisphere (see fig. 9.2). This lobe is concerned with the sensory perception and integration of taste; some visual information; and, primarily, *somatosensory* (body-wall and musculoskeletal) sensations such as touch, stretch, and pain.

3. The **occipital lobe** is at the rear of the head underlying the occipital bone. It is the principal visual center of the brain.

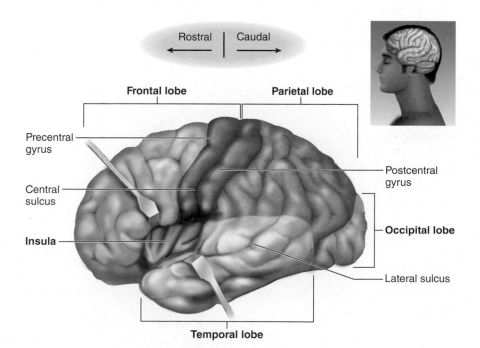

Figure 9.7 Lobes of the Cerebrum. The frontal and temporal lobes are retracted slightly to reveal the insula. AP|R

4. The **temporal lobe** is a lateral, horizontal lobe deep to the temporal bone, separated from the parietal lobe above it by a deep **lateral sulcus.** It is concerned with hearing, smell, learning, memory, visual recognition, and emotional behavior.

5. The **insula**[17] is a small mass of cortex deep to the lateral sulcus, made visible only by retracting or cutting away some of the overlying cerebrum (see figs. 9.1c and 9.4c). It is not as accessible to study in living people as other parts of the cortex and is therefore less-known territory. It apparently plays roles in understanding spoken language and speaking it, in the sense of taste, and in integrating sensory information from visceral receptors, such as visceral pain.

The Basal Nuclei

The **basal nuclei (basal ganglia)** are masses of cerebral gray matter buried deep in the white matter, lateral to the thalamus (see fig. 9.4). They receive input from the substantia nigra of the midbrain and motor areas of the cerebral cortex and send signals back to both of these locations. They are part of a feedback loop that serves in motor control and are further discussed in a later section on that topic.

The Limbic System

The **limbic**[18] **system** is a group of brain structures concerned primarily with learning, memory, and emotion. It forms a ring of tissue on the medial side of each cerebral hemisphere, encircling the corpus callosum and thalamus (fig. 9.8). It includes connections to slightly more distant parts of the brain, however, including portions of the frontal and temporal lobes, hypothalamus, thalamus, and the olfactory apparatus. The last of these may explain why certain odors can be associated with powerful memories and emotional responses. Two particularly important components of the limbic system are the **amygdala**[19] (ah-MIG-da-luh) and **hippocampus**[20]—both of which are parts of the temporal lobe.

---

[17] *insula* = island

[18] *limbus* = border

[19] *amygdala* = almond

[20] *hippocampus* = sea horse, named for its shape

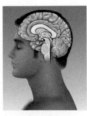

Figure 9.8 The Limbic System. This ring of structures (shown in violet) includes important centers of learning and emotion. In the frontal lobe, there is no sharp anterior boundary to limbic system components and there is no universal agreement among anatomists on how much to include in the limbic system. **AP|R**

• *What component of this system is especially associated with the sense of fear?*

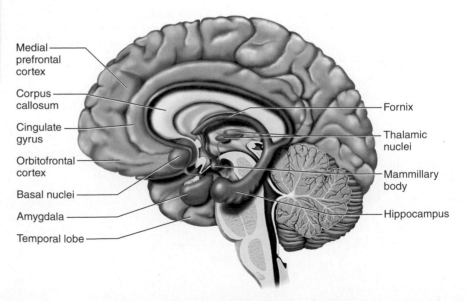

Medial prefrontal cortex

Corpus callosum

Cingulate gyrus

Orbitofrontal cortex

Basal nuclei

Amygdala

Temporal lobe

Fornix

Thalamic nuclei

Mammillary body

Hippocampus

The limbic system was originally thought to be associated mainly with smell, but later experiments revealed more significant roles in emotion and memory. It is involved in feelings that reinforce behaviors important to survival, such as the sense of reward associated with eating and sex, fear upon seeing a snake, and aversion to a food that has made one sick in the past. Most limbic structures have centers for both gratification and aversion. Stimulation of a gratification center produces a sense of pleasure or reward; stimulation of an aversion center produces unpleasant sensations such as sorrow, fear, or avoidance. Aversion centers dominate some limbic structures, such as the amygdala, whereas gratification centers dominate others. The roles of the amygdala in emotion and the hippocampus in memory are described in section 9.3.

Before You Go On

Answer these questions from memory. Reread the preceding section if there are too many you don't know.

5. List the four regions of the brainstem from inferior to superior, and list several functions of each.

6. Where are the thalami located, and what is their primary function?

7. Describe the location and functions of the hypothalamus.

8. Name two parts of the brain that are divided into right and left hemispheres and exhibit a convoluted surface of folds and grooves. What are the folds and grooves called?

9. State several roles of the cerebellum in addition to motor coordination.

10. List the five lobes of the cerebrum and some functions of each one.

11. State a location in the brain for each of the following functions: regulation of heart rate; formation of memories; visual awareness; blocking of pain signals; hearing; perception of touch; and initiation of body movements.

9.3 Multiregional Brain Functions

Expected Learning Outcomes

When you have completed this section, you should be able to:

a. identify the destinations of different types of sensory signals going to the brain;

b. identify some of the conscious, thinking (cognitive) areas of the brain;

c. identify areas of the brain involved in the creation of memories and in memory storage and retrieval;

d. describe how the brain controls the skeletal muscles;

e. describe the locations and functions of the language centers of the brain;

f. discuss the functional relationship between the right and left cerebral hemispheres; and

g. describe how electroencephalograms are recorded and why they are clinically useful.

Such brain functions as sensation, memory, language, emotion, motor control, and so forth are best considered as *multiregional* brain functions because they are not localized to one specific part of the brain, but involve many. There is no simple one-to-one relationship between these functions and the anatomical parts described up to this point. They are associated especially with the cerebral cortex, but not exclusively; they also involve such areas as the cerebellum, basal nuclei, and others. Some of these functions present the most difficult challenges for neurobiology, but they are the most intriguing functions of the brain and involve its largest areas.

Sensation

The human senses fall into two categories: (1) The **special senses** employ sense organs limited to the head and include vision, hearing, equilibrium, taste, and smell. (2) The **somatosensory**[21] or **general senses** are distributed over the entire body and include touch, itch, pain, heat, cold, pressure, stretch, and movement. The special senses, including regions of the brain that receive input for them, are (fig. 9.9a):

- vision—the posterior occipital lobe;
- hearing—the superior temporal lobe and nearby insula;
- equilibrium—mainly the cerebellum;
- taste—parts of the insula and lower parietal lobe; and
- smell—the medial surface of the temporal lobe and inferior surface of the frontal lobe.

These receiving areas are called **primary sensory areas.** They merely make one aware of a stimulus, but identifying and making sense of it are carried out by adjacent **sensory association areas** of the cerebrum. For example, a *visual association area* immediately anterior to the primary visual area enables one to identify what one sees and to relate effectively to one's visual world.

Signals for the general senses arrive in the **postcentral gyrus,** a vertical fold of the parietal lobe immediately posterior to the central sulcus. Signals coming from the spinal cord decussate along the way (see p. 274–275), so those arising on the right side of the body end in the left postcentral gyrus, and vice versa. Each half of the cerebrum therefore senses the opposite side of the body. Furthermore, each postcentral gyrus is like an upside-down sensory map of the body (fig. 9.9b). The lowermost part of the gyrus receives sensation from the face, for example, whereas sensation from the trunk and hips is perceived by the uppermost part of the gyrus, and sensation from the thigh through the foot is perceived by part of the gyrus in the longitudinal fissure between the cerebral hemispheres. This map, often called the *sensory homunculus,*[22] has an oddly distorted look because it shows body regions disproportionately large or small according to the relative amounts of cerebral cortex dedicated to them. Much more cortex is concerned with sensations from the face and hands, for example, than that concerned with the entire trunk of the body.

---

[21] *somato* = body
[22] *hom* = man; *uncul* = little

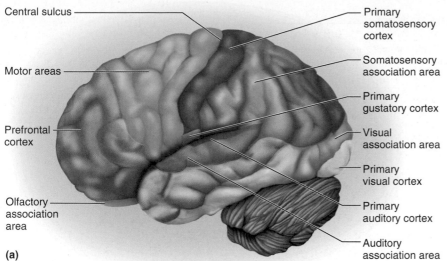

Central sulcus

Motor areas

Prefrontal cortex

Olfactory association area

Primary somatosensory cortex

Somatosensory association area

Primary gustatory cortex

Visual association area

Primary visual cortex

Primary auditory cortex

Auditory association area

(a)

Figure 9.9 Sensory Areas of the Cerebrum.
(a) Primary sensory areas and association areas.
(b) Sensory map of the postcentral gyrus. The human figure is distorted to represent the fact that larger areas of cortex are concerned with facial sensation, for example, than those concerned with the thigh and leg. Notice that higher regions of the gyrus, in general, are associated with lower regions of the body. **AP|R**

- *How do you predict a person would be affected by a stroke that destroyed much of the postcentral gyrus within the longitudinal fissure?*

Cognition

Cognition[23] refers to mental processes such as awareness, perception, thinking, knowledge, and memory. It is an integration of information between the routes of sensory input and motor output. Cognitive association areas of the cerebrum account for 75% of our brain tissue. They include the sensory association areas just described, but the earlier discussion also pointed out other, diverse cognitive functions in all five lobes of the cerebrum and even in the cerebellum. Lesions in these cognitive areas can produce a broad spectrum of effects such as inability to recognize familiar faces or to navigate within a familiar building; a variety of language deficits (aphasias) described later; or profound personality disorders and socially inappropriate behaviors.

Memory is one of the cognitive abilities we value most highly. It is a widely distributed function, not localized to any one area of the brain. The creation of new memories from ongoing sensory experience, called *memory consolidation,* is a function of the hippocampus, amygdala, and cerebellum, but these are not memory-storage depots. Different forms of memory are stored in very different regions of the cerebrum—for example, the memory of motor skills such as tying one's shoes or driving a car lies in the basal nuclei; a memory for faces and voices resides in the temporal lobe; a memory of languages resides in the posterior temporal to anterior occipital lobe; and memories of odors, tastes, and one's social position are associated with the frontal lobe.

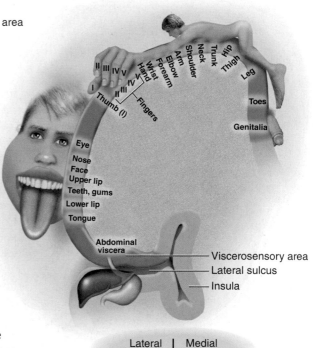

(b)

Viscerosensory area

Lateral sulcus

Insula

Lateral | Medial

Motor Control

Components of the nervous system that control the skeletal muscles are called the **somatic motor division.** This is distinguished from the *visceral motor (autonomic) division,* discussed later, which controls cardiac and smooth muscle. The somatic motor division is under voluntary control—we contract most skeletal muscles at will—whereas the visceral motor division is involuntary.

---

[23] *cognit* = to know

Figure 9.10 Motor Areas of the Cerebrum.
(a) Primary motor and motor association areas. (b) A frontal section through the precentral gyrus, mapping out areas of motor function. Like the sensory map in figure 9.9, the motor map shows an inverted arrangement with lower parts of the gyrus controlling muscles in higher areas of the body. More cerebral tissue is dedicated to fine motor control of the hands, face, and speech apparatus than to the trunk and lower limb. AP|R

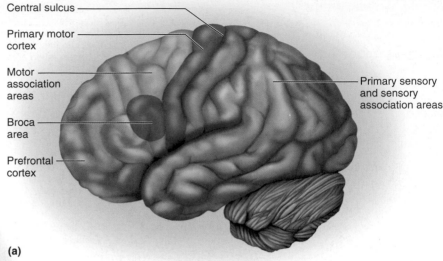

(a)

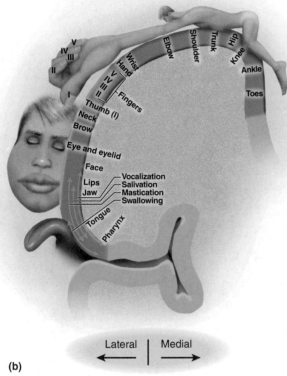

(b)

The decision to make a voluntary muscular movement originates in the **motor association area** of the frontal lobe (fig. 9.10a). A "plan of action" is then relayed to neurons in the **precentral gyrus**—the most posterior gyrus of the frontal lobe, immediately in front of the central sulcus. **Upper motor neurons** in this gyrus issue nerve fibers to the brainstem and spinal cord, where they synapse with **lower motor neurons.** Axons from the lower motor neurons then exit the CNS and go to the skeletal muscles. Most motor pathways bound for muscles below the neck cross over at a point called the **pyramidal decussation** near the inferior end of the pyramids of the medulla oblongata (see fig. 9.5a). Consequently, the right cerebrum controls movements on the left side of the body and vice versa. This is not true of most head–neck muscles, however, which are controlled by the cerebral hemisphere on the same side of the body.

The precentral gyrus, like the postcentral gyrus described earlier, has an upside-down map of the opposite side of the body (fig. 9.10b). That is, regions of the gyrus deep in the longitudinal fissure control muscles of the lower limb; the apex of the gyrus at the top of the head controls muscles of the hip, trunk, and shoulder; and as we descend the lateral part of the gyrus we find neurons that control the upper limb, a large area devoted to the hand, and finally a large area devoted to muscles of the face and tongue. This is depicted as a *motor homunculus* distorted to represent relative amounts of cerebrum dedicated to each region of the body. The amount of cerebral tissue dedicated to hands, face, and tongue reflects the importance of fine motor control in speech, facial expressions, and use of the hands.

Other areas of the brain important in muscle control are the basal nuclei (see fig. 9.4) and cerebellum. Among other functions, the basal nuclei assume control of highly practiced behaviors that one carries out with little thought—writing, typing, driving a car, or tying one's shoes, for example. They also control the onset and cessation of planned movements, and the repetitive movements at the shoulder and hip that occur during walking.

The cerebellum aids in learning motor skills, maintains muscle tone and posture, smooth muscle contractions, coordinates eye and body movements, and coordinates the motions of different joints with each other (such as the shoulder and elbow in pitching a baseball). When higher centers of the cerebrum tell the muscles what to do, a copy of this information goes to the cerebellum. When the muscles contract, they stimulate sense organs called **proprioceptors**[24] (PRO-pree-o-SEP-turs) in the muscles and joints, including

---

[24] *proprio* = one's own; *ceptor* = receiver

the muscle spindles described in chapter 8. Proprioceptors are specialized to monitor the position and movements of the body; they issue signals to the cerebellum to report what the muscles are actually doing. From the inner ears, the cerebellum also derives a sense of **equilibrium,** an awareness of the body's balance and movements. If there is a discrepancy between command and motor performance, the cerebellum sends a "progress report" to the cerebrum, which modifies its output to correct the muscle action.

Language

Language includes several cognitive and motor abilities—reading, listening, understanding words, writing, and speaking—involving multiple areas of the cerebrum (fig. 9.11). The cerebral control of these tasks is very complex, and we can consider only a few key centers here. One is the **Wernicke**[25] (WUR-ni-keh) **area,** usually found in the left cerebral hemisphere just posterior to the lateral sulcus. Immediately posterior to this is a fold of cerebrum called the **angular gyrus,** which some authorities consider part of the Wernicke area itself. These areas lie at a crossroad between the visual cortex of the occipital lobe and the auditory cortex of the temporal lobe. They receive input from both and interpret visual images and sounds in terms of learned rules of grammar and vocabulary. That is, they create linguistic meaning from what we see and hear, and they begin to formulate the phrases of what we will speak.

The Wernicke area transmits a plan of speech forward to the **Broca**[26] **area,** which lies low in the prefrontal cortex of the same hemisphere. The Broca area also plays a part in constructing grammatically sensible sentences, and it compiles a "plan" of motor neuron action that will soon govern the muscle contractions of the larynx, tongue, cheeks, and lips. It transmits this plan by way of the insula to the primary motor cortex (precentral gyrus), which then issues commands to the relevant muscles to carry out the speech. Injury to the Wernicke or Broca area from stroke or trauma can leave a person with language impairments called **aphasia**[27] (ah-FAY-zhee-uh) (see Clinical Application 9.2).

---

[25] Karl Wernicke (1848–1905), German neurologist

[26] Pierre Paul Broca (1824–80), French surgeon and anthropologist

[27] *a* = without; *phas* = speech

Figure 9.11 Language Centers of the Left Hemisphere.

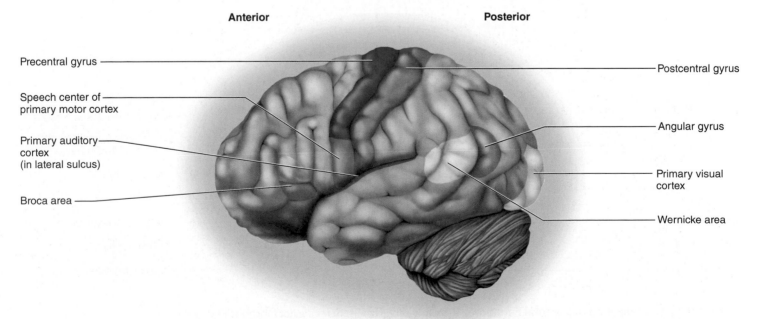

Anterior

Posterior

Precentral gyrus

Speech center of primary motor cortex

Primary auditory cortex (in lateral sulcus)

Broca area

Postcentral gyrus

Angular gyrus

Primary visual cortex

Wernicke area

Clinical Application 9.2

APHASIA

Aphasia takes many forms depending on which regions of cerebral tissue, or the white-matter connections between them, are damaged. Injury to the Wernicke area can make a person unable to recognize spoken or written words or remember the names of familiar objects *(anomic[28] aphasia)*. It often results in babbling incomprehensible speech filled with made-up words and nonsensical syntax *(fluent aphasia)*. Lesions of the Broca area, by contrast, leave a person knowing what to say but having great difficulty commanding the muscles to say it *(nonfluent aphasia)*. The result is typically a stammering, labored speech with partial sentences, or a frustrated tight-lipped reluctance to speak at all. Some strokes reduce a person's vocabulary to only a few words, sometimes the last few that he or she spoke before the stroke occurred.

Stranger still, damage to certain small areas of cerebral cortex can leave a person with impaired mathematical ability, a selective absence of vowels from one's writing, or difficulty understanding the second half of each word the person reads.

Speech therapy can help some individuals regain lost language skills, depending on the location and extent of tissue damage. Even sex can make a difference in recovery; men are three times as likely as women to suffer language deficits from left-sided cerebral damage, and women recover more easily than men.

[28] *a* = without; *nom* = names

Cerebral Lateralization

The two cerebral hemispheres look identical at a glance, but close examination reveals a number of differences. For example, in women the left temporal lobe is often longer than the right. In left-handed people, the left frontal, parietal, and occipital lobes are usually wider than those on the right. The two hemispheres also differ in some of their functions (fig. 9.12). Neither hemisphere is "dominant," but each is specialized for certain tasks. This difference in function is called **cerebral lateralization.**

One hemisphere, usually the left, is called the *categorical hemisphere.* It is specialized for spoken and written language and for the sequential and analytical reasoning employed in such fields as science and mathematics. This hemisphere seems to break information into fragments and analyze it in a linear way. The other hemisphere, usually the right, is called the *representational hemisphere.* It perceives information in a more integrated, holistic way. It is the seat of imagination and insight, musical and artistic skill, perception of patterns and spatial relationships, and comparison of sights, sounds, smells, and tastes.

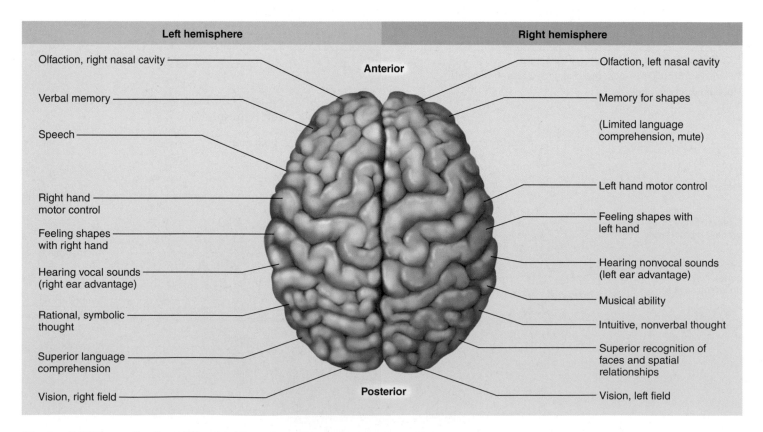

Figure 9.12 Lateralization of Cerebral Functions. The two cerebral hemispheres are not functionally identical.

The Electroencephalogram

Brain functions and disorders are often studied or diagnosed by placing an array of recording electrodes on the scalp and recording the subject's **brain waves**—rhythmic voltage changes resulting from the synchronized activity of cerebral neurons close to the surface. The recording is called an **electroencephalogram**[29] **(EEG).** It exhibits four types of brain waves, each of which dominates a different state of consciousness, wakefulness, or sleep (fig. 9.13). Normal EEGs are useful in studying brain functions such as sleep and consciousness, and abnormalities can help diagnose epilepsy, degenerative brain diseases, brain tumors, metabolic abnormalities, cerebral trauma, and so forth. The EEG changes with age and with states of consciousness ranging from high alert to deep sleep and coma. The complete and persistent absence of brain waves is often used as a clinical and legal criterion of brain death.

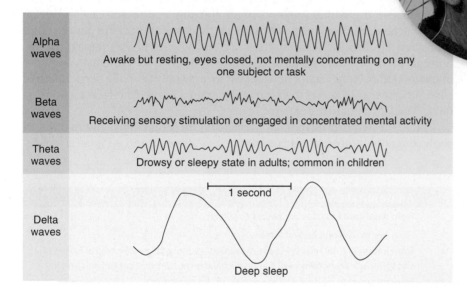

Figure 9.13 The Electroencephalogram (EEG).

Alpha waves — Awake but resting, eyes closed, not mentally concentrating on any one subject or task

Beta waves — Receiving sensory stimulation or engaged in concentrated mental activity

Theta waves — Drowsy or sleepy state in adults; common in children

1 second

Delta waves — Deep sleep

Before You Go On

Answer these questions from memory. Reread the preceding section if there are too many you don't know.

12. Describe the locations of the primary sensory areas for vision, hearing, taste, touch, and pain. How would a pain in the left foot differ from a pain in the right hand with respect to the area of cerebral cortex where these sensations are registered?

13. The hippocampus, amygdala, and cerebellum are all involved in memory, but how do their memory functions differ?

14. In what gyrus do motor commands to the skeletal muscles originate? How does this gyrus resemble and differ from the one that receives somatosensory information?

15. What are the roles of the Wernicke area, Broca area, and precentral gyrus in language?

---

[29] *electro* = electricity; *encephalo* = brain; *gram* = record

9.4 The Cranial Nerves

Expected Learning Outcomes

When you have completed this section, you should be able to:

a. list the 12 cranial nerve pairs by name and number;
b. identify where each cranial nerve originates and terminates; and
c. state the functions of each cranial nerve.

The brain communicates with the rest of the body through two routes: the spinal cord and 12 pairs of **cranial nerves.** The cranial nerves are numbered I to XII starting with the most anterior pair. Each nerve also has a descriptive name such as *optic nerve* or *vagus nerve.* Most of these nerves arise from the base of the brain, exit the skull through its foramina, and lead to muscles and sense organs located mainly in the head and neck. One exception to this pattern is the vagus nerve (X), which descends to reach many organs in the thoracic and abdominal cavities. Another is the accessory nerve (XI), once thought to originate in the medulla oblongata but now known to arise only from the cervical spinal cord. Strictly speaking, it is therefore not a cranial nerve, but neuroanatomists haven't kicked it out of the club yet.

Table 9.1 itemizes the cranial nerves and their functions, and figure 9.14 shows their points of origin and termination in the brain and peripheral organs.

Table 9.1 The Cranial Nerves

Number and Name	Function	Anatomical Course
I. Olfactory	*Sensory:* Smell	From nasal cavity to olfactory bulbs of brain
II. Optic	*Sensory:* Vision	From eye to thalamus and midbrain
III. Oculomotor	*Motor:* Eye movements	From midbrain to all muscles of eye movement except as noted for cranial nerves IV and VI; muscle that opens eyelid; and internal eye muscles that control lens and iris
IV. Trochlear	*Motor:* Eye movements	From midbrain to superior oblique muscle of eye movement
V. Trigeminal (three branches, V_1, V_2, V_3)	*Motor:* Chewing *Sensory:* Touch, temperature, and pain from face	*Motor:* From pons to masseter, temporalis, and pteryoid muscles of chewing; tensor tympani muscle of middle ear *Sensory:* From facial skin, surface of eye, nasal mucosa, paranasal sinuses, palate, teeth and gums, tongue (but not taste buds), floor of mouth, and dura mater to pons
VI. Abducens	*Motor:* Eye movements	From pons to lateral rectus muscle of eye movement
VII. Facial	*Motor:* Facial expression; secretion of tears, saliva, and nasal and oral mucus *Sensory:* Taste	*Motor:* From pons to muscles of facial expression, salivary glands, tear glands, and glands of nasal cavity and palate *Sensory:* From taste buds of anterior tongue to thalamus
VIII. Vestibulocochlear	*Motor:* Cochlear tuning *Sensory:* Hearing and equilibrium	*Motor:* From pons to cochlea of inner ear *Sensory:* From cochlea, semicircular ducts, and vestibule of inner ear, to medulla oblongata and medulla–pons junction
IX. Glossopharyngeal	*Motor:* Salivation, swallowing, gagging *Sensory:* Taste; sensations of touch, pressure, pain, and temperature in tongue and outer ear; regulation of blood pressure and respiration	*Motor:* From medulla oblongata to salivary glands, glands of tongue, and muscle of pharynx *Sensory:* From middle and outer ear, posterior one-third of tongue (including taste buds), and pharynx, to medulla oblongata
X. Vagus	*Motor:* Swallowing, speech, regulation of heart rate and bronchial airflow, gastrointestinal secretion and motility *Sensory:* Taste; sensations of hunger, fullness, and gastrointestinal discomfort	*Motor:* From medulla oblongata to tongue, palate, larynx, lungs, heart, liver, spleen, digestive tract, kidney, and ureter *Sensory:* From root of tongue, pharynx, larynx, epiglottis, outer ear, and several thoracic and abdominal viscera, to medulla oblongata
XI. Accessory	*Motor:* Swallowing; head, neck, and shoulder movements	From spinal cord segments C1 to C6, into the cranial cavity and back out to palate, pharynx, trapezius, and sternocleidomastoid muscles
XII. Hypoglossal	*Motor:* Tongue movements	From medulla oblongata to muscles of tongue

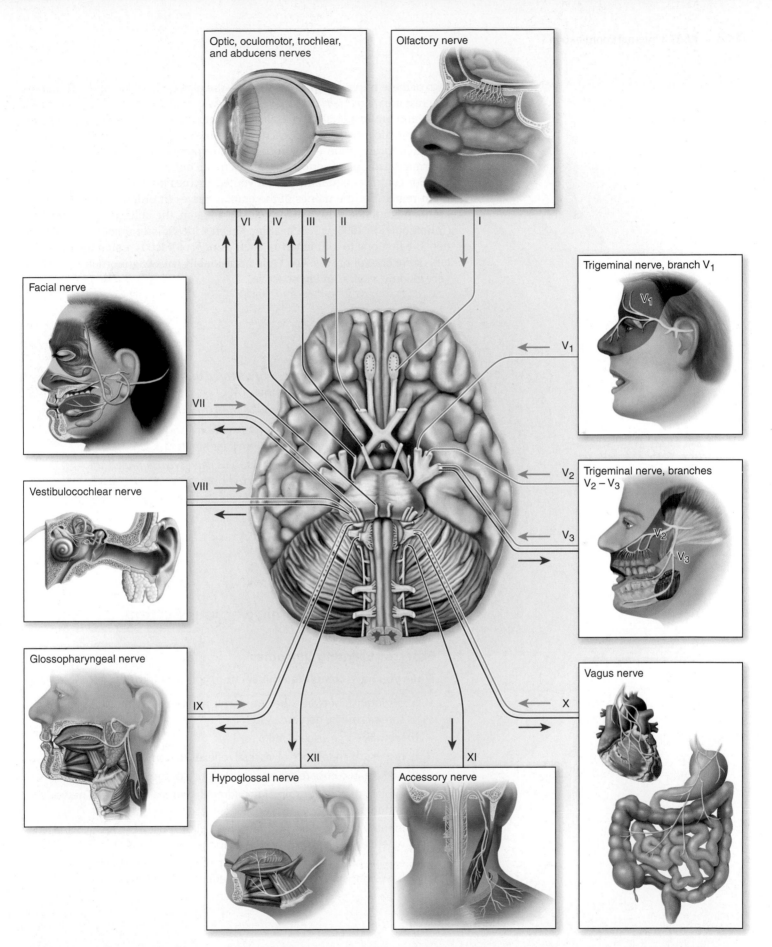

Figure 9.14 The Cranial Nerves. Pathways concerned with sensory function are shown in green, and those concerned with motor function are shown in red. Arrows indicate the direction in which the nerve signals travel. All cranial nerves except I and II carry sensory fibers, but these functions are omitted in cases where the nerve is essentially motor and its sensory function is limited to proprioception (muscle feedback). (n. = nerve, nn. = nerves) **AP|R**

Two of these nerves are purely sensory—the olfactory (I) and optic (II) nerves. The rest are *mixed nerves,* carrying a two-way traffic of sensory signals to the brain and motor signals to muscles and other organs. In cases where only the motor function of a nerve is described in table 9.2, the sensory function is limited to proprioception—monitoring the contraction of a muscle and sending information about its performance back to the brain.

For centuries, students have invented rhymes and other phrases to help them remember the names of the cranial nerves in order. One of the senior author's students, now a neurologist,[30] created the following memory aid (mnemonic) in 1993, which has since been widely disseminated on the Internet. The first one to four letters of each word (boldfaced) match the initial letters of the cranial nerves: **Old Op**ie **oc**casionally **tri**es **trig**onometry, **a**nd **f**eels **ve**ry **glo**omy, **vagu**e, and **hypo**active.

Before You Go On

Answer these questions from memory. Reread the preceding section if there are too many you don't know.

16. List the purely sensory cranial nerves and state the function of each.
17. What is the only cranial nerve that extends beyond the head–neck region? Name some organs in which it terminates.
18. If the oculomotor, trochlear, or abducens nerve was damaged, the effect would be similar in all three cases. What would that effect be?
19. Name two cranial nerves involved in the sense of taste and explain where their sensory fibers originate.

9.5 The Autonomic Nervous System

Expected Learning Outcomes
When you have completed this section, you should be able to:

a. distinguish between the autonomic nervous system (ANS) and the somatic motor nervous system;
b. identify some visceral reflexes;
c. discuss the relevance of visceral reflexes to homeostasis;
d. name and compare the two subdivisions of the ANS;
e. explain the relationship of the adrenal medulla and enteric nervous system to the ANS;
f. identify the two principal neurotransmitters used by the ANS and where each of them is employed in the system; and
g. explain how the sympathetic and parasympathetic divisions can have cooperative or antagonistic effects on various organs.

---

[30] Marti Haykin, M.D.

The **autonomic nervous system (ANS)** can be defined as a motor nervous system that controls glands, cardiac muscle, and smooth muscle. It is also called the **visceral motor division** to distinguish it from the somatic motor division that controls the skeletal muscles. The primary targets of the ANS are organs of the thoracic and abdominal cavities, such as the heart, lungs, digestive tract, and urinary tract; but it also innervates some structures of the body wall, including cutaneous blood vessels, sweat glands, and piloerector muscles.

The word *autonomic* literally means "self-governed." It refers to the fact that the ANS usually carries out its actions without one's conscious intent, awareness, or ability to control it at will. It is for this reason that autonomic responses are used as a basis for polygraph ("lie detector") tests. Visceral effectors do not depend on the autonomic nervous system to function, but only to adjust their activity to the body's changing needs. The heart, for example, goes on beating even if all nerves to it are severed, but the ANS adjusts the heart rate for such conditions as rest or exercise.

Visceral Reflexes

The ANS works through **visceral reflexes**—reflexes that regulate such primitive functions as blood pressure, heart rate, body temperature, digestion, energy metabolism, respiratory airflow, pupillary diameter, defecation, and urination. In short, the ANS quietly manages a myriad of unconscious processes responsible for the body's homeostasis; these are not specifically human but among our most basic animal functions. Many drug therapies are based on alteration of autonomic functions (see Perspectives on Health, p. 320).

A visceral reflex arc includes (1) receptors—nerve endings that detect stretch, tissue damage, blood chemicals, body temperature, and other internal stimuli; (2) afferent neurons leading to the CNS; (3) interneurons within the CNS; (4) efferent neurons carrying motor signals away from the CNS; and (5) effectors.

For example, high blood pressure activates a visceral **baroreflex**[31] involving the following components (fig. 9.15): ① It stimulates stretch receptors called *baroreceptors* in the carotid arteries and aorta. ② They issue signals through afferent fibers in the glossopharyngeal nerves to the medulla oblongata. ③ Interneurons of the medulla integrate this input with other information and issue signals back to the heart by way of efferent fibers in the vagus nerves. ④ The vagus nerves slow down the heart (the effector) and reduce blood pressure. This completes a homeostatic negative feedback loop and ideally restores blood pressure to normal. A separate autonomic reflex arc accelerates the heart when blood pressure drops below normal.

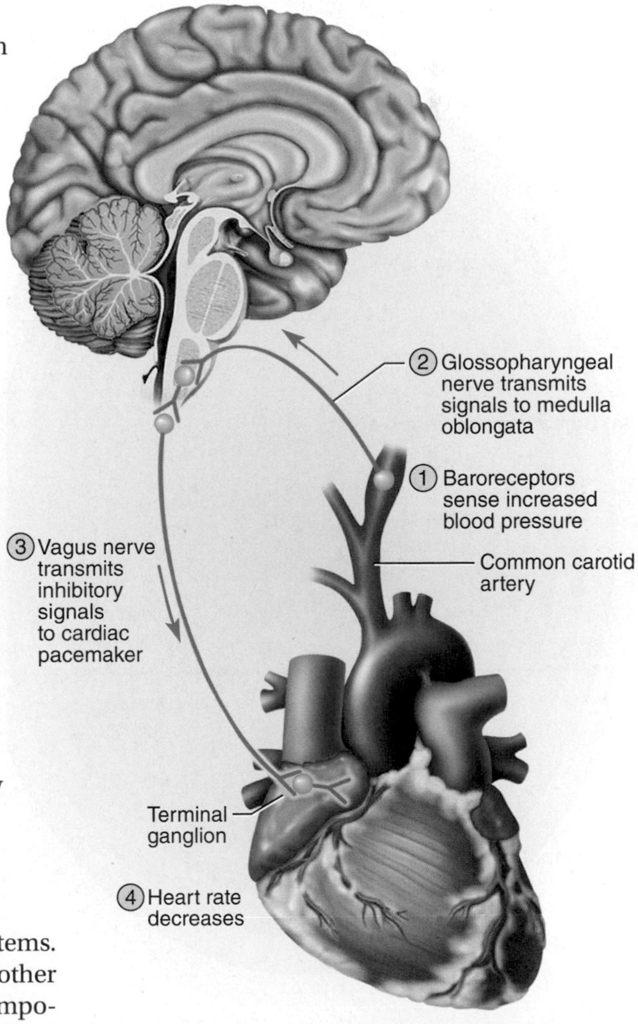

Neural Pathways

The ANS has components in both the central and peripheral nervous systems. In the CNS, there are autonomic control centers in the hypothalamus and other regions of the brainstem, and motor neurons in the spinal cord. PNS components include motor neurons in the ganglia and nerve fibers in the cranial and spinal nerves you have already studied.

The autonomic pathway to a target organ differs significantly from somatic motor pathways. In somatic pathways, a motor neuron in the brainstem or spinal cord issues a myelinated axon that reaches all the way to a skeletal

② Glossopharyngeal nerve transmits signals to medulla oblongata

① Baroreceptors sense increased blood pressure

Common carotid artery

③ Vagus nerve transmits inhibitory signals to cardiac pacemaker

Terminal ganglion

④ Heart rate decreases

Figure 9.15 An Autonomic Reflex Arc. In the arterial baroreflex, a rise in blood pressure is detected by baroreceptors in the carotid artery. The glossopharyngeal nerve transmits signals to the medulla oblongata, resulting in parasympathetic output through the vagus nerve that reduces the heart rate and lowers blood pressure.

---

[31] *baro*= pressure

Figure 9.16 Comparison of Somatic and Autonomic Efferent Pathways. (a) In the somatic nervous system, each motor nerve fiber extends all the way from the CNS to a skeletal muscle. (b) In the sympathetic nervous system, a short preganglionic fiber extends from the spinal cord to a sympathetic ganglion near the vertebral column. A longer postganglionic fiber extends from there to the target organ. (c) In the parasympathetic nervous system, a long preganglionic fiber extends from the CNS to a ganglion in or near the target organ, and a short postganglionic fiber completes the path to specific effector cells in that organ (smooth or cardiac muscle, or gland cells). (ACh = acetylcholine; NE = norepinephrine)

muscle (fig. 9.16a). In autonomic pathways, the signal must travel across two neurons to get to the target cells, and it must cross a synapse where these neurons meet in an autonomic ganglion (fig. 9.16b, c). The first neuron, called the **preganglionic neuron,** has a soma in the brainstem or spinal cord; its axon terminates in a ganglion outside the CNS. It synapses there with a **postganglionic neuron** whose axon extends the rest of the way to the target cells. The axons of these neurons are called the pre- and postganglionic fibers. The preganglionic fibers are myelinated, and the postganglionic fibers are not.

Subdivisions of the ANS

The ANS has two subdivisions: sympathetic and parasympathetic. The **sympathetic division** adapts the body in many ways for physical activity—it increases alertness, heart rate, blood pressure, pulmonary airflow, blood glucose concentration, and blood flow to cardiac and skeletal muscle. Extreme sympathetic responses are often called the "fight-or-flight" reaction because they come into play when an animal must attack, defend itself, or flee from danger. In our own lives, this reaction occurs in many situations involving arousal, competition, stress, danger, anger, or fear—ranging from a game of

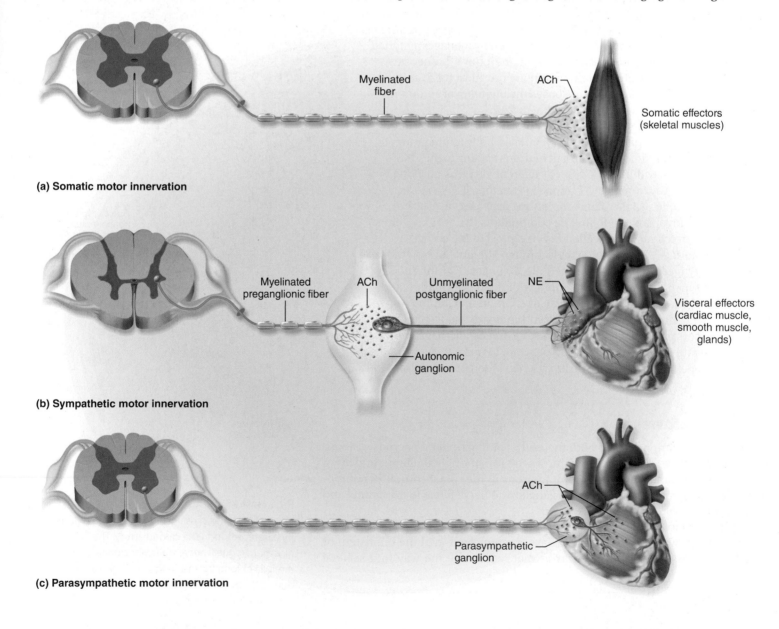

Myelinated fiber

ACh

Somatic effectors (skeletal muscles)

(a) Somatic motor innervation

Myelinated preganglionic fiber

ACh

Unmyelinated postganglionic fiber

NE

Autonomic ganglion

Visceral effectors (cardiac muscle, smooth muscle, glands)

(b) Sympathetic motor innervation

ACh

Parasympathetic ganglion

(c) Parasympathetic motor innervation

chess to defending oneself from an attacker. Ordinarily, however, the sympathetic division has more subtle effects that we notice barely, if at all.

The **parasympathetic division**, by comparison, has a calming effect on many body functions. It is associated with reduced energy expenditure and normal bodily maintenance, including such functions as digestion and waste elimination. This is sometimes called the "resting-and-digesting" state.

This does not mean that the body alternates between states where one division or the other is active and the opposite system is turned off. Normally both divisions are active simultaneously, but the balance between them shifts with the body's changing needs. Neither division has universally excitatory or calming effects. The sympathetic division, for example, excites the heart but inhibits digestive and urinary functions, while the parasympathetic division has the opposite effects. We will later examine how they achieve these contrasting effects.

Anatomy of the Sympathetic Division

All efferent, preganglionic fibers of the sympathetic division arise from the thoracic and first two lumbar segments of the spinal cord; hence this division is sometimes called the *thoracolumbar division*. These fibers travel a short distance to ganglia that lie just outside the vertebral canal alongside the vertebral column. Here, some of the fibers turn and travel up or down to higher and lower ganglia, uniting the ganglia into a string called the **sympathetic chain**. The chain extends upward into the cervical region and downward to the sacral and coccygeal levels (fig. 9.17a).

Usually, the preganglionic nerve fiber synapses with a postganglionic neuron somewhere in the sympathetic chain. Some of the fibers pass through the chain without synapsing, however, and go to three *collateral ganglia* wrapped around the abdominal aorta, the large artery on the posterior abdominal wall. From either the sympathetic chain or collateral ganglia, postganglionic nerve fibers then complete the path to the target organs. The general pattern in the sympathetic division is to have short preganglionic fibers and long postganglionic ones (fig. 9.16b).

Most preganglionic fibers branch before reaching their synapses. Typically, each preganglionic neuron synapses with 10 to 20 postganglionic neurons. This means that a localized output from the spinal cord can branch out and reach several target organs at once, such as the eyes, sweat glands, heart, and lungs, creating multiple effects of the fight-or-flight response (pupillary dilation, sweating, faster heart rate, and increased airflow).

> *Apply What You Know*
>
> Which of the types of neural circuits described in chapter 8 is represented by this branching of sympathetic pathways to multiple target organs?

Another component of the sympathetic nervous system is found in the **adrenal**[32] **glands,** which rest like hats atop the kidneys (see fig. 11.12, p. 381). Each adrenal has an outer rind, the **adrenal cortex,** which has no relation to sympathetic function, but its inner core, the **adrenal medulla,** is essentially a sympathetic ganglion. Preganglionic fibers penetrate through the cortex and terminate on cells of the medulla. They stimulate these cells to secrete epinephrine (adrenaline) and norepinephrine (noradrenaline) into the blood. These hormones accentuate and prolong the other effects of the sympathetic nervous system.

---

[32] *ad* = near; *ren* = kidney

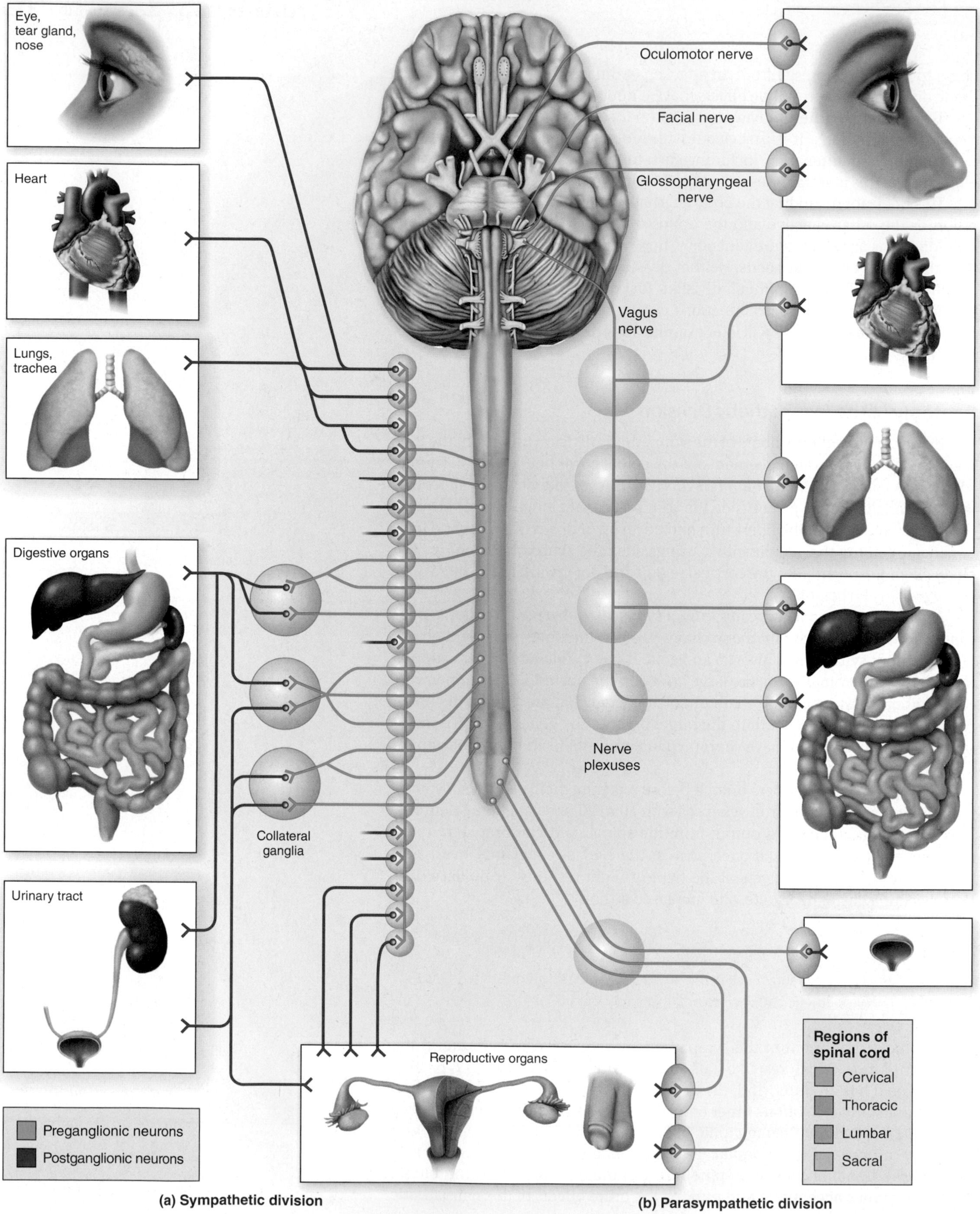
Eye,
tear gland,
nose
Heart
Lungs,
trachea
Digestive organs
Urinary tract
Oculomotor nerve
Facial nerve
Glossopharyngeal
nerve
Vagus
nerve
Nerve
plexuses
Collateral
ganglia
Reproductive organs
Preganglionic neurons
Postganglionic neurons
Regions of
spinal cord
Cervical
Thoracic
Lumbar
Sacral
(a) Sympathetic division
(b) Parasympathetic division

Figure 9.17 The Sympathetic and Parasympathetic Nerve Pathways. (a) Sympathetic pathways. (b) Parasympathetic pathways.

- *Suppose a person had a spinal cord injury that severed the cord at the cervical level. Which of these functions do you think would be lost, and why—acceleration of the heart in response to fear or the secretion of stomach juices in response to the smell of good food?* AP|R

Anatomy of the Parasympathetic Division

Preganglionic fibers of the parasympathetic division leave the CNS by way of four cranial nerves—the oculomotor (III), facial (VII), glossopharyngeal (IX), and vagus (X) nerves—and by way of spinal nerves from the sacral segments of the spinal cord (fig. 9.17b). Therefore, the parasympathetic division is also sometimes called the *craniosacral division*. In comparison to sympathetic preganglionic fibers, parasympathetic preganglionics travel relatively long distances to *terminal ganglia* in or near their target organs, and synapse there with short postganglionic fibers that complete the path to the target cells (figs. 9.16c, 9.17b). Parasympathetic fibers in cranial nerves III, VII, and IX supply target organs in the head, such as tear glands, salivary glands, nasal glands, and internal muscles of the eyeball (for control of the iris and lens). Cranial nerve X (the vagus nerve), however, carries 90% of all parasympathetic fibers and descends through the thorax and abdomen to reach numerous organs in these body cavities—heart, lungs, esophagus, stomach, intestines, liver, pancreas, and urinary tract. Parasympathetic fibers from the sacral region of the spinal cord supply the colon and rectum, urinary bladder, and reproductive organs. Fibers from the vagus and sacral nerves pass through various webs or *nerve plexuses* on their way to the target organs.

Neurotransmitters and Receptors

We have already seen a few cases in which the two divisions of the ANS have contrasting effects on the same organs. The sympathetic division accelerates the heartbeat, for example, and the parasympathetic division slows it down; the sympathetic division inhibits digestion and the parasympathetic division stimulates it. The key to such contrasting effects, and to many drug actions, depends on differences in the neurotransmitters employed by the two divisions and in the types of neurotransmitter receptors found in the target cells.

The two main neurotransmitters employed in the ANS are acetylcholine (ACh) and norepinephrine (NE). Nerve fibers that secrete ACh and receptors that bind it are called **cholinergic.**[33] Those that secrete or bind NE are called **adrenergic.**[34]

All preganglionic fibers of the ANS are cholinergic; they release ACh to stimulate the postganglionic neurons (table 9.2). Parasympathetic postganglionic fibers also are cholinergic, whereas most sympathetic postganglionic fibers are adrenergic. (A few of the latter are cholinergic—those that supply sweat glands and some blood vessels.)

The neurotransmitter is not the whole story, however, in how the ANS produces contrasting effects in different organs. For example, norepinephrine from sympathetic nerve fibers constricts most blood vessels but dilates the coronary vessels in the wall of the heart. Acetylcholine from parasympathetic fibers contracts the urinary bladder, yet relaxes the internal urethral sphincter that allows the urine to leave the bladder. The reason each division can produce opposite effects on different cells is that there are multiple classes of adrenergic and cholinergic receptors. The two major classes of adrenergic receptors are called α-*adrenergic* and β-*adrenergic* receptors, and there are subclasses of

---

[33] *cholin,* from *acetylcholine; erg* = work, action

[34] after *noradrenaline,* a synonym for *norepinephrine*

Table 9.2 Locations of Cholinergic and Adrenergic Fibers in the ANS

Division	Preganglionic Fibers	Postganglionic Fibers
Sympathetic	Always cholinergic	Mostly adrenergic; a few cholinergic
Parasympathetic	Always cholinergic	Always cholinergic

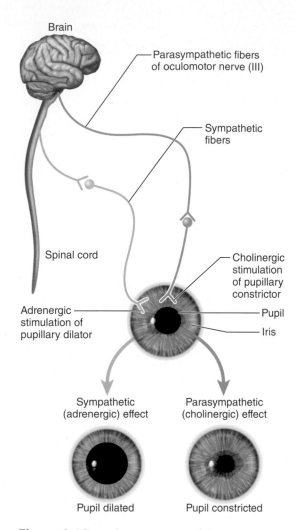

Figure 9.18 Dual Innervation of the Iris. Shows antagonistic effects of the sympathetic (yellow) and parasympathetic (blue) divisions on the iris.

both of these. The major classes of cholinergic receptors, again with subclasses of each, are called *nicotinic* and *muscarinic* receptors, after certain plant toxins used to discover them—nicotine from tobacco and muscarine, a mushroom poison. Certain receptor types excite the target cells and others inhibit them. Thus, a smooth muscle cell in one organ might have a nicotinic receptor and contract in response to ACh, whereas a smooth muscle cell in another organ might have a muscarinic receptor and relax in response to ACh.

Dual Innervation

Many organs receive both sympathetic and parasympathetic input. In the eye, for example, the iris receives fibers of both types (fig. 9.18). Sympathetic fibers lead to contractile cells arranged around the pupil like spokes of a wheel; when they contract, they dilate the pupil—a well-known sympathetic effect. Parasympathetic fibers lead to muscle cells that encircle the pupil. When they contract, they constrict the pupil. When the two divisions of the ANS have opposite effects on the same organ, they are said to have **antagonistic effects.**

In other cases, the two divisions have **cooperative effects.** Saliva, for example, consists of a watery solution of digestive enzymes plus slippery mucus that makes food easier to swallow. Sympathetic fibers stimulate the salivary glands to secrete the mucus, and parasympathetic fibers stimulate them to secrete the enzymes.

It is not always necessary to have dual innervation, however, for the ANS to produce opposite effects on an organ. Most blood vessels, for example, receive only sympathetic nerves. When the nerves increase their firing rate, they constrict the vessels, which raises the blood pressure. When the nerves decrease their firing rate, they allow the blood vessels to relax and dilate, which lowers the blood pressure. In some cases, such as sudden emotional shocks, the sympathetic firing rate decreases so much that the blood pressure falls too low and a person faints.

Table 9.3 presents several examples of the sympathetic and parasympathetic effects on various target organs and bodily processes.

Higher-Level Regulation of the ANS

Even though *autonomic* means "self-governed," the ANS is not an independent nervous system. All of its output originates in the CNS, and it is strongly influenced by the cerebral cortex, hypothalamus, medulla oblongata, and spinal cord.

There are many everyday examples of cerebral influences on the ANS: anger raises the blood pressure, fear makes the heart race, thoughts of good food make the stomach rumble, and anxiety or distraction inhibits sexual function. The limbic system is involved in many emotional responses and has extensive connections with the hypothalamus, thus linking the conscious centers of the brain with the autonomic control centers of the hypothalamus.

The hypothalamus contains nuclei for many primitive functions, including hunger, thirst, thermoregulation, emotions, and sexuality. Artificial stimulation of different regions of the hypothalamus can activate the fight-or-flight response typical of the sympathetic nervous system, or conversely, the calming effects typical of the parasympathetic. Output from the hypothalamus travels largely to nuclei in lower regions of the brainstem, and from there to the cranial nerves and the sympathetic neurons in the spinal cord.

The midbrain, pons, and medulla oblongata house numerous autonomic nuclei: centers for cardiac and vasomotor control, salivation, swallowing, sweating, digestive secretion, bladder control, pupillary construction and dilation, and other primitive functions. Autonomic output from these nuclei travels by way of the spinal cord and the oculomotor, facial, glossopharyngeal, and vagus nerves.

Finally, the spinal cord contains integrating centers for such autonomic reflexes as defecation, urination, erection, and ejaculation. Fortunately, the brain

Table 9.3 Effects of the Sympathetic and Parasympathetic Nervous Systems

Target	Sympathetic Effect	Parasympathetic Effect
Eye		
Iris	Pupillary dilation	Pupillary constriction
Lens	Relaxation for far vision	Contraction for near vision
Integumentary System		
Merocrine sweat glands	Secretion	No effect
Piloerector muscles	Hair erection	No effect
Circulatory System		
Heart rate and force	Increase	Decrease
Blood vessels of skeletal muscles	Dilation	No effect
Blood vessels of skin	Constriction	Dilation, blushing
Respiratory System		
Bronchioles	Dilation, increased airflow	Constriction, reduced airflow
Urinary System		
Kidneys	Reduced urine output	No effect
Bladder wall	No effect	Contraction
Internal urinary sphincter	Contraction, urine retention	Relaxation, urine release
Digestive System		
Salivary glands	Thick mucous secretion	Thin serous secretion
Gastrointestinal motility	Reduced	Increased
Liver	Glycogen breakdown	Glycogen synthesis
Reproductive System		
Penis and clitoris	No effect	Erection
Glandular secretions	No effect	Stimulation
Orgasm, smooth muscle roles	Stimulation	No effect

is able to inhibit defecation and urination consciously, but when injuries sever the spinal cord from the brain, the autonomic spinal reflexes alone control the elimination of urine and feces. Erection and ejaculation can occur through autonomic spinal reflexes alone even in men with spinal cord injuries who cannot feel the associated sensations.

Before You Go On

Answer these questions from memory. Reread the preceding section if there are too many you don't know.

20. Trace the pathway of a parasympathetic nerve fiber from the medulla oblongata to the small intestine.

21. What neurotransmitters are secreted by adrenergic and cholinergic fibers?

22. What are the two ways in which the sympathetic and parasympathetic systems can affect each other when they both innervate the same target organ? Give examples.

23. How can the sympathetic nervous system have contrasting effects in a target organ without dual innervation?

24. List some autonomic responses that are controlled by nuclei in the hypothalamus.

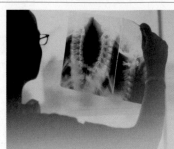

PERSPECTIVES ON H E A L T H

Drugs and the Nervous System

An understanding of neurotransmitter and receptor types in the brain and autonomic nervous system has given rise to the large field of *neuropharmacology*—the branch of science that deals with the design of drugs that mimic, enhance, or inhibit the action of these neurotransmitters.

A number of drugs work by stimulating adrenergic and cholinergic neurons or receptors. For example, phenylephrine, found in such cold medicines as Dimetapp and Sudafed PE, aids breathing by stimulating α-adrenergic receptors. It dilates the bronchioles and constricts nasal blood vessels, thus reducing swelling in the nasal mucosa. Some drugs work by binding to adrenergic receptors without stimulating them, thereby blocking the action of epinephrine and norepinephrine. Propanolol, for example, reduces high blood pressure by blocking β-adrenergic receptors; it is in a family of drugs called beta-blockers.

Other drugs target parasympathetic effects. Pilocarpine relieves glaucoma (excessive pressure in the eyeball) by mimicking an action of acetylcholine, dilating a vessel that drains fluid from the eye. Some drugs block ACh effects. For example, ACh normally stimulates constriction of the pupil and secretion of respiratory mucus. Atropine blocks muscarinic ACh receptors, so it is useful for dilating the pupils for eye examinations and drying the mucous membranes of the respiratory tract in preparation for inhalation anesthesia. It is an extract of the deadly nightshade plant, *Atropa belladonna*. Women of the Middle Ages used nightshade to dilate their pupils, which was regarded as a beauty enhancement.[35]

Many other drugs are designed to act on the central nervous system. Sigmund Freud predicted that psychiatry would eventually draw upon biology and chemistry to deal with emotional problems once treated only by counseling and psychoanalysis. A branch of neuropharmacology called *psychopharmacology* has fulfilled his prediction. This field dates to the 1950s when chlorpromazine, an antihistamine, was incidentally found to relieve the symptoms of schizophrenia.

The management of clinical depression is one example of how psychopharmacology has supplemented counseling approaches. A person's mood is strongly influenced by neurotransmitters in a family called the *monoamines,* including norepinephrine and serotonin. Some forms of depression result from monoamine deficiencies, and therefore yield to drugs that prolong the effects of the monoamines already present at the synapses. Fluoxetine (Prozac), for example, blocks the reabsorption of serotonin by the presynaptic neurons that release it. This causes serotonin to linger longer in the synapse, producing a mood-elevating effect. Fluoxetine is called a *selective serotonin reuptake inhibitor (SSRI).* It is also used to treat fear of rejection, excess sensitivity to criticism, lack of self-esteem, and inability to experience

---

[35] *bella* = beautiful, fine; *donna* = woman

Aging of the Nervous System

The nervous system reaches its peak development around age 30. The brain then begins to atrophy and loses, on average, about 56% of its weight by age 75. The cerebral gyri are narrower, the sulci are wider, the cortex is thinner, and there is more space between the brain and meninges. The remaining cortical neurons have fewer synapses, and for multiple reasons, synaptic transmission is less efficient: The neurons produce less neurotransmitter, they have fewer receptors, and the neuroglia around the synapses is more leaky and allows neurotransmitter to escape from the synapses. The degeneration of myelin sheaths with age also slows down nerve conduction.

Neurons exhibit less rough endoplasmic reticulum and Golgi complex with age, which indicates that their metabolism is slowing down. Old neurons accumulate more neurofibrillary tangles—dense mats of cytoskeletal elements in their cytoplasm. In the

pleasure, all of which were long handled only through counseling, group therapy, or psychoanalysis.

Our growing understanding of neurochemistry also gives us deeper insight into the action of addictive drugs of abuse such as amphetamines and cocaine. Amphetamines ("speed") chemically resemble norepinephrine and dopamine, two neurotransmitters associated with elevated mood. Dopamine is especially important in sensations of pleasure. Cocaine blocks dopamine reuptake and thus produces a brief rush of good feelings. But when dopamine is not reabsorbed by the neurons, it diffuses out of the synaptic cleft and is degraded elsewhere. Cocaine thus depletes the neurons of dopamine faster than they can synthesize it, so that finally there is no longer an adequate supply to maintain normal mood and the user develops dependency on the cocaine. The postsynaptic neurons make new dopamine receptors as if "searching" for the neurotransmitter—all of which leads ultimately to anxiety, depression, and the inability to experience pleasure without the drug.

Another drug from which many derive pleasure, but with far less harmful consequences, is caffeine. Caffeine works because of its similarity to adenosine (fig. 9.19). Adenosine, which you know as a component of DNA, RNA, and ATP, also acts in the brain to inhibit ACh release. One theory of sleepiness is that it results when prolonged metabolic activity breaks down so much ATP that the accumulated adenosine has a noticeably inhibitory effect. Caffeine resembles adenosine closely enough to bind to its receptors, but it does not produce the inhibitory effect. Thus, it blocks adenosine action, enhances ACh secretion, and makes a person feel more alert.

Adenosine   **Caffeine**

Figure 9.19 Adenosine and Caffeine. Adenosine, a breakdown product of ATP and other chemicals, inhibits ACh release and produces a sense of sleepiness. Caffeine is similar enough to block the action of adenosine by binding to its receptors. This results in increased ACh release and heightened arousal.

extracellular material, plaques of amyloid protein appear, especially in people with Down syndrome and Alzheimer disease (AD). AD is the most common nervous disability of old age (p. 272).

Not all functions of the central nervous system are equally affected by senescence. Motor coordination, intellectual function, and short-term memory decline more than language skills and long-term memory. Elderly people are often better at remembering things in the distant past than remembering recent events.

The sympathetic nervous system loses adrenergic receptors with age and becomes less sensitive to norepinephrine. This contributes to a decline in homeostatic control of such variables as body temperature and blood pressure. Many elderly people experience *orthostatic hypotension*—a drop in blood pressure when they stand, which sometimes results in dizziness, loss of balance, or fainting.

CAREER SPOTLIGHT

Electroneurodiagnostic Technologist

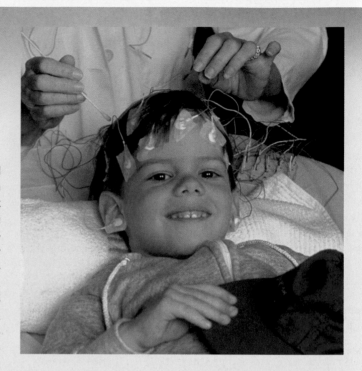

Electroneurodiagnostic (END) technologists are also called EEG technologists, although their duties are broader than recording electroencephalograms. END technologists work primarily in hospital neurology departments, operating equipment that records electrical activity of the brain, spinal cord, and peripheral nerves, thus obtaining data used by neurologists in diagnosing and monitoring a broad spectrum of neurological diseases: epilepsy, stroke, encephalitis, head trauma, and Alzheimer disease. END technologists are also involved in sleep studies and assessment of brain death.

There are more openings for END technologists than there are qualified people to fill them, so job opportunities are excellent. One must have good manual dexterity, excellent oral and written communication skills, and a basic ability to use computers and electronic equipment. Training programs require only a high school diploma for entry, and extend for 12 to 24 months leading to an associate degree or certification. The training includes electronics, computer skills, neuroanatomy, neuropathology, psychology, and other topics. See appendix B for links to further career information.

CONNECTIVE ISSUES

Ways in Which the Nervous System Affects Other Organ Systems

Integumentary System
Regulates piloerection, sweating, and cutaneous blood flow, thereby adjusting heat loss through the skin.

Skeletal System
Affects bone development by regulating the muscle tension exerted on the bones.

Muscular System
Indispensable to skeletal muscle contraction; maintains muscle tone.

Endocrine System
Stimulates secretion by the pituitary gland and adrenal medulla; brain synthesizes several hormones.

Circulatory System
Regulates heartbeat, blood vessel diameters, blood pressure, and routing of blood; sympathetic stimulation enhances blood clotting.

Lymphatic and Immune Systems
Stimulates lymphoid organs and thereby influences production of immune cells; regulates immune responses; emotional states influence disease susceptibility and severity.

Respiratory System
Regulates rate and depth of breathing and ease of pulmonary airflow.

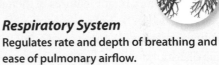

Urinary System
Regulates rate of urine production by adjusting renal blood flow and pressure; controls emptying of bladder.

Digestive System
Regulates appetite, feeding behavior, digestive tract motility and secretion, and defecation.

Reproductive System
Regulates sex drive, arousal, and orgasm; secretes a hormone (oxytocin) involved in sexual pair-bonding, childbirth, and lactation; stimulates pituitary to secrete other hormones that control sperm and egg production, menstrual cycle, pregnancy, and lactation.

323

Study Guide

Assess Your Learning Outcomes

To test your knowledge, discuss the following topics with a study partner or in writing, ideally from memory.

9.1 Overview of the Brain (p. 289)

1. Principal portions of the brain—cerebrum, cerebellum, and brainstem—and major features of their gross anatomy
2. Meninges of the brain and some ways in which they differ from the meninges of the spinal cord
3. The distribution of gray and white matter in the brain
4. Four ventricles of the brain
5. The functions, sources, flow, and reabsorption of cerebrospinal fluid
6. Arteries that supply the brain
7. The blood–brain barrier, how it protects the brain, and how it presents an obstacle to certain forms of medical treatment

9.2 Principal Divisions of the Brain (p. 296)

1. Locations, anatomical features, and functions of the medulla oblongata, pons, midbrain, thalamus, and hypothalamus
2. Gross anatomy, histology, and functions of the cerebellum
3. Gross anatomy of the cerebrum, including the names and boundaries of its five lobes
4. The distribution of gray and white matter in the cerebrum and the three types of tracts of white matter
5. Functions of the five lobes of the cerebrum
6. The location and basic functions of the basal nuclei of the cerebrum
7. Major components, locations, and functions of the limbic system, especially the amygdala and hippocampus

9.3 Multiregional Brain Functions (p. 303)

1. Areas of cerebrum that receive input for the special and general senses
2. The functional difference between primary sensory areas and sensory association areas
3. Sensory organization of the postcentral gyrus
4. Locations of some of the cognitive functions of the brain
5. Areas of the brain concerned with the creation and storage of memories

6. Areas of the brain concerned with voluntary motor control, especially organization of the precentral gyrus and the role of the cerebellum in motor coordination
7. Locations and functions of the language centers of the brain
8. Functional differences between the right and left cerebral hemispheres
9. Brain waves seen in the electroencephalogram, and the clinical uses of the EEG

9.4 The Cranial Nerves (p. 310)

1. Numbers and names of the cranial nerves
2. The origin, termination, and functions of each nerve

9.5 The Autonomic Nervous System (p. 312)

1. Differences between the autonomic and somatic motor divisions of the nervous system
2. Targets (effectors) of the autonomic nervous system (ANS)
3. Examples of visceral reflexes and a visceral reflex arc
4. Anatomical and functional differences between the sympathetic and parasympathetic divisions
5. Roles of the sympathetic chain ganglia and adrenal medulla in the sympathetic division
6. Which cranial and spinal nerves carry fibers of the parasympathetic nervous system, and the destinations of each
7. Two primary neurotransmitters of the ANS, and the distinction between and locations of adrenergic and cholinergic nerve fibers and receptors
8. Antagonistic and cooperative effects in the ANS
9. Examples of ANS regulation by the cerebrum, hypothalamus, lower brainstem, and spinal cord

Testing Your Recall

1. Which of these is caudal to the hypothalamus?
    a. the thalamus
    b. the optic chiasm
    c. the fourth ventricle
    d. the pituitary gland
    e. the corpus callosum

2. The blood–brain barrier is formed by
    a. Schwann cells.
    b. ependymal cells.
    c. epithelial cells of the capillaries.
    d. the pia mater.
    e. the arachnoid villi.

3. The right precentral gyrus controls muscles on the left side of the body because the motor nerve fibers decussate as they pass through
   a. the pyramids of the medulla oblongata.
   b. the posterior horn of the spinal cord.
   c. the corpus callosum between cerebral hemispheres.
   d. the white matter of the cerebral hemisphere.
   e. the corticospinal tracts of the spinal cord.

4. The medulla oblongata plays a direct role in all of the following *except*
   a. speech.
   b. respiration.
   c. swallowing.
   d. the heartbeat.
   e. memory.

5. Nearly all sensory signals pass through a "gateway to the cerebral cortex" called the ___ on their way to centers of consciousness in the brain.
   a. thalamus
   b. hypothalamus
   c. midbrain
   d. pons
   e. cerebellum

6. Movement disorders are most likely to result from injury to
   a. the Wernicke area.
   b. the cerebellum.
   c. the prefrontal cortex.
   d. the temporal lobe.
   e. the postcentral gyrus.

7. To smile, frown, or blow a kiss depends most on which of the following cranial nerves?
   a. accessory
   b. glossopharyngeal
   c. trigeminal
   d. facial
   e. trochlear

8. Most parasympathetic nerve fibers travel through
   a. the corpus callosum.
   b. thoracic spinal nerves.
   c. sacral spinal nerves.
   d. the vagus nerve.
   e. the trigeminal nerve.

9. Damage to the ___ nerve could result in defects of eye movement.
   a. optic
   b. vagus
   c. trigeminal
   d. facial
   e. abducens

10. All of the following *except* the ___ nerve begin or end in the orbit.
    a. optic
    b. oculomotor
    c. trochlear
    d. abducens
    e. accessory

11. The right and left cerebral hemispheres are connected to each other by a thick C-shaped tract of fibers called the _____.

12. The brain has four chambers called _____ filled with _____ fluid.

13. On a sagittal plane, the cerebellar white matter exhibits a branching pattern called the _____.

14. Acetylcholine binds to a category of neurotransmitter receptors called _____ receptors.

15. Cerebrospinal fluid is secreted partly by a mass of blood capillaries called the _____ in each ventricle.

16. The primary motor area of the cerebrum is the _____ gyrus of the frontal lobe.

17. Sense organs called _____ monitor stretch, movement, and tension in the muscles and joints, giving one a nonvisual awareness of the body's position and movement.

18. Any area of cerebral cortex that identifies or interprets sensory information is called a/an _____ area.

19. The awareness of visual input lies in the _____ lobe of the cerebrum.

20. The motor pattern for speech is generated in an area of cortex called the _____ and then transmitted to the primary motor cortex to be carried out.

Answers in appendix A

True or False

Determine which five of the following statements are false, and briefly explain why.

1. The two hemispheres of the cerebellum are separated by the longitudinal fissure.

2. The choroid plexuses produce some of the cerebrospinal fluid, but not all of it.

3. The parasympathetic nervous system slows down the heart.

4. Emotions are associated more with the limbic system than with the parietal lobe.

5. The sympathetic nervous system stimulates digestion.

6. Hearing is a function of the occipital lobe.

7. All cranial nerves except the olfactory and optic nerves contain both sensory and motor fibers.

8. In the parasympathetic nervous system, the preganglionic fibers are adrenergic and usually shorter than the postganglionic fibers.

9. Unlike other cranial nerves, the vagus nerve extends far beyond the head–neck region.

10. The optic nerve controls movements of the eye.

Answers in appendix A

Testing Your Comprehension

1. Which cranial nerve conveys pain signals to the brain in each of the following situations? (a) Sand blows into your eye. (b) You bite the rear of your tongue. (c) Your stomach hurts from eating too much.

2. A person can survive destruction of an entire cerebral hemisphere but cannot survive destruction of the hypothalamus, which is a much smaller mass of brain tissue. Explain this difference and describe some ways that destruction of a cerebral hemisphere would affect one's quality of life.

3. Suppose you were walking alone at night and suddenly heard a dog growling close behind you. Describe several ways in which your sympathetic nervous system would prepare you to deal with this situation.

The Sense Organs

Artistic conception of the anterior components of an eye looking toward the right. This depicts the cornea, iris, lens, and the muscles (orange) that focus images by adjusting the lens curvature.

Chapter Outline

BASE CAMP

Before ascending to the next level, be sure you're properly equipped with a knowledge of these concepts from earlier chapters:

- Basic structure of a neuron (p. 257)
- Action potentials in nerve fibers (p. 265)
- Lobes of the cerebrum (p. 301)
- Areas of primary sensory cortex (p. 304)

Anatomy & Physiology | REVEALED®
aprevealed.com

Module 7: Nervous System

Anyone who enjoys music, art, fine food, or a good conversation knows the value of the human senses. Indeed, sensory input seems necessary to the integrity of one's personality, intellectual function, and even sanity; extreme sensory deprivation can lead to hallucinations, incoherent thought patterns, deterioration of intellectual function, and sometimes morbid fear or panic. Such effects sometimes occur in burn patients who are immobilized and extensively bandaged (including the eyes). Sensory deprivation has often been used by oppressive governments for the torture of political prisoners.

Yet much of the information communicated by the sense organs never comes to our conscious attention—for example, information about blood pressure and composition, body temperature, and muscle tension. Despite our lack of awareness, such information is necessary for internal communication and control. The sense organs are therefore vital not just to our conscious awareness or enjoyment of the environment but also to homeostasis and our very survival in a ceaselessly changing and challenging world.

10.1 Receptors and Sensations

Expected Learning Outcomes

When you have completed this section, you should be able to:

a. define *receptor* and *sense organ*;
b. describe the types of information obtained from sensory receptors; and
c. classify the human sense organs into broad functional categories.

A sensory **receptor** is any structure specialized to detect a stimulus. Some receptors are simple, bare nerve endings, such as the receptors for heat and pain, while others are true sense organs. A **sense organ** is a structure composed of nervous tissue along with muscular, epithelial, or connective tissues that enhance the organ's response to a certain type of stimulus. Sense organs can be as complex as the eye and ear or as microscopic and simple as a dendrite wrapped in a little connective tissue.

Sensory signals to the brain sometimes produce a **sensation**—the subjective awareness of a stimulus. However, most sensory signals delivered to the CNS produce no conscious sensation at all. Some are filtered out in the brainstem before reaching the cerebral cortex, a valuable function that keeps us from being distracted by innumerable unimportant stimuli detected by the sense organs. Other nerve signals concern functions that do not require our conscious attention, such as monitoring blood pressure and pH.

Types of Sensory Information

Sensory receptors provide the CNS with four kinds of information about a stimulus—*type, location, intensity*, and *duration*:

1. **Type (modality)** refers to variations as different as vision, hearing, taste, and pain, but also to finer differences within these major categories—a red or blue color or a bitter or sweet taste, for example. The action potentials for vision are identical to the action potentials for taste or any other sensory type. So how can the brain tell a visual signal from a taste signal? It does this partly by "assuming" that if a signal comes from the retina, it must be visual; if it comes from a taste bud, it must be a taste; and so on. It's as if each nerve pathway from sensory cells to the brain is labeled to identify its origin, and the brain employs these labels to interpret what type of stimulus the nerve signal represents.

2. **Location** is also determined from which nerve fibers are issuing signals to the brain. Any sensory neuron detects stimuli within an area called its **receptive field.** In the sense of touch, for example, a single sensory neuron may cover an area of skin as great as 7 cm in diameter. No matter where the skin is touched within that field, the brain receives signals from that neuron (fig. 10.1a). One cannot determine whether the skin was touched at point A or at some other point 4 or 5 cm away, but in some areas of the body, such as the back, it isn't necessary to make finer distinctions. On the other hand, we must be able to more precisely localize touch sensations from the fingertips. Here, each sensory neuron may cover a receptive field as small as 1 mm in diameter, so two points of contact just 2 mm apart will be felt separately (fig. 10.1b). You can imagine the importance of this to reading braille, appreciating the fine texture of a fabric, or manipulating a small object such as a sesame seed.

Figure 10.1 The Receptive Field of a Nerve Ending. (a) A neuron with a large receptive field, as found in the skin of the back. If the skin is touched simultaneously at two closely spaced locations within the same receptive field, both touches stimulate the same nerve fiber and the brain perceives this as a single touch. (b) Neurons with small receptive fields, as found on the fingertips. Two closely spaced touches here are likely to stimulate different nerve fibers, enabling the brain to perceive them as separate touches.

- *Which of these do you think would be characteristic of your lips?*

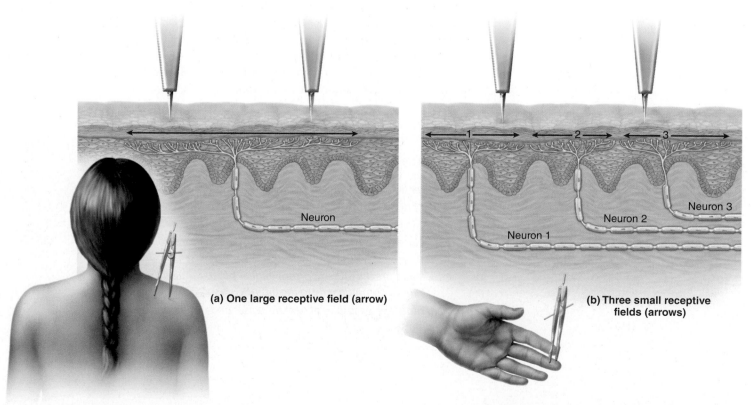

(a) One large receptive field (arrow)

(b) Three small receptive fields (arrows)

3. **Intensity** refers to whether a sound is loud or soft, a light is bright or dim, a pain is mild or excruciating, and so forth. One way of communicating stimulus intensity is for a nerve fiber to fire more rapidly in response to a high-intensity stimulus and more slowly in response to a low-intensity stimulus. Another is that intense stimuli stimulate a larger number of nerve fibers to fire than weak stimuli do. Yet a third is that weak stimuli activate only the most sensitive neurons, whereas strong stimuli also activate the less sensitive ones. Therefore, the brain can discern stimulus intensity by monitoring which nerve fibers are sending it signals, how many nerve fibers are doing so, and how fast those nerve fibers are firing.

4. **Duration,** or how long a stimulus lasts, is encoded by changes in the firing frequency of a nerve fiber with the passage of time. All receptors exhibit the property of **sensory adaptation.** If the stimulus is prolonged, the firing of the neuron gets slower over time; with it, so does our conscious sensation. Adapting to hot bathwater is an example of this. Sensory adaptation also explains why we may notice a suspicious odor (such as a gas leak) for several seconds and then the sensation fades in intensity even if the stimulus is still there. Some receptors adapt much more slowly, such as those for muscle tension and body position. The brain must always be aware of the body's position and movements, so it would be problematic if these receptors adapted too quickly and ceased to inform the brain of such things.

Interestingly, as diverse as this information is, all of it comes to the brain in one simple form—a chain of action potentials. We can think of all sense organs as **transducers**—devices for converting one form of energy (light, heat, chemicals, touch, sound) into another (action potentials). The process of converting stimulus energy to nerve energy is called *sensory transduction.* This chapter will explain how some receptors achieve that.

General and Special Senses

There are multiple ways of classifying the senses and receptors. The **general senses** are distributed over much or all of the body—in the skin, muscles, tendons, joints, and viscera. They include the senses of touch, pressure, pain, heat and cold, stretch, and others. The **special senses** are limited to the head and employ receptors that are innervated by the cranial nerves. They include taste, smell, hearing, equilibrium, and vision.

Before You Go On

Answer these questions from memory. Reread the preceding section if there are too many you don't know.

1. Not every sensory receptor is a sense organ. Explain.
2. Not every sensory signal results in a person's conscious awareness of a stimulus. Explain.
3. What are the four important qualities of a stimulus that sense organs communicate to the brain? Give some examples.
4. State two examples of general senses and two examples of special senses.

10.2 The General Senses

Expected Learning Outcomes

When you have completed this section, you should be able to:

a. describe some relatively simple and widespread sensory nerve endings and the sensations associated with them;

b. define and distinguish some types of pain;

c. identify some chemicals that stimulate pain receptors; and

d. explain the mysterious-seeming phenomena of referred pain, phantom pain, and the relative painlessness of some major injuries.

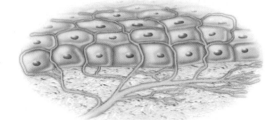

Free nerve endings

Receptors for the general senses are relatively simple in structure and physiology. They consist of one or a few sensory nerve fibers and, usually, a sparse amount of connective tissue. A few examples are illustrated in figure 10.2.

Simple Nerve Endings

The simplest receptors are bare dendrites with no connective tissue and include the following.

- **Free nerve endings** include *warm receptors* that respond to rising temperatures, *cold receptors* that respond to falling temperatures, and *nociceptors*[1] (NO-sih-sep-turs) that produce pain sensations in response to tissue injury. Free nerve endings are especially abundant in the skin and mucous membranes.

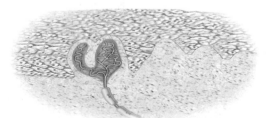

Tactile cell — — Nerve ending

Tactile disc

- **Tactile discs** are receptors for light touch that are employed for detecting textures, edges, and shapes. They are flattened nerve endings that terminate adjacent to a specialized *tactile cell* at the base of the epidermis.

- **Hair receptors** are dendrites that coil around a hair follicle and respond to movements of the hair. They are stimulated when, for example, an ant walks across one's skin, bending one hair after another. However, they adapt quickly, so we are not constantly irritated by the feel of clothing against the skin. Hair receptors are particularly important in the eyelids, where the slightest touch triggers a protective blink reflex.

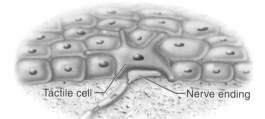

Hair receptor

Some other simple nerve endings have glial cells or connective tissue wrapped around the nerve fiber. Most of these are receptors for touch, pressure, or stretch. Some of them, such as the muscle spindles described in chapter 8, are considered **proprioceptors**[2] (PRO-pree-oh-SEP-turs)—receptors specialized to detect the position and movement of the body and its parts. Two other encapsulated nerve endings are as follows (several additional types are not considered here).

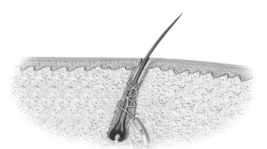

Tactile corpuscle

- **Tactile (Meissner**[3]**) corpuscles,** like tactile discs, are receptors for light touch and texture and are found in the dermal papillae at the boundary

---

[1] *noci* = pain

[2] *proprio* = of one's own; *ceptor* = sensory receptor

[3] Georg Meissner (1829–1905), German histologist

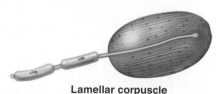

Lamellar corpuscle

Figure 10.2 Receptors for the General (Somatosensory) Senses.

between the dermis and epidermis. Each consists of two or three nerve fibers meandering upward through a tall mass of flattened Schwann cells. They are especially concentrated in sensitive hairless areas such as the fingertips, palms, eyelids, lips, nipples, and genitals. Drag a fingernail lightly across the back of one hand, then across the palm. The difference in sensation that you feel is due to the high concentration of tactile corpuscles in the palmar skin. Tactile corpuscles enable you to tell the difference between satin and sandpaper, for example, by light strokes of your fingertips.

- **Lamellar (pacinian[4]) corpuscles** are sensitive to deep pressure, stretch, tickle, and vibration. They are 1 or 2 mm long and visible to the naked eye. In cross section, they look like an onion slice, with the nerve fiber at the core surrounded by layers of flattened Schwann cells and fibroblasts. They are especially abundant in the periosteum of bone, joint capsules, the pancreas and some other viscera, and deep in the dermis, especially in the hands, feet, breasts, and genitals.

Pain

Pain is a discomfort caused by tissue injury or noxious stimulation; it typically leads to evasive action. Few of us enjoy pain and we may wish that no such thing existed, but it is one of our most important and purposeful senses and we would be far worse off without it. We see evidence of this in leprosy and diabetes mellitus, in which the sense of pain is often diminished by nerve damage. The absence of pain makes people unaware of minor injuries that, if not cared for, can become infected and grow worse. Victims may lose fingers, toes, or entire limbs. In short, pain is an adaptive sensation that enhances survival and well-being.

Pain is not simply an effect of overstimulation of nerve endings meant for other functions. It has its own specialized nerve fibers, the **nociceptors,** which respond to chemicals released by injured tissues. They are especially dense in the skin and mucous membranes and occur in nearly all organs, but are absent from the brain. During brain surgery, patients sometimes must be awake to communicate with the surgeon, but need only a local anesthetic for the skin and meninges; operation on the brain itself produces no pain.

Pain from the skin, muscles, and joints is called **somatic pain.** Pain from the internal organs of the body cavities is called **visceral pain.** The latter often results from stretch, chemical irritants such as alcohol and bacterial toxins, or a drop in blood flow, as in menstrual cramps. It is often accompanied by nausea.

Three fascinating aspects of pain are referred pain, phantom pain, and spinal gating of pain. **Referred pain** is a phenomenon in which pain from the viscera is mistakenly thought to come from the skin or other superficial sites—for example, when the pain of a heart attack is felt "radiating" along the left shoulder and arm. This results from the fact that pain fibers from the heart and skin converge on the same neurons in the central nervous system, so by the time these signals get to the pain centers of the brain, the brain cannot identify their source. Of these two, the skin is more frequently injured, so the brain acts as if it "assumes" that the pain most likely comes from the skin. Knowledge of the areas of referred pain (fig. 10.3) is important to a physician in diagnosing the true source of discomfort reported by a patient.

Phantom pain is the eerie sensation of pain coming from a limb that has been amputated. It may be associated with other illusory sensations as well, such as the bulk and weight of the absent limb or feelings of itching or movement. Any irritation of the stump of a limb can set off nerve impulses along the

---

[4]Filipo Pacini (1812–83), Italian anatomist

Clinical Application 10.1

MIGRAINE HEADACHE

About 5% of men and twice as many women in the United States suffer from *migraine headaches*. These differ from ordinary *tension headaches* and the *symptomatic headaches* that stem from underlying disorders such as brain tumors and infections. Migraines are felt as throbbing, moderate to severe pain, often behind one eye or ear. In slightly over half of migraine sufferers, these headaches are limited to one side of the head (the word *migraine* is a French corruption of the Latin *hemicrania*, "half-cranium"). They are commonly preceded by an *aura*, or subjective sensation that a headache is coming on, that may include visual disturbances such as light flashes and blind spots, and sometimes dizziness or ringing in the ears. The aura often lasts for 20 to 30 minutes before the onset of pain itself, and is associated with a wave of reduced cerebral blood flow starting in the occipital region and progressing forward in one hemisphere of the brain. A migraine headache commonly lasts from 4 hours to 3 days and may be accompanied by photophobia (aversion to light), scalp tenderness, nausea, and vomiting. The pain arises not from the brain tissue but from fibers of the trigeminal nerve that innervate the dura mater and cerebral arteries. The cause of migraine headaches is still not well known; possible culprits of special interest to medical physiologists are serotonin and other neurotransmitters that stimulate blood vessel dilation and constriction.

corresponding spinal pathway, and the brain does not perceive that the distal part of the limb is absent.

People are sometimes severely injured and yet feel little or no pain—for example, soldiers mortally wounded on a battlefield. The absence or minimal sense of pain results from a mechanism called **spinal gating.** Certain spinal and brainstem neurons secrete pain-blocking substances called *endorphins* and *enkephalins*, which stop pain signals at synapses in the cord or brainstem before they reach conscious levels of the brain. We make more everyday use of this phenomenon when, for example, we painfully bump our elbow and rub the area to ease the pain. The rubbing action activates the spinal gating mechanism. Unfortunately, efforts to employ endorphins or enkephalins for the clinical management of pain have been disappointing.

Figure 10.3 Referred Pain. (a) Pain from the viscera is often felt in specific regions of the skin. (b) Basis of referred pain in a heart attack. The brain cannot distinguish the source of pain if two or more origins feed into the same interneurons and CNS pathways to the sensory cortex.

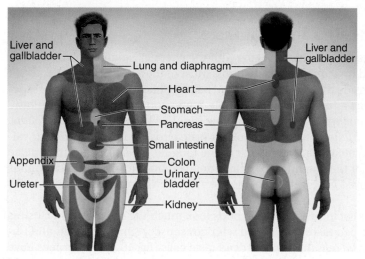

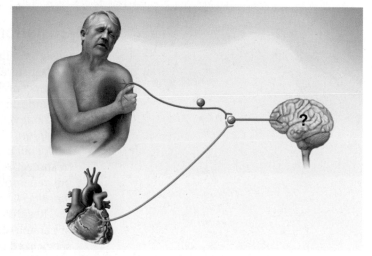

(a)

(b)

10.3 The Chemical Senses—Taste and Smell

Expected Learning Outcomes

When you have completed this section, you should be able to:

a. describe the receptor cells for taste and smell and identify their anatomical locations;

b. identify the five primary taste sensations and the chemicals that produce them;

c. discuss factors other than taste that contribute to the flavor of food; and

d. identify the brain regions that process gustatory and olfactory information.

Our world is made of chemicals, and many of these play an essential survival role in the selection of appropriate foods, avoidance of danger, and so forth. We chemically monitor our food and drink through the sense of taste and monitor the air through the sense of smell. These are called the **chemical senses.**

Gustation—The Sense of Taste

Gustation, the sense of taste, begins with the chemical stimulation of sensory cells clustered in about 4,000 **taste buds.** Most of these are on the tongue, but some occur inside the cheeks and on the palate, pharynx, and epiglottis, especially in infants and children. The visible bumps on the tongue are not taste buds but various types of **lingual papillae** (fig. 10.4), including *vallate, foliate, fungiform,* and *filiform* types. Most of our taste buds are in 7 to 12 circular vallate papillae that form a V at the rear of the tongue. The spiky filiform papillae (responsible for the roughness of a cat's tongue) have no taste buds in humans but are employed in perception of food texture, or what food technologists call *mouthfeel.*

Regardless of location, all taste buds look much the same—ovoid clusters of banana-shaped **taste cells** mixed with nonsensory *supportive cells* and unspecialized *basal cells* that replace dead taste cells (fig. 10.4d). Taste cells have

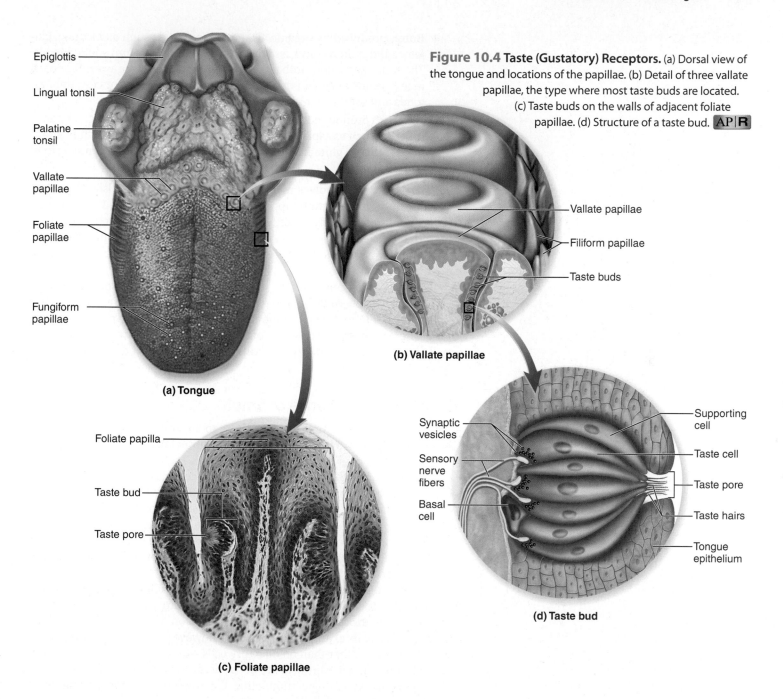

Figure 10.4 Taste (Gustatory) Receptors. (a) Dorsal view of the tongue and locations of the papillae. (b) Detail of three vallate papillae, the type where most taste buds are located. (c) Taste buds on the walls of adjacent foliate papillae. (d) Structure of a taste bud. AP|R

Epiglottis

Lingual tonsil

Palatine tonsil

Vallate papillae

Foliate papillae

Fungiform papillae

(a) Tongue

Vallate papillae

Filiform papillae

Taste buds

(b) Vallate papillae

Foliate papilla

Taste bud

Taste pore

(c) Foliate papillae

Synaptic vesicles

Sensory nerve fibers

Basal cell

Supporting cell

Taste cell

Taste pore

Taste hairs

Tongue epithelium

(d) Taste bud

microvilli called *taste hairs* that project into a pit called the *taste pore* at the apex of the taste bud. To stimulate the taste cells, food chemicals *(tastants)* must dissolve in the saliva and flow into the taste pore to reach the taste hairs. Taste cells are epithelial cells, not neurons, but they have synaptic vesicles at the base of the cell that release a neurotransmitter and stimulate an adjacent nerve fiber.

The five primary taste sensations are

1. **Salty,** produced by metal ions such as sodium and potassium. Electrolyte deficiencies can cause a person or animal to crave salty food.

2. **Sweet,** produced by sugars and many other organic compounds and usually associated with high-calorie foods.

3. **Sour,** produced by the acids in such foods as citrus fruit.

4. **Bitter,** produced by organic alkaloid compounds such as nicotine, caffeine, and quinine, and associated with plant toxins and spoiled foods. The bitter sensation probably served in evolution as a warning not to ingest a substance, but many people cultivate a liking for certain bitter foods and drinks such as capers and tonic water.

5. **Umami**[5] (pronounced "ooh-mommy"), a meaty taste produced by amino acids such as aspartic and glutamic acid and by monosodium glutamate (MSG), a glutamic acid salt widely used as a flavor-enhancing food additive.

The *flavor* of a food involves much more than its taste; it is also affected by aroma, texture, temperature, appearance, and state of mind. Without their aromas, cinnamon merely tastes faintly sweet and coffee and peppermint are bitter. The hot flavor of peppers and some other foods stems from stimulation of free sensory endings of the trigeminal nerve rather than stimulation of the taste buds.

The facial, glossopharyngeal, and vagus nerves (cranial nerves VII, IX, and X) convey taste signals to the medulla oblongata of the brain. Signals are relayed from here to regions of the insula and postcentral gyrus of the cerebrum, where we become conscious of a taste and integrate it with sensations of smell and vision to form our overall impression of the flavor and palatability of a food.

Olfaction—The Sense of Smell

Olfaction, the sense of smell, is a response to airborne chemicals *(odorants).* These are detected by a patch of sensory epithelium called the **olfactory mucosa** in the roof of the nasal cavity (fig. 10.5). The mucosa measures about 5 cm^2 and consists of 10 to 20 million **olfactory cells,** nonsensory supporting cells and basal cells, and mucus-secreting *olfactory glands.* Olfactory cells are true neurons, unlike taste cells. They have a swollen tip bearing 10 to 20 cilia called **olfactory hairs,** a bulbous body containing the nucleus, and a basal end that tapers to a thin axon leading to the brain. Being the only neurons directly exposed to the external environment, they live only about 60 days and are continually replaced by dividing basal cells.

The first step in olfaction is usually that an odor molecule diffuses through the mucus on the surface of the olfactory epithelium and binds to one of the hairs on a receptor cell. This leads to depolarization of the cell, generating action potentials in its axon. Some chemicals, however, act on nociceptors of the trigeminal nerve rather than on olfactory cells—for example, ammonia, menthol, chlorine, and the capsaicin of hot peppers. "Smelling salts" revive unconscious persons by strongly stimulating the trigeminal nerve endings with ammonia fumes.

The axons from multiple olfactory cells converge to form little bundles that collectively constitute the olfactory nerve (cranial nerve I). These penetrate through several pores in the cribriform plate of the ethmoid bone. Directly above this plate, the nerve fibers end in a pair of *olfactory bulbs* (see fig. 9.14, p. 311). From here, other neurons form bundles of fibers called *olfactory tracts* that relay signals to multiple destinations in the cerebrum and brainstem, especially the **primary olfactory cortex** in the temporal lobe. Higher brain centers interpret complex odors such as chocolate, coffee, and perfume by decoding a mixture of signals, like combining different primary colors to get all the colors of the light spectrum.

---

[5]*umami* = Japanese slang loosely meaning delicious or yummy

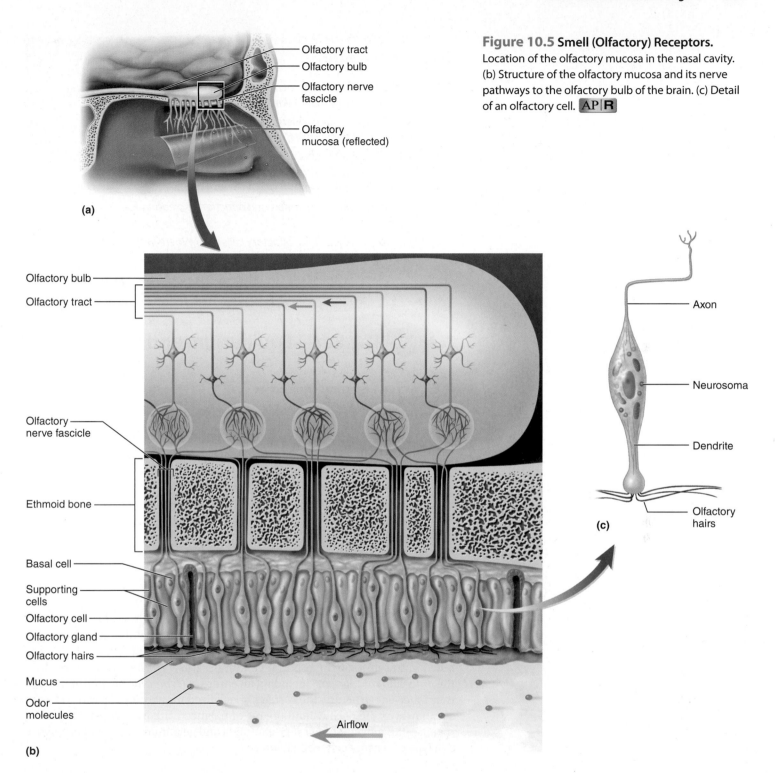

(a)

Figure 10.5 Smell (Olfactory) Receptors.
Location of the olfactory mucosa in the nasal cavity. (b) Structure of the olfactory mucosa and its nerve pathways to the olfactory bulb of the brain. (c) Detail of an olfactory cell. **AP|R**

(b)

(c)

Apply What You Know

Look up the meaning of *anosmia*. Explain why anosmia could endanger a person's health or life. Also describe a couple of your favorite foods or beverages and how your enjoyment of them would be affected by anosmia.

Before You Go On

Answer these questions from memory. Reread the preceding section if there are too many you don't know.

9. Other than the tongue, where are taste buds located? Are taste buds visible with the naked eye? Explain.

10. Which of these are neurons—taste cells or olfactory cells? Which cell type binds stimulant molecules with microvilli? Which one employs cilia?

11. What are the roles of gustation and olfaction in creating the sensation of flavor?

12. Where do the axons of the olfactory cells end? Where in the brain is the primary olfactory cortex located?

10.4 The Ear—Equilibrium and Hearing

Expected Learning Outcomes

When you have completed this section, you should be able to:

a. identify the anatomical components of the outer, middle, and inner ear;

b. explain how the vestibular system of the inner ear functions to give us a sense of the body's position and movements; and

c. explain how the cochlea of the inner ear converts the vibrations of a sound wave into a nerve signal, and how the brain interprets such signals.

The ear serves two very different human senses—equilibrium (the sense of balance) and hearing. The first of these was the original evolutionary function of the ear. Its role in hearing emerged only as vertebrates colonized land and benefited from sensitivity to airborne vibrations. This was when animals evolved the *outer* and *middle ears*, including an eardrum *(tympanic membrane)*, and a new inner-ear structure, the *cochlea*, specialized for hearing. As you will soon see, however, the cochlea and organs of balance work by surprisingly similar means—the movement of inner-ear fluids and gelatinous membranes relative to a type of sensory cells called *hair cells*.

General Anatomy of the Ear

The ear consists of three sections called the *outer, middle*, and *inner ear* (fig. 10.6). The outer ear collects sound waves; the middle ear relays them to the inner ear and has devices for protecting the ear from loud sounds; and the inner ear contains devices for converting vibrations and body movements into nerve signals.

The Outer Ear

The most prominent feature of the **outer ear** is the fleshy **auricle (pinna)** on the side of the head. The auricle is shaped and supported by elastic cartilage except

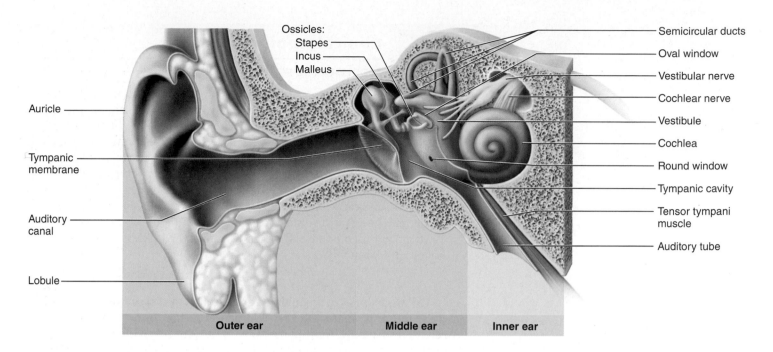

Ossicles:
Stapes
Incus
Malleus

Auricle

Tympanic membrane

Auditory canal

Lobule

Semicircular ducts
Oval window
Vestibular nerve
Cochlear nerve
Vestibule
Cochlea
Round window
Tympanic cavity
Tensor tympani muscle
Auditory tube

Outer ear   Middle ear   Inner ear

for the fatty *lobule* (earlobe), and has a pattern of whorls and recesses that direct sound into the auditory canal. The **auditory canal,** or **external acoustic meatus** (me-AY-tus), is a passage through the temporal bone leading to the eardrum. It follows a slightly S-shaped course for about 3 cm. It is lined with skin and possesses ceruminous and sebaceous glands. The secretions of these glands form **cerumen** (earwax), which waterproofs the canal, inhibits bacterial growth, and keeps the eardrum pliable. It also coats the *guard hairs* at the entrance to the canal, making them sticky and more effective at blocking parasites and debris from entering the canal.

The Middle Ear

What we colloquially call the eardrum is more properly known as the **tympanic membrane.** It marks the beginning of the **middle ear.** Medial to the membrane is a tiny air-filled chamber, the **tympanic cavity,** only 2 to 3 mm wide. It contains the body's three smallest bones and its two smallest skeletal muscles. The three bones, called **auditory ossicles,**[6] transfer vibrations from the tympanic membrane to the inner ear. From lateral to medial, they are the **malleus**[7] (attached to the inside of the tympanic membrane), **incus,**[8] and **stapes**[9] (STAY-peez).

The tympanic membrane would not vibrate freely if the air pressure on one side were much greater than pressure on the other side. Therefore, the middle ear is connected to the pharynx by the **auditory** (**pharyngotympanic** or **eustachian**[10]) **tube.** This tube is normally flattened and closed, but when we yawn or swallow, it opens and allows air to enter or leave the tympanic cavity. This equalizes air pressure on the two sides of the tympanic membrane, but unfortunately it also frequently allows throat infections to spread to the middle ear (see Clinical Application 10.2).

Figure 10.6 **General Anatomy of the Ear.**

- *Cranial nerves are numbered I through XII. Give the name and roman numeral of the thick yellow cranial nerve emerging at the right side of this figure.* **AP|R**

---

[6]*oss* = bone; *icle* = little

[7]*malleus* = hammer, mallet

[8]*incus* = anvil

[9]*stapes* = stirrup

[10]Bartholomeo Eustachio (1520–74), Italian anatomist

Clinical Application 10.2

MIDDLE-EAR INFECTION

Otitis[11] *media* (middle-ear infection) is especially common in children because their auditory tubes are relatively short and horizontal, providing easy passage for microbes to travel from the throat to the middle-ear cavity. Infection can lead to fluid accumulation in the middle ear, which inhibits the vibration of the auditory ossicles, impairs hearing, and causes pain. If untreated, the infection can spread and cause *meningitis*, a potentially deadly infection of the meninges around the brain. Chronic otitis media can also cause fusion of the ossicles, thus causing hearing loss. It is sometimes necessary to drain the fluid from the middle ear by puncturing the tympanic membrane and inserting a tiny drainage tube—a procedure called *tympanostomy*.

The two muscles of the middle ear are called the **stapedius** (sta-PEE-dee-us) and **tensor tympani** (TEN-sur TIM-pan-eye). They insert on the stapes and malleus, respectively. In response to loud but slowly building noises such as thunder, they contract and inhibit the movement of these bones, protecting sensory cells of the inner ear from destructive overstimulation. They are not adequately effective, however, at protecting the inner ear (or one's hearing) from constant loud noise such as amplified concert music or from sudden loud noises such as gunshots close to the ear. Such sounds can irreversibly destroy sensory cells of the inner ear and cause permanent hearing impairment.

The Inner Ear

The **inner ear** is a complex region that transforms mechanical energy (vibration) to nerve energy. It is housed in a maze of passages in the temporal bone called the *bony labyrinth* and consists mostly of a complex of fluid-filled chambers and tubes called the *membranous labyrinth* (fig. 10.7). The membranous labyrinth within the bony labyrinth is thus a tube-within-a-tube structure, like a bicycle inner tube within a tire. The membranous labyrinth is filled with a liquid called *endolymph*, similar to intracellular fluid, and surrounded by a liquid called *perilymph*, similar to cerebrospinal fluid.

The stapes of the middle ear has an elongated *base plate* suspended in an opening called the **oval window.** This window marks the beginning of the inner

[11]*ot* = ear; *itis* = inflammation

Figure 10.7 The Membranous Labyrinth of the Inner Ear. (a) Position and orientation of inner-ear structures within the temporal bone. (b) Structure of the membranous labyrinth and origins of the vestibulocochlear nerve. AP|R

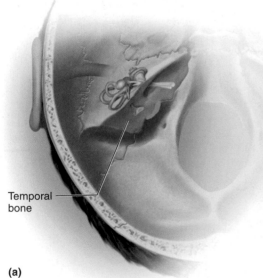

Temporal bone

(a)

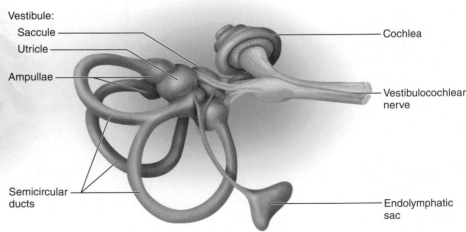

Vestibule:
Saccule
Utricle
Ampullae
Semicircular ducts

Cochlea
Vestibulocochlear nerve
Endolymphatic sac

(b)

ear. On the other side of it is a chamber, the **vestibule,** filled with endolymph. Within the vestibule are two pouches called the *saccule* and *utricle,* and arising from it are three loops called *semicircular ducts* and a snail-like coil called the *cochlea* (fig. 10.6). The saccule, utricle, and semicircular ducts are collectively called the **vestibular apparatus,** and are the organs of equilibrium. The cochlea is the organ of hearing.

Equilibrium

Loosely speaking, **equilibrium** is the sense of balance. More exactly, it is the perception of the orientation and movements of the head; such perception triggers muscular reflexes that maintain one's balance.

Two organs of equilibrium, the **saccule**[12] and **utricle,**[13] are pouches immediately medial to the middle ear. Each is filled with endolymph and contains a patch of epithelium called a **macula,**[14] composed of sensory **hair cells** and nonsensory supporting cells (fig. 10.8). A hair cell is a columnar to

---

[12]*saccule* = little sac

[13]*utricle* = little bag

[14]*macula* = spot, patch

Figure 10.8 The Saccule and Utricle.
(a) Locations of the maculae. (b) Structure of a macula. (c) Movement of the otolithic membrane when the head is tilted.

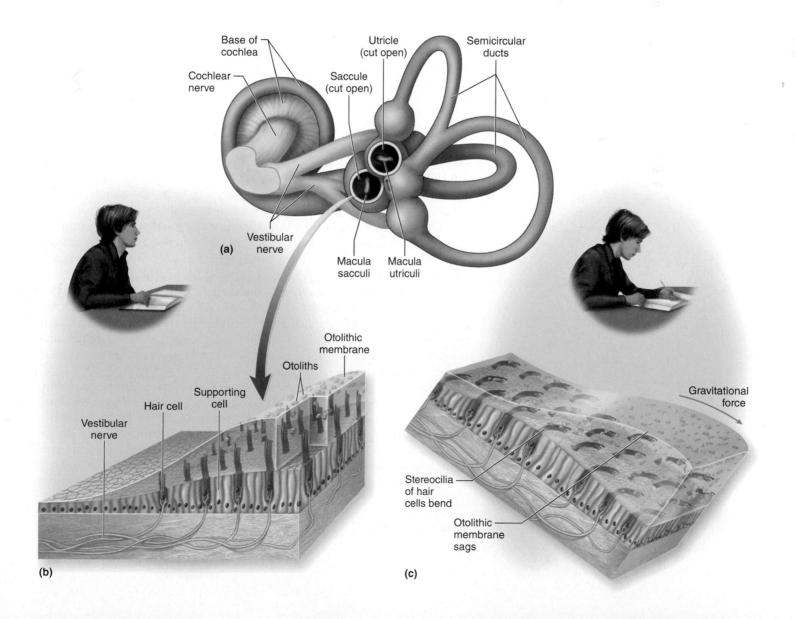

pear-shaped cell with hairlike microvilli called **stereocilia**[15] on its apical surface. Overlying the macula is a layer of gel called the **otolithic membrane,** containing granules of protein and calcium carbonate called **otoliths.**[16] The otoliths give the membrane added weight and inertia, which enhances its ability to stimulate the hair cells when the body moves.

To understand how this mechanism works, consider the macula that lies almost horizontally on the floor of the utricle. When you tilt your head down to tie your shoes, the heavy otolithic membrane sags downward and bends the stereocilia, stimulating the hair cells. These cells also respond when the body accelerates or decelerates horizontally. For example, if you are sitting in a car that begins to move, the macula moves at the same speed, but the inertia of the otolithic membrane makes it lag behind briefly. This bends the stereocilia backward and stimulates the hair cells. If you slow down and stop, the macula slows down at the same rate, but the otolithic membrane briefly keeps moving forward and bends the stereocilia the other way. By comparing input from the utricles of both ears, your brain can evaluate tilts of the head and changes in horizontal motion.

The saccule has an almost vertical macula and its hair cells therefore respond to vertical acceleration and deceleration. If, for example, you are standing in an elevator that begins to move up, the otolithic membrane pulls down on the stereocilia. As the elevator slows to a stop, the macula does too, but the membrane keeps moving upward for a moment. At both the start and stop, the hair cells are stimulated and your brain becomes aware of your vertical movements. The macula sacculi also senses more natural vertical movements of the head, as when you stand up and even as your head bobs up and down during walking and running.

The head also rotates, such as when you spin in a rotating chair, walk down a hall and turn a corner, or bend forward to pick something up from the floor. Such movements are detected by the three **semicircular ducts** (fig. 10.9). One of these lies on a plane about 30° from horizontal, and the other two are oriented vertically at right angles to each other. Each duct has a bulb at its base called the **ampulla.** Within the ampulla, there is a mound of sensory epithelium called the *crista ampullaris*, once again composed of hair cells and supporting cells. These are topped by a gelatinous cap called the **cupula,** but this gel has no otoliths. The ducts are filled with endolymph.

Suppose you're sitting in a rotating office chair and you spin around to get something from a desk behind you. The horizontal semicircular duct will spin with you, but the fluid within it will lag behind a bit. The moving cupula will push against the fluid and bend backward. This bends the stereocilia of the hair cells and stimulates them. Tilting your head forward or sideways—say, to pick up something on the floor or to lie down in bed—similarly engages the vertical semicircular ducts and stimulates their hair cells. The ducts are arranged to detect rotation in any of the *x, y,* and *z* geometric planes, and movements in a direction between any two of these planes create a mixture of signals from two or more of the ducts.

When a hair cell is stimulated, it releases a neurotransmitter from synaptic vesicles at its base and stimulates an adjacent nerve fiber. The nerve fibers from the utricle, saccule, and semicircular ducts form the *vestibular nerve,* which joins the *cochlear nerve* discussed later to form *cranial nerve VIII,* the *vestibulocochlear* nerve described in chapter 9. The nerve fibers lead mostly to the pons of the brainstem. The pons relays signals to several other centers you have studied in the last two chapters: (1) the primary somatosensory

---

[15]*stereo* = solid

[16]*oto* = ear; *lith* = stone

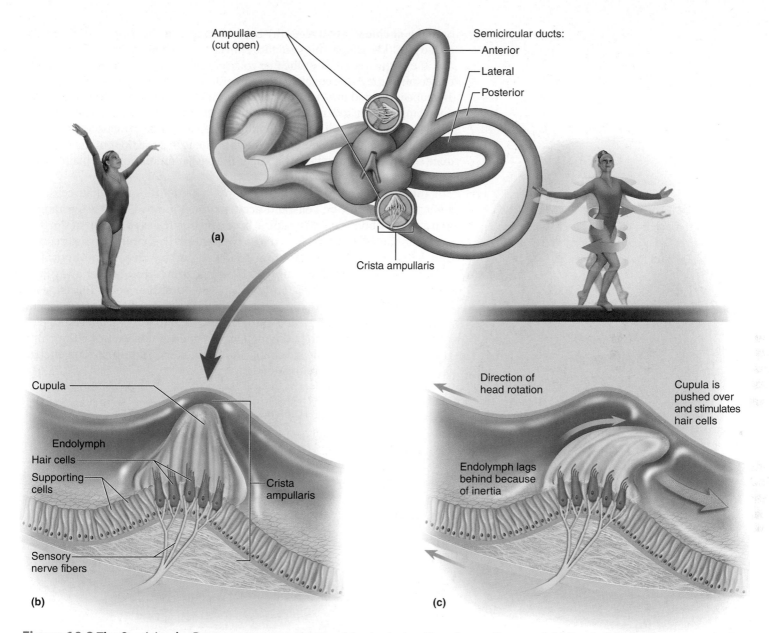

Figure 10.9 The Semicircular Ducts. (a) Structure of the semicircular ducts, with each ampulla opened. (b) Detail of the sensory receptor in the ampulla. (c) Action of the endolymph on the cupula and hair cells when the head rotates.

cortex, where one becomes conscious of head movements; (2) the spinal cord, where one generates signals to the muscles that maintain posture and balance; (3) the cerebellum, a center of motor coordination; and (4) midbrain nuclei that activate eye movements to compensate for head movements, keeping one's vision fixed on a given target in spite of head motion.

The Cochlea and Hearing AP|R

Hearing is the awareness of sound, waves of molecular motion created by vibrating objects. The outer and middle ear are concerned with collecting airborne sound waves and transferring vibration to the inner ear. In the inner ear, only the **cochlea**[17] (COC-lee-uh) is concerned with hearing. Named for its coiled snail-like shape, this structure converts vibrations to nerve signals.

The cochlea is a fleshy tube that coils around a screwlike bony core. A vertical section through the cochlea cuts the coil at each turn, as shown in figure 10.10. The tube is divided into a triangular space called the **cochlear duct,** filled with endolymph, and two larger spaces called *scalae* (SCALE-ee) above and below it, filled with perilymph. A thin membrane separates the upper scala from the cochlear duct and a thicker **basilar membrane** forms the floor of the cochlear duct, separating it from the lower scala. The basilar membrane is a platform for the sensory cells of the cochlea.

The sensory cells of hearing, like those of equilibrium, are hair cells. Together with supporting cells and accessory membranes, they are arranged into a coiled ribbon called the **spiral organ** (**acoustic organ, organ of Corti**[18]). Cochlear hair cells are organized into two groups. Everything we hear comes from a single row of about 3,500 **inner hair cells,** each with a row of about 50 or 60 stereocilia on the surface (fig. 10.11). Across from these, farther away from the bony core of the cochlea, are about 20,000 **outer hair cells.** These are arranged in three rows, and each cell bears a V-shaped array of stereocilia. Despite their greater number and arguably more interesting appearance, they do not generate the sounds we hear. They tune the cochlea to enable brain to better discriminate one pitch of sound from another.

---

[17]*cochlea* = snail

[18] Alfonso Corti (1822–88), Italian anatomist

Figure 10.10 The Cochlea. (a) A vertical section, which cuts through the cochlea at each turn of the coiled tube. (b) Detail of one section through the cochlea. (c) Detail of the spiral organ. AP|R

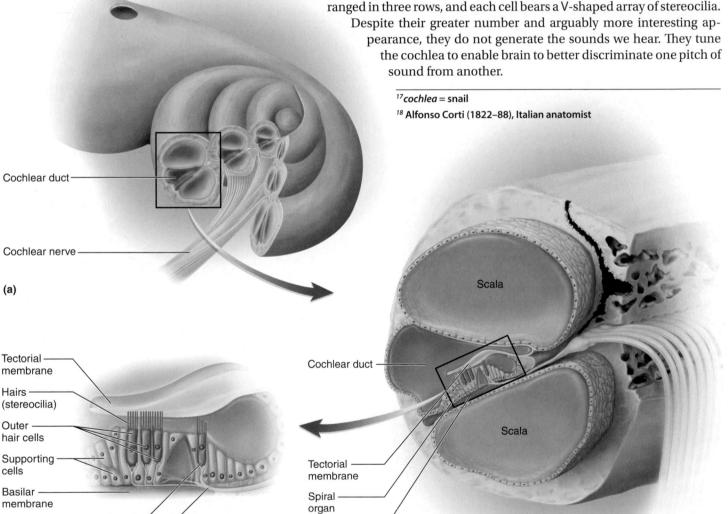

Cochlear duct

Cochlear nerve

(a)

Scala

Cochlear duct

Scala

Tectorial membrane

Spiral organ

Basilar membrane

(b)

Tectorial membrane

Hairs (stereocilia)

Outer hair cells

Supporting cells

Basilar membrane

Inner hair cell

Fibers of cochlear nerve

(c)

A gelatinous **tectorial**[19] **membrane** lies just above the hair cell stereocilia. Medially, this membrane is anchored to the core of the cochlea. Laterally, it is anchored by the tips of the outer hair cell stereocilia, which are embedded in the gel. The stereocilia of the inner hair cells stop just short of the tectorial membrane and lie free in the endolymph.

Figure 10.12 shows a very simplified model of how the ear responds to sound. Vibrating air molecules collide with the tympanic membrane and set it in motion. The ossicles of the middle ear transfer this vibration to the fluids of the inner ear. Each time the footplate of the stapes moves in and out at the oval window, it moves the perilymph of the upper scala, which in turn moves the basilar membrane up and down.

The key to hearing is the up-and-down motion of the basilar membrane, with the hair cells going along for the ride. With each upward movement, stereocilia of the inner hair cells are pushed against the tectorial membrane above, forcing them to bend over. As we saw in the vestibular apparatus, this bending excites the hair cells and stimulates them to release neurotransmitter from the synaptic vesicles at the base of each cell. The neurotransmitter excites adjacent fibers of the *cochlear nerve* and generates a nerve signal, which we will trace later. Thus, vibrations of the basilar membrane are translated to rhythmic bursts of nerve activity in the cochlear nerve.

Two features of a sound are particularly important to our sense of hearing—its loudness (amplitude) and its frequency (low-pitched or high). The basilar membrane is structured in such a way that it generates a code for each of these sound qualities in the nerve signal.

The louder a sound, the more vigorously the basilar membrane vibrates. This results in stronger signaling of the cochlear nerve fibers by the hair cells, and more action potentials per second in the cochlear nerve fibers to the brain. The brain interprets a high rate of nerve firing as a loud sound and a low firing rate as a quieter sound.

As for the frequency (pitch) of a sound, the basilar membrane tapers from a relatively stiff and narrow proximal end, anchored to the base of the cochlea, to a wide and limber distal end, with no anchorage, at the apex of the cochlea. High-pitched sounds cause the most vibration at the narrow basal end, much as the shortest strings of a piano produce the highest pitch

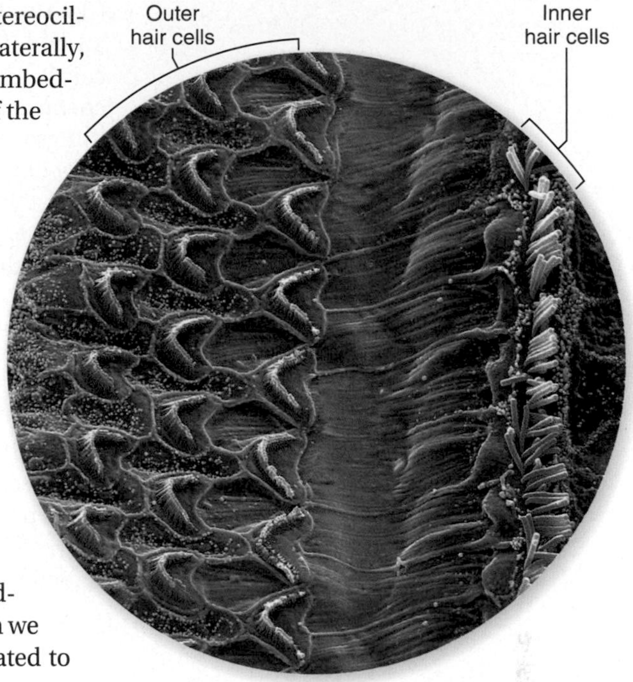

---

[19]*tect = roof*

Figure 10.11 Apical Surfaces of the Cochlear Hair Cells.

- *Which of the cells in this photo are the source of all sounds that one hears? What do the other cells do, if we cannot hear any of the nerve signals that come from them?*

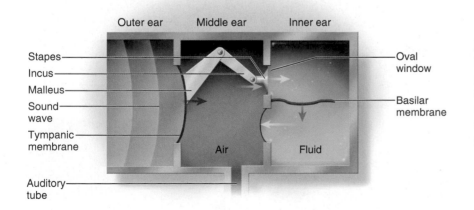

Figure 10.12 A Mechanical Model of Hearing. Each inward movement of the tympanic membrane pushes inward on the auditory ossicles and the fluids of the inner ear. Fluid movement pushes downward on the basilar membrane. Thus, the basilar membrane vibrates up and down in synchrony with the sound wave.

- *Why would high air pressure in the middle ear reduce the vibrations of the basilar membrane in the inner ear?*

(fig. 10.13). Low-pitched sounds produce the most vibration at the wide distal end, like the longest piano strings producing bass notes. The brain receives signals from a mixture of nerve fibers representing all areas of the cochlea, and decodes the mixture of signals into a perception of pitch. Most things we hear, of course, are not pure tones but complex sounds with a variety of pitches, stimulating all parts of the basilar membrane to varying degrees. The brain therefore has to carry out quite a phenomenal task of decoding the complex message. Neuroscientists have achieved a deep understanding of this mechanism, but the details are beyond the scope of this book.

The Auditory Projection Pathway

Nerve fibers from the basilar membrane feed into the *cochlear nerve*, which then becomes part of the already-discussed *vestibulocochlear* nerve leading to the pons (see fig. 10.7). The pons sends signals back to the cochlea for tuning by the outer hair cells, and back to the middle ear for the reflexes of the stapedius and tensor tympani muscles. It also compares input from the right

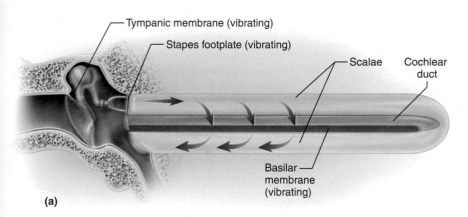

(a)

Figure 10.13 Frequency Response of the Basilar Membrane of the Cochlea. (a) The cochlea, uncoiled and laid out straight. (b) Sound waves produce a wave of vibration along the basilar membrane. The areas of greatest vibration vary with the frequency of the sound. The amount of vibration is greatly exaggerated in this illustration. (c) The taper of the basilar membrane and its correlation with sound frequencies. High frequencies are best detected by hair cells near the narrow proximal end of the membrane (left), and low frequencies by hair cells near the wide free end (right). (Hz = hertz, or cycles per second, a measure of the frequency of a sound wave)

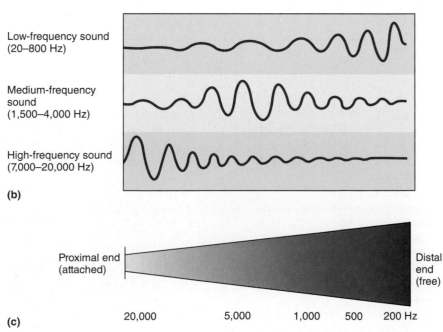

and left ears to give us a sense of *binaural*[20] *hearing*, the ability to localize sounds in space. The signals for hearing ascend to the midbrain, which further aids in binaural hearing and in processing fluctuations in pitch. Finally, signals ascend farther to be relayed through the thalamus to the **primary auditory cortex** in the temporal lobe of the cerebrum, which is where we become consciously aware of the sound.

Before You Go On

Answer these questions from memory. Reread the preceding section if there are too many you don't know.

13. What mechanism protects the inner ear from damage by loud sounds? What part of the brain controls this mechanism?

14. Name six places where hair cells can be found in the inner ear. How do hair cells resemble neurons? How do they differ from neurons?

15. Explain how a semicircular duct differs in function from the saccule and utricle.

16. Describe the events beginning with vibration of the tympanic membrane and ending with the release of neurotransmitter from a cochlear hair cell.

17. How can the brain tell a low-pitched sound from a high-pitched sound? A loud sound from a softer sound?

10.5 The Eye and Vision

Expected Learning Outcomes

When you have completed this section, you should be able to:

a. describe organs surrounding the eye that protect it and aid in its functions;

b. describe the anatomy of the eye itself;

c. explain how the eye controls light and focuses it on the retina;

d. explain how cells of the retina absorb light and generate nerve signals; and

e. describe the projection pathway that these signals take through the brain.

Vision is the ability to form a recognizable image of an object that emits or reflects light. It begins with focusing light on a membrane called the *retina*, the site of our sensory cells for vision. Here, the light creates a chemical reaction that leads to the generation of a nerve signal. The brain must receive and interpret this nerve signal for the process of vision to be complete.

---

[20]*bin* = two; *aur* = ears

Anatomy of the Orbital Region

The eyeball occupies a bony socket called the **orbit.** This general area of the face is the *orbital region*. It contains structures that protect and aid the eye (fig. 10.14):

- The **eyebrows** enhance facial expressiveness and perhaps shield the eye to some degree from glare and forehead perspiration.
- The **eyelids (palpebrae)** block foreign objects from the eye, screen out visual stimuli to facilitate sleep, and blink periodically to moisten the eye with tears and sweep debris from its surface. An eyelid consists mostly of *orbicularis oculi* muscle covered with skin. Along the margin of the eyelid are 20 to 25 **tarsal glands** that secrete a protective oily film onto the eye surface. The **eyelashes** serve as guard hairs to keep debris from the eye; touching them stimulates hair receptors and triggers the blink reflex.
- The **conjunctiva** is a transparent mucous membrane that covers the inner surface of the eyelid and anterior surface of the eyeball except for the region of the cornea. It produces a thin mucous film that prevents drying of the eye. It is richly innervated and very sensitive to pain, and well supplied with minute blood vessels. The appearance of "bloodshot" eyes is due to dilation of these vessels.
- The **lacrimal apparatus** is a group of structures that produce tears and drain them away from the eye. Its chief organ is the **lacrimal (tear) gland**—an almond-shaped gland in the superolateral corner of the orbit, between the eyeball and frontal bone. Approximately 12 short ducts lead from the gland and secrete tears onto the surface of the conjunctiva. Tears cleanse and lubricate the eye surface, supply oxygen and nutrients to the conjunctiva, and contain an antibacterial enzyme to prevent eye infections. After flowing across the eye, tears drain into a pore at the medial corner of each eyelid and from there into a **lacrimal sac** in the medial wall of the orbit. A **nasolacrimal duct** drains the tears from here into the nasal cavity; watery eyes can therefore lead to a runny nose.
- **Orbital fat** fills the space between the eyeball and bones of the orbit. It cushions the eye, allows it to move freely, and protects the blood vessels, nerves, and muscles of the orbit.

Figure 10.14 Accessory Structures of the Orbit. (a) Sagittal section of the orbit. (b) The lacrimal apparatus for secreting tears and draining them into the nasal cavity. AP|R

- *What would be the expected result of an infection that blocked the pores at the medial corner of the eyelids?*

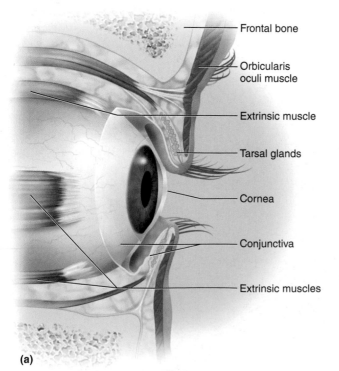

Frontal bone

Orbicularis oculi muscle

Extrinsic muscle

Tarsal glands

Cornea

Conjunctiva

Extrinsic muscles

(a)

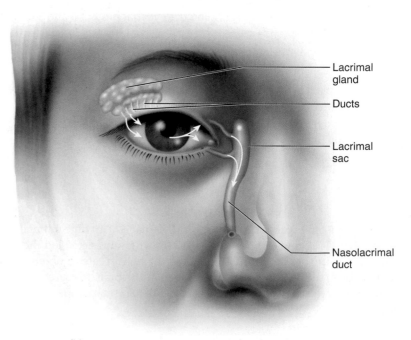

Lacrimal gland

Ducts

Lacrimal sac

Nasolacrimal duct

(b)

- Six **extrinsic muscles** arise from the walls of the orbit and insert on the surface of the eyeball. The term *extrinsic* distinguishes these from the *intrinsic* muscles inside the eyeball. The extrinsic muscles serve to move the eye when we look up, down, or to one side.

Anatomy of the Eyeball

The eyeball itself is a sphere about 24 mm (1 in.) in diameter. It is composed of three tissue layers, or *tunics,* that form the wall of the eye, optical components that admit and focus light, and neural components that absorb light and generate a nerve signal (fig. 10.15).

1. The outer **fibrous layer** is composed of the opaque white **sclera** ("white of the eye") over most of the eye surface, and the transparent **cornea** over the anterior central region. Only the cornea admits light into the eye.

2. The middle **vascular layer** consists of three regions:

   - The **iris,** an adjustable diaphragm that admits light through its central opening, the **pupil.** The iris contains pigment cells with variable amounts of melanin. If melanin is abundant, the eye color ranges from hazel to brown or black. If melanin is scanty, light reflects from a more posterior layer of the iris and gives it a blue, green, or gray color.

   - The **ciliary body,** a thick ring of muscular tissue that encircles and supports the iris and lens and adjusts lens shape for focusing.

   - The **choroid** (CORE-oyd), a deeply pigmented layer, rich in blood vessels, that underlies and nourishes the retina.

3. The inner **neural layer** consists of the *retina* and the beginning of the *optic nerve.*

Figure 10.15 The Eye (Sagittal Section). AP|R

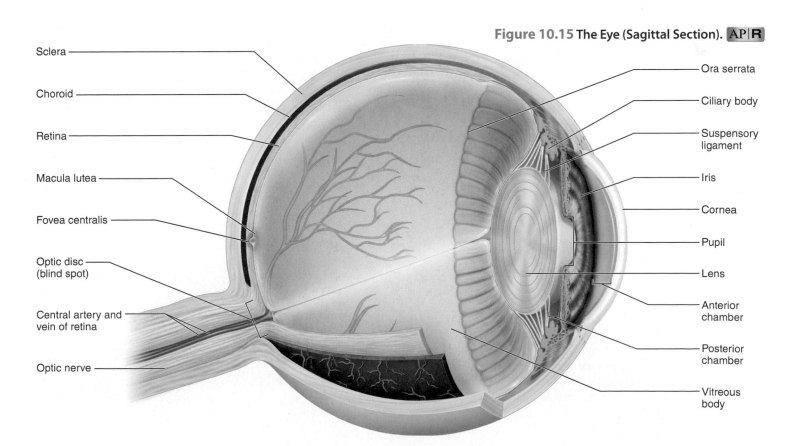

Sclera

Choroid

Retina

Macula lutea

Fovea centralis

Optic disc
(blind spot)

Central artery and
vein of retina

Optic nerve

Ora serrata

Ciliary body

Suspensory
ligament

Iris

Cornea

Pupil

Lens

Anterior
chamber

Posterior
chamber

Vitreous
body

Clinical Application 10.3

GLAUCOMA

Glaucoma[21] is a state of elevated pressure in the eye that occurs when aqueous humor is secreted faster than it is reabsorbed. It results from an obstruction of the scleral venous sinus. Pressure in the anterior and posterior chambers pushes the lens back and puts pressure on the vitreous body. The vitreous body then presses excessively on the retina and choroid and compresses the blood vessels that nourish the retina. Without a good blood supply, retinal cells die and the optic nerve may atrophy, leading to blindness. Often the damage is already irreversible by the time a person notices the first symptoms. Illusory light flashes are an early warning sign of glaucoma. In later stages, the sufferer may experience dimness of vision (origin of the word *glaucoma*), a narrow visual field, and colored halos around artificial lights. Glaucoma can be halted with drugs or surgery, but lost vision cannot be restored. The disease can be detected at an early stage in the course of regular eye examinations. The field of vision is checked, the retina and optic nerve are visually inspected with an ophthalmoscope, and pressure in the eye is checked with an instrument called a *tonometer,* which measures tension on the cornea.

The optical components of the eye are transparent elements that admit light rays, bend (refract) them, and focus images on the retina. They include the cornea (already discussed) and the following components:

1. The **aqueous humor,** a watery fluid that fills all the space between the cornea and lens. The space from lens to pupil is called the *posterior chamber*, and the space from pupil to cornea is called the *anterior chamber*. The ciliary body secretes aqueous humor into the posterior chamber, the fluid flows through the pupil into the anterior chamber, and here it is reabsorbed by a circular vein called the *scleral venous sinus* (fig. 10.16). Normally the rate of reabsorption balances the rate of secretion, but see Clinical Application 10.3 for an important exception.

2. The **lens** is suspended behind the iris by a ring of fibers that attach it to the ciliary body. Tension on these fibers somewhat flattens the lens so it is about 9.0 mm in diameter and 3.6 mm thick at the middle. When the

[21]*glauc* = grayness

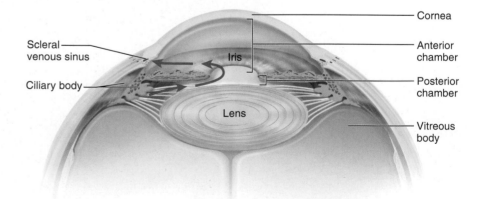

Figure 10.16 The Flow of Aqueous Humor.
Blue arrows indicate the flow of aqueous humor from the ciliary body into the posterior chamber behind the iris, its flow through the pupil into the anterior chamber, and its reabsorption by the scleral venous sinus anterior to the iris.

lens is removed from the eye and not under tension, it has a more spheroidal shape.

3. The **vitreous body** is a transparent jelly that fills the large space behind the lens. It maintains the spheroidal shape of the eyeball and keeps the retina smoothly pressed against the rear of the eyeball.

The neural components of the eye are the retina and optic nerve. The **retina** is a thin transparent membrane that lines the posterior two-thirds of the eye. It is attached to the rest of the eyeball at only two points—the **optic disc,** which is a posterior circular area where the **optic nerve** originates, and the **ora serrata,**[22] a scalloped margin that encircles the eye at the anterior edge of the retina, slightly posterior to the lens (fig. 10.15). The rest of the retina is unattached but is held firmly against the wall of the eyeball by the vitreous body.

The retina can be examined with an illuminating and magnifying instrument called an *ophthalmoscope* (off-THAL-mo-scope) (fig. 10.17). Directly posterior to the center of the lens, on the visual axis of the eye, is a patch of cells called the **macula lutea**[23] (MACK-you-la LOO-tee-ah), about 3 mm in diameter. In the center of the macula is a pit, the **fovea centralis,** which produces the most finely detailed visual images.

The optic disc lies about 3 mm medial to the fovea. Nerve fibers from all areas of the retina converge here and leave the eye to form the optic nerve. The optic disc is the one area of retina devoid of sensory cells; it forms a *blind spot* in the visual field. To compensate for this, the eye makes tiny, unnoticed scanning movements of the visual field, and the brain "fills in" the blind spot with

---

[22]*ora* = mouth; *serrata* = scalloped, serrated

[23]*macula* = spot; *lutea* = yellow

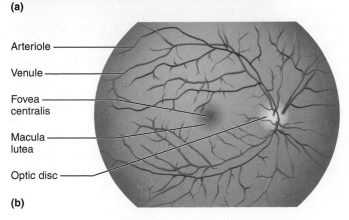

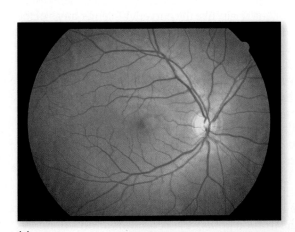

(a)

Arteriole

Venule

Fovea centralis

Macula lutea

Optic disc

(b)

(c)

Figure 10.17 The Fundus (Rear) of the Eye. (a) A typical ophthalmoscopic image. (b) Major features of the fundus. Note the blood vessels diverging from the optic disc, where they enter by way of the optic nerve. An eye examination also serves as an examination of the health of the blood vessels, and can reveal signs of such diseases as atherosclerosis or diabetes. (c) Use of the ophthalmoscope.

• *Is figure (a) the subject's right or left eye? How can you tell?*

essentially imaginary information so that we are not distracted by a patch of darkness in the field. Blood vessels also enter and leave at this point. Eye examinations serve for more than evaluating the visual system; they allow for direct, noninvasive examination of these vessels for signs of hypertension, diabetes mellitus, atherosclerosis, and other vascular diseases.

> **Apply What You Know**
>
> Some have suggested that the blind spot of one eye is compensated for by the other eye, with each eye seeing the area of the visual field that the other eye cannot. What simple experiment could you do that would prove this wrong?

Forming the Visual Image AP|R

The visual process begins when light rays enter the eye, focus on the retina, and produce a tiny inverted image. The pupil dilates or constricts to determine how much light enters the eye. Dilation is achieved by the **pupillary dilator,** a system of contractile cells that radiate from the pupil like the spokes of a wheel. These cells respond to the sympathetic nervous system and dilate the pupil when light intensity falls, when we look from a nearby to a more distant object, or when the body is in a general state of sympathetic arousal. Constriction is achieved by a circle of smooth muscle called the **pupillary constrictor,** innervated by the parasympathetic nervous system (see fig. 9.18, p. 318). Pupillary constriction occurs when light intensity rises and when we shift our focus to a relatively close object. The responses to light are called **photopupillary reflexes.**

Image formation depends on **refraction,** the bending of light rays (fig. 10.18). Light travels at different speeds through media such as air, water, and glass. When light rays strike a denser medium at a 90° angle, they merely slow down; but if they strike at an oblique angle, like the curved off-center areas of the cornea, they also bend. In vision, the greatest amount of refraction, and therefore most focusing, occurs at the air–cornea interface. The lens contributes additional refraction and fine-tunes the focusing process begun by the cornea. The aqueous humor and vitreous body have relatively little effect. An important difference between the cornea and lens is that the curvature of the cornea is fixed, and therefore so is the amount of refraction occurring at the air–cornea interface. The curvature of the lens, however, can be adjusted moment by moment.

When we look at objects more than 6 m away, our eyes are in a relatively relaxed state called **emmetropia** (EM-eh-TRO-pee-ah). Light rays entering the eyes are nearly parallel, the pupil is relatively dilated, and the lens is relatively thin, about 3.6 mm at the center (fig. 10.18a). When we look at something closer than 6 m, three things must happen: (1) **convergence**—the two eyes turn medially toward the fixation point; (2) **accommodation**—light rays are more divergent as they enter the eye, and the lens thickens to as much as 4.5 mm at the center to focus them; and (3) **pupillary constriction**—narrowing of the pupil to screen out divergent peripheral light rays that cannot be effectively focused (fig. 10.18b). Inadequate convergence can result from unequal strength of the extrinsic eye muscles and can cause double vision. You can simulate this effect by pressing gently on one eyelid as you view this page. The image will fall on noncorresponding regions of the two retinas and you will see double.

The closest an object can be and still come into focus is called the **near point** of vision. It depends on the flexibility of the lens. The lens stiffens with age, so a 10-year-old child can focus on objects only 9 cm away, but by age

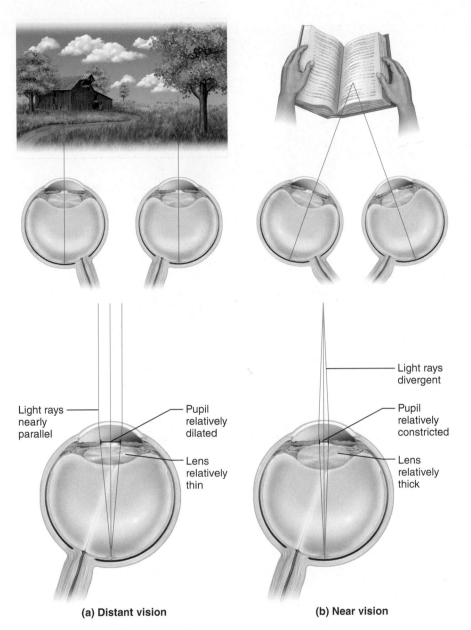

Figure 10.18 Focusing and Accommodation in Distance and Near Vision. (a) When we view a distant scene, light rays entering the eye are nearly parallel. The lens is relatively thin and the pupil relatively dilated. (b) When we view a nearby object, light rays are more divergent. The eyes converge, the lens thickens, and the pupil constricts. These reflexes are mediated through the midbrain.

Light rays nearly parallel

Pupil relatively dilated

Lens relatively thin

Light rays divergent

Pupil relatively constricted

Lens relatively thick

(a) Distant vision

(b) Near vision

60, the near point averages about 83 cm from the eye. That is, older people have more difficulty focusing on things close to the face—a condition called **presbyopia**[24]—hence the common joke among older people, "My arms have gotten shorter!"

Generating the Visual Nerve Signal

Sensory transduction, the conversion of light energy into nerve signals, occurs in the retina. To understand the process, we must begin with the retina's cellular layout (fig. 10.19). Its outermost layer is a dark *pigment epithelium* that absorbs excess light. Facing this is a layer of receptor cells called **rods** and **cones,** which absorb light and begin the process of sensory transduction.

Rods and cones are connected to a thick layer of neurons called **bipolar cells,** and these are connected, directly or indirectly, to a single layer of large

[24]*presby* = old age; *opia* = eye condition

Figure 10.19 The Retina. (a) Photomicrograph. The sclera and choroid are external to the retina; the pigment epithelium is the most superficial layer of the retina itself. (b) Diagram of the relationships of the retinal cells. **AP|R**

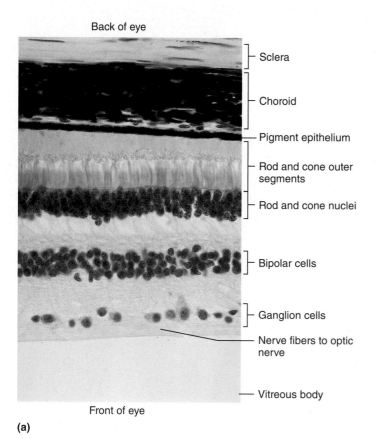

Back of eye

- Sclera
- Choroid
- Pigment epithelium
- Rod and cone outer segments
- Rod and cone nuclei
- Bipolar cells
- Ganglion cells
- Nerve fibers to optic nerve
- Vitreous body

Front of eye

(a)

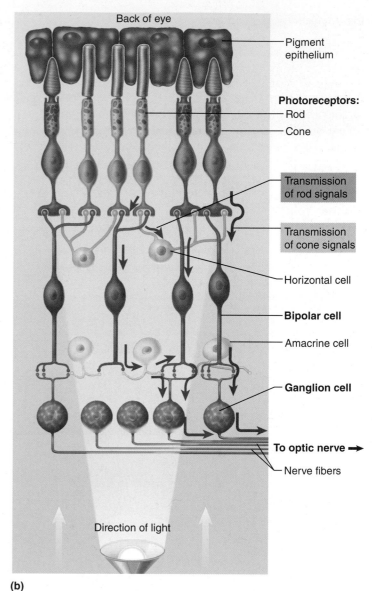

Back of eye

- Pigment epithelium

Photoreceptors:
- Rod
- Cone

- Transmission of rod signals
- Transmission of cone signals
- Horizontal cell
- **Bipolar cell**
- Amacrine cell
- **Ganglion cell**
- **To optic nerve →**
- Nerve fibers

Direction of light

(b)

round neurons called **ganglion cells.** Ganglion cells are the innermost cells of the retina, adjacent to the vitreous body. Their axons converge at the rear of the eye to form the **optic nerve** that leaves the eyeball and passes to the brain.

Photoreceptors are named for their shapes (fig. 10.20). The slender rods consist of a *cell body* containing the nucleus, an *inner segment* containing mitochondria and other organelles, and a cylindrical *outer segment* that faces the back of the eye. The outer segment contains a stack of membranous discs, like a roll of pennies in a paper wrapper. Each disc is packed with molecules of a pigment called *visual purple,* or **rhodopsin** (ro-DOP-sin), a derivative of vitamin A. (This is why a dietary deficiency of vitamin A results in defective vision.)

Cones are structurally similar to rods. The outer segment is shorter and more conical, and it has parallel infoldings of the plasma membrane rather than a stack of membranous discs. Cones work in a similar manner but in brighter light; we will examine their properties in the next section.

When a rod is in the dark, unstimulated, it releases a steady stream of inhibitory neurotransmitter from synaptic vesicles in the base of the cell. This inhibits the bipolar cells, the next cells in line. When light enters the eye, rhodopsin absorbs it and the rod stops releasing neurotransmitter. The bipolar cell, no longer inhibited, stimulates a ganglion cell. The ganglion cell is the only retinal cell type that produces action potentials. It responds to the foregoing changes with a burst of action potentials—a nerve signal that travels out the optic nerve and goes to the brain.

A substantial amount of information processing occurs in the retina before the resulting signal is sent to the brain. Other types of retinal cells, such as *horizontal cells* and *amacrine cells* (fig. 10.19b), are involved in detecting the boundaries of objects, movement, changes in light intensity, and other aspects of the visual scene.

The Roles of Rods and Cones

A human retina contains about 130 million rods and 6.5 million cones. These two receptor cells are concerned with distinctly different aspects of vision. All rods contain identical rhodopsin. As a result, they all respond to light in the same way and have no basis for distinguishing colors (different wavelengths of light) from each other; essentially, rods produce visual sensations of shades of gray (*monochromatic*, or black-and-white, vision). Furthermore, they are active only in dim light, producing **night (scotopic) vision** and grainy, low-resolution images.

In bright daylight and typical room lighting, rods are "bleached out" and don't respond at all. However, they are very sensitive to low intensities, responding to light even as dim as starlight reflected from white paper. Such dim light stimulates any one rod very weakly, but hundreds of rods converge to stimulate each bipolar cell, and many bipolar cells then collaborate to stimulate each ganglion cell. Although this extensive *neural convergence* (see p. 271) enables the eye to respond to very dim light, it also means that each optic nerve fiber going to the brain conveys information from a relatively large patch of retina with hundreds of rods. The resulting image is like a very coarse-grained, overenlarged newspaper photograph.

Cones, on the other hand, are somewhat responsive even at moonlight intensity, but function best at daylight and room light intensities; they are responsible for **day (photopic) vision.** Since they act in brighter light than rods do, they do not have to team up to stimulate a bipolar cell; they have far less neural convergence than the rod pathways. Indeed, in the fovea, there are about 4,000 tiny cones and no rods at all, and each cone has its own bipolar cell and ganglion cell—like a "private line" to the brain. As a result, each optic nerve fiber in this region sends a signal representing only a very tiny area of the retina, and the resulting image is a fine-grained, high-resolution one.

When we look directly at something, we aim each eye so the image falls on the fovea and we see the object as sharply as possible. If you fix your gaze on a few words of this book, for example, you will notice (don't move your eyes!) that the words to the left and right of that are blurry. That's because our peripheral vision is a relatively low-resolution system that serves mainly to alert us to objects or movements in that region, stimulating us to turn our eyes or head to look that way and identify what is there.

Another well-known aspect of cones is that they provide *trichromatic color vision.* The sensory pigment of cone cells, called **photopsin** (fo-TOP-sin),

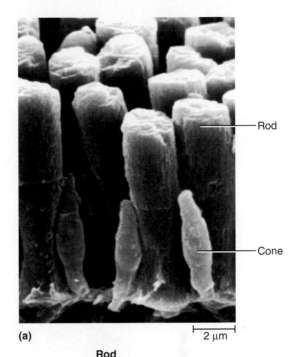

(a)   2 μm

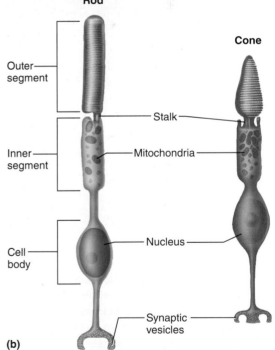

(b)

Figure 10.20 Rod and Cone Cells. (a) Rods and cones of a salamander retina (electron micrograph). (b) Structure of human rod and cone cells.

- *Vision requires a steady supply of ATP. Would you expect this to be made in the outer segment, inner segment, or cell body? Explain.*

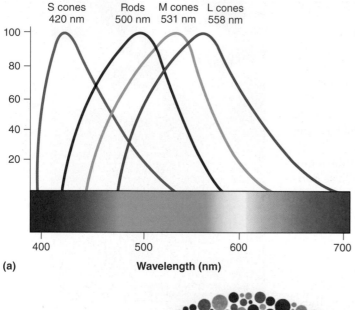

(a)

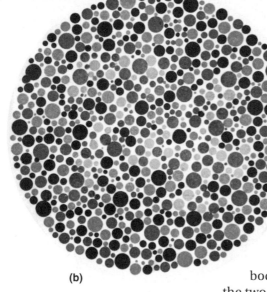

(b)

Figure 10.21 Color Vision. (a) Absorption curves of rods and the three types of cones. (b) What number do you see? People with normal trichromatic vision should see 74, but people with red–green color blindness see no discernible number.

exists in three varieties (fig. 10.21): So-called **short-wavelength (S) cones** have a photopsin that best absorbs light in the violet part of the spectrum; **medium-wavelength (M) cones** peak in the green part of the spectrum; and although **long-wavelength (L) cones** peak in the yellow-green wavelengths, they are the only cones whose absorption extends into the red wavelengths, so they bear sole responsibility for our ability to see red. The brain distinguishes colors from each other based on the mixture of signals it gets from these three types of cones.

People with a genetic lack of any one of these three photopsins have **color blindness.** Those with *red–green color blindness,* the most common form, have difficulty distinguishing shades of green, orange, and red from each other (fig. 10.21b). This is a sex-linked genetic trait occurring in about 8% of males but only about 0.5% of females. Such males inherit the gene for red–green color blindness on the X chromosome received from their mothers. Females can also inherit the gene from their mothers, but they also usually inherit a gene for normal color vision on the X chromosome contributed by their fathers, which masks the effect of the maternal gene. The father contributes no X chromosome to his sons, so he cannot contribute a gene to mask the maternal effect on a son's color vision.

The Visual Projection Pathway

Once nerve signals are generated in the retina, where do they go? The two optic nerves enter the cranial cavity, converge, and form an X called the **optic chiasm**[25] (fig.10.22). Here, half of the nerve fibers from each eye cross over to the opposite side of the brain. The right cerebral hemisphere thus receives input from the medial (nasal) side of the left eye and the lateral side of the right eye. The left cerebral hemisphere receives the medial fibers from the right eye and lateral fibers from the left. Consequently, the right brain sees things on the left side of the body, the left brain sees things on the right side of the body, and the two overlap in the middle. Recall that the right brain controls most voluntary motor responses on the left side of the body, and vice versa, so each cerebral hemisphere visually monitors the side of the body where it exercises its primary motor control.

Slightly posterior to the optic chiasm, most optic nerve fibers terminate at synapses in the thalamus. New nerve fibers begin here and continue the rest of the way to the **primary visual cortex** in the occipital lobe at the rear of the cerebrum. This is where one becomes aware of the visual stimulus. A stroke that destroys occipital lobe tissue can leave a person blind in one-half of the visual field even if the eyes still function normally. Neurons of the primary visual cortex communicate with neurons of the nearby **visual association area,** just anterior to the primary cortex. The visual association area integrates visual input with memory and enables us to identify and interpret what we are seeing at the moment, as well as to form new short- or long-term visual memories.

[25]chi = X

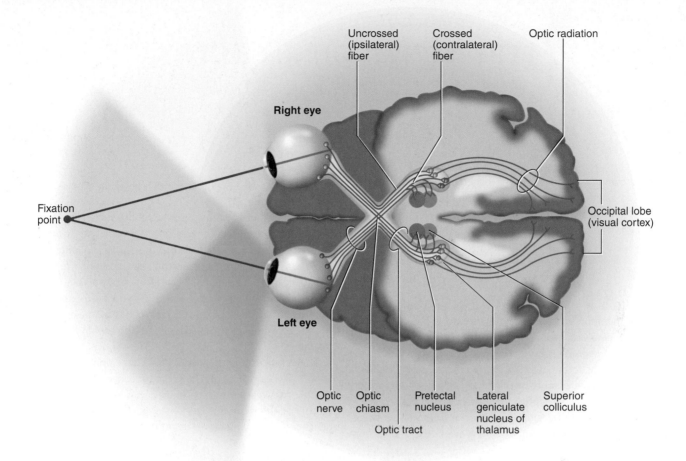

Right eye

Uncrossed (ipsilateral) fiber

Crossed (contralateral) fiber

Optic radiation

Fixation point

Left eye

Optic nerve

Optic chiasm

Pretectal nucleus

Optic tract

Lateral geniculate nucleus of thalamus

Superior colliculus

Occipital lobe (visual cortex)

A few optic nerve fibers take a different course. Instead of going to the thalamus, they lead to two midbrain nuclei—one that controls accommodation of the lens and the photopupillary reflex, and one that controls reflexes of the muscles that move the eye (for example, causing us to look toward something we see moving in our peripheral vision).

Figure 10.22 The Visual Projection Pathway to the Brain. Note that half of the nerve fibers from each retina (red and violet) cross over to the opposite side of the brain when passing through the optic chiasm. The right occipital lobe thus monitors the left half of the visual field and vice versa.

- *If a cerebral hemorrhage destroyed the optic radiation in just one cerebral hemisphere, where would you expect blindness to be experienced, and why—total blindness in both eyes; total blindness in just one eye; blindness in the medial field of vision of both eyes; blindness in the peripheral field of both eyes; or blindness in the peripheral field of just one eye?*

Before You Go On

Answer these questions from memory. Reread the preceding section if there are too many you don't know.

18. *Briefly state the function of each of these orbital structures: tarsal glands, lacrimal gland, conjunctiva, and extrinsic muscles.*
19. *How does the aqueous humor differ from the vitreous body in composition, location, and function?*
20. *What specific region of the retina produces the most high-resolution images? Why?*
21. *What component of the eye has the greatest role in refracting light rays to form an image on the retina?*
22. *Explain why cone cells produce high-resolution images and rod cells do not; why cones provide day vision and rods do not; and why cones provide color vision and rods do not.*

PERSPECTIVES ON HEALTH

The most common health problems related to the senses are in vision and hearing. Some of these are commonplace changes that occur almost inevitably with age, whereas others are caused by disease or trauma that can occur at any age.

Hearing Disorders

The most common complaint about hearing is partial deafness. *Deafness* means any hearing loss, whether temporary or permanent. *Conductive deafness* results from conditions that reduce the transfer of vibrations to the inner ear. Such conditions may include a damaged tympanic membrane, otitis media, blockage of the auditory canal with earwax or other matter, and otosclerosis. *Otosclerosis*[26] is fusion of the auditory ossicles to each other or fusion of the stapes footplate to the oval window. Either way, it prevents the ossicles from vibrating freely.

Sensorineural (nerve) deafness results from death of hair cells or any of the nervous elements concerned with hearing. It is a common occupational disease of factory and construction workers, musicians, and other people exposed to frequent and sustained loud sounds. Deafness leads some people to delusions of being talked about, disparaged, or cheated. Beethoven said his deafness nearly drove him to suicide.

Sensorineural deafness is often accompanied by *tinnitus* (ti-NITE-us), a persistent sense of ringing, whistling, clicking, or buzzing in the ear. The physiological basis of tinnitus is not well known, but it may represent an illusion of sensory input from cochlear hair cells that are no longer present, much like the illusion of phantom pain from a limb that has been amputated.

Visual Disorders

The most common visual disorders concern inability to focus properly. *Myopia* (nearsightedness) is an inability to focus on distant objects. It results from eyeballs that are a little too long, so light rays come to a focus before they reach the retina instead of focusing on the retina. The rays have begun to diverge again by the time they reach the retina, so the resulting image is blurry. Myopia is treated with corrective lenses that make the rays diverge slightly before entering the eye, so they converge again at a greater distance behind the cornea—on the retina.

Hyperopia (farsightedness) is much the opposite—the eyeball is abnormally short, light rays have not yet converged to a focal point by the time they reach the retina, and one has difficulty focusing on nearby objects. This is corrected by lenses that begin the process of convergence before the light rays enter the eye, so they reach their focal point sooner, on the retina.

Many people suffer *astigmatism,* a condition in which focusing on objects in one plane, such as the vertical, makes objects in a different plane, such as the horizontal, go out of focus. This is the result of a cornea that is somewhat ovoid, like the back of a spoon, instead of circular. Astigmatism is also treated with corrective lenses.

More serious problems are glaucoma (see Clinical Application 10.3) and cataracts. A *cataract* is a cloudiness of the lens, making the vision dim or milky, as if looking at a scene from behind a waterfall—hence the name.[27] The lenses become more predisposed to cataracts as they thicken with age. Their degeneration is often accelerated and worsened by diabetes mellitus. Heavy smoking and excessive exposure to the ultraviolet rays of sunlight are additional risk factors for cataracts. Cataracts can be treated by replacing the lenses with plastic ones. The artificial lenses improve vision almost immediately, but glasses may still be needed for near vision.

---

[26]*oto* = ear; *scler* = hardening, stiffening; *osis* = condition, process

[27]*cataract* = waterfall

Aging of the Sense Organs

Some sensory functions begin to decline shortly after adolescence, or even earlier. Many schoolchildren and adolescents already experience the beginning of presbyopia, the stiffening of the lenses mentioned earlier, and thus find it harder to focus without corrective lenses. Later in life, people often notice a deficiency of night vision, as more and more light is needed to stimulate the retina. This has several causes: There are fewer receptor cells in the retina, the vitreous body becomes less transparent, the pupil becomes narrower as the pupillary dilators atrophy, and the lenses tend to yellow with age. Dark adaptation takes longer because enzymatic reactions in the rod cells are slower. Elderly people also have an elevated risk of glaucoma (see Clinical Application 10.3) and cataracts (see Perspectives on Health)—both of which can cause serious visual impairment, even blindness. Having to give up reading and driving can be among the most difficult changes of lifestyle for elderly people.

Auditory sensitivity peaks in adolescence and declines afterward. The tympanic membrane and the joints between the auditory ossicles become stiffer, so vibrations are transferred less effectively to the inner ear, creating a degree of *conductive deafness. Nerve deafness* occurs as the numbers of cochlear hair cells and auditory nerve fibers decline. (See conductive and nerve deafness in Perspectives on Health.) The greatest hearing loss occurs in the frequency range of conversation. The death of receptor cells in the semicircular ducts, utricle, and saccule and of nerve fibers in the vestibular nerve and neurons in the cerebellum results in poor balance and dizziness—a factor in falls that sometimes lead to severely disabling bone fractures.

The senses of taste and smell are blunted as the numbers of taste buds, olfactory cells, and second-order neurons in the olfactory bulbs decline. Elderly people often find food less appealing, so declining sensory function can be a factor in malnutrition.

CAREER SPOTLIGHT

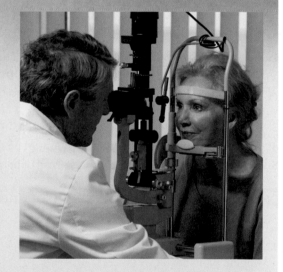

Optician

An optician fills prescriptions for eyeglasses or contact lenses. Whereas opthalmologists or optometrists perform eye tests and determine the appropriate strength of corrective lenses, opticians ensure that the prescription is filled accurately. Opticians assist patients with selection of eyeglasses or contact lenses to ensure that they fit with maximum comfort and functionality. They also educate the patient about proper cleaning and care of the eyewear.

Opticians are typically employed by chain stores with optical departments or by hospitals, health clinics, or private doctors' offices that have their own stores. In addition to customer interaction, the job often entails business management. The optician does not require the same level of advanced training as optometrists or ophthalmologists. Most opticians obtain a 2-year associate degree from a college with a program in optometric technology. Several states require that opticians be licensed. For links to further career information, see appendix B.

Study Guide

Assess Your Learning Outcomes

To test your knowledge, discuss the following topics with a study partner or in writing, ideally from memory.

10.1 Receptors and Sensations (p. 328)

1. The meaning of *receptor* and distinction between a receptor and a sense organ
2. Why not all sensory signals result in a sensation
3. Four properties of a stimulus—type, location, intensity, and duration—and how the brain is able to interpret each on the basis of the nerve signals it receives
4. Why sense organs are described as transducers
5. The difference between general and special senses, and which senses belong in each category

10.2 The General Senses (p. 331)

1. Five types of simple nerve endings, their structural differences, and the senses for which each is responsible
2. The importance of pain and the harmful consequences of lacking a sense of pain
3. Distinctions between somatic and visceral pain
4. The reasons for referred pain and phantom pain
5. How spinal gating alters a person's sense of pain

10.3 The Chemical Senses—Taste and Smell (p. 334)

1. The locations of taste buds and their relationship to the lingual papillae
2. The cellular organization of a taste bud and the relationship of taste cells to sensory nerve fibers
3. The five primary taste sensations and the types of tastants that produce each sensation
4. The difference between taste and flavor; nongustatory factors that influence flavor
5. The location and cellular structure of the olfactory mucosa
6. The mechanism by which odorants trigger an olfactory nerve signal
7. The cranial nerves that carry gustatory and olfactory stimuli, and the brain regions that receive and integrate these signals

8. The olfactory regions of the brain and the route that olfactory nerve signals travel to get there

10.4 The Ear—Equilibrium and Hearing (p. 338)

1. The three principal regions of the ear and general function of each one
2. Anatomy of the auricle and auditory canal, and functions of cerumen
3. Anatomy of the middle ear, especially its tympanic membrane, auditory tube, auditory ossicles, and muscles; the function of each of these
4. The relationship between the bony labyrinth, membranous labyrinth, perilymph, and endolymph of the inner ear
5. The function and anatomical components of the vestibular apparatus
6. The structure of hair cells and how it relates to the generation of sensory signals
7. The locations and cellular organization of the saccule and utricle, how they generate sensations of motion and body orientation, and how they differ
8. Anatomy of the semicircular ducts, the type of sensation they produce, and how the three ducts in each ear differ from each other in orientation and function
9. The destinations of signals for equilibrium in the brain, and the pathways taken to these destinations
10. Gross anatomy of the cochlea, and the anatomical relationships of its fluid-filled spaces, basilar membrane, and spiral organ
11. Structure of the cochlear hair cells, their relationship to the basilar and tectorial membranes, and the spatial arrangement and functional difference between inner and outer hair cells
12. How sound waves in the auditory canal ultimately result in movement of the basilar membrane and hair cells
13. How hair cell movements result in auditory nerve signals
14. How the brain discriminates between sounds of different loudness and pitch
15. The destinations of signals for hearing in the brain, the pathway taken to these destinations, and types of information processing that occur at different points on this route

10.5 The Eye and Vision (p. 347)

1. Structure and function of components of the orbital region external to the eyeball
2. Anatomy of the lacrimal apparatus and the pathway that tears follow from their origin to their drainage into the nasal cavity

3. The three layers of the eyeball and the components of each one

4. Optical components of the eyeball, their locations, and their functions

5. Neural components of the eyeball and specialized regions of the retina

6. Locations of the pupillary dilator and constrictor, their respective nerve supplies, and the conditions under which they alter the diameter of the pupil

7. Relative contributions of the cornea and lens to the refraction of light rays

8. Responses of the eye to shifting one's focus from a distant object to a nearby one

9. Principal cell types of the retina and their relationships to each other

10. Parts of a rod or cone cell, location of the visual pigment, and how light absorption by the pigment leads to the generation of a nerve signal

11. How and why the rod and cone system differ with respect to light sensitivity, visual resolution, and capacity for color vision

12. Why vision is sharpest at the fovea

13. How the retina generates signals that the brain can interpret as differences in color

14. The destination of visual signals in the brain and the pathways taken to get there

Testing Your Recall

1. All of the following are classified as special senses except
   a. taste.
   b. touch.
   c. vision.
   d. hearing.
   e. equilibrium.

2. A nociceptor is a nerve ending responsible for the sense of
   a. touch.
   b. vibration.
   c. heat and cold.
   d. taste or smell.
   e. pain.

3. The occipital lobe is concerned mainly with the sense of
   a. vision.
   b. proprioception.
   c. pain.
   d. taste.
   e. hearing.

4. A stomachache is classified as
   a. fast pain.
   b. phantom pain.
   c. visceral pain.
   d. somatic pain.
   e. referred pain.

5. Which of these is *not* one of the five primary taste sensations detected by the taste buds?
   a. sweet
   b. spicy
   c. bitter
   d. salty
   e. umami

6. Olfactory cells are found in
   a. the olfactory tract.
   b. the olfactory bulb.
   c. the primary olfactory cortex.
   d. the olfactory mucosa.
   e. the temporal lobe.

7. The malleus, incus, and stapes are components of
   a. the outer ear.
   b. the middle ear.
   c. the vestibular system.
   d. the bony labyrinth.
   e. the inner ear.

8. The malleus, incus, and stapes have their closest anatomical and functional association with
   a. the stapedius and tensor tympani.
   b. the auditory canal.
   c. the auditory tube.
   d. the semicircular ducts.
   e. the cochlear hair cells.

9. A single sensory cell type is associated with all of the following *except*
   a. low-resolution vision.
   b. low-light vision.
   c. black-and-white vision.
   d. photopic vision.
   e. rhodopsin.

10. Which lobe of the cerebrum is concerned with receiving and interpreting visual input?
    a. the insula
    b. the frontal lobe
    c. the parietal lobe
    d. the temporal lobe
    e. the occipital lobe

11. Pressure on a tissue is detected by a simple sense organ called a _____, which has a nerve fiber surrounded by onion-like layers of flattened Schwann cells and fibroblasts.

12. Our sharpest vision is in a retinal pit called the _____ directly behind the central axis of the lens.

13. The optic nerves cross at a point called the _____, where half of the nerve fibers from each eye are routed to each cerebral hemisphere.

14. A person might have dry eyes if the ducts of the _____ glands were overly constricted or obstructed.

15. The sense of vertical movement, as when rising from a chair or riding in an elevator, originates in a patch of sensory cells in a chamber called the _____.

16. _____ is a change in the thickness of the lens to enable the eye to focus on nearby objects.

17. Nerve fibers for the sense of smell originate in the olfactory mucosa and terminate in a part of the brain called the _____.

18. The sensory cells of hearing, called _____, are arranged in four spiral rows in the cochlea.

19. A person will have color blindness if he or she has a defective gene for one of the three types of _____ cells.

20. The position and movements of the body and its limbs are detected by sense organs called _____.

Answers in appendix A

True or False

Determine which five of the following statements are false, and briefly explain why.

1. The usual cause of pain is an undesirable overstimulation of receptors whose normal function is touch.

2. Olfactory hairs are true cilia, but the "hairs" of a cochlear hair cell are not.

3. The intensity of a stimulus is communicated to the brain in terms of the firing frequency of a sensory nerve fiber.

4. *Hair receptor* is another name for a hair cell of the inner ear.

5. In vision, most light refraction occurs as light passes through the lens of the eye.

6. The right occipital lobe receives visual input from both the right and left eyes.

7. Deep bass tones stimulate different cochlear hair cells than high treble tones.

8. The auditory (pharyngotympanic) tube lets air into the inner ear to equalize pressure on both sides of the tympanic membrane.

9. After passing across the surface of the eye, tears ultimately drain into the nasal cavity.

10. The taste buds, also known as lingual papillae, can be seen on the surface of the tongue with a magnifying glass.

Answers in appendix A

Testing Your Comprehension

1. You are sitting in a chair attending lecture, your head tilted forward as you take notes. When the lecture ends, you raise your head from your notebook, rise from your chair and leave the room, walk down the hall, and turn a corner headed for your next class. Describe all the actions that would occur in your utricle, saccule, and semicircular ducts to make your brain aware of these movements.

2. Although it is unusual for a female to be color-blind, Julie is one such case. Do you think her father is color-blind? Her mother? Explain.

3. Frank has worked for years as a stage manager for a traveling rock band, and never took precautions to protect his hearing. By the age of 30, he finds he can no longer hear the high notes on an electronic keyboard. What region of his cochlea do you believe is damaged? Name the specific cells in that region that have likely been destroyed.

Chapter 11

The Endocrine System

Artistic conception of the insulin-secreting beta cells of the pancreas. Insulin (yellow dots) enters the bloodstream through the capillaries that permeate the pancreatic islets.

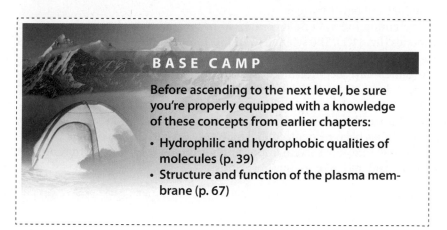

BASE CAMP

Before ascending to the next level, be sure you're properly equipped with a knowledge of these concepts from earlier chapters:

- Hydrophilic and hydrophobic qualities of molecules (p. 39)
- Structure and function of the plasma membrane (p. 67)

Chapter Outline

Anatomy & Physiology REVEALED®
aprevealed.com

Module 8: Endocrine System

If the body is to function as an integrated whole, its cells must be able to communicate and coordinate their functions. Even simple organisms of just a few cells have mechanisms for intercellular communication. Complex animals such as human beings achieve this through the nervous and endocrine systems.

Hormones are critical for maintaining homeostasis. They regulate growth and development, metabolism and energy balance, water balance, all aspects of reproduction, and stress responses. Perhaps you have heard of the pituitary and thyroid glands, hormones such as estrogen and insulin, and endocrine diseases such as goiter and diabetes mellitus. We will examine the structure of hormones and how they work, and provide a survey of endocrine glands and tissues, the hormones they produce, and the functions that these hormones perform.

11.1 Overview of the Endocrine System

Expected Learning Outcomes
When you have completed this section, you should be able to:

a. define *hormone* and *endocrine system;* and
b. explain the major functional differences between the nervous and endocrine systems.

Hormones and Endocrine Glands

Hormones[1] are chemical messengers secreted by endocrine glands or tissues into the blood. They travel everywhere the blood goes, but they affect only cells that have receptors for them—called their **target cells.** Hormones produced in one part of the body may have an effect on very distant organs or cells. For example, those produced by the pituitary gland in the head affect organs in the abdominal and pelvic cavities.

The **endocrine system** is composed of the glands and cells that secrete hormones (fig. 11.1). The science of **endocrinology** embraces the study of this system and the diagnosis and treatment of its disorders. Traditionally, endocrine glands have been distinguished from exocrine glands (see chapter 4). Exocrine glands typically release their secretions to a tissue surface by way of ducts, whereas endocrine glands are ductless and release their secretions (hormones) into the blood—as reflected in the word *endocrine.*[2] In recent decades, however, it has been recognized that many organs such as the brain, heart, liver, and small intestine secrete important hormones, even though they are not usually considered to be endocrine glands.

---

[1] *hormone* = to excite, set in motion

[2] *endo* = internal; *crin* = to secrete

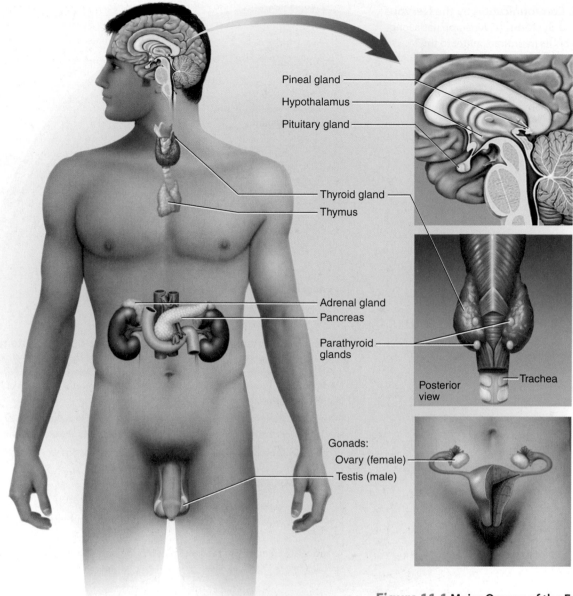

Pineal gland
Hypothalamus
Pituitary gland

Thyroid gland
Thymus

Adrenal gland
Pancreas

Parathyroid glands

Trachea

Posterior view

Gonads:
Ovary (female)
Testis (male)

Figure 11.1 Major Organs of the Endocrine System. This system also includes gland cells in many other organs not shown here. AP|R

- *After reading this chapter, name at least three hormone-secreting organs that are not shown in this illustration.*

Comparison of the Nervous and Endocrine Systems

One way to better understand the endocrine system is to compare it to the nervous system (fig. 11.2). Both systems serve for internal communication, but they are not redundant; they differ in significant ways (table 11.1). For example, the nervous system communicates by means of both electrical and chemical signals, whereas the endocrine system relies solely on chemical messengers. The nervous system responds very quickly to a stimulus and stops responding quickly when stimulation ceases; the endocrine system is slower to respond and stop. The nervous system generally targets its signals to a specific organ, in contrast to the endocrine system whose hormones go everywhere the blood goes.

Despite their differences, the nervous and endocrine systems have much in common. For example, some chemicals such as norepinephrine and dopamine function as both hormones and neurotransmitters. Some hormones are secreted by specialized neurons called *neuroendocrine cells,*

Figure 11.2 Communication by the Nervous and Endocrine Systems. (a) A neuron has a long fiber that delivers its neurotransmitter to the immediate vicinity of its target cells. (b) Endocrine cells secrete a hormone into the bloodstream (left). At a point often remote from its origin, the hormone leaves the bloodstream and enters or binds to its target cells (right). **AP|R**

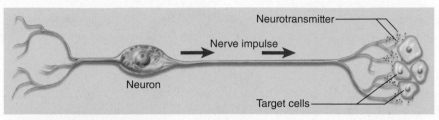

(a) Nervous system

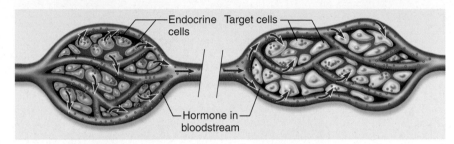

(b) Endocrine system

Table 11.1 Comparison of the Nervous and Endocrine Systems

Nervous System	Endocrine System
Communicates by means of electrical impulses and neurotransmitters	Communicates by means of hormones
Releases neurotransmitters at synapses at specific target cells	Releases hormones into bloodstream for general distribution throughout body
Usually has relatively local, specific effects	Sometimes has very general, widespread effects
Reacts quickly to stimuli, usually within 1–10 ms	Reacts more slowly to stimuli, often taking seconds to days
Stops quickly when stimulus stops	May continue responding long after stimulus stops
Adapts relatively quickly to continual stimulation	Adapts relatively slowly; may respond for days to weeks

which functionally belong to both the endocrine and nervous systems. In addition, hormones and neurotransmitters may have similar effects on the same target cells; for example, norepinephrine (a neurotransmitter) and glucagon (a hormone) both target liver cells and cause the release of glucose. Neurons sometimes trigger hormone secretion, and hormones may stimulate or inhibit neurons. Thus, the two systems regulate each other as they coordinate activities of other body systems.

Before You Go On

Answer these questions from memory. Reread the preceding section if there are too many you don't know.

1. *Define* endocrine system, hormone, *and* target cell.
2. *Compare and contrast how the nervous system and endocrine system coordinate human physiology, and describe some ways in which these two systems interact.*

11.2 Endocrine Physiology

Expected Learning Outcomes

When you have completed this section, you should be able to:

a. characterize the three chemical classes of hormones and state examples of each;

b. describe how hormones act on their target cells and receptors;

c. discuss some types of interaction between multiple hormones;

d. explain how the body regulates rates of hormone secretion; and

e. explain how the body rids itself of a hormone whose job is done.

Hormone Structure

Hormones are classified into three groups on the basis of their chemical structure (fig. 11.3). (1) **Steroid hormones** are lipids synthesized from cholesterol, and share its basic structure of four organic rings with varied side groups. Some familiar steroid hormones are testosterone and estradiol, one of the estrogens. (2) **Monoamines** are small molecules synthesized from the amino acids tyrosine and tryptophan, such as thyroxine (a thyroid hormone) and epinephrine. They are named for the fact that they have an amino group ($-NH_2$). (3) **Peptide hormones** are chains ranging from 3 to more than 200 amino acids long; those 50 amino acids or longer are considered proteins, and some are glycoproteins—proteins with carbohydrate attached. Oxytocin is an example of a small peptide hormone, composed of only 9 amino acids. Insulin is composed of 51 amino acids, large enough to be regarded as a protein.

Figure 11.3 The Chemical Classes of Hormones. (a) Two steroid hormones, defined by their four-membered rings derived from cholesterol. (b) Two monoamines, derived from amino acids and defined by their -NH- or -NH$_2$ (amino) groups. (c) A small peptide hormone, oxytocin, and a protein hormone, insulin, defined by their chains of amino acids (the yellow circles).

Testosterone

Estradiol

(a) Steroids

Thyroxine

Epinephrine

(b) Monoamines

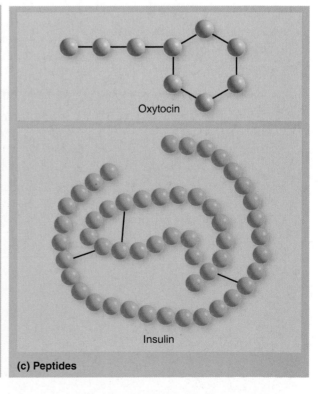

Oxytocin

Insulin

(c) Peptides

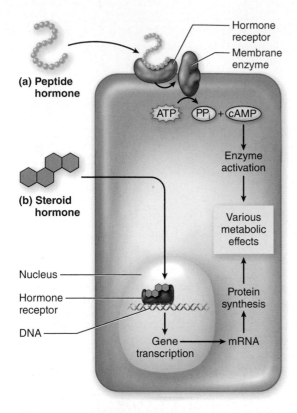

Figure 11.4 Hormone Actions on a Target Cell.
(a) Peptide hormones cannot enter their target cells;
they bind to surface receptors and activate
intracellular processes, working through second
messengers such as cAMP. (b) Steroid hormones freely
enter their target cells and usually bind to receptors in
the nucleus associated with specific genes in the DNA.
Either process can lead to a great variety of metabolic
effects on the target cell. **AP | R**

- *Would testosterone act by mechanism (a) or
  mechanism (b)? Explain.*

Hormone Receptors and Effects

The function of a hormone is to stimulate a physiological change in a target
cell. In order to do this, the hormone must first bind to a receptor in or on that
cell. A given hormone fits its receptor much like a key fits only one lock. A
"key" such as insulin will fit only an insulin receptor ("lock"), for example; it
will not fit an estrogen receptor. Many hormones are in the blood at one time,
but any given organ or cell responds only to those for which it has the proper
receptors—just as there are many radio broadcast signals in the air at once,
but you hear only the one to which you tune your own radio.

The location of the lock varies with the chemical nature of the hormone.
Peptide hormones and most monoamines (such as epinephrine) are hydrophilic and cannot pass into the target cell. They must bind to a receptor at the
cell surface (fig. 11.4a). The hormone acts as a *first messenger* "knocking at the
door" of the target cell. Its receptor is closely associated with a membrane enzyme that responds by producing a *second messenger* inside the cell. Often, this
is *cyclic adenosine monophosphate (cAMP),* made by removing two phosphate
groups (PP$_i$) from ATP. Cyclic AMP generally activates or deactivates cytoplasmic enzymes. This can lead to various metabolic effects in the cell. For example, an activated enzyme can turn certain metabolic pathways on; it can "wake
up" certain genes, leading to the production of new enzymes; or it can change
the permeability of the plasma membrane, thereby altering the transport of
materials into or out of the target cell. Glucagon, for example, acts on liver cells
to activate metabolic pathways leading to the release of blood sugar (see
fig. 11.11).

Steroid hormones, in contrast, are hydrophobic and readily diffuse
through the phospholipids of the target-cell membrane. Most of them pass
directly into the cell's nucleus and bind to a receptor associated with the DNA
(fig. 11.4b). This typically causes a gene to be transcribed, leading through
mRNA to the production of a specific protein. That protein can then alter cell
metabolism in various ways. For example, when estrogen enters a cell of the
uterus during the first half of a woman's menstrual cycle, it binds to a nuclear
receptor and activates a gene for progesterone receptors. The cell produces
those receptors, preparing itself to respond to the progesterone that will arrive in the second half of the cycle. The progesterone itself will activate other
genes that cause the lining of the uterus to thicken and prepare for the possibility of pregnancy.

Hormone Interactions

We never have just a single hormone in circulation; the blood plasma is always
a "cocktail" of many hormones traveling together. Often, two or more of these
act on the same target organ and alter each other's effects. There are three types
of hormone interactions:

1. **Permissive effects,** in which one hormone enhances the target organ's response to another hormone secreted later. As described in
   the previous section, for example, estrogen stimulates the uterus to
   produce progesterone receptors. When progesterone is secreted later
   in the menstrual cycle, the uterus is very responsive to it (see chapter 19), but it would not respond at all if estrogen had not primed the
   uterus for it.

2. **Antagonistic effects,** in which one hormone opposes the action of
   another. For example, insulin stimulates the liver to absorb glucose and
   lower its blood concentration, whereas glucagon stimulates the liver to
   release glucose and raise the blood sugar level.

Clinical Application 11.1

HORMONE RECEPTORS AND THERAPY

In treating endocrine disorders, it is essential to understand the role of hormone receptors. For example, pituitary dwarfism is now rare because it is easily treated in childhood with growth hormone, and genetic engineering has now made human growth hormone freely available. However, there is another form of dwarfism, *Laron dwarfism,* which results from a defect in the receptor for growth hormone. Administering growth hormone has no benefit because the body doesn't respond to it. Similarly, *type 2 diabetes mellitus,* the most common form of diabetes, is thought to result from an insulin receptor defect or deficiency (among other possible causes). No amount of insulin replacement can correct this. Another implication of hormone–receptor physiology for therapy is that estrogen stimulates the growth of some malignant tumors with estrogen receptors. For this reason, estrogen replacement therapy should not be used for women with estrogen-dependent cancer or at risk for it.

3. **Synergistic effects,** in which two hormones acting together produce a much stronger response than the sum of their separate effects. In men, for example, testosterone and follicle-stimulating hormone would have very little effect on sperm production if they each worked alone, but when they work together, the testes produce 300,000 sperm per minute.

Hormonal Control of Homeostasis

The endocrine system precisely regulates numerous variables in the body through feedback control. Most often, hormonal control relies on negative feedback, in which the secretion of a hormone counteracts (negates) a change in a regulated variable. In a negative feedback loop, the body senses a change in a variable, an endocrine gland responds by altering the rate of hormone secretion, and target cells respond in a way that will ultimately bring the variable back to its original state.

An example of this is the regulation of blood glucose. When the blood glucose level rises during digestion of a meal, endocrine cells of the pancreas respond by secreting insulin. Among other effects, insulin targets liver, muscle, and other cells and causes them to remove glucose from the blood and store it for later use. The feedback mechanism thus reverses the original change in blood glucose levels and maintains homeostasis.

Both positive and negative feedback mechanisms also regulate the endocrine glands themselves, as section 11.3 will show with regard to the pituitary gland.

Hormone Clearance

After they are secreted, hormones do not linger in the blood forever. Most are taken up by the liver and kidneys, chemically degraded, and excreted in the urine or bile. Some hormones are removed from the blood very quickly; for example, growth hormone levels decline by 50% within just 6 to 20 minutes. Others linger longer; thyroid hormone, for example, maintains an effective level in the blood for as long as 2 weeks after its secretion ceases (as when the thyroid gland has been surgically removed to treat thyroid cancer).

Before You Go On

Answer these questions from memory. Reread the preceding section if there are too many you don't know.

3. Define the three chemical classes of hormones and give an example of each.
4. Explain what a hormone receptor is and state two places where receptors are located in the target cells.
5. Describe three ways in which hormones can influence each other's actions.
6. Explain how the endocrine system can maintain homeostasis through negative feedback. Give an example.

11.3 The Hypothalamus and Pituitary Gland

Expected Learning Outcomes

When you have completed this section, you should be able to:

a. describe the anatomical relationship between the pituitary gland and hypothalamus;
b. explain how the hypothalamus controls pituitary functions; and
c. name the pituitary gland hormones and explain their effects.

This section and the next surveys the major glands of the endocrine system. We begin our exploration with the pituitary and a nearby region of the brain that controls it, the hypothalamus. You may remember that the hypothalamus regulates a wide range of body functions, including water balance, energy balance, growth, and reproduction (see p. 299). It fulfills these roles in large part by controlling the production and secretion of hormones from the pituitary gland. Thus, the pituitary is an important link in the neural and hormonal regulation of homeostasis.

Pituitary Anatomy

The **pituitary gland (hypophysis**[3]**)** controls more bodily functions than any other endocrine gland, yet despite this functional significance, it is only about as big as a kidney bean; it grows about 50% larger, however, during pregnancy. It is located just below the hypothalamus (see fig. 11.1), sheltered in the *sella turcica,*[4] a saddlelike enclosure in the sphenoid bone (see p. 165).

---

[3] *hypo* = below; *physis* = growth
[4] *sella* = saddle; *turcica* = Turkish

The pituitary appears to be a single structure, but functionally and embryologically, it is divided into **anterior** and **posterior lobes.** The anterior pituitary gland develops from a pouch in the roof of the embryonic pharynx (throat) (fig. 11.5). Meanwhile, the posterior pituitary gland arises as a downgrowth from the hypothalamus, and retains its connection to the brain throughout life. The two parts of the pituitary gland come to lie side by side, and are so closely joined together that they appear to be a single gland.

The hypothalamus is attached to the posterior pituitary gland by a **stalk** containing bundles of nerve fibers (fig. 11.6a). These nerve fibers originate in two pairs of nuclei (right and left) in the hypothalamus, represented in blue and green in the figure. Those nuclei synthesize two hormones, *oxytocin* and *antidiuretic hormone,* which we will examine shortly, and transport them down the nerve fibers for storage in the posterior pituitary. Under appropriate conditions later on, the hypothalamus sends electrical signals down the same nerve fibers to stimulate the posterior pituitary to release those hormones into the blood.

The hypothalamus connects to the anterior pituitary gland by a network of small blood vessels called the **hypophyseal portal system** (fig. 11.6b). A *portal system* is one in which blood flows from one capillary bed to another before returning to the heart. In this specific case, the hypothalamus secretes chemical signals into the capillaries at its end of the portal system; these signals travel a short distance to the anterior pituitary. Here they leave the bloodstream through a second capillary network and tell the anterior pituitary what to do. We will examine these signals shortly in more detail.

Figure 11.5 Embryonic Development of the Pituitary Gland. (a) Sagittal section of the head at 4 weeks, showing the neural and pharyngeal origins of the pituitary gland. (b) Pituitary development at 8 weeks. The anterior pituitary has now separated from the pharynx. (c) Development at 16 weeks. The two lobes are now encased in bone and so closely associated they appear to be a single gland.

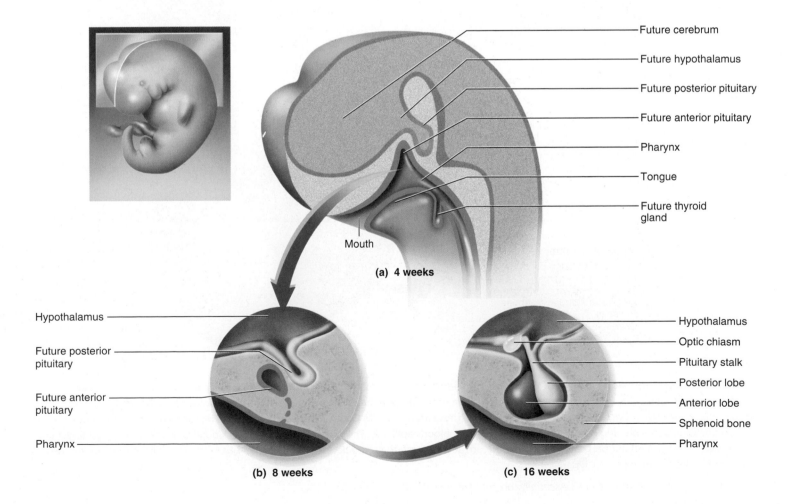

Future cerebrum
Future hypothalamus
Future posterior pituitary
Future anterior pituitary
Pharynx
Tongue
Future thyroid gland

Mouth

(a) 4 weeks

Hypothalamus
Future posterior pituitary
Future anterior pituitary
Pharynx

(b) 8 weeks

Hypothalamus
Optic chiasm
Pituitary stalk
Posterior lobe
Anterior lobe
Sphenoid bone
Pharynx

(c) 16 weeks

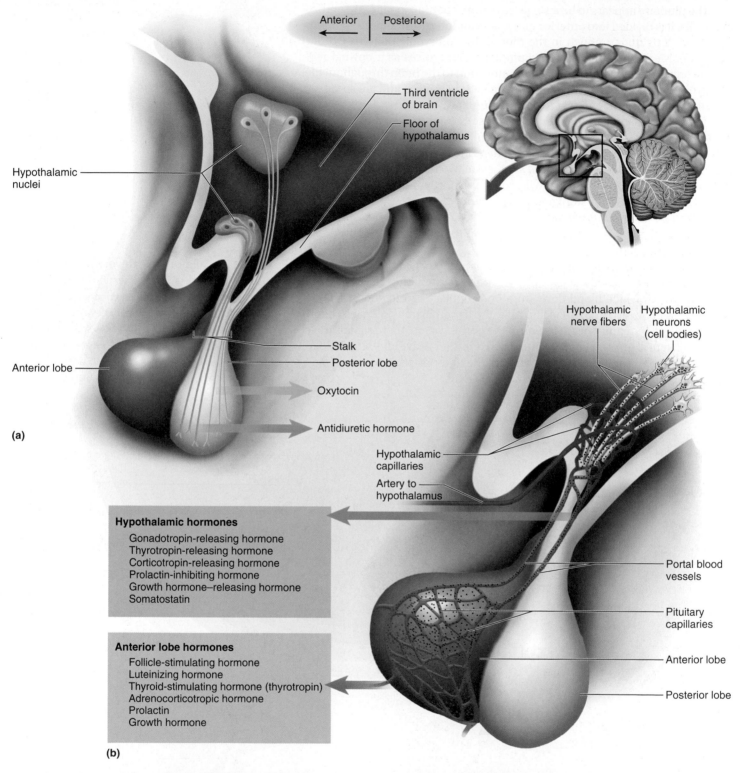

Anterior | Posterior

Third ventricle of brain

Floor of hypothalamus

Hypothalamic nuclei

Hypothalamic nerve fibers

Hypothalamic neurons (cell bodies)

Anterior lobe

Stalk

Posterior lobe

Oxytocin

Antidiuretic hormone

Hypothalamic capillaries

Artery to hypothalamus

(a)

Hypothalamic hormones
Gonadotropin-releasing hormone
Thyrotropin-releasing hormone
Corticotropin-releasing hormone
Prolactin-inhibiting hormone
Growth hormone–releasing hormone
Somatostatin

Anterior lobe hormones
Follicle-stimulating hormone
Luteinizing hormone
Thyroid-stimulating hormone (thyrotropin)
Adrenocorticotropic hormone
Prolactin
Growth hormone

Portal blood vessels

Pituitary capillaries

Anterior lobe

Posterior lobe

(b)

Figure 11.6 Anatomy and Control of the Pituitary Gland. (a) Control of the posterior lobe. The nuclei in blue and green synthesize oxytocin and antidiuretic hormone and transport them down nerve fibers in the stalk for storage in the posterior lobe. Later, nerve signals passing down the stalk trigger the posterior lobe to release these hormones. (b) Control of the anterior lobe. Neurons in the hypothalamus secrete releasing and inhibiting hormones into hypothalamic blood capillaries. These hormones (small black dots) travel down the portal system to capillaries in the anterior lobe. Leaving the bloodstream at that point, they stimulate anterior lobe cells to secrete their hormones, or in other cases inhibit them from doing so. AP|R

• *Which lobe of the pituitary gland is essentially composed of brain tissue?*

Anterior Pituitary Hormones

The anterior pituitary gland synthesizes six peptide hormones. The first two are called **gonadotropins**[5] because they target the gonads (ovaries and testes).

1. **Follicle-stimulating hormone (FSH).** FSH stimulates the development of gametes (sex cells) in women and men. Once females reach reproductive age, it stimulates the maturation of bubblelike follicles that contain the eggs. In the process, it stimulates the follicles to secrete estrogen. Therefore, FSH plays an important role in the regulation of monthly ovarian and menstrual cycles. In males, it stimulates the production of sperm.

2. **Luteinizing hormone (LH).** LH also targets the ovaries and testes. In females, it stimulates ovulation, the release of the egg from the follicle. The follicle left behind after ovulation develops into a yellowish structure called the **corpus luteum.**[6] Under the influence of LH, the corpus luteum secretes progesterone, which serves to maintain the lining of the uterus should pregnancy occur. In males, LH stimulates the testes to secrete testosterone.

3. **Thyroid-stimulating hormone (TSH),** also called **thyrotropin.** TSH stimulates the thyroid gland to secrete thyroid hormone, which is necessary for regulation of metabolism.

4. **Adrenocorticotropic hormone (ACTH),** also called **corticotropin.** ACTH stimulates the adrenal cortex, the outer layer of the adrenal gland near the kidney, to secrete hormones called *glucocorticoids.* These are important in glucose, fat, and protein metabolism.

5. **Prolactin**[7] **(PRL).** Pituitary cells that produce PRL increase greatly in size and number during pregnancy. After a woman gives birth, PRL stimulates the mammary glands to secrete milk. In males, PRL makes the testes more sensitive to LH.

6. **Growth hormone (GH),** also called **somatotropin.** The pituitary gland produces at least a thousand times as much GH as any other hormone. In general, GH promotes tissue growth by mobilizing energy from fat, raising levels of calcium and other electrolytes, and stimulating protein synthesis, mitosis, and cellular differentiation. Rather than targeting one or a few organs, GH has widespread effects on the body. They are most conspicuous during childhood and adolescence, but important for tissue maintenance and repair throughout life.

Posterior Pituitary Hormones

The posterior pituitary gland stores and releases two small peptide hormones produced by the hypothalamic nuclei.

1. **Oxytocin**[8] **(OT)** (see fig. 11.3c). The most obvious effect of oxytocin is stimulation of uterine smooth muscle, leading to contractions during labor and delivery. A synthetic form of oxytocin, pitocin, is sometimes administered to women to induce labor. During lactation, OT targets cells in the mammary glands that cause milk to flow to the nipple. In both men and women, oxytocin surges during sexual arousal and orgasm; its

---

[5] *trop* = to turn, change

[6] *corpus* = body; *lute* = yellow

[7] *pro* = favoring; *lact* = milk

[8] *oxy* = sharp, quick; *toc* = birth

release may help propel sperm through the male reproductive tract and assist in transporting sperm up the female reproductive tract. Recent evidence indicates that oxytocin also plays a role in emotional bonding between mothers and offspring, and between sexual partners.

2. **Antidiuretic hormone (ADH), also called vasopressin.** ADH has the same structure as oxytocin except for two of its amino acids. One function of the hypothalamus is to regulate fluid balance in the body. It does this in part by synthesizing ADH, which stimulates the kidneys to retain water and reduce urine output. Alcohol inhibits ADH release, which partially explains the increased urine production when one consumes alcoholic beverages.

Control of the Pituitary Gland

The hypothalamus controls the anterior pituitary gland by secreting several releasing and inhibiting hormones into the portal system. They travel to the anterior pituitary, leave the blood vessels, and either stimulate or suppress the output of the corresponding pituitary hormones. The hypothalamus thereby regulates the numerous body functions that are affected by the anterior pituitary.

The hypothalamic hormones are listed in the violet box in figure 11.6b. *Releasing hormones* stimulate the anterior pituitary to secrete its hormones (pink box) into the blood. For example, *thyrotropin-releasing hormone (TRH)* stimulates the pituitary to secrete *thyrotropin (thyroid-stimulating hormone, TSH)* into the blood. Hypothalamic inhibiting hormones suppress pituitary output—such as *prolactin-inhibiting hormone* suppressing prolactin secretion when a woman is not pregnant or nursing, and *somatostatin* reducing the pituitary secretion of growth hormone.

The hypothalamus controls the posterior pituitary by sending nerve signals down the stalk, commanding it to release ADH when there is a need to reduce urinary water loss, or OT in such conditions as childbirth and breastfeeding.

Figure 11.7 summarizes the "chain of command" from hypothalamus to anterior pituitary and from the pituitary to its target organs. Superficially, this may seem like a military chain of command from hypothalamus ("the general") to pituitary ("the lieutenant") to the target organs ("the privates"). But unlike a military chain of command, the endocrine system does not work strictly from the top down. Here, the privates also give commands to the lieutenants and even to the general, through communication loops of both negative and positive feedback. For example, we have already seen that the hypothalamus secretes TRH to stimulate the pituitary, the pituitary secretes TSH to stimulate the thyroid, and the thyroid responds by secreting thyroid hormone (TH). But TH then feeds back to both the hypothalamus and anterior pituitary to reduce TRH and TSH output, as if saying, "That's enough for now." This keeps TH secretion from becoming excessive. This particular chain of responses is called **negative feedback inhibition** of the pituitary.

By contrast, we saw in chapter 1 that during childbirth, pituitary oxytocin (OT) stimulates labor contractions in the uterus, and stretch receptors in the uterus issue nerve signals back to the hypothalamus to say, "Send more oxytocin!" The hypothalamus commands the posterior pituitary to release more OT. This is a case of **positive feedback control** of the pituitary.

Apply What You Know

Explain the phrase, "The pituitary gland is a link between the neural regulation and hormonal regulation of homeostasis."

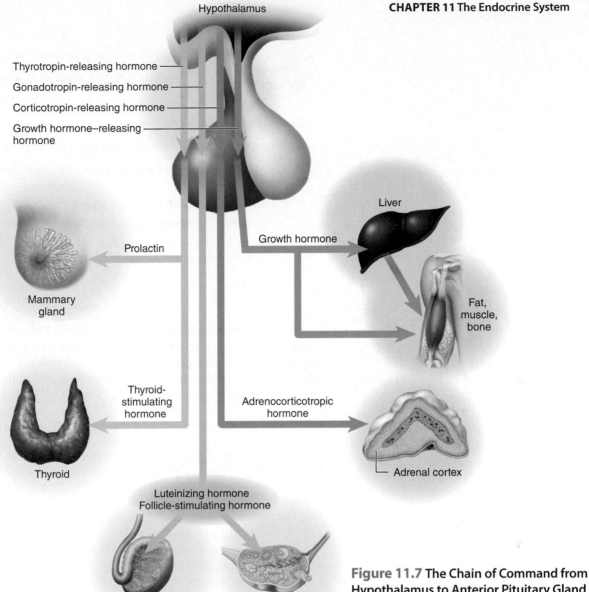

Hypothalamus

Thyrotropin-releasing hormone

Gonadotropin-releasing hormone

Corticotropin-releasing hormone

Growth hormone–releasing hormone

Prolactin

Mammary gland

Thyroid-stimulating hormone

Thyroid

Luteinizing hormone
Follicle-stimulating hormone

Testis

Ovary

Liver

Growth hormone

Fat, muscle, bone

Adrenocorticotropic hormone

Adrenal cortex

Figure 11.7 The Chain of Command from Hypothalamus to Anterior Pituitary Gland to Target Organs. AP|R

Before You Go On

Answer these questions from memory. Reread the preceding section if there are too many you don't know.

7. Explain why the anterior and posterior pituitary glands are considered to be separate glands, despite their physical proximity.

8. Prepare a study aid for yourself by constructing a table with three columns. In the left column, list the six hormones that are synthesized and secreted by the anterior pituitary gland; in the middle column, identify the target organ of each hormone; and in the right column, list the effects of each hormone.

9. How does the hypothalamus control the anterior pituitary gland? How does it control the posterior pituitary gland?

10. List the posterior pituitary gland hormones and explain their functions.

11. Describe the influence of the pituitary target organs over hypothalamic and pituitary function.

11.4 Other Endocrine Glands and Tissues

Expected Learning Outcomes

When you have completed this section, you should be able to:

a. describe the structure and locations of the remaining endocrine glands;

b. name the hormones that the glands produce;

c. name the targets and explain the effects of the hormones; and

d. discuss hormones produced by endocrine cells in tissues and organs other than the classic endocrine glands.

Numerous hormones are produced throughout the body. We will begin our discussion with the most superior gland, the pineal gland, and will work our way inferiorly through glands in the neck and trunk. Finally, we will discuss hormones produced by tissues and organs other than those traditionally regarded as endocrine glands.

The Pineal Gland

The **pineal gland** (PIN-ee-ul) is anatomically associated with the brain. It attaches to the roof of the third ventricle, near the posterior end of the corpus callosum (see fig. 9.2, p. 291, and fig. 11.1). The pineal gland secretes **melatonin,** a monoamine, primarily during hours of darkness. Melatonin plays a role in sleep, in the basic daily cycles of activity known as *circadian rhythms,* and possibly in timing of the onset of puberty.

Intriguing recent studies show that increased melatonin levels are associated with decreased incidence of cancer. This finding has implications for our modern lifestyle because exposure to light suppresses the production of melatonin. This may explain the finding that blind women, whose melatonin production is not suppressed by artificial light, have a lower incidence of breast cancer. In contrast, women who work during the night under artificial light (and hence have lower levels of melatonin) have a greater incidence of breast cancer.

The pineal gland is large in children but begins to atrophy at about the age of 7. By adulthood, it is no more than a fibrous, shrunken mass. As it shrinks, granules of calcium phosphate and calcium carbonate called *pineal sand* appear. These tiny grains are visible on X-rays, and enable radiologists to determine the position of the pineal gland. This can be clinically useful if the gland has been displaced by a brain tumor or other structural abnormality.

The Thyroid Gland AP|R

The **thyroid gland** lies adjacent to the trachea, near the shieldlike *thyroid*[9] *cartilage* of the larynx. It is the largest endocrine gland in adults, weighing 20 to 25 g, and has a dark reddish brown color due to its unusually great blood flow. It is composed of two lobes that are bulbous at their inferior end and taper superiorly. In most people, a narrow bridge of tissue, the *isthmus,* crosses the front of the trachea and joins the two lobes (fig. 11.8a).

The thyroid gland is composed of sacs called *thyroid follicles* (fig. 11.8b). These are lined with simple cuboidal epithelium composed of **follicular cells,**

---

[9] *thyr* = shield; *oid* = like, resembling

Clinical Application 11.2

THYROID DISORDERS

Thyroid disorders result from too little thyroid hormone *(hypothyroidism)* or too much *(hyperthyroidism)*. They occur more frequently in women than in men. The most common cause of hypothyroidism in adults is *Hashimoto disease,* an autoimmune disorder that destroys the thyroid gland. Symptoms include thinning hair and scaly skin, cold intolerance, weight gain, and mental sluggishness. *Graves[10] disease (toxic goiter)* is a form of hyperthyroidism in which dysfunctional antibodies mimic the effect of thyroid-stimulating hormone, causing the thyroid gland to secrete excessive thyroid hormone. The resulting signs and symptoms include rapid weight loss, increased sweating, irritability, heart palpitations, and protruding eyeballs *(exopthalmia)*. Graves disease is sometimes treated with radioactive iodine that kills the thyroid cells, or the gland may be surgically removed. Either treatment requires replacement of thyroid hormone for the rest of one's life.

which synthesize **thyroid hormone (TH).** TH is a double-ringed monoamine made from two molecules of the amino acid tyrosine. About 90% of the daily output of TH is a form of TH called **thyroxine,** also known as T_4 or **tetraiodothyronine** (TET-ra-EYE-oh-doe-THY-ro-neen) because it has four iodine atoms. The other 10% is called T_3 or **triiodothyronine** (try-EYE-oh-doe-THY-ro-neen) because it has only three iodine atoms.

When TH reaches a target cell, T_4 is converted to T_3. T_3 enters the nucleus and binds to receptors associated with the DNA. Its primary effect is to stimulate an increase in metabolic rate. There is a resulting rise in the body's ATP, oxygen, and fuel consumption (and therefore in appetite) and in heat production. TH secretion rises in cold weather to compensate for the body's heat loss. TH also stimulates growth of the bones, skin, and other tissues; increases alertness; and is crucial to development of the fetal nervous system.

In addition, the thyroid gland has small clusters of **C (clear) cells,** or **parafollicular cells,** nestled between the follicles. They produce a peptide hormone called **calcitonin** in response to elevated blood calcium levels.

[10] Robert James Graves (1769–1853), Irish physician

Figure 11.8 The Thyroid Gland. Gross anatomy, anterior view. Major blood vessels are shown only on the anatomical right. (b) Histology. AP|R

- *What is the function of the C cells in part (b)?*

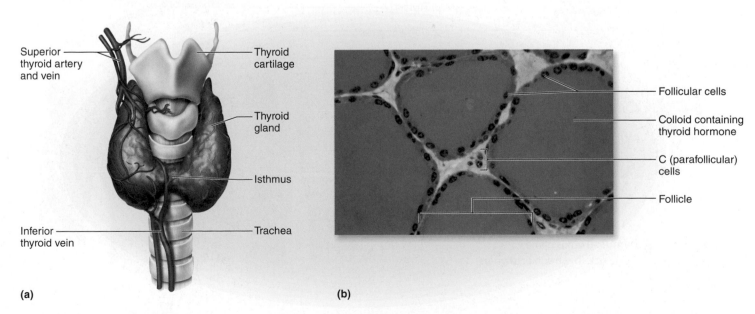

(a)   (b)

Calcitonin inhibits the bone-resorbing activity of osteoclasts and stimulates bone deposition by osteoblasts, thus promoting bone deposition. Calcitonin is significant in children but seems to have no significant effect in adults. No disease results from an excess or deficiency of calcitonin.

The Parathyroid Glands AP|R

Four small, ovoid **parathyroid glands** are usually located on the posterior surface of the thyroid (see fig. 11.1), although their location varies. They secrete **parathyroid hormone (PTH),** which is the dominant calcium-regulating hormone in adults. In response to a blood calcium deficiency, it stimulates osteoclasts and inhibits osteoblasts, thereby dissolving bone and raising the blood calcium level. Note that these effects are opposite from those of calcitonin. PTH also stimulates the kidneys to produce calcitriol, which promotes absorption of calcium in the digestive tract. If the parathyroids are accidentally removed in the course of neck surgery, immediate hormone replacement therapy is necessary; a person can otherwise die within a few days of blood calcium deficiency.

The Thymus

The **thymus** is a bilobed gland in the mediastinum superior to the heart, behind the sternal manubrium. It is the site of maturation for certain white blood cells called T lymphocytes (*T* for *thymus-dependent*), which play a critical role in immunity. The thymus secretes hormones called **thymosin** and **thymopoietin,** which regulate the development and activity of T lymphocytes and stimulate the development of other lymphatic organs. The thymus is discussed more thoroughly in chapter 14 in relation to its immune function.

In the fetus and infant, the thymus is enormous compared to adjacent organs, sometimes extending from near the diaphragm to the neck (fig. 11.9a). It continues to grow until age 5 or 6, but after the age of 14 it shrinks dramatically (fig. 11.9b). By old age, it is a small fibrous and fatty remnant barely distinguishable from surrounding tissue.

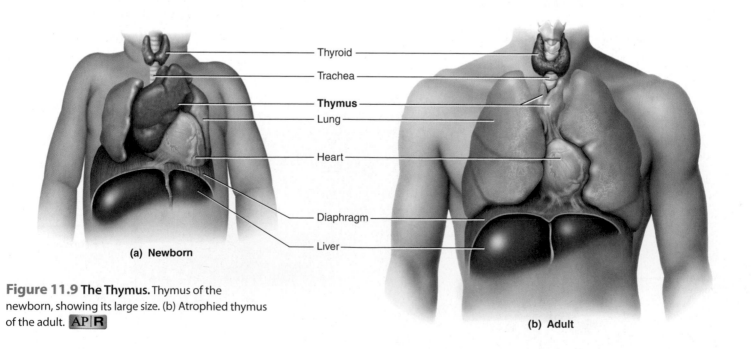

Thyroid
Trachea
Thymus
Lung
Heart
Diaphragm
Liver

(a) **Newborn**

(b) **Adult**

Figure 11.9 The Thymus. Thymus of the newborn, showing its large size. (b) Atrophied thymus of the adult. AP|R

The Pancreas AP|R

The **pancreas** is an elongated, spongy organ located below and behind the stomach (fig. 11.10a). It is approximately 15 cm long and 2.5 cm wide. One end of it nestles in the curve of the first part of the small intestine, the duodenum. Most pancreatic tissue functions as an exocrine digestive gland, but scattered throughout the exocrine tissue are 1 to 2 million little endocrine cell clusters called **pancreatic islets (islets of Langerhans[11])** (fig. 11.10b, c).

Insulin and **glucagon** are the most important pancreatic hormones. They regulate the blood concentration of glucose, a vital fuel for most of the body's tissues. Neurons especially depend on glucose and quickly die if it is not available.

Insulin is produced by **beta (β) cells** of the islets in response to a rise in blood glucose during and after a meal. Insulin promotes the uptake of glucose by liver, muscle, and fat cells, among others, and promotes glycogen synthesis *(glycogenesis[12])* and fat storage (fig. 11.11a). These processes lower the blood glucose concentration.

Glucagon is produced by **alpha (α) cells** of the islets in response to falling levels of blood glucose. In contrast to insulin, glucagon promotes the mobilization of fuels. Its primary action is on the liver, where it stimulates the breakdown of glycogen *(glycogenolysis[13])* and the synthesis of glucose from proteins (or their amino acids) *(gluconeogenesis[14])* (fig. 11.11b). These processes release glucose into the blood, raising its concentration and making it available to other tissues.

---

[11] Paul Langerhans (1847–88), German anatomist

[12] *glycogen* + *genesis* = creation, formation of

[13] *glycogen* + *lysis* = splitting, breaking apart

[14] *gluco* = glucose, sugar; *neo* = new; *genesis* = creation, formation of

Figure 11.10 The Pancreas. (a) Gross anatomy of the pancreas and its relationship to the duodenum. (b) Cell types of a pancreatic islet. (c) Light micrograph of a pancreatic islet amid the darker exocrine acini. AP|R

- *What is the function of the exocrine cells in this gland?*

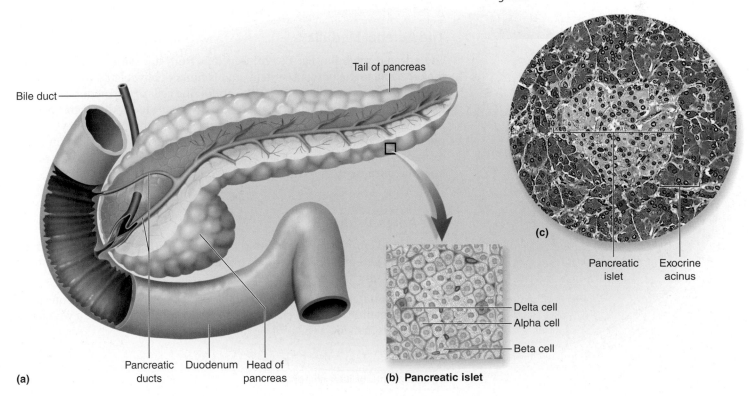

Tail of pancreas

Bile duct

Pancreatic ducts Duodenum Head of pancreas

(a)

(b) Pancreatic islet

Delta cell
Alpha cell
Beta cell

(c)

Pancreatic islet   Exocrine acinus

Figure 11.11 Effects of Insulin and Glucagon.
(a) Insulin promotes the uptake of blood glucose by cells and its polymerization to make glycogen, an energy-storage carbohydrate. In doing so, it lowers the blood sugar level. (b) Glucagon stimulates some cells, especially in the liver, to break down glycogen and release glucose to the bloodstream, thereby raising blood sugar level. Glycogenesis is the synthesis of glycogen; glycogenolysis is its hydrolysis (breakdown to glucose); and gluconeogenesis is the synthesis of glucose from noncarbohydrates, especially fats and proteins.

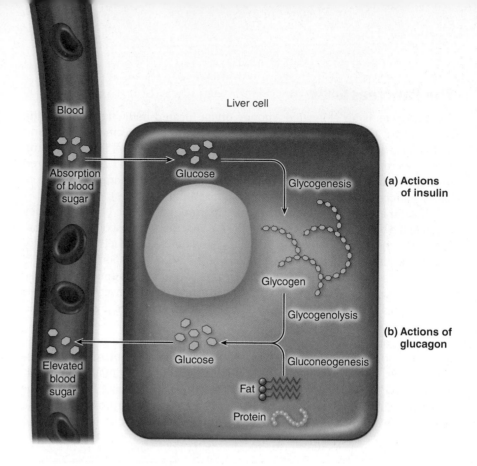

The Adrenal Glands AP|R

Each kidney is topped by a hatlike **adrenal gland** (fig. 11.12a). These are relatively large in the fetus and infant but shrink about 50% by the age of 2 years. The adrenal gland has two distinct portions that differ in embryonic origin and the hormones produced—an inner *medulla* surrounded by an outer *cortex*.

The Adrenal Medulla

The **adrenal medulla** (fig. 11.12a) develops from the same embryonic cells that give rise to the nervous system, and is functionally related to the sympathetic nervous system. Essentially, the endocrine cells of the medulla are sympathetic postganglionic neurons (see p. 315) that are modified to have no axons or dendrites but secrete their products into the blood. They secrete mainly **epinephrine** and **norepinephrine,** two monoamines involved in stress responses. They are discussed later under stress physiology. As you may recall, norepinephrine is also a sympathetic neurotransmitter.

The Adrenal Cortex

The **adrenal cortex** produces steroid hormones *(corticosteroids),* all of which are synthesized from cholesterol. It has three tissue layers that differ in their hormone output (fig. 11.12b).

- The **zona glomerulosa**[15] (glo-MER-you-LO-suh) is the thin, outermost layer of the cortex. Its cells are arranged in rounded clusters for which this zone is named.

---

[15] *glomerul* = little ball; *osa* = full of

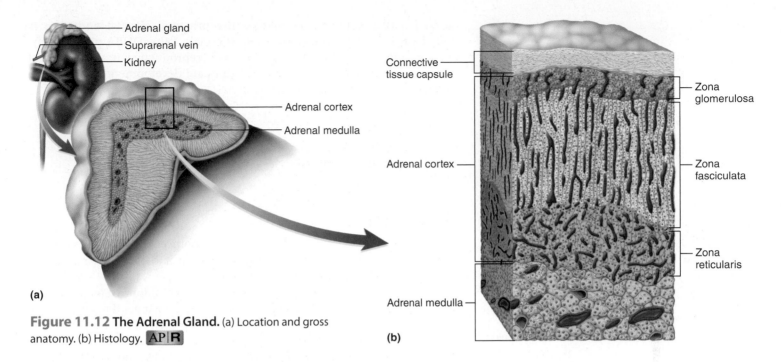

Figure 11.12 The Adrenal Gland. (a) Location and gross anatomy. (b) Histology. **AP|R**

(a)

(b)

- The **zona fasciculata**[16] (fah-SIC-you-LAH-ta) is the middle and thickest layer of the cortex. Its cells are arranged in parallel cords (fascicles) perpendicular to the gland surface. They have a foamy look stemming from their high lipid content.
- The **zona reticularis**[17] (reh-TIC-you-LAR-iss), is a narrow, innermost layer whose cells form a branching network (reticulum).

One of the corticosteroids is **aldosterone,** produced only in the zona glomerulosa. Nicknamed "the salt-retaining hormone," aldosterone targets kidney cells and stimulates them to return sodium ions to the blood instead of allowing them to be lost in the urine. Water follows the sodium by osmosis, so aldosterone also slows down the body's rate of urinary water loss and helps to maintain blood volume and pressure. Aldosterone and some similar, minor hormones are collectively called *mineralocorticoids.*

Another corticosteroid is **cortisol,** produced in the zona fasciculata and zona reticularis. It is secreted in response to adrenocorticotropic hormone (ACTH) from the anterior pituitary gland. Its most noticeable effects are on adipose tissue, muscle, and the liver. It raises the levels of blood glucose and other fuels, providing the body with the energy to respond to stress. Cortisol stimulates muscles to break down their own proteins and release the amino acids into circulation; it stimulates the liver to synthesize glucose from these free amino acids; and it stimulates adipose tissue to break down fats, releasing fatty acids into circulation as a supplemental fuel. These effects are important to the body's ability to adapt to stress and will be discussed later under stress physiology. Cortisol and similar, lesser hormones are called *glucocorticoids.*

The third group of corticosteroids are sex hormones—**androgens** and smaller amounts of **estrogens**—produced in both the zona fasciculata and zona reticularis. Androgens stimulate many aspects of male development and reproductive physiology. In men, the testes secrete most of the male hormones, so their adrenal androgens are of relatively minor importance. Androgens are also important in female physiology, however, and the

---

[16] *fascicul* = little bundle; *ata* = possessing

[17] *reticul* = little network; *aris* = pertaining to

adrenal cortex is their sole source. The principal adrenal androgen is **dehydroepiandrosterone (DHEA).** In the male fetus, DHEA plays an important role in development of the male reproductive tract. At sexual maturity, DHEA acts in both sexes to stimulate development of pubic and axillary hair and apocrine sweat glands.

> **Apply What You Know**
>
> The zona fasciculata grows significantly thicker in pregnant women. What do you think could be the benefit of this?

The Gonads

The **gonads** (ovaries and testes) function as both endocrine and exocrine glands. Their exocrine products are eggs and sperm, and their endocrine products are gonadal hormones, mostly steroids.

The ovaries secrete **estrogen, progesterone,** and **inhibin.** In females of reproductive age, the ovaries undergo monthly cyclical changes that involve hormonal production. In the first part of the ovarian cycle, a bubblelike follicle develops, consisting of the egg (oocyte) surrounded by *follicular cells* that pile atop each other to form layers (fig. 11.13a). Follicular cells produce estrogen. In the middle of the cycle, the egg is released by ovulation. The remains of the follicle become the corpus luteum, a structure that secretes **progesterone.**

Estradiol and progesterone regulate the ovarian and menstrual cycles, prepare for and maintain pregnancy, and contribute to the development of the reproductive system. Inhibin is also secreted by the follicle and corpus luteum. It is an inhibitory signal to the anterior pituitary gland that regulates the secretion of follicle-stimulating hormone (FSH). The ovarian and menstrual cycles and associated hormones are discussed in more detail in chapter 19.

The **testes** secrete **testosterone** and inhibin. Most of the testis is composed of microscopic *seminiferous tubules* that produce sperm. The wall of the tubules is formed of **sustentacular cells,** the source of inhibin. By limiting FSH secretion, inhibin regulates the rate of sperm production. Nestled between the tubules are clusters of **interstitial cells (cells of Leydig[18]),** the source of testosterone (fig. 11.13b). Testosterone sustains sperm production and the sex drive throughout adolescent and adult life. It stimulates development of the male reproductive system in the fetus, and the development of the masculine physique in adolescence.

Endocrine Functions of Other Organs and Tissues

In recent decades, physiologists have discovered several hormones secreted by organs and tissues that have not traditionally been thought of as endocrine glands.

- **The heart.** Rising blood pressure stretches the heart wall and stimulates muscle cells in the atria to secrete **natriuretic[19] peptides.** These hormones increase urine output and sodium excretion and oppose the actions of angiotensin II, described shortly. The effect of natriuretic peptides is to lower blood pressure.

- **The liver.** The liver is involved in the production of at least five hormones: (1) It collaborates with the skin and kidneys to produce **calcitriol (vitamin D).** (2) It secretes a blood protein called *angiotensinogen,* which

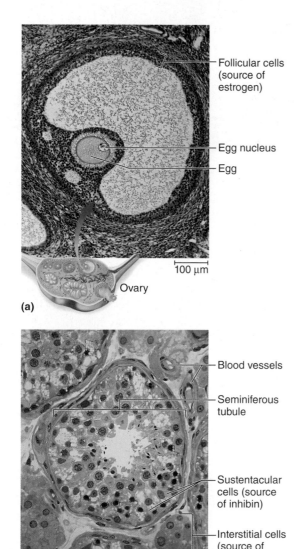

Follicular cells (source of estrogen)

Egg nucleus

Egg

100 μm

Ovary

(a)

Blood vessels

Seminiferous tubule

Sustentacular cells (source of inhibin)

Interstitial cells (source of testosterone)

50 μm

Testis

(b)

Figure 11.13 The Gonads. (a) Histology of an ovarian follicle. (b) Histology of the testis. AP|R

[18] Franz von Leydig (1821–1908), German histologist

[19] *natri* = sodium; *uretic* = pertaining to urine

the kidneys, lungs, and other organs convert to the hormone **angiotensin II,** a regulator of blood pressure. (3) It secretes **erythropoietin (EPO)** (er-RITH-ro-POY-eh-tin), a hormone that stimulates red blood cell production. (4) It secretes **hepcidin,** a regulator of iron homeostasis. (5) It secretes **insulin-like growth factor I (IGF-I),** a hormone that aids the action of growth hormone.

- **The kidneys.** The kidneys, like the liver, produce erythropoietin and perform certain steps in both calcitriol and angiotensin II synthesis. Angiotensin II has numerous effects. It constricts blood vessels and increases cardiac output and blood pressure. It also stimulates the adrenal cortex to secrete aldosterone, discussed earlier. In multiple ways, angiotensin II therefore elevates the blood pressure.

- **The stomach and intestines.** These organs secrete **enteric**[20] **hormones** that coordinate different regions and glands of the digestive system with each other. For example, the stomach secretes **gastrin** when food arrives. Gastrin stimulates the production of hydrochloric acid. Other hormones regulate feelings of hunger and satiety, and are the focus of much interest because of the prevalence of obesity. For example, **ghrelin,** a "hunger hormone," sends signals to the hypothalamus to stimulate the appetite. In contrast, **peptide YY (PYY),** secreted by the small and large intestines, creates a feeling of satiety and has an appetite-suppressing effect.

- **Adipose tissue**. Fat cells secrete the protein hormone **leptin,** which has long-term effects on appetite-regulating centers of the hypothalamus. A low level of leptin, signifying a low level of body fat, increases appetite and food intake, whereas a high level of leptin tends to blunt the appetite. Early-onset obesity may sometimes be related to defective leptin receptors. Leptin also serves as a signal for the onset of puberty, which is delayed in persons with abnormally low levels of body fat.

- **Osseous tissue.** The bones secrete **osteocalcin,** a hormone that stimulates the pancreatic islets and adipose tissue, increasing the secretion and effectiveness of insulin and reducing fat deposition.

Before You Go On

Answer these questions from memory. Reread the preceding section if there are too many you don't know.

12. Make a chart that lists the glands from the preceding section, the hormones they secrete, the target tissues, and the actions of the hormones.

13. What hormone increases the body's heat production in cold weather? What other functions does this hormone have?

14. What hormones are produced by the adrenal cortex and what are their functions?

15. Name one hormone produced by each of the following organs—the heart, kidney, stomach, and ovary—and state the function of each hormone.

---

[20] *entero* = intestine

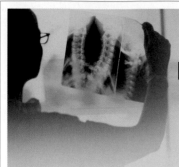

PERSPECTIVES ON HEALTH

Medical problems with the endocrine system center mostly on three dysfunctions: hyposecretion (hormone deficiency), hypersecretion (hormone excess), and hormone inaction, in which a normal amount of hormone is secreted but the body does not respond normally to it. Hypo- and hypersecretion disorders have specific names that often refer to the specific endocrine gland, such as *hypo-* and *hyperthyroidism.*

One example of a hyposecretion disorder is *diabetes insipidus,* in which the posterior pituitary puts out little or no antidiuretic hormone. The kidneys, as a result, fail to conserve water; one therefore produces enormous quantities of urine, although the urine is glucose-free (unlike the sugary urine in diabetes mellitus). Hormone hyposecretion is often treated with *hormone replacement therapy.* An example of a hypersecretion disorder is *pheochromocytoma*[21] (FEE-oh-CRO-mo-sy-TO-muh), a tumor of the adrenal medulla that secretes up to 27 times the normal amounts of epinephrine and norepinephrine. This often causes drenching sweats, crises of severe headaches and hypertension, and sudden death from stroke or cardiac arrhythmia, often before the disease is diagnosed. An example of target-cell insensitivity is *androgen-insensitivity syndrome (AIS),* in which the body lacks testosterone receptors. Even though testosterone is present, it has no effect. In males with AIS, the genitalia show female anatomy, and at puberty, the small amount of estrogen produced by the testes causes males to develop enlarged breasts and other feminine features.

Diabetes Mellitus

Diabetes mellitus (DM) is the world's most common metabolic disease and occurs in forms that exemplify both hyposecretion and hormone inaction. It is the leading cause of adult blindness, renal failure, gangrene, and the necessity for limb amputations.

Classic signs and symptoms of DM are "the three polys": *polyuria*[22] (excessive urine output), *polydipsia*[23] (intense thirst), and *polyphagia*[24] (ravenous hunger). These are common patient complaints when first presenting to a physician, and diabetes can be confirmed by three clinical findings in blood and urine tests: *hyperglycemia*[25] (elevated blood glucose), *glycosuria*[26] (glucose in the urine), and *ketonuria* (ketones in the urine).

There are two types of diabetes mellitus: type 1 and type 2. *Type 1 diabetes,* formerly called *insulin-dependent* or *juvenile-onset diabetes,* accounts for about 10% of cases. It is an autoimmune condition that results from destruction of pancreatic beta cells by one's own antibodies. Type 1 diabetes is most often diagnosed around the age of 12 but it can also appear later in life. Treatment includes meal planning, exercise, self-monitoring of blood glucose levels, and periodic injections of insulin or delivery of insulin by a pump worn on the body.

Type 2 diabetes mellitus, formerly called *non-insulin-dependent* or *adult-onset diabetes,* accounts for 90% of cases, including nearly 7% of the U.S. population. Type 2 diabetics may have normal or even elevated insulin levels; the problem is *insulin resistance*—a failure of target cells to respond to it. The three major risk factors for type 2 diabetes are heredity, age, and obesity. It tends to run in families, and obesity and physical inactivity increase insulin resistance. Lifestyle changes over recent decades mean that many of us eat too much and do not exercise enough. The incidence of type 2 diabetes has therefore increased, and more and more people are developing it at a young age. The relationship between obesity and diabetes mellitus is complex, but one factor is that adipocytes produce a hormonelike secretion that indirectly interferes with glucose transport into most kinds of cells. Type 2 diabetes can often be managed with a combination of weight loss, careful attention to diet, and exercise. Some patients receive medications that improve insulin secretion or target-cell sensitivity, and some take insulin injections to supplement insulin-sensitizing drugs.

When cells cannot absorb glucose, they rely on fat and protein for energy. Fat and protein breakdown results in muscular atrophy and weakness. Rapid fat breakdown elevates blood concentrations of free fatty acids and their breakdown products, ketones. As acids, ketones lower the blood pH and produce a condition called *ketoacidosis.* Ketoacidosis depresses the nervous system, produces a deep gasping breathing typical of terminal diabetes, and leads to diabetic coma and death.

In the long term, diabetes mellitus can lead to degenerative neurological and cardiovascular diseases, including *diabetic neuropathy* (nerve damage) and *atherosclerosis* (blockage of blood vessels with lipid deposits). Diabetic neuropathy can make a patient dangerously unaware of minor injuries, which can thus fester from neglect and contribute to gangrene and the necessity of amputation. Many diabetics lose their toes, feet, or legs to the disease. Atherosclerosis leads to poor circulation, which contributes to gangrene, kidney failure, and blindness.

---

[21]*pheo* = dusky; *chromo* = colored; *cyt* = cells; *oma* = tumor

[22] *poly* = much, excessive; *uri* = urine

[23] *poly* = much, excessive; *dipsia* = drinking

[24] *poly* = much, excessive; *phagia* = eating

[25] *hyper* = excess; *glyc* = sugar, glucose; *emia* = blood condition

[26] *glyco* = glucose, sugar; *uria* = urine condition

11.5 Stress Physiology

Expected Learning Outcomes

When you have completed this section, you should be able to:

a. define *stress* in physiological terms;

b. identify the principal hormonal responses to stress and explain their roles in adapting to it.

Stress is defined as any situation that upsets homeostasis and threatens one's physical or emotional well-being. Physical causes of stress *(stressors)* include injury, surgery, hemorrhage, infection, intense exercise, pain, and malnutrition, among others. Emotional causes include anger, grief, depression, anxiety, and guilt. Whatever the cause, we react to stress in a fairly consistent way called the **stress response,** mediated by the endocrine system and sympathetic nervous system. Three stages of stress are recognized, each with its own characteristic hormonal responses.

1. The **alarm reaction** is mediated mainly by norepinephrine from the sympathetic nerves and epinephrine from the adrenal medulla. These and other hormones raise the heart rate, blood pressure, and blood glucose level; promote pulmonary ventilation; and prepare the body for action such as fighting or escaping danger. In most cases, we recover from stress during this stage. If a stressful situation persists, however, the increased consumption of liver and muscle glycogen can push one to the next stage.

2. The **stage of resistance** develops as stored glycogen is depleted but the demand for glucose still must be met. ACTH and cortisol are secreted to mobilize alternative fuels—breaking down stored fat and protein into fatty acids, glycerol, and amino acids. The liver utilizes glycerol and amino acids to synthesize more glucose. Fatty acids and glucose are oxidized to make ATP. Sometimes to our misfortune, long-term cortisol increases inhibit the immune system. Chronic stress therefore increases one's susceptibility to infections, ulcers, and even cancer.

3. The **stage of exhaustion** sets in as even one's reserve alternative fuels are depleted. When fat stores are gone, the body begins breaking down more and more protein to meet its energy needs. Thus there is a progressive wasting away of muscles and weakening of the body. Elevated aldosterone secretion can cause excessive salt and water retention, hypertension, and depletion of potassium from the body. Death can result in this stage from heart failure, kidney failure, or overwhelming infection.

Before You Go On

Answer these questions from memory. Reread the preceding section if there are too many you don't know.

16. *Identify a variety of physical and emotional causes of stress.*

17. *Identify the dominant hormones of each stage of stress, and explain how the pathological progression of stress relates to the use or depletion of various body fuels.*

Aging of the Endocrine System

The endocrine system shows less degenerative change in old age than any other organ system. Although the reproductive hormones drop sharply in old age and growth hormone and thyroid hormone secretion decline steadily after adolescence, other hormones continue to be secreted at fairly stable levels even into old age. Target-cell sensitivity declines, however, so some hormones have less effect. For example, the pituitary gland is less sensitive to negative feedback inhibition by adrenal glucocorticoids; consequently, ACTH secretion and the response to stress are more prolonged than usual. Diabetes mellitus is more common in old age, largely because target cells have fewer insulin receptors. In part, this is an effect of increased adiposity in old age. The more fat there is at any age, the less sensitive other cells are to insulin. In addition, most people lose muscle mass in old age, and muscle is the body's most significant glucose-buffering tissue.

CAREER SPOTLIGHT

Diabetes Educator

Diabetes mellitus is the world's most widespread metabolic disease, and people with diabetes often express frustration and dissatisfaction with the care they receive. Patients who cannot adequately manage their diabetes suffer needless disability and premature death. A diabetes educator is a professional who assists, educates, and provides emotional support for diabetes patients and their caregivers. Diabetes educators teach techniques of blood glucose monitoring, insulin injection, and the care and use of insulin pumps. They counsel patients on diet, exercise, taking their medication regularly, coping emotionally with the disease, and other skills—all aimed at behavioral changes that will produce better health outcomes. They work in hospitals, physicians' offices, outpatient clinics, pharmacies, home health-care organizations, and other settings. They are usually able to spend more time with a patient than the physician can, and to provide more emotional support.

A candidate for diabetes educator must have very good communication skills; a sound knowledge of biological, clinical, and social sciences; and an ability to relate comfortably, patiently, and effectively with people of all kinds. Many diabetes educators are registered nurses who specialize and become diabetes nurse educators (DNEs). Diabetes educators, however, also come from a wide variety of other backgrounds and include dietitians, pharmacists, social workers, exercise physiologists, and mental health professionals. With experience and self-directed online coursework, one may become a Certified Diabetes Educator (CDE) or sit for board exams to become board-certified in advanced diabetes management (BC-ADM). See appendix B for further information on becoming a diabetes educator.

CONNECTIVE ISSUES

Ways in Which the Endocrine System Affects Other Organ Systems

All Systems
Growth hormone, insulin, insulin-like growth factors, thyroid hormone, and glucocorticoids affect the development and metabolism of most tissues.

Integumentary System
Sex hormones affect skin pigmentation, development of body hair and apocrine glands, and subcutaneous fat deposition.

Skeletal System
Many hormones affect bone development and homeostasis, including growth hormone, parathyroid hormone, calcitonin, calcitriol, estrogen, and progesterone.

Muscular System
Growth hormone and testosterone stimulate muscular growth; insulin regulates carbohydrate metabolism in muscle; other hormones affect electrolyte balance, which is critical to muscular function.

Nervous System
The endocrine system exerts negative feedback inhibition on the hypothalamus; several hormones affect nervous system development, mood, and behavior; other hormones affect electrolyte balance, which is critical to neuron function.

Circulatory System
Angiotensin II, aldosterone, ADH, and natriuretic peptides regulate blood volume and pressure; EPO regulates production of red blood cells; epinephrine, thyroid hormone, and other hormones affect heart rate and contraction force.

Lymphatic and Immune Systems
Hormones stimulate lymphoid organs and thereby influence production of immune cells; regulate immune responses and thus influence disease susceptibility and severity.

Respiratory System
Epinephrine and norepinephrine increase pulmonary airflow.

Urinary System
ADH regulates water excretion; PTH, calcitriol, and aldosterone regulate electrolyte excretion.

Digestive System
Insulin and glucagon regulate nutrient storage and metabolism; enteric hormones regulate gastrointestinal secretion and motility.

Reproductive System
Gonadotropins, sex steroids, and other hormones regulate sexual development, sperm and egg production, sex drive, the menstrual cycle, pregnancy, fetal development, and lactation.

387

Study Guide

Assess Your Learning Outcomes

To test your knowledge, discuss the following topics with a study partner or in writing, ideally from memory.

11.1 Overview of the Endocrine System (p. 364)

1. The meaning of *hormone* and the components of the endocrine system
2. The reason why only some organs respond to a hormone even though it travels throughout the body
3. The distinction between exocrine glands and endocrine glands
4. Functional similarities and differences between the endocrine system and nervous system and examples of how the systems overlap

11.2 Endocrine Physiology (p. 367)

1. The three chemical classes of hormones, with examples of each
2. Why the relationship between a hormone and receptor is described as a "lock-and-key" mechanism
3. How hydrophilic and hydrophobic hormones differ in their actions at a target cell
4. Three ways in which hormones influence each other's action when two or more hormones act on the same target organ
5. How hormone secretion is regulated by negative and positive feedback loops
6. How a hormone's actions are halted once its purpose has been served

11.3 The Hypothalamus and Pituitary Gland (p. 370)

1. The structural relationship and embryonic origins of the two lobes of the pituitary gland

2. How the hypothalamus is functionally connected to each of the lobes
3. Hormones of the anterior pituitary and their target organs and functions
4. Hormones of the posterior pituitary and their target organs and functions
5. How the hypothalamus controls secretion by each of the pituitary lobes
6. How positive and negative feedback from target organs regulate pituitary function

11.4 Other Endocrine Glands and Tissues (p. 376)

1. The location of the pineal gland and the action of melatonin
2. The location and histology of the thyroid gland
3. The chemical nature of thyroid hormone and the differences between T_3 and T_4
4. The effects of thyroid hormone
5. The source and effects of calcitonin
6. The location of the parathyroid glands, their hormone, and its function
7. The location and endocrine function of the thymus
8. Why the pancreas is considered to be both an exocrine and endocrine gland
9. The cells that produce insulin and glucagon and the functions of these hormones
10. The location and two main components of the adrenal glands
11. Hormones of the adrenal medulla and their effects
12. Tissue layers of the adrenal cortex
13. Hormones of the adrenal cortex, their functions, and the tissue layers that produce each one
14. Hormones of the female and male gonads, their cellular sources, and their functions
15. Names and functions of the hormones produced by the heart, liver, kidneys, stomach, intestines, adipose tissue, and bones

11.5 Stress Physiology (p. 385)

1. The physiological definition of stress
2. Physical and emotional causes of stress
3. Three stages of the stress response, the major controlling hormones of each, and the events that mark transitions from one stage to the next

Testing Your Recall

1. Which hormone relies on a second messenger to cause a cellular response?
   a. testosterone
   b. glucagon
   c. progesterone
   d. aldosterone
   e. cortisol

2. Which of the following does not influence blood pressure?
   a. aldosterone
   b. natriuretic peptide
   c. renin
   d. cortisol
   e. angiotensin II

3. _____ leads to increased osteoclast activity and elevates blood calcium concentration.
   a. Parathyroid hormone
   b. Calcitonin
   c. Thyroxine
   d. Aldosterone
   e. ACTH

4. Where are receptors for insulin located?
   a. in the pancreatic beta cells
   b. in the blood plasma
   c. on the target-cell membrane
   d. in the target-cell cytoplasm
   e. in the target-cell nucleus

5. Which hormone is *not* released in response to stress?
   a. ACTH
   b. cortisol
   c. leptin
   d. epinephrine
   e. norepinephrine

6. Which hormone would be released if a person was dehydrated?
   a. glucagon
   b. ADH
   c. ACTH
   d. insulin
   e. natriuretic peptide

7. Which gland is derived from an embryonic pouch in the roof of the pharynx?
   a. pineal gland
   b. posterior pituitary gland
   c. thyroid gland
   d. parathyroid gland
   e. anterior pituitary gland

8. Which hormone is secreted by alpha cells?
   a. calcitonin
   b. glucagon
   c. insulin
   d. growth hormone
   e. epinephrine

9. Which hormone is *not* synthesized and secreted by the anterior pituitary gland?
   a. TSH
   b. ACTH
   c. prolactin
   d. FSH
   e. oxytocin

10. What endocrine disorder is characterized by polyuria, polyphagia, and polydipsia?
    a. hypothyroidism
    b. Graves disease
    c. diabetes insipidus
    d. diabetes mellitus
    e. Cushing syndrome

11. In response to stress, _____ stimulates glucose synthesis in liver cells and the release of fatty acids from adipose tissue.

12. Antidiuretic hormone (ADH) is synthesized in the _____ and released by the _____.

13. The hypothalamus can monitor the body's fat content by the blood concentration of _____, a hormone secreted by adipocytes.

14. The stomach secretes _____, a hormone that stimulates appetite.

15. The hypophyseal portal system is a means for the brain to communicate with the _____.

16. In males, testosterone is mainly secreted by the _____ cells of the testes.

17. Follicle-stimulating hormone (FSH) targets the _____ and _____.

18. The _____ gland secretes melatonin.

19. The _____ produces hormones that stimulate T lymphocyte maturation.

20. Muscular contractions of the uterus during labor and delivery are stimulated by _____.

Answers in appendix A

True or False

Determine which five of the following statements are false, and briefly explain why.

1. Insulin sensitivity declines when the percentage of body fat rises.

2. All hormones are secreted by endocrine glands.

3. If fatty plaques of atherosclerosis blocked the arteries of the hypophyseal portal system, the ovaries and testes would probably malfunction.

4. The pineal gland and thymus become larger as one ages.

5. The adrenal cortex secretes monoamines.

6. Unlike neurotransmitters, hormones cannot be selectively delivered to just one target organ.

7. The second stage of the stress response is dominated by the hormone cortisol.

8. Oxytocin and antidiuretic hormone (ADH) are secreted through a duct leading to the posterior pituitary.

9. Thyroxine decreases metabolic rate and reduces heat production.

10. A steroid hormone can freely enter a target cell, but a peptide hormone cannot.

Answers in appendix A

Testing Your Comprehension

1. Suppose you were browsing in a health-food store and saw a product that advertised: "Put an end to heart disease. This herbal medicine will totally rid your body of cholesterol!" Would you buy it? Why or why not? If the product were as effective as claimed, what are some other effects that it would produce?

2. Tom Harley is a motorcycle enthusiast who goes out riding every weekend. One day he rear-ends a car that stops unexpectedly in front of him. Tom falls hard on his face and badly fractures his nose and sphenoid bone. He recovers, but ever since then he has had polyuria—he produces enormous amounts of urine each day and he is continually dehydrated and intensely thirsty. In view of your knowledge of the pituitary gland, explain how his injury could have led to this condition.

3. We saw in this chapter that a given enzyme fits only one specific receptor, like a lock that accepts only one key. We also saw in chapter 2 that enzyme–substrate interactions are like a lock and key. (a) Discuss the similarities between hormone–receptor and enzyme–substrate interactions, and (b) identify the most important difference between them. One of these systems requires only very tiny amounts of either hormone or substrate, whereas the other uses a much greater quantity. (c) Identify which is which, and (d) explain the reason for this difference in view of your answer to part (b).

Chapter

12

The Circulatory System I

BLOOD

The formed elements of blood. This shows lymphocytes (blue), erythrocytes (red), and platelets (violet) in a mesh of fibrin, a sticky blood-clotting protein.

Chapter Outline

BASE CAMP

Before ascending to the next level, be sure you're properly equipped with a knowledge of these concepts from earlier chapters:

- Osmosis (p. 74)
- Blood as a connective tissue (p. 115)
- Red bone marrow (p. 153)

Anatomy & Physiology | REVEALED®
aprevealed.com

Module 9: Circulatory System

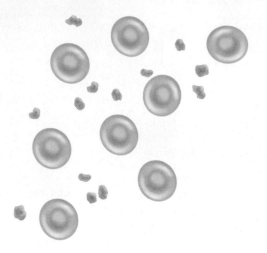

Blood has always had a special mystique. Since ancient times, people have seen blood flow from the body and with it, the life of the individual. It is no wonder that blood was seen as the source of a mysterious "vital force." It was long thought that hereditary and even moral traits were transmitted through the blood, and people still use such unfounded or metaphorical expressions as "I have one-quarter Cherokee blood."

Even though it is an especially accessible tissue, little was known about the specific functions of blood before the twentieth century. From ancient Egypt to the 1800s, physicians believed they could treat almost any malady, from pneumonia to menstrual cramps, by draining "bad blood" from their patients. Indeed, bloodletting was a likely factor in the death of George Washington, who had a probable streptococcus infection and insisted that his doctors bleed him.

Over the past several decades, technology has ushered in a new era for **hematology**,[1] the study of blood. Diseases such as cancer and AIDS have given added impetus to hematology research, and recent discoveries have saved and improved the lives of countless people who otherwise would have suffered and died.

12.1 Introduction

Expected Learning Outcomes
When you have completed this section, you should be able to:

a. identify the major components of the circulatory system;
b. explain the functions of the circulatory system;
c. describe the components and the physical properties of blood; and
d. summarize the composition of blood plasma.

Functions of the Circulatory System AP|R

The **circulatory system** consists of the heart, blood vessels, and blood. The term **cardiovascular system**[2] refers only to the heart and blood vessels, which are discussed in chapter 13.

Single-celled organisms and even the simplest multicellular animals can meet their needs by diffusion of substances from the environment through the plasma membranes of their cells. However, as animals evolved larger bodies, some cells were too far from the surface to have their needs met so simply. Larger animals developed a system of internal passages—blood vessels—to distribute the necessities of life to all cells of the body. A pump, the heart, was

---

[1] *hem, hemato* = blood; *ology* = study of
[2] *cardio* = heart; *vas* = vessel

necessary to move materials along these vessels. And of course there had to be blood itself, the liquid medium of transport for those materials. But the functions of blood are more diverse than that. In humans, they are as follows.

Transport

- The blood carries oxygen from the lungs to all of the body's tissues, while it picks up carbon dioxide from those tissues and carries it to the lungs to be removed from the body.
- It picks up nutrients from the digestive tract and distributes them to all tissues.
- It carries metabolic wastes to the kidneys for removal.
- It carries hormones from endocrine cells to their target cells.

Protection

- White blood cells destroy harmful microorganisms and cancer cells.
- Antibodies and other blood proteins neutralize toxins and help to destroy pathogens.
- Blood plays several roles in inflammation, a mechanism for limiting the spread of infection and promoting the repair of injured tissue.
- Platelets and other blood-borne agents initiate blood clotting and other processes for minimizing blood loss.

Regulation

- Blood helps maintain optimal fluid balance and distribution in the body by absorbing or giving off fluid under different conditions.
- Blood proteins stabilize the pH of extracellular fluids by buffering acids and bases.
- Shifts in blood flow help to regulate body temperature by routing blood to the skin for heat loss or routing blood deeper to retain heat in the body.

In sum, the blood plays numerous roles in maintaining homeostasis. Considering its vital functions, it is easy to understand why an excessive loss of blood is fatal, and why the circulatory system evolved mechanisms for minimizing such loss.

Components and General Properties of Blood

Table 12.1 lists several properties of blood. It is a liquid connective tissue composed of plasma and formed elements. **Plasma** is a clear, light-yellow fluid that

Table 12.1 General Properties of Blood

Characteristic	Typical Values for Healthy Adults*
Mean fraction of body weight	8%
Volume in adult body	Female: 4–5 L; male: 5–6 L
Mean temperature	38°C (100.4°F)
pH	7.35–7.45
Hematocrit (packed cell volume)	Female: 37% to 48%
	Male: 45% to 52%
Hemoglobin	Female: 12–16 g/dL
	Male: 13–18 g/dL
Mean RBC count	Female: 4.2–5.4 million/mL
	Male: 4.6–6.2 million/mL
Platelet count	130,000–360,000/L
Total WBC count	5,000–10,000/L

*Values vary slightly depending on the testing methods used.

Figure 12.1 The Formed Elements of Blood.

- *What do erythrocytes and platelets lack that the other formed elements have?*

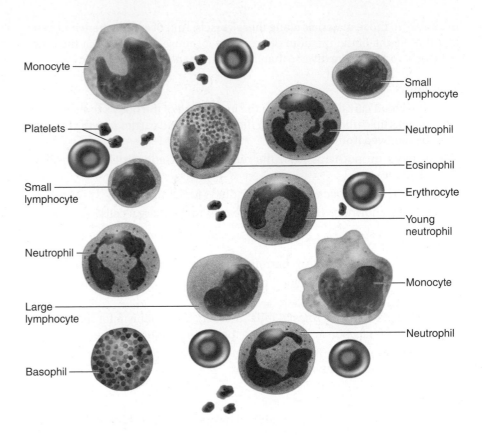

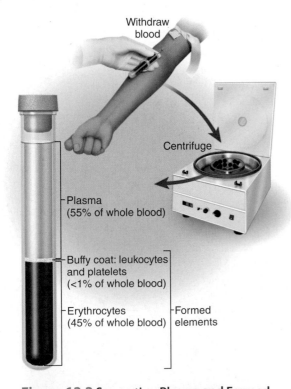

Figure 12.2 Separating Plasma and Formed Elements of Blood. Centrifuging a sample of blood separates the erythrocytes from the leukocytes and platelets (buffy coat) and plasma. The percentage of the volume composed of erythrocytes is called the hematocrit.

forms the extracellular matrix of the tissue. **Formed elements** are cells and cell fragments—*red blood cells, white blood cells,* and *platelets* (fig. 12.1). The term *formed elements* denotes that these are membrane-enclosed entities with a specific structure; they cannot all be called cells because platelets are merely fragments of certain bone marrow cells.

Most adults have 4 to 6 L of blood. It is typically about 55% plasma and 45% formed elements by volume. These proportions are determined by taking a sample of blood in a tube and spinning it for a few minutes in a centrifuge. The relatively dense formed elements are forced to the bottom of the tube and the plasma rises to the top (fig. 12.2). The red blood cells are the densest elements and constitute anywhere from about 35% to 50% of total blood volume depending on one's sex, age, diet, and degree of physical activity. This value, called the *hematocrit,* is commonly used in clinical assessment. The white blood cells and platelets settle in a narrow cream- or buff-colored zone called the *buffy coat* just above the red blood cells. At the top of the tube is the plasma, which accounts for the rest of the volume.

Blood Plasma

Plasma is a complex mixture of water, proteins, nutrients, nitrogenous wastes, hormones, and gases (table 12.2). **Serum** is the fluid that remains when the blood clots and the solids are removed.

Protein is the most abundant plasma solute by weight, totaling 6 to 9 grams per deciliter (g/dL). Plasma proteins play a variety of roles including clotting, defense, and transport of other solutes such as iron, lipids, and hydrophobic hormones. The three major categories of proteins are albumin, globulins, and fibrinogen. Other plasma proteins are essential for life but are found in far smaller quantities.

Albumin is the smallest and most abundant plasma protein. It transports various solutes and buffers the pH of the plasma. It also makes a major

Table 12.2 Composition of Blood Plasma

Water	92% by weight
Proteins	6–9 g/dL
Nutrients	Glucose, amino acids, lipids (cholesterol, fatty acids, triglycerides, and phospholipids), vitamins, iron, and trace elements
Electrolytes	Sodium, potassium, magnesium, calcium, and others
Nitrogenous wastes	Urea, uric acid, creatine, creatinine, bilirubin, ammonia
Hormones	Insulin, thyroid hormone, estrogen, and all other hormones
Blood gases	Oxygen, carbon dioxide, and nitrogen

contribution to two important physical properties of blood: viscosity and osmolarity. **Viscosity** refers to the thickness or stickiness of a fluid that results from cohesion of its particles. Whole blood is 4.5 to 5.5 times as viscous as water, mainly because of the red blood cells. Plasma alone, however, is 2.0 times as viscous as water, mainly because of albumin. Viscosity is important in circulation because it is one determinant of blood flow through the vessels. If the blood is too viscous, it flows sluggishly, whereas if viscosity is reduced, blood flows too easily. Either condition puts a strain on the heart and may lead to cardiovascular problems.

The **osmolarity** of blood is the concentration (molarity) of particles that cannot pass through the walls of the blood vessels. As blood reaches the tissues, fluid moves through the capillary wall, carrying nutrients and oxygen to the surrounding cells. The water must then be reabsorbed to maintain appropriate blood volume and fluid balance. Reabsorption into the bloodstream by osmosis is governed by the relative osmolarity of the blood versus the tissue fluid. If the blood osmolarity is too high, the bloodstream absorbs too much water, and blood volume and pressure are elevated, placing a potentially dangerous strain on the heart and vessels. If osmolarity is low, excess water remains in the tissues, they become edematous (swollen), and blood pressure may drop. Albumin is essential to maintaining optimal osmolarity and, therefore, optimal fluid balance.

The other main categories of protein are globulins and fibrinogen. **Globulins** play important roles in immunity, clotting, and transport. **Fibrinogen** is a soluble precursor of *fibrin,* a sticky protein that forms the framework of a blood clot. The liver produces all the major plasma proteins except one class of globulins, the antibodies.

Apply What You Know

How could a disease such as liver cancer or hepatitis result in impaired blood clotting?

Production of Blood AP|R

Blood must be continually produced to replace blood cells that grow old and die and plasma components that are consumed or excreted. Plasma is formed mainly by water absorption from the digestive tract. Electrolytes and nutrients also come from the digestive tract. The liver supplies nearly all plasma proteins and secretes as much as 4 g of protein per hour.

Production of formed elements is called **hemopoiesis**[3] (HE-mo-poy-EE-sis). An adult typically produces 400 billion platelets, 200 billion red blood cells,

---

[3] *hemo* = blood; *poiesis* = formation

and 10 billion white blood cells every day. From infancy onward, the red bone marrow is the site of most blood cell production. **Hemopoietic stem cells (HSCs)** found in red bone marrow have the potential to give rise to all the different kinds of formed elements. Later sections of this chapter explain the formation of specific types.

Before You Go On

Answer these questions from memory. Reread the preceding section if there are too many you don't know.

1. Describe the transport, protective, and regulatory functions of the blood.

2. What are the two principal components of blood? List the different kinds of formed elements and their primary functions.

3. What are the three major classes of plasma proteins? Briefly explain their functions.

4. Define viscosity *and* osmolarity *and explain how each contributes to normal function of the circulatory system.*

5. Why must blood be constantly replaced? What is hemopoiesis and where does it occur?

12.2 Erythrocytes

Expected Learning Outcomes

When you have completed this section, you should be able to:

a. describe the structure and explain the function of red blood cells;

b. characterize the structure and function of hemoglobin;

c. define some clinical measurements of RBC and hemoglobin quantities and give some typical values for each;

d. discuss the life cycle of erythrocytes; and

e. explain the molecular basis of blood types and their clinical significance.

Erythrocytes [4] **(red blood cells, RBCs)** have two main functions: (1) to pick up oxygen from the lungs and transport it to body tissues, and (2) to pick up carbon dioxide from the tissues and unload it in the lungs. Although severe deficiency of leukocytes or platelets can be fatal within a few days, a severe deficiency of erythrocytes can be fatal within minutes. Death ensues rapidly in cases of massive hemorrhaging, for example, because of the loss of erythrocytes and the oxygen they carry.

---

[4] *erythro* = red; *cyte* = cell

Quantity and Structure

Erythrocytes are the most abundant formed elements. Viewed with a microscope, the round, pillowlike RBCs crowd the field (see fig. 4.23, p. 115). Their number is critically important to health, because it determines the amount of oxygen the blood can carry. One of the most routine measurements in hematology is **hematocrit (packed cell volume, PCV)**,[5] the percentage of blood volume composed of RBCs (fig. 12.2). In women it normally ranges between 37% and 48%; in men, between 45% and 52%. The RBC count is normally 4.2 to 5.4 million/mL in women and 4.6 to 6.2 million/mL in men. This is often expressed as cells per cubic millimeter (mm³); 1 mL = 1 mm³. The sex difference occurs because male hormones, androgens, stimulate red blood cell production, whereas estrogens do not. In addition, women of reproductive age may have lower numbers of RBCs due to blood loss during menstruation or because of fetal demands for iron during pregnancy.

Erythrocytes are discoidal cells with a thick rim and a thin sunken center (fig. 12.3). The cells have a diameter of about 7.5 μm. The interior of the cell is curiously devoid of distinguishing features because, unlike most cells, RBCs have few organelles. The lack of a nucleus and DNA renders them incapable of mitosis and protein synthesis, resulting in an inability for self-repair. The advantage of losing the nucleus during development, however, is that cells acquire their distinctive biconcave shape by caving in at the center where the nucleus used to be. The primary benefit of this shape is that it enables the dense slurry of RBCs to flow smoothly through blood vessels with minimal turbulence, and allows the RBCs to bend and pass through the tiniest blood vessels.

---

[5] *hemato* = blood; *crit* = to separate

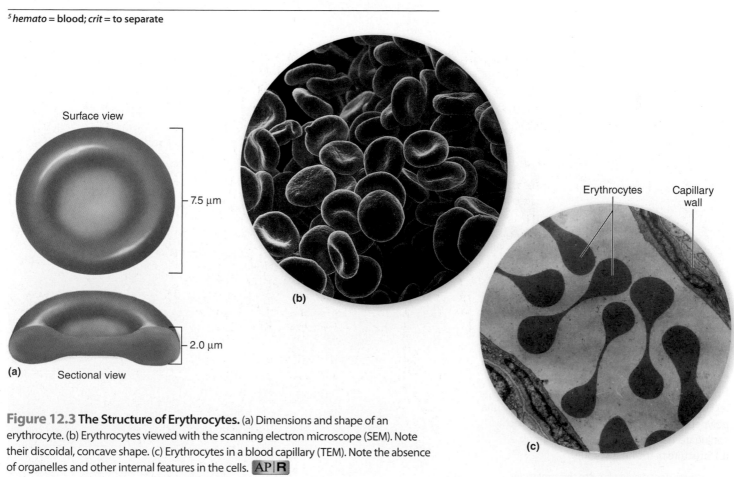

Figure 12.3 The Structure of Erythrocytes. (a) Dimensions and shape of an erythrocyte. (b) Erythrocytes viewed with the scanning electron microscope (SEM). Note their discoidal, concave shape. (c) Erythrocytes in a blood capillary (TEM). Note the absence of organelles and other internal features in the cells. AP|R

- *Why are erythrocytes caved in at the center?*

Hemoglobin

The color of blood is due to **hemoglobin (Hb)**, a red, iron-containing protein. Hemoglobin consists of four polypeptide chains called **globins** (fig. 12.4). Two of these, the *alpha (α) chains*, are 141 amino acids long, and the other two, the *beta (β) chains*, are 146 amino acids long. Each globin has a nonprotein component bound to it called the **heme** group, a carbon–nitrogen ring with an iron atom (Fe) at the center. An O_2 molecule binds to each Fe; therefore, each hemoglobin can carry up to four O_2. Hemoglobin also transports about 5% of the CO_2 in the blood, but this is bound to the globin component rather than to the heme. A hemoglobin molecule can thus transport both gases simultaneously.

RBCs are designed to transport oxygen very efficiently. Lacking organelles, they have more space to pack with hemoglobin. The cytoplasm of an RBC consists mainly of a 33% solution of hemoglobin, with about 280 million molecules per cell. The total **hemoglobin concentration** of whole blood is normally 13 to 18 g/dL in men and 12 to 16 g/dL in women. RBCs lack mitochondria, and therefore rely exclusively on anaerobic fermentation to meet their own ATP requirement. If they made their ATP with mitochondria and aerobic respiration, they would consume their oxygen cargo instead of delivering it to other cells, like a bakery truck driver who's supposed to deliver doughnuts to the grocery stores but eats half of them along the way.

Apply What You Know

How many molecules of O_2 can be carried in one red blood cell?

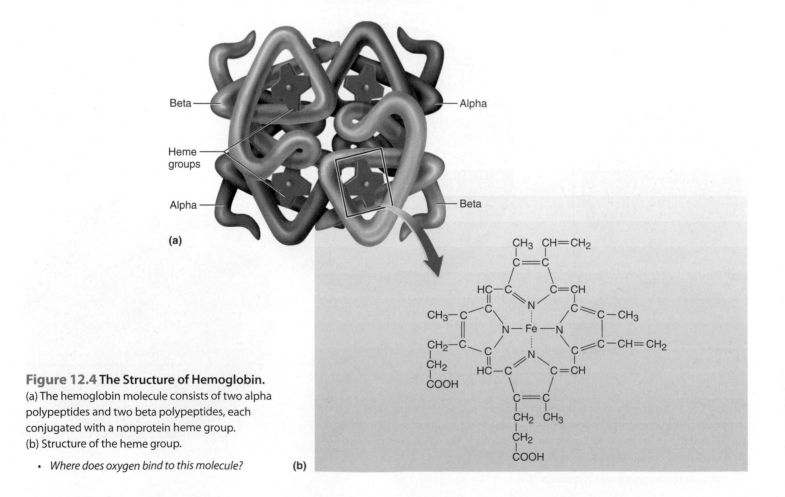

Figure 12.4 The Structure of Hemoglobin.
(a) The hemoglobin molecule consists of two alpha polypeptides and two beta polypeptides, each conjugated with a nonprotein heme group.
(b) Structure of the heme group.

• *Where does oxygen bind to this molecule?*

Clinical Application 12.1

SICKLE-CELL DISEASE

Sickle-cell disease is a hereditary hemoglobin defect occurring mostly among people of African descent; its symptoms occur in about 1.3% of African-Americans, and about 8.3% are asymptomatic carriers with the potential to pass the genetic trait to their children. The disease is caused by a defective gene that codes for the amino acid valine instead of glutamic acid at one position in each beta hemoglobin chain. Manifestation of the disease requires that an individual inherit defective copies of the gene from both parents. The abnormal hemoglobin (HbS) turns to gel at low oxygen levels, as when blood passes through the oxygen-hungry skeletal muscles. The RBCs become elongated, stiffened, and pointed (sickle-shaped) (fig. 12.5). These deformed, inflexible cells adhere to each other and clog the tiny blood capillaries, shutting down the blood flow to tissues downstream. This produces severe pain and can lead to heart failure or stroke, among other effects. The spleen removes defective RBCs faster than they can be replaced, thus leading to anemia. Without treatment, a child has little chance of living to age 2, and even with the best treatment, few victims live to age 50.

Sickle-cell disease originated in areas of Africa where millions of lives have been lost to malaria over thousands of years. Malarial parasites normally invade and reproduce in RBCs but cannot survive in RBCs with HbS hemoglobin. Thus the sickle-cell gene confers resistance to malaria, even in individuals who carry only one copy of the gene (are heterozygous for it) and do not have sickle-cell disease. Natural selection has favored the persistence of the gene in certain areas in Africa where the number of lives saved by resistance to malaria has far outweighed the deaths due to sickle-cell disease. In North America, where malaria is rare, the incidence of the gene among those of African descent has declined because there is little evolutionary benefit from possessing it and so many of those who do possess it die young, without reproducing.

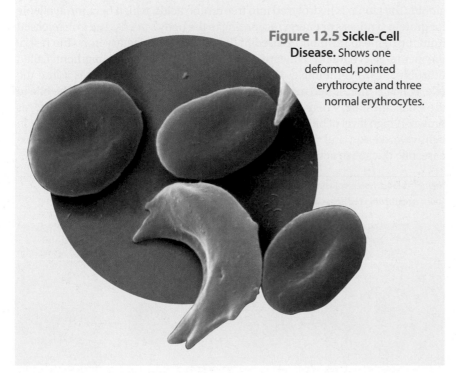

Figure 12.5 Sickle-Cell Disease. Shows one deformed, pointed erythrocyte and three normal erythrocytes.

The Erythrocyte Life Cycle

An erythrocyte lives for an average of 120 days from the time it is produced in the red bone marrow until it dies. In a state of balance, the birth and death of RBCs amount to over 1 million cells per second.

The production of red blood cells is called **erythropoiesis** (eh-RITH-ro-poy-EE-sis) (fig. 12.6). It is one aspect of the more general process of hemopoiesis, the production of all formed elements of the blood. Like all formed elements, RBCs trace their origin to a bone marrow stem cell, the hemopoietic stem cell (HSC). HSCs give rise to more specialized cells called **colony-forming units (CFUs)**, each type destined to produce one class of formed elements.

The transformation from an HSC to a mature RBC takes 3 to 5 days and involves four major developments—a reduction in cell size, an increase in cell number, the synthesis of hemoglobin, and the loss of the nucleus and other organelles. Once the nucleus and other organelles have been expelled, the cell is called a *reticulocyte*, and is ready to leave the bone marrow and enter the circulating blood. Reticulocytes are named for cytoplasmic clusters of ribosomes visible as a fine network in young RBCs. Even this disappears within the first day or two in circulation, but an elevated reticulocyte count can suggest a blood disease with abnormally rapid RBC production. Production of RBCs is stimulated by the hormone **erythropoietin (EPO)**, which is produced in the kidneys and liver and targets bone marrow. EPO levels rise when blood oxygen levels fall below normal, thus stimulating a corrective rise in RBC count.

As the RBC ages, its membrane proteins deteriorate. Without a nucleus or ribosomes, it cannot synthesize new proteins and its membrane becomes increasingly fragile. Eventually, the cell ruptures as it tries to flex its way through tight capillaries. Many RBCs die in the spleen, which has been called an "erythrocyte graveyard." Its narrow channels severely challenge the ability of old, fragile RBCs to squeeze through the organ and they become trapped, broken up, and destroyed.

When the RBCs rupture, or undergo **hemolysis**[6] (he-MOLL-ih-sis), hemoglobin is released and an empty plasma membrane is left behind. Macrophages digest the membrane fragments and hemoglobin, separating heme from the globin. The globin is hydrolyzed into free amino acids, which become available for protein synthesis or energy, according to the body's needs. The iron released from the heme to the blood is used in the same way as dietary iron. The rest of the heme is eventually converted to a vivid yellow–green pigment called **bilirubin.**[7] The liver removes bilirubin from the blood and secretes it into the bile, which is then stored in the gallbladder. The gallbladder releases the bile to the small intestine, where bacteria convert it to a pigment that colors the feces brown. A high level of bilirubin in the blood causes *jaundice,* a yellowish cast in light-colored skin and the whites of eyes. Jaundice may be a sign of liver disease, bile duct obstruction, or rapid RBC destruction.

---

[6] *hemo* = blood; *lysis* = splitting, breaking
[7] *bili* = green; *rub* = red; *in* = substance

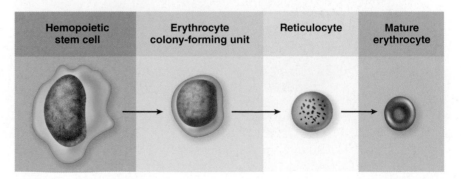

| Hemopoietic stem cell | Erythrocyte colony-forming unit | Reticulocyte | Mature erythrocyte |

Figure 12.6 Erythropoiesis. Stages in the development of erythrocytes.

Blood Types

Transfusions of blood from person to person have saved countless lives. Even in ancient Rome, physicians began to experiment with transfusions, but had mixed results; some lives were saved, but many died of transfusion reactions. It was not until 1900 that a young researcher in Vienna, Karl Landsteiner, discovered blood types A, B, and O. He showed that plasma from one group of people (group A) caused the red cells of another group (B) to clump, or **agglutinate**. This explained why a transfusion might have disastrous consequences—if a person received a blood type different from his own, agglutination would occur and the person probably would not survive. This concept of compatibility laid the groundwork for successful therapeutic use of transfusions.

Blood types are based on interactions between large molecules called *antigens* and *antibodies*. **Antigens** occur on the surface of cells and enable the body to distinguish its own cells from foreign matter. When the body detects an antigen of foreign origin, it activates an immune response. One part of the response is the production of **antibodies**, proteins produced by certain white blood cells. One method of antibody action is agglutination, in which antibody molecules bind to antigen molecules and stick them together.

Human blood groups are determined by genetic variation in antigenic glycolipids on the RBC surface. A person can have the type A antigen, type B, both, or neither, making his or her blood type A, B, AB, or O, respectively (fig. 12.7). One who has type A antigens on the RBCs has anti-B antibodies in the plasma and vice versa. A type O person has both anti-A and anti-B, and a type AB person has neither of them. That is, one has antibodies against any ABO antigen *except* the antigens on one's own RBCs. Consequently, if a type A patient mistakenly received a transfusion with type B blood, the patient's anti-B antibodies would attack the type B erythrocytes. The transfused RBCs would agglutinate, block small blood vessels, burst open (**hemolyze**), and release their hemoglobin. Free hemoglobin can block kidney tubules and cause death from acute renal failure. Another consideration in transfusion compatibility is that antibodies in a mismatched donor's plasma can attack RBCs in the recipient—for example, if type O blood was transfused into a type B patient. In short, a transfusion patient should always receive blood of his or her own type; strictly speaking, there is no such thing as a universal donor or universal recipient, despite the former popularity of these expressions.

Apply What You Know

A person who has type AB blood needs a transfusion. What type or types of blood can he or she receive?

In 1940, the **Rh blood group** was discovered in rhesus monkeys, for which it is named. *Rh-positive* people have Rh antigens on their RBCs, whereas these are lacking in *Rh-negative* individuals. The Rh blood type is often combined with the ABO type in a single expression such as A+ for type A, Rh-positive, or O– for type O, Rh-negative.

Clinical problems can arise for an Rh+ fetus carried by a woman who is Rh–. During the first pregnancy, there is usually no problem because maternal and fetal blood do not mix until birth, when shearing of the placenta from the uterine wall exposes the mother to the Rh+ fetal blood. She then begins producing antibodies against it that may jeopardize her future pregnancies. If she becomes pregnant with another Rh+ fetus, those antibodies can diffuse through the placenta into the fetal blood and attack its RBCs. The baby may be born with a severe anemia called **hemolytic disease of the newborn,** or **erythroblastosis fetalis.** Pregnant women who are Rh– may be given an Rh immune globulin that binds to the fetal RBC antigens so they cannot stimulate

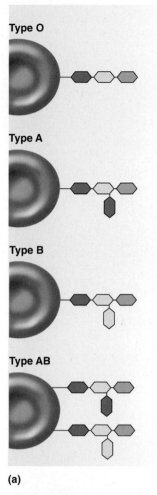

Type O

Type A

Type B

Type AB

(a)

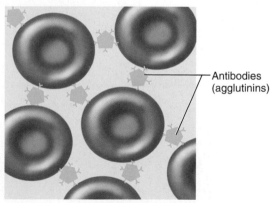

Antibodies (agglutinins)

(b)

Figure 12.7 Chemical Basis of the ABO Blood Types. (a) The four blood types in the ABO group differ in the terminal sugars of the carbohydrate chains attached to the RBC plasma membranes. Each color represents a different sugar. Note that type AB red cells have both the type A and type B antigens. (b) RBC agglutination in a mismatched transfusion. Five-sided antibodies in the recipient blood plasma bind A or B antigens on the RBCs of the donor blood.

her immune system to produce the antibodies that would cause agglutination. In severe cases of erythroblastosis fetalis, an *exchange transfusion* may be given to completely replace the infant's Rh+ blood with Rh– blood.

Before You Go On

Answer these questions from memory. Reread the preceding section if there are too many you don't know.

6. Define hematocrit *and give the normal ranges for women and men.*
7. Describe the size, shape, and contents of an erythrocyte. How does the lack of mitochondria relate to the function of red blood cells?
8. What is the function of hemoglobin? Describe its structure. What is the role of iron? Define hemoglobin concentration *and provide units of measurement.*
9. List the stages in the production of an RBC. Why do RBCs die? What happens to their components when they die and disintegrate?
10. Describe the antigens that determine the ABO blood types. What are antibodies and antigens and how do they interact to cause a transfusion reaction? What is the clinical significance of the Rh blood group?

12.3 Leukocytes

Expected Learning Outcomes

When you have completed this section, you should be able to:

a. discuss the general function of leukocytes and the specific functions of each individual type;
b. characterize the appearance and relative abundance of each type of leukocyte; and
c. describe the life cycle of leukocytes.

Leukocytes[8] (also called **white blood cells** or **WBCs**) protect us from infectious microorganisms and other pathogens, and we would soon succumb to disease without their defense. When viewed with a microscope, they stand out as vividly stained islands among a sea of the more numerous and pale erythrocytes. They are much more abundant in the body than their low numbers in blood films would suggest because they spend only a few hours in the bloodstream, then migrate through the walls of capillaries and spend the rest of their lives in connective tissues. The bloodstream is like the subway that the WBCs take to work; in the blood films, we see only the ones on their commute, not those already at work in the tissues.

---

[8] *leuko* = white; *cyte* = cell

Structure and Function

Leukocytes are not uniform but vary in form and function. In contrast to erythrocytes, they retain their organelles throughout life; thus, when viewed with a transmission electron microscope, they show a complex internal structure (fig. 12.8). Among their organelles are the usual instruments of protein synthesis—the nucleus, rough endoplasmic reticulum, ribosomes, and Golgi complex—for leukocytes must synthesize proteins to carry out a variety of functions. For example, phagocytic WBCs have digestive enzymes packaged into lysosomes that enable the cell to digest pathogens.

Types of Leukocytes

The five types of leukocytes are distinguished from each other by their relative size and abundance, the size and shape of their nuclei, the presence or absence of certain cytoplasmic granules, the coarseness and staining properties of those granules, and most importantly their functions.

　　All WBCs have organelles that appear as granules under the light microscope. Some of these are lysosomes called **nonspecific granules** because they occur in all five WBC types. Others are various kinds of **specific granules** that occur only in three WBC types—neutrophils, eosinophils, and basophils—which are therefore called **granulocytes.** Specific granules stain conspicuously and distinguish each cell type from the others. They contain enzymes and other chemicals employed in defense. The two remaining WBC types—monocytes and lymphocytes—are called **agranulocytes** because they lack specific granules.

　　Bear in mind that the color of WBC nuclei (usually violet) and the colors described below for their granules are not natural colors. They result from the dyes used to stain blood films, and they can vary depending on what stain was used on any blood films you may study.

Granulocytes

1. **Neutrophils** (NEW-tro-fills) (fig. 12.9) are the most abundant WBCs, constituting 60% to 70% of the circulating leukocytes. Not only are they easy to find in blood films, they are also easily recognized by their violet-staining, multilobed nucleus—usually three or four lobes connected by thin strands. The cytoplasm is lightly stippled with tiny, pale granules. Neutrophils are aggressive antibacterial cells. They are quickly attracted to areas of infection or inflammation, where they crawl about phagocytizing and digesting bacteria they encounter. They also secrete a cloud of toxic chemicals (hydrogen peroxide and bleachlike hypochlorite) that kill even greater numbers. The chemicals

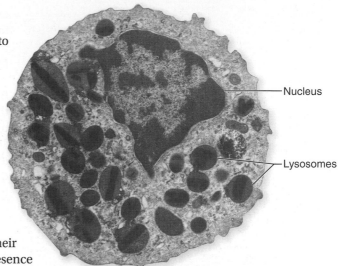

Figure 12.8 The Structure of a Leukocyte. This example is an eosinophil. The lysosomes seen here are the coarse pink granules seen in the cytoplasm of the eosinophil in Figure 12.9.

Figure 12.9 The Five Kinds of Leukocytes. Note the size of each WBC type compared to the RBCs around it. AP|R

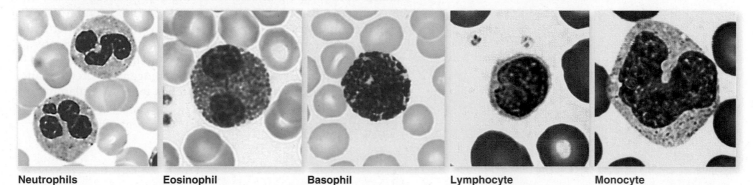

Neutrophils　　　**Eosinophil**　　　**Basophil**　　　**Lymphocyte**　　　**Monocyte**

are lethal to the neutrophils themselves, which typically have a short life span. Neutrophils that have died in combat are a major component of pus and are responsible for its milky color. An elevated neutrophil count in the blood is clinically important because it may indicate an infection.

2. **Eosinophils** (EE-oh-SIN-oh-fills) (fig. 12.9) represent only 2% to 4% of the WBC total, but are also easy to recognize. They usually have a large, bilobed nucleus, like two balloons tied together by a short string. The cytoplasm contains an abundance of large, rosy pink to pinkish orange specific granules. Eosinophils secrete chemicals that weaken or destroy large parasites such as hookworms or tapeworms, which are too large to be destroyed by any one WBC. They also help to dispose of antigen–antibody complexes formed in immune reactions (see chapter 14). Their numbers rise in parasitic infections, allergies, and diseases of the spleen and central nervous system, and fluctuate between day and night and over the course of the menstrual cycle.

3. **Basophils** (BASE-oh-fills) (fig. 12.9) are the rarest of WBCs, comprising from less than 0.5% to about 1% of the total. They are recognizable because of their abundant, very coarse granules, usually stained dark violet (sometimes pink). The granules hide the nucleus from view, but the nucleus is large, pale, and typically S- or U-shaped. Basophils secrete **histamine,** a vasodilator that widens blood vessels, speeds the flow of blood to injured tissue, and makes the blood vessels more permeable so that agents such as neutrophils and clotting proteins can get into the connective tissues more quickly. They also secrete **heparin,** an anticoagulant that inhibits blood clotting and thus promotes the mobility of other WBCs in the area. They are important in allergy, inflammation, and attracting neutrophils and other WBCs to sites of infection. An elevated basophil count is commonly seen in diabetes, chickenpox, and various other diseases.

Agranulocytes

4. **Lymphocytes** (LIM-fo-sites) (fig. 12.9) are second to neutrophils in abundance at 25% to 33% of the WBC count, and are quickly spotted on a blood film. Their dark violet nucleus fills nearly the entire cell. The cytoplasm stains a clear light blue color and forms a narrow and sometimes barely detectable rim around the nucleus. The subclasses of lymphocytes are indistinguishable under the light microscope but have distinct immune functions. The lymphocyte count rises in diverse infectious diseases and immune responses. The different types of lymphocytes and their functions are explored further in chapter 14.

5. **Monocytes** (MON-oh-sites) (fig. 12.9) are the largest WBCs and comprise about 3% to 8% of the WBC count. The nucleus is large and typically ovoid, kidney-shaped, or horseshoe-shaped. The cytoplasm contains sparse, fine granules. After migrating from the blood into the connective tissues, monocytes transform into even larger cells called **macrophages**[9] (MAC-ro-fay-jez). They destroy dead or dying host and foreign cells, microorganisms, and other foreign material, consuming up to 25% of their own volume per hour. They also aid in alerting the immune system to invading pathogens or foreign matter (see chapter 14). The monocyte count tends to rise in inflammation and viral infections.

---

[9] *macro* = big; *phage* = eater

The Leukocyte Life Cycle

Leukopoiesis (LOO-co-poy-EE-sis), the production of white blood cells, begins with the same hemopoietic stem cells as erythropoiesis. Some HSCs differentiate into distinct types of colony-forming units (CFUs), which then go on to produce any of three cell lines—one line committed to producing the three kinds of granulocytes *(myeloblasts)*, one to monocytes *(monoblasts)*, and one producing lymphocytes *(lymphoblasts)* (fig. 12.10).

Granulocytes and monocytes stay in the red bone marrow until they are needed; the marrow contains 10 to 20 times as many of these cells as the circulating blood does. Lymphocytes begin developing in the bone marrow but do not stay there. Some migrate to the thymus and mature there; then from either the bone marrow or thymus, lymphocytes disperse and colonize the lymph nodes, spleen, tonsils, and mucous membranes. Their life histories and migrations are further discussed in chapter 14.

Circulating leukocytes do not stay in the blood for very long. Granulocytes circulate for 4 to 6 hours and then migrate into the tissues, where they live another 4 or 5 days. Monocytes travel in the blood for 10 to 20 hours, then migrate into the tissues and differentiate into macrophages, which can live for as long as a few years. Lymphocytes are responsible for long-term immunity and therefore may survive for lengths of time ranging from a few weeks to decades. They leave the bloodstream for the tissues and eventually enter the lymphatic system, which empties them back into the blood. They are continually cycled between the blood, lymph, and tissue fluid.

Figure 12.10 Leukopoiesis. Stages in the development of leukocytes.

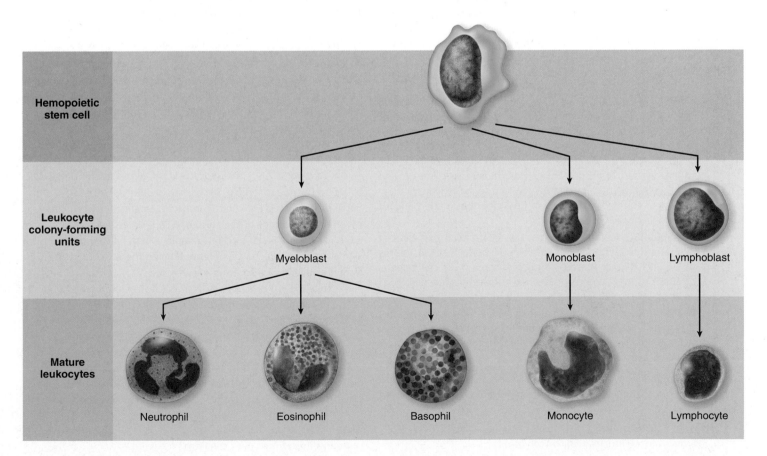

Hemopoietic stem cell

Leukocyte colony-forming units

Myeloblast   Monoblast   Lymphoblast

Mature leukocytes

Neutrophil   Eosinophil   Basophil   Monocyte   Lymphocyte

PERSPECTIVES ON HEALTH

Given the vital functions of blood, it is no surprise that disorders can have disastrous consequences for health. A deficiency or excess of a specific cell type underlies many blood diseases. Misshapen or oddly colored cells may also indicate certain underlying conditions.

Erythrocyte Disorders

The two most common RBC disorders are **anemia**[10] (an RBC or hemoglobin deficiency) and **polycythemia**[11] (POL-ee-sih-THEME-ee-uh) (an RBC excess). There are three fundamental causes of anemia:

1. **Reduced erythropoiesis.** When hemoglobin or erythrocytes are produced too slowly to keep up with the rate of RBC death, an individual becomes anemic. One cause of suppressed RBC manufacture or hemoglobin synthesis is dietary deficiency of minerals or vitamins needed for their production. For example, a lack of iron suppresses erythropoiesis and leads to *iron-deficiency anemia*. Elderly people may experience *pernicious anemia*, an immune disorder that destroys stomach tissue. Normally the stomach produces a protein called *intrinsic factor*, needed for absorbing vitamin B_{12} from the food. Vitamin B_{12} is required for erythropoiesis. The immune attack on the stomach thus results in a vitamin B_{12} deficiency. Gastric-bypass patients also experience a similar anemia because of the surgical removal of stomach tissue or disconnection of the stomach from the small intestine, where the intrinsic factor is needed. Radiation, viruses, and some poisons can cause anemia by destroying bone marrow. Kidney disease can lead to inadequate production of *erythropoietin* (EPO), a hormone that is the principal stimulus for RBC production.

2. **Rapid RBC destruction.** Excess RBC destruction that exceeds the rate of erythropoiesis results in *hemolytic anemia*. Causes include various poisons, drug reactions, sickle-cell disease, snake venoms, and blood-destroying parasitic infections such as malaria.

3. **Hemorrhage.** Blood loss may lead to *hemorrhagic anemia*. This can be a consequence of trauma such as gunshot, automobile, or battlefield injuries; hemophilia; ruptured aneurysms; or heavy menstruation.

Anemia, whatever its fundamental cause, may lead to **hypoxia**, a deprivation of oxygen in the tissues. In severe cases, the affected person is lethargic and becomes short of breath upon physical exertion. The skin may be pallid because of the lack of hemoglobin.

Polycythemia can result from cancer of the bone marrow or from a multitude of other conditions. Dehydration, for example, is characterized by loss of water from the blood while erythrocytes remain and become highly concentrated. Abnormally high oxygen demand (as in people who engage in overzealous aerobic exercise) or low oxygen supply (as occurs with smoking, emphysema, or ascent to high altitude) stimulates erythropoiesis. RBC counts can rise as high as 11 million RBCs/μL and hematocrit as high as 80%. The thick blood "sludges" in the vessels, increases blood pressure, and puts dangerous strain on the cardiovascular system that can lead to heart failure or stroke.

Leukocyte Disorders

A deficiency of WBCs is called **leukopenia**[12] (LOO-co-PEE-nee-uh). A low WBC count may be due to poisoning from lead, mercury, or arsenic, or to radiation exposure. Infectious diseases such as AIDS, measles, mumps, chickenpox, influenza, and poliomyelitis also deplete WBCs. Certain drugs such as glucocorticoids, anticancer drugs, and immunosuppressant drugs given to transplant patients lower the WBC count. A patient with insufficient WBCs cannot effectively fight disease and is susceptible to infections and cancer.

An excess of WBCs is **leukocytosis**.[13] A higher-than-normal WBC count can be due to infection or allergy, or even dehydration or emotional disturbances. For clinical purposes, a *differential WBC count*, in which the percentage of each type of WBC is determined, is more useful than a total WBC count. For example, a high neutrophil count is a sign of bacterial infection, whereas a high eosinophil count usually indicates an allergy or a parasitic infection such as worms.

Leukemia is a cancer that originates from mutated hemopoietic cells. Proliferation of the malignant cells results in an extraordinarily high number of circulating leukocytes and their precursors (fig. 12.11). Leukemia is classified as myeloid or lymphoid, acute or chronic. **Myeloid leukemia** is characterized by uncontrolled granulocyte production, whereas **lymphoid leukemia** involves uncontrolled lymphocyte or monocyte production. **Acute leukemia** appears suddenly, progresses rapidly, and causes death within a few months if left

---

[10] *an* = without; *emia* = blood

[11] *poly* = many; *cyt* = cells; *hemia* = blood condition

[12] *leuko* = white; *penia* = deficiency

[13] *leuko* = white; *cyt* = cell; *osis* = condition

untreated. **Chronic leukemia** develops more slowly and may be undetected for many months.

As the cancerous cells proliferate uncontrollably, they replace normal bone marrow and a person suffers from a deficiency of normal WBCs, RBCs, and platelets. The numerous cancerous leukocytes that are produced are immature and are incapable of fulfilling their roles as defenders against disease. The lack of functional WBCs renders the patient vulnerable to infection; the paucity of RBCs leads to anemia and fatigue; and the platelet deficiency results in hemorrhaging and impaired blood clotting. Although the immediate cause of death is usually hemorrhage or opportunistic infection, cancerous cells may metastasize from the bone marrow to other organs of the body, where they displace or compete with normal cells. Widespread metastasis is not easily treated by either surgery or radiation therapy and may lead to death of the individual.

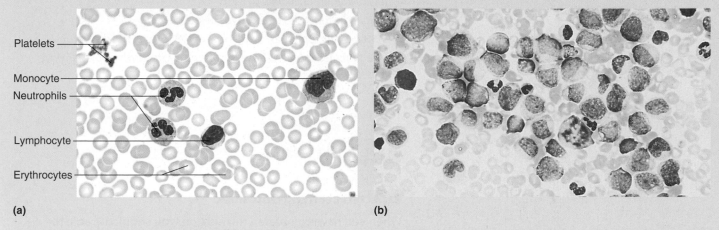

Platelets
Monocyte
Neutrophils
Lymphocyte
Erythrocytes

(a)   (b)

Figure 12.11 Normal and Leukemic Blood. (a) A normal blood smear. (b) Blood from a patient with acute monocytic leukemia. Note the abnormally high number of WBCs, especially monocytes.

- *With all these extra white cells, why isn't the body's infection-fighting capability increased in leukemia?*

Clinical Application 12.2

THE COMPLETE BLOOD COUNT

The *complete blood count (CBC)* is a very common clinical test that affords a wealth of information about a patient's health. It includes hematocrit; hemoglobin concentration; total RBC, WBC, and platelet counts; RBC size; and a *differential WBC count*, among other values. The differential WBC count quantifies what percentage of the WBCs consists of each cell type. The relative proportions of one WBC type to another can be valuable indications of various disease states such as infections, inflammation, allergies, cancer, AIDS, and certain metabolic diseases.

Blood counts used to be done by visual examination with a microscope, but now are usually done with an electronic cell counter. Cell counters force a blood sample through a tiny orifice and count cells as voltage pulses that vary with cell type and size. They give much faster and more accurate results than visual inspection. However, cell counters still misidentify some cells, and a medical technologist must review the results for suspicious data and visually identify cells that the instrument cannot. If a CBC does not provide enough information or if it suggests certain disorders, additional tests may be done such as coagulation time or a bone marrow biopsy.

12.4 Platelets

Expected Learning Outcomes

When you have completed this section, you should be able to:

a. describe the structure and functions of blood platelets;

b. describe platelet production; and

c. describe blood clotting and other mechanisms for controlling bleeding.

Circulatory systems developed very early in animal evolution, and with them evolved mechanisms for stopping leaks, which are potentially fatal. Platelets are the major players in cessation of bleeding and are the focus of this section.

Structure and Function

Platelets are small circulating fragments of bone marrow cells called **megakaryocytes** and serve various roles in blood clotting and other functions listed below. They are among the formed elements of blood, but they are not cells, since they never have a nucleus of their own at any time in their development. They are the second most abundant formed element after erythrocytes. They are so much smaller than RBCs, however (2 to 4 μm in diameter), that they contribute even less than WBCs to total blood volume.

Platelets have a complex internal structure that includes lysosomes, mitochondria, a cytoskeleton, granules filled with platelet secretions, and a system of channels that open onto the platelet surface (the **open canalicular system**) (fig. 12.12a). When activated, they form pseudopods and are capable of ameboid movement.

Platelets have a variety of functions:

1. They secrete *vasoconstrictors*, chemicals that cause spasmodic constriction of broken vessels and thus help reduce blood loss.

2. They stick together to form temporary *platelet plugs* to seal small breaks in injured blood vessels.

3. They secrete *clotting factors* that promote blood clotting.

4. They initiate the formation of a clot-dissolving enzyme that dissolves blood clots that have outlasted their usefulness.

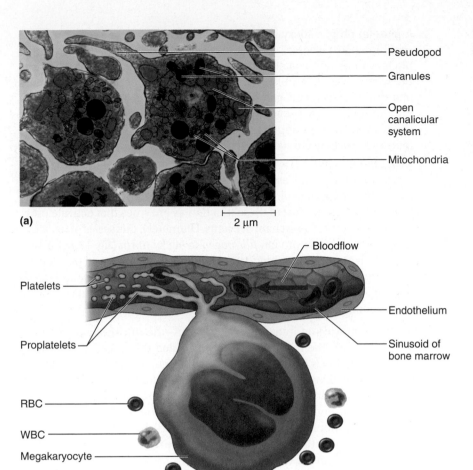

Pseudopod

Granules

Open
canalicular
system

Mitochondria

(a)

2 μm

Bloodflow

Platelets

Proplatelets

Endothelium

Sinusoid of
bone marrow

RBC

WBC

Megakaryocyte

(b)

Figure 12.12 Platelets. (a) Structure of blood platelets. (b) Platelet production. The megakaryocyte extends tendrils into the sinusoids of the red bone marrow, where blood flow shears off pieces called proplatelets. These break up into platelets as they travel in the blood stream, especially while passing through the lungs. Note the size of the megakaryocyte compared to that of the mature RBCs and WBCs around it. AP|R

5. They secrete *growth factors* that stimulate mitosis in fibroblasts and smooth muscle and thus help maintain and repair blood vessels.

Platelet Production

Production of platelets, called **thrombopoiesis,** is one aspect of hemopoiesis. Some hemopoietic stem cells differentiate into **megakaryocytes**[14] (meg-ah-CAR-ee-oh-sites), gigantic cells up to 150 μm in diameter, large enough to be visible to the naked eye. Each has a huge multilobed nucleus and multiple sets of chromosomes (fig. 12.12b).

Platelets arise from long, beaded processes that extend from the surface of the megakaryocyte into the lumen of blood sinusoids in the bone marrow. The force of the blood flow shears the tips of these processes off, releasing them as *proplatelets* that further fragment into individual platelets, especially as they pass through the lungs. About 25% to 40% of the platelets are stored in the spleen and released as needed. The rest circulate freely in the blood and live for 5 or 6 days.

Hemostasis and Coagulation

The circulatory system has evolved overlapping mechanisms to seal itself after injury. **Hemostasis,**[15] the cessation of bleeding, occurs in three stages:

1. **Vascular spasm,** the prompt constriction of a broken vessel. This narrows the opening and reduces blood loss. It is triggered in part by **serotonin,** a vasoconstrictor secreted by the platelets.

[14]*mega* = giant; *karyo* = nucleus; *cyte* = cell
[15]*hemo* = blood; *stasis* = stability

2. A **platelet plug,** a sticky mass of platelets that acts as a stopper to close small breaks in a vessel. In an undamaged blood vessel, the endothelium is smooth and coated with a platelet repellent to prevent adhesion. But when a vessel is broken, platelets adhere to the roughened surfaces and now-exposed collagen fibers of the vessel wall. The platelets then put out long, spiny pseudopods that adhere to other platelets and the vessel wall. The pseudopods contract and pull the wall together, and the platelet mass may reduce or stop minor bleeding.

3. **Coagulation (clotting)** is the last but most effective defense against bleeding. It is a complex process involving more than 30 chemical reactions and a multitude of protein **clotting factors (procoagulants)** produced by the liver and platelets. Most of these function as enzymes that activate other enzymes in a cascading chain of events. Ultimately, this series of reactions converts the plasma protein *fibrinogen* to sticky *fibrin* (fig. 12.13). Fibrin adheres to the wall of the blood vessel, and as blood cells and platelets arrive, many of them stick to it like insects in a spider web. The resulting mass of fibrin, platelets, and blood cells forms a clot that ideally seals the break in the blood vessel long enough for the vessel to heal.

There are two ways of initiating coagulation—the intrinsic and extrinsic pathways (fig. 12.13). The **intrinsic pathway** is called this because everything needed for coagulation is contained in the blood itself. It begins when platelets adhere to a damaged vessel and secrete a procoagulant named *factor XII.* It continues through a series of three enzymatic reactions with more and more product at each step, ending with *factor VIII.* This is also called *antihemophiliac factor A* because hemophilia results from a hereditary inability to produce factor VIII. The **extrinsic pathway** is so named because it employs a procoagulant from sources other than the blood—*factor III (thromboplastin),* released from damaged tissues surrounding a blood vessel, as when you cut yourself or hit your thumb with a hammer.

Factor III and factor VIII both activate another procoagulant called *factor X,* the start of a **common pathway** that completes the coagulation process. Thus you can see that the extrinsic pathway gets to factor X in just one step, whereas the intrinsic pathway requires four steps. The extrinsic pathway therefore clots the blood more quickly.

From factor X to the end, coagulation is completed by the same mechanism (the common pathway) regardless of how it started. Factor X activates an enzyme called *prothrombin activator.* This enzyme activates *prothrombin,* a protein from the liver, and converts it to *thrombin.* Thrombin is an enzyme that converts another liver protein, *fibrinogen,* into *fibrin.* Fibrin is the sticky mesh that traps platelets and blood cells and forms the vessel-sealing clot.

Several steps in all three pathways—intrinsic, extrinsic, and common—also require calcium as a cofactor. Blood clotting can therefore be blocked by binding all the calcium in a blood sample, making it unavailable to these enzymes. In drawing blood for clinical testing or blood banking, this can be done with calcium-binding salts of oxalate, citrate, and others, or with the chemical *heparin,* which works by blocking thrombin action and other mechanisms.

It is important to distinguish coagulation from agglutination. In both reactions, blood cells or platelets aggregate and stick together, so they may seem to be superficially similar processes. However, the mechanisms are entirely different. In coagulation, formed elements are stuck together by fibrin. In the agglutination of blood, RBCs are stuck together by antibody molecules (see p. 401).

Once a leak is sealed and the crisis has passed, platelets secrete *platelet-derived growth factor (PDGF),* a substance that stimulates fibroblasts and smooth muscle to proliferate and replace the damaged tissue of the blood vessel. When tissue repair is completed and the blood clot is no longer necessary, the clot must be disposed of. Platelets then secrete a protein that initiates a series of reactions leading to a fibrin-digesting enzyme called *plasmin.* Plasmin dissolves the old blood clot.

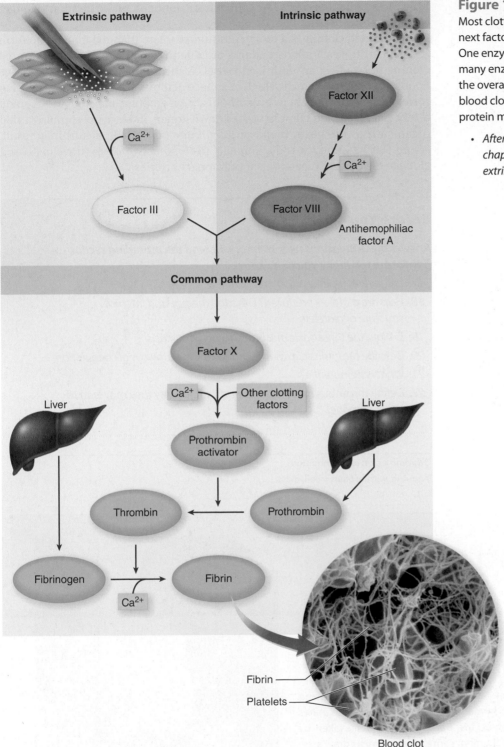

Figure 12.13 The Pathways of Coagulation.
Most clotting factors act as enzymes that convert the next factor from an inactive form to an active form. One enzyme molecule at any given level activates many enzyme molecules at the next level down, so the overall effect becomes amplified at each step. A blood clot with platelets (orange) trapped in a sticky protein mesh is shown at the bottom.

- *After you read about hemophilia later in this chapter, explain whether it would affect the extrinsic mechanism, intrinsic mechanism, or both.*

A deficiency of any clotting factor can shut down the coagulation cascade. In **hemophilia,** for example, a hereditary lack of a single protein in the cascade results in an inability to clot and therefore an inability to control bleeding. Most hemophiliacs suffer from a lack of factor VIII; before purified factor VIII became available in the 1960s, more than half of those with hemophilia died before age 5 and only 10% lived to age 21. Although purified factor VIII enabled hemophiliacs to quickly control bleeding and engage in typical activities, it came at a steep price. Many hemophiliacs became infected with hepatitis or HIV carried in the donated plasma. Factor VIII is now obtained from recombinant bacteria rather than from donor blood, so there is no longer any risk of infection to patients.

Unwanted clotting is a more common problem than the inability to form clots. Most strokes and heart attacks are due to **thrombosis**[16]—the abnormal clotting of blood in an unbroken vessel. A **thrombus** (clot) may grow large enough to obstruct a small vessel, or a piece of it may break loose and begin to travel in the bloodstream as an **embolus.**[17] An embolus may lodge in a small artery and block blood flow to tissues downstream. If the vessel supplies a vital organ such as the heart, brain, or lung, *infarction* (tissue death) may result. About 650,000 Americans die annually of *thromboembolism* (traveling blood clots) in the coronary, cerebral, or pulmonary arteries.

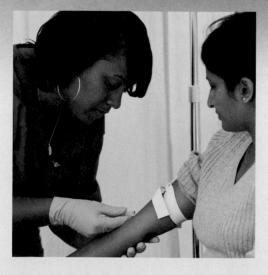

Before You Go On

Answer these questions from memory. Reread the preceding section if there are too many you don't know.

17. List several functions of blood platelets.
18. How are platelets produced? Describe the size and internal structure of platelets.
19. Explain the three basic mechanisms of hemostasis.
20. Describe the intrinsic, extrinsic, and common pathways of coagulation.
21. What is hemophilia?
22. Define thrombus *and* embolus *and explain how they may lead to tissue damage.*

---

[16] *thromb* = clot; *osis* = condition
[17] *em* = in, within; *bolus* = ball, mass

CAREER SPOTLIGHT

Phlebotomist

A phlebotomist[18] draws blood for the purposes of clinical testing, blood banking, or transfusion. Although this is often done by nurses and physicians, phlebotomists are employed to reduce the workload on these other professionals, especially in hospitals and blood drives. Some phlebotomists travel to nursing homes, outpatient clinics, or patients' homes to collect samples; some states employ phlebotomists as dialysis technicians. Samples are usually taken by puncturing a vein, but with specialized training, some phlebotomists draw arterial samples, usually from the radial or ulnar arteries at the wrist.

Training and certification requirements vary from state to state. One must have a high school diploma or general educational development (GED) certificate, then may undertake training ranging from 6 weeks at a trade school or career center to a college associate degree. Training typically includes anatomy, legal aspects of blood collection, patient interaction skills, and blood-collecting techniques. A state may require certification by such agencies as the National Phlebotomy Association or the American Society of Clinical Pathologists. For further information on careers in phlebotomy, see appendix B.

---

[18]*phlebo* = vein; *tom* = cut; *ist* = one who

CONNECTIVE ISSUES

Ways in Which the Blood Affects Other Organ Systems

All Systems
Delivers oxygen and nutrients to all systems and carries away their metabolic wastes for elimination; conveys hormones to their target organs; plays a role in fluid balance.

Integumentary System
Nourishes the dermis and deep layers of epidermis; delivers WBCs to dermis that phagocytize foreign invaders.

Skeletal System
Transports nutrients to bone cells; delivers hormones that regulate skeletal growth.

Muscular System
Delivers oxygen and nutrients; removes lactic acid and thus reduces its fatigue-producing effect.

Nervous System
Cerebrospinal fluid is derived from blood plasma.

Endocrine System
Transports all hormones.

Lymphatic and Immune Systems
Source of tissue fluid, which becomes lymph; provides WBCs involved in immunity.

Respiratory System
Transports respiratory gases.

Urinary System
Kidneys modify a blood filtrate to form urine.

Digestive System
Transports digested nutrients.

Reproductive System
Distributes sex hormones.

413

Study Guide

Assess Your Learning Outcomes

To test your knowledge, discuss the following topics with a study partner or in writing, ideally from memory.

12.1 Introduction (p. 392)

1. The components of the circulatory system; the differences between the cardiovascular system and circulatory system
2. The diverse functions of the blood; contributions of blood to homeostasis
3. The two main components of whole blood; the relative amounts of formed elements and their plasma in whole blood; and the three main types of formed elements
4. The components of blood plasma and their functions
5. The basis of blood viscosity and osmolarity and their importance in cardiovascular function
6. The formation of blood plasma
7. The definition of *hemopoiesis* and description of where it occurs (after infancy)
8. The name of the bone marrow cell type that ultimately gives rise to all formed elements

12.2 Erythrocytes (p. 396)

1. Ways of quantifying erythrocytes and hemoglobin levels, and typical values for women and men
2. The structure and function of erythrocytes, and the reason for their unusual structure
3. The structure and function of hemoglobin and how it binds blood gases

4. The process of erythropoiesis; the hormone that stimulates it
5. The life span, death, and disposal of RBCs and hemoglobin
6. The molecular basis of blood types A, B, AB, and O, and definitions of *antigen* and *antibody*
7. The cause and mechanism of a transfusion reaction and why it can lead to death; the meanings of *agglutination* and *hemolysis*
8. The two blood types of the Rh group
9. The cause of hemolytic disease of the newborn; why it is less common in a first pregnancy than in later ones

12.3 Leukocytes (p. 402)

1. The general function of leukocytes (WBCs)
2. The difference between granulocytes and agranulocytes
3. The name, physical characteristics, and functions of each WBC type
4. The cell lines that give rise to WBCs
5. The life history of WBCs after their release from the bone marrow

12.4 Platelets (p. 408)

1. The structure of platelets; why they are not considered to be cells
2. Functions of platelets
3. The site and process of platelet production
4. The general term for cessation of bleeding, and the three stages of the process
5. The intrinsic, extrinsic, and common pathways of coagulation and the ultimate product of these reactions
6. Differences between coagulation and agglutination
7. The process of breaking down a clot that is no longer needed
8. The hereditary and molecular bases of hemophilia
9. The differences between *thrombosis, thrombus,* and *embolus;* the danger of thromboembolism

Testing Your Recall

1. Which contribute(s) most to viscosity of the blood?
   a. albumin
   b. fibrinogen
   c. globulins
   d. sodium
   e. erythrocytes

2. What is the normal hematocrit for a male?
   a. between 45% and 52%
   b. between 37% and 48%
   c. 4.6 to 6.2 million/μL
   d. between 60% and 70%
   e. about 55%

3. Which is *not* a cause of leukocytosis?
   a. dehydration
   b. AIDS
   c. allergy
   d. leukemia
   e. bacterial infection

4. What blood type is a person who has only A antigens on his or her erythrocytes?
   a. type B
   b. type AB
   c. type O
   d. type A
   e. Rh+

5. Which of these is a granulocyte?
   a. a monocyte
   b. an erythrocyte
   c. a lymphocyte
   d. a macrophage
   e. an eosinophil

6. Platelets have all of the following functions *except*
   a. coagulation.
   b. plugging broken blood vessels.
   c. stimulating vasoconstriction.
   d. transporting oxygen.
   e. repairing blood vessels after an injury.

7. Histamine and heparin are secreted by
   a. lymphocytes.
   b. basophils.
   c. monocytes.
   d. eosinophils.
   e. platelets.

8. Order the formed elements from most to least abundant.
   1. erythrocytes
   2. basophils
   3. neutrophils
   4. lymphocytes
   5. platelets

   a. 1, 2, 4, 5, 3
   b. 5, 1, 4, 3, 2
   c. 3, 4, 2, 1, 5
   d. 1, 5, 3, 4, 2
   e. 1, 5, 4, 3, 2

9. Which is *not* true of erythrocytes?
   a. They have no nucleus.
   b. They have no mitochondria.
   c. Their cytoplasm consists mostly of a 33% solution of hemoglobin.
   d. They live for about 60 days.
   e. Their main function is to transport O_2 and CO_2.

10. What is the first event that occurs in hemostasis?
    a. vascular spasm
    b. platelet plug formation
    c. coagulation
    d. conversion of fibrinogen to fibrin
    e. secretion of platelet-derived growth factor

11. Production of all the formed elements of blood is _____.

12. The percentage of blood volume composed of RBCs is called the _____.

13. After monocytes emigrate from the blood into the connective tissues, they become _____.

14. The cessation of bleeding is called _____.

15. _____ is the fluid that remains if all the formed elements and fibrinogen are removed from the blood.

16. The hereditary lack of factor VIII causes a disease called _____.

17. An abnormally low WBC count is called _____.

18. The smallest and most abundant protein, _____, plays a role in maintaining blood osmolarity and viscosity.

19. An excess of RBCs is called _____.

20. The production of red blood cells is called _____.

Answers in appendix A

True or False

Determine which five of the following statements are false, and briefly explain why.

1. By volume, the blood usually contains more plasma than formed elements.

2. An increase in the albumin concentration of the blood would tend to increase blood pressure.

3. Anemia is caused by a low oxygen concentration in the blood.

4. The most important WBCs in combating a bacterial infection are basophils.

5. Platelets and erythrocytes lack nuclei.

6. Lymphocytes are the most abundant WBCs in the blood.

7. Platelet count is often depressed in people with leukemia.

8. All formed elements of the blood come ultimately from hemopoietic stem cells.

9. Since RBCs have no nuclei, they do not live as long as granulocytes do.

10. Leukemia is a severe deficiency of WBCs.

Answers in appendix A

Testing Your Comprehension

1. A patient is found to be seriously dehydrated and to have an elevated RBC count. Does the RBC count necessarily indicate a disorder of erythropoiesis? Why or why not?

2. Patients suffering from renal failure are typically placed on hemodialysis and erythropoietin (EPO) replacement therapy. Explain the reason for giving EPO, and predict what the consequences would be of not including this in the treatment regimen.

3. Explain why athletes sometimes illegally inject the hormone erythropoietin (EPO) in an effort to boost performance. What is a possible health hazard associated with this form of "blood doping?"

Chapter 13

The Circulatory System II

HEART AND BLOOD VESSELS

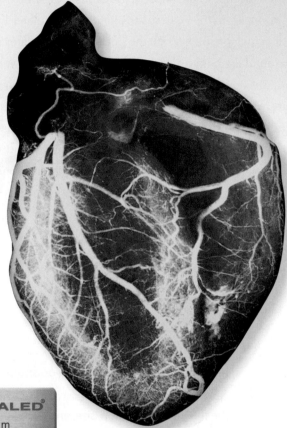

A colorized image of coronary blood vessels that have been injected with a dye to enhance their visibility in an X-ray. Such an image is called a coronary angiogram.

Anatomy & Physiology | **REVEALED**®
aprevealed.com

Module 9: Circulatory System

BASE CAMP

Before ascending to the next level, be sure you're properly equipped with a knowledge of these concepts from earlier chapters:

- Thoracic cavity anatomy (p. 19)
- Desmosomes and gap junctions (p. 72)
- Simple squamous epithelium (p. 103)
- Microscopic structure of skeletal muscle (p. 206)
- Resting membrane potentials and action potentials (p. 264)

We are more conscious of our heart than we are of most organs, and more wary of its failure. The Greek scholar Aristotle thought the heart served primarily as the seat of emotion, a view that persisted through the Middle Ages. Even today, the heart is associated with emotions, as is reflected in its prominence on Valentine's Day and in common phrases such as "heartfelt," "dart to the heart," and "heartbroken" to describe various feelings. It was not until the Renaissance that the anatomy and physiology of the heart were systematically studied. Later, the use of microscopes revealed the existence of tiny vessels, the capillaries, which connect the blood flowing from the heart to the blood that circulates back to it, thus clinching the theory that the same blood circulates repeatedly around the body and is not consumed by the organs.

The heart and blood vessels are of great interest in medical research and practice due to the prevalence of cardiovascular disease. In recent decades, advances such as coronary bypass surgery, valve replacements, and artificial pacemakers have prolonged countless lives. It is certain that cardiology will long remain one of the most dramatic and attention-getting fields of medicine.

13.1 Overview of the Cardiovascular System

Expected Learning Outcomes

When you have completed this section, you should be able to:

a. distinguish between the pulmonary and systemic circulations;

b. describe the general location, size, and shape of the heart; and

c. describe the pericardium, which encloses the heart.

The **cardiovascular system** consists of the heart and the blood vessels, whereas the circulatory system includes these and the blood. The heart functions as a muscular pump that keeps blood flowing through the vessels. The vessels deliver blood to nearly all tissues of the body and then return it to the heart. **Cardiology** is the study of the heart, clinical evaluation of its functions and disorders, and treatment of cardiac diseases.

The Pulmonary and Systemic Circuits

The cardiovascular system has two major divisions: a **pulmonary circuit**, which carries blood to the lungs for gas exchange and returns it to the heart, and a **systemic circuit**, which supplies blood to all organs of the body, including the lungs and the wall of the heart itself (fig. 13.1).

The right side of the heart supplies the pulmonary circuit. It receives oxygen-poor blood that has circulated through the body and pumps it into a large artery, the *pulmonary trunk*. From there, the oxygen-poor blood is distributed to the lungs, where it unloads carbon dioxide and picks up a fresh load of oxygen. It then returns this oxygen-rich blood to the left side of the heart by way of *pulmonary veins*.

The left half of the heart supplies the systemic circuit. It pumps blood into the largest artery, the *aorta*, which gives off branches that ultimately deliver oxygen to every organ of the body. After the blood has picked up carbon dioxide from the tissues, it returns to the right heart by way of the two largest veins—the *superior vena cava*, which drains the upper body, and the *inferior vena cava*, which drains everything below the diaphragm.

Position, Size, and Shape of the Heart

The heart is located in the thoracic cavity in the mediastinum, between the lungs and deep to the sternum (fig. 13.2). Tilted slightly toward the left, about two-thirds of the heart lies to the left of the median plane. The broad superior portion of the heart, the **base**, is the point of attachment for the pulmonary trunk, pulmonary veins, and aorta—the so-called *great vessels*. The inferior end tapers to a blunt point, the **apex**, immediately above the diaphragm.

The normal adult heart weighs about 300 g (10 ounces) and measures about 9 cm (3.5 in.) wide at the base, 13 cm (5 in.) from base to apex, and 6 cm (2.5 in.) from anterior to posterior at its thickest point. Whatever one's body size, the healthy heart is roughly the same size as the fist.

The Pericardium

The heart is enfolded in a double-walled sac called the **pericardium** (fig. 13.2c). The outer wall, the **pericardial sac (parietal pericardium)**, has a tough, superficial *fibrous layer* of dense irregular connective tissue, and a deep *serous layer*. The serous layer turns inward at the base of the heart and forms the **epicardium (visceral pericardium)**, which covers the heart surface.

The space between the parietal and visceral pericardium is the **pericardial cavity**. The heart is not inside the pericardial cavity; all the cavity contains is a small amount (5–30 mL) of **pericardial fluid**, exuded by the serous layer of the pericardium. The fluid lubricates the membranes and allows the heart to beat with minimal friction. The pericardium not only reduces friction but also isolates the heart from other thoracic organs and anchors it within the thoracic cavity.

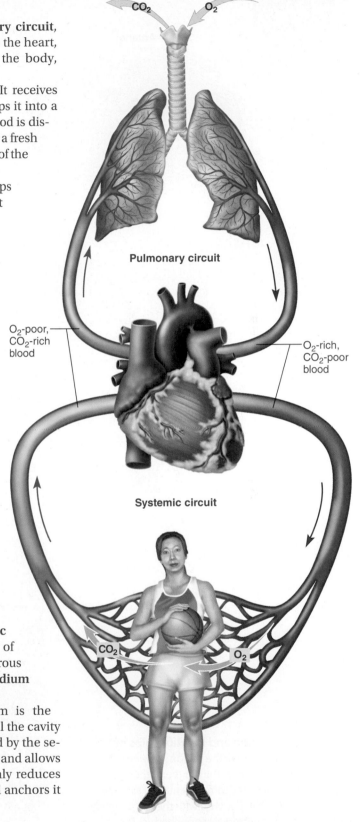

CO_2 O_2

Pulmonary circuit

O_2-poor, CO_2-rich blood

O_2-rich, CO_2-poor blood

Systemic circuit

CO_2 O_2

Figure 13.1 General Schematic of the Cardiovascular System. AP|R

- *Are the lungs supplied by the pulmonary circuit, the systemic circuit, or both? Explain.*

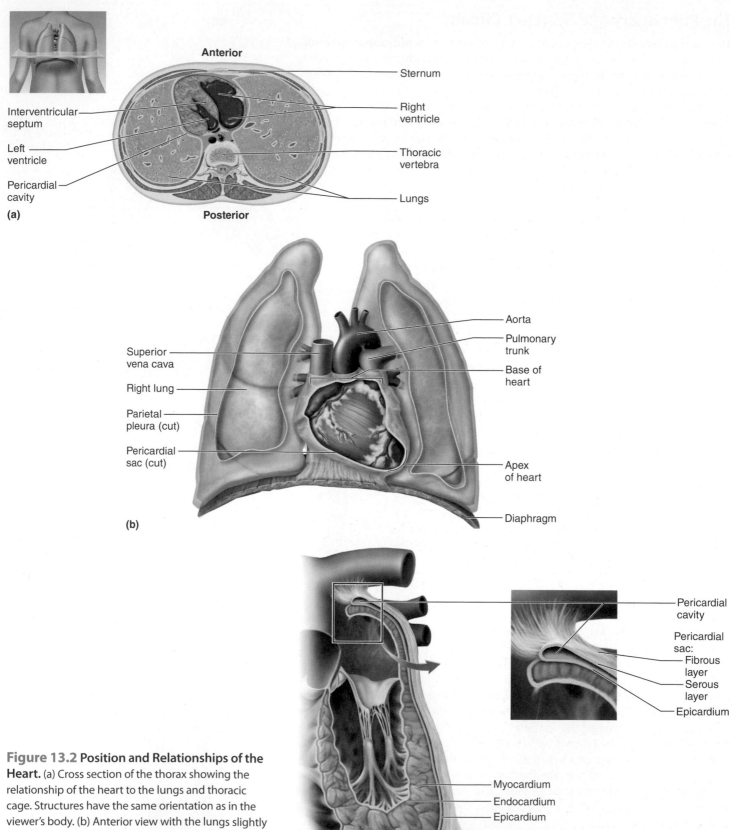

Figure 13.2 Position and Relationships of the Heart. (a) Cross section of the thorax showing the relationship of the heart to the lungs and thoracic cage. Structures have the same orientation as in the viewer's body. (b) Anterior view with the lungs slightly retracted and the pericardial sac opened. (c) The pericardium and heart wall. **AP|R**

• *Does most of the heart lie to the right or left of the median plane?*

Pericarditis is inflammation of the pericardium, most often caused by viral infection, in which membranes may become roughened and produce a painful *friction rub* with each heart beat.

Before You Go On

Answer these questions from memory. Reread the preceding section if there are too many you don't know.

1. Distinguish between the pulmonary and systemic circuits and state which part of the heart supplies each one.

2. Make a two-color sketch of the pericardium, using one color for the pericardial sac and another for the epicardium. For the pericardial sac, label both the fibrous and serous layers. Show the relationship between the pericardium, pericardial cavity, and heart wall.

13.2 Gross Anatomy of the Heart

Expected Learning Outcomes

When you have completed this section, you should be able to:

a. describe the three layers and four chambers of the heart;

b. describe the surface features of the heart;

c. explain the structure and function of the valves;

d. trace the flow of blood through the chambers; and

e. describe the vessels that supply blood to the heart wall.

The Heart Wall

The heart wall consists of three layers—a thick muscular *myocardium* sandwiched between two thin serous membranes, the *epicardium* and *endocardium* (fig. 13.2). The **epicardium**,[1] as we have already seen, covers the surface of the heart. It consists of a simple squamous epithelium overlying a layer of areolar tissue; in most places it is as thin as tissue paper, and translucent so that the underlying muscle shows through. In some areas, it also has a layer of adipose tissue. Some individuals have a thick blanket of fat around the entire heart, but this is not a healthy condition.

The **endocardium**[2] lines the interior of the heart chambers, covers the valve surfaces, and is continuous with the inner lining *(endothelium)* of the blood vessels. Like the epicardium, it is composed of simple squamous epithelium overlying a thin layer of areolar tissue, but it contains no adipose tissue.

The **myocardium**,[3] composed of cardiac muscle, lies between those two layers and makes up most of the mass of the heart. It does the contractile work of the heart, and its thickness varies greatly according to the workload of

---

[1] *epi* = upon; *cardi* = heart

[2] *endo* = internal; *cardi* = heart

[3] *myo* = muscle; *cardi* = heart

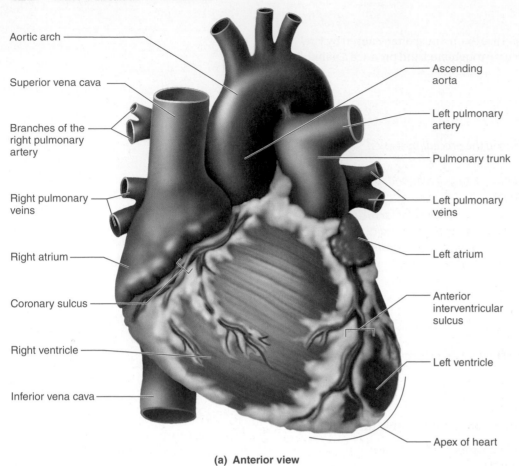

Aortic arch

Superior vena cava

Branches of the
right pulmonary
artery

Right pulmonary
veins

Right atrium

Coronary sulcus

Right ventricle

Inferior vena cava

Ascending
aorta

Left pulmonary
artery

Pulmonary trunk

Left pulmonary
veins

Left atrium

Anterior
interventricular
sulcus

Left ventricle

Apex of heart

(a) Anterior view

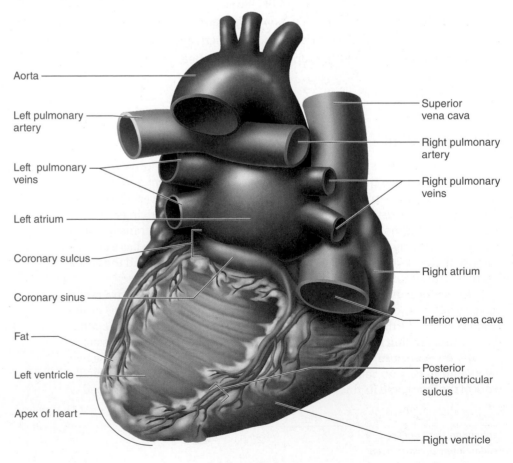

Aorta

Left pulmonary
artery

Left pulmonary
veins

Left atrium

Coronary sulcus

Coronary sinus

Fat

Left ventricle

Apex of heart

Superior
vena cava

Right pulmonary
artery

Right pulmonary
veins

Right atrium

Inferior vena cava

Posterior
interventricular
sulcus

Right ventricle

(b) Posterior view

**Figure 13.3 External Anatomy of
the Heart.** The unlabeled coronary
blood vessels on the heart surface are
identified in figure 13.5. AP|R

individual chambers. This muscle coils around in such a way as to create a twisting action when it contracts, "wringing" blood out of the heart like wringing out a wet towel.

The heart also has a connective tissue framework called the **fibrous skeleton**, concentrated especially in the walls between the chambers and in rings around the valve orifices. It has several roles: (1) It provides attachment for the valves and holds the valve orifices open, yet prevents them from being excessively stretched when blood surges through. (2) It anchors the cardiac muscle cells, giving them something to pull on when they contract. (3) It serves as an electrical insulator between the atria and ventricles, and thus prevents the atria from stimulating the ventricles directly; this is important in the timing and coordination of cardiac excitation and contraction.

The Chambers

The heart has four chambers that receive and eject blood (figs. 13.3 and 13.4). The two smaller, superior chambers are the **right** and **left atria** (AY-tree-uh; singular *atrium*[4]). They are thin-walled receiving chambers for blood returning to the heart by way of the great veins. Most of the mass of each atrium is on the posterior side of the heart, so only a small portion is visible from the anterior view (fig. 13.3).

---

[4] *atrium* = entryway

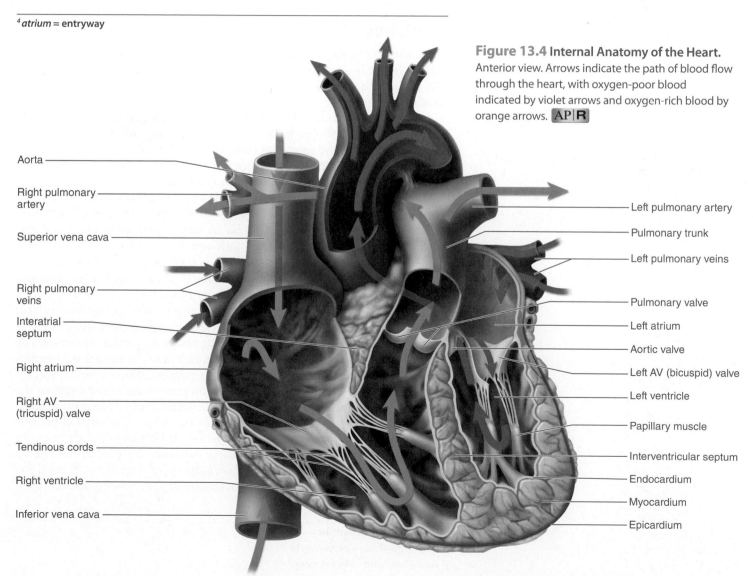

Figure 13.4 **Internal Anatomy of the Heart.** Anterior view. Arrows indicate the path of blood flow through the heart, with oxygen-poor blood indicated by violet arrows and oxygen-rich blood by orange arrows. AP|R

Aorta

Right pulmonary artery

Superior vena cava

Right pulmonary veins

Interatrial septum

Right atrium

Right AV (tricuspid) valve

Tendinous cords

Right ventricle

Inferior vena cava

Left pulmonary artery

Pulmonary trunk

Left pulmonary veins

Pulmonary valve

Left atrium

Aortic valve

Left AV (bicuspid) valve

Left ventricle

Papillary muscle

Interventricular septum

Endocardium

Myocardium

Epicardium

The two inferior chambers, the **right** and **left ventricles**,[5] compose most of the bulk of the heart. They are strong pumps that eject blood into the arteries and keep it flowing around the body. The right ventricle constitutes most of the anterior portion of the heart, whereas the left ventricle forms the apex and inferoposterior portion.

The boundaries of the four chambers are marked on the surface by three sulci (grooves), which are largely filled with fat and coronary blood vessels. The **coronary**[6] **sulcus** encircles the heart near the base and separates the atria from the ventricles. It is a deep groove that is easily seen when one lifts the edges of the atria (fig. 13.3a). The **anterior** and **posterior interventricular sulci** run obliquely on the front and back of the heart from the base toward the apex. They are useful landmarks for distinguishing the boundary between the right and left ventricles because they overlie the *interventricular septum*.

The thickness of the walls varies among the chambers. The atria have thin flaccid walls, corresponding to their light workload—all they do is pump blood to the ventricles immediately below. The walls of the ventricles are much thicker, especially the left, which is two to four times as thick as the right because it pumps blood through the entire body, whereas the right ventricle pumps blood only to the lungs and back. The **interventricular septum** is a thick wall between the two ventricles; the **interatrial septum** lies between the atria.

The Valves

Four heart valves ensure a one-way flow of blood—one between each atrium and its ventricle and one at the exit from each ventricle to its great artery (fig. 13.4). Each valve consists of two or three flaps of thin tissue called **cusps (leaflets)**.

The **atrioventricular (AV) valves** control the opening between each atrium and the ventricle below it. They ensure that blood cannot regurgitate back into the atria when the ventricles contract. The **right AV valve** is often called the **tricuspid valve** because it has three cusps. The **left AV valve** is also known as the **mitral valve** for its resemblance to a miter, the headdress of a church bishop. (It was formerly known also as the bicuspid valve from a mistaken belief that it had only two cusps.) Each cusp is anchored to conical **papillary muscles** on the floor of the ventricle by way of stringy **tendinous cords (chordae tendineae)** (KOR-dee ten-DIN-nee-ee). When the ventricles contract, the papillary muscles tense the cords and prevent the AV valves from bulging back into the atria or flipping inside out like windblown umbrellas. Excessive bulging due to slack tendinous cords is called *valvular prolapse*.

The pulmonary and aortic valves (collectively called **semilunar**[7] **valves**) lie between the ventricles and great arteries. The **pulmonary valve** controls the opening from the right ventricle into the pulmonary trunk, and the **aortic valve** controls the opening from the left ventricle into the aorta. Each semilunar valve has three cusps shaped like shirt pockets. When blood is ejected from the ventricles, it opens these valves and presses their cusps against the arterial walls. When the ventricles relax and expand, arterial blood flows backward toward the ventricles, but quickly fills the cusps. The inflated pockets meet at the center and quickly seal the opening, so little blood is able to return to the ventricles. Because of the way these cusps are attached to the arterial wall, they cannot prolapse any more than a shirt pocket turns inside out if you jam your hand into it. Thus, they do not require or possess tendinous cords.

[5] *ventr* = belly, lower part; *icle* = little

[6] *coron* = crown; *ary* = pertaining to

[7] *semi* = half; *lunar* = like the moon

Blood Flow Through the Chambers AP|R

Until the sixteenth century, anatomists thought that blood flowed directly from the right ventricle to the left through invisible pores in the septum. In actuality, blood in the right and left chambers is kept entirely separate in the adult. Thus, deoxygenated blood returning from the tissues and destined for the lungs does not mix with oxygenated blood that has returned from the lungs. Figure 13.4 shows the pathway of blood as it travels from the right atrium through the body and back to the starting point.

Blood that has been through the systemic circuit returns by way of the superior and inferior venae cavae to the right atrium. It flows directly from the right atrium, through the right AV valve, into the right ventricle. When the right ventricle contracts, it ejects this blood through the pulmonary valve into the pulmonary trunk, on its way to the lungs to exchange carbon dioxide for oxygen.

Blood returns from the lungs by way of two pulmonary veins on the left and two on the right; all four of these empty into the left atrium. Blood flows through the left AV valve into the left ventricle. Contraction of the left ventricle ejects this blood through the aortic valve into the ascending aorta, on its way to another trip around the systemic circuit.

Coronary Circulation

If your heart lasts for 80 years and beats an average of 75 times a minute, it will beat more than 3 billion times and pump more than 200 million liters of blood. It is a remarkably hardworking organ that requires an abundant supply of oxygen and nutrients.

The energetic demands of the heart mean that it receives a disproportionate amount of the body's blood supply. At rest, the coronary blood vessels supply the heart wall with about 250 mL of blood per minute. This is about 5% of the circulating blood going to meet the metabolic needs of the heart alone, even though the heart is only 0.5% of the body's weight. It receives 10 times its "fair share" to sustain its strenuous workload. Of course, the amount increases dramatically during exercise. The blood vessels devoted to supplying the heart wall constitute the **coronary circulation** (fig. 13.5).

Arterial Supply

Coronary arteries are a focus of attention by health professionals and laypeople alike because their blockage by atherosclerosis can lead to heart attacks (fig. 13.5d, and see p. 427). The pattern of vessel branching is one of the most variable aspects of anatomy; the following description describes the pattern seen in about 70% to 85% of the population.

Immediately after the aorta leaves the left ventricle, it gives off a right and left coronary artery. The openings of the two arteries lie behind the cusps of the aortic valve. When the ventricles relax, a small amount of blood backflows from the aorta and fills the cusps, allowing flow into the coronary arteries. Thus, unlike most arteries in the body, coronary artery flow peaks during ventricular relaxation rather than ventricular contraction.

- The **left coronary artery (LCA)** curves under the left atrium and travels a short distance before dividing into two branches: the **anterior interventricular** and **circumflex branch**. The anterior interventricular branch is also known clinically as the *left anterior descending (LAD) branch*. It travels down the anterior interventricular sulcus to the apex of the heart, rounds the bend, and continues a short distance up the posterior side. It supplies both ventricles and the anterior two-thirds of their septum. The circumflex branch continues around the left side of the heart in the coronary sulcus. It gives off branches to the left ventricle and the left atrium.

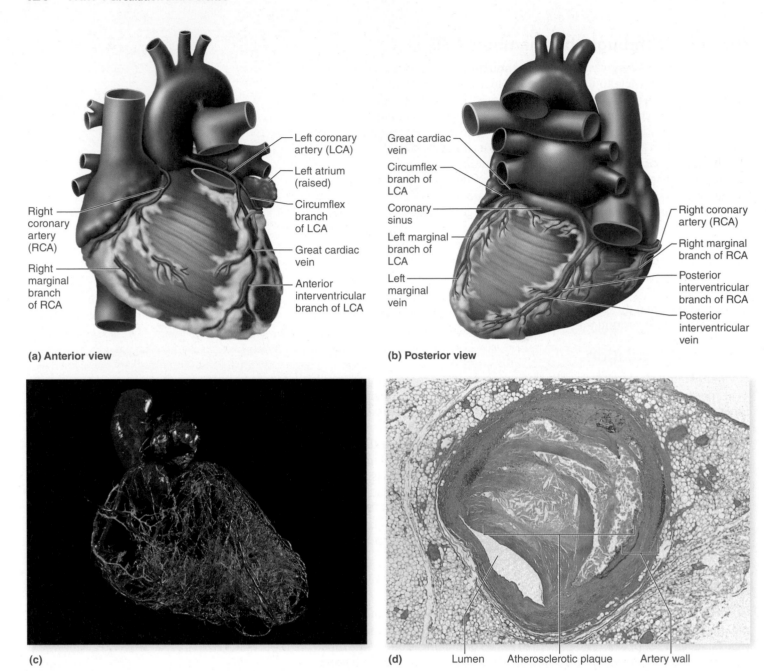

(a) Anterior view

Left coronary artery (LCA)
Left atrium (raised)
Circumflex branch of LCA
Great cardiac vein
Anterior interventricular branch of LCA

Right coronary artery (RCA)
Right marginal branch of RCA

(b) Posterior view

Great cardiac vein
Circumflex branch of LCA
Coronary sinus
Left marginal branch of LCA
Left marginal vein

Right coronary artery (RCA)
Right marginal branch of RCA
Posterior interventricular branch of RCA
Posterior interventricular vein

(c)

(d)   Lumen   Atherosclerotic plaque   Artery wall

Figure 13.5 The Major Coronary Blood Vessels. (a) Anterior view. (b) Posterior view. (c) A polymer cast of the coronary circulation. (d) Cross section of a coronary artery with advanced atherosclerosis. Most of the lumen is obstructed by a plaque of calcified scar tissue. The small remaining space can easily be blocked by thrombosis, embolism, or vasoconstriction, causing a myocardial infarction.

AP|R

- The **right coronary artery (RCA)** supplies the right atrium, continues along the coronary sulcus, and gives off two branches. The **right marginal branch** runs toward the apex and supplies portions of the right atrium and ventricle. The **posterior interventricular branch** runs down the posterior interventricular sulcus and supplies the posterior walls of both ventricles as well as the posterior portion of the interventricular septum. It ends where it joins the anterior interventricular branch of the LCA.

Venous Drainage

Venous drainage refers to the route by which blood leaves an organ. In the heart, after the blood flows through the coronary arteries and passes through capillaries, it drains into several veins that return the blood to the right atrium. The four most significant are the following.

- The **coronary sinus** is a large, horizontal, venous tunnel in the coronary sulcus on the posterior side of the heart (fig. 13.5b). It collects blood from following veins, among others, and then empties into the right atrium.

PERSPECTIVES ON HEALTH

Coronary Artery Disease

Coronary artery disease (CAD) is a degenerative change in the coronary arteries associated with atherosclerosis (see Clinical Application 13.2). It is the most common cause of heart failure and one of the leading killers in the United States. An early warning symptom of CAD is *angina pectoris*, a transient substernal chest pain that occurs when obstructed coronary arteries are not delivering enough oxygen to some areas of the myocardium. Angina is often a forerunner to myocardial infarction.

One of the treatments for CAD is *balloon angioplasty*. A cardiologist inserts a catheter into the femoral artery at the groin and threads it all the way up to the heart and into the diseased coronary artery. A tiny balloon at the tip of the catheter is then inflated, which crushes the atheroma and plasters it against the vessel wall. This opens the vessel, restores blood flow, and relieves pain, but it is typically a temporary solution and often has to be repeated. It does not reduce the risk of heart attack or death in patients under medical treatment regimens.

A more invasive procedure is *coronary bypass surgery*, in which the chest is opened, the patient is put on a heart–lung bypass machine, and undamaged vessels (usually from the leg or chest) are attached above and below the damaged portion of the coronary artery. This provides a detour around the obstruction so blood can reach downstream tissues.

Prevention of CAD is, of course, preferable to treatment. Regular exercise, a healthy body weight, and a diet rich in fruits and vegetables and low in saturated fat and cholesterol all protect against CAD. Avoiding smoking is also important. Chapter 18 discusses the relationship of cardiovascular health and disease to so-called "good" and "bad cholesterol" (more correctly known as high- and low-density lipoproteins, or HDLs and LDLs).

While lifestyle plays a role, some risk factors for CAD are unavoidable. One of these is heredity. In a pair of identical twins, if one suffers CAD or dies of a resulting MI, the other is much more likely to do so at a similar age than is a nonidentical sibling with a similar lifestyle. Males and older people are also more at risk of CAD than females and younger people. Drugs that control hypertension and lower cholesterol levels are commonly used to reduce the progression of cardiovascular disease.

- The **posterior interventricular (middle cardiac) vein** leads up the posterior interventricular sulcus to the sinus.
- The **left marginal vein** passes from the apex up the left margin of the heart to the sinus.
- The **great cardiac vein** passes up the anterior aspect of the heart from the apex toward the coronary sulcus, then arcs around the left side of the heart and empties into the sinus.

Before You Go On

Answer these questions from memory. Reread the preceding section if there are too many you don't know.

3. Which layer of the heart wall is thickest? How do the other two resemble and differ from each other?
4. What are structural and functional differences between the atrioventricular and semilunar valves?
5. Trace the route of the blood through the heart, naming each chamber and valve in order.
6. Trace the alternative routes of blood flow through the coronary arteries and veins that supply the heart wall.

13.3 Physiology of the Heart

Expected Learning Outcomes

When you have completed this section, you should be able to:

a. describe the structure of cardiac muscle cells and their attachments to each other;

b. explain why the heart needs such a high capacity for aerobic respiration;

c. describe the heart's electrical system, how it coordinates the beating of the heart chambers, and the unusual action potentials of the pacemaker and myocardium;

d. describe the nerve supply to the heart;

e. diagram the electrocardiogram and label and explain each of its waves;

f. describe the key events that occur through one cycle of cardiac contraction and relaxation;

g. explain what causes the "lubb-dupp" sounds of the heartbeat; and

h. define *cardiac output* and explain what determines it and what chemical agents can modify it.

Despite our justifiable fear of heart failure, the heart is a remarkably hardworking and reliable organ. Two of its most important physiological properties are its rhythmicity and its resistance to fatigue. In this section, we examine the bases for those two properties and the operation of the heart through one cycle of contraction and relaxation.

Cardiac Muscle

The work of the heart is, of course, done by its muscle cells, called **cardiomyocytes**.[8] Like skeletal muscle fibers, they are striated, but they differ in several other ways. They are short stocky cells rather than slender fibers—typically 50 to 100 μm long and 10 to 20 μm wide (fig. 13.6). Each has a single centrally placed nucleus, often surrounded with glycogen.

The ends of each cardiomyocyte are slightly branched, so the cell somewhat resembles a short log with notches in the ends. Each branch reaches out to connect end to end with another cardiomyocyte. The cells stimulate each other through these connections and therefore act as a unit rather than independently. Their synchronized activity is necessary for the coordinated contractions of the heart chambers.

The ends where two cardiomyocytes meet are marked by **intercalated** (in-TUR-ku-LAY-ted) **discs**, which appear through the microscope as dark

---

[8] *cardio* = heart; *myo* = muscle; *cyte* = cell

Striations

Nucleus

Intercalated discs

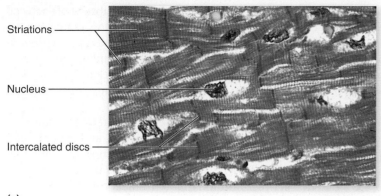

(a)

Figure 13.6 Cardiac Muscle. (a) Light micrograph. (b) Structure of a cardiomyocyte and its relationship to adjacent cardiomyocytes. All of the colored area is a single cell. Note that it is notched at the ends and typically linked to two or more neighboring cardiomyocytes by the mechanical and electrical junctions of the intercalated discs. (c) Structure of an intercalated disc. **AP|R**

Striated myofibril   Glycogen   Nucleus   Mitochondria   Intercalated discs

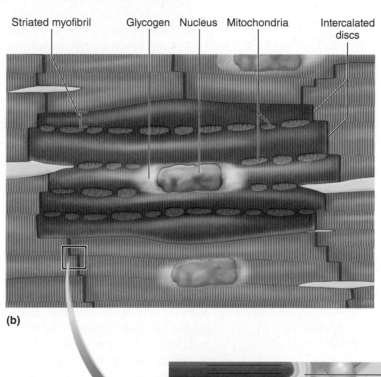

(b)

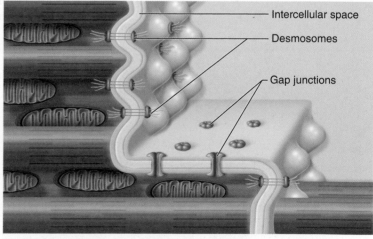

Intercellular space

Desmosomes

Gap junctions

(c)

transverse lines, heavier than the striations. The plasma membrane of each cell at this point is folded somewhat like the bottom of an egg carton, increasing the surface area of cell-to-cell contact. The cells are joined at the discs by *desmosomes* and other mechanical junctions that keep them from pulling apart when the myocardium contracts, and by electrical *gap junctions* (see p. 73) that allow ions to flow from one cell directly into the cytoplasm of the next, thus enabling electrical excitation to spread rapidly from cell to cell.

The heart requires a great deal of energy to beat so steadily, year after year without fail. Although it weighs only 300 g, it produces and uses about 5 kg of ATP per day. Fatty acids and glucose provide most of the energy for ATP production. Cardiomyocytes make ATP almost exclusively by aerobic respiration, and are therefore very resistant to fatigue. To meet this ATP demand, they have exceptionally large mitochondria constituting about 25% of the cell volume, in contrast to smaller mitochondria making up only 2% of a skeletal muscle cell. Cardiomyocytes need an abundant, reliable supply of oxygen, which makes the coronary circulation especially critical, but they also are rich in the oxygen-storing pigment myoglobin.

The Cardiac Conduction System

The heart is described as **autorhythmic** because it beats at its own rhythm without need of stimuli from the nervous system. This results from the action of modified cardiomyocytes that have lost the ability to contract, and are instead specialized to depolarize spontaneously at regular time intervals. Such cells are concentrated in two masses called *sinoatrial* and *atrioventricular nodes*. Other noncontractile cardiomyocytes behave much like nerves; together with the nodes, they form the **cardiac conduction system**. Electrical signals travel rapidly through the system in the following order, ultimately to stimulate the contractile cardiomyocytes and generate the heartbeat (fig.13.7):

① The **sinoatrial (SA) node**. This is the heart's pacemaker, a patch of cells in the right atrium near the superior vena cava. These cells depolarize every 0.8 seconds or so at rest, setting off each heartbeat and determining the heart rate (about 70 beats/min. at that rate of pacemaker firing, and faster if the pacemaker fires more often).

② **Internodal conduction**. Signals from the SA node spread through both atria, causing them to contract before the ventricles.

Figure 13.7 The Cardiac Conduction System. Electrical signals travel along the pathways indicated by the arrows. **AP|R**

- *One atrium begins contracting slightly before the other. Which do you think begins first, and why?*

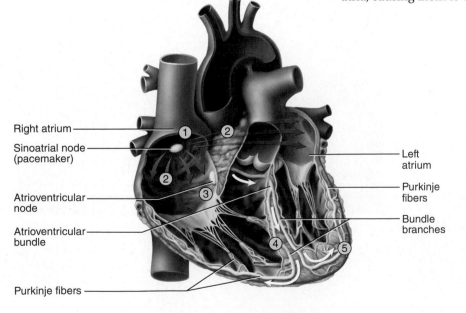

Right atrium

Sinoatrial node (pacemaker)

Atrioventricular node

Atrioventricular bundle

Purkinje fibers

Left atrium

Purkinje fibers

Bundle branches

① SA node fires.

② Excitation spreads through atrial myocardium.

③ AV node fires.

④ Excitation spreads down AV bundle.

⑤ Purkinje fibers distribute excitation through ventricular myocardium.

③ The **atrioventricular (AV) node**. The AV node is a patch of cells similar to the SA node but located in the interatrial septum just above the tricuspid valve. It acts as the gateway for all electrical signals headed for the ventricles. The signals must pass through this node because the fibrous skeleton blocks them from traveling to the ventricles by any other route. The AV node delays the signal somewhat, allowing the ventricles time to fill with blood before the signal travels on and stimulates them to contract. The signal travels like a car that has to slow down on a state highway at a small town with one traffic light (the AV node) before speeding up again.

④ The **atrioventricular (AV) bundle (bundle of His**[9]**)**. This is a cord of modified cells that leaves the AV node and travels to the interventricular septum. In the septum, it forks into the **right** and **left bundle branches**, which descend toward the apex of the heart. Like a car on the highway at the far side of that little town, here the signal speeds up again.

⑤ **Purkinje**[10] (pur-KIN-jee) **fibers**. Purkinje fibers are large cells that arise from the bundle branches. They pass down the interventricular septum just under the endocardium, then turn upward at the apex of the heart and branch into little vinelike tips throughout the ventricular myocardium. They conduct action potentials very rapidly, and excitation spreads almost immediately throughout the entire ventricular mass, making it contract in near unison.

In further comparison to the highway analogy, cardiac signals travel at a velocity of about 1 meter per second (m/s) through the muscle of the atria. When they reach the "small town" of the AV node, they slow down to 0.05 m/s and they're delayed for about 100 milliseconds before reaching the AV bundle on "the other side of town." In the bundle and Purkinje fibers, the signal speeds up to 4 m/s. As small as the heart is, signals traveling at such a speed can reach all cells of the ventricular myocardium in a small fraction of a second.

Electrical Activity and Contraction

We saw in chapter 8 that neurons have a stable voltage across the plasma membrane called a *resting membrane potential*, as they "wait" for excitation from an external source. Stimulation opens membrane channels, allowing ions to flow through, and this changes that resting potential into an *action potential*. Cells of the cardiac SA node, however, behave differently. They do not rely on an external source of stimulation and they do not have a stable resting potential. Their membranes are leaky, continually allowing sodium ions into the cell. Each time this leakage raises the membrane potential to its threshold voltage, it sets off an action potential; the pacemaker fires and initiates another heartbeat.

The contractile cardiomyocytes, however, do have a stable resting potential, waiting at a voltage of about –90 mV to receive a signal from the Purkinje fibers or from each other. The arrival of a signal opens ion channels in the cell and causes an action potential, which triggers contraction of the cell much like we saw in skeletal muscle (chapter 7). Unlike the action potentials of neurons and skeletal muscle, the action potential in cardiac muscle is not simply

---

[9] Wilhelm His Jr. (1863–1934), German physiologist
[10] Johannes E. Purkinje (1798–1869), Bohemian physiologist

a quick up-and-down voltage spike. In skeletal muscle, an action potential falls back to the resting level in about 2 milliseconds (ms). In cardiac muscle, however, the voltage remains elevated for about 200 ms (one-fifth of a second). This produces a prolonged contraction of the myocardium, necessary to ensure effective ejection of blood from the heart chambers. It also prevents the cardiomyocytes from being stimulated again too soon. Premature restimulation could create a state of tetanus (see p. 216), causing the heart to "seize up" and cease pumping.

Nerve Supply to the Heart

Even though the heart has its own pacemaker, it obviously doesn't always beat at the same rate. A typical resting heart rate is about 70 to 75 beats/min. (bpm), whereas vigorous exercise can raise it to as high as 230 bpm. The heart rate is modified by the autonomic nervous system. Sympathetic fibers travel through *cardiac nerves* to the SA and AV notes and the myocardium. They release norepinephrine, which stimulates more rapid firing of the SA node and stronger contractions in the cardiomyocytes. Parasympathetic fibers travel to the heart through the two vagus nerves. In the heart wall itself, they synapse with short postganglionic neurons that lead to the SA and AV nodes; the myocardium receives little or no parasympathetic input. Parasympathetic fibers secrete acetylcholine, which slows the heartbeat by decreasing the speed of depolarization in the pacemaker cells.

The Electrocardiogram

Next to listening to the heart sounds with a stethoscope, the most common clinical method of evaluating heart function is the **electrocardiogram**[11] (**ECG** or **EKG**[12]). The depolarization and repolarization of the atrial and ventricular myocardium generate electrical currents that are detectable with electrodes on the skin. These signals can be amplified and displayed on either a paper chart or a video monitor. The ECG is the record of these signals (fig. 13.8).

Three major events are seen in the ECG: the **P wave, QRS complex**, and **T wave**. The letters do not stand for words, but were used to designate the waves by the German physiologist Willem Einthoven. He developed the technique and made the first successful ECG in 1895, and later was awarded the Nobel Prize in Physiology or Medicine. The P wave reflects depolarization of the atria. The QRS complex represents depolarization of the ventricles. It is the largest wave of the ECG because the ventricles constitute the largest muscle mass in the heart and generate the strongest electrical current. The atria repolarize during the QRS complex but their effect is largely masked by depolarization of the ventricles. The T wave represents ventricular repolarization. These waves are not action potentials of individual cardiomyocytes but composite voltage changes produced by many cells.

Heart rate can be determined by measuring the time between successive R peaks. We will see in the next section how the waves of the ECG are correlated with the mechanical pumping actions of the heart.

The electrocardiogram is useful because it provides a noninvasive, if indirect, snapshot of heart function and a tool for rapid preliminary diagnosis of a variety of cardiac conditions (see Clinical Application 13.1).

---

[11] *electro* = electricity; *cardio* = heart; *gram* = record of

[12] EKG is from the German spelling, *Elektrokardiogramm*

Clinical Application 13.1

CARDIAC ARRHYTHMIA

Cardiac arrhythmia means any deviation from the normal heart rhythm. It can occur in either the atria or the ventricles. The most common form in the atria is *atrial flutter*, in which these chambers show weak chaotic contractions at about 250 to 350 bpm. The ECG shows a "sawtooth" pattern with multiple small atrial peaks between QRS waves. In itself, it is not life-threatening and it can come and go, but it may cause fatigue, pain, dizziness, and breathlessness and it can sometimes lead to more serious conditions such as blood clots in the heart.

The most serious ventricular arrhythmia is *ventricular fibrillation* (*VF* or *V-fib*), an uncoordinated squirming contraction of the ventricles often brought on by myocardial infarction (tissue death) or drugs of abuse such as amphetamines. A heart in VF is often said to feel like a squirming "bag of worms." The ECG shows grossly irregular waves of depolarization with no recognizable P, QRS, and T waves. There is no effective pumping of blood from such weak, uncoordinated ventricles, and without treatment, death is imminent. VF can be halted with a strong electrical shock from a *defibrillator*, but this is only a temporary life-prolonging measure. The patient may require a pacemaker, coronary bypass surgery, or other measures to manage the underlying problem.

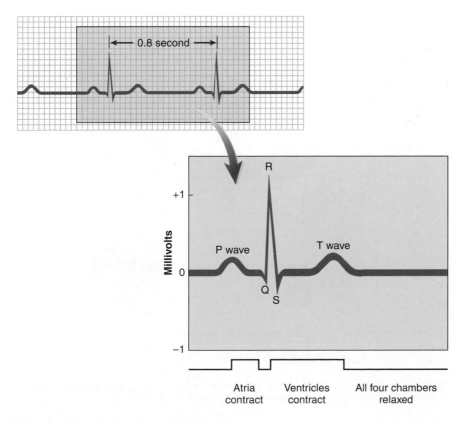

Figure 13.8 The Electrocardiogram.

The Cardiac Cycle

A **cardiac cycle** consists of one complete contraction and relaxation of all four heart chambers, encompassing events from the beginning of one heartbeat to the beginning of the next. Any chamber is said to be in **systole** (SIS-toe-lee) when it is contracting and in **diastole** (dy-ASS-toe-lee) when it is relaxed; when these words are used without specifying a chamber, however, they usually refer to the ventricles. Excitation of a heart chamber depolarizes its myocardium and leads to systole; repolarization of the myocardium relaxes its muscle and leads to the diastole of a chamber. If the relevant valves are open and allow blood to flow through, systole generally expels blood from a chamber and diastole allows a chamber to refill.

Figure 13.9 illustrates the mechanical events of the cardiac cycle correlated with the electrocardiogram and heart sounds. Closely follow the figure as you study the text.

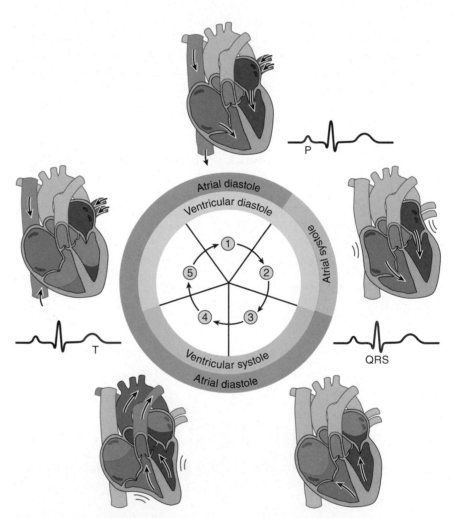

① All four chambers relaxed; AV valves open; ventricles fill passively.

② SA node fires; atria contract; ventricular filling completed.

③ AV node fires; ventricles contract; atria relax; AV valves are forced closed; heart sound S_1 occurs; with all valves closed, no blood is ejected yet.

④ Ventricular pressure forces semilunar valves open; ventricles eject blood.

⑤ Ventricles repolarize and relax; semilunar valves close; heart sound S_2 occurs; with all valves closed, no blood enters ventricles yet.

Figure 13.9 The Cardiac Cycle. Transitions from phases 1 to 2, 2 to 3, and 4 to 5 correspond to the electrocardiogram waves indicated in red. **AP|R**

• *The ventricles begin contracting at phase 3 but cannot change volume until phase 4. Why?*

① Initially, all four chambers are relaxed, in diastole. The AV valves are open, and without any effort from the heart, blood flows passively through these valves into the ventricles. This alone accounts for more than two-thirds of the ventricular filling.

② The SA node fires, exciting the atrial myocardium, and creating the P wave in the ECG. This initiates atrial systole and contributes the last 30% or so to ventricular filling. At the end of atrial systole, each ventricle contains about 130 mL of blood.

③ The AV node fires and electrical excitation spreads throughout the ventricles, producing the QRS complex in the ECG. This sets off ventricular systole. Meanwhile, the atria relax. Pressure in the ventricles rises sharply. Blood surges against the AV valves and forces them shut; the surge against this obstacle creates a vibration heard with the stethoscope as the **first heart sound** (S₁). At first, the ventricles cannot eject any blood because pressure in the aorta and pulmonary trunk is still higher than ventricular pressure and it holds the two semilunar valves closed.

④ When ventricular pressure exceeds pressure in the arteries, the semilunar valves open and blood spurts into the aorta and pulmonary trunk. Ventricular ejection lasts about 200 to 250 ms, the length of time described earlier when the cardiomyocyte membrane voltage remains elevated. At rest, the ventricles expel about 70 mL of blood each. This is called the **stroke volume (SV)**. Another 60 mL, almost half of what was in the ventricle, remains behind to mix with fresh blood on the next cycle. In vigorous exercise, however, stroke volume increases and the ventricles may eject as much as 90% of their blood. Diseased hearts may eject much less than 50%.

⑤ The myocardium repolarizes and produces the T wave of the ECG. The ventricles now relax and expand. All four chambers are again in diastole. Very briefly, blood in the aorta and pulmonary trunk surges backward, filling the pocketlike cusps of those two valves. The three valve cusps come together in the middle of the artery and close that opening. The surge of aortic and pulmonary blood against the closed valves creates another bump, heard as the **second heart sound** (S₂). As the ventricles expand, the AV valves reopen and the ventricles begin to refill with blood pouring down from the atria. The cycle begins anew.

Listening to the heart sounds with a stethoscope is a common part of a physical examination. It is an example of **auscultation** (AWS-cul-TAY-shun), listening to the sounds the body makes. The first and second heart sounds are often described as sounding like a "lubb-dupp." The first heart sound, S₁ ("lubb"), is louder and longer, and the second, S₂ ("dupp"), is softer and shorter.

Heart murmurs are whooshing or swishing sounds that are not part of the "lubb-dupp" sounds. Most heart murmurs are harmless, but some of them indicate valve problems. For example, blood may be regurgitated from the ventricles into the atria during ventricular systole because of structural problems with the AV valves. If the defect is great enough to interfere with the normal flow of blood, valve replacements may be an option. Since the 1960s, there has been a great deal of success in replacing defective valves with either mechanical valves or biological material such as heart valves from pigs. These surgical procedures have prolonged numerous lives.

Cardiac Output

The purpose of the cardiac cycle is to eject blood into the great arteries. The volume of blood ejected by each ventricle per minute (mL/min.) is called the **cardiac output**. Mathematically, it is the product of stroke volume, *SV* (mL/beat), and heart rate, *HR* (beats/min.)—that is, $SV \times HR = CO$. Typical adult resting values are

$$70 \text{ mL/beat } (SV) \times 75 \text{ beats/min. } (HR) = 5{,}250 \text{ mL/min. } (CO)$$

Thus, the body's entire volume of blood (usually 4–6 L) passes through the heart every minute. Vigorous exercise can raise *CO* to as much as 21 L/min., and world-class athletes can pump up to 35 L/min.

Given that $SV \times HR = CO$, there are obviously only two ways to increase cardiac output: Increase the stroke volume or increase the heart rate (or both). Agents that increase the heart rate include the sympathetic nervous system and its neurotransmitter, norepinephrine; the similar hormone epinephrine; and thyroid hormone, nicotine, and caffeine. Agents that slow down the heart include parasympathetic stimulation with acetylcholine, as well as excessive levels of calcium or potassium.

Other agents increase cardiac output by increasing the contraction strength of the cardiomyocytes. These include epinephrine and norepinephrine, calcium, the hormone glucagon, and the drug digitalis. Yet other agents weaken myocardial contractions and reduce cardiac output, including potassium or calcium excesses, oxygen deficiency, and abnormally low blood pH *(acidosis)*. Note that some agents affect both the heart rate and contraction strength; epinephrine increases both, and potassium excess (hyperkalemia) reduces both.

> **Apply What You Know**
>
> Physical exercise obviously increases cardiac output. Do you think it achieves this through heart rate, contraction strength, or both? Explain.

Before You Go On

Answer these questions from memory. Reread the preceding section if there are too many you don't know.

7. Describe the structure of intercalated discs and explain the functional importance of each feature.

8. Why must the signal for each heartbeat slow down at the AV node?

9. Why is it so important that the action potential of a cardiomyocyte remain elevated for an extended time, instead of being a quick up-and-down voltage change like it is in skeletal muscle?

10. Describe the major events in each stage of the cardiac cycle, beginning when all chambers are in diastole.

11. What is represented by the P, QRS, and T waves of the ECG? Draw a simple diagram of an ECG trace and label the waves.

12. Itemize some agents that speed up and slow down the heart, and that strengthen and weaken its contractions.

13.4 General Anatomy of Blood Vessels

Expected Learning Outcomes
When you have completed this section, you should be able to:

a. describe the structure of a blood vessel;

b. compare and contrast arteries, capillaries, and veins; and

c. explain how portal systems and anastomoses differ from the most common route in which blood flows from the heart and back again.

There are three categories of blood vessels—arteries, veins, and capillaries. The three differ in direction of blood flow, the pressure they must withstand, and the corresponding histological structure of their walls.

The Vessel Wall

The walls of arteries and veins are composed of the following three tissue layers (fig. 13.10).

1. The **tunica interna**[13] (TOO-ni-ca in-TER-nuh) lines the inside of the vessel and is exposed to the blood. It consists of a simple squamous epithelium called the **endothelium,** overlying a basement membrane

[13] *tunica* = coat; *interna* = internal

Figure 13.10 Histology of Blood Vessels. Note the changes in wall structure as we progress around the circuit from high-pressure arteries, to blood capillaries (exchange vessels), to low-pressure veins.

- *Why do the arteries need so much more elastic tissue than the veins do?*

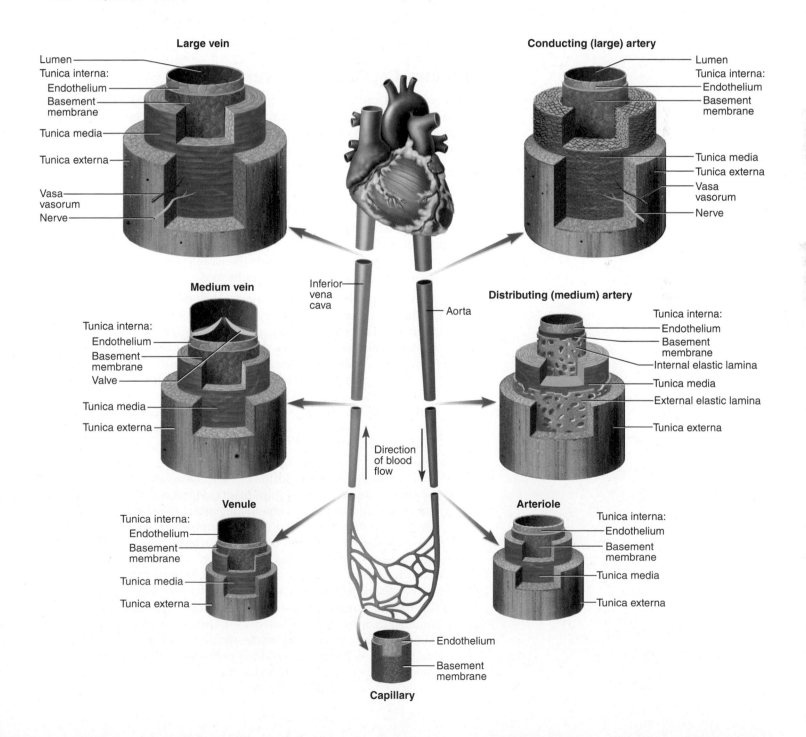

and a sparse layer of loose connective tissue. The endothelium acts as a selectively permeable barrier to materials entering or leaving the bloodstream. It also secretes chemicals that prevent blood cells and platelets from adhering to the vessel wall, and others that dilate or constrict blood vessels and thereby regulate the flow of blood.

2. The **tunica media,** the middle layer, is usually thickest. It consists of smooth muscle, collagen, and elastic tissue. It strengthens the vessels and prevents the blood pressure from rupturing them, and its muscle carries out vasoconstriction and vasodilation.

3. The **tunica externa** is the outermost layer. It consists of loose connective tissue that often blends into surrounding tissues, anchors the blood vessel to them, and is penetrated by small nerves, lymphatic vessels, and small blood vessels that supply the larger vessel.

In contrast to these, capillaries are very thin-walled vessels with only a tunica interna.

Arteries, Capillaries, and Veins

All vessels that carry blood away from the heart are classified as **arteries,** whereas all vessels that carry blood back to the heart are classified as **veins.** **Capillaries** are microscopic vessels that connect the smallest arteries to the smallest veins.

Arteries

Each beat of the heart creates a surge of pressure in the arteries as blood is ejected into them. Arteries withstand these surges because of their resilient structure. They are more muscular than veins, and retain their round shape even when empty. They are divided into three categories by size, but of course there is a gradual transition from one to the next.

1. **Conducting (elastic** or **large) arteries** include the largest arteries, such as the aorta and pulmonary trunk, and may be up to 2.5 cm (1 in.) in diameter. The tunica media is dominated by layers of perforated elastic sheets like slices of Swiss cheese; smooth muscle and collagen are present but less visible. The abundance of elastic tissue enables conducting arteries to expand during ventricular systole and recoil during diastole. Their expansion protects smaller downstream arteries by reducing the pressure surge during ejection of blood from the heart, and their recoil prevents blood pressure from dropping too low when the heart relaxes.

2. **Distributing (muscular** or **medium) arteries** are smaller branches that distribute blood to specific organs. These arteries are dominated by smooth muscle in the tunica media, with less abundant elastic tissue. You could compare a conducting artery to an interstate highway and distributing arteries to the exit ramps and state highways that serve individual towns. All conducting arteries and most distributing arteries are named and can be traced in a careful dissection or surgery. Examples include the brachial and femoral arteries in the arm and thigh, and the renal artery that supplies blood to the kidney.

3. **Resistance (small) arteries** are usually too variable in location and number to be given individual names. Their tunica media is composed almost entirely of smooth muscle, with little elastic tissue. The smallest of these, called **arterioles,** have only one to three layers of smooth muscle and a very thin tunica externa.

Clinical Application 13.2

ATHEROSCLEROSIS AND HARDENING OF THE ARTERIES

As we get older, our arteries become less distensible and less able to protect smaller downstream arteries from blood pressure surges. The increasing stiffness of the arteries is called *arteriosclerosis* ("hardening of the arteries"). It results mainly from cumulative damage by free radicals (see p. 35), which cause gradual deterioration of elastic and other tissue in the arterial wall, much like old rubber bands become less stretchy.

Another contributing factor in arterial stiffness is *atherosclerosis*,[14] the growth of lipid deposits in the arterial walls. Atherosclerosis begins when a vessel is damaged by hypertension, diabetes, or other factors. Macrophages invade the damaged tissue and accumulate cholesterol and fat from the blood. In time, the lesion grows into a fatty *plaque (atheroma)*. Platelets adhere to these plaques and secrete growth factors that stimulate the proliferation of smooth muscle and connective tissue. The plaque grows into a bulging mass of lipid, fiber, and smooth muscle that blocks blood flow through the vessel (see fig. 13.5). Blood clots can easily develop on these plaques and finish off the obstruction of the vessel, or break free and travel as *emboli* to block smaller vessels downstream.

Degradation and blockage of vessels by atheromas often lead to kidney failure, heart attack, or stroke. Blockage occurs in both arteries and veins, but is more crucial in arteries because this is where it has the potential to shut off blood flow to a vital tissue. As the arteries harden, they become increasingly unable to expand and relieve the pressure surges generated by the heart, and more delicate arteries downstream are more vulnerable to rupturing.

Capillaries

For the blood to serve any purpose, materials such as nutrients, wastes, and hormones must pass between the blood and tissue fluids, through the walls of the vessels. This occurs mostly in the capillaries (fig. 13.11a), which are therefore often called the *exchange vessels* of the cardiovascular system. The number of capillaries has been estimated at a billion. Scarcely any cell in the body is more than four to six cell widths from the nearest capillary.

Capillaries are composed of only an endothelium and basement membrane. Their walls are as thin as 0.2 to 0.4 µm, and they range about 5 to 9 µm in diameter. Since erythrocytes are about 7.5 µm in diameter, they often have to stretch into elongated shapes to squeeze through the smallest capillaries.

Continuous capillaries, the most common type, are tubes of endothelial cells rolled up like burritos and held together by tight junctions (fig. 13.11b). Water and small solutes pass through the wall via narrow gaps called **intercellular clefts** between the cells, or by diffusing through the cells themselves. Continuous capillaries sometimes have cells called *pericytes* associated with them. Among other functions, these can differentiate into new endothelial and smooth muscle cells and contribute to the repair of damaged capillaries and other vessels.

In organs such as the kidneys where rapid filtration is important, or pancreatic islets where it is necessary to secrete large molecules (insulin and

[14] *athero* = fatty; *scler* = hard; *osis* = condition

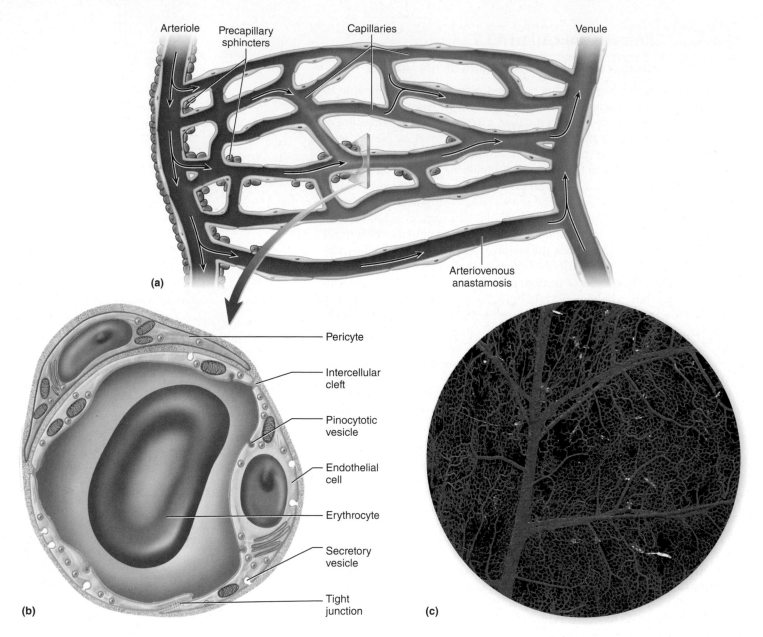

Figure 13.11 Blood Capillaries. (a) A capillary bed arising from an arteriole and draining into a venule. Arrows indicate the direction of blood flow. (b) Cross section of a continuous capillary. (c) Photograph of a dye-injected capillary bed.

glucagon) into the blood, there are special **fenestrated capillaries** with large *filtration pores* in them, somewhat like a sheet of cookie dough with a lot of holes punched in it by a small cookie cutter, then rolled into a tube. Intercellular clefts are only about 4 nm wide, but filtration pores range from 20 to 100 nm wide and thus allow much larger molecules to pass through.

Apply What You Know

Considering the functions of the skeletal muscles and the anterior lobe of the pituitary gland, which would you expect to have continuous capillaries and which to have fenestrated capillaries? Explain.

Capillaries are organized into **capillary beds,** webs of 10 to 100 vessels arising from a single arteriole (fig. 13.11). At the beginning of each capillary is a ring of smooth muscle cells called a **precapillary sphincter,** which can dilate to let blood into a capillary or constrict to shut down local blood flow.

Blood flow through the capillary beds is regulated to match the metabolic needs of tissues. There is not enough blood in the body to fill the entire vascular system at once; consequently, about three-quarters of the body's capillaries are shut down at any given time. For example, about 90% of capillaries in skeletal muscle have little or no blood flow during rest. In contrast, during exercise, they receive an abundant flow, while capillary beds elsewhere—for example, in the intestines—shut down to compensate.

Veins

Veins are relatively thin-walled and flaccid, and expand easily to accommodate a greater volume of blood than arteries do. Being farther from the heart, veins have a much lower blood pressure than arteries do, and their blood flow is steady, rather than pulsating with the heartbeat as it does in arteries. Therefore, veins do not require such muscular or elastic walls to withstand pressure surges.

Small veins merge to form larger and larger ones as they approach the heart. In examining the types of veins, we will follow the direction of blood flow, working from smallest to largest vessels.

1. **Venules** receive blood from capillaries. They range up to 1 mm in diameter. The smallest are quite porous, and this is where most white blood cells leave the bloodstream to wander among the connective tissues.
2. **Medium veins** range up to 10 mm in diameter and have a thicker tunica media and externa; they are more muscular than venules but less so than medium arteries. Most veins with individual names are in this category, such as the brachial veins in the arm. Many medium veins in the limbs have **venous valves** that ensure a one-way flow of blood toward the heart, against the pull of gravity. These are later discussed under the subject of venous blood flow.
3. **Large veins** have diameters greater than 10 mm. These include veins that empty into the heart—the two venae cavae and four pulmonary veins.

Variations in Circulatory Routes

The most common pattern of systemic blood flow is for blood to leave the heart via the arteries, pass through one capillary bed (often quite distant from the heart), then return to the heart via the veins. However, there are important variations on this pattern (fig. 13.12). In a **portal system,** the blood flows through two capillary beds in a row before returning to the heart. We have seen one of these already between the hypothalamus and anterior pituitary gland (chapter 11), and we will encounter others in the intestine–liver relationship and in the kidneys. Portal systems are found where a substance is to be picked up by one capillary bed and immediately given off by another, or vice versa.

In contrast, **anastomoses** (ah-NASS-tuh-MO-seez), or **shunts,** are routes in which the blood bypasses capillaries, going directly from an artery to a vein (an *arteriovenous anatomosis*), from one vein to another (a *venous anastomosis*), or from one artery to another (an *arterial anastomosis*). Arteriovenous anastomoses are found at the bases of the fingers, for example, where they can allow blood to detour and avoid the fingers in cold weather in order to reduce heat loss. The resulting lack of warming of the fingers, however, is the reason why frostbite so commonly occurs here. (The same is true of the

Figure 13.12 Variations in the Systemic Circuit. (a) The most common pathway, in which blood passes through only one capillary bed before returning to the heart. (b) A portal system, in which blood passes through two capillary beds in series before returning to the heart. (c) An arteriovenous anastomosis (shunt), in which blood bypasses the capillaries by flowing directly from an artery to a vein.

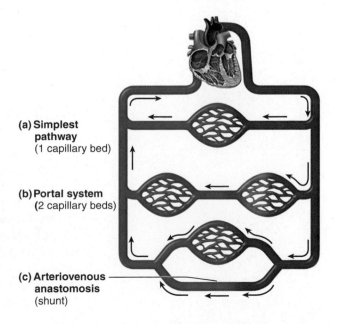

(a) Simplest pathway (1 capillary bed)

(b) Portal system (2 capillary beds)

(c) Arteriovenous anastomosis (shunt)

toes and ears.) Venous anastomoses are very common in such places as the hand and forearm, where they provide alternative routes of venous return in case one route is blocked. One place to find arterial anastomoses is in the coronary blood vessels, allowing blood to bypass an arterial blockage and nourish the heart tissue by an alternative route. They are also common at joints such as the knee and elbow, where they allow alternative routes of blood flow if flexion of a joint temporarily compresses one of the arteries.

Before You Go On

Answer these questions from memory. Reread the preceding section if there are too many you don't know.

13. Name the three tunics of a typical blood vessel and explain how they differ from each other.

14. Contrast the tunica media of a conducting artery, arteriole, and medium vein, and explain how the histological differences are related to the functional differences between these vessels.

15. Describe the structural and functional differences between continuous and fenestrated capillaries.

16. Define portal system *and* anastomosis, *and explain their special functional purposes.*

13.5 Physiology of Circulation

Expected Learning Outcomes

When you have completed this section, you should be able to:

a. define *blood pressure* and explain how it is commonly measured;

b. explain the relationship between blood pressure, resistance, and flow;

c. describe three factors that determine resistance to blood flow;

d. discuss local, neural, and hormonal control of blood pressure;

e. explain how blood pressure and osmotic pressure interact in capillary fluid exchanges; and

f. describe the mechanisms for returning venous blood to the heart.

We have addressed how the heart pumps blood, and we turn now to how the blood vessels deliver it to the organs that need it—that is, to blood **flow**. The flow of blood, like that of any other fluid, is governed by fundamental principles of pressure and resistance. In turn, blood flow through the vessels governs **perfusion,** the amount of blood received by a given tissue or organ in a given span of time, measured in such terms as milliliters of blood per gram of tissue per minute.

Blood Pressure

Blood pressure (BP) is the force exerted by blood on a vessel wall. Contraction of the heart initiates a wave of pressure that sharply decreases as the blood flows farther and farther away from the ventricle of origin. Blood always flows down a gradient from a point of high pressure to a point of lower pressure. The greater the pressure difference, the greater the flow. Think of a garden hose by analogy. If you slightly open the tap, you get a trickle of water from the end of the hose. Open the tap wider, and the water gushes out. Flow increases because you have increased the pressure difference between the beginning and end of the hose.

Arterial blood pressure is expressed as a ratio of the **systolic pressure** generated by contraction (systole) of the left ventricle, to **diastolic pressure,** the minimum to which the BP falls when the ventricle is in diastole. Both are expressed in millimeters of mercury (mm Hg) for a reason explained shortly. Measured at a point close to the heart, such as the brachial artery of the arm, a typical healthy young adult pressure would be 120/80 mm Hg (120 mm Hg systolic, 80 mm Hg diastolic) or lower.

The method of obtaining these values is a familiar part of a routine physical examination, using a device called a **sphygmomanometer** (SFIG-mo-meh-NOM-eh-tur). This consists of an inflatable cuff connected to a rubber bulb for pumping air into it, and a dial gauge or a calibrated mercury column for measuring air pressure in the cuff. To take a patient's blood pressure, the examiner wraps the cuff snugly around the patient's arm and inflates it with the bulb until it exceeds the systolic blood pressure. By squeezing the brachial muscles, this procedure collapses the brachial artery deep within. Even during systole, the heart cannot force blood through the artery, and there is no blood flow distal to that point. The examiner now listens with a stethoscope at the bend of the elbow (cubital region) while slowly releasing air from the cuff.

At first there is no sound, but as soon as the cuff pressure equals the systolic BP, each heartbeat forces the brachial artery open and allows a brief jet of blood to pass through. The vessel collapses again at diastole. This jet of blood and the subsequent surge of blood against the recollapsed artery cause turbulence that the examiner hears as a faint "bump" sound. The cuff pressure at the instant of the first bump is noted as the systolic BP. The examiner continues to bleed air from the cuff. During this time, a bump is heard every time the artery collapses—that is, once in each heartbeat. But soon the cuff pressure falls to a point that the brachial artery remains open even during diastole, and no further sounds are heard. The point at which the last bump is heard is noted as the diastolic BP.

Blood pressures considered to be healthy vary with age, and pressure typically rises in and beyond middle age as the arteries become less resilient. A persistent resting blood pressure above 140/90 mm Hg is considered to be **hypertension** (high blood pressure). This does not include temporary elevations of BP due to exercise or emotional responses. A persistent low BP is called **hypotension**. It may result from blood loss, dehydration, or inability to regulate blood pressure fluctuation in old age, and it is routinely seen in patients approaching the moment of death. There is no particular numerical criterion for hypotension.

Blood pressure is determined by three principal variables: (1) cardiac output, discussed earlier in the chapter; (2) blood volume, which is regulated mainly by the kidneys; and (3) resistance to flow, discussed in the next section.

Peripheral Resistance

Resistance is a measure of hindrance to blood flow through a vessel caused by friction between the moving fluid and stationary vessel walls. **Peripheral resistance** is opposition to flow in vessels away from the heart, as opposed to resistance encountered in the heart itself. When resistance increases, flow decreases unless the heart pumps harder to compensate for it.

Resistance is determined by three principal variables:

1. **Viscosity**, the "thickness" of the blood. Higher viscosity increases resistance and impedes flow. (Imagine the difference between trying to suck molasses compared to water into a syringe.) The most significant determinants of blood viscosity are the concentrations of erythrocytes and albumin. A deficiency of RBCs or albumin reduces viscosity, and an excess of either increases it.

> **Apply What You Know**
>
> How can dehydration raise the viscosity of the blood even if it involves no loss of RBCs or albumin from the bloodstream?

2. **Vessel length**. The farther a liquid travels through a vessel, the more cumulative resistance it encounters; pressure and flow both decline with distance. This is why, in a reclining person, arteries of the feet have less flow and a weaker pulse than arteries near the heart.

3. **Vessel radius**, the most important variable in flow. Vessel lengths and blood viscosity do not change from one minute to the next, so the only way to control peripheral resistance from moment to moment is by **vasomotion**—adjusting the radius of the blood vessels. This includes **vasoconstriction,** the narrowing of the vessel, and **vasodilation,** the widening of a vessel. These are achieved by contraction or relaxation of the muscle in the tunica media.

   Radius *(R)* is a very potent factor in blood flow *(F)* because flow is proportional to the *fourth power* of radius *($F \propto r^4$);* in other words, their relationship is exponential. Doubling the radius of a vessel increases flow by 16 times ($r^4 = 2 \times 2 \times 2 \times 2 = 16$). Arterioles can change their radius as much as threefold, which would result in an 81-fold (3^4) change in blood flow. Going back to our earlier garden hose analogy, you could compare the effect of vasoconstriction on flow to what would happen if you backed your car over the hose.

Regulation of Blood Pressure and Flow

The perfusion of each organ is not constant but varies depending on its ever-changing demand for blood. During exercise, for example, vasodilation of vessels supplying skeletal muscle and the heart results in a greater percentage of blood being directed to those organs. Perfusion is controlled by local, neural, and hormonal mechanisms.

Local Control

A tissue with a very high metabolic rate needs more blood, oxygen, and nutrients, but such a tissue also produces metabolic by-products such as CO_2, lactic acid, adenosine, and others that relax the smooth muscle of arterioles. The vessels dilate and perfusion increases to meet the demands of the tissue. Other local chemicals released by platelets, endothelial cells, and connective tissue cells—such as histamine, prostaglandins, and nitric oxide—also dilate vessels and increase perfusion.

Long-term metabolic demand also stimulates new blood vessels to grow into a tissue—a process called **angiogenesis**.[15] A denser capillary network then meets the elevated needs of the tissue. This occurs in such cases as muscular conditioning and in the monthly regrowth of the uterine lining after a woman's menstrual period. Cancerous tumors also induce angiogenesis to feed their growth, and one line of cancer research is to develop drugs that may starve a tumor by blocking angiogenesis.

Neural Control

Vasomotion is also regulated by the autonomic nervous system. Arteries near the heart—especially the aortic arch and internal carotid arteries—have sensory nerve endings called **baroreceptors** that monitor blood pressure (BP) and transmit signals to the medulla oblongata of the brainstem (fig. 13.13). The medulla has a group of neurons called the **vasomotor center** that issues signals to the blood vessels. When BP rises above normal, the vasomotor center dilates blood vessels to lower it; when BP falls below normal, the center constricts blood vessels to raise it. Such reflexes are important in adjusting BP to postural changes. When you rise quickly from bed, for example, gravity draws the blood downward and BP in the head and neck falls. The baroreceptors and medulla respond quickly to constrict blood vessels and accelerate the heart, maintaining blood flow to the brain and preventing you from fainting. If the reflex is not quick enough, you may briefly feel dizzy. The vasomotor center also receives input from brain centers concerned with thought and emotion. This is why stress, anger, or excitement raises the blood pressure.

Hormonal Control AP|R

Some hormones affect blood pressure by constricting or dilating blood vessels.

- **Epinephrine** from the adrenal medulla and **norepinephrine** from sympathetic nerves dilate some vessels such as the coronary arteries and arteries of the muscles, but constrict other vessels such as those of the digestive tract and skin. During exercise, this gives priority to perfusion of the organs where it is needed most, routing blood away from other organs that can temporarily do with less. Epinephrine secreted during exercise or stress can also constrict vessels throughout the body and raise overall blood pressure.

- **Angiotensin II** is a potent vasoconstrictor that raises blood pressure. It is produced through the collaborative action of the liver, kidneys, and lungs (see fig. 16.9, p. 539). An enzyme required for its synthesis is *angiotensin-converting enzyme (ACE)*. Hypertension is often treated with drugs called ACE inhibitors, which block this enzyme, thus lowering angiotensin II levels and blood pressure.

Certain other hormonal mechanisms can lower blood pressure by promoting urinary loss of water, thus lowering blood volume.

- **Natriuretic peptides**, secreted by the heart when blood pressure is too high, stimulate the kidneys to excrete more sodium. Water follows by osmosis and is lost from the body, thereby lowering blood volume and pressure. Thus the more natriuretic peptide secreted, the lower the BP will be.

---

[15] *angio* = vessels; *genesis* = production of

Figure 13.13 Neural Control of Blood Pressure. Baroreceptors in the aorta and internal carotid artery monitor blood pressure and transmit signals to the vasomotor center of the medulla oblongata. The vasomotor center integrates this input and issues signals to the blood vessels via sympathetic nerves. Blood vessels may dilate or constrict to adjust blood pressure. AP|R

- *If the blood vessels dilated, would that raise or lower the blood pressure? Explain.*

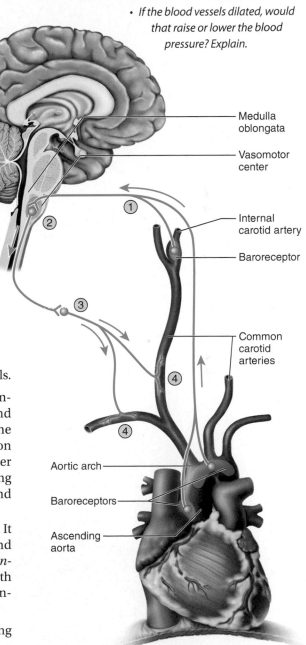

Medulla oblongata

Vasomotor center

Internal carotid artery

Baroreceptor

Common carotid arteries

Aortic arch

Baroreceptors

Ascending aorta

① Baroreceptors detect degree of stretch in arteries and transmit signals to vasomotor center.

② Vasomotor center integrates input.

③ Outgoing signals are carried by sympathetic nerves to blood vessels.

④ Blood vessels constrict or dilate to adjust blood pressure.

- **Aldosterone** promotes sodium and water retention and thus supports BP, but one way of lowering BP is for the adrenal glands to secrete less aldosterone and allow greater elimination of water from the body.
- **Antidiuretic hormone** also promotes water retention, and here again BP can be lowered by reducing ADH secretion and producing more urine.

The actions of these hormones on the kidneys will be further described in chapter 16.

Capillary Fluid Exchange

Capillaries are "the business end" of the cardiovascular system. The system would be useless if the blood could not release some materials to the tissues and pick up other materials from them. These exchange processes occur across the thin walls of the capillaries and smallest venules, but capillaries greatly outnumber venules and are the site of most fluid exchange.

Some ways for substances to pass across a vessel wall follow.

1. **Diffusion.** If a substance is more concentrated in the blood than in the surrounding tissue fluid and it is capable of crossing the wall, it will leave by diffusion; if it is more concentrated in the tissue fluid, it will tend to enter the blood by diffusion. Substances leaving the systemic blood in this manner include glucose, oxygen, and steroid hormones; substances picked up by the systemic blood in this way include carbon dioxide and other wastes. Oxygen and steroids can diffuse through the plasma membranes and cytoplasm of the endothelial cells; hydrophilic substances such as glucose and electrolytes can diffuse through the intercellular clefts between the cells.

2. **Filtration.** This is a process in which the blood pressure forces fluid through the capillary wall, carrying solutes with it.

Figure 13.14 Capillary Fluid Exchange. (a) At the arterial end of a capillary, blood pressure overrides osmosis and fluid filters from the capillary into the tissues. (b) At the venous end, blood pressure is lower and is now overridden by osmosis, which results in net fluid uptake by the capillary. (c) Excess fluid not reabsorbed by the blood capillary is absorbed from the tissues by a lymphatic capillary. AP|R

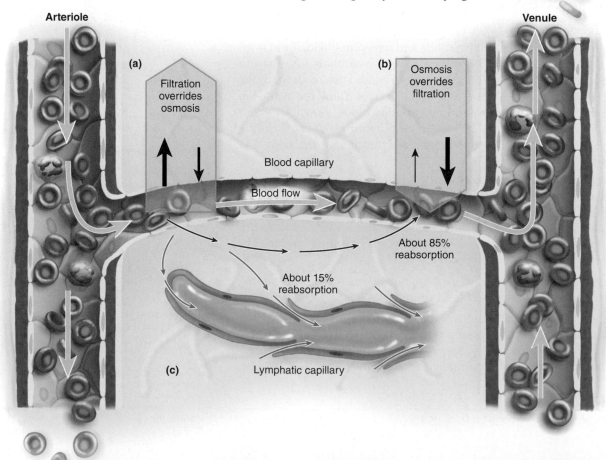

Arteriole

Venule

(a) Filtration overrides osmosis

(b) Osmosis overrides filtration

Blood capillary

Blood flow

About 85% reabsorption

About 15% reabsorption

(c)

Lymphatic capillary

3. **Osmosis**. Because of the high concentration of sodium, protein, and erythrocytes in the blood, capillaries have a strong tendency to absorb water from surrounding tissues by osmosis.

Near the arterial end, a typical capillary tends predominantly to give off fluid to the tissues, delivering vital materials to the cells such as oxygen and nutrients. At the venous end, it tends predominantly to absorb fluid, picking up and carrying away such materials as metabolic wastes and excess water (fig. 13.14). How can a single capillary do both?

When blood enters a capillary from an arteriole, it is under relatively high pressure (typically about 30 mm Hg). This tends to force fluid out through the capillary wall. At the same time, however, the capillary tends to absorb water by osmosis. But since the outward-acting blood pressure overrides the inward-acting osmotic pressure, the effect in most capillaries is for more fluid to leave the capillary at this end than to reenter it. But blood pressure drops along the length of the capillary because of resistance to flow, and by the time blood reaches the venous end of a capillary, it is about 13 mm Hg. Now osmosis overrides the outward blood pressure, so the venous end of a capillary tends to absorb more fluid from the tissues than it gives off.

Typically, though, a capillary reabsorbs only about 85% as much fluid as it filters into the tissue. The excess 15% is reabsorbed by lymphatic vessels, discussed in chapter 14. If the amount of fluid leaving the capillaries is greater than the total amount reabsorbed by the blood capillaries and lymphatics, however, then fluid builds up in the tissue, causing swelling or **edema**.

Venous Return

After the blood has done its job of exchanging material with the tissues, it has to return to the heart to be repressurized. Flow back to the heart is called **venous return**. Even though blood pressure in the veins is much lower than it is in the arteries, the pressure gradient generated by the heartbeat is still the most important means of returning blood to the heart. For a person in a sitting or standing position, gravity is also a significant force aiding or opposing venous return. It aids blood in the head and neck—anywhere above the heart—to flow down the superior vena cava into the heart. However, it opposes the venous return of blood from points below the heart such as the abdomen and lower limbs. The pressure gradient generated by the heart isn't sufficient to force all this blood to flow uphill back to the heart, against the pull of gravity. The heart needs help, and it gets this in two main ways.

One is the **skeletal muscle pump** (fig. 13.15). As muscles of the limbs contract and relax, they squeeze the blood vessels that lie among them, promoting a flow of blood. Medium-size vessels in the limbs are equipped with one-way valves as we saw earlier. These valves point upward, so as muscles compress a vein, blood can flow upward toward the heart; but when the muscles relax, the valves prevent it from flowing back down toward the hands and feet. Muscular activity therefore causes blood to flow upward, bit by bit, with the valves acting as checkpoints to prevent backflow. However, in many people—especially those who must stand for long periods on their jobs—blood accumulates in the veins and stretches them until the valve cusps cannot meet in the middle. The valves then allow blood to flow back down. This distension of the veins can then become permanent, creating **varicose veins** that are often visible through the skin.

The other aid to venous return is the **thoracic pump**. When you begin to inhale, your chest expands and pressure in the thoracic cavity drops below the pressure in your abdominal cavity.

Figure 13.15 Venous Valves and the Skeletal Muscle Pump. (a) Muscle contraction squeezes the deep veins and forces blood through the next valve in the direction of the heart. Valves below the point of compression prevent backflow. (b) When the muscles relax, blood flows back downward under the pull of gravity but can flow only as far as the nearest valve.

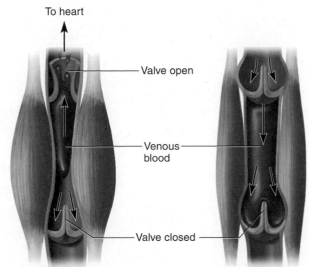

To heart

Valve open

Venous blood

Valve closed

(a) Contracted skeletal muscles   **(b) Relaxed skeletal muscles**

Visualize your inferior vena cava (IVC) as a soft tube of blood extending from your pelvic region to the thoracic cavity, and you can see how these respiratory pressure changes would squeeze the abdominal portion of the IVC while reducing pressure on its thoracic portion. Like squeezing toothpaste from a tube, this pressure difference helps blood flow upward through the IVC to the heart. This is one reason why exercise and deep breathing are good for circulation.

When the heart fails to pump enough blood to meet the demands of the organs, a person enters a state of **circulatory shock**. There are two forms of this: one in which cardiac output is poor because of heart failure, called **cardiogenic shock**, and one on which output is low because there isn't enough blood returning to refill the heart between beats, called **low venous return (LVR) shock**. Some causes of LVR shock include loss of fluid volume by hemorrhaging or profuse sweating; tumors or other factors compressing veins and blocking the flow of blood; and **venous pooling**, in which blood accumulates in the lower parts of the body instead of returning to the heart. There are many causes of venous pooling shock including immune reactions (allergies) causing widespread vasodilation; standing still for too long (as in choir or the military) so the skeletal muscle pump is not working, often leading to fainting (**syncope**); or sitting still for too long, as in a cramped airliner seat on a long flight. In many cases the body can recover from mild shock on its own, as when one recovers from a faint, but in more severe cases a person can die from circulatory shock.

Before You Go On

Answer these questions from memory. Reread the preceding section if there are too many you don't know.

17. *How does a sphygmomanometer measure blood pressure? In a BP such as 120/80 mm Hg, what do the top and bottom numbers represent and what does "mm Hg" mean?*

18. *List the three main determinants of blood pressure.*

19. *How do viscosity, vessel length, and vessel radius influence resistance and flow? Why is radius the most important of these?*

20. *Describe the mechanisms of local, neural, and hormonal control of blood flow. Identify the hormones that affect blood pressure, and how they work.*

21. *Draw a negative feedback loop that illustrates blood pressure homeostasis. Use a drop in blood pressure as the stimulus and include baroreceptors, nerves, and the medulla oblongata in your loop. (See fig. 1.4 for an example of how to diagram a negative feedback loop.)*

13.6 Circulatory Routes and Blood Vessels

Expected Learning Outcomes

When you have completed this section, you should be able to:

a. trace the route of blood through the pulmonary circuit;

b. identify the principal systemic arteries and veins of the body; and

c. trace the flow of blood from the heart to any major organ or body region and back to the heart, naming the vessels through which it would travel.

We conclude this chapter with a survey of the major arterial and venous circulatory pathways, identifying the major arteries and veins and the routes taken by circulating blood. This discussion is confined to adult circulation; special features of fetal circulation and how it changes after birth are described in chapter 20.

The Pulmonary Circuit

The pulmonary circuit (fig. 13.16) serves only to exchange CO_2 for O_2. The lungs receive a separate systemic blood supply, via the *bronchial arteries*, to nourish the pulmonary tissues. The pulmonary circuit begins with the **pulmonary trunk** arising from the right ventricle. The pulmonary trunk branches into right and left **pulmonary arteries**, and these give off further branches just before or after entering each lung. Ultimately these arteries lead to a web of capillaries around each pulmonary air sac, or alveolus. Their relationship to the alveolus and the exchange of O_2 and CO_2 are detailed in chapter 15. Leaving the alveolar capillaries, the blood flows into pulmonary venules and larger veins until finally, two **pulmonary veins** from each lung enter the left atrium of the heart (see fig. 13.3).

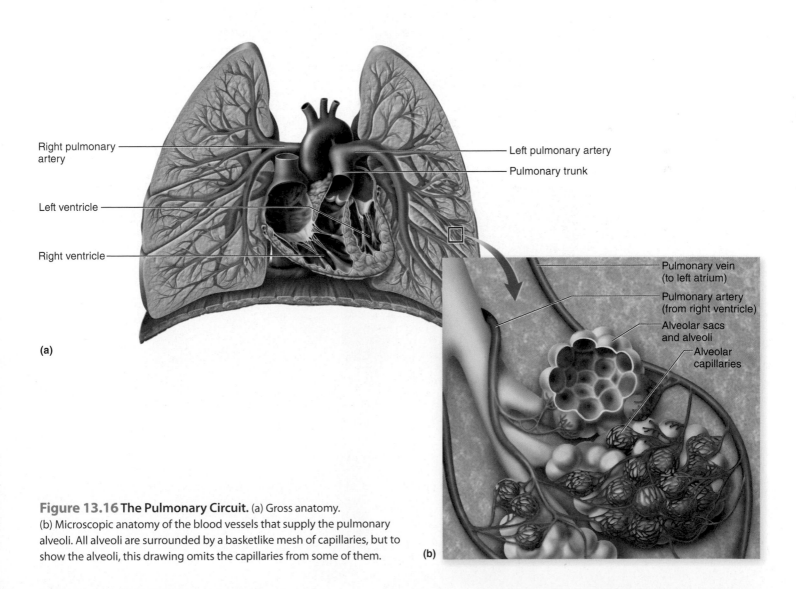

Right pulmonary artery

Left pulmonary artery

Pulmonary trunk

Left ventricle

Right ventricle

(a)

Pulmonary vein (to left atrium)

Pulmonary artery (from right ventricle)

Alveolar sacs and alveoli

Alveolar capillaries

Figure 13.16 The Pulmonary Circuit. (a) Gross anatomy. (b) Microscopic anatomy of the blood vessels that supply the pulmonary alveoli. All alveoli are surrounded by a basketlike mesh of capillaries, but to show the alveoli, this drawing omits the capillaries from some of them.

(b)

The Systemic Circuit

The systemic circuit (figs. 13.17 and 13.18) supplies oxygen and nutrients to all organs and removes their metabolic wastes. Part of it, the *coronary circulation*, was described earlier in this chapter. This section surveys the other major adult arteries and veins region by region, from head to foot. It must be

Figure 13.17 The Major Systemic Arteries.
Different arteries are illustrated on the left than on the right for clarity, but nearly all of them occur on both sides. (a. = artery)

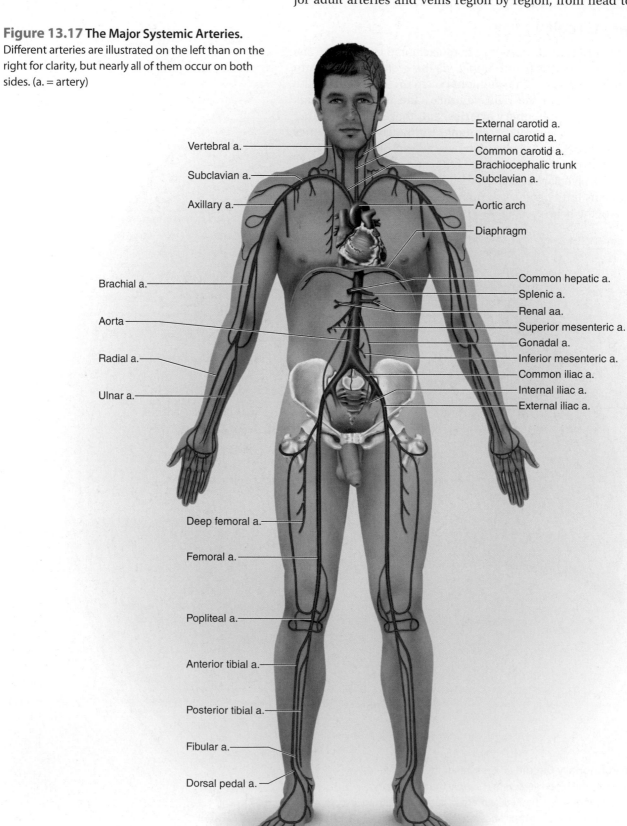

noted that there is a great deal of anatomical variation in the circulatory system; this description outlines only the most common circulatory pathways.

It is important that you consult the illustrations as you read the text, just as you use a map to visualize the branches of highways and city streets. A verbal description alone does not allow effective navigation of the circulatory

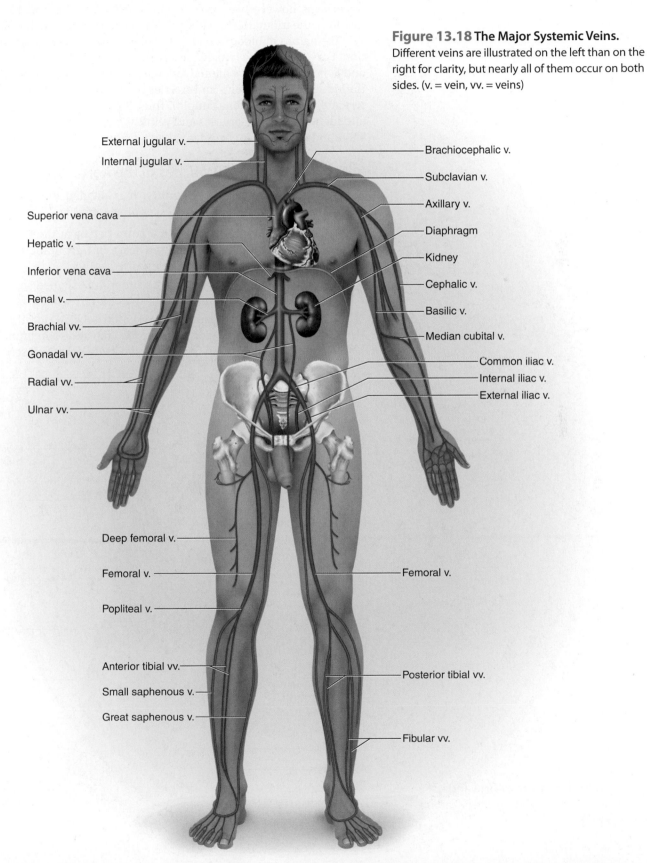

Figure 13.18 The Major Systemic Veins.
Different veins are illustrated on the left than on the right for clarity, but nearly all of them occur on both sides. (v. = vein, vv. = veins)

External jugular v.
Internal jugular v.
Superior vena cava
Hepatic v.
Inferior vena cava
Renal v.
Brachial vv.
Gonadal vv.
Radial vv.
Ulnar vv.
Deep femoral v.
Femoral v.
Popliteal v.
Anterior tibial vv.
Small saphenous v.
Great saphenous v.

Brachiocephalic v.
Subclavian v.
Axillary v.
Diaphragm
Kidney
Cephalic v.
Basilic v.
Median cubital v.
Common iliac v.
Internal iliac v.
External iliac v.
Femoral v.
Posterior tibial vv.
Fibular vv.

landscape. As we trace the arteries and veins, we will do so in the direction the blood flows. That is, in describing arteries, we begin near the heart and trace them distally, whereas in describing veins, we will begin distally and trace them approaching the heart. In describing the smaller divisions of larger arteries, we speak of them as *branches*. In describing the smaller divisions of larger veins, however, we speak of them as *tributaries*, because like the streams that form the tributaries of a river, they contribute blood to the larger vessel.

The names of blood vessels often give clues to their location or destination. For example, the *brachial artery* runs through the arm region; the *ulnar* and *femoral arteries* are named for adjacent bones; and the *renal vein* drains the kidney. Adjacent veins and arteries often have similar names (*femoral artery* and *femoral vein*, for example). Thus, the names may help you remember and identify principal arteries and veins. When an artery or vein is extremely short and quickly branches, it is called a *trunk*.

Initial Segments and Branches of the Aorta

The aorta is the body's largest artery. It is divided into three portions—ascending aorta, aortic arch, and descending aorta (fig. 13.19).

- **Ascending aorta.** The ascending aorta begins at the left ventricle and rises for a short distance from the heart. Its only branches are the *coronary arteries* to the heart wall, described earlier.
- **Aortic arch.** The *aortic arch* is the superior ∩-shaped segment, like the handle on a cane. It gives off three major arteries in the following order:
  - **Brachiocephalic**[16] **trunk** (BRAY-kee-oh-seh-FAL-ic). This short vessel rises from the right side of the aortic arch and soon branches into the *right common carotid* and *right subclavian arteries* described later.

---

[16] *brachio* = arm; *cephal* = head

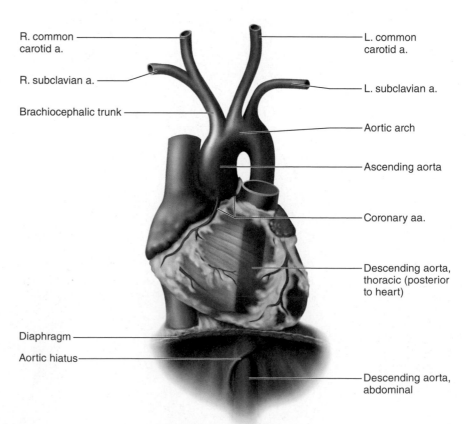

R. common carotid a.

L. common carotid a.

R. subclavian a.

L. subclavian a.

Brachiocephalic trunk

Aortic arch

Ascending aorta

Coronary aa.

Descending aorta, thoracic (posterior to heart)

Diaphragm

Aortic hiatus

Descending aorta, abdominal

Figure 13.19 The Thoracic Aorta and Its Branches. (L. = left, R. = right, a. = artery, aa. = arteries) **AP|R**

- **Left common carotid artery** (cah-ROT-id). Unlike the right common carotid, the left one arises directly from the aortic arch, as its middle branch.
- **Left subclavian**[17] **artery** (sub-CLAY-vee-un). Unlike the right subclavian, the left one arises directly from the aortic arch, since there is no trunk on the left. This is the third and last artery to arise from the arch before the aorta begins its descent.

- **Descending aorta**. The *descending aorta* passes downward behind the heart, near the vertebral column, through the thoracic and abdominal cavities. It is called the *thoracic aorta* above the diaphragm and the *abdominal aorta* below it. It ends in the lower abdominal cavity by forking into the lower limbs.

Arteries of the Head and Neck

The head and neck receive blood primarily from the following two pairs of arteries (fig. 13.20a), but also from several smaller arteries in the shoulder that we will not discuss.

- **Common carotid arteries**. The *right common carotid artery* splits from the brachiocephalic trunk, whereas the *left common carotid artery* arises directly from the aortic arch. The common carotids pass up the front of the neck, alongside the trachea, where their pulse can be palpated.
- **Vertebral arteries**. These arise from the right and left subclavian arteries and travel up the back of the neck through the transverse foramina of vertebrae C1 through C6. They enter the cranial cavity through the foramen magnum, bound for the brain.

---

[17] *sub* = below; *clavi* = clavicle, collarbone

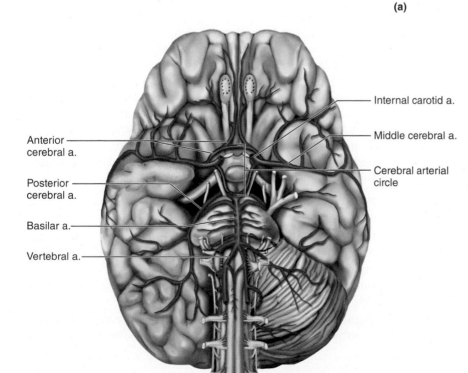

Internal carotid a.

External carotid a.

Carotid sinus

Vertebral a.

Thyroid gland

Common carotid a.

Subclavian a.

Axillary a.

Brachiocephalic trunk

(a)

Internal carotid a.

Middle cerebral a.

Cerebral arterial circle

Anterior cerebral a.

Posterior cerebral a.

Basilar a.

Vertebral a.

(b)

Figure 13.20 Arteries of the Head and Neck.
(a) Superficial arteries. (b) Base of the brain showing the blood supply to the brainstem, cerebellum, and cerebral arterial circle. (a. = artery) **AP|R**

- *List the arteries, in order, that an erythrocyte must travel to get from the left ventricle to the skin of the left side of the forehead.*

The common carotid arteries have the most extensive distribution of all the head–neck arteries. Near the *laryngeal prominence* ("Adam's apple"), each common carotid splits into external and internal branches.

- **External carotid artery.** This branch ascends the side of the head external to the cranium. It provides branches to the thyroid gland, larynx, tongue, teeth, scalp, facial skin and muscles, and chewing muscles.
- **Internal carotid artery.** This branch passes medial to the angle of the mandible and enters the cranial cavity through the carotid canal of the temporal bone. It supplies the orbits and about 80% of the cerebrum. Compressing the internal carotids near the mandible can therefore cause loss of consciousness.

The vertebral arteries give rise to branches that supply the cervical vertebrae, spinal cord and its meninges, and deep muscles of the neck. They then enter the foramen magnum and continue as follows (fig. 13.20b).

- **Basilar artery.** Within the cranial cavity, the vertebral arteries converge to form a single median *basilar artery* along the anterior surface of the brainstem. The basilar artery gives off branches to the cerebellum, pons, and inner ear, then divides and flows into the cerebral arterial circle.
- **Cerebral arterial circle (circle of Willis**[18]**).** The basilar and internal carotid arteries flow into this loop of vessels surrounding the pituitary gland. Most people lack one or more components of this circle; only 20% have a complete arterial circle. The circle gives off *anterior, posterior,* and *middle cerebral arteries* that provide the most significant blood supplies to the cerebrum. Blood supply to the brain is so critical that it is furnished by several arterial anastomoses, so that if blood flow along one route is cut off, there are alternative routes through the arterial circle to reach the same brain tissue. Knowledge of the distribution of the arteries arising from the circle is crucial for understanding the effects of blood clots, aneurysms, and strokes on brain function.

Veins of the Head and Neck

- **Dural sinuses.** After blood circulates through the brain tissue, it collects in large thin-walled veins called *dural sinuses*—blood-filled spaces between the layers of the dura mater (see fig. 9.3, p. 293). The most superficial of these is the *superior sagittal sinus*, which courses along the top of the head just beneath the cranium along the midsagittal line; students often see this vessel in animal brain dissections if the dura mater is present. At the rear of the head, a pair of *transverse sinuses* arches around toward the ears just under the cranial bone (fig. 13.21b). They collect blood from all the other dural sinuses and then exit the cranium through the *jugular foramen* just posterior to each ear.

Blood flows down the neck mainly through three veins on each side, all of which empty into the subclavian vein:

- **Internal jugular**[19] **veins** (JUG-you-lur). These are continuations of the transverse sinuses after the vessels exit the cranium. They travel deep to the sternocleidomastoid muscles. They also pick up blood from the face, temporal region, and thyroid gland before emptying into the subclavian vein.
- **External jugular veins.** These are smaller veins on the superficial surface of the sternocleidomastoids. They too pass down the neck, draining

---

[18] Thomas Willis (1621–75), English anatomist
[19] *jugul* = neck, throat

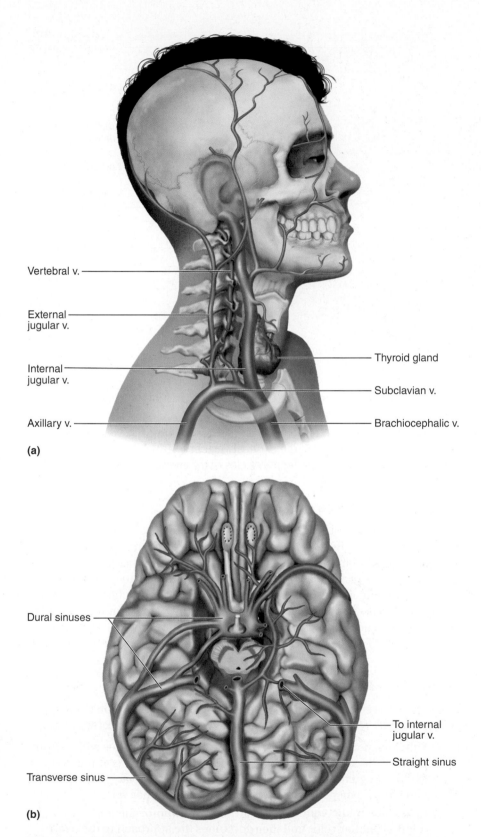

Vertebral v.

External jugular v.

Internal jugular v.

Axillary v.

Thyroid gland

Subclavian v.

Brachiocephalic v.

(a)

Dural sinuses

To internal jugular v.

Straight sinus

Transverse sinus

(b)

Figure 13.21 Veins of the Head and Neck. (a) Superficial veins. (b) Base of the brain showing some of the dural sinuses and other drainage from the cerebrum. (Not all of these veins and sinuses are discussed in the text.) (v. = vein) AP|R

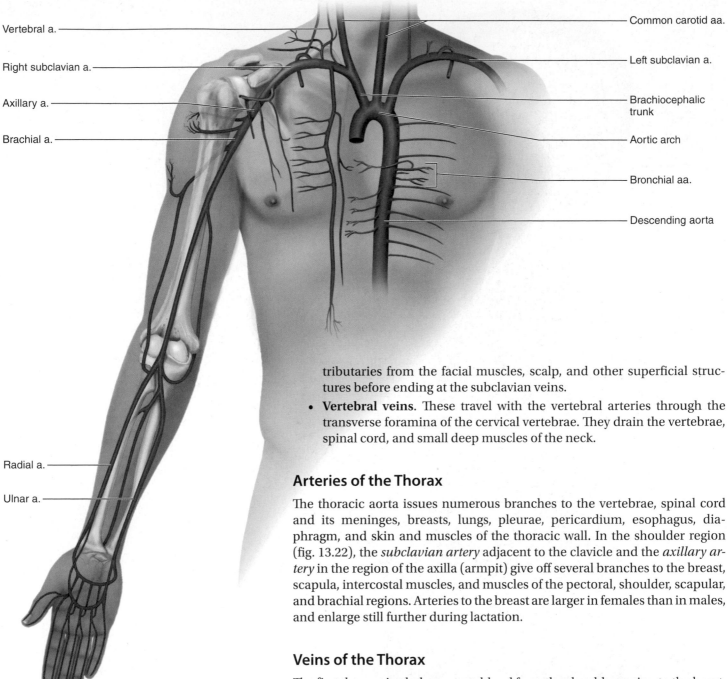

Vertebral a.

Right subclavian a.

Axillary a.

Brachial a.

Radial a.

Ulnar a.

Common carotid aa.

Left subclavian a.

Brachiocephalic trunk

Aortic arch

Bronchial aa.

Descending aorta

Figure 13.22 Arteries of the Thorax and Upper Limb. (a. = artery, aa. = arteries) AP|R

tributaries from the facial muscles, scalp, and other superficial structures before ending at the subclavian veins.

- **Vertebral veins.** These travel with the vertebral arteries through the transverse foramina of the cervical vertebrae. They drain the vertebrae, spinal cord, and small deep muscles of the neck.

Arteries of the Thorax

The thoracic aorta issues numerous branches to the vertebrae, spinal cord and its meninges, breasts, lungs, pleurae, pericardium, esophagus, diaphragm, and skin and muscles of the thoracic wall. In the shoulder region (fig. 13.22), the *subclavian artery* adjacent to the clavicle and the *axillary artery* in the region of the axilla (armpit) give off several branches to the breast, scapula, intercostal muscles, and muscles of the pectoral, shoulder, scapular, and brachial regions. Arteries to the breast are larger in females than in males, and enlarge still further during lactation.

Veins of the Thorax

The first three veins below return blood from the shoulder region to the heart, and the azygos system drains the wall and viscera of the thorax (fig. 13.23).

- **Subclavian vein.** This vessel drains the upper limb. It runs with its companion, the subclavian artery, posterior to the clavicle. It receives the external jugular vein and vertebral vein along the way, then ends (in name only) where it merges with the internal jugular vein.
- **Brachiocephalic vein.** This is formed by union of the subclavian and internal jugular veins and then continues the approach to the heart.
- **Superior vena cava.** This large vein is formed by the union of the right and left brachiocephalic veins. It travels inferiorly for about 7 cm and empties into the right atrium of the heart. It drains all structures superior to the diaphragm except the pulmonary circuit and coronary circulation. It also receives drainage from the abdominal cavity by way of the azygos system.

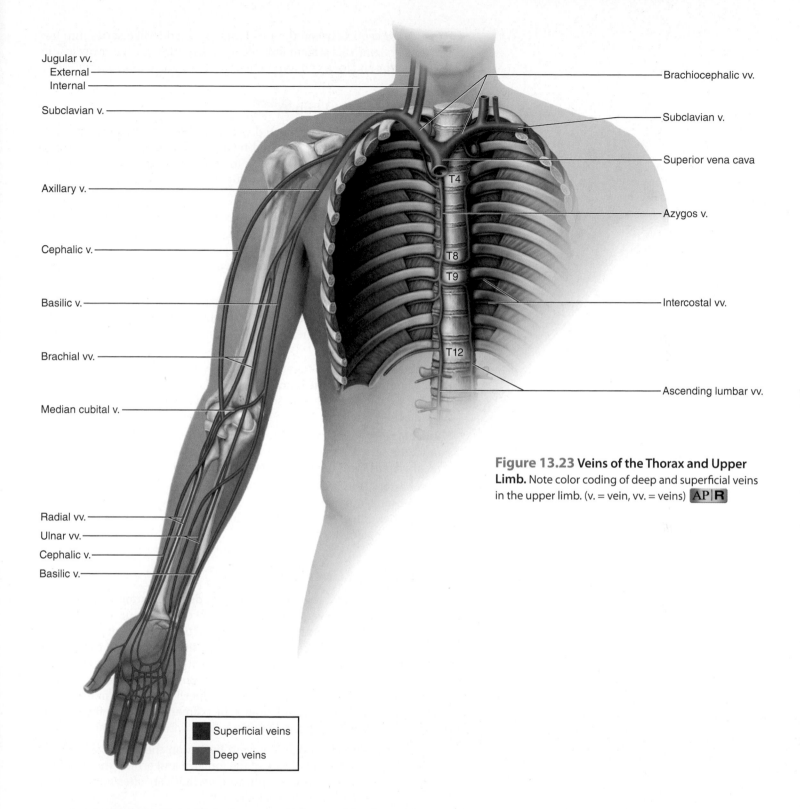

Jugular vv.
External
Internal
Subclavian v.
Axillary v.
Cephalic v.
Basilic v.
Brachial vv.
Median cubital v.
Radial vv.
Ulnar vv.
Cephalic v.
Basilic v.

Brachiocephalic vv.
Subclavian v.
Superior vena cava
Azygos v.
Intercostal vv.
Ascending lumbar vv.

T4
T8
T9
T12

Figure 13.23 Veins of the Thorax and Upper Limb. Note color coding of deep and superficial veins in the upper limb. (v. = vein, vv. = veins) **AP|R**

Superficial veins
Deep veins

- **Azygos system.** The principal venous drainage of the thoracic organs is by way of the *azygos* (AZ-ih-goss) *system*. The most prominent vein of this system is the *azygos*[20] *vein*, an unpaired vessel that ascends the right side of the posterior thoracic wall. The azygos system receives numerous tributaries from many thoracic organs and tissues. A series of

[20] unpaired; from *a* = without; *zygo* = union, mate

intercostal veins between the ribs empty in a ladderlike series into the azygos system. This system also receives blood from as far away as the lower limbs and lower abdomen by way of two *ascending lumbar veins*.

Arteries of the Upper Limb AP|R

In the limbs (appendicular region), the arteries are usually deep and well protected. The upper limb is supplied by a prominent artery that changes name along its course from *subclavian* to *axillary* to *brachial*, then issues branches to the arm, forearm, and hand (fig. 13.22).

- **Subclavian artery.** The *left subclavian artery* arises from the brachiocephalic trunk and the *right subclavian artery* arises directly from the aortic arch, as described earlier. Each subclavian artery arches as high as the base of the neck slightly superior to the clavicle. It then passes posterior to the clavicle, downward over the first rib, and ends in name only at the rib's lateral margin.

- **Axillary artery.** The subclavian changes name to *axillary artery* as soon as it passes the first rib. It continues through the axillary region and ends, again in name only, at the neck of the humerus. Beyond this, it is called the brachial artery.

- **Brachial artery** (BRAY-kee-ul). This artery continues down the medial and anterior sides of the humerus and ends just distal to the elbow, supplying the anterior flexor muscles of the arm along the way. Just distal to the elbow, it branches into the next two arteries.

- **Radial artery.** This branch descends the forearm laterally, alongside the radius, nourishing the lateral forearm muscles. The most common place to take a pulse is at the radial artery just proximal to the thumb.

- **Ulnar artery.** This branch descends medially through the forearm, alongside the ulna, nourishing the medial forearm muscles. The radial and ulnar arteries join at the wrist to form arterial arches that issue smaller arteries to the palmar region and fingers.

Veins of the Upper Limb AP|R

Veins of the appendicular region occur in both deep and superficial groups; you may be able to see several of the superficial ones in your forearms and hands. Veins of the upper limb lead ultimately to the axillary and subclavian veins (fig. 13.23). The superficial veins are larger in diameter and carry more blood than the deep veins. Beginning distally with the veins that drain the hand, and following the direction of blood flow toward the heart, the most prominent superficial veins are as follows. A superficial venous network, often visible on the back of the hand, drains the palmar region and fingers and gives rise to the cephalic and basilic veins.

- **Cephalic**[21] **vein** (sef-AL-ic). This vein begins on the lateral side of the hand, travels up the lateral side of the forearm and arm to the shoulder, and joins the *axillary vein* there. Intravenous fluids are often administered through the distal end of the cephalic vein.

- **Basilic**[22] **vein** (bah-SIL-ic). This vein begins on the medial side of the hand and travels up the posterior side of the forearm, continuing into the arm. It turns deeper about midway up the arm and joins the *brachial vein* near the axilla. As an aid to remembering which vein is cephalic and

---

[21] *cephalic* = related to the head

[22] *basilic* = royal, prominent, important

which is basilic, visualize your arm held straight away from the torso (abducted) with the thumb up. The cephalic vein runs along the upper side of the arm closer to the head (as suggested by *cephal*, "head"), and the name *basilic* is suggestive of the lower (basal) side of the arm (although not named for that reason).

- **Median cubital vein.** This is a short connection between the cephalic and basilic veins that obliquely crosses the elbow in the cubital fossa. It is often clearly visible through the skin and is the most common site for drawing blood.

Deeper venous arches in the hand also receive blood from the fingers and palmar region and give rise to the first two veins below.

- **Radial veins.** A pair of *radial veins* arises from the lateral side of the hand and courses up the forearm alongside the radius. They join near the elbow and form one of the brachial veins.

- **Ulnar veins.** A pair of *ulnar veins* arises from the medial side of the palmar arches and course up the forearm alongside the ulna. They unite near the elbow to form the other brachial vein.

- **Brachial veins.** The two *brachial veins* continue up the brachium, flanking the brachial artery, and converge into a single vein just before the axillary region.

- **Axillary vein.** This vein arises by the union of the brachial and basilic veins. It passes through the axillary region, picking up the cephalic vein along the way. At the lateral margin of the first rib, its name changes to the *subclavian vein*. We have already traced the course of the subclavian vein in discussion of veins of the thorax.

Arteries of the Abdominopelvic Region

The descending aorta passes through an opening in the diaphragm called the *aortic hiatus*, then descends through the abdominal cavity and ends at the level of vertebra L4, where it branches into right and left *common iliac arteries* (fig. 13.24a). Along its descent, it gives rise to the following branches in the order listed, and a few others not listed. Those indicated in the

Figure 13.24 Arterial Supply to the Abdominopelvic Region. (a) The abdominal aorta and its branches. (b) Branches of the celiac trunk arising from the upper aorta. (c) Branches of the superior mesenteric artery supplying the intestines. (a. = artery) AP|R

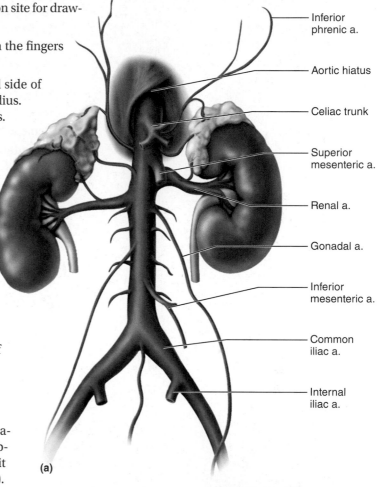

- Inferior phrenic a.
- Aortic hiatus
- Celiac trunk
- Superior mesenteric a.
- Renal a.
- Gonadal a.
- Inferior mesenteric a.
- Common iliac a.
- Internal iliac a.

(a)

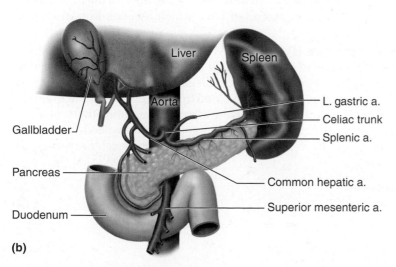

Liver
Spleen
L. gastric a.
Celiac trunk
Splenic a.
Aorta
Gallbladder
Pancreas
Common hepatic a.
Duodenum
Superior mesenteric a.

(b)

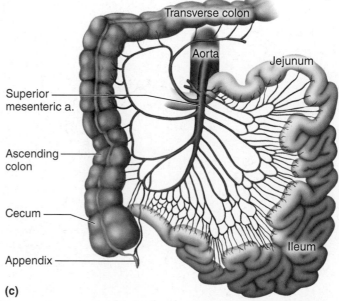

Transverse colon
Aorta
Jejunum
Superior mesenteric a.
Ascending colon
Cecum
Appendix
Ileum

(c)

plural are paired right and left, and those indicated in the singular are solitary median arteries.

- **Inferior phrenic**[23] **arteries.** These supply blood to the inferior surface of the diaphragm.
- **Celiac trunk.** The short, stubby *celiac*[24] (SEE-lee-ac) trunk supplies the upper abdominal viscera. It goes for only about 1 cm before dividing into three branches: the *common hepatic artery* to the liver, *splenic artery* to the spleen, and *left gastric artery* to the stomach. These arteries and their branches also supply the gallbladder, lower esophagus, and greater omentum (fig. 13.24b).
- **Superior mesenteric artery.** This is the most significant intestinal blood supply. It arises medially from the upper abdominal aorta, below the celiac trunk, and fans out in the mesentery to supply nearly all of the small intestine and the proximal half of the large intestine (fig. 13.24c).
- **Renal arteries.** Right and left *renal arteries* supply the kidneys.
- **Gonadal arteries.** More specifically called *ovarian arteries* in the female and *testicular arteries* in the male, these are long, slender arteries that descend along the posterior body wall to the female pelvic cavity or male scrotum.
- **Inferior mesenteric artery.** This artery arises medially from the lower aorta and fans out within the mesentery to supply the distal end of the large intestine.
- **Common iliac arteries.** Finally, the aorta forks into these two short arteries at its inferior end.
- **Internal iliac artery.** The common iliac artery descends for about 5 cm along the posterior wall of the pelvic cavity, then divides in two. One division, the *internal iliac artery*, supplies mainly the pelvic wall, gluteal muscles, hip joint, urinary bladder, rectum, the uterus and part of the vagina in women, and the prostate gland in men. The other division, the *external iliac artery*, is discussed later with the lower limb.

Veins of the Abdominopelvic Region

We will trace the veins in ascending order, following the blood flow back toward the heart.

- **Internal iliac vein.** This vein drains the gluteal muscles, the medial aspect of the thigh, the urinary bladder and rectum, the prostate of the male, and the uterus and vagina of the female.
- **Common iliac vein.** This vein is formed by the union of the internal iliac and an external iliac, described later, which ascends from the lower limb.
- **Inferior vena cava (IVC).** The IVC is the body's largest blood vessel, having a diameter of about 3.5 cm. It is a single median vein formed by the union of the right and left common iliac veins at the level of vertebra L5 (fig. 13.25a). The IVC picks up blood from several veins that drain the lumbar body wall, and the following tributaries, among others. At its superior end, it penetrates the diaphragm and immediately enters the right atrium of the heart from below. It does not receive any thoracic drainage.

---

[23] *phren* = diaphragm
[24] *celi* = belly, abdomen

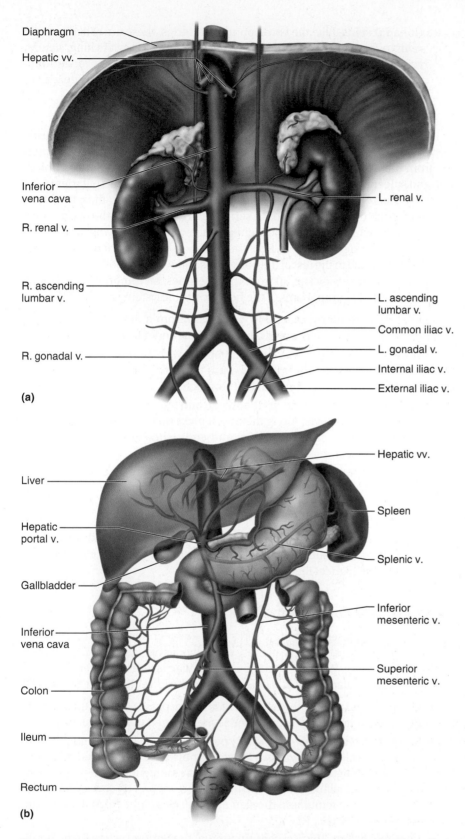

Diaphragm

Hepatic vv.

Inferior
vena cava

R. renal v.

R. ascending
lumbar v.

R. gonadal v.

(a)

L. renal v.

L. ascending
lumbar v.

Common iliac v.

L. gonadal v.

Internal iliac v.

External iliac v.

Hepatic vv.

Liver

Hepatic
portal v.

Gallbladder

Inferior
vena cava

Colon

Ileum

Rectum

Spleen

Splenic v.

Inferior
mesenteric v.

Superior
mesenteric v.

(b)

Figure 13.25 Venous Drainage of the Abdominopelvic Region. (a) Anatomy of
the inferior vena cava and its tributaries. (b) The hepatic portal system, which drains blood
from the major digestive organs and others into the liver. (v. = vein, vv. = veins) AP|R

- **Gonadal veins.** Like the corresponding arteries, these are called *ovarian veins* in the female and *testicular veins* in the male. They are long, slender veins that originate in the ovary or testis and climb the posterior abdominal wall. The right gonadal vein empties into the IVC whereas the left gonadal vein empties higher, into the left renal vein.
- **Renal veins.** These veins drain the kidneys into the IVC.

The venous drainage of much of the digestive system forms the **hepatic**[25] **portal system** (fig. 13.25b). This network of veins receives all of the blood draining from the abdominal digestive tract as well as from the pancreas, gallbladder, and spleen. It is called a *portal system* because it connects capillaries of the intestines and other digestive organs to capillaries in the liver; thus, the blood passes through two capillary beds in series before it returns to the heart. Intestinal blood is richly laden with nutrients for a few hours following a meal. The hepatic portal system gives the liver first claim to these nutrients before the blood is distributed to the rest of the body. It also allows the blood to be cleansed of bacteria and toxins picked up from the intestines, an important function of the liver. The main components of the hepatic portal system are: AP|R

- **Inferior mesenteric vein.** This vein arises from a fanlike array of small veins in the mesentery that drain the rectum and distal part of the colon.
- **Superior mesenteric vein.** This vein begins with a similar fanlike array of veins in the mesentery, but these drain the entire small intestine, the ascending and transverse colon, and stomach.
- **Splenic vein.** This vein drains the spleen and travels across the abdominal cavity toward the liver. Along the way, it picks up veins from the pancreas, then the inferior mesenteric vein, and ends where it meets the superior mesenteric vein.
- **Hepatic portal vein.** This vein is formed by convergence of the splenic and superior mesenteric veins. It enters the inferior surface of the liver and leads ultimately to capillaries called the *hepatic sinusoids,* detailed in chapter 17.
- **Hepatic veins.** Blood from the liver drains into a pair of short hepatic veins that exit its superior surface and pass only a very short distance before ending high on the inferior vena cava.

Arteries of the Lower Limb AP|R

As we have already seen, the aorta forks at its lower end into the right and left *common iliac arteries*, and each of these soon divides again into an *internal* and *external iliac artery.* We discussed the internal iliac artery earlier, and we now trace the external iliac artery as it descends into the lower limb (see fig. 3.17).

- **External iliac artery.** The *external iliac artery* passes behind the *inguinal ligament* at the top of the thigh and is then named the femoral artery.
- **Femoral artery.** The *femoral artery* passes superficially through the groin in a region called the *femoral triangle.* Its pulse can be palpated in the upper medial thigh, and this is an important emergency pressure point for stopping arterial bleeding in the lower limb. It gives off several branches that supply the femur and the skin and muscles of the thigh before it descends to the knee.
- **Popliteal artery.** This is a continuation of the femoral artery in the popliteal fossa of the posterior side of the knee. Its branches supply the knee joint, patella, fibula, and some calf muscles. Just below the knee, the popliteal artery branches into the next two arteries.

[25] *hepat* = liver; *ic* = pertaining to

- **Posterior tibial artery**. This is a continuation of the popliteal artery that passes down the posterior side of the leg, supplying flexor muscles along the way. Inferiorly, it passes behind the medial malleolus of the ankle and into the plantar region of the foot.

- **Anterior tibial artery**. This artery immediately penetrates the interosseous membrane between the upper tibia and fibula and emerges on the anterior side of the membrane. It travels to the ankle, supplying the extensor muscles of the leg along the way.

- **Dorsal pedal artery**. The inferior end of the anterior tibial artery gives rise to the *dorsal pedal artery* on the upper side of the foot. This is a useful site for palpating the pulse in infants and judging the adequacy of lower limb circulation (from strength of the pulse) in adults.

Veins of the Lower Limb AP|R

As in the upper limb, the lower limb has both superficial and deep veins. The two most prominent superficial veins are the *saphenous*[26] (sah-FEE-nus) *veins*, which both arise from superficial veins on the dorsal side of the foot (see fig. 13.18). These two are among the most common sites of varicose veins.

- **Small (short) saphenous vein**. This arises from the lateral side of the foot, passes up that side of the leg as far as the knee, and drains into the popliteal vein.

- **Great (long) saphenous vein**. This is the longest vein in the body. It arises from the medial side of the foot and travels all the way up the leg and thigh to the inguinal region, where it empties into the *femoral vein* described shortly. It is commonly used as a site for the long-term administration of intravenous fluids; it is a relatively accessible vein in infants and in patients in shock whose veins have collapsed. Portions of this vein are often used as grafts in coronary bypass surgery.

Following are the deep veins of the lower limb.

- **Fibular (peroneal) veins**. Blood draining the foot enters a pair of *fibular veins* on the lateral side of the ankle. These veins run side by side, ascending the back of the leg, and converge like an inverted Y into a single vein.

- **Posterior tibial veins**. Similarly, blood from the foot drains into a pair of *posterior tibial veins* on the medial side of the ankle. These also ascend the leg side by side, embedded deep in the calf muscles. They converge like an inverted Y into a single vein about two-thirds of the way up the tibia.

- **Popliteal vein**. The two inverted Ys just described converge on each other near the knee to form the *popliteal vein*, which passes through the popliteal fossa.

- **Anterior tibial veins**. Two *anterior tibial veins* arise from the superomedial aspect of the foot and travel up the anterior compartment of the leg between the tibia and fibula. They converge just distal to the knee and then flow into the popliteal vein.

- **Femoral vein**. This is a continuation of the popliteal vein into the thigh. It drains blood from the deep thigh muscles and femur.

- **External iliac vein**. This vein is formed by the union of the femoral and great saphenous veins near the inguinal ligament. We already traced the course of the internal iliac vein of the pelvic region. The external and internal iliacs unite to form the *common iliac vein*, also described earlier.

---

[26] *saphen* = standing

Before You Go On

Answer these questions from memory. Reread the preceding section if there are too many you don't know.

22. Contrast the pathways and destinations of the external and internal carotid arteries.

23. Briefly state the organs or parts of organs that are supplied with blood by (a) the cerebral arterial circle, (b) the celiac trunk, (c) the superior mesenteric artery, and (d) the internal iliac artery.

24. If you were dissecting a cadaver, where would you look to locate the following vessels? (a) cephalic vein, (b) dorsal pedal artery, (c) femoral artery and vein.

25. If you were dissecting a cadaver, where would you look for the internal and external jugular veins? What muscle would help you distinguish one from the other?

26. Trace one possible path of a red blood cell from the left ventricle to the foot.

27. State two reasons why the great saphenous vein is clinically significant.

Aging of the Cardiovascular System

Cardiovascular disease (CVD) is a leading cause of death in old age. As cardiac muscle atrophies with age, the heart wall becomes thinner and weaker, resulting in reduced cardiac output. The fibrous components of the heart stiffen, limiting cardiac expansion and weakening its contractions. Degenerative changes in the nodes and conduction pathways increase the risk of cardiac arrhythmias. In addition, coronary atherosclerosis is, to varying degrees, a universal feature of the aging heart and increases the risk of myocardial infarction (see Clinical Application 13.2 and Perspectives on Health).

Systemic arteries stiffened by arteriosclerosis cannot expand effectively to accommodate the pressure surges of cardiac systole. Consequently, blood pressure rises steadily with age. Atherosclerosis also narrows the arteries and reduces blood flow to such vital organs as the brain, kidneys, and the heart itself, weakening these organs. Reduced flow to skeletal muscles reduces physical endurance. Atherosclerotic plaques can become focal points of spontaneous blood clotting *(thrombosis),* especially in the lower limbs, where blood flow is relatively slow and blood clots more easily. About 25% of people over 50 experience venous blockage by blood clots—especially people who do not exercise regularly to stimulate optimal blood flow. Degenerative changes in the veins are most evident in the lower limbs, where the venous valves become weaker and less able to stop the backflow of blood. Chronic stretching of the veins makes varicose veins and hemorrhoids more common in and beyond middle age.

CAREER SPOTLIGHT

Electrocardiographic Technician

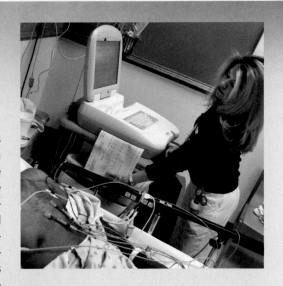

An *electrocardiographic (ECG or EKG) technician* prepares electrocardiograms (ECGs) for diagnostic, exercise testing, and other purposes. The ECG technician prepares the patient for the test by attaching electrodes to specific sites on the chest and limbs and monitors the equipment while results are recorded. One can become a certified ECG technician through programs at community colleges or vocational colleges. A typical course of training entails 4 months beyond high school and includes anatomy and physiology, medical terminology, interpretation of cardiac rhythms, patient-care techniques, cardiovascular medication, and medical ethics. Many people, however, become ECG technicians through on-the-job training rather than formal programs. Most employers prefer to train people who are already in a health-care profession, such as nurses' aides. With more advanced training, one may become a cardiovascular technologist and assist physicians in diagnosis, cardiac catheterization, echocardiography, and other more specialized skills and for correspondingly better salaries. For further information on a career as an ECG technician or cardiovascular technologist, see appendix B.

CONNECTIVE ISSUES

Ways in Which the Cardiovascular System Affects Other Organ Systems

All Systems
Delivers oxygen and nutrients to all systems and carries away their metabolic wastes for elimination; controls body temperature by carrying heat from deep organs to the body surface from which it is lost; conveys hormones to their target organs.

Integumentary System
Dermal blood flow affects sweat production.

Skeletal System
Provides minerals for bone deposition; delivers erythropoietin to bone marrow; delivers hormones that regulate skeletal growth.

Muscular System
Removes heat generated by exercise; removes lactic acid and thus reduces its fatigue-producing effect.

Nervous System
Endothelial cells of the cerebral blood vessels produce the blood–brain barrier and help to produce cerebrospinal fluid.

Endocrine System
Transports all hormones.

Lymphatic and Immune Systems
Produces tissue fluid, which becomes lymph; provides the leukocytes and plasma proteins (antibodies) involved in immunity.

Respiratory System
Transports respiratory gases; low capillary blood pressure prevents fluid accumulation in the lungs.

Urinary System
High blood pressure in certain capillaries of the kidneys is the basis for the first step in urine production; bloodstream carries away water reabsorbed by the kidneys.

Digestive System
Picks up and transports digested nutrients; helps to reabsorb and recycle bile acids and minerals from the intestines.

Reproductive System
Distributes sex hormones; vasodilation produces erection.

Study Guide

Assess Your Learning Outcomes

To test your knowledge, discuss the following topics with a study partner or in writing, ideally from memory.

13.1 Overview of the Cardiovascular System (p. 418)

1. Definitions of *cardiovascular system* and *cardiology*
2. The purpose of the pulmonary and systemic circuits and what side of the heart supplies each
3. Description of the location and position of the heart, using precise anatomical terms
4. The relationship between the epicardium, pericardial sac, and pericardial cavity
5. The function of pericardial fluid

13.2 Gross Anatomy of the Heart (p. 421)

1. The three layers of the heart wall and the histology of each
2. Location and functions of the connective tissue of the heart
3. Surface features of the heart and how they relate to its internal structure
4. The reason for variations in wall thickness of different heart chambers.
5. Similarities and differences in the structure and function of the atrioventricular valves compared to the semilunar valves
6. The flow of blood through the heart, from the venae cavae to the aorta, naming all chambers, vessels, and valves through which the blood passes
7. Why the heart receives a disproportionate amount of blood relative to its percentage of body mass
8. Names and pathways of the coronary blood vessels

13.3 Physiology of the Heart (p. 428)

1. Structurally distinctive properties of cardiac muscle and the reasons for its special features
2. Components of the cardiac conducting system and their locations
3. The origin and propagation of electrical excitation through the heart
4. The anatomical and functional relationship between the SA node and AV node
5. How and why the electrical activity of the SA node and cardiomyocytes differ from that of skeletal muscle

6. Sympathetic and parasympathetic innervation of the heart
7. The waves of the ECG and the electrical events associated with each
8. Definitions of *systole* and *diastole*
9. The key events in each phase of the cardiac cycle: electrical events, chamber contractions and relaxations, valve opening and closing, sources of the two heart sounds, correlations with the waves of the ECG, and movements of the blood
10. The meaning of *cardiac output* and the two factors that determine it
11. Neural and chemical factors that increase or reduce cardiac output and how they do so

13.4 General Anatomy of Blood Vessels (p. 436)

1. The layers of a blood vessel wall and their structural differences
2. The definitions of *arteries, capillaries,* and *veins*
3. Types or subclasses of arteries, capillaries, and veins; how they differ structurally, and the functional relevance of those differences
4. Functional significance of capillary beds and precapillary sphincters
5. The structure and functional significance of portal systems and anastomoses

13.5 Physiology of Circulation (p. 442)

1. Definitions of *flow* and *perfusion*
2. The method of measuring blood pressure, and a typical healthy adult value
3. How peripheral resistance influences blood flow; the three variables that govern resistance; and why vessel radius is the most important of these variables
4. How a tissue can regulate its own blood supply independently of nerves and hormones
5. How baroreceptors and the vasomotor center regulate blood flow
6. Hormones that can raise and lower blood pressure, and how they do so
7. Mechanisms of capillary fluid exchange
8. How a capillary can give off fluid at one end and reabsorb it at the other end

9. Mechanisms that allow blood to return to the heart even against the pull of gravity

10. Types and causes of circulatory shock

13.6 Circulatory Routes and Blood Vessels (p. 448)

1. The path of a blood cell from pulmonary trunk to left atrium, naming the blood vessels

2. Segments of the thoracic aorta, the arteries arising from each, and destinations of the blood in those arteries

3. Major arteries of the head and neck, including the routes of blood flow to the brain

4. Veins of the head, neck, and thorax, and the route that blood travels in these veins to the heart

5. Major arteries and veins of the upper limb

6. Major branches of the abdominal aorta and the structures they supply

7. Major abdominal veins and the organs or regions they drain

8. The general function and circulatory routes of the hepatic portal system

9. Major arteries and veins of the lower limb

Testing Your Recall

1. The cardiac conduction system includes all of the following *except*
   a. the SA node.
   b. the AV node.
   c. the bundle branches.
   d. the tendinous cords.
   e. the Purkinje fibers.

2. The most prominent vessel draining blood from the anterior ventricular myocardium is
   a. the great saphenous vein.
   b. the anterior interventricular branch.
   c. the middle cardiac vein.
   d. the great cardiac vein.
   e. the not-so-great cardiac vein.

3. The outermost layer of the heart wall is known as
   a. the pericardial sac.
   b. the epicardium.
   c. the parietal pericardium.
   d. the tunica externa.
   e. the fibrous layer.

4. Intestinal blood flows into the liver by way of
   a. the superior mesenteric vein.
   b. the hepatic portal vein.
   c. the abdominal aorta.
   d. the hepatic sinusoids.
   e. the hepatic veins.

5. Ventricular contraction is associated with
   a. opening of the AV valves.
   b. the P wave.
   c. the T wave.
   d. the QRS complex.
   e. the second heart sound.

6. Janet is allergic to bee stings. One day she was stung and her blood pressure dropped so low that she lost consciousness and required emergency treatment. Janet had experienced
   a. myocardial infarction.
   b. cerebral vascular accident.
   c. ventricular fibrillation.
   d. low venous return shock.
   e. cardiogenic shock.

7. Which of the following is most responsible for speeding up the heart rate during exercise?
   a. increased outflow of Na^+ from pacemaker cells in the SA node
   b. norepinephrine released from sympathetic nerves
   c. acetylcholine released from vagus nerves
   d. outward leakage of K^+ from pacemaker cells
   e. release of histamine

8. Michael has a tumor that grows to the size of a grapefruit and compresses his external iliac artery. This will result in greatly reduced blood flow to
   a. his lower limb.
   b. his brachial region.
   c. his cerebrum.
   d. his kidneys.
   e. his small and large intestines.

9. What event in the heart corresponds to the first heart sound?
   a. the P wave
   b. the T wave
   c. diastole
   d. closing of the semilunar valves
   e. closing of the AV valves

10. Arthur has a diseased AV bundle that often fails to conduct impulses from the AV node to the bundle branches and Purkinje fibers. This is most likely to result in
    a. two or more P waves between QRS waves.
    b. two or more QRS waves between P waves.
    c. a doubling of the second heart sound.
    d. mitral valve prolapse.
    e. ventricular fibrillation.

11. The contraction of any heart chamber is called _____ and its relaxation is called _____.

12. The instrument that measures blood pressure is the _____.

13. New blood vessels grow through a process called _____.

14. The _____ nerves innervate the heart and serve to reduce heart rate.

15. Most of the blood supply to the brain comes from a complex of arteries around the pituitary gland called the _____.

16. The clinical term for high blood pressure is _____.

17. Mechanoreceptors that respond to pressure changes in major arteries near the heart are called _____.

18. Death of cardiac tissue from lack of blood flow is commonly known as a heart attack, but clinically called _____.

19. Repolarization of the ventricles produces the _____ of the electrocardiogram.

20. The _____ is an artery whose branches supply parts of both ventricles and the anterior portion of the interventricular septum.

Answers in appendix A

True or False

Determine which five of the following statements are false, and briefly explain why.

1. The blood supply to the myocardium is the coronary circulation; everything else is called the systemic circuit.

2. The body's longest blood vessel is the great saphenous vein.

3. Blood pressure can be lowered by increasing the secretion of aldosterone.

4. High blood levels of CO_2 and lactic acid stimulate an increase in heart rate.

5. Arteries have a series of valves that ensure a one-way flow of blood.

6. If the radius of a blood vessel doubles and all other factors remain the same, blood flow through that vessel will also double.

7. In a portal system, blood passes through two capillary beds in series before the blood returns to the heart.

8. An electrocardiogram is a tracing of the action potential of a cardiomyocyte.

9. The first heart sound occurs at the time of the QRS wave of the electrocardiogram.

10. The tendinous cords extend from the cusps of the AV valves to the papillary muscles in the ventricles.

Answers in appendix A

Testing Your Comprehension

1. Mr. Jones, 78, dies of a massive myocardial infarction triggered by coronary thrombosis. Upon autopsy, necrotic myocardium is found in the lateral and posterior right ventricle and posterior interventricular septum. Based on the information in this chapter, where in the coronary circulation do you think the thrombosis occurred?

2. Why would a choke hold (a tight grip around the neck) cause a person to pass out? What arteries would be involved?

3. Diuretics are drugs that increase urine output, and are often used to treat hypertension. Explain why they would have the desired effect.

The Lymphatic System and Immunity

Two natural killer (NK) cells (orange) attacking a cancer cell (red). NK cells are an important defense against cancer, often destroying malignant cells before they can seed the growth of a tumor.

BASE CAMP

Before ascending to the next level, be sure you're properly equipped with a knowledge of these concepts from earlier chapters:

- Leukocyte types and their functions, especially lymphocytes (p. 403)
- Structure of capillaries (p. 439)
- Structure of veins (p. 441)

Anatomy & Physiology | **REVEALED**®
aprevealed.com

Module 10: Lymphatic System

Y ou may be surprised to know that the human body harbors at least 10 times as many bacterial cells as it does human cells. We provide them with a warm, wet, nutritious environment where all of their needs are met. We are also constantly exposed to new invaders through the food we eat, the water we drink, the air we breathe, and the surfaces we touch. Most of these are harmless or even beneficial, but others have the potential to be pathogenic, or disease-causing. It is a wonder that the body isn't overrun and consumed by microbes—which indeed quickly happens when one dies and homeostasis ceases.

What prevents these microbes from taking over? In 1882, Russian zoologist Elie Metchnikoff discovered mobile, phagocytic cells that attacked and seemed to be devouring foreign matter introduced into animals such as starfish larvae. He named the attacking cells *phagocytes* and coined the term *phagocytosis* for the cell process we know so well today. His work set the stage for the science of *immunology*, the study of the *immune system* that protects us from so many diseases; he justifiably received a Nobel Prize in 1908 for this work.

The immune system is not an organ system in itself, but a population of cells that inhabit all of our organs and defend the body against disease. Immune cells are especially concentrated in a true organ system, the *lymphatic system,* which plays crucial roles not only in immunity but also in body fluid distribution. This chapter explores and integrates the functions of the lymphatic and immune systems.

14.1 The Lymphatic System

Expected Learning Outcomes
When you have completed this section, you should be able to:
a. list the functions and basic components of the lymphatic system;
b. discuss how lymph is formed;
c. explain how lymph is returned to the bloodstream; and
d. identify the major lymphatic tissues and organs, and describe their location, structure, and functions.

Our environment is populated by innumerable **pathogens**—bacteria, viruses, fungi, and other microbes that can cause disease. One of our defenses against these is the **lymphatic system** (fig. 14.1), a network of tissues, organs, and vessels that recover tissue fluid, inspect it and cleanse it of pathogens, activate immune responses, and return the fluid to the bloodstream. Components of the system include: (1) **lymph**, the fluid that it collects from the tissues and returns to the bloodstream; (2) **lymphatic vessels** (also simply called *lymphatics*),

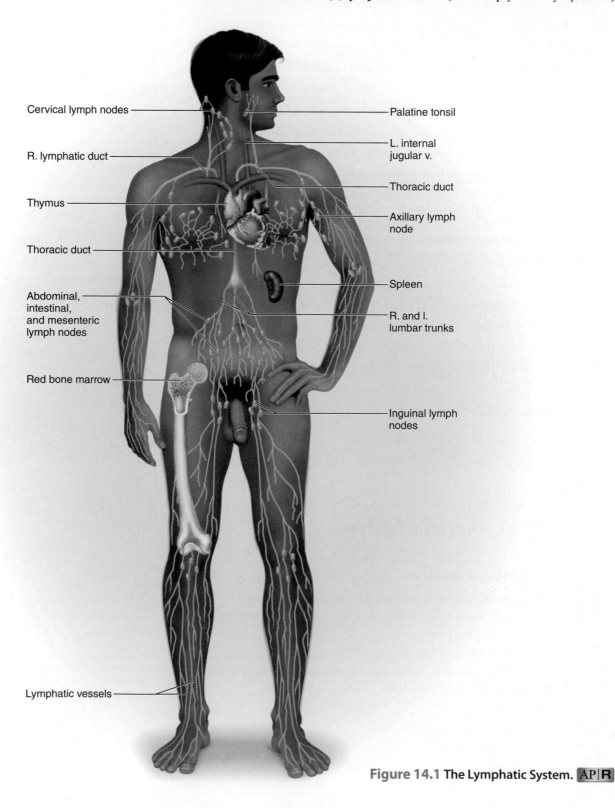

Cervical lymph nodes

R. lymphatic duct

Thymus

Thoracic duct

Abdominal,
intestinal,
and mesenteric
lymph nodes

Red bone marrow

Lymphatic vessels

Palatine tonsil

L. internal
jugular v.

Thoracic duct

Axillary lymph
node

Spleen

R. and l.
lumbar trunks

Inguinal lymph
nodes

Figure 14.1 The Lymphatic System. AP|R

which resemble veins and transport the lymph; (3) **lymphatic tissue**, composed of loose aggregations of lymphocytes in the connective tissues of various organs such as the digestive and respiratory tracts; and (4) **lymphatic organs**, structures enclosed in a fibrous capsule and containing organized masses of lymphatic tissue. The principal lymphatic organs we will discuss are the tonsils, thymus, spleen, and lymph nodes.

Functions of the Lymphatic System

The lymphatic system has three functions:

1. **Fluid recovery.** Fluid continually seeps from the blood capillaries into the tissue spaces. The blood capillaries reabsorb about 85% of this fluid and the lymphatic system absorbs the other 15%. If not for the lymphatic system, this 15% would amount to a loss of 2 to 4 L of water and one-quarter to one-half of the protein from the blood plasma per day. One would quickly die from the loss of blood volume and resulting circulatory failure.

2. **Immunity.** Fluid recovered from body tissues is checked by the lymphatic system for toxins, microbes, and other threats. The lymphatic system also guards the openings of the digestive, respiratory, and other tracts. When disease agents are detected, immune cells of the lymphatic system are quickly mobilized to fight them off.

3. **Lipid absorption.** Special lymphatic vessels called *lacteals* in the small intestine absorb dietary lipids. The lipids travel through lymphatic vessels that ultimately empty into the large left subclavian vein. From here, the bloodstream can distribute these lipids throughout the body for storage or immediate use.

Lymphatic Vessels and Lymph

The lymphatic system recovers excess tissue fluid by means of innumerable microscopic vessels called **lymphatic capillaries** that permeate almost all of the body's tissues. These are similar to blood capillaries, but with important differences. They are closed at one end (fig. 14.2) and are essentially sacs of

Figure 14.2 Lymphatic Capillaries.
(a) Relationship of the lymphatic capillaries to a bed of blood capillaries. (b) Uptake of tissue fluid by a lymphatic capillary.

- *Why can metastasizing cancer cells get into the lymphatic system more easily than they can enter the bloodstream?*

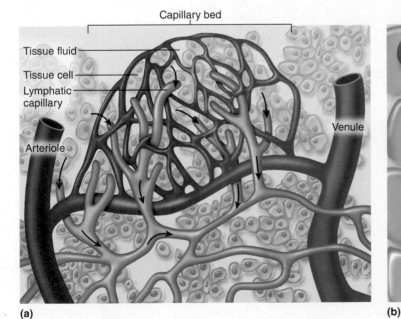

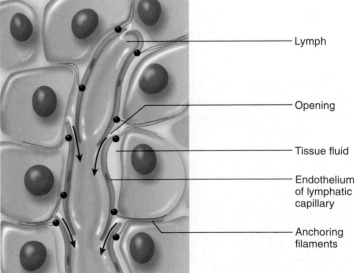

(a)

(b)

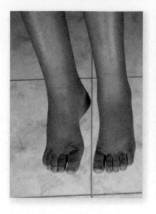
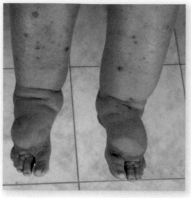

Figure 14.3 Edema. The person on the right shows severe edema of the legs and feet compared to a person without edema on the left. Blockage of lymphatic vessels is one of several causes of edema.

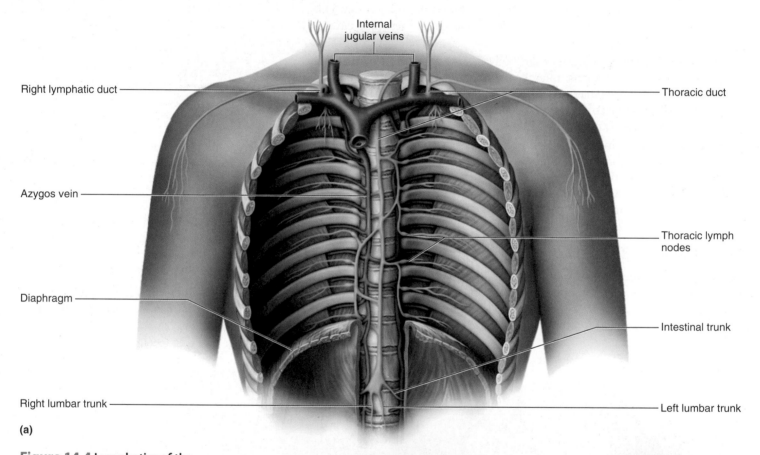

Internal jugular veins

Right lymphatic duct

Thoracic duct

Azygos vein

Thoracic lymph nodes

Diaphragm

Intestinal trunk

Right lumbar trunk

Left lumbar trunk

(a)

Figure 14.4 Lymphatics of the Thoracic Region. (a) Lymphatics of the thorax and upper abdomen and their relationship to the subclavian veins, where the lymph returns to the bloodstream. (b) Lymphatic drainage of the right mammary and axillary regions. (c) Regions of the body drained by the right lymphatic duct and thoracic duct. AP|R

• *Why are the axillary lymph nodes often biopsied in cases of suspected breast cancer?*

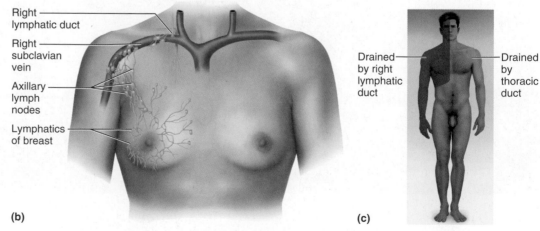

Right lymphatic duct

Right subclavian vein

Axillary lymph nodes

Lymphatics of breast

Drained by right lymphatic duct

Drained by thoracic duct

(b)

(c)

thin, loosely overlapping endothelial cells. The endothelial cells are tethered to surrounding tissue by protein filaments that keep the sac from collapsing. Unlike endothelium of blood capillaries, the cells are not joined by tight junctions but have large gaps between them for passage of proteins, bacteria, and white blood cells.

Overlapping edges of the endothelial cells act as valves that open and close depending on pressure in the surrounding tissue fluid. When tissue fluid pressure is high, the flaps are pushed inward (open) and fluid enters the lymphatic vessel. Inadequate fluid recovery from the tissues can result in **edema**, a swelling and puffiness due to fluid accumulation (fig.14.3). Once recovered tissue fluid enters the lymphatic vessels, it is called **lymph**.

As lymphatic capillaries converge, they become larger vessels with different names. Lymphatic capillaries converge to form thin **collecting vessels**. These converge to form six **lymphatic trunks**, each of which drains a major portion of the body. For example, the *lumbar trunks* drain the lumbar region as well as the lower limbs. All lymphatic trunks converge to form just two **collecting ducts**, the largest lymphatic vessels (fig. 14.4). The lesser of these, the **right lymphatic duct**, receives lymphatic drainage from the right side of the head and chest and from the right upper limb. It empties into the right subclavian vein. The other one, called the **thoracic duct**, is larger and longer. It receives several lymphatic trunks that drain all of the body below the diaphragm; the left side of the head, neck, and chest; and the left upper limb. It empties into the left subclavian vein. Thus, both ducts return lymph to the bloodstream at the subclavians. In summary, the route from the tissue fluid back to the bloodstream is lymphatic capillaries → collecting vessels → lymphatic trunks → two collecting ducts → subclavian veins. There is a continual recycling of fluid from blood to tissue fluid to lymph and back to the blood.

The larger lymphatic vessels are similar to veins in their structure, with a three-layered wall and valves that ensure a one-way flow of fluid (fig. 14.5). In contrast to blood, the lymph does not flow in response to a pump like the heart. As a result, it flows at even lower pressure and speed than venous blood, propelled by four mechanisms: (1) Lymphatic vessels contract rhythmically when flowing lymph stretches their walls. (2) Contracting skeletal muscles compress the lymphatics and move the lymph, like the skeletal muscle pump that moves venous blood. (3) Breathing creates pressure differences between the abdominal and thoracic cavities that cause lymph to flow upward into the thoracic lymphatics, in the same manner as the thoracic pump that aids venous blood flow. (4) Finally, at the point where the collecting ducts empty into the subclavian veins, the rapidly flowing bloodstream draws lymph into it. Given these mechanisms, it is easy to see why physical exercise significantly increases the rate of lymphatic return.

Lymph is similar to blood plasma but with less protein. Usually clear and colorless, its composition and characteristics vary somewhat from place to place. For example, lymph draining from the small intestine may have a milky appearance because it contains dietary lipids. In addition to proteins, lymph contains lymphocytes, macrophages, bacteria, viruses, cellular debris, and sometimes even traveling cancer cells.

Lymphatic Tissues

Lymphatic tissues are aggregations of lymphocytes in the connective tissue of mucous membranes and various organs. Lymphatic tissue is particularly abundant in body passages open to the exterior—the respiratory, digestive,

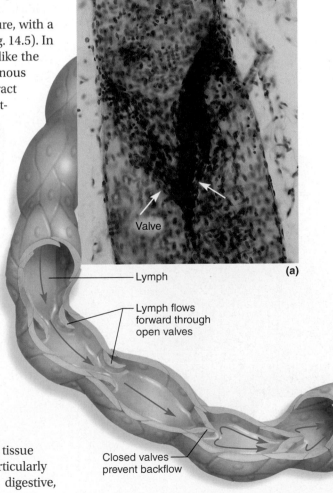

Figure 14.5 Valves in the Lymphatic Vessels. (a) Photograph of a lymphatic valve. (b) Operation of the valves to ensure a one-way flow of lymph.

- *What would be the consequence if these valves did not exist?*

Valve

Lymph   (a)

Lymph flows forward through open valves

Closed valves prevent backflow

(b)

urinary, and reproductive tracts—where it is called **mucosa-associated lymphatic tissue (MALT)**. It guards the body against environmental pathogens entering the body orifices such as the nose, mouth, anus, vagina, and urethra. In other places, lymphocytes and macrophages congregate in dense masses called **lymphatic nodules**, which come and go as pathogens invade the tissues and the immune system answers the challenge. Abundant lymphatic nodules are a relatively constant feature of lymph nodes, tonsils, and the appendix.

Lymphatic Organs

Lymphatic organs have well-defined anatomical sites and at least partial connective tissue capsules that separate the lymphatic tissue from neighboring tissues. The principal lymphatic organs are the tonsils, thymus, spleen, and lymph nodes.

The Tonsils

The **tonsils** are patches of lymphatic tissue located at the entrance of the pharynx, where they guard against inhaled and ingested pathogens (fig. 14.6). Numerous **lingual tonsils** are concentrated in a patch on each side of the root of the tongue; a pair of **palatine tonsils** lies on the lateral walls of the posterior oral cavity; and a single **pharyngeal tonsil** is found medially at the rear of the nasal cavity. Each tonsil has deep pits *(tonsillar crypts)* surrounded by lymphatic nodules. A partial capsule underlies each tonsil and separates it from deeper tissues, whereas the superficial surface of the tonsil is covered by the epithelium of the tongue or pharynx. Inflammation of the tonsils—most often the palatine—is called *tonsillitis*, and is characterized by redness, swelling, and pain. Tonsillitis used to be routinely treated by surgical removal of the tonsils *(tonsillectomy)* but is now usually treated with antibiotics.

Figure 14.6 The Tonsils. (a) Locations of the tonsils. (b) Histology of the palatine tonsil. AP|R

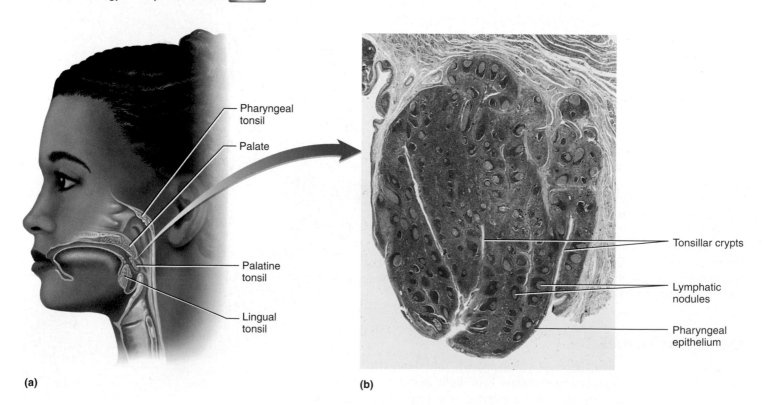

(a)

(b)

The Thymus

The **thymus** (fig. 14.7) is located between the heart and the base of the neck. It houses developing lymphocytes and secretes hormones that regulate their activity. The thymus is completely enclosed in a fibrous capsule. Extensions of the capsule form *septa* (walls) that divide the interior of the thymus into compartments called *lobules*. Each lobule has a relatively light center called the *medulla*, surrounded by a darker cortex (fig. 14.7b), both heavily populated by lymphocytes called *T cells*. *Reticular epithelial cells* seal off the cortex from the medulla and surround the blood vessels. They secrete *thymosin* and other hormones that promote the development and action of T cells, and form a *blood–thymus barrier* that isolates developing T cells from blood-borne antigens. The importance of thymic signaling molecules is demonstrated by the fact that if the thymus is removed from newborn mammals, they waste away and never develop immunity. Later in this chapter, we examine the role of the thymus in maturation of the T cells.

Figure 14.7 The Thymus. (a) Gross anatomy. (b) Histology. (c) Cellular architecture of a lobule. The reticular epithelial cells are interconnected to form a blood–thymus barrier. AP|R

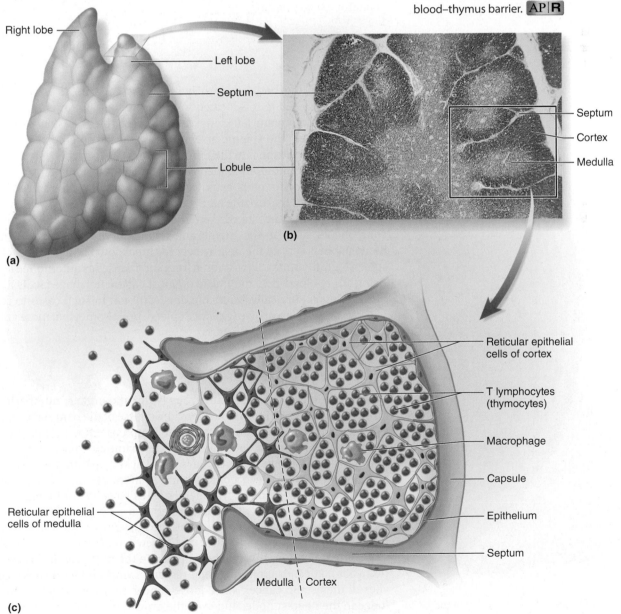

Right lobe

Left lobe

Septum

Lobule

(a)

Septum

Cortex

Medulla

(b)

Reticular epithelial cells of cortex

T lymphocytes (thymocytes)

Macrophage

Capsule

Epithelium

Septum

Reticular epithelial cells of medulla

Medulla   Cortex

(c)

Clinical Application 14.1

LYMPH NODES AND METASTATIC CANCER

Metastasis is the phenomenon in which cancer cells break free of the original *primary tumor,* travel to other sites of the body, and establish new tumors. Gaps in the walls of lymphatic capillaries allow cancer cells to enter and travel in the lymph. The abnormal cells eventually lodge in a lymph node, multiply there, and destroy the node. Cancerous lymph nodes are swollen, relatively firm, and usually painless. Cancer of a lymph node is called *lymphoma.*

Once a tumor is established in one node, cells may emigrate and travel to the next node. If the cancer is detected early enough, it may be eradicated by removing not only the primary tumor, but the lymph nodes downstream from that point. For example, breast cancer is often treated with a combination of lumpectomy (removing the primary tumor and surrounding tissue) or mastectomy (removing the breast) and removal of nearby axillary nodes.

The Spleen

The **spleen** is the body's largest lymphatic organ—up to 12 cm long and weighing up to 160 g. It is located in the upper left abdomen, tucked snugly between the diaphragm, left kidney, and stomach (fig. 14.8). It has a medial slit, the **hilum**, where the **splenic artery** enters and the **splenic vein** leaves. The spleen's major function is to filter the blood and cleanse it of bacteria and other foreign matter. It is densely stuffed with lymphocytes and macrophages, which monitor the blood for pathogens. Lymphocytes can leave the spleen by way of lymphatic vessels that exit the hilum, eventually to find their way into the blood. The spleen is also the site where old erythrocytes die and break down; RBC components such as the membrane fragments are disposed of; and free hemoglobin is sent to the nearby liver for breakdown and iron recycling. The spleen is quite vulnerable to trauma from such causes as automobile accidents. Being a very blood-rich and fragile organ, it can hemorrhage fatally and is difficult to repair surgically. Therefore, it is common in such cases to remove it *(splenectomy).* One can live without a spleen, but is somewhat more vulnerable to infections.

Lymph Nodes

Lymph nodes are the most numerous lymphatic organs, numbering in the hundreds (see fig. 14.1). They lie along the course of the lymphatic vessels and serve essentially to filter and cleanse the lymph as it flows on its way back to the bloodstream, and to detect and activate an immune response to microbes or other disease agents that may have been picked up from the tissue fluid. In order to do so, they need *both* incoming (afferent) and outgoing (efferent) lymphatic vessels; all other lymphatic organs have efferent lymphatics only. Lymph nodes are especially concentrated in the neck, axillary region, groin, mediastinum, and mesenteries around the intestines, with smaller clusters in the bend of the elbow and knee and elsewhere. Infections often result in swollen lymph nodes that can be palpated in the neck (cervical region), armpit (axilla), knee, or groin (inguinal region). Lymph nodes also are common sites of metastatic cancer, since loose cancer cells can travel in the lymph, lodge in the nearest node, and seed the growth of a new tumor (see Clinical Application 14.1).

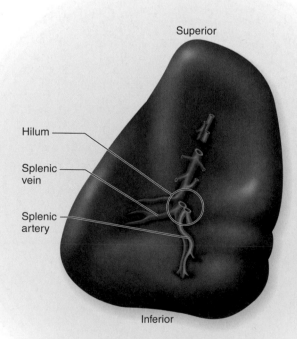

Superior

Hilum

Splenic vein

Splenic artery

Inferior

Figure 14.8 The Spleen. Gross anatomy of the medial surface. AP|R

A lymph node (fig. 14.9) is a small elongated or bean-shaped structure, usually less than 3 cm long, embedded in fat or other connective tissue. About two-thirds of the node has a convex surface that receives several afferent lymphatics. The opposite concave surface has a slit called the *hilum* where the efferent lymphatics, usually one to three in number, leave the node. Small arteries and veins enter and leave the node at the hilum as well.

The node is enclosed in a fibrous capsule with septa that penetrate and loosely compartmentalize its interior. The interior is divided into a relatively light-staining and loosely organized *medulla* just inside the hilum, encircled by a darker, C-shaped *cortex* in the convex regions of the node. The lymphocytes in the cortex are arranged in ovoid *lymphatic nodules*, whereas in the medulla they are arranged in elongated *medullary cords* with lymph-filled sinusoids between them. The dense concentration of lymphocytes in the node, especially in the cortex, give the tissue a granular appearance. Lymphocytes multiply profusely in the nodes, which are a major source of lymphocytes entering the bloodstream.

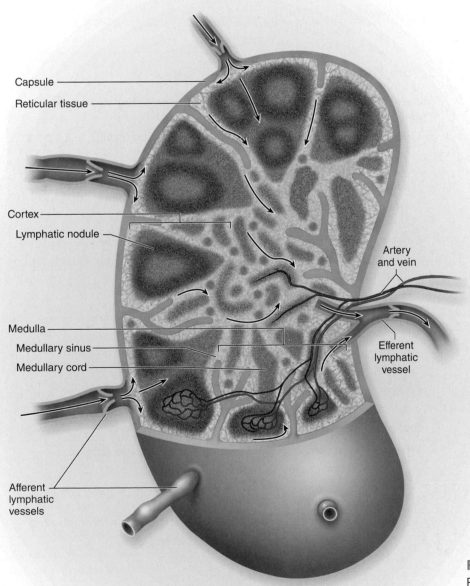

Figure 14.9 Anatomy of a Lymph Node.
Partially bisected lymph node showing pathway of lymph flow. AP|R

Flowing lymph typically passes through one node after another as it finds its way to the subclavian veins, where it reenters the bloodstream. In addition to lymphocytes, the node is populated by phagocytic macrophages and reticular cells. They remove about 99% of the impurities that enter the node (microbes, tissue debris, and other such matter), so by the time the lymph has flowed through a series of nodes and empties into the blood, it is essentially devoid of foreign matter but has numerous lymphocytes that it contributes to the blood. In addition to simple filtration and phagocytosis, the lymph nodes monitor the collected tissue fluid for foreign **antigens**—macromolecules that can activate an immune response. If a macrophage detects and phagocytizes an antigen, it digests it and displays molecular fragments of it bound to proteins on its surface. Lymphocytes detect these displayed fragments and begin an immune response detailed later in this chapter. The macrophages and certain other cells are called **antigen-presenting cells (APCs)** for this behavior. More about their role in immunity will be said later.

Apply What You Know

Why is it important that afferent and efferent lymphatic vessels ensure that lymph only flows one way in the lymph node?

Before You Go On

Answer these questions from memory. Reread the preceding section if there are too many you don't know.

1. List the primary functions and four major components of the lymphatic system.
2. How does fluid get into the lymphatic system? What prevents it from draining back out?
3. What is the difference between a lymphatic tissue and a lymphatic organ? Give an example of each.
4. List the lymphatic organs and describe the location, structure, and function of each.
5. What two functions are performed by the lymphocytes, macrophages, and reticular cells of a lymph node?

14.2 Nonspecific Resistance

Expected Learning Outcomes

When you have completed this section, you should be able to:

a. contrast nonspecific and adaptive immunity;

b. explain the types and importance of physical barriers in defense against pathogens;

c. enumerate the defensive functions of each kind of white blood cell;

d. describe the role of the complement system;

e. explain the process of inflammation; and

f. discuss the benefits and drawbacks of fever.

Our environment is heavily populated with pathogens as well as nonliving disease agents such as toxins and radiation. We have three lines of defense against such disease agents: (1) a system of external barriers, especially the skin and mucous membranes, that are impenetrable to most pathogens; (2) leukocytes, inflammation, fever, and other defenses that deal with pathogens that break through those barriers; and finally (3) the immune system, which not only defeats a pathogen but leaves our body with a memory of it so we can react more quickly and efficiently to it the next time.

The first two are called **nonspecific (innate) defenses** because they are present from birth and act equally well against a broad spectrum of pathogens. The third is called **adaptive (specific) immunity** because it adapts the body to its experiences with pathogens, and immunity to one pathogen does not give us immunity to another one. Immunity to measles, for example, does not make us immune to influenza. AP|R

Physical Barriers

The body is especially vulnerable to invasion anyplace that an internal passage opens to the exterior environment—the nose, mouth, anus, urethra, and vagina. The mucous membranes that line our digestive, reproductive, urinary, and respiratory tracts, and the skin that covers the body, are physical barriers that microbes cannot easily breach.

Several features make the skin an especially effective barrier. The cells are joined by tight junctions and are composed mainly of keratin, a tough protein that is difficult to penetrate. Furthermore, the skin surface is inhospitable for microbial growth. It is mostly dry, whereas microbes prefer a moist environment, and it lacks nutrients conducive to their survival. A thin film of acidic sweat and antimicrobial chemicals also retards pathogen growth and survival. Mucous membranes are hard to penetrate because of tight junctions between the epithelial cells; because the mucus itself is sticky and entraps microbes; and because it contains *lysozyme*, an enzyme that destroys bacteria.

Leukocytes and Macrophages

Nevertheless, the foregoing barriers are not completely impenetrable. Wounds or insect bites may breach the skin, for example, and the mucous membranes of the internal surfaces may be torn or irritated. If microbes get past the physical barrier, they are attacked by **phagocytes** that have a voracious appetite for foreign matter, and are assaulted by toxic chemicals produced by various white blood cells (fig. 14.10). The five types of leukocytes and details of their functions are described in chapter 12, so the following will serve only as brief reminders.

- **Neutrophils** destroy bacteria, especially, by phagocytosis and secretion of a chemical *killing zone* of hypochlorite and hydrogen peroxide.
- **Eosinophils** weaken or destroy parasites too large for phagocytosis, respond to allergens, and phagocytize and remove antigen–antibody complexes from the tissues.
- **Basophils** secrete histamine and heparin—a vasodilator and anticoagulant, respectively—which enhance the delivery of blood and immune cells to sites of infection or inflammation.
- **Lymphocytes** are of several types with varied roles explained in the ensuing discussion, including secretion of antibodies, direct attacks on pathogenic microbes, and destruction of infected and cancerous cells.
- **Monocytes** transform to **macrophages** in connective tissues, where they function as phagocytes and antigen-presenting cells and carry out cell signaling in inflammation and other defense processes.

Figure 14.10 Macrophage Phagocytizing Bacteria. Filamentous pseudopods of the macrophages snare the rod-shaped bacteria and draw them to the cell surface, where they are phagocytized.

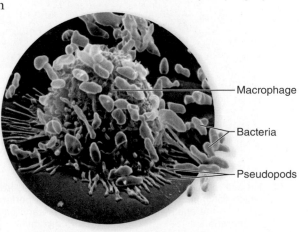

Natural Killer Cells

Natural killer (NK) cells are lymphocytes that continually patrol the body "on the lookout" for pathogens or diseased host cells (fig. 14.11). They attack and destroy bacteria, cells infected with viruses, cancer cells, and cells of transplanted tissues and organs. Upon recognition of an enemy cell, the NK cell binds to it and releases proteins called *perforins*. This is a "kiss of death"—it creates a hole in the enemy cell membrane that allows a rapid inflow of water and salts. This alone may kill the cell, but in addition, the NK cell secretes protein-degrading enzymes that enter through the hole, destroy the enemy's enzymes, and induce apoptosis (programmed cell death). This is a nonspecific defense; it does not rely on prior exposure to a pathogen, and it has no memory of prior exposures. This distinguishes the action of NK cells from those of other lymphocytes discussed later (T and B cells).

Defensive Proteins

Multiple types of proteins provide short-term, nonspecific resistance to pathogens. **Interferons** are a family of proteins secreted by cells in response to infection by viruses. They signal nearby cells to produce other proteins that inhibit viral replication. In addition, they activate natural killer (NK) cells, which destroy infected cells before they can liberate a swarm of newly replicated viruses.

The **complement system** is a group of 30 or more proteins synthesized mainly by the liver. They circulate in the blood in an inactive form; when activated by pathogens, they initiate a complex cascade of chemical events, which ultimately leads to the demise of invading microbes. They were named for the fact that they complement the action of the antibodies that we will study shortly, but they also aid in nonspecific defenses. Their effects are to stimulate inflammation, to promote phagocytosis of pathogens by neutrophils and macrophages, and to attack and rupture bacterial cells.

Inflammation

When you suffer a local infection, a tissue irritation such as blisters from tight shoes, a splinter, or just an itchy mosquito bite, you may notice redness, swelling, heat, and pain in the affected tissue. Those signs and symptoms have traditionally been called the four *cardinal signs* of the defensive process called **inflammation**. Words ending in the suffix *–itis* denote inflammation of specific organs and tissues—for example, *arthritis* is inflammation of the joints, *dermatitis* pertains to the skin, and *gingivitis* is inflammation of the gums. The purposes of inflammation are to (1) attract phagocytes and plasma proteins

Figure 14.11 The Action of a Natural Killer Cell.

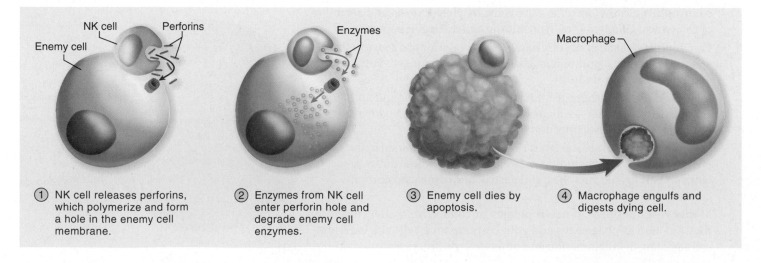

① NK cell releases perforins, which polymerize and form a hole in the enemy cell membrane.

② Enzymes from NK cell enter perforin hole and degrade enemy cell enzymes.

③ Enemy cell dies by apoptosis.

④ Macrophage engulfs and digests dying cell.

that prevent the spread of pathogens and promote their destruction, (2) remove the debris of damaged tissue, and (3) initiate tissue healing and repair.

Almost immediately after invasion by a microorganism, blood vessels dilate and blood flow to the area increases. Tissue damage stimulates basophils of the blood and similar *mast cells* in the connective tissue to release the vasodilator histamine. The cardinal signs of inflammation result from these early responses: (1) heat and (2) redness result from the increased blood flow; (3) swelling (edema) is due to increased fluid filtration from the capillaries; and (4) pain results from direct injury to the nerves, pressure on the nerves from the edema, and stimulation of pain receptors by a type of lipids called *prostaglandins*.

The increased blood flow rapidly delivers an army of defensive WBCs. Endothelial cells of the blood vessels interact with leukocytes—they produce cell-adhesion molecules that make their membranes sticky and snag the cells as they pass in the bloodstream, causing them to slowly tumble along the endothelium. Adhesion to the vessel wall gives WBCs time to check for signals from nearby injured or infected tissues (fig. 14.12). In the meantime, chemicals cause the endothelial cells of the blood capillaries and venules to separate a little, widening the intercellular gaps and permitting the WBCs to crawl through. The increased permeability also permits

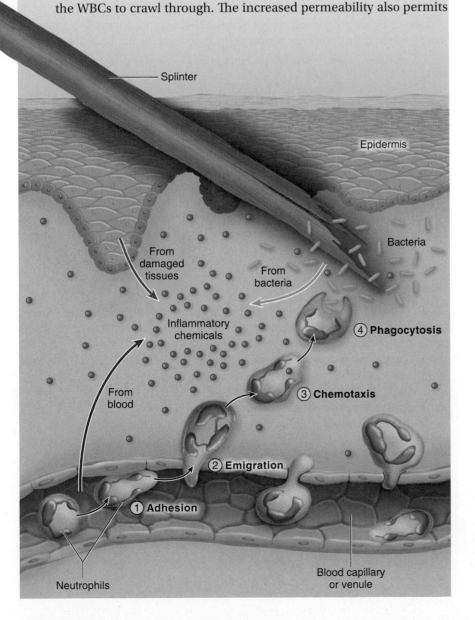

Figure 14.12 Neutrophil Behavior in Inflammation. Tissue injury and infection cause release of inflammatory chemicals from bacteria, the blood, and the damaged tissues themselves. Inflammatory chemicals stimulate neutrophils to adhere to the capillary wall, emigrate from the bloodstream through gaps between the capillary cells, follow the chemical trail by chemotaxis, and phagocytize (engulf) bacteria.

fluid, antibodies, and clotting factors and complement proteins to move into the surrounding tissue.

One priority of inflammation is to prevent pathogens from spreading throughout the body. Pathogens are physically trapped by clots that form in areas adjacent to the injury; a sticky mesh of fibrin walls off the area, so bacteria and other pathogens are essentially trapped in a fluid pocket surrounded by a gelatinous, clotted capsule. They are unable to escape attacks by phagocytes, antibodies, and other defenses.

The leukocytes wreak destruction upon pathogens that have invaded the body. Neutrophils are the chief enemies of bacteria. The major stages of their action are summarized in figure 14.12. Within an hour of injury, they begin to accumulate in inflamed tissue owing to **chemotaxis,** a process in which they are attracted by chemicals released by the tissues and bacteria. As they encounter bacteria, neutrophils avidly phagocytize and digest them. They also secrete chemical signals that recruit more WBCs, especially monocytes, to the scene.

The battle leaves an aftermath of tissue destruction. Monocytes are the major agents of tissue cleanup and repair, arriving within 8 to 12 hours of injury. They emigrate from the bloodstream and turn into macrophages. Macrophages engulf and destroy bacteria and other pathogens, damaged host cells, and dead and dying neutrophils.

As the battle progresses, all the neutrophils and most of the macrophages die. The dead cells, tissue debris, and tissue fluid form a milky pool of **pus.** Pus is usually absorbed, but sometimes it accumulates in a cavity called an **abscess**[1] that may have to be drained.

Edema also contributes to tissue cleanup. Swelling compresses veins and reduces venous drainage, while it forces open valves of lymphatic capillaries and promotes lymphatic drainage. The lymphatic vessels collect and remove bacteria, dead cells, proteins, and tissue debris better than blood capillaries can.

Vasodilation promotes healing by speeding the delivery of oxygen and amino acids necessary for protein synthesis. Endothelial cells and platelets release chemicals that stimulate tissue regrowth. Meanwhile, the heat of inflamed tissue increases the metabolic rate and the speed of mitosis and tissue repair. Pain also contributes to recovery. It is an alarm signal that calls our attention to the injury and causes us to limit the use of a body part so it has a chance to rest and heal.

Fever

Fever is an abnormal elevation of body temperature resulting from such causes as infection, trauma, and drug reactions. It was long regarded as an undesirable side effect of illness, and efforts were (and still are) made to reduce it for the sake of comfort. It is now recognized, however, as a defensive mechanism that, in moderation, does more good than harm. People recover from colds and other infectious diseases more quickly when they allow a fever to run its course rather than using fever-reducing medications such as aspirin.

When neutrophils and macrophages phagocytize bacteria, they secrete a **pyrogen,**[2] or fever-producing agent, that stimulates the hypothalamus to secrete a *prostaglandin*. The prostaglandin, in turn, raises the hypothalamic set point for body temperature—say to 39°C (102°F) instead of 37°F. Aspirin and ibuprofen reduce fever by inhibiting prostaglandin synthesis.

[1] *ab* = away; *scess* (from *cedere*) = to go
[2] *pyro* = fire, heat; *gen* = producing

Figure 14.13 The Course of a Fever.

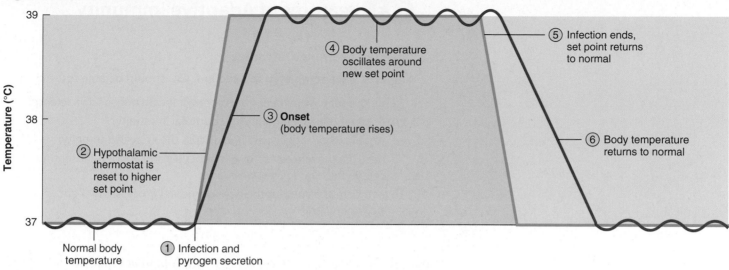

When the set point rises, the cutaneous blood vessels constrict to reduce heat loss and a person may shiver to generate heat. During the onset of a fever, one experiences chills, feels cold and clammy to the touch, and has a rising temperature (fig. 14.13). In the next stage, the body temperature hovers around the higher set point for as long as the pathogen is present. The elevated temperature inhibits reproduction by bacteria and viruses, promotes interferon activity, and increases metabolic rate and speeds tissue repair. When the infection is defeated, pyrogen secretion ceases and the hypothalamic thermostat is set back to normal. This activates heat-losing mechanisms, especially cutaneous vasodilation and sweating. The skin is warm and flushed during this stage. The result is falling temperature until it drops back to normal.

Even though most fevers are beneficial, excessively high temperature can be dangerous. It speeds up different enzymatic pathways, and may lead to metabolic discoordination and cellular dysfunction. Fevers above 40.5°C (105°F) can make a person delirious. Convulsions and coma ensue at higher temperatures, and death or irreversible brain damage commonly results from fevers that range from 44°C to 46°C (111°F to 115°F).

Before You Go On

Answer these questions from memory. Reread the preceding section if there are too many you don't know.

6. List several features that make the skin an effective barrier against pathogens.

7. What are macrophages? What roles do they play in defense?

8. How do interferons and complement proteins protect against disease?

9. Explain the causes of the four cardinal signs of inflammation.

10. How does inflammation promote the destruction of pathogens?

11. List three benefits of fever.

14.3 Features of Adaptive Immunity

Expected Learning Outcomes

When you have completed this section, you should be able to:

a. define *adaptive immunity* and contrast it with nonspecific defense;

b. contrast cellular immunity and humoral immunity;

c. explain what antigens are and discuss their role in immunity;

d. describe the role played by lymphocytes and antigen-presenting cells in the immune response; and

e. compare and contrast the developmental processes of T and B lymphocytes.

The rest of this chapter is concerned with **adaptive (specific) immunity**, which relies on the body's ability to recognize foreign substances individually, respond to them selectively, and remember them so it can defeat them more quickly in future exposures. Two characteristics distinguish adaptive immunity from nonspecific defense:

1. **Specificity.** Adaptive immunity is directed against a particular pathogen. Immunity to one pathogen does not usually confer immunity to others. A person may be immune to measles but still susceptible to polio, for example.

2. **Memory.** When reexposed to the same pathogen, the body reacts so quickly that there is no noticeable illness. The reaction time for inflammation and other nonspecific defenses, by contrast, is just as long for later exposures as for the initial one.

There are two forms of adaptive immunity: cellular and humoral. **Cellular (cell-mediated) immunity** employs lymphocytes that directly attack and destroy foreign cells or diseased host cells. It is a means to rid the body of pathogens that reside inside our cells, where they are inaccessible to attack by antibodies—for example, intracellular bacteria, viruses, yeasts, and protozoans. Cellular immunity also acts against parasitic worms, cancer cells, and cells of transplanted organs and tissues.

Humoral (antibody-mediated) immunity employs antibodies, which do not directly destroy a pathogen but tag it for destruction by other mechanisms. The term *humoral* comes from the fact that many of the antibodies are dissolved in body fluids, once called "humors." Humoral immunity is effective against extracellular microbes and noncellular entities such as toxins, venoms, and allergens.

Both forms rely on the body's ability to recognize specific foreign invaders and mount a targeted response to them. How this occurs is the focus of the following sections. Before we address those mechanisms (the "plot"), let us look at the "cast of characters" that play various roles in the drama of immunity.

Antigens

An **antigen**[3] is any molecule that triggers an immune response. Some antigens are components of plasma membranes and bacterial cell walls; others are venoms or toxins. Most antigens are complex molecules unique to each organism—usually proteins, polysaccharides, glycoproteins, or glycolipids.

---

[3] acronym from *antibody generating*

Clinical Application 14.2

HYPERSENSITIVITY, ALLERGIES, AND ANAPHYLACTIC SHOCK

Hypersensitivity is an excessive, harmful immune reaction to antigens. It includes reactions to tissues and organs transplanted from another person, abnormal reactions to one's own tissues (**autoimmunity**), and **allergies,**[4] which are reactions to environmental antigens. Such antigens, called **allergens,** occur in dust; mold; animal dander; toxins from poison oak and other plants; and foods such as nuts, eggs, shellfish, and gluten, a protein in wheat flour and some other grain products. Drugs such as penicillin are allergenic to some people.

Hypersensitivity may be characterized by a very rapid, acute response, or may exhibit a slower onset (1–3 hours after exposure) and last longer (10–15 hours). The most common allergies are acute in nature and begin seconds after exposure. They usually subside after 30 minutes but can be severe or even fatal. Allergens bind to the membranes of basophils and stimulate secretion of histamine and other inflammatory chemicals that trigger glandular secretion, vasodilation, increased capillary permeability, smooth muscle spasms, and other effects. The clinical signs include local edema, mucus hypersecretion, watery eyes, runny nose, hives (red itchy skin), and sometimes breathing problems, cramps, diarrhea, and vomiting. Examples of acute sensitivity are food allergies and asthma (see Clinical Application 15.1, p. 512).

Anaphylaxis[5] (AN-uh-fih-LAC-sis) is an immediate and sometimes severe acute reaction. Local anaphylaxis can be relieved with antihistamines. **Anaphylactic shock,** however, is a widespread hypersensitivity characterized by dyspnea (labored breathing), widespread vasodilation, circulatory shock, and sometimes, sudden death. Antihistamines are inadequate to counter anaphylactic shock, but epinephrine relieves symptoms by dilating bronchioles, increasing cardiac output, and restoring blood pressure.

Their uniqueness enables the body to distinguish its own ("self") molecules from those of any other individual or organism ("nonself"). The immune system "learns" to distinguish self- from nonself-antigens prior to birth; thereafter, it normally attacks only nonself-antigens.

Small molecules such as simple sugars and amino acids are not generally antigenic. They are too universal to all individuals (and species) to help the immune system distinguish self from nonself; if the immune system did attack them, it would be attacking substances on which our lives depend. However, certain small molecules called **haptens**[6] *become* antigenic by binding to proteins of one's own body, creating an antigenic complex. After the first exposure, the hapten alone can activate an allergic response (see Clinical Application 14.2). Haptens include penicillin and some other drugs, and various chemicals in cosmetics, detergents, industrial chemicals, animal dander, and poison ivy.

Lymphocytes

The two broad categories of lymphocytes involved in adaptive immunity are *T* and *B lymphocytes.* T lymphocytes are named for the thymus, where they spend time as part of their maturation process. B lymphocytes are named for

[4] *allo* = altered; *erg* = action, reaction

[5] *ana* = against; *phylax* = protection

[6] *hapt* = grasp, seize

an organ *(bursa)* in chickens where they were discovered, but the *B* can serve as a reminder that they remain in the bone marrow throughout their maturation process. Once mature, both types populate the lymph nodes, mucous membranes, and other lymphatic tissues and organs throughout the body, and remain on constant surveillance for pathogens. They have the ability to recognize and respond to an almost limitless variety of foreign agents. In addition, T cells specialize in recognizing and destroying body cells that have gone awry and become cancerous.

T Lymphocytes

T lymphocytes (T cells) have a life history involving three stages and anatomical stations in the body. We can loosely think of these as their "birth," their "training" or maturation, and finally their "deployment" to locations where they will carry out their immune functions.

T cells are "born" in the red bone marrow as descendants of the hemopoietic stem cells described in chapter 12. Immature T cells travel from the bone marrow to the thymus, where they mature into fully functional T cells (fig. 14.14). Their "training" in the thymus is a complex multistep process. The essence of it is that young T cells develop antigen receptors on the cell surface and are then tested to see if they will respond to one's self-antigens. About 98% of them fail this test (and you thought your final exams were hard!); they respond to self-antigens, signifying that they could be harmful to one's own tissues. These self-reactive cells are destroyed or rendered inactive, leaving only 2% with the ability to respond selectively to foreign antigens. This creates a state of **self-tolerance** in which the immune system normally will not attack one's own body. These 2% of the cells multiply and build up a huge population of self-tolerant T cells. Now considered to be **immunocompetent**, they "graduate from school" and disperse to the lymph nodes and other organs, ready to do battle.

T cells mature mostly during fetal life and early childhood. The thymus gradually atrophies as a person ages, but it continues to produce *thymosin*, a hormone that promotes the proliferation of T cells throughout the body.

Figure 14.14 The Life History and Migrations of B and T Cells. T cells are represented by the red pathways and B cells by the violet. (a) T stem cells emigrate from the bone marrow and attain immunocompetence in the thymus. (b) Immunocompetent T cells emigrate from the thymus and recolonize the bone marrow or colonize various lymphatic organs (right). (c) B cells achieve immunocompetence in the red bone marrow (left), and many emigrate to lymphatic tissues and organs, including the lymph nodes, tonsils, and spleen (right).

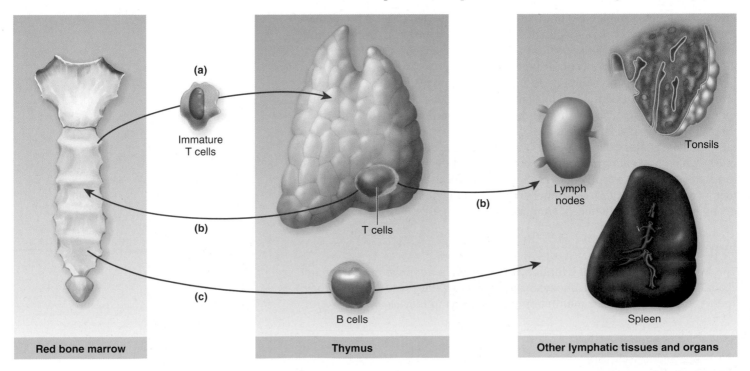

| Red bone marrow | Thymus | Other lymphatic tissues and organs |

B Lymphocytes

B lymphocytes (B cells) mature in the bone marrow, but they undergo a similar process of testing and selection, then likewise disperse and colonize lymphatic tissues and organs throughout the body. B cells are abundant in lymph nodes, mucous membranes, and the spleen. They differ from T cells in their mode of attack against foreign invaders, as we will soon see in more detail.

Antigen-Presenting Cells

T lymphocytes cannot recognize unprocessed antigens; they must first be introduced to them by antigen-presenting cells (APCs) that process and display antigen fragments. Antigen-presenting cells include *macrophages* of various kinds, *dendritic cells* of the skin and mucous membranes, and even B lymphocytes, which can play an antigen-presenting role in addition to their other functions.

When an APC encounters an antigen, it internalizes it by phagocytosis, digests it with lysosomal enzymes, and displays the relevant (antigenic) fragments attached to proteins on its surface (fig. 14.15). This is like an invitation to the immune system to launch an attack on the antigenic invader. We shall see shortly how the immune cells respond in different forms of adaptive immunity.

With so many cell types involved in immunity, it is no surprise that chemical messengers are required to coordinate their activities. Lymphocytes and APCs communicate with a great variety of chemical signals called **interleukins.**[7]

---

[7] *inter* = between; *leuk* = leukocytes

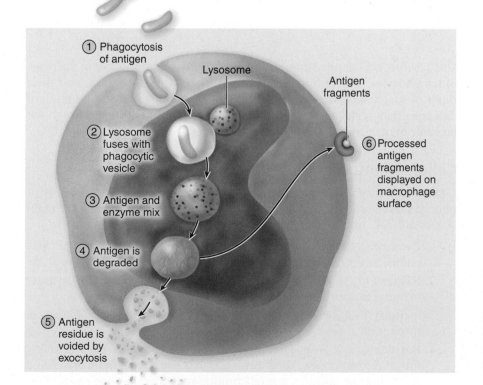

1. Phagocytosis of antigen
Lysosome
Antigen fragments
2. Lysosome fuses with phagocytic vesicle
3. Antigen and enzyme mix
4. Antigen is degraded
5. Antigen residue is voided by exocytosis
6. Processed antigen fragments displayed on macrophage surface

Figure 14.15 The Action of Antigen-Presenting Cells (APCs). Stages in the processing and presentation of an antigen by an APC such as a macrophage. **AP|R**

14.4 Cellular and Humoral Immunity

Expected Learning Outcomes

When you have completed this section, you should be able to:

a. list the types of lymphocytes involved in cellular immunity and describe the roles they play;

b. describe the process of antigen presentation and T cell activation;

c. describe how T cells destroy enemy cells;

d. explain how B cells recognize and respond to an antigen;

e. describe the structure and actions of antibodies;

f. explain the role of memory cells in immunity; and

g. compare and contrast cellular and humoral immunity.

The two forms of adaptive immunity, cellular and humoral, differ in the types of lymphocytes involved, the types of pathogens they recognize, and their mode of attack. Thus, they are not redundant but complementary methods of defeating pathogens and remembering them so those pathogens are much less able to make one ill in the future.

Cellular Immunity

Cellular (cell-mediated) immunity is a form of specific defense in which T lymphocytes directly attack and destroy diseased or foreign cells. The four main classes of T cells involved in cellular immunity follow.

1. **Cytotoxic T (T_C) cells** are the "effectors" of cellular immunity that carry out the attack on enemy cells.

2. **Helper T (T_H) cells** promote the action of T_c cells, but also play key roles in humoral immunity and nonspecific defense.

3. **Memory T (T_M) cells** are descended from the cytotoxic T cells and are responsible for memory in cellular immunity.

4. **Regulatory T (T_R) cells** inhibit the responses of other T cells, enabling the imune response to wind down after a pathogen has been eradicated and helping to prevent harmful immune attacks against one's own body.

Cellular immunity occurs in three stages: recognition, attack, and memory (or "the three *R*s of immunity"—recognize, respond, remember).

① **Recognition.** Cytotoxic and helper T cells patrol the lymph nodes and other tissues, inspecting the APCs as though looking for trouble. APCs use two different sorts of surface proteins to display antigens—one for antigens of foreign origin and another for antigens that come from one's own cells *(host cells)* with signs of disease, such as cells infected by viruses or that have turned cancerous. T_H cells respond only to displayed bits of foreign antigen, whereas T_C cells respond only to antigens of diseased host cells. When a T cell encounters an APC displaying an antigen that it's programmed to recognize, it initiates an immune response (fig. 14.16). In a process called **clonal selection,** the T cell multiplies rapidly and builds a huge population of cells programmed against the same antigen. Some cloned cells become effector cells that carry out the immune attack, while others become memory T cells.

② **Attack.** The attack phase requires both helper and cytotoxic T cells. Helper T cells play a central coordinating role in cellular immunity

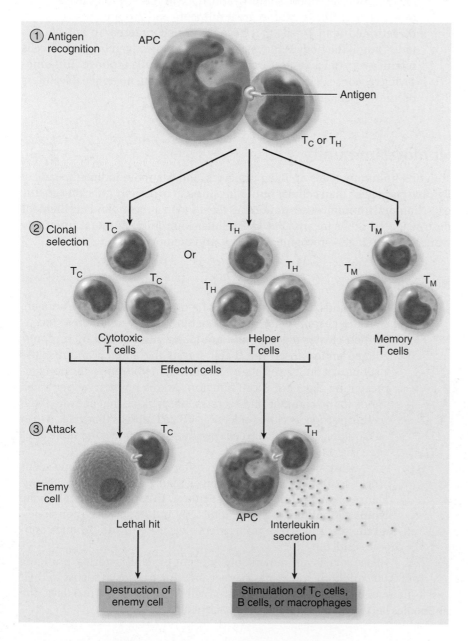

Figure 14.16 T Cell Activation.

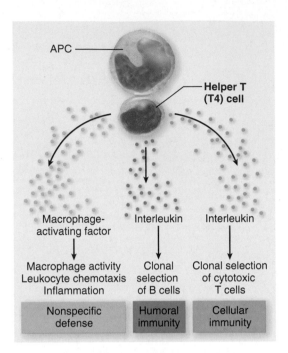

Figure 14.17 The Role of Helper T Cells in Defense and Immunity. AP|R

- *Why does AIDS reduce the effectiveness of all three defenses listed across the bottom of the figure? (See Perspective on Health, p. 497.)*

Figure 14.18 Destruction of a Cancer Cell by Cytotoxic T Cells. Four T$_c$ cells (pink) have bound to this cancer cell and delivered a lethal hit of cytotoxic chemicals that will destroy it. AP|R

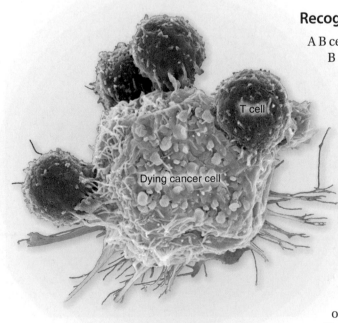

(fig. 14.17); indeed, they serve to regulate most immune functions. When a helper T cell recognizes an antigen displayed on an APC, it secretes interleukins that exert three effects: (1) they attract macrophages and stimulate their phagocytic activity; (2) they stimulate the activation of B cells and trigger humoral immunity; and (3) they stimulate the maturation and mitosis of cytotoxic T cells.

Cytotoxic T cells are the only T lymphocytes that directly attack and kill other cells (fig. 14.18). When a T$_C$ cell detects an enemy antigen on a diseased cell, it "docks" on the cell and delivers a **lethal hit** of cytotoxic chemicals that destroy it. These include *interferons,* which inhibit viral replication; *perforins,* which make holes in an enemy cell in the same manner as described earlier for natural killer cells; and other chemicals that enter and degrade the enemy cell's enzymes, activate macrophages, and kill cancer cells. Having delivered its lethal hit, the T$_C$ cell goes off in search of other diseased cells, leaving these chemicals behind to perform the assassination.

③ **Memory.** The foregoing response is followed by immune memory. Following clonal selection, some T$_C$ and T$_H$ cells become long-lived memory (T$_M$) cells. These exist in great numbers and require fewer steps to be activated than T cells that have never encountered an antigen before. Thus, if the body is reexposed to the same antigen later, the T cells mount a very quick response called the **T cell recall response.** This time-saving response destroys a pathogen so quickly that no noticeable illness occurs—a person is immune to the disease.

Humoral Immunity

Humoral immunity relies on B lymphocytes, and is a more indirect means of fighting pathogens than cellular immunity. Instead of directly attacking enemy cells, humoral immunity uses antibodies that bind to antigens and tag them for destruction by other means. But like cellular immunity, humoral immunity works in three stages—recognition, attack, and memory.

Recognition

A B cell has thousands of surface receptors for one particular type of antigen. B cell activation begins when an antigen binds to several of these receptors, which cluster together and are taken into the cell by receptor-mediated endocytosis. The B cell digests the antigen and displays portions of it on the surface of the plasma membrane—the antigen-presenting aspect of B cell function. At this point, a helper T cell binds to the complex and secretes interleukins that activate the B cell. This triggers clonal selection—B cell mitosis that gives rise to a battalion of identical B cells programmed against the same antigen (fig. 14.19).

Most cells of the clone differentiate into **plasma cells** (fig. 14.20). Plasma cells are larger than B cells due to an abundance of rough endoplasmic reticulum. They develop mainly in the lymphatic nodules of lymph nodes. About 10% stay in lymph nodes, but the rest leave, take up residence elsewhere, and produce antibodies until they die.

Plasma cells are remarkably efficient protein factories, pumping out 2,000 antibody molecules per second. This comes at a cost to their self-maintenance, however, and a plasma cell dies after only 4 to 5 days. The antibodies lay the groundwork for the attack phase.

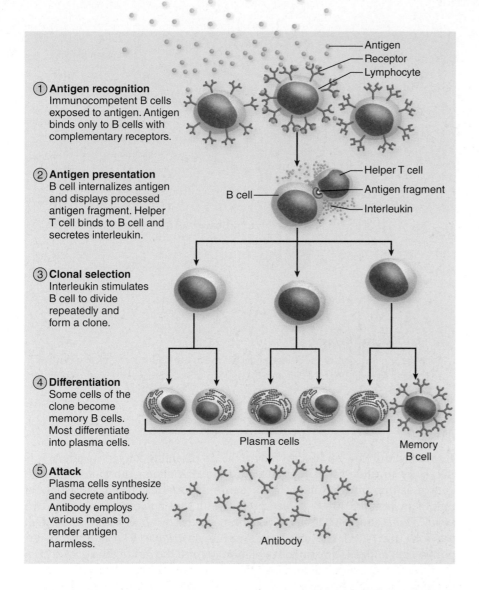

① **Antigen recognition**
Immunocompetent B cells exposed to antigen. Antigen binds only to B cells with complementary receptors.

Antigen
Receptor
Lymphocyte

② **Antigen presentation**
B cell internalizes antigen and displays processed antigen fragment. Helper T cell binds to B cell and secretes interleukin.

Helper T cell
B cell
Antigen fragment
Interleukin

③ **Clonal selection**
Interleukin stimulates B cell to divide repeatedly and form a clone.

④ **Differentiation**
Some cells of the clone become memory B cells. Most differentiate into plasma cells.

Plasma cells
Memory B cell

⑤ **Attack**
Plasma cells synthesize and secrete antibody. Antibody employs various means to render antigen harmless.

Antibody

Figure 14.19 Clonal Selection and Ensuing Events of the Humoral Immune Response.

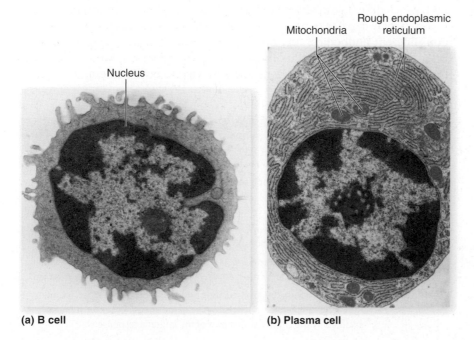

Nucleus

Rough endoplasmic reticulum
Mitochondria

(a) B cell

(b) Plasma cell

Figure 14.20 B Cell and Plasma Cell. (a) B cells have little cytoplasm and scanty organelles. (b) A plasma cell, which differentiates from a B cell, has an abundance of rough endoplasmic reticulum.

• *What does this endoplasmic reticulum do in the plasma cell?*

Figure 14.21 Antibody Structure.
A molecule of IgG.

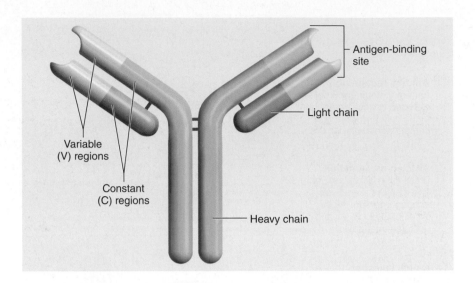

> **Apply What You Know**
>
> Consider the function of rough endoplasmic reticulum and the chemical nature of antibodies. Explain how the difference between a B cell and plasma cell relates to plasma cell function.

Attack

To understand how antibodies play a role in disabling foreign invaders, it is necessary to understand their structure and function. Antibodies are also known as **immunoglobulins (Ig)** or **gamma globulins.** The basic structural unit of an antibody is a Y-shaped protein with four polypeptide chains—two *heavy chains* about 400 amino acids long and the two *light chains* about half that long (fig. 14.21). The arms have a terminal *variable (V) region* that determines what antigens an antibody will bind to, and the rest of the antibody is a *constant (C) region* that determines the functional class to which an antibody belongs. There are five classes of antibodies named **IgA, IgD, IgE, IgG, and IgM,** which differ in their functions (Table 14.1). The human immune system is believed to be capable of producing at least 10 billion and perhaps up to 1 trillion different antibodies. Any one individual has a much smaller subset of these, but nevertheless, such diversity helps to explain why we can deal with the staggering diversity of antigens that exists in our environment.

Table 14.1 Antibody Classes

Class	Location and Function
IgA	Found in blood plasma, mucus, tears, milk, saliva, and intestinal secretions. Provides passive immunity to the newborn.
IgD	A membrane protein of B cells; thought to function in activation of B cells by antigens.
IgE	A membrane protein of basophils and mast cells. Stimulates them to release histamine and other chemical mediators of inflammation and allergy; important in hypersensitivity reactions; attracts eosinophils to sites of parasitic infection.
IgG	Constitutes about 80% of circulating antibodies in blood plasma. IgG and IgM are responsible for most adaptive immune responses against invaders.
IgM	Constitutes about 10% of circulating antibodies in plasma. When found in the membrane of B cells, it acts as part of the antigen receptor. It is also secreted by the cell in the first part of the immune response.

Antibodies can use any of the following means of rendering antigens harmless and marking them for destruction.

- **Neutralization.** Only certain regions of an antigen are pathogenic—for example, the parts of a toxin molecule or virus that enable them to bind to human cells. Antibodies can neutralize an antigen by binding to these active regions and masking their effects.

- **Agglutination.** The basic Y-shaped antibody has antigen-binding sites at the tip of each arm of the Y. An antigen may have more than one of these Ys; for example, each IgM molecule has 5 Ys, and therefore 10 antigen-binding sites. If one site binds to one antigen molecule and another site binds to another antigen, those two antigen molecules are stuck together. Repeating this over and over, antibodies can clump *(agglutinate)* antigens into large masses called *antigen–antibody complexes* (fig. 14.22). This renders the antigens powerless to harm the body. The antigen–antibody complexes can be removed and degraded by eosinophils, or they can bind to red blood cells and travel with them to the liver and spleen, where they are stripped off the RBC and disposed of. This process of ridding the body of agglutinated antigens is called *immune clearance.*

- **Activation of the complement system.** Activation of the complement system is the most important way antibodies provide defense against foreign cells such as bacteria. When an appropriate antigen binds to an antibody of the IgG or IgM class, complement proteins bind to receptors on the antibody tail. This initiates the cascade of events that leads to destruction of the invader by the mechanisms described earlier for the complement system (p. 482).

Memory

When a person is exposed to a particular antigen for the first time, the immune reaction is called the **primary response.** The appearance of protective antibodies is delayed for 3 to 6 days while B cells multiply and differentiate into plasma cells. As the plasma cells begin secreting antibody, the antibody level in the blood plasma and tissue fluids rises. It drops to low levels within a month.

The primary response, however, leaves one with an immune memory of the antigen. During clonal selection, some members of the clone become **memory B cells** rather than plasma cells. Memory B cells, found mainly in the lymph nodes, mount a very quick **secondary response** if reexposed to the same antigen (fig. 14.23). Plasma cells form within hours; the antibody level (titer) rises much higher than before and peaks within a few days. The response is so rapid that the antigen has little chance to exert a noticeable effect on the body, and no illness results. The antibody level remains elevated for weeks to years, conferring lasting protection. Memory does not last as long in humoral immunity, however, as it does in cellular immunity.

Active and Passive Immunity

Up to this point, we have examined humoral immunity only as a process that occurs in response to unintentional but natural exposure to environmental antigens. However, there are other varieties of humoral immunity bestowed by intentional antigen exposure or by acquiring antibodies produced by another individual.

Any case in which the body produces its own antibodies is called **active immunity.** What we have studied so far is considered *natural active immunity,* because it results from natural

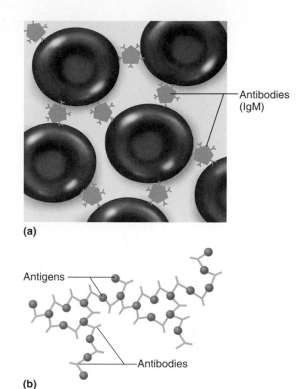

(a)

(b)

Figure 14.22 Agglutination by Antibodies.
(a) Agglutination of foreign erythrocytes by IgM.
(b) An antigen–antibody complex involving an antigen and an antibody such as IgG.

Figure 14.23 The Primary and Secondary Responses in Humoral Immunity. The individual is exposed to antigen on day 0 in both cases. Note the differences in the speed of response, the height of the antibody titer, and the rate of decline in antibody titer.

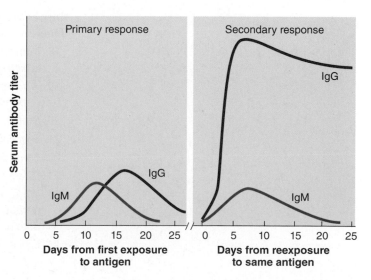

exposure to antigens. In contrast, *artificial active immunity* is a process in which we deliberately subject ourselves to antigens in order to stimulate antibody production and produce an immune memory that protects us against future disease—in other words, **vaccination.**

Because of vaccination, certain diseases that were once dreaded harbingers of death are now only ghostly memories. Vaccinations rely on injecting dead or disabled pathogens that stimulate the production of antibodies but do not make the person ill. If one then encounters that antigen later in life, a rapid secondary response occurs and prevents it from causing disease. Spectacular success has been achieved with certain vaccines; for example, the World Health Organization launched a campaign in the 1960s to eradicate smallpox worldwide, and no cases have been known since 1978. Although smallpox is the only disease to be eradicated, other diseases such as polio are virtually unknown in the United States, although cases have been increasing in India and parts of Africa where vaccine is either unavailable or people refuse to have children vaccinated.

At one time, optimists hoped that vaccines would wipe out infectious disease. We now know that infectious diseases are so complex that we will never completely conquer them—we are locked in an arms race that has been carried out for millions of years. For example, despite great effort, scientists have yet to develop effective vaccines against global diseases such as malaria, tuberculosis, or the virus HIV, which causes AIDS.

We can also acquire disease resistance by **passive immunity,** bestowed by either natural or artificial means. *Natural passive immunity* is the process in which a fetus or infant acquires antibodies from the mother through the placenta before birth or via the breast milk after. It provides important early protection of the infant until it can build its own immunity by exposure to environmental antigens. *Artificial passive immunity* can be acquired at any age by injecting blood serum from another person or from animals (such as horses) that have made antibodies against a specific antigen. Such *immune serum* is used for emergency treatment of snakebites, botulism, tetanus, rabies, and other diseases. The immunity conferred by serum is short-lived, lasting only until the body degrades the injected antibodies.

Before You Go On

Answer these questions from memory. Reread the preceding section if there are too many you don't know.

18. Name four types of lymphocytes involved in cellular immunity. Which of these is also essential to humoral immunity?
19. What are the three phases of cellular and humoral immune responses?
20. Explain how helper and cytotoxic T cells are activated.
21. Describe how cytotoxic T cells destroy target cells.
22. Compare and contrast a B cell and a plasma cell.
23. Discuss three ways antibodies act against antigens.
24. Why does the secondary immune response prevent a pathogen from causing disease, yet the primary response does not?
25. Explain how vaccination works.

PERSPECTIVES ON HEALTH

AIDS: A Global Epidemic

AIDS (acquired immunodeficiency syndrome) is a modern plague that affects every part of the world, but incredibly, was unknown before the early 1980s. Globally, 7,500 people are infected every day; the number of infected people is 33 million. Each year 2 million people die of the disease. Sub-Saharan Africa is the most heavily affected region, with as many as 25% of adults infected in some areas. The disease has had profound demographic, social, and economic impacts in many parts of the world, and a global effort is required to address it.

AIDS is caused by *HIV (human immunodeficiency virus),* which is spread by contact with body fluids such as blood, semen, vaginal secretions, and breast milk. It is most frequently passed from person to person during sexual activity, but may also be spread through infected needles, or from mothers to their infants in breast milk. HIV primarily infects helper T cells. Figure 14.24a shows components of the virus. Its surface molecules bind to corresponding receptors on the plasma membrane of the helper T (T_H) cells. Once inside the cell, the virus may remain latent for anywhere from 2 to 15 years. This is why most people have no clinical symptoms for a period of time after infection. Once the virus becomes active in T_H cells, it replicates and the cells eventually die. As the cells die, millions of new infectious viral particles spew forth into the bloodstream and infect more T cells (fig. 14.24b). A healthy T_H count is 600 to 1,200 cells/μL; a criterion for AIDS is a count less than 200/μL. The scarcity of T_H cells disables the specific immune response and the patient succumbs to opportunistic infections such as pneumonia or tuberculosis or dies from cancer, diseases that are normally kept under control by cell-mediated immunity.

There is currently no cure for AIDS. The disease can be prevented by avoiding contact with body fluids such as blood or semen that might be infected. In recent years, medications that disable the virus and prevent its replication have dramatically improved the survival of HIV-positive persons. However, the disease is spreading at an alarming rate (mostly through heterosexual sexual activity) in less-developed countries, where the cost of these drugs means they are out of reach for most people. The AIDS epidemic has triggered unprecedented effort to develop a vaccine or cure, but the challenges to doing so are daunting. The virus is one of the fastest-evolving microbes that exist. It reproduces incredibly rapidly—a single virus can produce billions of copies in a day. In addition, it accumulates numerous mutations as it copies its genetic material, and therefore is constantly reinventing itself. It thus presents a moving target that frustrates attempts to develop a vaccine with an effective antigen.

HIV and AIDS dramatically illustrate the importance of understanding the fundamental mechanisms of the immune system. They also highlight the vulnerability of the human species to disease, even in this era of sophisticated medical intervention.

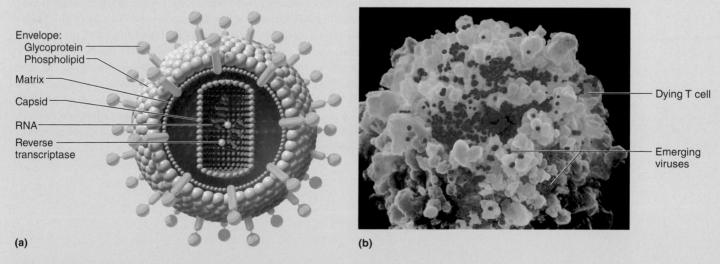

Envelope:
Glycoprotein
Phospholipid
Matrix
Capsid
RNA
Reverse transcriptase

Dying T cell

Emerging viruses

(a)   (b)

Figure 14.24 The Human Immunodeficiency Virus (HIV). (a) Structure of the virus. (b) Viruses emerging from a dying helper T cell. Each virus can now invade a new helper T cell and produce a similar number of descendants.

Aging of the Lymphatic and Immune Systems

The amounts of lymphatic tissue and red bone marrow decline with age. Consequently, there are fewer hemopoetic stem cells, disease-fighting leukocytes, and antigen-presenting cells (APCs). Also, the lymphocytes produced in these tissues often fail to mature or become immunocompetent. Both humoral and cellular immunity depend on APCs and helper T cells, and therefore both types of immune response are blunted. As a result, an older person is less protected against cancer and infectious diseases. It becomes especially important in old age to be vaccinated against influenza and other acute seasonal infections.

CAREER SPOTLIGHT

Public Health Nurse

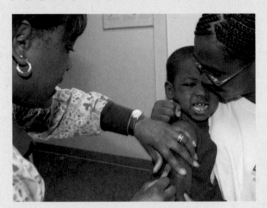

Public health nurses are registered nurses (RNs) who usually work in government, for private nonprofit agencies, or in clinics. They focus on the overall health of communities, working to educate individuals, families, and other groups about health-care issues, disease prevention, nutrition, and child care. They collaborate with community leaders, teachers, parents, and physicians in ensuring the health of the community. The tasks of a public health nurse focus on prevention and education and might include vaccinating children against childhood diseases, administering flu shots, or screening populations for hypertension. Education about, prevention of, and treatment of communicable diseases are also in the realm of a public health nurse. Qualifications vary from state to state—some require a bachelor's degree in nursing whereas in other states an associate degree from a community college suffices. See appendix B for further information on careers in public health nursing.

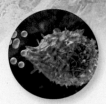

CONNECTIVE ISSUES

Ways in Which the Lymphatic and Immune Systems Affect Other Organ Systems

All Systems
Lymphatics drain excess tissue fluid and remove cellular debris and pathogens from tissues. The immune system defends other systems from infection and cancer.

Integumentary System
Lymphatic drainage prevents cutaneous edema; dendritic cells of the immune system densely populate the skin and guard against invasion.

Skeletal System
Interleukins affect the rate of hemopoiesis in the red bone marrow.

Muscular System
Lymphatics prevent muscular edema.

Nervous System
Microglia of the immune system protect tissues of the central nervous system from agents of disease.

Endocrine System
Lymph transports some hormones.

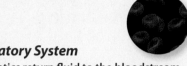

Circulatory System
Lymphatics return fluid to the bloodstream and help to sustain blood volume; spleen and other lymphatic organs prevent accumulation of debris and foreign matter in blood; spleen disposes of old RBCs.

Respiratory System
Alveolar macrophages remove debris from lungs.

Urinary System
Lymphatics absorb fluid and proteins in the kidneys, helping the kidneys to concentrate the urine and conserve water.

Digestive System
Lymph absorbs and transports dietary lipids; hepatic macrophages play a defensive role in the liver.

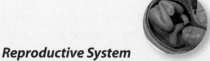

Reproductive System
Ovaries and testes require immune barriers to protect haploid germ cells against immune attack.

499

Study Guide

Assess Your Learning Outcomes

To test your knowledge, discuss the following topics with a study partner or in writing, ideally from memory.

14.1 The Lymphatic System (p. 471)

1. Definitions of *lymphatic system, immune system,* and *immunology*
2. The types of pathogens
3. Major components and functions of the lymphatic system
4. Types and structure of lymphatic vessels
5. Origin and composition of lymph and the route of its flow
6. Lymphatic tissues, their locations and histology, and how they differ from lymphatic organs
7. The types, locations, structure, and function of tonsils
8. The location, anatomy, and functions of the thymus
9. The location, anatomy, and functions of the spleen
10. Anatomy, histology, and functions of lymph nodes, and areas of lymph node concentration in the body
11. The relationship of lymph nodes to lymphatic vessels and how this relates to their lymph-cleansing function

14.2 Nonspecific Resistance (p. 480)

1. The three lines of defense against agents of disease
2. Major differences between nonspecific defense and adaptive immunity
3. The role of skin and mucous membranes as a first line of defense
4. How each leukocyte type functions in the second line of defense
5. The role of natural killer (NK) cells in recognizing and destroying enemy cells
6. The role of interferons in defense
7. The definition of *complement system* and ways in which it leads to the destruction of pathogens
8. The definition of *inflammation,* and its four cardinal signs
9. The mechanisms by which inflammation combats a pathogen and promotes healing
10. Benefits and dangers of fever
11. The roles of pyrogens, prostaglandins, and the hypothalamus in fever

14.3 Features of Adaptive Immunity (p. 486)

1. Two characteristics of adaptive immunity that distinguish it from nonspecific defense
2. The two types of adaptive immunity, and their similarities and differences
3. Definition of *antigen* and why antigens are fundamental to the specific immune response
4. How haptens differ from other antigens, how they become antigenic, and where one might encounter haptens in everyday life
5. Similarities and differences between T and B lymphocytes
6. The life history of a T cell from its "birth" until it is ready to do battle
7. Similarities and differences between the life history of a B cell and that of a T cell
8. Some types of antigen-presenting cells and their role in the immune response
9. Definition of *interleukins* and their role in immunity

14.4 Cellular and Humoral Immunity (p. 490)

1. The four types of T cells associated with cellular immunity
2. Differences between T_H cells and T_C cells in the types of antigens they recognize
3. How antigen-presenting cells help T cells recognize enemy antigens
4. The roles of clonal selection and interleukins in cellular and humoral immunity
5. The nature of a T_C cell's lethal hit on an enemy cell
6. The mechanism of memory and the T cell recall response in cellular immunity
7. The process of antigen recognition and B cell activation in humoral immunity
8. The relationship between a B cell and a plasma cell, and their structural and functional differences
9. The basic structure and varied classes of antibodies, and how antibody structure relates to the ability to recognize and bind a specific antigen
10. Why we say only that antibodies indirectly promote the destruction of antigens rather than destroying them directly
11. Three ways in which antibodies promote antigen destruction
12. The mechanism of memory and the secondary response in humoral immunity; similarities and differences between memory in the two forms of adaptive immunity
13. The difference between active and passive immunity and the reason artificial immunity is associated with vaccination

Testing Your Recall

1. Classic signs of inflammation include all of the following *except*
   a. edema.
   b. fever.
   c. redness.
   d. pain.
   e. heat.

2. Which of the following cells are involved in nonspecific but not adaptive immunity?
   a. helper T cells
   b. cytotoxic T cells
   c. natural killer cells
   d. B cells
   e. plasma cells

3. The lethal hit is used by _____ to kill enemy cells.
   a. neutrophils
   b. basophils
   c. mast cells
   d. NK cells
   e. cytotoxic T cells

4. Which of these lymphatic organs has a cortex and a medulla: (I) spleen; (II) lymph node; (III) thymus; (IV) red bone marrow?
   a. II only
   b. III only
   c. II and III only
   d. III and IV only
   e. I, II, and III

5. All of the following are important in nonspecific defense *except*
   a. histamine.
   b. clonal selection.
   c. inflammation.
   d. pyrogens.
   e. interferons.

6. Where do T cells attain immunocompetence?
   a. in the red bone marrow
   b. in the lymph nodes
   c. in the thymus
   d. in the spleen
   e. in the mucosa-associated lymphatic tissue

7. The only lymphatic organ with both afferent and efferent lymphatic vessels is
   a. the spleen.
   b. a lymph node.
   c. a palatine tonsil.
   d. the pharyngeal tonsil.
   e. the thymus.

8. Antibodies cannot get inside a host cell to destroy a viral infection. Therefore, we use _____ to combat viral infections.
   a. interleukins
   b. complement
   c. humoral immunity
   d. haptens
   e. cytotoxic T cells

9. Which statement is *not* true of antibodies?
   a. They directly kill infected cells by using a lethal hit.
   b. They are proteins.
   c. They have constant regions that define the antibody class.
   d. They are produced by plasma cells.
   e. They activate the complement system.

10. All lymph ultimately reenters the bloodstream at what point?
    a. the right atrium
    b. the common carotid artery
    c. the internal iliac veins
    d. the subclavian veins
    e. the inferior vena cava

11. Any organism capable of causing disease is called a/an _____.

12. The lymphatic vessels of the small intestine that absorb dietary lipids are called _____.

13. Cells that inhibit T cell responses once a pathogen has been eradicated are called _____.

14. _____ are small proteins, synthesized in response to viral infections, that inhibit viral replication and activate natural killer (NK) cells.

15. B cells become _____ cells before they begin to secrete antibodies.

16. Cells that process antigens and display fragments of them to activate the adaptive immune response are called _____.

17. Tonsillitis most often involves inflammation of the _____.

18. The _____ is located in the mediastinum, and is large in childhood but atrophies after about age 15.

19. _____ are phagocytic cells that wander in connective tissues and kill bacteria by releasing a lethal cloud of chemicals, including hydrogen peroxide and hypochlorite.

20. Basophils release a chemical called _____ that causes vasodilation.

Answers in appendix A

True or False

Determine which five of the following statements are false, and briefly explain why.

1. B cells play roles in both innate and adaptive immunity.

2. Perforins have a role in both nonspecific defense and adaptive immunity.

3. Lymphatic capillaries are more permeable than blood capillaries.

4. T lymphocytes are involved exclusively in cell-mediated immunity.

5. Obstruction of a major lymphatic vessel is likely to cause edema.

6. Lymph nodes are populated by B cells but not T cells.

7. HIV infects primarily B cells.

8. Most plasma cells form in the lymph nodes.

9. Plasma cells secrete about 20 antibody molecules per second.

10. Lymphatic nodules are a prominent feature of the tonsils but are not found in the thymus.

Answers in appendix A

Testing Your Comprehension

1. Severe combined immunodeficiency disease (SCID) is a genetic disorder that results in scarcity or absence of both T and B cells from birth. Explain why children with SCID must live in protective enclosures, such as sterile plastic chambers and suits. How are SCID and AIDS similar? How do they differ?

2. In treating a woman for malignancy in the right breast, a surgeon removes some of her axillary lymph nodes. Following the surgery, the patient experiences edema of her right arm. Explain why.

3. A burn research center uses mice for studies of skin grafting. To prevent rejection of grafted tissues, the mice are thymectomized at birth (their thymus is removed). Even though B cells do not develop in the thymus, the mice show no humoral response and are very susceptible to infection. Explain why the removal of the thymus would improve the success of skin grafts but adversely affect humoral immunity.

Chapter

15

The Respiratory System

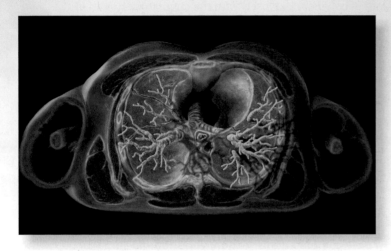

Branching of the bronchial trees into the lungs, seen from below in this artistic conception. Each lung receives one main bronchus, which ultimately branches into 8 million microscopic alveolar sacs and 150 million alveoli, where the bloodstream picks up oxygen.

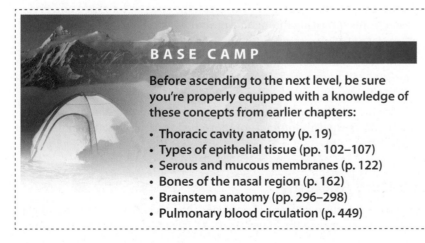

BASE CAMP

Before ascending to the next level, be sure you're properly equipped with a knowledge of these concepts from earlier chapters:

- Thoracic cavity anatomy (p. 19)
- Types of epithelial tissue (pp. 102–107)
- Serous and mucous membranes (p. 122)
- Bones of the nasal region (p. 162)
- Brainstem anatomy (pp. 296–298)
- Pulmonary blood circulation (p. 449)

Anatomy & Physiology | REVEALED®
aprevealed.com

Module 11: Respiratory System

B reath represents life. The first breath of a baby and the last gasp of a dying person are two of the most dramatic moments of human experience. But why do we breathe? It comes down to the fact that most of our metabolism directly or indirectly requires ATP. Most ATP synthesis requires oxygen and generates carbon dioxide—thus driving the need to breathe in order to supply the former and eliminate the latter. The respiratory system consists essentially of tubes that deliver air to the lungs, where oxygen diffuses into the blood and carbon dioxide diffuses out. It collaborates closely with the circulatory system in providing oxygen to the tissues and removing carbon dioxide from them.

15.1 Functions and Anatomy of the Respiratory System

Expected Learning Outcomes

When you have completed this section, you should be able to:

a. explain the functions of the respiratory system;

b. name and describe the structures of the respiratory system; and

c. trace the passage of air from the nose to the alveoli.

Respiratory Functions

The obvious function of the respiratory system is to obtain oxygen (O_2) from the atmosphere and to expel carbon dioxide (CO_2). However, the respiratory system has a broad range of functions:

- **Gas exchange.** Exchange of oxygen and carbon dioxide between the atmosphere and blood occurs in the lungs.
- **Communication.** Speech and vocalizations such as laughing and crying result from air passing through respiratory passageways.
- **Sense of smell.** Chemoreceptors in the nose allow us to perceive odors.
- **Acid–base balance.** Exhalation rids the body of CO_2. If respiration cannot keep pace with CO_2 production, the pH of body fluids falls abnormally low.
- **Expulsion of abdominal contents.** A breath-holding Valsalva maneuver helps to expel abdominal contents during urination, defecation, and childbirth.

Respiratory Anatomy

The **respiratory system** consists of passages that serve for airflow and of gas-exchange surfaces deep in the lungs (fig. 15.1). It is often divided into the **upper**

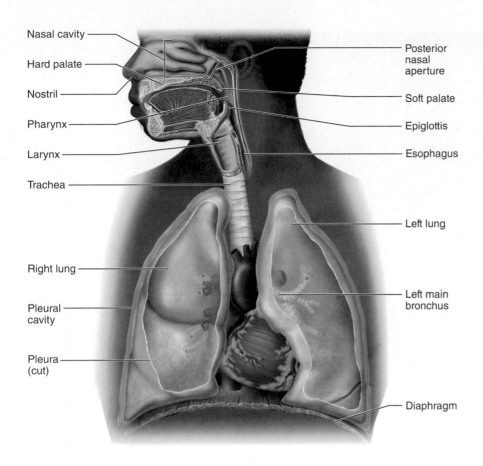

Nasal cavity

Hard palate

Nostril

Pharynx

Larynx

Trachea

Right lung

Pleural
cavity

Pleura
(cut)

Posterior
nasal
aperture

Soft palate

Epiglottis

Esophagus

Left lung

Left main
bronchus

Diaphragm

Figure 15.1 The Respiratory System. AP|R

respiratory tract, which includes structures in the head and neck, and the **lower respiratory tract,** which extends from the trachea through the lungs (the thoracic structures). Functionally, the respiratory tract is divided into: (1) a **conducting zone,** which extends from the nose to the *terminal bronchioles* and serves only to conduct air to the finer branches of the tract; and (2) a **respiratory zone,** which extends from terminal bronchioles to alveoli and has very thin walls that allow for gas exchange with the blood.

The Nose

The nose is shaped by hyaline cartilage at the tip and bone at the bridge. You can easily find the boundary between bone and cartilage by palpating your own nose. During normal breathing, air is drawn through the **nostrils** or **nares** (NAIR-eze), the anterior boundary of the nose (fig. 15.2). The nostrils have stiff **guard hairs,** or **vibrissae** (vy-BRISS-ee), that prevent insects and large airborne particles from entering. The nose extends posteriorly to the **posterior nasal apertures,** or **choanae**[1] (co-AH-nee), that empty into the throat. The ethmoid and sphenoid bones compose the roof and walls of the nasal cavity, and the **palate** forms its floor. The palate separates the nasal cavity from the oral cavity and allows you to breathe while chewing food. It is divided into a bony *hard palate* anteriorly and a fleshy *soft palate* posteriorly.

    The internal **nasal cavity** is divided into right and left halves by a wall, the **nasal septum,** composed of bone and hyaline cartilage. Folds of the lateral wall called the *nasal conchae* occupy most of the nasal cavity and cause the airflow to be quite turbulent. As the air contacts the mucous membrane covering the conchae, it is warmed and humidified and dust sticks to the mucus.

---

[1]*choana* = funnel

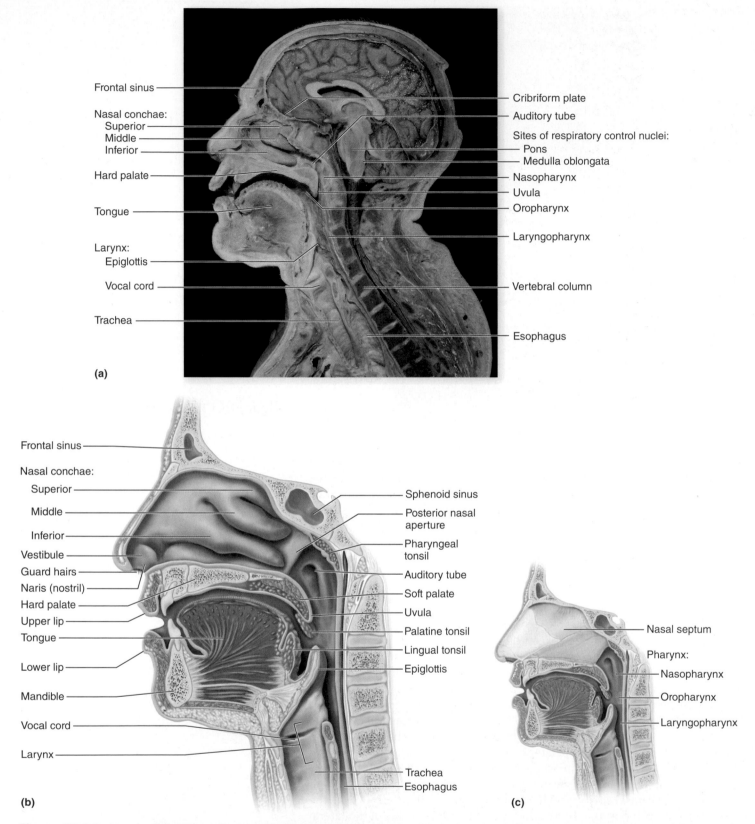

Figure 15.2 Anatomy of the Upper Respiratory Tract. (a) Median section of the cadaver head. (b) Internal anatomy. (c) The nasal septum and regions of the pharynx. AP|R

- *Draw a line across part (b) of this figure to indicate the boundary between the upper and lower respiratory tract.*

Odors are detected by the **olfactory epithelium** located in a small area of the roof of the nasal cavity. Most of the nasal cavity is lined with **respiratory epithelium** of the pseudostratified ciliated columnar type. Abundant *goblet cells* secrete mucus that is swept by the cilia toward the pharynx. Additional mucus is produced by glands beneath the epithelium. The *paranasal sinuses* described in chapter 6 also drain mucus into the nasal cavity. Inhaled dust, pollen, bacteria, and other foreign matter stick to the mucus and are swallowed, passing to the digestive system without harming the lungs.

The Pharynx

The **pharynx** (FAIR-inks), or throat, is a muscular funnel that extends from the posterior nasal apertures to the larynx. It consists of three regions: the *nasopharynx, oropharynx,* and *laryngopharynx* (fig. 15.2c). The nasopharynx is a space superior to the soft palate that serves as a passageway for air. It houses the pharyngeal tonsil, a patch of lymphoid tissue that is well positioned to respond to airborne pathogens. The **auditory (eustachian) tube** from the middle ear opens into the nasopharynx. This explains why an infection in the nose and throat may cause a middle-ear infection *(otitis media).*

The oropharynx lies between the soft palate and root of the tongue. It contains the palatine and lingual tonsils and serves as a passageway for both food and air. The laryngopharynx (la-RIN-go-FAIR-inks) extends from the union of the nasopharynx and oropharynx to the opening of the esophagus. The nasopharynx is lined with pseudostratified columnar epithelium whereas the oropharynx and laryngopharynx are lined with stratified squamous epithelium. Muscles of the pharynx facilitate swallowing and contribute to speech.

The Larynx

The **larynx** (LAIR-inks), or "voice box," is a chamber composed mainly of cartilage and muscle (fig. 15.3). Its main function is to prevent food and drink from entering the airway, but it has evolved to have the additional role of sound production.

Figure 15.3 Anatomy of the Larynx. Most muscles are removed in order to show the cartilages.

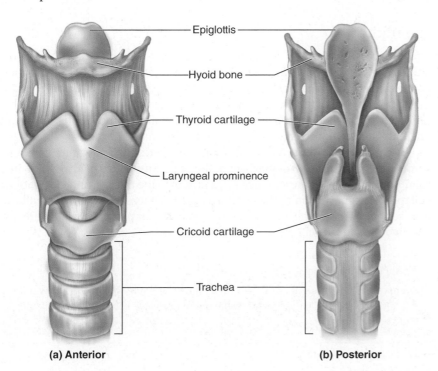

(a) Anterior (b) Posterior

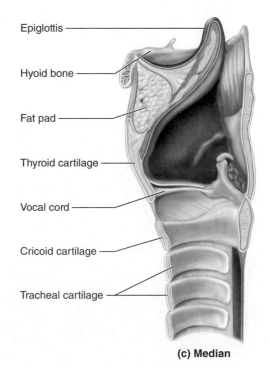

(c) Median

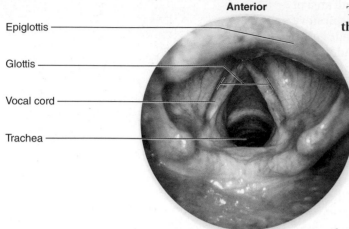

Anterior

Epiglottis

Glottis

Vocal cord

Trachea

Posterior

Figure 15.4 Endoscopic View of the Larynx.
Looking into the glottis with a laryngoscope.

The larynx is framed by nine cartilages. The largest of these is the **thyroid**[2] **cartilage,** named for its shieldlike shape; it has an anterior peak *(laryngeal prominence)* commonly called the "Adam's apple." It is especially visible in men because testosterone promotes its growth during puberty. It forms most of the "box" of the voice box and provides anterior anchorage for the vocal cords. A large *cricoid cartilage* and its ligaments attach the inferior end of the larynx to the trachea. Other smaller cartilages provide support to soft tissues, operate the vocal cords, and shape the epiglottis above.

The **epiglottis**[3] is a spoon-shaped flap that seals the laryngeal opening during swallowing to keep food and liquid out of the airway. When you swallow, the root of the tongue pushes the epiglottis downward while muscles attached to the thyroid cartilage pull the larynx up to meet it. The epiglottis thus caps the opening like a lid and diverts food and drink to the esophagus, behind the larynx. If food or liquid gets past the epiglottis, it triggers a *cough reflex* to expel it and, ideally, prevent it from getting as far as the trachea or lungs. The epiglottis and smaller cartilages of the larynx serve as important landmarks for the placement of a breathing tube by emergency personnel, a procedure called *endotracheal intubation.*

Within the larynx, the **vocal cords** stretch like a V from the midpoint of the thyroid cartilage in front to two small cartilages posteriorly (fig. 15.4). The vocal cords and the opening between them are collectively called the *glottis.* Airflow through the glottis vibrates the vocal cords and produces sound. Under the control of several small muscles, the cartilages at the posterior ends of the cords pivot, altering the tension on the cords and the angle between them. This alters the pitch of the voice. The larynx alone produces only crude sounds that some have compared to a hunter's duck call; the oral cavity, cheeks, tongue, and lips shape these sounds into intelligible words. Longer and less taut vocal cords produce a lower-pitched sound than shorter or tauter cords; for this reason, men generally have lower-pitched voices than women, and adults lower than children.

The Trachea

The **trachea,** or "windpipe," is about 12 cm (4.5 in.) long and 2.5 cm (1 in.) in diameter (fig. 15.5). At its inferior end, it forks into right and left *bronchi.* It is lined with a ciliated epithelium rich in mucus-producing goblet cells. Cilia sweep the mucus and foreign particles upward toward the throat, away from the lungs. This **mucociliary escalator** effectively removes potentially harmful debris, including microorganisms that could otherwise cause pulmonary infections.

The trachea is supported by 16 to 20 C-shaped rings of hyaline cartilage, some of which can be felt at the front of the neck. Similar to the wire spiral in a vacuum cleaner hose, the cartilage rings reinforce the trachea and keep it from collapsing when we inhale. The open part of the C faces posteriorly and allows the esophagus to expand and push into the muscular posterior wall of the trachea when food is swallowed.

Apply What You Know

Chemicals in cigarette smoke paralyze cilia of the respiratory epithelium. How does this affect the mucociliary escalator? What are the possible consequences for a person who smokes?

[2] *thyr* = shield; *oid* = resembling
[3] *epi* = above, upon

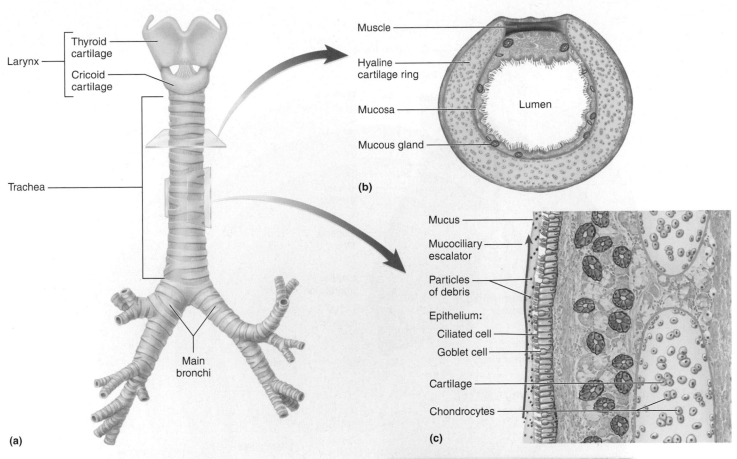

Larynx
- Thyroid cartilage
- Cricoid cartilage

Trachea

Main bronchi

(a)

Muscle

Hyaline cartilage ring

Mucosa

Mucous gland

Lumen

(b)

Mucus

Mucociliary escalator

Particles of debris

Epithelium:
  Ciliated cell
  Goblet cell

Cartilage

Chondrocytes

(c)

Figure 15.5 Anatomy of the Lower Respiratory Tract.
(a) Anterior view. (b) Cross section of the trachea showing the
C-shaped cartilage rings. (c) Longitudinal section of the
trachea showing the action of the mucociliary escalator.
(d) Tracheal epithelium showing ciliated cells and goblet cells.
AP|R

- *Why do inhaled objects more often go into the right main
bronchus than into the left?*

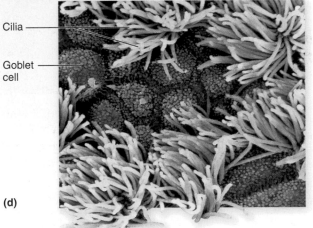

Cilia

Goblet cell

(d)

The Lungs, Pleurae, and Bronchial Tree

The **lungs** are soft, spongy organs that occupy most of the thoracic cavity.
They are roughly cone-shaped with a broad **base** that rests on the diaphragm;
a superior peak, the **apex,** that projects slightly above the clavicle; and a
broad, curved **costal surface** that lies against the rib cage (fig. 15.6). The
lungs are separated from each other by a partition called the **mediastinum,**
which contains the heart, esophagus, and major blood vessels. Each lung
has a slit *(hilum)* on its medial surface through which it receives bronchi,
blood vessels, and nerves. The two lungs are not symmetrical. The right lung
is divided into three lobes while the left has two. The left lung has an indentation called the **cardiac impression** to accommodate the left-leaning apex of
the heart.

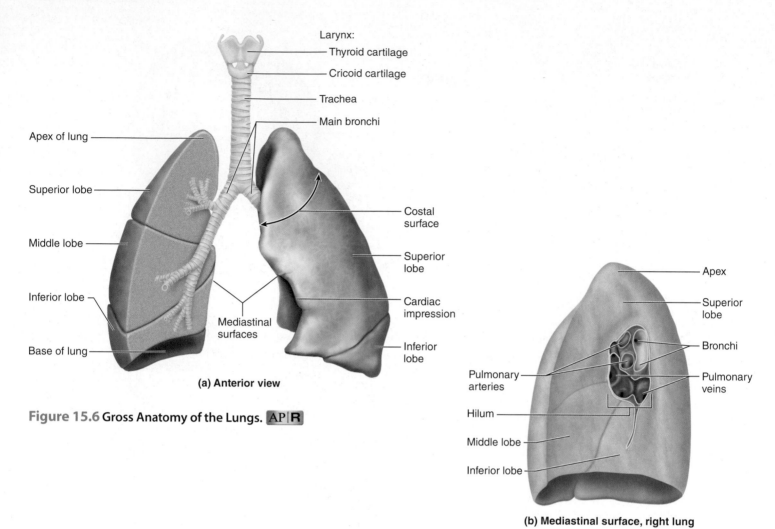

Figure 15.6 Gross Anatomy of the Lungs. AP|R

(a) Anterior view

(b) Mediastinal surface, right lung

A serous membrane, the **pleura,** lines the thoracic wall and adheres to the lungs. Where it forms the lung surface, the membrane is called the **visceral pleura.** The **parietal pleura** clings to the walls of the rib cage and the superior surface of the diaphragm (fig. 15.7). The space between the parietal and visceral pleurae is the **pleural cavity.** However, the parietal and visceral layers normally adhere to each other like two sheets of wet paper, with nothing in the pleural cavity but a thin film of lubricating *pleural fluid.* The pleural cavity does not *contain* the lung, but *wraps around it* much like the pericardial cavity wraps around the heart (p. 419). Under abnormal conditions such as chest wounds, however, air can enter and separate the two membranes—a condition called *pneumothorax.* This is accompanied by collapse of the lung, and may require insertion of a chest tube to draw the air out.

Pleural fluid acts as a lubricant that reduces friction as the lungs expand and contract. Infection of pleural membranes produces a condition called *pleurisy* where the membranes may become roughened and rub together, making each breath a painful experience.

The **bronchial tree** consists of a series of branching passages that carry air from the trachea to the exchange surfaces of the lungs. It begins when the trachea splits into two **main (primary) bronchi** (BRONK-eye) that lead to the right and left lungs (fig. 15.6a). The right main bronchus is slightly wider and more vertical than the left one.

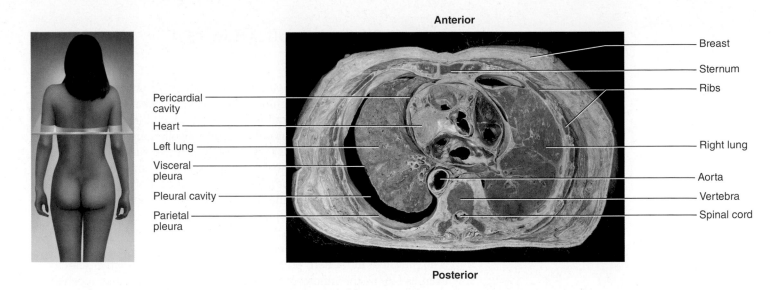

Anterior

Pericardial cavity
Heart
Left lung
Visceral pleura
Pleural cavity
Parietal pleura

Breast
Sternum
Ribs
Right lung
Aorta
Vertebra
Spinal cord

Posterior

Like the trachea, the main bronchi are supported by rings of hyaline cartilage and have a layer of smooth muscle in the wall. Progressing distally, the bronchi branch into smaller and smaller passageways supported by cartilage plates rather than rings. These lead eventually to tubes called **bronchioles,** less than 1 mm in diameter and with no cartilage at all (fig. 15.8). Smooth muscle is the main supportive tissue of the bronchioles, and unhindered by cartilage, the bronchioles have more capacity to dilate and constrict than the bronchi do. Such changes in diameter are called **bronchodilation** and **bronchoconstriction,** respectively.

The bronchioles branch into smaller divisions called *terminal bronchioles,* which are the end of the conducting division. Beyond this, the air passages have such thin walls they can exchange gases with the blood, and constitute

Figure 15.7 Lungs in Relation to the Thoracic Cavity. This cross section of the cadaver is oriented the same way as the reader's body. The pleural cavity is especially evident where the left lung has shrunken away from the thoracic wall, but in a living person the lung fully fills this space, the parietal and visceral pleurae are pressed together, and the pleural cavity is only a potential space between the membranes, as on the right side of this photograph.

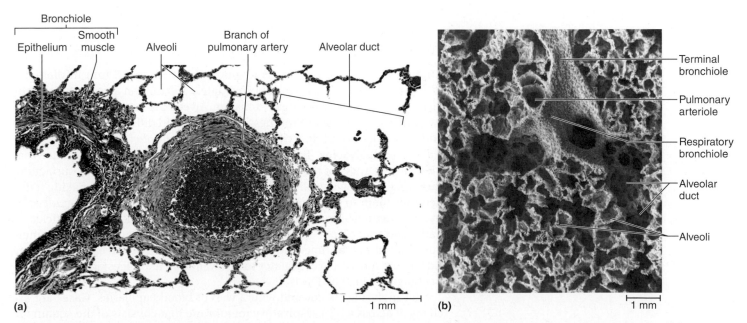

Bronchiole
Smooth
Epithelium   muscle   Alveoli
Branch of pulmonary artery   Alveolar duct

Terminal bronchiole
Pulmonary arteriole
Respiratory bronchiole
Alveolar duct
Alveoli

(a)   1 mm   (b)   1 mm

Figure 15.8 Histology of the Lung. (a) Light micrograph. (b) Scanning electron micrograph. Note the spongy texture of the lung. **AP|R**

• *Histologically, how can we tell that the passage at the top of part (a) is a bronchiole and not a bronchus?*

Clinical Application 15.1

ASTHMA

Asthma sufferers are familiar with the symptoms of an attack—shortness of breath, wheezing, and coughing. One person in 10 has asthma in the United States, and the most common sufferers are children. Exposure to a trigger such as cat dander or pollen causes swollen and inflamed airways, and release of chemicals that cause violent and sudden spasms of the bronchi. In the short term, asthma can be medically managed with inhaled drugs called *β-adrenergic agonists* that mimic the effect of epinephrine (adrenaline) and dilate the bronchioles. Such inhalers treat acute symptoms but not the underlying causes. In more severe cases, patients receive inhaled corticosteroids on a daily basis to suppress inflammation. Recently, researchers have tried to answer the question of why asthma is so prevalent in modern populations, in contrast to its relative scarcity even as late as the nineteenth century. A hypothesis that has received some support is that enhanced hygiene in modern societies has led to reduced exposure to bacteria and parasitic worms in childhood. The immune system may not develop as robustly in individuals who lack such exposure to pathogens. In addition, women in industrialized countries have the choice of whether to breast-feed or bottle feed. Individuals who are not breast-fed may lack stimulants to the developing immune system that are present in breast milk. This evolutionary point of view provides us with insight into why the disease has become one of the most prevalent conditions in the modern world.

the *respiratory division.* These passages, in order, are elongated *respiratory bronchioles,* some of which are still ciliated; then *alveolar ducts,* which have a simple squamous epithelium and no cilia; and finally, *alveolar sacs,* which are grapelike clusters of alveoli arranged around a central space.

The Alveoli

An **alveolus** (AL-vee-OH-lus) is a pouch about 0.2 to 0.5 mm in diameter (fig. 15.9). Each lung contains approximately 150 million alveoli, which provide a large surface area for gas exchange. Alveoli are lined mostly with *squamous alveolar cells* whose thinness allows for rapid diffusion between the air and the blood. In addition, more rounded *great alveolar cells* are found in the walls of each alveolus. They secrete *pulmonary surfactant,* a substance that prevents the walls of the alveoli from sticking together. Without surfactant, the surfaces of the alveoli would cling together like wet paper, making it difficult for them to inflate during inhalation. A third cell type is the *alveolar macrophages,* which wander the alveoli and phagocytize dust particles or pathogens that escape the mucociliary escalator. They are the most numerous of all cells in the lung, and as many as 100 million perish each day as they ride up the mucociliary escalator to be swallowed and digested, thus ridding the lungs of a load of debris.

Each alveolus is covered with a web of blood capillaries. Gases are exchanged across a thin **respiratory membrane** that consists of the squamous alveolar cell, the squamous endothelial cell that lines the capillary, and their shared thin basement membrane. The total thickness of the respiratory membrane is only 0.5 µm—a very minimal barrier to gas exchange. By comparison, a red blood cell has a diameter 15 times as great, about 7.5 µm.

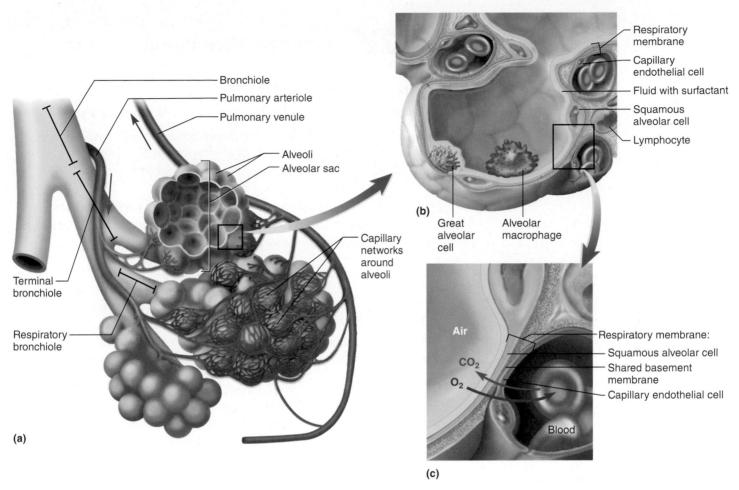

(a)

Bronchiole
Pulmonary arteriole
Pulmonary venule
Alveoli
Alveolar sac
Terminal bronchiole
Respiratory bronchiole
Capillary networks around alveoli

(b)

Respiratory membrane
Capillary endothelial cell
Fluid with surfactant
Squamous alveolar cell
Lymphocyte
Great alveolar cell
Alveolar macrophage

(c)

Air
CO₂
O₂
Blood
Respiratory membrane:
Squamous alveolar cell
Shared basement membrane
Capillary endothelial cell

Figure 15.9 Pulmonary Alveoli. (a) Clusters of alveoli and their blood supply. (b) Structure of an alveolus. (c) Structure of the respiratory membrane.

Blood Supply to the Lungs

The lungs receive two blood supplies—pulmonary and systemic. The pulmonary arteries serve only for exchanging CO_2 for O_2, and lead only to the capillaries surrounding the alveoli. The pulmonary veins lead immediately back to the heart (see p. 449). The systemic supply comes from about five *bronchial arteries* arising from the thoracic aorta. They serve to nourish the lung tissues and pleurae, leading to capillaries in the bronchi and other lung tissues but not to the alveoli. Bronchial veins leaving the lungs drain indirectly, via the azygos system (see p. 457), into the superior vena cava.

Before You Go On

Answer these questions from memory. Reread the preceding section if there are too many you don't know.

1. *A dust particle is inhaled and gets into an alveolus without being trapped along the way. Describe the path it takes, naming all air passages from the external nares to the alveolus. What would happen to it after arrival in the alveolus?*

2. *Explain how the structure of the larynx prevents choking and enables the production of sound.*

3. *Describe the relationship between the visceral pleura, parietal pleura, lungs, and thoracic wall. Why can we say the lungs are "surrounded by" the pleural cavity but not "in" the pleural cavity?*

4. *List the cell types found in the alveoli and explain their functions.*

15.2 Pulmonary Ventilation

Expected Learning Outcomes

When you have completed this section, you should be able to:

a. explain how differences in pressure between the atmosphere and lungs account for the flow of air into and out of the lungs;

b. name the muscles of respiration and describe their roles in inspiration and expiration;

c. define the clinical measurements of pulmonary ventilation; and

d. describe the brainstem centers that control breathing and the inputs they receive from other parts of the nervous system.

The Mechanics of Breathing

Breathing, or **pulmonary ventilation,** consists of **inspiration** (inhaling) and **expiration** (exhaling). One complete inspiration and expiration is called a **respiratory cycle.**

Airflow into and out of the lungs is driven by differences in air pressure between the atmosphere and the lungs. Like blood and other fluids, air flows from a point of high pressure to a point of lower pressure. If **atmospheric (barometric) pressure** is greater than **intrapulmonary pressure** (pressure inside the lungs), air flows into the lungs; if intrapulmonary pressure exceeds atmospheric pressure, air flows out. A person cannot, of course, change the atmospheric pressure at will, so the only way one can breathe is to raise and lower the intrapulmonary pressure so that it cyclically falls below and rises above the atmospheric pressure (fig. 15.10).

This is achieved by contracting and relaxing the diaphragm and other muscles to alter the pressure in the thoracic cavity. For any container of a given volume, if the volume increases, the pressure falls, and vice versa. Consider a syringe, for example. If you pull back the plunger, you increase the volume within the syringe, causing the internal pressure to drop. When it falls below the surrounding air pressure, air flows into the syringe. Conversely, if you push the plunger in, you reduce the volume and increase the pressure within the syringe. When its internal pressure rises above the surrounding air pressure, air flows out. The respiratory muscles of the thoracic cavity have the same effect on volume and pressure within the thorax, and on airflow into and out of the lungs.

The principal muscle (prime mover) of pulmonary ventilation is the **diaphragm,** the muscular dome that separates the thoracic and abdominal cavities (fig. 15.11). It alone accounts for about two-thirds of the pulmonary airflow. When relaxed, it bulges upward and presses against the base of the lungs, and the lungs are at their minimum volume. When the diaphragm contracts, it flattens, dropping about 1.5 cm in relaxed inspiration and as much as 7 cm in deep breathing. This enlarges the thoracic cavity (including the lungs), reduces its pressure, and produces an inflow of air. Other muscles assist the diaphragm. The **external intercostal muscles** pull the ribs upward and the sternum forward, increasing the diameter of the chest. In quiet breathing, the dimensions of the thoracic cage increase by only 2 to 5 mm in each direction, but combined with the movement of the diaphragm, this increases thoracic volume by approximately 500 mL. Thus, during a typical inspiration, about 500 mL of air flows into the respiratory tract.

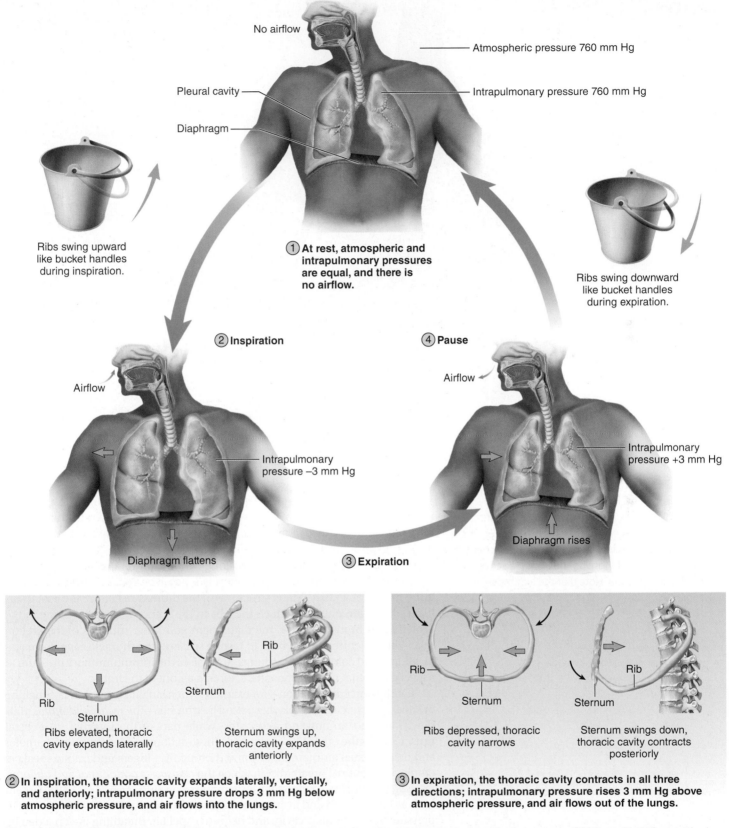

No airflow

Atmospheric pressure 760 mm Hg

Pleural cavity

Intrapulmonary pressure 760 mm Hg

Diaphragm

Ribs swing upward like bucket handles during inspiration.

① **At rest, atmospheric and intrapulmonary pressures are equal, and there is no airflow.**

Ribs swing downward like bucket handles during expiration.

② **Inspiration**

④ **Pause**

Airflow

Airflow

Intrapulmonary pressure –3 mm Hg

Intrapulmonary pressure +3 mm Hg

Diaphragm flattens

Diaphragm rises

③ **Expiration**

Rib

Sternum

Ribs elevated, thoracic cavity expands laterally

Sternum

Rib

Sternum swings up, thoracic cavity expands anteriorly

② **In inspiration, the thoracic cavity expands laterally, vertically, and anteriorly; intrapulmonary pressure drops 3 mm Hg below atmospheric pressure, and air flows into the lungs.**

Rib

Sternum

Ribs depressed, thoracic cavity narrows

Sternum

Rib

Sternum swings down, thoracic cavity contracts posteriorly

③ **In expiration, the thoracic cavity contracts in all three directions; intrapulmonary pressure rises 3 mm Hg above atmospheric pressure, and air flows out of the lungs.**

Figure 15.10 The Respiratory Cycle. The pressures given here are based on an assumed atmospheric pressure of 760 mm Hg. Values with plus and minus signs are relative to atmospheric pressure. Like bucket handles, each rib is attached at both ends (to the spine and sternum) and swings up and down during inspiration and expiration. **AP|R**

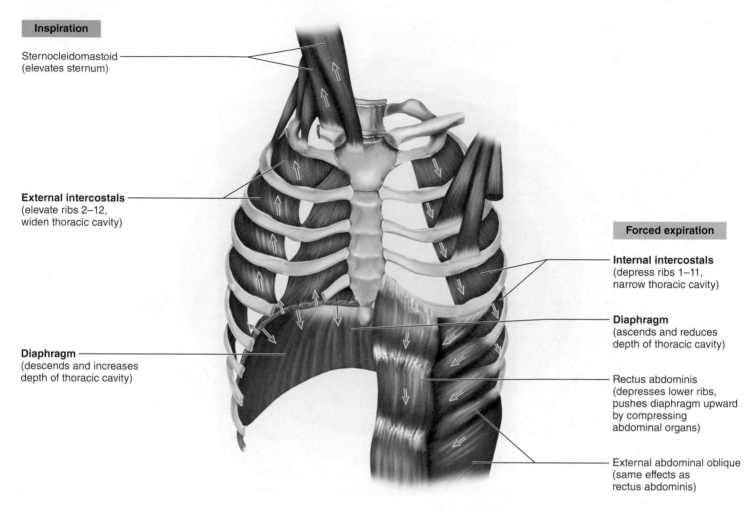

Inspiration

Sternocleidomastoid
(elevates sternum)

External intercostals
(elevate ribs 2–12,
widen thoracic cavity)

Diaphragm
(descends and increases
depth of thoracic cavity)

Forced expiration

Internal intercostals
(depress ribs 1–11,
narrow thoracic cavity)

Diaphragm
(ascends and reduces
depth of thoracic cavity)

Rectus abdominis
(depresses lower ribs,
pushes diaphragm upward
by compressing
abdominal organs)

External abdominal oblique
(same effects as
rectus abdominis)

Figure 15.11 The Respiratory Muscles.
Boldface indicates the principal respiratory muscles; the others are accessory. Arrows indicate the direction of muscle action. Muscles listed on the left are active during inspiration and those on the right are active during forced expiration. Note that the diaphragm is active in both phases.

Lung expansion does not depend on muscle action alone. The parietal pleura adheres to the wall of the rib cage and clings to the visceral pleura, which forms the outer surface of the lung. When the ribs swing upward and outward during inspiration, the pleurae follow them, carrying the lung surface with them. Thus, the entire lung expands along with the thoracic cage.

Unlike inspiration, quiet expiration is a passive process that does not require muscular activity. Many structures of the thorax are elastic: the costal cartilages, the ligaments that attach the ribs to the spine, and elastic tissue in the bronchi and bronchioles. As the diaphragm and other muscles of inspiration relax, the structures spring back to resting size and the thoracic cage decreases in volume and compresses the lungs. This raises the intrapulmonary pressure to a level above that of the atmosphere, forcing air out of the lungs.

During forced expiration—for example, when singing or shouting, coughing or sneezing, exercising, or playing a wind instrument—the *rectus abdominis* pulls down on the sternum and lower ribs, while part of the **internal intercostal muscles** pulls the other ribs downward. These actions reduce the chest diameter and help to expel air more rapidly and thoroughly. Singers and public speakers employ "abdominal breathing," in which the *transverse abdominal* and *abdominal oblique muscles* compress the abdominal cavity and push some of the viscera such as the stomach and liver upward against the diaphragm. This increases the pressure in the thoracic cavity and helps to expel air. Breathing is also aided by several other muscles of the neck, abdomen, lower back, and even the pelvic floor.

Not only does abdominal pressure affect thoracic pressure, the opposite is also true. Depression of the diaphragm raises abdominal pressure and helps to expel the contents of certain abdominal organs, thus aiding in childbirth, urination, defecation, and vomiting. During such actions, we often consciously or unconsciously employ the **Valsalva**[4] **maneuver.** This consists of taking a deep breath, holding it by closing the glottis, and then contracting the abdominal muscles to raise abdominal pressure and push the organ contents out.

Measurements of Ventilation

Measurements of pulmonary ventilation are used to assess the severity of respiratory disease or to monitor a patient's improvement or disease progression. To measure pulmonary function, a subject breathes into a device called a **spirometer,**[5] which recaptures the expired breath and records the rate of breathing and volumes of air that are inspired or expired. The measurement of pulmonary function is called **spirometry.** Figure 15.12 shows values for a healthy adult. These values are proportional to body size; consequently, they are generally higher for men than for women.

One value obtained with a spirometer is **tidal volume,** the amount of air inhaled or exhaled in one breath during quiet breathing. The average tidal volume is about 500 mL. Other volumes include the **inspiratory reserve volume,** the amount of air that can be drawn into the lungs beyond the normal quiet inspiration (typically about 3,000 mL), and **expiratory reserve volume,** the amount of air that can be expelled beyond the normal quiet expiration (typically about 1,200 mL). **Vital capacity** is the maximum amount of air that can be exhaled after the deepest possible breath. It is the sum of tidal volume, inspiratory reserve volume, and expiratory reserve volume. It represents the maximum ability to ventilate lungs in one breath, and is an especially important measure of pulmonary health. Even when one exhales as much as possible, an amount of air called the **residual volume** remains in the lungs and mixes with fresh air on the next inspiration. The residual volume prevents collapse of the alveoli and allows gas exchange to continue even between inspirations.

Spirometry is especially valuable for the diagnosis and assessment of *restrictive* and *obstructive* lung disorders. **Restrictive disorders** are conditions associated with poor pulmonary *compliance*—the ability of the lungs to expand when pressure changes. For example, one person's lungs may expand more readily than another person's in the face of an identical pressure change in the thorax. Compliance is compromised when the lungs are stiffened by a buildup of scar tissue in diseases such as tuberculosis and black lung disease. Low compliance results in reduced vital capacity. **Obstructive disorders** interfere with airflow by narrowing or blocking airways, making it harder to inhale or exhale. Examples include asthma, emphysema, and chronic bronchitis (see Clinical Application 15.1 and Perspectives on Health, p. 519). Obstructive disorders are assessed by having a person exhale as rapidly as possible into a spirometer. A healthy adult should be able to expel 75% to 85% of the vital capacity in 1.0 second. A value lower than 75% indicates the patient probably has an obstructive disorder.

Figure 15.12 Spirometry. The wavy line indicates inspiration when it rises and expiration when it falls.

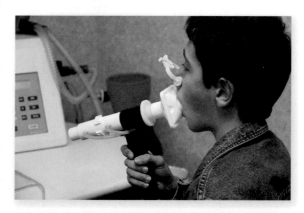

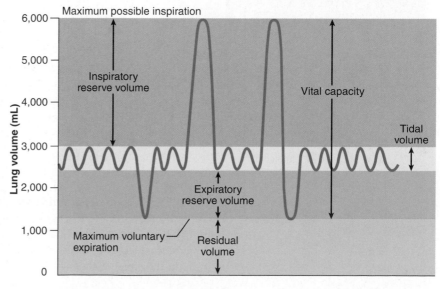

Apply What You Know

Suppose a person had a tidal volume of 500 mL, an inspiratory reserve volume of 2,800 mL, and an expiratory reserve of 1,000 mL. Calculate vital capacity. If this person can exhale 2,500 mL in 1.0 second with maximum effort, what percentage of the vital capacity is this? What does the percentage indicate about the person's health?

---

[4] Antonio Maria Valsalva (1666–1723), Italian anatomist

[5] *spiro* = breath; *meter* = measuring device

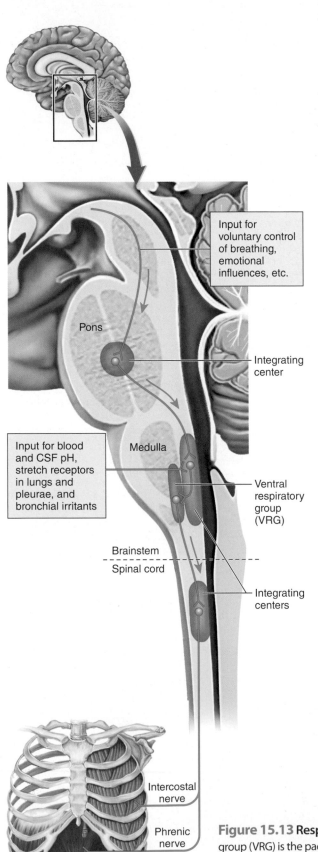

Neural Control of Breathing

Breathing, like the heartbeat, is automatic; we breathe without thinking about it, which is a good thing or we would fear going to sleep. However, in contrast to the heart, the lungs do not have a built-in pacemaker. Instead, they rely on messages from the nervous system to coordinate the complex skeletal muscle actions required to sustain ventilation. The unconscious cycle of breathing is controlled by respiratory centers in the brain, specifically in the medulla oblongata and pons (fig. 15.13).

The principal pacemaker for breathing is the **ventral respiratory group (VRG)** of neurons in the medulla oblongata. It contains *inspiratory neurons* that, at rest, typically fire for about 2 seconds and issue nerve signals to the respiratory muscles via the spinal cord, paired *phrenic nerves* to the diaphragm, and paired *intercostal nerves* to the intercostal muscles. The VRG also contains *expiratory neurons,* which then begin to fire and inhibit the inspiratory neurons for about 3 seconds. The typical resting breathing rhythm of about 12 breaths/min. results from these neuron groups in the VRG "taking turns" in this manner.

Of course, we do not always breathe at that relaxed rate. The medulla and pons receive input from several sources that influence one's respiratory rate and depth. *Chemoreceptors* monitor the pH of the blood and cerebrospinal fluid as well as blood CO_2 and O_2; *stretch receptors* in the pleurae and bronchial tree monitor expansion of the lungs; *irritant receptors* in the bronchial tree inhibit respiration in response to smoke, noxious fumes, and other agents; centers in the hypothalamus and limbic system can alter one's respiration in response to emotions; and cerebral centers enable us to voluntarily modify our respiration, as in speaking, singing, laughing, swimming, or crying.

Before You Go On

Answer these questions from memory. Reread the preceding section if there are too many you don't know.

5. *Explain pressure changes that occur in the thorax during inspiration and expiration. Why does contraction of the diaphragm cause inspiration but contraction of the abdominal muscles cause expiration?*

6. *What is the prime mover of respiration? What other muscles act as synergists during inspiration and forced expiration?*

7. *Define tidal volume, inspiratory reserve volume, and expiratory reserve volume. How is vital capacity calculated? Explain why vital capacity is useful for assessing pulmonary function.*

8. *Name the principal brainstem center that sets the respiratory rhythm and explain how it does so.*

9. *Identify several sources of input that the brainstem center receives and uses to modify the respiratory rhythm under special circumstances.*

Figure 15.13 Respiratory Control Centers in the Central Nervous System. The ventral respiratory group (VRG) is the pacemaker that generates the rhythm of breathing. Its output goes to a spinal integrating center that gives rise to the intercostal and phrenic nerves to the intercostal muscles and diaphragm. A brainstem integrating center processes information from multiple sources and then issues output to the VRG to adjust the respiratory rhythm.

PERSPECTIVES ON HEALTH

Smoking and Respiratory Health

Smoking cigarettes is a relatively recent cultural practice that has profound impacts on respiratory health. Tobacco use contributes to premature death due to emphysema, cancer, and cardiovascular disease. The history of respiratory diseases is linked to the use of cigarettes, which were not manufactured until the end of the nineteenth century. At the beginning of the twentieth century, lung cancer was uncommon, but by the 1990s, it had become the leading cause of cancer deaths in the United States. Despite decades of research that demonstrates the link between smoking and myriad diseases, approximately 20% of the U.S. population still smokes.

Chronic Obstructive Pulmonary Disease

Chronic obstructive pulmonary disease (COPD) is a collective term for chronic bronchitis and emphysema. It is among the few leading causes of disability and death in the United States, and is almost always caused by cigarette smoking.

Chronic bronchitis begins with cellular abnormalities brought on by chronic exposure to cigarette smoke. Many ciliated cells of the respiratory tract transform into nonciliated squamous cells. Even the cilia on cells that do retain them are paralyzed by toxins in the smoke. The mucociliary escalator becomes less efficient in transporting debris and microorganisms to the throat. Attempts to clear the airway results in a persistent cough, the "smoker's hack." Inhaled particulates in smoke bypass the damaged cilia and travel as far as the alveoli. Alveolar macrophages ingest these, especially tar, and become filled with black material, leading to the black color of smokers' lungs. The immune function of the macrophages is also impaired. Stagnation of the respiratory mucus and immune suppression of the macrophages invite recurring infections and inflammation, leading to shortness of breath (dyspnea) and oxygen deficiency in the blood and tissues (hypoxia).

In emphysema, the alveoli become hyperinflated and many of them burst. The lungs can become permeated with grape-sized and larger spaces taking the place of functional alveoli. As the disease progresses, symptoms include dyspnea, extreme fatigue, inability to tolerate even slight physical exertion such as walking across a room, and mental sluggishness due to inadequate oxygenation of the brain.

Chronic hypoxia from COPD stimulates the kidneys to secrete erythropoietin, which stimulates the production of more erythrocytes (see p. 400). A sharp rise in RBC count increases blood viscosity and pressure, puts a strain on the heart, and thus leads to cardiac disease as a further consequence of smoking.

Lung Cancer

Cigarette smoke contains over 60 carcinogens, substances known to cause cancer. Accumulation of mutations results in the formation of lung tumors (fig. 15.14), over 90% of which develop in the mucous membranes of the large bronchi. In the most common form of lung cancer, *squamous-cell carcinoma*, the bronchial epithelium changes from the ciliated pseudostratified type to the stratified squamous type. Proliferating cells invade the tissues under the epithelium, producing bleeding lesions. Dense masses of fibrous tissue (collagen) replace functional lung tissue. In the most dangerous form of lung cancer, *small-cell (oat-cell) carcinoma*, malignant cells quickly spread beyond the lungs and establish secondary (metastatic) tumors in the mediastinum and other organs.

Degeneration of the lungs produces a cough, but coughing is such an everyday occurrence among smokers that it seldom causes much alarm. The first sign of serious trouble may be coughing up blood. By the time lung cancer is diagnosed, it usually has already metastasized to other organs such as the heart, bones, liver, lymph nodes, and brain. The chance of recovery is poor, with only 7% of patients surviving for 5 years after diagnosis.

(a) Healthy lung, mediastinal surface

Tumors

(b) Smoker's lung with carcinoma

Figure 15.14 Effect of Smoking.

15.3 Gas Exchange and Transport

Expected Learning Outcomes

When you have completed this section, you should be able to:

a. define *partial pressure* and explain how it affects diffusion of gases across the respiratory membrane;

b. describe gas exchange in the lungs and systemic capillaries;

c. describe how O_2 and CO_2 are transported in the blood; and

d. explain the effect of blood gases and pH on the respiratory rhythm.

Composition of Air AP|R

Ultimately, respiration is about gases, especially oxygen and carbon dioxide. The respiratory structures and mechanisms discussed so far only set the stage for delivering oxygen to the tissues and returning carbon dioxide to the lungs where it can be expelled. To grasp how these processes occur, it is necessary to begin with the composition of the air we inhale.

The air we breathe consists of 78.6% nitrogen, 20.9% oxygen, 0.04% carbon dioxide, traces of several other gases, and variable amounts of water vapor (0.5% on a typical cool clear day). Atmospheric pressure is the sum of the pressures exerted separately by each of the gases. The separate contribution of each gas is called its **partial pressure,** and is symbolized with a *P* followed by the formula of the gas, such as P_{O_2}. For example, oxygen accounts for 20.9% of the volume of air. At the average sea-level atmospheric pressure of 760 mm Hg, the P_{O_2} is therefore:

$$0.209 \times 760 \text{ mm Hg} = 159 \text{ mm Hg}$$

When there is a difference between the partial pressure of a gas at one point and its partial pressure at another point, we say there is a **partial-pressure gradient.** Movement of a gas from the point of high partial pressure to the point of low partial pressure is movement *down the gradient.*

Alveolar Gas Exchange AP|R

Alveolar gas exchange (external respiration) refers to the *loading* of O_2 and *unloading* of CO_2 in the pulmonary alveoli. It depends on the tendency of gases to diffuse down their partial-pressure gradients. Air breathed into the alveoli has a P_{O_2} of about 104 mm Hg, compared to 40 mm Hg in the blood of the surrounding capillaries. Oxygen therefore diffuses down its pressure gradient across the respiratory membrane from the alveolar air into the blood. In contrast, carbon dioxide has a P_{CO_2} of about 46 mm Hg in the capillary blood and 40 mm Hg in the inhaled air, so it diffuses from the capillary into the alveolus.

Alveolar gas exchange is subject to several factors that affect diffusion through the respiratory membrane:

1. **Partial-pressure gradient.** The greater the difference in partial pressure between one point and another, the faster the gas will diffuse. This is why we can improve a patient's blood oxygen level by giving oxygen-enriched air.

Clinical Application 15.2

HIGH-ALTITUDE PHYSIOLOGY

One of the most thrilling adventures one can have is to climb mountains, but climbing at high altitudes poses physiological challenges. In the past several decades, many people have attempted to ascend Mount Everest, which at 8,848 m (29,029 ft) is the highest point on the earth's surface. The atmospheric pressure at the peak is approximately 255 mm Hg, or about one-third of the pressure at sea level. Oxygen is still approximately 21% of the mix of gases in the air; thus, the P_{O_2} is only 53 mm Hg. At very high altitudes, the P_{O_2} gradient between the alveolus and the blood capillaries is not great enough to drive the diffusion of O_2 to the blood.

At high altitude, the body responds in the short term by increasing ventilation and heart rate, resulting in enhanced oxygen delivery to the tissues. Climbers attempt to offset the lack of oxygen by a process called *acclimatization;* they ascend to increasingly higher elevations over a period of several days or even weeks. The slow ascent leads to physiological changes that help one cope with the "thin air." For example, within a few days at high altitude, red blood cell production increases due to increased levels of erythropoietin (EPO) secreted by the kidneys. As a result, the hematocrit, normally about 40% to 45%, can rise to over 55%. In addition, the mitochondria become more efficient at using O_2 to produce ATP. Despite acclimatization, ascent to very high altitudes (over 20,000 ft) often leads to symptoms such as fatigue, headache, nausea, and a decline in cognitive ability.

2. **Membrane thickness.** Normally the respiratory membrane is very thin, but in conditions such as pneumonia, it can be thickened by fluid accumulation (pulmonary edema). Gases then have to travel farther from the alveolar air space to the blood, and as a result, the blood can become oxygen-deficient (a state called *hypoxemia*).

3. **Membrane surface area.** Each lung has about 70 m^2 of respiratory membrane, normally a more-than-adequate surface area for gas exchange. Degenerative pulmonary diseases such as emphysema, however, dramatically reduce the surface area and result in low blood P_{O_2}. This is why persons with emphysema often must use supplemental oxygen—to compensate for the reduced alveolar surface area by accentuating the pressure gradient in point 1 above.

Gas Transport

Gas transport refers to the manner in which the blood carries oxygen and carbon dioxide between the lungs and systemic tissues. As we saw earlier, oxygen has a higher P_{O_2} in the alveoli than in the blood arriving at an alveolus, so it diffuses into the blood. About 98.5% of it enters the erythrocytes (RBCs) and binds to the iron at the center of each heme group in the hemoglobin molecule (see fig. 12.4, p. 398). Hemoglobin with oxygen bound to it is called **oxyhemoglobin.** Since the hemoglobin molecule has four heme groups, and each heme can bind one O_2 molecule, each hemoglobin can carry up to four O_2. The remaining 1.5% of the transported O_2 is dissolved gas in the blood plasma. Carbon monoxide also binds to the iron, and much more tightly than oxygen does. It prevents hemoglobin from binding oxygen, and this is precisely why breathing carbon monoxide in the air can be fatal.

The blood also transports carbon dioxide. The P_{CO_2} in systemic tissue fluid is typically about 46 mm Hg and in the arriving arterial blood it is about 40 mm Hg. Therefore, CO_2 diffuses from the tissue fluid into the blood. About 5% of it binds to proteins such as hemoglobin and plasma albumin, and 5% is transported as dissolved gas much like the CO_2 in champagne and carbonated beverages. Ninety percent, however, reacts with water to produce **carbonic acid** (H_2CO_3), which then dissociates into bicarbonate and hydrogen ions:

$$CO_2 + H_2O \rightarrow H_2CO_3 \rightarrow HCO_3^- + H^+$$

This reaction occurs primarily within the RBCs, which have an enzyme called **carbonic anhydrase** that greatly accelerates the process. The H^+ binds to hemoglobin and thus does not greatly affect the blood pH; the HCO_3^- diffuses out of the RBC into the blood plasma. Carbon dioxide binds to a different site on hemoglobin (the globin) than the oxygen does (the heme), so hemoglobin can transport both gases simultaneously.

Systemic Gas Exchange

Systemic gas exchange (internal respiration) means the unloading of O_2 and loading of CO_2 at the systemic capillaries. Blood arriving here typically has a P_{O_2} of about 95 mm Hg and the surrounding tissue fluid has about 40 mm Hg. Oxygen therefore *dissociates* from the oxyhemoglobin molecule and diffuses out of the blood into the tissue fluid. Typically, the blood gives up about 22% of its oxygen in one pass through a systemic capillary bed, so it still has 78% of its load as it returns to the heart and lungs.

The amount of oxygen unloaded, however, is not constant in all tissues or at all times. Active tissues need more oxygen, and fortunately, hemoglobin adjusts its unloading to variations in local demand. In highly active tissue such as exercising muscle, it can unload up to 80% of its oxygen. There are multiple mechanisms by which hemoglobin adjusts O_2 unloading to the metabolic state of a local tissue. A few of these (not a complete list) follow.

1. Active tissues consume O_2 rapidly, lowering the P_{O_2} of the tissue fluid. If there is a greater difference between the P_{O_2} of the blood and tissue fluid, then O_2 diffuses more rapidly into the tissue. This is a case of the partial-pressure gradient issue explained on page 520.

2. Active tissues produce more CO_2 than less active ones. CO_2 generates acid through the carbonic acid reaction shown above, lowering the pH of the tissue fluid. Hemoglobin responds to the lower pH by releasing more O_2—a phenomenon called the *Bohr effect*.

3. Active tissues generate more heat than less active ones. Hemoglobin releases O_2 more easily at higher temperatures, thus unloading more O_2 to the warmest, most active tissues.

Blood Gases and the Respiratory Rhythm

We breathe more heavily at some times (during exercise, for example) than at others. Pulmonary ventilation varies as it adjusts to the body's rate of CO_2 production and demand for O_2. The effect is to maintain a stable blood pH (7.40 ± 0.05) and systemic arterial P_{O_2} and P_{CO_2} of 95 and 40 mm Hg, respectively. Regulation of these values depends on input to brainstem respiratory centers from the previously described chemoreceptors that monitor the composition of blood and cerebrospinal fluid. The most potent stimulus for breathing is pH, followed by CO_2. Perhaps surprisingly, the least significant is O_2.

A blood pH below 7.35 is known as **acidosis.** The corrective homeostatic response to acidosis is an increase in the rate and depth of breathing, "blowing off" CO_2 faster than the body produces it. As CO_2 is eliminated from the body, the carbonic acid reaction shifts to the left:

$$CO_2 + H_2O \leftarrow H_2CO_3 \leftarrow HCO_3^- + H^+$$

Thus, the H^+ on the right is consumed, raising the pH and ideally returning it to normal.

A pH greater than 7.45 is **alkalosis.** The response to this is a reduced respiratory rate and depth, allowing some CO_2 to accumulate and lower the pH to normal. You can see that P_{CO_2} and pH are very closely linked in regulating respiration. Oxygen normally has little effect because hemoglobin is almost saturated with O_2 anyway when it leaves the lungs, and little can be gained from heavier breathing. Oxygen does become more influential, however, in conditions involving high altitude (aviation and mountain climbing) and in diseases such as emphysema and pneumonia.

In summary, the main chemical stimulus to pulmonary ventilation is a change in pH that stimulates chemoreceptors of the brain and arteries. A pH change derives from changes in P_{CO_2} of the arterial blood. Therefore, changes in levels of carbon dioxide in the blood indirectly influence pulmonary ventilation. Ventilation is adjusted to maintain arterial pH at about 7.40. Normally, arterial P_{O_2} has relatively little effect on respiration.

Before You Go On

Answer these questions from memory. Reread the preceding section if there are too many you don't know.

10. *Identify three factors that affect the efficiency of alveolar gas exchange.*

11. *How is most oxygen transported in the blood?*

12. *What are the three ways in which blood transports CO_2? Which one is most significant?*

13. *Give three reasons why highly active tissues extract more oxygen from the blood than less active tissues.*

14. *What is the most potent chemical stimulus to respiration?*

15. *Explain how changes in pulmonary ventilation can correct pH imbalances.*

Aging of the Respiratory System

Respiratory failure is one of the leading causes of death in elderly people, but the gradual decline of respiratory function begins as early as the 30s. This decline is one of several factors in the gradual loss of physical endurance. The costal cartilages and joints of the thoracic cage become less flexible, and the lungs have less elastic tissue and fewer alveoli. Vital capacity, forced expiratory volume, and other values fall. Elderly people are also less capable of clearing the lungs of irritants and pathogens and are therefore increasingly vulnerable to respiratory infections. Pneumonia causes more deaths than any other infectious disease and is often contracted in hospitals and nursing homes. The chronic obstructive pulmonary diseases (COPDs)—emphysema and chronic bronchitis—are more common in old age since they represent the cumulative effects of a lifetime of degenerative change. COPDs also contribute to cardiovascular disease, hypoxemia (low blood Po_2), and hypoxic degeneration in all organ systems.

CAREER SPOTLIGHT

Respiratory Therapist

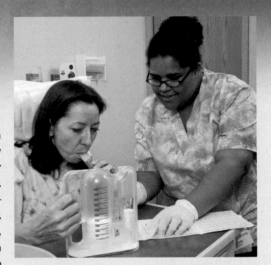

The respiratory therapist (RT) is an essential part of medical teams that deal with respiratory health. An RT specializes in assessment and treatment of respiratory conditions such as asthma or emphysema, and conditions that stem from other causes but affect breathing, such as strokes, heart attack, and spinal cord trauma. An RT is skilled in the use of equipment such as mechanical ventilators, which assist with breathing, and spirometers, devices that measure respiratory volumes. He or she may work in a variety of settings including hospital intensive care units, emergency rooms, or neonatal units; on emergency flight teams; or in home health care. To become an RT, students usually must complete either an associate degree or certificate program at a community college, or a 4-year bachelor's degree. One must pass national exams to become either a certified respiratory therapist or registered respiratory therapist. See appendix B for further information on careers in respiratory therapy.

CONNECTIVE ISSUES

Ways in Which the Respiratory System Affects Other Organ Systems

All Systems
Provides oxygen to all tissues and removes their carbon dioxide.

Integumentary System
Respiratory dysfunctions can cause such skin discolorations as cyanosis (blueness) or erythema (redness).

Skeletal System
Any respiratory disorder that causes hypoxemia stimulates accelerated erythropoiesis in the red bone marrow.

Muscular System
Acid–base imbalances caused by respiratory dysfunction can affect neuromuscular function. Acidosis, for example, makes muscular reflexes more sluggish.

Nervous System
Respiratory control is necessary in carrying out vocalizations initiated by the nervous system; acid–base balances resulting from respiratory dysfunction can make neurons insufficiently or excessively excitable.

Endocrine System
Lungs contain angiotensin-converting enzyme (ACE), which converts angiotensin I to the hormone angiotensin II; hypoxemia stimulates the secretion of erythropoietin.

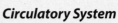

Circulatory System
By participation in angiotensin II production, lungs contribute to hormonal control of blood flow and blood pressure; proplatelets break up into platelets while passing through lungs; variations in respiratory rate stabilize blood pH; thoracic pump aids in venous return of blood to the heart; obstruction of pulmonary circulation can lead to right-sided heart failure.

Lymphatic and Immune Systems
Thoracic pump aids lymph flow.

Urinary System
Valsalva maneuver aids in emptying urinary bladder; respiratory and urinary systems collaborate in regulating acid–base balance.

Digestive System
Valsalva maneuver aids in vomiting and defecation.

Reproductive System
Valsalva maneuver aids in childbirth.

525

Study Guide

Assess Your Learning Outcomes

To test your knowledge, discuss the following topics with a study partner or in writing, ideally from memory.

15.1 Functions and Anatomy of the Respiratory System (p. 504)

1. Five functions of the respiratory system
2. Distinctions between the upper and lower respiratory tract and the conducting and respiratory zones of the tract
3. Function of the vibrissae (guard hairs)
4. Anterior and posterior boundaries of the nose
5. Structure and functional significance of the nasal conchae
6. Histology of the respiratory epithelium and the functional significance of its ciliated cells and goblet cells
7. Three divisions of the pharynx
8. Major structural features of the larynx and the function of the epiglottis
9. How the vocal cords produce sound; how their sound is transformed into intelligible words; and why men, women, and children differ in the pitch of the voice
10. Structure of the trachea and the functional significance of its cartilage, smooth muscle, and epithelial type
11. The nature and functional importance of the mucociliary escalator
12. Gross anatomy of the lungs and their spatial relationships with other thoracic structures
13. Structure and function of the pleurae
14. Anatomy and histology of the bronchial tree from the main bronchi through the bronchioles
15. Components of the respiratory zone of the airway and the relationship of their special histology to their function
16. Cellular organization of the alveoli, the function of each cell type, and the structure of the respiratory membrane
17. The source and function of pulmonary surfactant
18. The two sources of blood to the lungs and their functional difference

15.2 Pulmonary Ventilation (p. 514)

1. Definitions of *inspiration, expiration,* and *respiratory cycle*
2. How respiratory airflow is governed by atmospheric and intrapulmonary pressures
3. The principal muscles of inspiration and how their actions result in an inflow of air
4. Role of the pleurae in inspiration
5. The normal mechanism of relaxed expiration, and why it does not require a muscular effort
6. Muscles that assist with forced expiration
7. How the Vasalva maneuver is performed and what bodily functions it aids
8. Definitions of the lung volumes and capacities measured with a spirometer, and why this is done
9. Meanings and examples of restrictive and obstructive lung disorders
10. Locations of the principal pacemaker of breathing, how its neurons control the respiratory cycle, and the inputs it receives to modify respiratory rate and depth

15.3 Gas Exchange and Transport (p. 520)

1. The composition of the air we breathe
2. The meaning of *partial pressure* and how the partial pressures of gases such as O_2 and CO_2 are calculated
3. The meaning of *alveolar gas exchange* and how it is governed by gradients of partial pressure
4. Three factors that can alter the efficiency of alveolar gas exchange
5. How O_2 and CO_2 are transported in the blood; which of the methods depend on hemoglobin and how they relate to hemoglobin structure
6. The chemical equation for the formation of carbonic acid from water and CO_2; how it relates to CO_2 transport and blood pH
7. The meaning of *systemic gas exchange* and how it relates to partial-pressure gradients
8. How hemoglobin is able to release more oxygen to some tissues than to others in accordance with the metabolic state of different tissues
9. How respiration relates to the pH of the blood
10. Definitions of *acidosis* and *alkalosis* and how the respiratory system responds to each
11. Why P_{O_2} normally has little effect on pulmonary ventilation

Testing Your Recall

1. Which enzyme is responsible for catalyzing the conversion of CO_2 and water to H_2CO_3?
   a. surfactant
   b. carbonic anhydrase
   c. erythropoietin
   d. bicarbonate oxidase
   e. oxyhemoglobin

2. The structures responsible for warming and humidifying air as it travels through the nasal cavity are
   a. choanae.
   b. conchae.
   c. nares.
   d. vibrissae.
   e. goblet cells.

3. In speech, crude sounds are converted to recognizable words primarily by
   a. the trachea.
   b. the thyroid cartilage.
   c. the vocal cords.
   d. the epiglottis.
   e. the lips and tongue.

4. A restrictive lung disorder is most likely to result in
   a. slower airflow.
   b. a reduced vital capacity.
   c. alkalosis.
   d. pulmonary scar tissue.
   e. a COPD.

5. The source of pulmonary surfactant is
   a. the visceral pleura.
   b. tracheal glands.
   c. alveolar capillaries.
   d. squamous alveolar cells.
   e. great alveolar cells.

6. Which of the following are fewest in number but largest in diameter?
   a. alveoli
   b. terminal bronchioles
   c. alveolar ducts
   d. bronchi
   e. respiratory bronchioles

7. Which of the following is/are *not* lined with pseudostratified columnar epithelium?
   a. nasal cavity
   b. trachea
   c. nasopharynx
   d. alveolar ducts
   e. main bronchi

8. What is the average P_{O_2} at sea level?
   a. 160 mm Hg
   b. 100 mm Hg
   c. 0.03 mm Hg
   d. 40 mm Hg
   e. 46 mm Hg

9. Which of these values would normally be the highest?
   a. tidal volume
   b. inspiratory reserve volume
   c. expiratory reserve volume
   d. vital capacity
   e. It would differ from one person to another.

10. During inspiration, most of the airflow results from the action of
    a. the external intercostal muscles.
    b. the internal intercostal muscles.
    c. the diaphragm.
    d. the rectus abdominis muscle.
    e. the abdominal oblique muscles.

11. The process of measuring pulmonary function is called _____.

12. The _____ is a respiratory pacemaker in the medulla oblongata.

13. The largest cartilage of the larynx is the _____ cartilage.

14. The last line of defense against inhaled particles is phagocytic cells called _____.

15. The superior opening into the larynx is guarded by a tissue flap called the _____.

16. In the blood, CO_2 combines with H_2O to form _____.

17. Nerve endings that monitor the pH and chemical composition of the blood are called _____.

18. The mechanism that moves debris-laden mucus up the bronchi and trachea to the pharynx, where it can be swallowed, is called the _____.

19. Each lung receives a main bronchus, blood vessels, and nerves through a slit called the _____.

20. _____ is a condition caused by bronchospasm that results in shortness of breath, wheezing, and coughing.

Answers in appendix A

True or False

Determine which five of the following statements are false, and briefly explain why.

1. An average tidal volume is 1,000 mL.

2. Restrictive lung diseases result in a reduced inspiratory reserve volume.

3. Most of the CO_2 transported by the blood is in the form of dissolved gas.

4. The nasopharynx serves as a passageway for air only.

5. Each lung contains approximately 150 million alveoli.

6. Squamous alveolar cells are the most numerous cell type in the lung.

7. The most important factor that influences breathing rate and depth is P_{O_2}.

8. The lungs contain more respiratory bronchioles than terminal bronchioles.

9. Spirometry is a clinical technique for measuring a person's pulmonary functions.

10. Filling of the lungs during inspiration is what makes the chest expand.

Answers in appendix A

Testing Your Comprehension

1. In emphysema, there is a substantial loss of pulmonary alveoli and gas-exchange surface area, resulting in a low blood P_{O_2}. Explain why emphysema would also result in a high blood P_{CO_2} and acidosis.

2. Why is it more sensible for the hilum of each lung to be on its medial surface than it would be for it to be located at its apex, base, or costal surface?

3. The rhythmic beating of the heart is driven by a pacemaker located in the heart itself, yet the rhythmic expansion and contraction of the lungs are controlled by a pacemaker in the brainstem. Why do you think this is necessary, rather than the lungs having their own pacemaker?

The Urinary System

Each kidney has about 1 million of these microscopic, blood-filtering structures called glomeruli. An arteriole (yellow) feeds into the glomerulus (pink), a ball of blood capillaries that filter the blood and begin the process of urine formation.

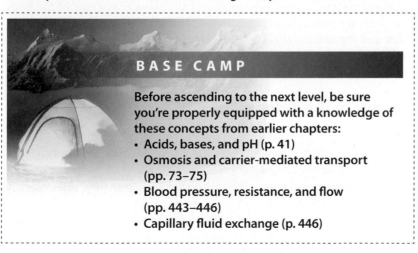

BASE CAMP

Before ascending to the next level, be sure you're properly equipped with a knowledge of these concepts from earlier chapters:
- Acids, bases, and pH (p. 41)
- Osmosis and carrier-mediated transport (pp. 73–75)
- Blood pressure, resistance, and flow (pp. 443–446)
- Capillary fluid exchange (p. 446)

Anatomy & Physiology | REVEALED®
aprevealed.com

An inevitable effect of metabolism is the production of waste products. These can poison the organism itself, so every living thing from bacterium to plant to human must have a way of eliminating them. For us, the most significant way of doing so is the urinary system. Yet the kidneys do much more than this, as you will see. They have close functional relationships with the endocrine, circulatory, and respiratory systems that we have covered in recent chapters. For this reason, patients with kidney failure can also exhibit anemia, hypertension, and a host of other problems.

Anatomically, the urinary system is closely associated with the reproductive system. The two systems have a shared embryonic development, and in many animals, the eggs and sperm are emitted through the urinary tract. In human males, the urethra continues to serve as a passage for both urine and sperm. Thus, the urinary and reproductive systems are often collectively called the *urogenital (U-G) system,* and *urologists* treat both urinary and male reproductive disorders.

16.1 Functions of the Urinary System

Expected Learning Outcomes
When you have completed this section, you should be able to:

a. name and locate the organs of the urinary system;
b. list several functions of the kidneys in addition to urine formation;
c. define *excretion* and identify the systems that excrete wastes; and
d. name the major nitrogenous wastes and identify their sources.

The **urinary system** is one of the anatomically simpler systems of the body, having only six principal organs: two **kidneys,** two **ureters,** the **urinary bladder,** and the **urethra** (fig. 16.1). Most of this chapter concerns the kidneys, the organs of urine production. The others—the organs of urine elimination—are treated near the end of the chapter.

Functions of the Kidneys AP|R

The kidneys play more roles than are commonly realized:

- They filter the blood, separate wastes and excrete them, and return useful material to the bloodstream.
- They regulate blood volume and pressure by eliminating or conserving water as necessary.
- They regulate the osmolarity of body fluids by controlling the relative amounts of water and solutes eliminated.

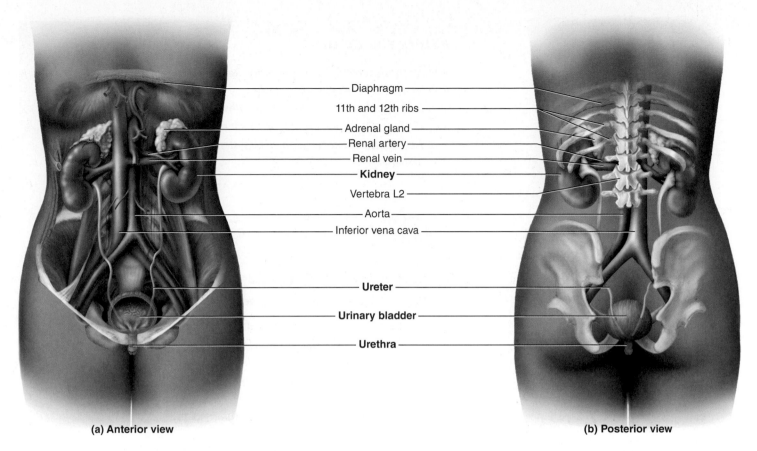

Diaphragm
11th and 12th ribs
Adrenal gland
Renal artery
Renal vein
Kidney
Vertebra L2
Aorta
Inferior vena cava

Ureter

Urinary bladder

Urethra

(a) Anterior view

(b) Posterior view

Figure 16.1 The Urinary System. AP|R

- They secrete the enzyme *renin,* which activates hormonal mechanisms that control blood pressure and electrolyte balance.
- They secrete the hormone *erythropoietin,* which stimulates the production of red blood cells and thus supports the oxygen-carrying capacity of the blood.
- They carry out the final step in synthesizing the hormone *calcitriol* (p. 158) and thereby contribute to calcium homeostasis.
- They detoxify some drugs.

Metabolic Wastes and Excretion

Excretion is the process of separating wastes from body fluids and eliminating them. A **waste** is any substance that is useless or present in excess of the body's needs. A **metabolic waste,** more specifically, is a waste substance produced by the body—for example, as a by-product of ATP production or protein catabolism. The urinary system excretes a variety of metabolic wastes, toxins, and drugs in the urine. Other systems also excrete wastes; for example, the respiratory system excretes carbon dioxide, and the digestive and integumentary systems excrete water and salts.

The kidneys are responsible for ridding the body of toxic metabolic wastes, especially small nitrogen-containing compounds called **nitrogenous wastes.** About 50% of the nitrogenous waste is **urea,** a by-product of protein metabolism. Protein breakdown produces ammonia, which is exceedingly toxic. The liver quickly converts ammonia to urea, which is somewhat less toxic. The urea is carried in blood plasma to the kidneys where it is filtered into the urine and excreted.

Other nitrogenous wastes in the urine include **uric acid** and **creatinine** (cree-AT-ih-neen), produced by the catabolism of nucleic acids and creatine phosphate, respectively. Although less toxic than ammonia and less abundant than urea, these too are far from harmless.

Before You Go On

Answer these questions from memory. Reread the preceding section if there are too many you don't know.

1. State four functions of the kidneys other than forming urine.
2. List four nitrogenous wastes and their metabolic sources.
3. Name some wastes eliminated by systems other than the urinary system.

16.2 Anatomy of the Kidney

Expected Learning Outcomes

When you have completed this section, you should be able to:

a. describe the location and general appearance of the kidney, and its relationship to neighboring organs;
b. identify the major external and internal features of the kidney;
c. trace the flow of blood through the kidney;
d. describe the components of a nephron; and
e. trace the flow of fluid through the renal tubules.

The kidneys lie against the posterior abdominal wall at the level of vertebrae T12 to L3 (fig. 16.1). About half of the left kidney is within the rib cage, as rib 12 crosses it. The right kidney is slightly lower because the space above it is occupied by the large right lobe of the liver. The kidneys are retroperitoneal, along with the ureters, urinary bladder, renal artery and vein, and adrenal glands.

Gross Anatomy AP|R

Each kidney weighs about 150 g and measures about 11 cm long, 6 cm wide, and 3 cm thick—roughly the size of a bar of bath soap. The lateral surface is convex, and the medial surface is concave and has a slit, the **hilum,** where it receives the renal blood vessels, nerves, lymphatics, and ureter (fig. 16.2).

Connective tissue layers anchor the kidney and protect it from trauma. A thick layer of adipose tissue, the *perirenal fat capsule,* cushions the kidney and holds it in place. The *fibrous (renal) capsule* forms the outer layer of the kidney, enclosing the glandular tissue like a cellophane wrapper.

The glandular, urine-forming tissue of the kidney is called the *parenchyma.* It has a C shape enclosing a medial space called the **renal sinus.** Its outer rind, the **renal cortex,** is about 1 cm thick in most places. Its inner tissue, the **renal medulla,** is divided into 6 to 10 cones called **renal pyramids,** separated from each other by extensions of the cortex. The broad base of each pyramid faces the overlying cortex, and its blunt apex, the **renal papilla,** faces the sinus. One pyramid and the overlying cortex constitute one *lobe* of the kidney.

The papilla of each pyramid is nestled in a cup called a **minor calyx**[1] (CAY-lix), which collects its urine. Two or three minor calyces (CAY-lih-seez) converge to form a **major calyx,** and two or three major calyces converge in the sinus to form the **renal pelvis.**[2] The funnel-like pelvis converges on the ureter, a tubular continuation leading to the urinary bladder.

---

[1] *calyx* = cup
[2] *pelvis* = basin

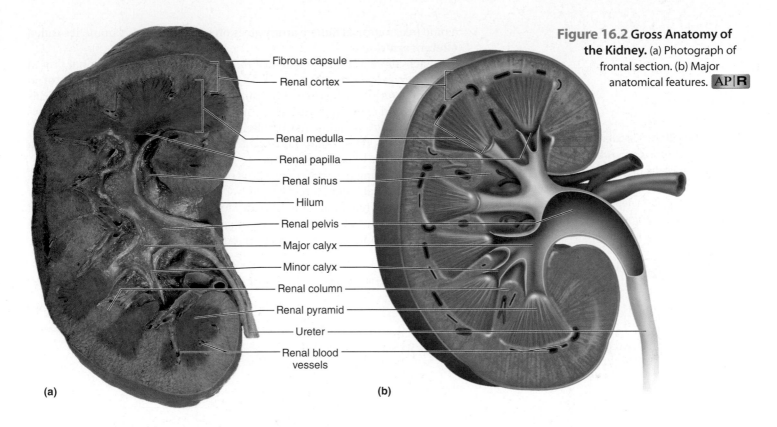

Figure 16.2 Gross Anatomy of the Kidney. (a) Photograph of frontal section. (b) Major anatomical features. **AP|R**

- Fibrous capsule
- Renal cortex
- Renal medulla
- Renal papilla
- Renal sinus
- Hilum
- Renal pelvis
- Major calyx
- Minor calyx
- Renal column
- Renal pyramid
- Ureter
- Renal blood vessels

(a)

(b)

Renal Circulation

Although the kidneys account for only 0.4% of the body weight, they receive about 21% of the cardiac output. This is a reflection of how important they are in eliminating toxic waste and regulating blood volume and composition.

Each kidney is supplied by a **renal artery** arising from the aorta. Branches of the renal artery lead to *arcuate arteries* that arch over the base of each renal pyramid, and from here, slender *cortical radiate arteries* penetrate into the cortex and ascend toward the kidney surface (fig. 16.3). Like the trunk of

Figure 16.3 Renal Circulation. (a) The larger blood vessels of the kidney. (b) Flowchart of renal circulation (omitting some minor vessels).

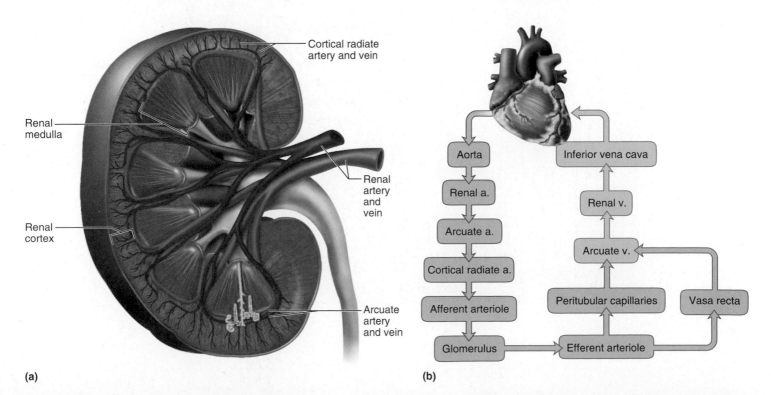

- Cortical radiate artery and vein
- Renal medulla
- Renal artery and vein
- Renal cortex
- Arcuate artery and vein

(a)

Aorta → Renal a. → Arcuate a. → Cortical radiate a. → Afferent arteriole → Glomerulus → Efferent arteriole → Peritubular capillaries → Vasa recta → Arcuate v. → Renal v. → Inferior vena cava

(b)

a pine tree, a cortical radiate artery gives off a series of lateral branches called **afferent arterioles.**

Each afferent arteriole supplies the capillaries of one functional unit of the kidney, called a *nephron* (fig. 16.4). Blood leaves the nephron by way of an **efferent arteriole,** which usually leads to a web of **peritubular capillaries** that surround the winding tubule of the nephron. From here, the blood flows through a series of veins that parallel the arteries and have similar names, ultimately leading to the **renal vein.** This vein exits the kidney at the hilum and leads to the inferior vena cava.

The foregoing description pertains only to the blood supply of the renal cortex. The renal medulla receives only 1% to 2% of the total blood flow, yet this is crucial to renal function. It is supplied by relatively straight, small blood vessels called the **vasa recta,**[3] which arise from the efferent arterioles of the deepest nephrons and then drain into the *arcuate veins.*

---

[3] *vasa* = vessels; *recta* = straight

Figure 16.4 Blood Circulation in the Nephron. For clarity, vasa recta are shown only on the left and peritubular capillaries only on the right. In nephrons close to the medulla (left), the efferent arteriole gives rise to the vasa recta. In nephrons farther out on the cortex (right), the nephron loop barely dips into the renal medulla and the efferent arteriole gives rise to peritubular capillaries. (DCT = distal convoluted tubule; PCT = proximal convoluted tubule)

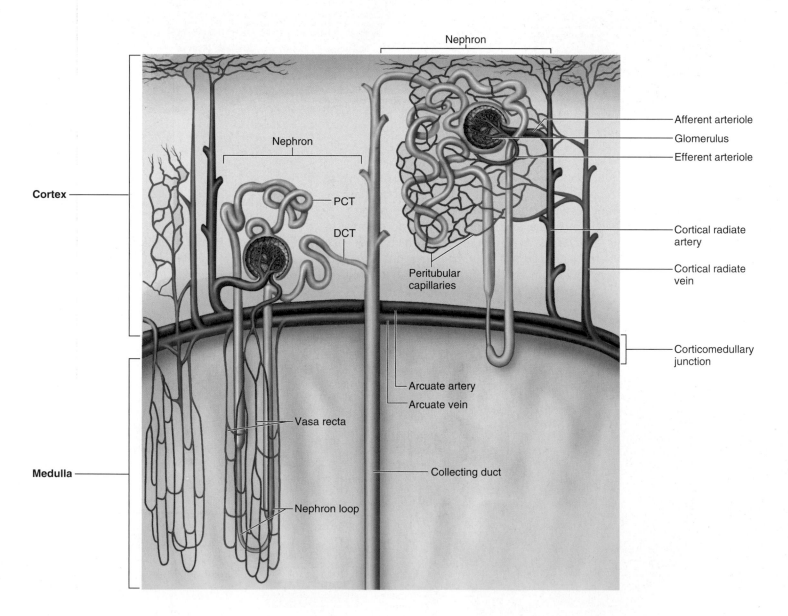

The Nephron

Each kidney has about 1.2 million urine-forming functional units called **nephrons**[4] (NEF-rons). This is where the blood plasma is filtered and processed to form urine. To understand how just one of these works is to understand nearly everything about how the whole kidney works. Each nephron is composed of two principal parts: a *renal corpuscle,* which filters the blood plasma, and a long coiled *renal tubule,* which converts the filtrate to urine (figure 16.4). The renal tubule extends beyond one nephron, however, to include a *collecting duct* shared by multiple nephrons.

The Renal Corpuscle

The **renal corpuscle** is composed of a ball of capillaries called a **glomerulus**[5] (glo-MERR-you-lus) and a two-layered **glomerular (Bowman**[6]**) capsule** that encloses it like a cup (fig. 16.5). The outer layer of the capsule is a simple squamous epithelium, and the inner layer consists of elaborate cells called **podocytes**[7] wrapped around the capillaries of the glomerulus (see fig. 16.8). The two layers are separated by a filtrate-collecting **capsular space.**

The afferent arteriole delivers blood to the glomerulus and the efferent arteriole carries it away. The two arterioles lie snugly side by side, penetrating the capsule at a single opening on one side. The afferent arteriole is significantly larger than the efferent arteriole. Thus, the glomerulus has a large inlet and a small outlet—resulting in unusually high blood pressure in the glomerulus, which is crucial to the kidney's task of blood filtration.

The filtrate enters the capsular space and then passes to the renal tubule, described next.

[4] *nephro* = kidney
[5] *glomer* = ball; *ulus* = little
[6] Sir William Bowman (1816–92), British physician
[7] *podo* = foot; *cyte* = cell

Figure 16.5 The Renal Corpuscle. (a) Anatomy of the corpuscle. (b) A resin cast of the glomerulus and nearby arteries. Note the difference in diameters between the afferent and efferent arterioles. [Part (b) © Dr. Richard Kessel & Dr. Randy Kardon/Visuals Unlimited, Inc.]

- *Which is larger, the afferent or efferent arteriole? How does the difference affect glomerular blood pressure?*

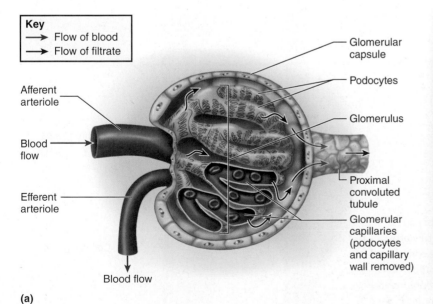

(a)

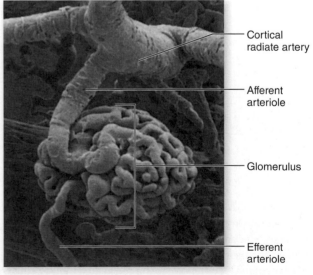

(b)

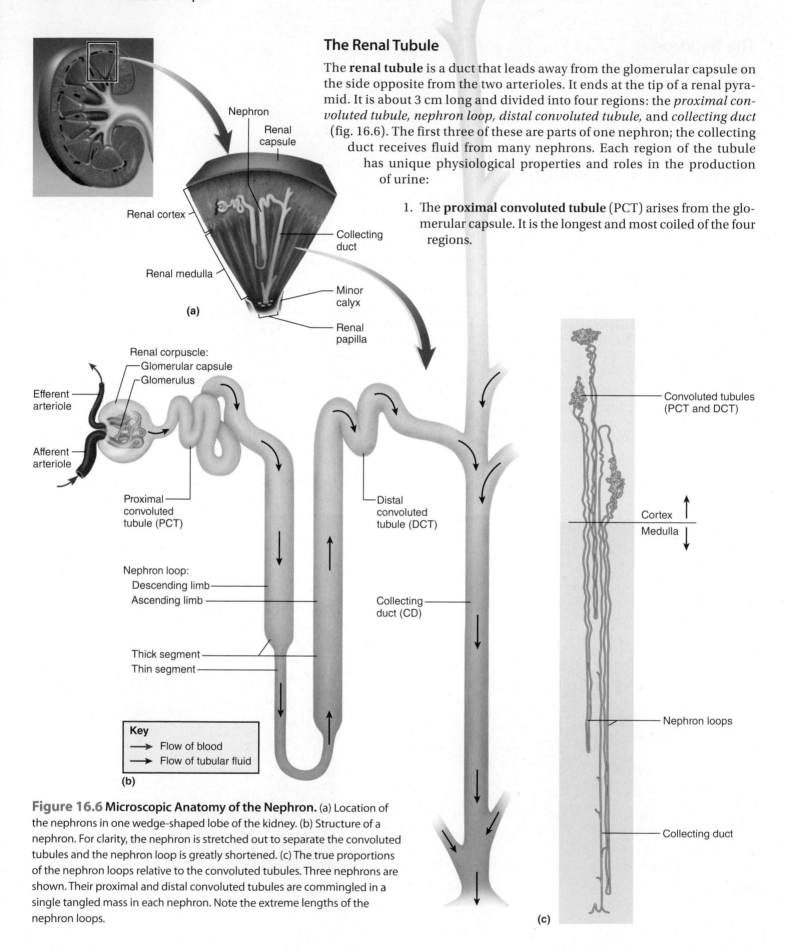

The Renal Tubule

The **renal tubule** is a duct that leads away from the glomerular capsule on the side opposite from the two arterioles. It ends at the tip of a renal pyramid. It is about 3 cm long and divided into four regions: the *proximal convoluted tubule, nephron loop, distal convoluted tubule,* and *collecting duct* (fig. 16.6). The first three of these are parts of one nephron; the collecting duct receives fluid from many nephrons. Each region of the tubule has unique physiological properties and roles in the production of urine:

1. The **proximal convoluted tubule** (PCT) arises from the glomerular capsule. It is the longest and most coiled of the four regions.

Figure 16.6 Microscopic Anatomy of the Nephron. (a) Location of the nephrons in one wedge-shaped lobe of the kidney. (b) Structure of a nephron. For clarity, the nephron is stretched out to separate the convoluted tubules and the nephron loop is greatly shortened. (c) The true proportions of the nephron loops relative to the convoluted tubules. Three nephrons are shown. Their proximal and distal convoluted tubules are commingled in a single tangled mass in each nephron. Note the extreme lengths of the nephron loops.

2. The **nephron loop (loop of Henle[8])** is a long U-shaped portion of the renal tubule found mostly in the medulla. It begins where the PCT straightens out and dips toward the medulla, forming the *descending limb* of the loop. At its deep end, the loop turns 180° and forms the *ascending limb*, which returns to the cortex, traveling parallel and close to the descending limb. Some portions of the loop have a simple cuboidal epithelium and therefore look thicker than others; thus, they are called the *thick segments.* The rest of the loop has a simple squamous epithelium and is called the *thin segment.* The thin segment is very permeable to water, whereas the rest is less so. The thick segments are heavily engaged in the active transport of ions, so they require a lot of ATP and the cells here are packed with mitochondria, accounting for their thickness. The functional significance of the thick and thin segments will be discussed later.

3. The **distal convoluted tubule (DCT)** begins shortly after the ascending limb reenters the cortex. It is shorter and less coiled than the proximal convoluted tubule. The DCT is the end of the nephron.

4. The **collecting duct** receives fluid from the DCTs of several nephrons as it passes through the cortex before descending into the medulla. Numerous collecting ducts converge toward the apex of a renal pyramid and end in pores at the conical tip of the papilla. Urine drains from these pores into the minor calyx that encloses the papilla.

The flow of fluid from the point where it is filtered from the blood to the point where urine leaves the body is glomerular capsule → proximal convoluted tubule → nephron loop → distal convoluted tubule → collecting duct → minor calyx → major calyx → renal pelvis → ureter → urinary bladder → urethra.

Before You Go On

Answer these questions from memory. Reread the preceding section if there are too many you don't know.

4. *Define the two parts of a renal corpuscle and explain what distinguishes a renal corpuscle from a nephron.*

5. *Trace the path taken by one red blood cell from the renal artery, through the renal cortex, to the renal vein.*

6. *Trace a drop of urine on its route from the glomerular capsule to the point where it leaves the body.*

16.3 Glomerular Filtration

Expected Learning Outcomes

When you have completed this section, you should be able to:

a. describe the process by which the kidney filters the blood plasma;

b. describe how the sympathetic nervous system, hormones, and the kidney itself regulate filtration.

---

[8] Friedrich G. J. Henle (1809–85), German anatomist

The kidney converts blood plasma to urine in four stages: glomerular filtration, tubular reabsorption, tubular secretion, and water conservation (fig. 16.7), which will be explored in this order. As it is transformed to urine, the fluid is referred to by different names that reflect its changing composition: (1) The fluid in the capsular space, called **glomerular filtrate,** is similar to blood plasma except that it has almost no protein. (2) The fluid from the proximal convoluted tubule through the distal convoluted tubule will be called **tubular fluid.** It differs from the glomerular filtrate because of substances removed and added by the tubule cells. (3) The fluid will be called **urine** once it enters the collecting duct, since it undergoes little alteration beyond that point except for a change in concentration.

Glomerular filtration, the first step of urine formation, is a process in which water and some solutes in the blood plasma are filtered from the capillaries of the glomerulus into the capsular space of the nephron. The fluid passes through three barriers that constitute the **filtration membrane** (fig. 16.8):

1. **The endothelium of the capillary.** The glomerulus is composed of *fenestrated capillaries* perforated by large filtration pores (see pp. 439–440). Like fenestrated capillaries elsewhere, these are highly permeable, although their pores are small enough to exclude blood cells from the filtrate.

2. **The basement membrane.** This is a gel that holds back most protein and other particles too large to pass through its fine mesh of glycoproteins. Passing large molecules through it would be like trying to force sand through a kitchen sponge; a little plasma protein gets through, but not much.

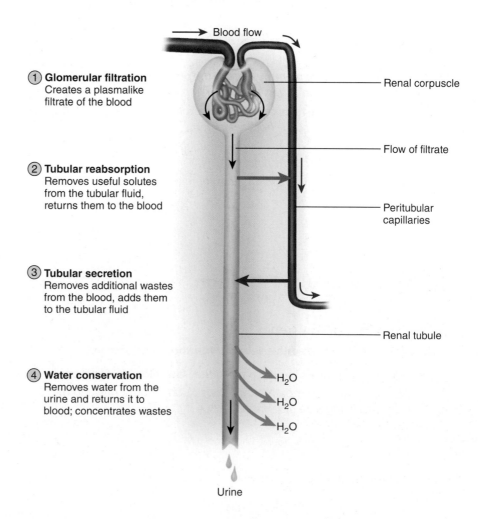

Blood flow

① **Glomerular filtration**
Creates a plasmalike filtrate of the blood

Renal corpuscle

Flow of filtrate

② **Tubular reabsorption**
Removes useful solutes from the tubular fluid, returns them to the blood

Peritubular capillaries

③ **Tubular secretion**
Removes additional wastes from the blood, adds them to the tubular fluid

Renal tubule

④ **Water conservation**
Removes water from the urine and returns it to blood; concentrates wastes

H_2O

H_2O

H_2O

Urine

Figure 16.7 Basic Stages of Urine Formation.
AP|R

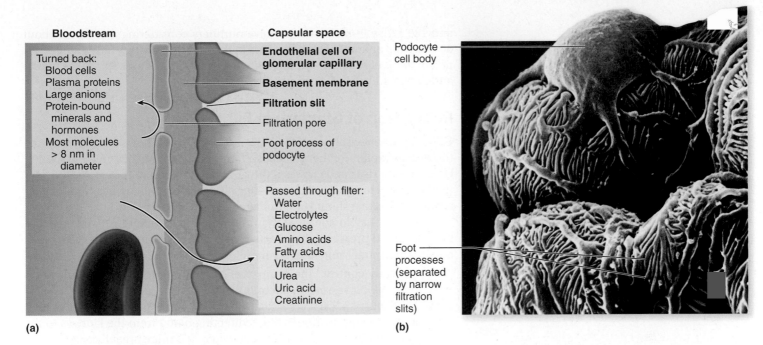

Bloodstream

Capsular space

Turned back:
  Blood cells
  Plasma proteins
  Large anions
  Protein-bound
    minerals and
    hormones
  Most molecules
    > 8 nm in
    diameter

**Endothelial cell of
glomerular capillary**

Basement membrane

Filtration slit

Filtration pore

Foot process of
podocyte

Passed through filter:
  Water
  Electrolytes
  Glucose
  Amino acids
  Fatty acids
  Vitamins
  Urea
  Uric acid
  Creatinine

(a)

Podocyte
cell body

Foot
processes
(separated
by narrow
filtration
slits)

(b)

**Figure 16.8 The Glomerular Filtration
Membrane.** (a) Components of the filtration
membrane. (b) Blood capillaries in the glomerulus
closely wrapped in spidery podocytes.

3. **Filtration slits.** The podocytes of the glomerular capsule are shaped somewhat like octopi, with bulbous cell bodies and several thick arms. Each arm has numerous little extensions called *foot processes (pedicels*[9]*)* that wrap around the capillaries and interdigitate with each other, like wrapping your hands around a pipe and lacing your fingers together.

Small molecules can pass freely through the filtration membrane into the capsular space. These include water, electrolytes, glucose, fatty acids, amino acids, nitrogenous wastes, and vitamins. Such substances have about the same concentration in the glomerular filtrate as in the blood plasma. Larger particles are left behind in the blood.

Filtration Pressure

Glomerular filtration is driven by blood pressure in the glomerular capillaries. Blood pressure is much higher here than in most other capillaries—about 60 mm Hg compared with 10 to 15 mm Hg elsewhere. This results from the fact that the afferent arteriole is substantially larger than the efferent arteriole, giving the glomerulus a large inlet and small outlet. Two forces oppose the blood pressure in the glomerulus: (1) fluid pressure in the capsular space, and (2) osmotic pressure of the blood (see p. 447). Normally, the glomerular blood pressure remains high enough to override these, so these capillaries engage solely in filtration. They reabsorb little or no fluid.

The high blood pressure in the glomeruli makes the kidneys especially vulnerable to hypertension, which can have devastating effects on renal function. Hypertension ruptures glomerular capillaries and leads to scarring of the kidneys *(nephrosclerosis)*. It promotes atherosclerosis of the renal blood vessels, just as it does elsewhere in the body, and thus diminishes renal blood supply. Over time, hypertension often leads to renal failure.

Glomerular Filtration Rate

Glomerular filtration rate (GFR) is the amount of filtrate formed per minute by the two kidneys combined. In adult females, the GFR is about 105 mL/min.; in males, about 125 mL/min. These rates are equivalent to 150 L/day in females

---

[9] *pedi* = foot; *cel* = little

and 180 L/day in males—impressive numbers considering that this is about 60 times the amount of blood plasma in the body. Obviously, only a small portion of this is eliminated as urine. An average adult reabsorbs 99% of the filtrate and excretes 1 to 2 L of urine per day.

Regulation of Glomerular Filtration

GFR must be precisely controlled. If it is too high, fluid flows through the renal tubules too rapidly for them to reabsorb the usual amount of water and solutes. Urine output rises and creates a threat of dehydration and electrolyte depletion. If GFR is too low, fluid flows sluggishly through the tubules, and they reabsorb wastes that should be eliminated in the urine.

The only way to adjust GFR from moment to moment is to change glomerular blood pressure. This is achieved by three homeostatic mechanisms: nervous, hormonal, and local control.

1. **Nervous control.** The sympathetic nervous system richly innervates renal blood vessels. In strenuous exercise or acute conditions such as circulatory shock, it constricts the afferent arterioles. This reduces GFR and urine production, while redirecting blood from the kidneys to the heart, brain, and skeletal muscles where it is more urgently needed. Under such conditions, GFR may be as low as a few milliliters per minute. Urine output may be reduced during participation in a sporting event for this reason.

2. **Hormonal control.** GFR is regulated by a hormonal chain reaction that responds to deviations in blood pressure. When blood pressure (BP) drops, sympathetic nerves stimulate the kidneys to secrete the enzyme **renin** (REE-nin). Renin converts *angiotensinogen,* a blood plasma protein, into *angiotensin I.* An enzyme in the lungs and kidneys, *angiotensin-converting enzyme (ACE),* further converts this to **angiotensin II,** a hormone that acts in several ways to restore fluid volume and blood pressure (fig. 16.9). Among other effects, angiotensin II makes one thirsty and encourages water intake; it stimulates the adrenal cortex to secrete **aldosterone,** which promotes retention of sodium and water in the body; and it stimulates the posterior pituitary to secrete **antidiuretic hormone (ADH),** which promotes water retention. Aldosterone and ADH reduce urinary water loss from the body, while thirst, of course, builds fluid volume and raises BP.

   Angiotensin II is also a potent vasoconstrictor, and generalized vasoconstriction raises blood pressure. In the kidneys, it constricts the efferent arteriole of each nephron more than the afferent arteriole. By narrowing the glomerular outlet more than its inlet, it sustains glomerular blood pressure and prevents a drop in blood pressure from proportionately lowering the GFR. By stabilizing GFR, angiotensin II ensures that the kidneys continue filtering waste from the body even when systemic blood pressure is abnormally low.

3. **Local control.** Local control, or **renal autoregulation,** is the ability of the nephrons to adjust their own blood flow and GFR even without external (nervous or hormonal) control. It enables them to maintain a relatively stable GFR in spite of changes in arterial blood pressure. Because of renal autoregulation, urine output increases very little even under circumstances that substantially raise blood pressure. There are two mechanisms of autoregulation: the myogenic mechanism and tubuloglomerular feedback.

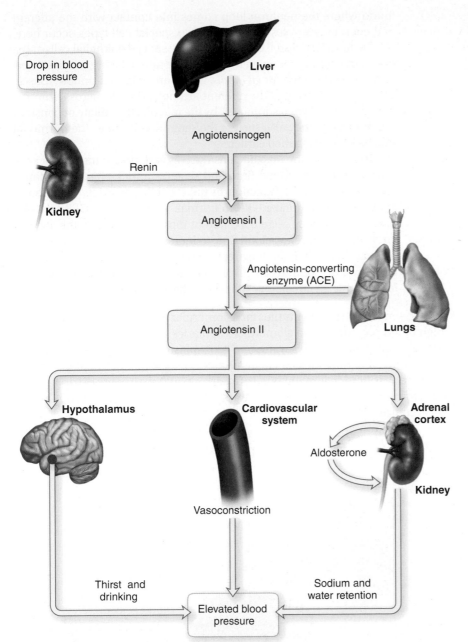

Figure 16.9 The Renin–Angiotensin–Aldosterone Mechanism. This chain of events is activated by a drop in blood pressure and acts to raise it again.

- **The myogenic** [10] **mechanism** is named for the fact that smooth muscle (*myo-*) has a tendency to contract when stretched. A rise in arterial blood pressure stretches the afferent arteriole. In response, the arteriole constricts and prevents blood flow into the glomerulus from changing very much. Conversely, when blood pressure falls, the afferent arteriole relaxes and allows blood to flow more easily into the glomerulus. Either way, GFR remains fairly stable.

- **Tubuloglomerular feedback** involves communication between the glomerulus and renal tubule through a complex of structures called the **juxtaglomerular** (JUX-tuh-glo-MER-you-lur) **apparatus.** This is

---

[10] *myo* = muscle; *genic* = produced by

found where the nephron loop comes into contact with the afferent and efferent arterioles (fig. 16.10). Two special cell types occur here: (1) The *macula densa*,[11] a patch of closely spaced epithelial cells at the end of the nephron loop on the side of the tubule facing the arterioles; and (2) *juxtaglomerular (JG) cells,* which are enlarged smooth muscle cells located mostly in the afferent arteriole, directly across from the macula densa. When stimulated by the macula, they dilate or constrict the arterioles. They also secrete the renin that initiates the hormonal mechanisms described earlier.

If GFR rises, the macula densa senses variation in the flow or concentration of the tubular fluid and secretes a messenger that stimulates the JG cells. Contraction of the JG cells constricts the afferent arteriole, thus reducing GFR to normal. Conversely, if GFR falls, the macula densa may secrete a different messenger causing the afferent arteriole cells to relax, blood flow to increase, and GFR to rise back to normal. In conditions of extreme blood pressure changes, however, GFR and urine output can fall precipitously or even cease. This is seen, for example, in circulatory shock from hemorrhage.

To summarize the events thus far: Glomerular filtration occurs because the high blood pressure of the glomerular capillaries overrides reabsorption. The filtration membrane allows water and most plasma solutes into the capsular space while retaining formed elements and most protein in the bloodstream. Glomerular filtration is maintained at a fairly steady rate in spite of variations in systemic blood pressure. This stability is achieved by nervous, hormonal, and local control mechanisms.

---

[11] *macula* = spot, patch; *densa* = dense

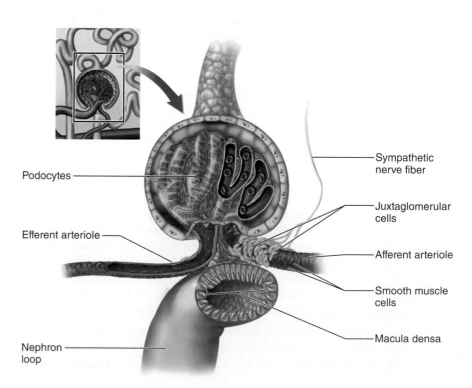

Podocytes

Efferent arteriole

Nephron loop

Sympathetic nerve fiber

Juxtaglomerular cells

Afferent arteriole

Smooth muscle cells

Macula densa

Figure 16.10 The Juxtaglomerular Apparatus.

Before You Go On

Answer these questions from memory. Reread the preceding section if there are too many you don't know.

7. Name the four stages in which blood plasma is converted to urine.

8. Explain why, in view of the structure of the glomerular filtration membrane, the glomerular filtrate has a different composition from the blood.

9. Assume a person is exercising, sweating profusely, and experiences a significant drop in body fluid and blood volume. Describe the homeostatic mechanisms that would help the kidneys maintain a normal GFR.

16.4 Tubular Reabsorption and Secretion

Expected Learning Outcomes

When you have completed this section, you should be able to:

a. describe how the renal tubules reabsorb useful solutes from the glomerular filtrate and return them to the blood;

b. describe how the tubules secrete solutes from the blood into the tubular fluid; and

c. describe how the nephron regulates water excretion.

The next two steps of converting glomerular filtrate to urine are *tubular reabsorption* and *tubular secretion,* the removal of some chemicals from the tubular fluid and addition of other chemicals to it. These two processes occur simultaneously, so we discuss them together in this section. We will trace the course of the tubular fluid through the nephron, from proximal convoluted tubule through distal convoluted tubule, and see how it is modified at each point along the way. Refer to figure 16.7 to put these processes into perspective.

The Proximal Convoluted Tubule

Tubular reabsorption is the process of reclaiming water and solutes from the tubular fluid and returning them to the blood. The proximal convoluted tubule (PCT) reabsorbs about 65% of the tubular fluid, while it also removes some substances from the blood and secretes them into the tubule for disposal in the urine. The importance of the PCT is reflected in its relatively great length and a shaggy fringe *(brush border)* of tall microvilli that greatly increase its absorptive surface area. It requires a great deal of energy to transport materials across the epithelial cells; the PCTs alone account for about 6% of one's resting ATP and calorie consumption.

The PCT reabsorbs a greater variety of chemicals than any other part of the nephron, including the following substances and many others.

- **Electrolytes.** Sodium ions (Na^+) are reabsorbed by transport proteins in the tubule cell membranes. Sodium is the key to everything else, because it creates an osmotic and electrical gradient that drives the reabsorption of water and the other solutes. Chloride ions (Cl^-) follow Na^+ by electrical attraction; potassium, magnesium, and phosphate ions are reabsorbed by membrane transport proteins and by leaking through the spaces between epithelial cells.
- **Glucose.** Glucose is cotransported with Na^+. Normally, all glucose in the tubular fluid is reabsorbed and there is none in the urine.
- **Nitrogenous wastes.** Some urea in the tubular fluid is reabsorbed along with water in the PCT, but about half of it passes in the urine. Thus, the kidneys do not completely clear the blood of this waste, but keep its concentration down to a safe level.
- **Water.** The kidneys reduce about 180 L of glomerular filtrate to 1 or 2 L of urine each day, so obviously water reabsorption is a significant function. About two-thirds of the water is reabsorbed by osmosis in the PCT.

After water and solutes cross the tubule epithelium, they are reabsorbed into the blood by the peritubular capillaries.

There is a limit to the amount of solute that the renal tubule can reabsorb because there are limited numbers of transport proteins in the plasma membranes. If all the transporters are occupied as solute molecules pass through, we reach a point called the **transport maximum (T_m),** where no more solute can be reabsorbed and the excess passes through the tubule and into the urine. Each organic solute reabsorbed by the renal tubule has its own T_m. Glucose, for example, normally enters the renal tubule at a rate well within the T_m; thus, all of it is reabsorbed. It does not saturate all of its transporters. But at any blood glucose level above 220 mg/dL or so, glucose is filtered faster than the renal tubule can reabsorb it, and the excess passes in the urine—a condition called **glycosuria**[12] (GLY-co-soo-ree-uh). In untreated diabetes mellitus, the plasma glucose concentration may exceed 400 mg/dL, so glycosuria is one of the classic signs of this disease.

The Nephron Loop

The primary function of the nephron loop is to generate an osmotic gradient that enables the collecting duct to concentrate the urine and conserve water, but in addition, the loop reabsorbs water and salts. The thin segment of the loop is very permeable to water but not to solutes. As tubular fluid flows down the descending limb, about 15% of the water is reabsorbed and returned to the bloodstream. Cells in the thick segment of the ascending limb are actively engaged in salt reabsorption, but are impermeable to water. Therefore, the ascending limb selectively reabsorbs salts from the fluid flowing through it. The more important function of the loop in supporting the work of the collecting duct is explained later.

The Distal Convoluted Tubule

Fluid arriving in the distal convoluted tubule (DCT) still contains about 20% of the water and 7% of the salts from the glomerular filtrate. If this were all passed

---

[12] *glycos* = sugar; *uria* = urine condition

as urine, it would amount to 36 L/day and one would never be able to leave the bathroom! Obviously, a great deal of fluid reabsorption still has to occur. The DCT reabsorbs variable amounts of water and salts, but much less than the PCT. In keeping with its lesser (yet not unimportant) role in reabsorption, it is much shorter than the PCT and does not have such a prominent brush border of microvilli. The DCT is regulated by several hormones—particularly aldosterone and natriuretic peptides.

Aldosterone, the "salt-retaining hormone," is secreted by the adrenal cortex when the blood pressure or Na^+ concentration falls or its K^+ concentration rises (see chapter 11). It stimulates the ascending limb of the nephron loop, the DCT, and the collecting duct to reabsorb more NaCl and secrete more K^+. Its net effect is that the body retains NaCl and water, urine volume is reduced, and the urine has an elevated K^+ concentration. The retention of salt and water helps to maintain blood volume and pressure.

Natriuretic peptides are secreted by the heart in response to high blood pressure. Their effect is to increase excretion of salt and water in the urine, thus reducing blood volume and pressure.

Tubular Secretion

Tubular secretion is a process in which the renal tubule extracts chemicals from blood of the peritubular capillaries and secretes them into the tubular fluid (see fig. 16.7). In the proximal convoluted tubule and nephron loop, it serves three purposes: (1) acid–base balance, through secretion of varying proportions of hydrogen (H^+) to bicarbonate (HCO_3^-) ions; (2) waste removal, through extracting urea, uric acid, bile acids, and ammonia from the blood; and (3) clearance of drugs and contaminants from the body, such as morphine, penicillin, and aspirin. One reason so many drugs must be taken three or four times a day is to keep pace with this rate of clearance and maintain a therapeutically effective drug concentration in the blood.

In summary, the PCT reabsorbs about 65% of the glomerular filtrate and returns it to the blood of the peritubular capillaries. Much of this reabsorption occurs by osmosis linked to the active transport of sodium ions. The nephron loop reabsorbs another 25% of the filtrate. The DCT reabsorbs more sodium, chloride, and water, but its rates of reabsorption are subject to control by hormones, especially aldosterone and natriuretic peptides. Drugs, wastes, and other solutes from the blood are secreted into the tubular fluid. The DCT essentially completes the process of determining the chemical composition of the urine. The principal function left to the collecting duct is to conserve water.

Before You Go On

Answer these questions from memory. Reread the preceding section if there are too many you don't know.

10. List five substances that are reabsorbed in the proximal convoluted tubule. Which of these does the most to drive reabsorption of the others? Explain how it does that.

11. Explain why glucose appears in the urine of a person with diabetes.

12. How do the thin and thick segments of the nephron loop differ in their absorptive functions?

13. Contrast the effects of aldosterone and natriuretic peptides on the distal convoluted tubule and on blood volume and pressure.

16.5 Water Conservation

Expected Learning Outcomes

When you have completed this section, you should be able to:

a. explain how the collecting duct and antidiuretic hormone regulate the volume and concentration of urine; and

b. explain how the kidney maintains an osmotic gradient in the renal medulla that enables the collecting duct to function.

The kidney serves not just to eliminate metabolic waste from the body but also to prevent excessive water loss, and thus to support the body's fluid balance. As the kidney returns water to the tissue fluid and bloodstream, the fluid remaining in the renal tubule, and ultimately passed as urine, becomes more and more concentrated. In this section, we examine the kidney's mechanism for conserving water and concentrating the urine.

The Collecting Duct

The collecting duct (CD) begins in the renal cortex, where it receives tubular fluid from numerous nephrons. As it passes through the medulla, it usually reabsorbs water and concentrates the urine. When urine enters the upper end of the CD, it is isotonic with blood plasma—300 milliosmoles per liter (mOsm/L) (see appendix C for explanation of this unit of measurement). By the time it leaves the lower end, it can be up to four times as concentrated—that is, highly hypertonic to the blood plasma.

Two facts enable the collecting duct to produce such hypertonic urine: (1) The osmolarity of the tissue fluid surrounding it is four times as high in the lower medulla as it is in the cortex. (2) The medullary portion of the CD is more permeable to water than to solutes. Therefore, as urine passes down the CD through the increasingly salty medulla, water leaves the tubule by osmosis, most NaCl and other wastes remain behind, and the urine becomes more and more concentrated (fig. 16.11).

It may seem surprising that the tissue fluid is so hypertonic in the deep medulla compared to the cortex and upper medulla. We would expect the salt to diffuse toward the cortex until it was evenly distributed through the kidney. However, there is a mechanism that overrides this—the nephron loop. Through a complex mechanism beyond the scope of this book, it acts as a *countercurrent multiplier* that continually recaptures salt and returns it to the deep medullary tissue. It is called a *multiplier* because it multiplies the salinity deep in the medulla and a *countercurrent* mechanism because it is based on fluid flowing in opposite directions in the descending and ascending limbs of the loop. The different permeabilities of the thin segment (to water only) and thick segment (to salts only) are also important to this mechanism. Without the action of this countercurrent multiplier, the collecting duct would be unable to absorb water and produce such concentrated urine.

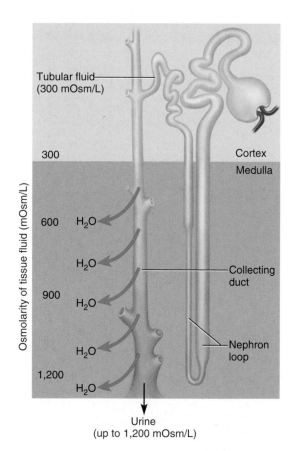

Figure 16.11 Water Reabsorption by the Collecting Duct. Note that the osmolarity of the tissue fluid increases fourfold from 300 mOsm/L in the cortex to 1,200 mOsm/L deep in the medulla. Urine concentration increases proportionately as water leaves the duct through its aquaporins.

Control of Water Loss

Just how concentrated the urine becomes depends on the body's state of hydration. If you are dehydrated, you produce scanty and very concentrated urine. Your high blood osmolarity stimulates the release of **antidiuretic hormone (ADH)** from the posterior pituitary gland. ADH induces the cells of the collecting duct to synthesize *aquaporins* (water-channel proteins) and install them in the plasma membrane, so more water can pass through the epithelium. More water is reabsorbed and is carried away by the surrounding blood capillaries *(vasa recta)*. Urine output is consequently reduced, and its osmolarity can be as high as 1,200 mOsm/L—four times the osmolarity of the blood plasma.

By contrast, if you drink a large volume of water, you soon produce a large volume of hypotonic urine. Hydration inhibits ADH secretion, and the tubule cells remove aquaporins from the plasma membrane. The cortical portion of the CD reabsorbs NaCl, but the CD is less permeable to water. More water passes through as abundant, dilute urine, with an osmolarity as low as 50 mOsm/L. This response is called *water diuresis*[13] (DY-you-REE-sis).

To summarize what we have studied in this section, the collecting duct can adjust water reabsorption to produce urine as hypotonic (dilute) as 50 mOsm/L or as hypertonic (concentrated) as 1,200 mOsm/L, depending on the body's need for water conservation or removal. In a state of hydration, ADH is not secreted and the cortical part of the CD reabsorbs salt without reabsorbing water; the water remains to be excreted in the dilute urine. In a state of dehydration, ADH is secreted, the medullary part of the CD reabsorbs water, and the urine is more concentrated. The CD is able to do this because it passes through a salinity gradient in the medulla from 300 mOsm/L near the cortex to 1,200 mOsm/L near the papilla. This gradient is produced by a countercurrent multiplier of the nephron loop, which concentrates NaCl in the lower medulla. The vasa recta are arranged in such a way that they can supply blood to the medulla without subtracting from its salinity gradient.

---

[13] *diuresis* = passing urine

Before You Go On

Answer these questions from memory. Reread the preceding section if there are too many you don't know.

14. *Explain why the osmotic gradient of the renal medulla is necessary for the ability of the collecting duct to concentrate the urine.*

15. *Discuss the role of the nephron loop in maintaining that osmotic gradient.*

16. *Predict how a lack of ADH would affect the volume of urine, and explain why.*

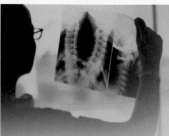

PERSPECTIVES ON HEALTH

Renal Insufficiency and Dialysis

Renal insufficiency is a state in which the kidneys cannot maintain homeostasis due to extensive destruction of their nephrons. Some causes of nephron destruction include the following:

- Hypertension
- Atherosclerosis of the renal arteries
- Chronic or repetitive kidney infections
- Trauma from such causes as blows to the lower back or continual vibration from machinery
- Prolonged ischemia and hypoxia, as in long-distance runners and swimmers
- Poisoning by heavy metals and solvents
- Blockage of renal tubules with proteins small enough to be filtered by the glomerulus—for example, myoglobin released by skeletal muscle damage and hemoglobin released by a transfusion reaction
- Glomerulonephritis, an autoimmune disease of the glomerular capillaries

Nephrons can regenerate and restore kidney function after short-term injuries. Even when some of the nephrons are irreversibly destroyed, others hypertrophy and compensate for their lost function.

Indeed, a person can survive on as little as one-third of one kidney. When 75% of the nephrons are lost, however, urine output may be as low as 30 mL/h compared with the normal rate of 50 to 60 mL/h. This is insufficient to rid the body of toxic substances, and can progress to a dangerous state called *uremia*. Convulsions, coma, and death can follow within a few days. Renal insufficiency also tends to cause anemia because the diseased kidneys produce too little erythropoietin (EPO), the hormone that stimulates red blood cell formation.

Hemodialysis is a procedure for artificially clearing wastes from the blood when the kidneys are not adequately doing so. Blood is pumped from the radial artery to a *dialysis machine* (artificial kidney) and returned to the patient by way of a vein. In the dialysis machine, the blood flows through a semipermeable cellophane tube surrounded by dialysis fluid. Urea, potassium, and other solutes that are more concentrated in the blood than in the dialysis fluid diffuse through the membrane into the fluid, which is discarded. Glucose, electrolytes, and drugs can be administered by adding them to the dialysis fluid so they will diffuse through the membrane into the blood. People with renal insufficiency accumulate substantial amounts of body water between treatments, and dialysis serves also to remove this excess water. Patients are typically given EPO to compensate for its lack of production by the kidneys.

16.6 Urine Storage and Elimination

Expected Learning Outcomes

When you have completed this section, you should be able to:

a. describe the functional anatomy of the ureters, urinary bladder, and male and female urethra;

b. explain how the nervous system and urethral sphincters control the voiding of urine; and

c. describe some physical and chemical properties of urine.

Urine is produced continually, but fortunately it does not drain continually from the body. Urination is episodic—occurring when we allow it. This is made possible by an apparatus for storing urine and by neural controls for its timely release.

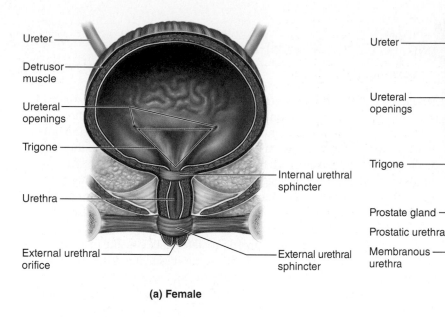

(a) Female

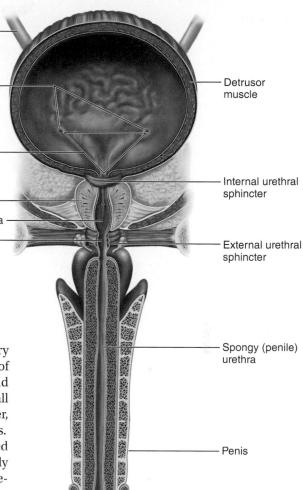

(b) Male

Figure 16.12 The Urinary Bladder and Urethra. Frontal sections. **AP|R**

- *Why are women more susceptible than men to bladder infections?*

The Ureters

The renal pelvis funnels urine into the ureter, a tube that extends to the urinary bladder. The ureter is about 25 cm long and reaches a maximum diameter of about 1.7 cm near the bladder. The ureters pass posterior to the bladder and enter it from below, pierce its muscular wall, and open onto its floor. A small flap of mucosa acts as a valve at the opening of each ureter into the bladder, preventing urine from backing up into the ureter when the bladder contracts.

The ureter consists mostly of a thick tube of smooth muscle, surrounded externally by a sheath of loose fibrous connective tissue that binds it to the body wall, and lined internally by a mucosa with a transitional epithelium, an epithelial type unique to the urinary tract (see p. 107). When urine enters the ureter and stretches it, the muscular wall contracts and initiates a peristaltic wave that milks the urine from the renal pelvis down to the bladder. These contractions occur every few seconds to every few minutes, proportional to the rate at which urine enters from above. The lumen of the ureter is very narrow and is easily obstructed or injured by kidney stones (see Clinical Application 16.1).

The Urinary Bladder

The urinary bladder (fig. 16.12) is a muscular sac on the floor of the pelvic cavity posterior to the pubic symphysis. Its muscle layer, called the **detrusor**[14] (deh-TROO-zur), consists of three layers of smooth muscle. The openings of the two ureters and the urethra mark a smooth triangular area called the *trigone*[15] (TRY-goan) on the bladder floor. This is a common site of bladder infection (see Clinical Application 16.2).

The bladder is highly distensible. As it fills, it expands superiorly, the wrinkled mucosa flattens, and the epithelium thins from five or six cell layers to only two or three. A moderately full bladder contains about 500 mL of urine and extends about 12.5 cm from top to bottom. The maximum capacity is 700 to 800 mL.

The Urethra

The urethra conveys urine from the urinary bladder to the **external urethral orifice,** the point of exit from the body. In the female, it is a tube 3 to 4 cm long

---

[14] *de* = down; *trus* = push

[15] *tri* = three; *gon* = angle

Clinical Application 16.1

KIDNEY STONES

A *renal calculus*[16] (kidney stone) is a hard granule composed usually of calcium phosphate or calcium oxalate. Renal calculi form in the renal pelvis and are usually small enough to pass unnoticed in the urine flow. Some, however, grow as large as several centimeters and block the renal pelvis or ureter, which can lead to the destruction of nephrons as pressure builds in the kidney. A large, jagged calculus passing down the ureter stimulates strong contractions that can be excruciatingly painful. It can also damage the ureter and cause *hematuria* (blood in the urine). Causes of renal calculi include hypercalcemia, dehydration, pH imbalances, frequent urinary tract infections, or an enlarged prostate gland causing urine retention. Calculi are sometimes treated with stone-dissolving drugs, but often they require surgical removal. A nonsurgical technique called *lithotripsy*[17] uses ultrasound to pulverize the calculi into fine granules easily passed in the urine.

bound to the anterior wall of the vagina by fibrous connective tissue. Its orifice lies between the vaginal orifice and clitoris. The male urethra is about 18 cm long and passes through the penis. It has three regions—*prostatic, membranous,* and *spongy urethra*—detailed in relation to sexual function in chapter 19.

In both sexes, the detrusor is thickened near the urethra to form an **internal urethral sphincter,** which compresses the urethra and retains urine in the bladder. Since this sphincter is composed of smooth muscle, it is under involuntary control. Where the urethra passes through the pelvic floor, it is encircled by an **external urethral sphincter** of skeletal muscle, which provides voluntary control over the voiding of urine.

Voiding Urine AP|R

When the bladder is filling, the sympathetic nervous system normally prevents premature leakage of urine. It issues nerve fibers from the spinal cord to the detrusor to *relax* that muscle, preventing bladder contraction, and fibers to the internal urethral sphincter to *contract* that muscle and hold the urine in. The external urethral sphincter, being composed of skeletal muscle, is innervated by somatic rather than sympathetic motor nerve fibers. They, too, hold the sphincter closed.

The act of urinating, called **micturition**[18] (MIC-too-RISH-un), is initiated by an autonomic spinal *micturition reflex.* Filling of the bladder excites stretch receptors in the bladder wall. They issue signals to the spinal cord, which returns signals to the bladder via parasympathetic nerves. Parasympathetic stimulation contracts the detrusor and relaxes the internal urethral sphincter. In very young children and in people with spinal cord injuries that disconnect the brain from the lower spinal cord, this alone results in emptying of the bladder.

Normally, however, one also has voluntary control over urination. Stretch signals from the bladder ascend the spinal cord to a nucleus in the pons called the **micturition center.** This nucleus integrates information about bladder tension with information from other brain centers such as the cerebral cortex.

[16] *calc* = calcium, stone; *ul* = little

[17] *litho* = stone; *tripsy* = crushing

[18] *mictur* = to urinate

Clinical Application 16.2

URINARY TRACT INFECTION

Infection of the urinary bladder is called *cystitis*.[19] It is especially common in females because bacteria such as *Escherichia coli* can travel easily from the perineum up the short urethra. Because of this risk, young girls should be taught never to wipe the anus in a forward direction. Cystitis is frequently triggered in women by sexual intercourse ("honeymoon cystitis"). If cystitis is untreated, bacteria can spread up the ureters and cause *pyelitis*,[20] infection of the renal pelvis. If it reaches the renal cortex and nephrons, it is called *pyelonephritis*. Kidney infections can also result from invasion by blood-borne bacteria. Urine stagnation due to renal calculi or prostate enlargement increases the risk of infection.

Thus, urination can be inhibited by knowledge that the circumstances are inappropriate. If it is desirable and appropriate to urinate, signals descend from the pons to inhibit the spinal somatic neurons that keep the external urethral sphincter constricted. That sphincter then relaxes, and micturition proceeds. If the bladder is not full enough to trigger the micturition reflex, but one wishes to "go" anyway because of a long drive or lecture coming up, the Valsalva maneuver (p. 517) can be used to compress the bladder and excite the stretch receptors early, thereby getting the reflex started.

Urine Volume and Properties

An average adult produces 1 to 2 L of urine per day. An output in excess of 2 L/day is called **diuresis** or **polyuria**[21] (POL-ee-YOU-ree-uh). Fluid intake and some drugs can temporarily increase output to as much as 20 L/day. Diseases such as diabetes may be characterized by chronic polyuria. Low urine output, called **oliguria**,[22] can result from kidney disease, dehydration, circulatory shock, and prostate enlargement, among other causes. If urine output drops to less than 400 mL/day, unsafe concentrations of wastes accumulate in the blood plasma. **Uremia** is a condition of dangerously high nitrogenous waste levels in the blood, usually indicating kidney failure.

The basic composition and properties of urine are as follows:

- **Appearance.** Urine varies from almost colorless to deep amber, depending on the body's state of hydration. The yellow color of urine is due to a pigment produced by the breakdown of hemoglobin from expired erythrocytes. Other unusual colors can be imparted to it by certain foods, vitamins, drugs, and metabolic diseases. **Pyuria**[23] (pus in the urine) can indicate a kidney infection. **Hematuria** (blood in the urine) may be due to a urinary tract infection, trauma, or kidney stones.

- **Odor.** Fresh urine has a distinctive but not repellent odor. As it stands, however, bacteria multiply, degrade urea to ammonia, and produce the pungent odor typical of stale wet diapers. Asparagus, other foods, and some metabolic diseases can impart distinctive aromas to the urine. Diabetes mellitus gives it a sweet, "fruity" odor of acetone.

---

[19] *cyst* = bladder; *itis* = inflammation

[20] *pyel* = pelvis; *itis* = inflammation

[21] *poly* = many, much; *ur* = urine; *ia* = condition

[22] *olig* = few, a little; *ur* = urine; *ia* = condition

[23] *py* = pus; *ur* = urine; *ia* = condition

Table 16.1 Properties and Composition of Urine

Physical Properties		
Specific gravity	1.001–1.028	
Osmolarity	50–1,200 mOsm/L	
pH	6.0 (range 4.5–8.2)	

Solutes	Concentration (mg/dL)*	Output (g/day)**
Urea	1,800	21
Creatinine	150	1.8
Ammonia	60	0.68
Chloride	533	6.4
Sodium	333	4.0
Potassium	166	2.0
Phosphate	83	1
Calcium	17	0.2
Magnesium	13	0.16

* Typical values for a young adult male
** Assuming a urine output of 1.2 L/day

- **Specific gravity.** This is a ratio of the density (g/mL) of a substance to the density of distilled water. Distilled water has a specific gravity of 1.000 by definition, and urine ranges from 1.001 when it is very dilute to 1.028 when it is very concentrated.
- **Osmolarity.** Urine can have an osmolarity as low as 50 mOsm/L in a very hydrated person or as high as 1,200 mOsm/L in a dehydrated person. Compared with the osmolarity of blood (300 mOsm/L), then, urine can be either hypotonic or hypertonic under different conditions.
- **pH.** The pH of urine ranges from 4.5 to 8.2 but is usually about 6.0 (mildly acidic).
- **Chemical composition.** Urine averages 95% water and 5% solutes by volume; table 16.1 lists several of the major solutes. It is abnormal to find glucose, free hemoglobin, albumin, or bile pigments in the urine; their presence can serve as important indicators of disease.

Apply What You Know

Predict the effect of dehydration on the specific gravity of urine.

Before You Go On

Answer these questions from memory. Reread the preceding section if there are too many you don't know.

17. *Where does the ureter begin and end? What prevents contraction of the bladder from forcing urine back up to the kidneys?*

18. *Compare and contrast the functions and neural control of the internal and external urethral sphincters.*

19. *Define polyuria, hematuria, and uremia and identify one possible cause of each.*

16.7 Fluid, Electrolyte, and Acid–Base Balance

Expected Learning Outcomes

When you have completed this section, you should be able to:

a. name the major fluid compartments;

b. list the body's sources of water and routes of water loss;

c. describe mechanisms of regulating water intake and output;

d. list the functions of sodium and potassium;

e. explain how electrolyte balance is regulated; and

f. describe three ways the body regulates pH.

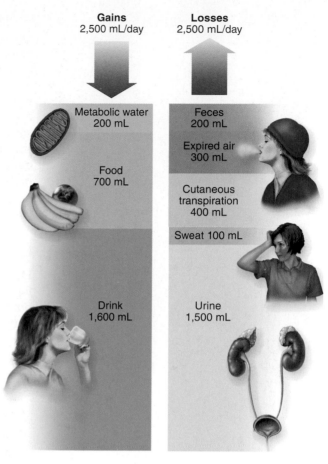

Figure 16.13 Typical Water Gains and Losses in a State of Fluid Balance.

- *How would the right side of this figure change if the person had an abnormally high rate of ADH secretion?*

In the course of producing urine and eliminating metabolic wastes, the kidneys play a vital role in the homeostatic regulation of the body's fluid, electrolyte, and acid–base balance. Imbalances in these variables can easily have life-threatening consequences. They are closely regulated by the collaborative effort of the urinary, respiratory, and digestive systems, coordinated by nervous and hormonal mechanisms.

Fluid Balance

Water makes up 55% to 60% of body mass in the average adult. Variation is related to amounts of lean body mass and fat; lean mass is mostly skeletal muscle, which contains more water than adipose tissue. Women on average have more adipose tissue than men, and therefore usually have less water relative to total body mass.

Fluids are distributed between two main areas, or **fluid compartments.** About two-thirds of the body water is in the **intracellular compartment,** within the cells *(intracellular fluid).* This compartment is separated from the **extracellular compartment** *(extracellular fluid)* by plasma membranes. The latter includes all the fluids outside cells—mainly *interstitial fluid* within the tissue spaces, *plasma* in the blood vessels, and *lymph* in lymphatic vessels.

> ### Apply What You Know
> Review synovial fluid of the joints (p. 187), cerebrospinal fluid of the brain (p. 295), and aqueous humor of the eye (p. 350) if necessary, and explain whether (and why) you would classify those as belonging to the intracellular or extracellular compartment.

Fluid balance means that on average, our daily water gains and losses are equivalent (about 2,500 mL/day for the average adult) and body water is properly distributed among the fluid compartments (not abnormally pooled somewhere, as in edema). A little of our body water comes from metabolic processes such as aerobic respiration, but for the most part, fluid balance is a matter of the amount we ingest in food and drink, and the amount we lose in various ways (fig. 16.13). Much of our water loss is through urine, fecal moisture, and the breath. We lose water through the skin in two ways—sweating and a nonglandular evaporative loss called *cutaneous transpiration.*

Fluid intake is governed mainly by thirst, which is controlled by mechanisms shown in figure 16.14. Dehydration raises blood osmolarity and reduces blood volume. The rise in osmolarity stimulates neurons called *osmoreceptors* in the hypothalamus, which leads to the sensation of thirst and a desire to drink. Normally, we drink long before there is any significant deficit in blood volume or osmolarity. In a more significant state of dehydration, however, the hormonal mechanisms of angiotensin II and antidiuretic hormone, detailed earlier, come into play to accentuate the thirst and reduce urinary loss of water. Contrary to popular myth, it is not at all necessary to drink 8 glasses (64 ounces) of water per day, or to drink even if you don't feel thirsty, to maintain proper hydration.

Fluid output, as we have seen, is governed mainly by varying urine volume and osmolarity, and that is regulated mainly by antidiuretic hormone.

Electrolyte Balance

Electrolytes are salts that dissociate into free ions in water. They are essential for a tremendous variety of cellular functions, such as nerve and muscle action potentials, bone building, enzyme activation, pH balance, osmotic effects on fluid balance, and more.

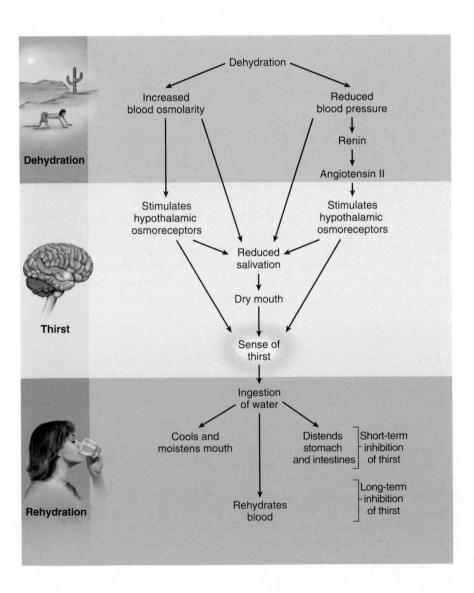

Figure 16.14 Dehydration, Thirst, and Rehydration.

The major cations from electrolytes are sodium (Na^+), potassium (K^+), and calcium (Ca^{2+}), and the major anions are chloride (Cl^-), bicarbonate (HCO_3^-), and phosphates (P_i). In a state of **electrolyte balance,** each of these is within its own normal range of concentration in the body fluids. Electrolyte balance is achieved mainly through equivalent daily rates of absorption by the intestines and excretion by the kidneys. Calcium homeostasis was explained in chapter 6, and anion homeostasis tends to come about automatically as an effect of cation regulation, so here we will focus on sodium and potassium.

Electrolyte concentrations differ in the intracellular and extracellular compartments. Sodium is the most abundant extracellular cation, and potassium is the most abundant in the intracellular fluid. These two ions play pivotal roles in resting and action potentials, and sodium is the most significant solute in determining the distribution of water among fluid compartments, because water follows sodium through the process of osmosis.

Both sodium and potassium balance are regulated by aldosterone. Aldosterone promotes Na^+ reabsorption and K^+ excretion by the kidneys. A rise in aldosterone secretion thus results in less Na^+ but more K^+ in the urine. When the aldosterone level is especially high, the urine can be sodium-free. Aldosterone has little effect on plasma sodium *concentration,* however, because when Na^+ is reabsorbed, a proportionate amount of water is reabsorbed along with it; more fluid is retained, but its concentration is unchanged. Natriuretic peptides, on the other hand, promote sodium and water excretion and lower the blood volume.

Electrolyte imbalances are named with the prefixes *hyper-* and *hypo-* to indicate concentrations above or below their normal ranges. *Hyponatremia,* a sodium deficiency, often arises among athletes who lose proportionate amounts of water and sodium through copious sweating but compensate by drinking plain water. The resulting dilution of sodium in the body fluids can have consequences ranging from weakness and confusion to death. Such deaths happen all too often among high-school and college athletes whose coaches are insufficiently vigilant about electrolyte balance. *Hypernatremia,* a sodium excess, is less common but can result from badly managed administration of I.V. fluids. Potassium imbalances are even more dangerous. *Hypokalemia,* a potassium deficiency, can result from chronic vomiting or diarrhea, excessive use of laxatives, or aldosterone hypersecretion. Symptoms include muscle weakness and cardiac arrhythmia. *Hyperkalemia,* a potassium excess, can set in rapidly when a crush injury (such as getting an arm caught in farm or factory machinery) ruptures large numbers of cells and releases intracellular K^+ into the extracellular fluids, or it can have a slower onset in cases of aldosterone hyposecretion or renal failure. Muscle cramps, diarrhea, vomiting, and atrial fibrillation are associated with hyperkalemia. Cardiac arrest can result from either condition. Indeed, deliberate cardiac arrest from hyperkalemia is part of the method used by some states for execution of criminals by lethal injection. Clinical administration of I.V. potassium must be done with extreme care for such reasons.

Acid–Base Balance

Our life-sustaining metabolic pathways are directly or indirectly dependent on enzymes, and enzymes, being proteins, are very pH-sensitive. Conditions slightly too acidic or basic can cause life-threatening disruption of enzyme function. Consequently, the pH of our blood and tissue fluid is maintained within a very narrow margin of safety—7.35 to 7.45. **Acid–base balance,** or **pH balance,** means the regulation of blood pH within this range.

A pH above this range is called **alkalosis,** and a pH below this range is called **acidosis.** Alkalosis causes nerve and muscle cells to be hyperexcitable and can lead to consequences as severe as convulsions and respiratory paralysis. Acidosis depresses the nervous system and can lead to coma and death. Among diabetics who do not manage their condition well, acidosis is a common cause of death. Our homeostasis is constantly challenged by acids of metabolic origin—lactic acid from anaerobic fermentation, carbonic acid from carbon dioxide, and fatty acids from fat catabolism. Acidosis is therefore more common than alkalosis.

The body regulates pH in three ways: chemical buffers, the respiratory system, and the urinary system.

Chemical Buffers

A chemical buffer is a mixture that binds excess H^+ and removes it from solution where there is an excess (low pH), or releases H^+ into solution to correct a deficit (high pH). Buffer solutions therefore tend to have a stable pH even if challenged by the addition of more acid or base. For example, bicarbonate ions (HCO_3^-) and the amino groups ($-NH_2$) of proteins can bind and neutralize H^+, thereby raising pH, as follows:

$$H^+ + HCO_3^- \rightarrow H_2CO_3$$

$$H^+ + -NH_2 \rightarrow -NH_3^+$$

Conversely, if pH rises too high, carbonic acid (H_2CO_3) and the carboxyl groups ($-COOH$) of proteins can liberate hydrogen ions and lower the pH:

$$H_2CO_3 \rightarrow HCO_3^- + H^+$$

$$-COOH \rightarrow -COO^- + H^+$$

Bicarbonate, proteins, and phosphates are the most significant chemical buffers in the body. Chemical buffers can restore and stabilize normal pH in a fraction of a second, but they are relatively limited in the amount of excess acid or base they can neutralize.

The Respiratory Buffer

The respiratory system takes longer than chemical buffers to act—a few minutes—but it can neutralize more acid or base than the chemical buffers alone can do. It depends on the relationship between carbon dioxide and pH, and the ability to vary the breathing rhythm so we retain more CO_2 in the body or expel it faster.

As we saw in chapter 15, carbon dioxide reacts with water to release acid and lower the pH:

$$CO_2 + H_2O \rightarrow H_2CO_3 \rightarrow HCO_3^- + H^+$$

A CO_2 excess in the body therefore tends to cause acidosis, a common occurrence in lung diseases such as emphysema where the body cannot expel CO_2 as fast as it is made. On the other hand, if there is a CO_2 deficiency in the body fluids, these reactions happen in reverse to raise the CO_2 level but lower the H^+ concentration, thus raising pH:

$$HCO_3^- + H^+ \rightarrow H_2CO_3 \rightarrow CO_2 + H_2O$$

We also saw in chapter 15 that the brainstem respiratory centers are very sensitive to pH changes. Acidosis stimulates one to breathe faster to expel

CO_2 faster than it is produced, thus raising the pH to normal. Alkalosis stimulates one to breathe more slowly, thus accumulating CO_2 and lowering the pH to normal.

The Renal Buffer

Of all buffering mechanisms, the kidneys are slowest to act and require several hours or even days to correct a pH imbalance, but they can rid the body of more acid than the chemical or respiratory buffer system can. The mechanisms are complex and beyond the scope of this book, but the essence of it is this: When there is an excess of acid (H^+) in the blood, it reacts with blood bicarbonate ions to form carbonic acid, which breaks down to H_2O and CO_2, like the last of the reactions shown in the previous section. These move from the blood into the cells of the renal tubule. These epithelial cells have an enzyme that reverses this reaction and generates HCO_3^- and H^+ again. The HCO_3^- is returned to the blood and the H^+ is secreted into the tubular fluid—a case of the *tubular secretion* phenomenon described earlier in this chapter. Thus, H^+ is removed from the blood and appears in the tubular fluid, although indirectly through the carbonic acid reactions. In the tubular fluid, some of this H^+ reacts with bicarbonate ions to again produce H_2O and CO_2, and some remains as free H^+, giving the urine its characteristic mildly acidic pH, about 6.0.

Ammonia (NH_3) in the tubular fluid also can react with H^+ and chloride ions to produce ammonium chloride, NH_4Cl. This mechanism can come to the aid of acid–base balance when the body is under a particularly heavy acid challenge, and the urine then shows up to 10 times as much ammonium chloride as usual. A high urine ammonium chloride level can be a sign of an acid-generating disease such as diabetes mellitus.

Before You Go On

Answer these questions from memory. Reread the preceding section if there are too many you don't know.

20. What are the major fluids of the extracellular compartment?

21. What is the most important mechanism for regulating fluid intake? What hormone is most important in regulating fluid output?

22. What two organ systems are most important in regulating electrolyte balance?

23. What are the terms for sodium and potassium excess and deficiency? Which of the four conditions has the potential to be most quickly lethal, and by what mechanism?

24. What hormones are most important in regulating sodium balance? Describe their actions.

25. What are the terms for excessively high and low pH? State some causes and effects of each.

26. Describe the essence of how chemical buffers, the respiratory system, and the kidneys resist or compensate for shifts in body fluid pH.

Aging of the Urinary System

Nephrons are abundant early in life, but the number declines markedly with age. By an age of 85 to 90, there can be as much as a 40% reduction in the number of functional nephrons and size of the kidneys. In addition, the renal blood supply and GFR are significantly reduced. The result is that although the aged kidneys function adequately under normal conditions, they cope less efficiently with illness and in clearance of medications. In men, an enlarged prostate may compress the urethra and make it more difficult to void urine. In women, strain on the pelvic floor muscles during childbirth makes incontinence more likely later in life.

CAREER SPOTLIGHT

Dialysis Technician

Dialysis technicians support the care of patients with kidney failure who require dialysis. The job entails setting up and operating hemodialysis machines. In addition, the technician prepares the patient for dialysis and monitors both the patient and the machine during the process. The technician must learn the scientific principles of dialysis, understand the equipment involved, and respond to physical and emotional needs of the person undergoing treatment. Dialysis technicians work closely with a team of nurses and physicians who care for the patient. They work at hemodialysis facilities in hospitals and outpatient facilities, or may assist patients with in-home dialysis. Training is provided on the job or through vocational schools or community colleges. Dialysis technicians must pass a state exam to become certified. See appendix B for further information on dialysis technician careers.

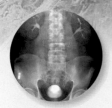

CONNECTIVE ISSUES

Ways in Which the Urinary System Affects Other Organ Systems

All Systems
The urinary system serves all other systems by eliminating metabolic wastes and maintaining fluid, electrolyte, and acid–base balance.

Integumentary System
Renal control of fluid balance is essential for sweat secretion; the epidermis is normally a barrier to fluid loss—thus, the skin and kidneys collaborate to maintain fluid balance.

Skeletal System
Renal control of calcium and phosphate balance and role in calcitriol synthesis are essential for bone deposition.

Muscular System
Renal control of Na^+, K^+, and Ca^{2+} balance is important for muscle contraction.

Nervous System
The nervous system is very sensitive to fluid, electrolyte, and acid–base imbalances that may result from renal dysfunction.

Endocrine System
Renin secretion by kidneys leads to angiotensin synthesis and aldosterone secretion; the kidneys produce erythropoietin.

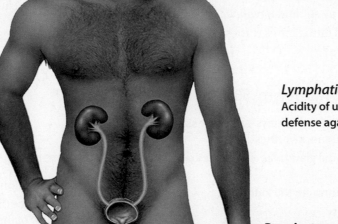

Circulatory System
Kidneys control blood pressure more than any other organ; erythropoietin from kidneys regulates hematocrit; kidneys regulate plasma composition.

Lymphatic and Immune Systems
Acidity of urine provides nonspecific defense against infection.

Respiratory System
Urinary and respiratory systems collaborate to regulate acid–base balance.

Digestive System
Kidneys excrete toxins absorbed by digestive tract; kidneys excrete hormones and metabolites after liver deactivates them; calcitriol synthesized by kidneys regulates Ca^{2+} absorption by small intestine.

Reproductive System
Urethra serves as common passageway for urine and semen in males; the urinary system of pregnant woman eliminates metabolic wastes of fetus.

Study Guide

Assess Your Learning Outcomes

To test your knowledge, discuss the following topics with a study partner or in writing, ideally from memory.

16.1 Functions of the Urinary System (p. 530)

1. The principal organs of the urinary system
2. Seven functions of the kidneys
3. Definitions of *excretion, waste,* and *metabolic waste*
4. The types and sources of nitrogenous wastes

16.2 Anatomy of the Kidney (p. 532)

1. The anatomical position of the kidneys, and their size, shape, and connective tissue coverings
2. Internal organization of the glandular and collecting tissues of the kidney
3. The route of blood flow through the kidney
4. Structure of the nephron
5. The route of fluid flow through the nephron as it becomes urine

16.3 Glomerular Filtration (p. 537)

1. The four stages in urine formation, and the changing names of the fluid at various stages
2. The meaning of *glomerular filtration* and the site of its occurrence
3. Components of the filtration membrane and how they affect the composition of the filtrate
4. Factors that determine filtration pressure and glomerular filtration rate (GFR)
5. The mechanism of sympathetic nervous control of glomerular filtration
6. Hormones that influence glomerular filtration, and their mechanisms of action
7. Mechanisms of local control by which a nephron regulates its own glomerular filtration
8. Structure of the juxtaglomerular apparatus and how it regulates GFR

16.4 Tubular Reabsorption and Secretion (p. 543)

1. Definitions of *tubular reabsorption* and *tubular secretion*
2. Structural adaptations of the proximal convoluted tubule (PCT) for its role in absorption
3. Substances reabsorbed by the PCT, the mechanisms of reabsorption, and how sodium reabsorption drives the others
4. Where the reabsorbed materials go after they leave the renal tubule
5. The concept of the transport maximum and how it relates to diabetes mellitus
6. Structure of the nephron loop and how it relates to tubular reabsorption
7. Substances that are reabsorbed by the distal convoluted tubule (DCT)
8. Hormones that influence DCT function, and their modes of action
9. The role of tubular secretion in the composition of the urine

16.5 Water Conservation (p. 546)

1. The basic function of the collecting duct
2. The significance of tissue fluid osmolarity and selective permeability of the collecting duct in urine formation and water conservation
3. How the nephron loop creates the osmotic gradient of the renal medulla necessary for water conservation
4. Hormonal control of collecting duct function and urine volume and concentration

16.6 Urine Storage and Elimination (p. 548)

1. Anatomy and peristaltic function of the ureters
2. Anatomy of the urinary bladder, its spatial relationship with the ureters and urethra, and the changes it undergoes as it fills and stretches
3. Structure of the female and male urethra
4. The two urethral sphincters, their locations, and differences in their muscular composition and nervous control
5. Components and actions of the spinal micturition reflex
6. The mechanism of voluntary control over micturition
7. Typical daily output of urine, factors that influence it, and consequences of insufficient output
8. Physical and chemical properties of normal urine

16.7 Fluid, Electrolyte, and Acid–Base Balance (p. 553)

1. Typical body water content and reasons for its variability
2. The body's fluid compartments and how they differ
3. The meaning of *fluid balance*
4. Major sources of water and means of water loss
5. The principal factors that regulate water intake and output
6. The meanings of *electrolytes* and *electrolyte balance*
7. The most physiologically important ions that arise by electrolyte dissociation in water
8. Hormonal control of sodium and potassium balance
9. The names and pathological consequences of sodium and potassium deficiencies and excesses
10. The meaning of *acid–base balance* and why it is crucial to one's metabolic pathways
11. The normal range of blood pH and terms for deviations above and below this range
12. Three major chemical buffer systems in the body and examples of how they stabilize pH
13. How variations in respiration can adjust body pH
14. How the kidneys excrete acid and adjust body pH
15. Relative importance, speed, and effectiveness of the chemical, respiratory, and renal buffer systems

Testing Your Recall

1. Micturition occurs when the _____ contracts.
   a. detrusor
   b. internal urethral sphincter
   c. external urethral sphincter
   d. muscular wall of the ureter
   e. muscular wall of the urethra

2. The compact ball of capillaries in a nephron is called
   a. the nephron loop.
   b. the peritubular plexus.
   c. the renal corpuscle.
   d. the glomerulus.
   e. the vasa recta.

3. Which of these is the most abundant nitrogenous waste in the blood?
   a. uric acid
   b. urea
   c. ammonia
   d. creatinine
   e. albumin

4. Which of these is not a part of the nephron?
   a. proximal convoluted tubule
   b. collecting duct
   c. distal convoluted tubule
   d. nephron loop
   e. glomerulus

5. What percentage of glomerular filtrate is reabsorbed by the proximal convoluted tubule?
   a. 6%
   b. 25%
   c. 45%
   d. 65%
   e. The amount varies depending on a person's state of hydration.

6. A glomerulus and glomerular capsule make up one
   a. renal capsule.
   b. renal corpuscle.
   c. kidney lobule.
   d. kidney lobe.
   e. nephron.

7. Each nephron receives its blood supply directly from
   a. a renal artery.
   b. an interlobar artery.
   c. a segmental artery.
   d. an efferent arteriole.
   e. an afferent arteriole.

8. Determine the correct order of flow through the renal tubule, if we let
   I = collecting duct,
   II = distal convoluted tubule,
   III = glomerular capsule,
   IV = proximal convoluted tubule, and
   V = nephron loop.

   a. IV–III–V–II–I
   b. I–II–IV–III–V
   c. IV–II–V–I–III
   d. III–IV–V–II–I
   e. III–I–IV–V–II

9. Which of the following is/are secreted into the renal tubule?
   a. urea
   b. glucose
   c. bicarbonate ions
   d. sodium chloride
   e. albumin

10. Increased ADH secretion should change the urine in what way?
   a. reduced volume
   b. increased volume
   c. reduced osmolarity
   d. reduced urea concentration
   e. reduced potassium concentration

11. The _____ reflex is an autonomic reflex activated by pressure in the urinary bladder.

12. _____ is the amount of filtrate formed per minute by the two kidneys combined.

13. The two _____ carry urine from the kidneys to the urinary bladder.

14. The _____ is a group of epithelial cells of the nephron loop that monitors the flow or composition of the tubular fluid.

15. To enter the capsular space, filtrate must pass between foot processes of the _____, cells that form part of the filtration membrane in the glomerular capsule.

16. The _____ is the functional unit of the kidney where urine is produced.

17. _____ hormone regulates the amount of water reabsorbed by the collecting duct.

18. The _____ sphincter is under involuntary control and relaxes during the micturition reflex.

19. Aldosterone stimulates the ascending limb of the nephron loop, the DCT, and the collecting duct to reabsorb more _____ and secrete more _____.

20. The _____ tightly wraps the kidney and protects it from trauma and infection.

Answers in appendix A

True or False

Determine which five of the following statements are false, and briefly explain why.

1. Aldosterone has no effect on the proximal convoluted tubule.

2. Sodium is the most abundant solute in the urine.

3. The kidney has more distal convoluted tubules than collecting ducts.

4. Sodium is reabsorbed only in the proximal convoluted tubule.

5. Natriuretic peptides increase blood volume by promoting sodium reabsorption in the kidney.

6. If all other conditions remain the same, constriction of the afferent arteriole reduces the glomerular filtration rate.

7. Angiotensin II raises blood pressure.

8. The minimum osmolarity of urine is 300 mOsm/L, equal to the osmolarity of the blood.

9. The parasympathetic nervous system sends signals that relax the detrusor and stimulate constriction of the internal urethral sphincter.

10. Micturition depends on contraction of the detrusor.

Answers in appendix A

Testing Your Comprehension

1. Animals that live in freshwater, such as muskrats and beavers, have very short nephron loops. In view of the function of the nephron loop in humans, explain why it is less necessary in such aquatic mammals. As a result of this difference, would you expect their urine to be hypotonic or hypertonic to ours? Why?

2. In view of how the human bladder changes as it fills, why is it important that the ureters enter it from below instead of from above?

3. Cholera is a bacterial infection common in impoverished countries and areas stricken by warfare and natural disasters. It causes profuse diarrhea and often death, especially in children. In view of section 16.7 of this chapter, why do you think these children could die in a few days from unrelenting diarrhea? Be as detailed as you can about the physiological mechanisms.

The Digestive System

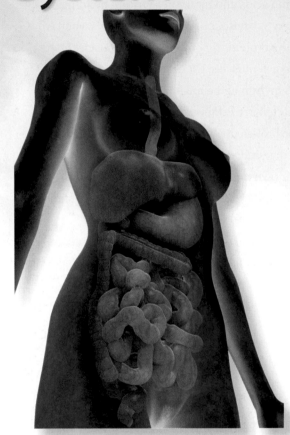

The digestive tract is our longest organ system, extending for about 5 m (16 feet) from end to end. It is a disassembly line that breaks food down into molecules that can be used for human structure and function.

Module 12: Digestive System

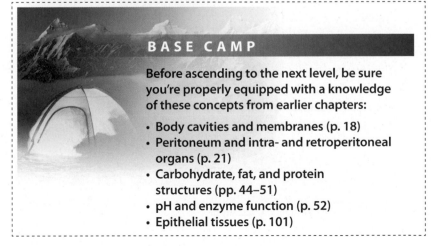

BASE CAMP

Before ascending to the next level, be sure you're properly equipped with a knowledge of these concepts from earlier chapters:

- Body cavities and membranes (p. 18)
- Peritoneum and intra- and retroperitoneal organs (p. 21)
- Carbohydrate, fat, and protein structures (pp. 44–51)
- pH and enzyme function (p. 52)
- Epithelial tissues (p. 101)

Chapter Outline

ating is a fundamental human behavior—required for survival and woven into the fabric of every culture, where celebrations and everyday social life revolve around food. For the first time in the history of a species, our main problem as modern humans is that most of us have access to an overabundance of food. For our ancestors, in contrast, the focus was procuring adequate food to maintain health.

The digestive system breaks down food we eat into smaller components, such as amino acids and monosaccharides, building blocks used by cells of every species. Consider what happens when we eat a steak, for example. The proteins that make up beef muscle differ very little from that of our own muscles. Even if they were identical, however, proteins such as myosin in the meat we eat could not be directly absorbed, transported in the blood, and incorporated into our muscles. Proteins must be broken down into amino acids before our cells can use them. Once digested, the amino acids in the beef proteins might be used to build our own muscle myosin, but could equally wind up in our insulin, collagen, or any other protein. The digestive system is essentially a disassembly line—its primary purpose is to break nutrients down into forms that can be used by the body and to absorb them so they can be distributed to the tissues. The study of the digestive tract and the diagnosis and treatment of its disorders is called **gastroenterology**.[1]

17.1 Overview of the Digestive System

Expected Learning Outcomes

When you have completed this section, you should be able to:

a. list the functions and major physiological processes of the digestive system;

b. distinguish between mechanical and chemical digestion;

c. list the regions of the digestive tract and the accessory organs of the digestive system;

d. describe the histological layers of the digestive tract;

e. describe the special nervous network of the digestive tract; and

f. describe the relationship between the digestive tract, peritoneum, and mesenteries.

---

[1] *gastro* = stomach; *entero* = intestine; *logy* = study of

Digestive System Functions

The **digestive system** processes food, extracts nutrients from it, and eliminates the residue. It does this in five stages:

1. **Ingestion**, the selective intake of food
2. **Digestion**, the mechanical and chemical breakdown of food into a form usable by the body
3. **Absorption**, the uptake of nutrient molecules into the epithelial cells of the digestive tract and then into the blood or lymph
4. **Compaction**, absorption of water and consolidation of the indigestible residue into feces
5. **Defecation**, the elimination of feces

   Mechanical digestion is the physical breakdown of food into smaller particles achieved by the cutting and grinding action of the teeth and the churning movements of the stomach and small intestine. Mechanical digestion exposes more food surface to the action of digestive enzymes. **Chemical digestion** employs enzymes to break dietary macromolecules into their basic building blocks: polysaccharides into monosaccharides, proteins into amino acids, fats into monoglycerides and fatty acids, and nucleic acids into nucleotides. All chemical digestion is by the process of enzymatic hydrolysis, breaking covalent bonds by adding water to them (see p. 43). The digestive enzymes that carry this out are produced by the salivary glands, stomach, pancreas, and small intestine. Some nutrients are already present in usable form in the ingested food and are absorbed without being broken down by digestion: vitamins, minerals, cholesterol, and water.

General Anatomy

The two anatomical subdivisions of the digestive system are the digestive tract and the accessory organs (fig. 17.1). The **digestive tract** (*alimentary*[2] *canal*) is a tube that extends from mouth to anus, and measures about 5 m (16 feet) long in a living person (but longer in the cadaver). It includes the mouth, pharynx, esophagus, stomach, small intestine, and large intestine. Part of the digestive tract, the stomach and intestines, constitutes the *gastrointestinal (GI) tract*. The **accessory organs** are the teeth, tongue, salivary glands, liver, gallbladder, and pancreas. We will first explore the structure of the system as a whole and then we will further explore the structure and functions of each region.

   The digestive tract is essentially a tube open to the environment at both ends. With slight variations from one region to another, most of it follows the same basic structural plan—a wall composed of four main tissue layers, from inner to outer surface (fig. 17.2):

1. The **mucosa (mucous membrane)** lines the lumen. It consists of an inner epithelium, a loose connective tissue layer called the *lamina propria*, and a thin layer of smooth muscle. Simple columnar epithelium

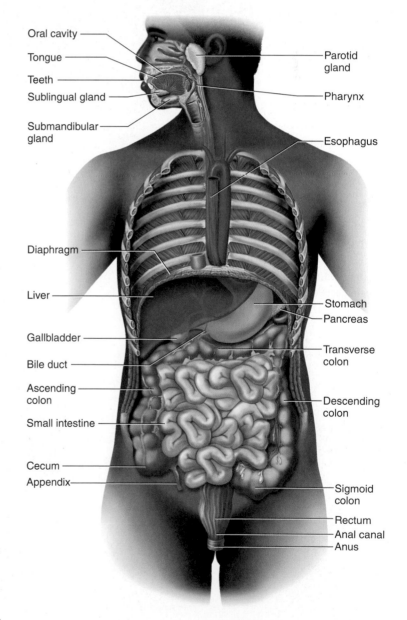

Oral cavity

Tongue

Teeth

Sublingual gland

Submandibular gland

Parotid gland

Pharynx

Esophagus

Diaphragm

Liver

Gallbladder

Bile duct

Ascending colon

Small intestine

Cecum

Appendix

Stomach

Pancreas

Transverse colon

Descending colon

Sigmoid colon

Rectum

Anal canal

Anus

Figure 17.1 The Digestive System. AP|R

---

[2] *aliment* = food

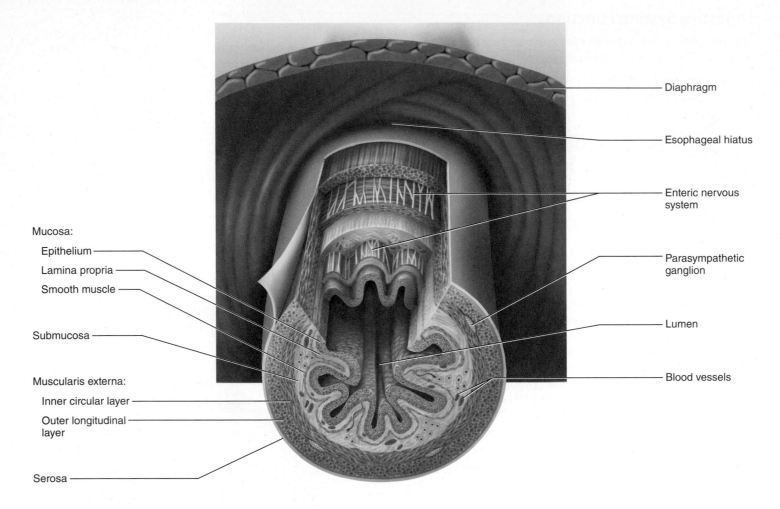

Diaphragm

Esophageal hiatus

Enteric nervous system

Parasympathetic ganglion

Lumen

Blood vessels

Mucosa:

  Epithelium

  Lamina propria

  Smooth muscle

Submucosa

Muscularis externa:

  Inner circular layer

  Outer longitudinal layer

Serosa

Figure 17.2 Tissue Layers of the Digestive Tract. Cross section of the esophagus just below the diaphragm. The layers shown here are typical of most of the digestive tract.

lines the digestive tract from the stomach through the large intestine; stratified squamous epithelium is found from the oral cavity through the esophagus and in the lower anal canal, where the tract is subject to more abrasion. The mucosa often exhibits an abundance of lymphocytes and lymphatic nodules—the *mucosa-associated lymphatic tissue (MALT)* (see p. 476).

2. The **submucosa** is a layer of loose connective tissue containing blood vessels, lymphatic vessels, and nerves.

3. The **muscularis externa** consists of smooth muscle, usually arranged in two layers: a deep layer whose muscle fibers encircle the tract and a more superficial layer whose fibers run longitudinally. The muscularis externa is responsible for the motility that propels food and residue through the digestive tract.

4. A thin **serosa (serous membrane)** covers the outer surface of most of the digestive tract. It consists of areolar tissue topped by a simple squamous epithelium. The serosa begins in the lower 3 to 4 cm of the esophagus and ends just before the rectum. The pharynx, most of the esophagus, and the rectum are covered instead by a fibrous connective tissue layer called the **adventitia** that blends into the connective tissues of adjacent organs.

An independent network of neurons called the **enteric[3] nervous system** lies among the layers of the digestive tract of the esophagus, stomach, and

---

[3] *enter* = intestine

intestines. It regulates digestive tract blood flow, secretion, and movements such as swallowing and intestinal peristalsis. This system is thought to have more neurons than the spinal cord! It can function completely independently of the central nervous system, although the CNS usually exerts a significant influence on it.

The stomach and intestines undergo such strenuous contractions as they process food that they need freedom to move in the abdominal cavity. Thus, they are not tightly bound to the abdominal wall, but rather are loosely suspended from it by connective tissue sheets called **mesenteries** (fig. 17.3; see also fig. 1.13, p. 20). The *parietal peritoneum*, a serous membrane that lines the wall of the abdominal cavity, turns inward and forms the double-layered mesenteries. In organs such as the stomach and small intestine, the two layers of the mesentery separate and pass around opposite sides of the organ, forming the serosa. The mesenteries hold the abdominal viscera in their proper relationship to each other and prevent the small intestine from becoming twisted and tangled by changes in body position and by the intestine's own contractions. They also provide passage for blood vessels and nerves that supply the digestive tract, and contain many lymph nodes and lymphatic vessels.

A mesentery called the *lesser omentum* extends from the right superior margin of the stomach to the liver. A larger mesentery, the *greater omentum*, hangs from the inferior margin of the stomach and loosely covers the small intestine like an apron. The omenta have a loosely organized, lacy appearance due partly to gaps in the membranes and partly to an irregular distribution of fatty tissue.

Some digestive organs such as the stomach and liver are *intraperitoneal*—enclosed by serosa on all sides. Others such as the pancreas and duodenum are *retroperitoneal*—lying against the posterior body wall and covered by peritoneum only on the anterior surface.

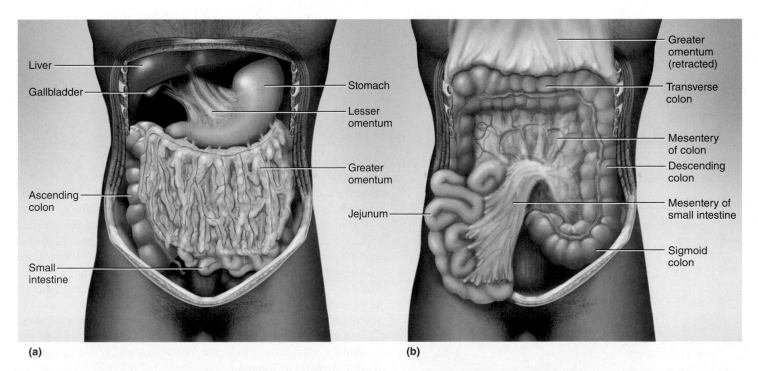

(a)　　　　　　　　　　　　　　　　　　　　(b)

Figure 17.3 Mesenteries of the Digestive Tract. (a) Shown with apron-like greater omentum overhanging the small intestine in its natural position. (b) Greater omentum and small intestine retracted to expose other mesenteries of the small and large intestines. See figures 1.13 and 1.14 for other mesenteries. **AP|R**

17.2 The Mouth Through Esophagus

Expected Learning Outcomes

When you have completed this section, you should be able to:

a. describe the gross anatomy of the digestive tract from the mouth through esophagus;

b. describe the general structure of a tooth, and identify the types of teeth and their respective positions;

c. list the functions of saliva and describe its composition; and

d. describe the process of swallowing.

The Mouth

The mouth is also known as the **oral,** or **buccal** (BUCK-ul), **cavity.** Its functions include ingestion (food intake), taste, chewing, mechanical and chemical digestion, swallowing, speech, and respiration. The mouth is lined with stratified squamous epithelium.

The cheeks, lips, palate, and tongue enclose the mouth (fig. 17.4). The anterior opening between the lips is the *oral fissure* and the posterior opening into the throat is the *fauces*[4] (FAW-seez). The **vestibule** is the space that separates the teeth from the cheeks and lips—the space where you insert your toothbrush to brush the outer surfaces of the teeth. The fleshiness of the lips and cheeks is due mainly to subcutaneous fat and the orbicularis oris and buccinator muscles (see p. 232). The lips and cheeks retain food and are essential for articulate speech and for sucking and blowing actions, including suckling by infants.

The **tongue,** although muscular and bulky, is a remarkably agile and sensitive organ. It manipulates food between the teeth while it

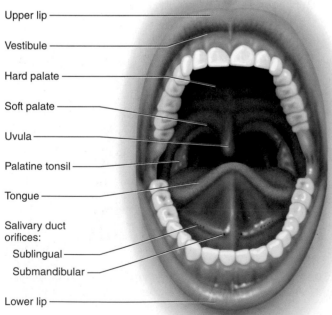

Upper lip

Vestibule

Hard palate

Soft palate

Uvula

Palatine tonsil

Tongue

Salivary duct orifices:

    Sublingual

    Submandibular

Lower lip

Figure 17.4 The Oral Cavity. AP|R

---

[4] *fauces* = throat

avoids being bitten, and it is sensitive enough to feel a stray hair in a bite of food. Its surface is covered with nonkeratinized stratified squamous epithelium and exhibits bumps and projections called *lingual papillae*, the site of the taste buds (see fig. 10.4, p. 335). Its anterior two-thirds, called the *body*, occupies the oral cavity; the posterior one-third, the *root*, occupies the first part of the pharynx. The body is attached to the floor of the mouth by a median fold called the *lingual frenulum.*

The **palate** separates the oral cavity from the nasal cavity, making it possible to breathe while chewing food. Its anterior portion, the *hard (bony) palate*, consists of horizontal plates of the maxillae and palatine bones covered with a thin mucous membrane. The posterior portion, the *soft palate*, has no bone but is more spongy, containing glandular tissue and skeletal muscle. A conical median projection, the *uvula,*[5] is clearly visible at the rear of the mouth.

The teeth are collectively called the **dentition**. They serve to *masticate* the food, breaking it into smaller pieces, making it easier to swallow and speeding up chemical digestion by exposing more surface area for enzyme action. Adults normally have 16 teeth in the mandible and 16 in the maxilla. From the midline to the rear of each jaw, there are two incisors, a canine, two premolars, and up to three molars (fig. 17.5). The chisel-like **incisors** bite off a piece of food, and

---

[5] *uvula* = little grape

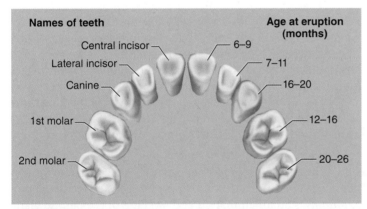

Names of teeth	Age at eruption (months)
Central incisor	6–9
Lateral incisor	7–11
Canine	16–20
1st molar	12–16
2nd molar	20–26

(a) Deciduous (baby) teeth

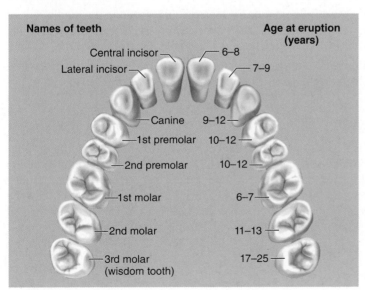

Names of teeth	Age at eruption (years)
Central incisor	6–8
Lateral incisor	7–9
Canine	9–12
1st premolar	10–12
2nd premolar	10–12
1st molar	6–7
2nd molar	11–13
3rd molar (wisdom tooth)	17–25

(b) Permanent teeth

Figure 17.5 The Dentition and Structure of a Tooth and Its Alveolus. (a) The deciduous (baby) teeth and ages at which they emerge from the gums (erupt). (b) The adult teeth and ages of eruption. Each figure shows only the upper teeth. The ages at eruption are composite ages for the corresponding upper and lower teeth. Generally, the lower (mandibular) teeth erupt somewhat earlier than their upper (maxillary) counterparts. (c) Structure of a tooth and periodontal tissues. This particular example is a molar. **AP|R**

- *Which teeth are absent from a 3-year-old child?*

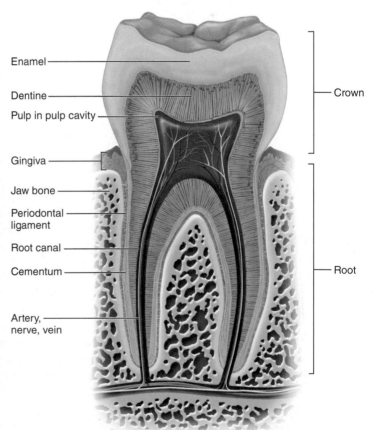

Enamel

Dentine

Pulp in pulp cavity

Gingiva

Jaw bone

Periodontal ligament

Root canal

Cementum

Artery, nerve, vein

Crown

Root

(c)

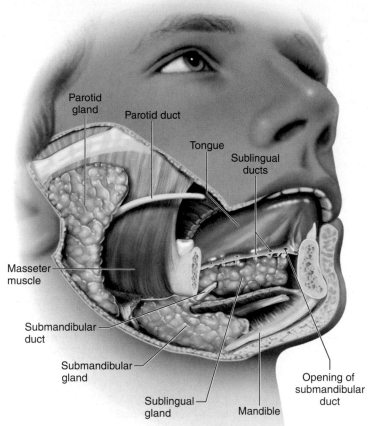

Parotid gland

Parotid duct

Tongue

Sublingual ducts

Masseter muscle

Submandibular duct

Submandibular gland

Sublingual gland

Mandible

Opening of submandibular duct

Figure 17.6 The Salivary Glands. Part of the mandible has been removed to expose the sublingual gland medial to it. **AP|R**

the slightly pointed **canines** puncture and shred it. The relatively broad surfaces of the **premolars** and **molars** are adapted for crushing, shredding, and grinding food.

Regions of a tooth are defined by their relationship to the gum, or **gingiva** (JIN-jih-vuh): the **crown** is the portion above the gum and the **root** is the portion inserted into the tooth socket below the gum. Internally, the **pulp cavity** in the crown and the narrow **root canal** in the root are occupied by **pulp**—a mass of loose connective tissue, blood and lymphatic vessels, and nerves.

Most of a tooth consists of hard yellowish tissue called **dentine**, covered with **enamel** in the crown and neck and **cementum** in the root. Enamel is a nonliving, hardened secretion, whereas dentine and cementum are living connective tissues with cells or cell processes embedded in a calcified matrix. Damaged dentine and cementum can regenerate, but damaged enamel cannot—it must be artificially repaired.

Saliva and Salivary Glands

Saliva is a watery solution of mucus, enzymes, and electrolytes. It moistens the mouth, cleanses the teeth, inhibits bacterial growth, and dissolves molecules so they can stimulate the taste buds. It lubricates food and binds particles together into a soft, slippery, easily swallowed mass. Its *salivary amylase* digests some of the starch in our food as we chew, and some of the fat after we swallow it and the stomach acid activates its fat-digesting *lingual lipase*.

Some of the saliva is produced by glandular tissue embedded in the tongue, lips, and palate, but most of it comes from three pairs of salivary glands outside the oral cavity (fig. 17.6). They produce about 1.0 to 1.5 liters of saliva per day.

1. The **parotid**[6] **glands** are located just beneath the skin anterior to the earlobes. Their ducts open into the mouth opposite each second upper molar tooth. *Mumps* is an inflammation and swelling of the parotid gland caused by a virus.

2. The **submandibular glands** are located halfway along the medial side of the body of the mandible. Their ducts empty into the mouth at a papilla beneath the tongue, near the lower central incisors.

3. The **sublingual glands** are located in the floor of the mouth. They have multiple ducts that empty into the mouth beneath the tongue.

The Pharynx

The pharynx is a muscular funnel that connects the oral cavity to the esophagus. It also connects the nasal cavity to the larynx; thus, it is a point where the digestive and respiratory tracts intersect. It has a deep layer of longitudinally oriented skeletal muscle and a superficial layer of circular skeletal muscle. The circular muscle is divided into superior, middle, and inferior **pharyngeal constrictors**, which force food downward during swallowing.

---

[6]*par* = next to; *ot* = ear

The Esophagus

The **esophagus** is a straight muscular tube 25 to 30 cm long (see fig. 17.1). It begins posterior to the trachea near the inferior border of the larynx. It passes through the mediastinum, penetrates the diaphragm at an opening called the *esophageal hiatus,* then continues another 3 to 4 cm to meet the stomach at the **cardiac orifice** (named for its proximity to the heart). A constriction called the **lower esophageal sphincter** prevents stomach contents from regurgitating into the esophagus, thus protecting the esophageal mucosa from the erosive effect of stomach acid. "Heartburn" has nothing to do with the heart, but is the burning sensation produced by acid reflux into the esophagus.

The wall of the esophagus is organized into the tissue layers described earlier, with some regional specializations (see fig. 17.2). The mucosa has a nonkeratinized stratified squamous epithelium. The submucosa contains mucous glands that aid in swallowing. When the esophagus is empty, the mucosa and submucosa are deeply folded into longitudinal ridges, giving the lumen a starlike shape in cross section. Most of the esophagus is covered with a connective tissue adventitia; the short segment below the diaphragm is covered by a serosa.

Swallowing

Swallowing, or **deglutition** (DEE-glu-TISH-un), is a complex action involving over 22 muscles in the mouth, pharynx, and esophagus. It is coordinated by the **swallowing center** in the medulla oblongata, and occurs in three phases (fig. 17.7).

1. The **oral phase** is under voluntary control. The tongue collects food, presses it against the palate to form a soft cohesive mass called a **bolus**, and pushes it posteriorly. Food accumulates in the oropharynx in front of the blade of the epiglottis. When the bolus reaches a critical size, the epiglottis tips posteriorly and the bolus slides around it through a space on each side, into the laryngopharynx.

2. The **pharyngeal phase** is involuntary and is initiated when food contacts touch receptors of the laryngopharynx. The soft palate and root of the tongue block food from entering the nasal cavity or reentering the mouth. The vocal cords adduct (come together) in the middle, and the larynx is pulled up against the epiglottis, preventing food from entering the larynx and trachea. These actions also widen the upper end of the esophagus to receive the bolus. The pharyngeal constrictors contract in order from superior to inferior to drive the bolus downward into the esophagus. Breathing is automatically suspended for a moment during this phase.

3. The **esophageal phase** is a wave of involuntary contraction called **peristalsis**, controlled jointly by the brainstem swallowing center and the enteric nervous system in the esophageal wall. The esophagus constricts above the bolus and relaxes below it, propelling the bolus along. When you are standing or sitting upright, most food and liquid drop through the esophagus by gravity faster than the peristaltic wave can catch up to it. Peristalsis, however, propels larger food pieces and ensures that you can swallow regardless of the body's position—even standing on your head! A second wave of peristalsis follows if the first wave is insufficient to clear food from the esophagus. Liquid normally reaches the stomach in 1 to 2 seconds and a food bolus in 4 to 8 seconds. As a bolus reaches the lower end of the esophagus, the lower esophageal sphincter relaxes to let it pass into the stomach.

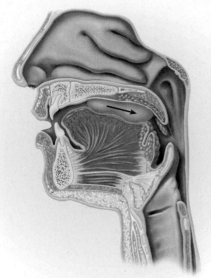

1 **Oral phase.** The tongue forms a food bolus and pushes it into the laryngopharynx.

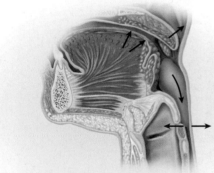

2 **Pharyngeal phase.** The palate, tongue, vocal cords, and epiglottis block the oral and nasal cavities and airway while pharyngeal constrictors push the bolus into the esophagus.

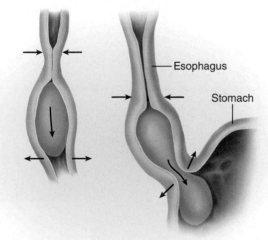

Esophagus

Stomach

3 **Esophageal phase.** Peristalsis drives the bolus downward, and relaxation of the lower esophageal sphincter admits it into the stomach.

Figure 17.7 Swallowing. See numbered steps in the text for further explanation. **AP|R**

• *What actions prevent the pharynx from forcing food back into the mouth or nose?*

Before You Go On

Answer these questions from memory. Reread the preceding section if there are too many you don't know.

6. Identify the diverse functions of the mouth and which parts perform each function.

7. Describe the spatial relationship of dentine, enamel, and cementum in a tooth. Name the four kinds of teeth and state how many of each a person has. Name the major salivary glands, state their locations, and describe the composition of the saliva.

8. Where is the cardiac orifice? Name the sphincter found here and discuss its function.

9. Explain the three stages of swallowing. Which is voluntary and which are involuntary?

17.3 The Stomach

Expected Learning Outcomes

When you have completed this section, you should be able to:

a. describe the gross and microscopic anatomy of the stomach;

b. identify the cell types of the gastric mucosa and their functions;

c. identify the secretions of the stomach and their functions;

d. describe how the nervous system and hormones regulate the stomach; and

e. explain why the stomach doesn't digest itself.

The stomach is a muscular sac in the upper left abdominal cavity immediately inferior to the diaphragm. It functions primarily as a food-storage organ, and holds 1.0 to 1.5 L after a typical meal. When extremely full, it may hold up to 4 L and extend nearly as far as the pelvis.

Well into the nineteenth century, authorities regarded the stomach as essentially a grinding chamber, fermentation vat, or cooking pot. We now know that it mechanically breaks up food particles, liquefies the food, and begins the chemical digestion of proteins and fat. This produces a pasty mixture of semidigested food called **chyme**[7] (pronounced "kime"). Most chemical digestion occurs after the chyme passes on to the small intestine.

Anatomy

The stomach is J-shaped and divided into four regions (fig. 17.8). (1) The *cardiac region* is a small area immediately inside the cardiac orifice. (2) The *fundic region (fundus)* is the superior dome that nestles against the diaphragm. (3) The

---

[7] *chyme* = juice

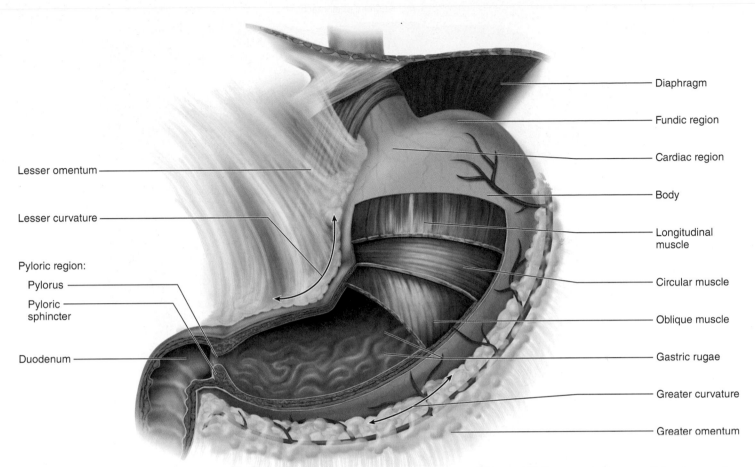

Labels: Lesser omentum, Lesser curvature, Pyloric region:, Pylorus, Pyloric sphincter, Duodenum, Diaphragm, Fundic region, Cardiac region, Body, Longitudinal muscle, Circular muscle, Oblique muscle, Gastric rugae, Greater curvature, Greater omentum

Figure 17.8 Gross Anatomy of the Stomach
AP|R

- *How does the muscularis externa of the stomach differ from that of the esophagus? Why?*

body makes up the greatest part of the stomach distal to the cardiac orifice. (4) The *pyloric region* is a slightly narrower pouch at the inferior end that terminates at the **pylorus**,[8] a narrow passage into the duodenum. The pylorus is surrounded by a thick ring of smooth muscle, the **pyloric (gastroduodenal) sphincter**, which regulates the passage of chyme into the duodenum.

The stomach wall has the layers typical of the digestive tract with some variations. The mucosa is covered with simple columnar epithelium (fig. 17.9). The mucosa and submucosa are flat and smooth when the stomach is full, but as it empties, these layers form conspicuous longitudinal wrinkles called *gastric rugae*[9] (ROO-jee). The lamina propria is almost entirely occupied by glands, to be described shortly. The muscularis externa has three layers rather than two: an outer longitudinal, middle circular, and inner oblique layer.

Apply What You Know

Contrast the epithelium of the esophagus with that of the stomach. Why is each epithelial type best suited to the function of its respective organ?

The gastric mucosa is pocked with depressions called *gastric pits* lined with the same columnar epithelium as the surface. Cells near the bottom of the

[8] *pylorus* = gatekeeper
[9] *rugae* = folds, creases

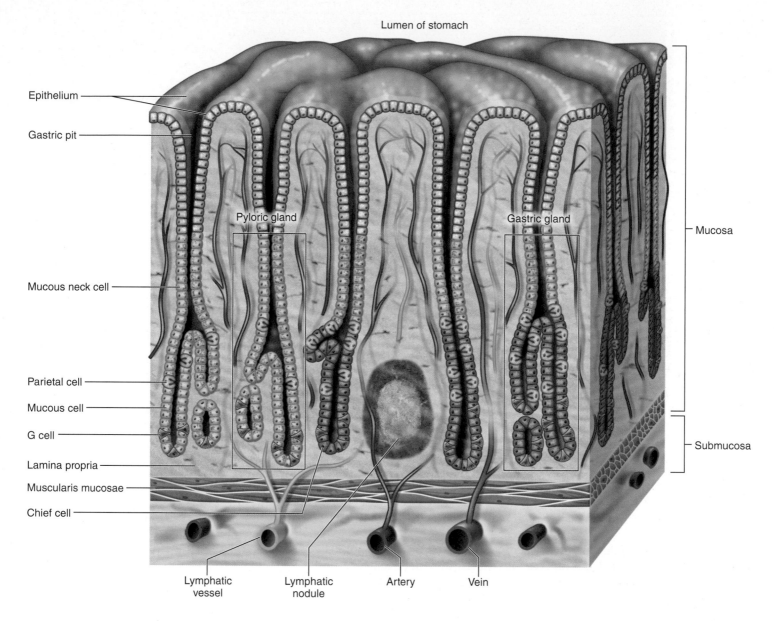

Lumen of stomach

Epithelium

Gastric pit

Pyloric gland

Gastric gland

Mucosa

Mucous neck cell

Parietal cell

Mucous cell

G cell

Lamina propria

Muscularis mucosae

Chief cell

Submucosa

Lymphatic vessel

Lymphatic nodule

Artery

Vein

Figure 17.9 Microscopic Anatomy of the Gastric Mucosa. Both gastric and pyloric glands are shown, although pyloric glands are found in the lower end of the stomach and gastric glands in the upper to middle stomach (fundus and body). AP|R

gastric pits divide and produce new epithelial cells that continually migrate upward and replace old epithelial cells that are sloughed off into the chyme.

Glands open into the bottom of each gastric pit. Near the esophagus and pylorus, the stomach has *cardiac* and *pyloric glands,* respectively; their main secretion is mucus. But in most of the stomach, we find **gastric glands,** with a greater variety of cell types and secretions:

- **Mucous cells** secrete mucus.
- **Regenerative cells**, found in the base of the pit and neck of the gland, divide rapidly and produce a continual supply of new cells to replace cells that die.
- **Parietal cells**, found mostly in the upper half of the gland, secrete *hydrochloric acid* and *intrinsic factor.*
- **Chief cells**, so named because they are the most numerous, secrete the enzymes *pepsinogen* and *gastric lipase.*
- **Enteroendocrine cells**, concentrated especially in the lower end of a gland, secrete hormones and local chemical messengers that regulate digestion. These include, for example, the *G cells* in figure 17.9, which secrete the regulatory hormone *gastrin.*

Gastric Secretions

The gastric glands produce 2 to 3 L of *gastric juice* per day, composed mainly of water and the following solutes:

- **Hydrochloric acid (HCl)** is produced by the parietal cells. The gastric juice can have a pH as low as 0.8, low enough to cause a serious chemical burn on the skin. Stomach acid has several functions: (1) It activates the enzyme *pepsin*; (2) it converts dietary iron to a form that the body can absorb and use; (3) it breaks up connective tissues and plant cells, helping to liquefy food and form chyme; and (4) it contributes to nonspecific disease resistance by destroying most ingested pathogens.

- **Pepsin** is an enzyme that digests proteins to shorter peptide chains, which then pass to the small intestine where their digestion is completed. Chief cells secrete an inactive precursor called *pepsinogen,* and hydrochloric acid converts it to pepsin.

- **Gastric lipase**, also secreted by chief cells, digests a small percentage of the dietary fat. Most fat, however, is digested in the small intestine.

- **Intrinsic factor** is a glycoprotein secreted by the parietal cells. It is essential for the absorption of vitamin B_{12} by the small intestine. Without vitamin B_{12}, hemoglobin cannot be synthesized and anemia develops. The secretion of intrinsic factor is the only indispensable function of the stomach. Digestion can continue following removal of the stomach (*gastrectomy*), but a person must then take vitamin B_{12} by injection, or vitamin B_{12} and intrinsic factor orally.

- **Hormones and local chemical messengers** are produced by enteroendocrine cells. Most of them travel in the bloodstream to distant target cells, but some of them diffuse only a short distance to stimulate nearby cells in the gastric mucosa. Surprisingly, several are peptides identical to those produced in the central nervous system, and have thus been called **gut–brain peptides**. These include *gastrin, serotonin, histamine, somatostatin*, and many others. Their roles in digestion and appetite regulation are discussed in later sections of this chapter and in chapter 18. Several of the gastric secretions are summarized in table 17.1.

Table 17.1 Major Secretions of the Gastric Glands

Secretory Cells	Secretion	Function
Mucous cells	Mucus	Protects mucosa from HCl and enzymes
Parietal cells	Hydrochloric acid	Activates pepsin; helps liquefy food; converts dietary iron to usable form; destroys ingested pathogens
	Intrinsic factor	Enables small intestine to absorb vitamin B_{12}
Chief cells	Pepsinogen	Converted to pepsin, which digests protein
	Gastric lipase	Digests fat
Enteroendocrine cells	Gastrin	Stimulates gastric glands to secrete HCl and enzymes; stimulates intestinal motility; relaxes ileocecal valve
	Serotonin	Stimulates gastric motility
	Histamine	Stimulates HCl secretion
	Somatostatin	Inhibits gastric secretion and motility; delays emptying of stomach; inhibits secretion by pancreas; inhibits gallbladder contraction and bile secretion; reduces blood circulation and nutrient absorption in small intestine

Gastric Motility

As you begin to swallow, the swallowing center of the medulla oblongata signals the stomach to relax, thus preparing it to receive food. The arriving food also causes smooth muscle to relax and the stomach stretches to accommodate more food.

Peristalsis

Soon, the stomach begins rhythmic peristaltic contractions that churn the food, mix it with gastric juice, and promote its physical breakup and chemical digestion. The contractions are governed by pacemaker cells in the muscularis externa. When these waves reach the pyloric sphincter, they squeeze it shut. The stomach thus squirts only about 3 mL of chyme into the duodenum at a time. Receiving such small amounts of chyme enables the duodenum to neutralize the stomach acid and digest nutrients little by little. A typical meal is emptied from the stomach in about 4 hours, but it takes less time if the meal is more liquid and as long as 6 hours if the meal is high in fat.

Vomiting

Vomiting is the forceful ejection of stomach and intestinal contents from the mouth. It involves multiple muscular actions integrated by the *emetic*[10] *center* of the medulla oblongata. Vomiting is commonly induced by overstretching of the stomach, chemical irritants such as alcohol and bacterial toxins, visceral trauma (especially to the pelvic organs), intense pain, or psychological and sensory stimuli that activate the emetic center (repugnant sights, smells, and thoughts). Vomiting is usually preceded by nausea, and is often accompanied by tachycardia, profuse salivation, and sweating.

Chronic vomiting can cause dangerous fluid, electrolyte, and acid–base imbalances. In cases of frequent vomiting, as in the eating disorder *bulimia*, the tooth enamel becomes severely eroded by the hydrochloric acid in the chyme. Aspiration (inhalation) of this acid is very destructive to the respiratory tract. Many have died from aspiration of vomit when they were unconscious or semiconscious. This is the reason that surgical anesthesia, which may induce nausea, must be preceded by fasting until the stomach and duodenum are empty.

Regulation of Gastric Function

Gastric secretion and motility are divided into three stages called the *cephalic, gastric,* and *intestinal phases,* based on whether the stomach is being controlled by the brain, by itself, or by the small intestine, respectively. These phases overlap and all three can occur simultaneously.

1. The **cephalic phase** (fig. 17.10a) is the stage in which the stomach responds to the mere sight, smell, taste, or thought of food. These sensory and mental inputs converge on the hypothalamus, which relays signals to the medulla oblongata. Vagus nerve fibers from the medulla stimulate the enteric nervous system, which, in turn, stimulates gastric activity.

---

[10]*emet* = vomiting

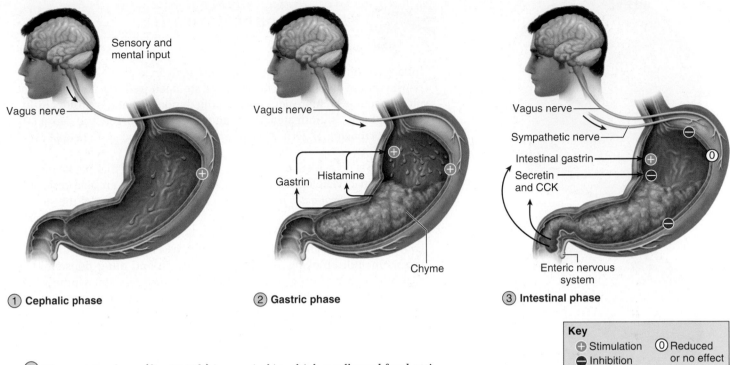

① **Cephalic phase**

② **Gastric phase**

③ **Intestinal phase**

Key
⊕ Stimulation   ⓪ Reduced or no effect
⊖ Inhibition

Figure 17.10 Neural and Hormonal Control of the Stomach. See numbered steps in the text for further explanation.

② The **gastric phase** (fig. 17.10b) is a period in which swallowed food activates gastric activity. About two-thirds of gastric secretion occurs during this phase. Ingested food stimulates gastric activity both by stretching the stomach and by raising the pH of its contents. Three chemical signals activate HCl secretion by the parietal cells: **acetylcholine** from the vagus nerves and **histamine** and **gastrin** from the enteroendocrine cells of the gastric glands. Gastrin and acetylcholine also signal the chief cells to secrete pepsinogen.

③ The **intestinal phase** occurs as chyme enters the duodenum and triggers local nervous reflexes and the secretion of regulatory hormones (fig. 17.10c). Acid and semidigested fats in the duodenum trigger inhibitory signals, sent to the stomach by way of the enteric nervous system. In addition, signals from the duodenum to the medulla oblongata inhibit vagal stimulation of the stomach and activate sympathetic neurons, which inhibit gastric activity. Furthermore, duodenal enteroendocrine cells release the hormones **secretin** and **cholecystokinin (CCK),** which suppress gastric activity. Collectively, all these neural and hormonal signals from the duodenum are like a message to the stomach to slow down and give the duodenum time to process the chyme it already contains before sending it more.

Protection of the Stomach

One might think that the stomach would be its own worst enemy; it is, after all, made of meat. Some people enjoy haggis and tripe, dishes made from animal stomachs, and have no difficulty digesting those. Why, then, doesn't the human stomach digest itself? The stomach is protected in three ways from the harsh acidic and enzymatic environment it creates:

1. The mucosa is coated with thick, highly alkaline mucus that neutralizes HCl and resists the enzymes.

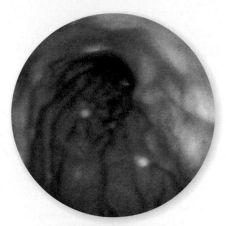

(a) Normal

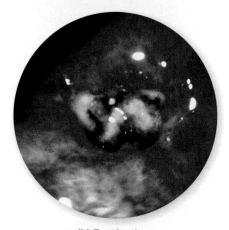

(b) Peptic ulcer

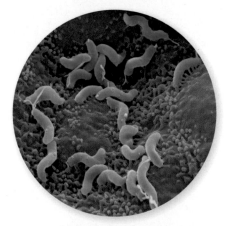

(c) *Helicobacter pylori*

Figure 17.11 Peptic Ulcer. (a) A healthy mucosa in the lower esophagus where it opens into the stomach. The small white spots are reflections of light from the endoscope. (b) A bleeding peptic ulcer. A peptic ulcer typically has an oval shape and yellow-white color. Here the yellowish floor of the ulcer is partially obscured by black blood clots, and fresh blood is visible around the margin of the ulcer. (c) *Helicobacter pylori,* the bacterium responsible for most peptic ulcers.

Clinical Application 17.1

PEPTIC ULCER

Inflammation of the stomach, called *gastritis,* can lead to a *peptic ulcer* as pepsin and hydrochloric acid erode the stomach wall (fig. 17.11). Peptic ulcers occur even more commonly in the duodenum and occasionally in the esophagus. If untreated, they can perforate the organ and cause fatal hemorrhaging or peritonitis. Most such fatalities occur in people over age 65.

There is no evidence to support the popular belief that peptic ulcers result from psychological stress. Most ulcers involve an acid-resistant bacterium, *Helicobacter pylori,* which invades the mucosa of the stomach and duodenum and opens the way to chemical damage to the tissue. Ulcers are often successfully treated with antibiotics against *Helicobacter* combined with bismuth suspensions such as Pepto-Bismol. Other risk factors include smoking and the use of aspirin and other nonsteroidal anti-inflammatory drugs (NSAIDs). NSAIDs suppress the synthesis of prostaglandins, which normally stimulate the secretion of protective mucus and acid-neutralizing bicarbonate. Aspirin itself is an acid that directly irritates the gastric mucosa.

2. The epithelial cells are joined by tight junctions that prevent gastric juice from seeping between them and digesting the underlying tissue.

3. In spite of these other protections, the stomach's epithelial cells live only 3 to 6 days and are then sloughed off into the chyme and digested with the food. They are replaced just as rapidly, however, by cell division in the gastric pits.

The breakdown of these protective mechanisms can result in inflammation and peptic ulcer (see Clinical Application 17.1).

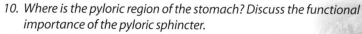

Before You Go On

Answer these questions from memory. Reread the preceding section if there are too many you don't know.

10. *Where is the pyloric region of the stomach? Discuss the functional importance of the pyloric sphincter.*

11. *List five kinds of cells in gastric glands and state their functions.*

12. *Name the chemical messengers that regulate gastric function and state the effect of each.*

13. *Compare and contrast the three phases of gastric regulation. State whether gastric secretion and motility increase or decrease in each phase. Summarize the roles of sympathetic, parasympathetic, and enteric nervous stimulation over the course of these three phases.*

14. *Describe three ways the stomach is protected from self-digestion.*

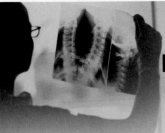

PERSPECTIVES ON HEALTH

Acute Gastrointestinal Illness

In recent years, headlines about contaminated apple juice, hamburger, and spinach, and outbreaks of illness on cruise ships, have focused public attention on contagious gastrointestinal illnesses. Acute gastrointestinal illness is caused by a variety of bacteria and viruses, and is marked by diarrhea, nausea, vomiting, and abdominal pain. *Diarrhea* occurs when the large intestine is irritated by pathogens, and feces pass through too quickly for adequate water reabsorption. In most cases, the person recovers after a few unpleasant days; but in severe cases, fluid loss can lead to dehydration, electrolyte imbalance, shock, and even death.

Pathogens that cause gastrointestinal disease are transmitted via contaminated water or food. Before widespread efforts in the nineteenth century to improve sanitation and provide clean drinking water, gastrointestinal infections such as cholera were devastatingly common. Gastrointestinal infections continue to cause illness and death throughout the world, especially in developing countries. As many as 17% of all deaths worldwide in children less than 5 years of age are due to acute gastrointestinal illness.

Three types of bacteria are responsible for many cases of gastrointestinal disease in the United States—*Salmonella, Campylobacter,* and *Shigella. Salmonella* is usually transmitted by eating foods that are contaminated with animal feces, such as poultry or eggs. Thorough cooking kills the bacteria. *Campylobacter* is often contracted by eating or coming into contact with raw or undercooked poultry or

meat, as when one cuts a chicken on a cutting board and then prepares other food such as a salad on the same board without washing it thoroughly. Some people acquire *Campylobacter* through contact with feces of an ill pet dog or cat. *Shigella* is transmitted from the stool or soiled fingers of one person to the mouth of another. This may occur, for example, when food handlers forget to wash their hands with soap after they use the bathroom. Flies may breed in infected feces and transmit the pathogen when they land on food. *Shigella* outbreaks sometimes occur in day-care centers when caretakers do not carefully clean their hands or surfaces where diapers are changed.

Rotavirus is the most common cause of severe diarrhea in infants and young children in the United States and, indeed, throughout the world. Almost all children in the United States are likely to be infected with rotavirus before their fifth birthday, and most recover without medical intervention. Severe cases require hospitalization, where fluids are replaced intravenously. The virus spreads rapidly. It can spread through feces, hand-to-mouth contact, or physical contact with infected persons or toys. Rotavirus is life threatening in infants, who can suffer massive loss of fluid. There is no specific drug treatment for rotavirus infection; viruses do not respond to antibiotics. Recently, a rotavirus vaccine has been developed, and public health officials hope that it will prevent severe disease among children in both developed and poorer nations.

17.4 The Liver, Gallbladder, and Pancreas

Expected Learning Outcomes

When you have completed this section, you should be able to:

a. describe the gross and microscopic anatomy of the liver, gallbladder, and pancreas;

b. discuss the functions of each; and

c. explain how hormones regulate secretions of the liver and pancreas.

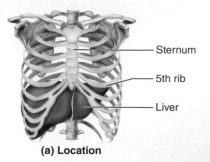

- Sternum
- 5th rib
- Liver

(a) Location

The small intestine receives not only chyme from the stomach but also secretions from the liver and pancreas, which enter the digestive tract near the junction of the stomach and small intestine. These secretions are so important to the digestive processes of the small intestine that it is necessary to understand them before we continue with intestinal physiology.

Figure 17.12 The Liver. (a) Location; note that most of the liver lies within the rib cage. (b) Anterior vew. (c) Inferior view. (d) Histology. Note the relationship of the hepatic lobules to the blood vessels and bile passages. **AP|R**

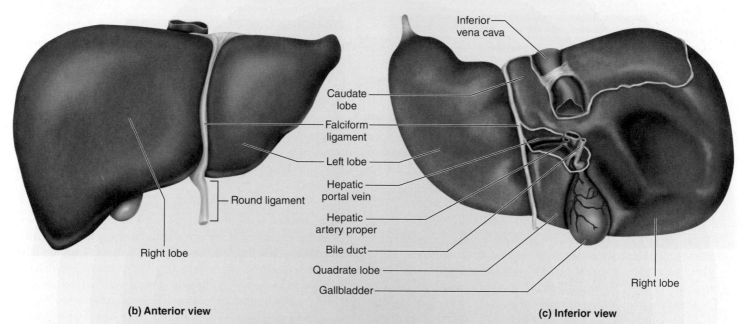

- Inferior vena cava
- Caudate lobe
- Falciform ligament
- Left lobe
- Hepatic portal vein
- Hepatic artery proper
- Bile duct
- Quadrate lobe
- Gallbladder
- Round ligament
- Right lobe
- Right lobe

(b) Anterior view

(c) Inferior view

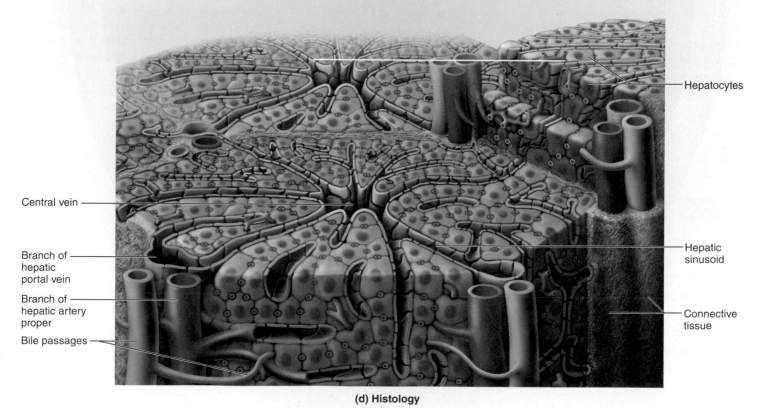

- Hepatocytes
- Central vein
- Branch of hepatic portal vein
- Hepatic sinusoid
- Branch of hepatic artery proper
- Bile passages
- Connective tissue

(d) Histology

The Liver AP|R

The **liver** (fig. 17.12) is a reddish brown gland located immediately inferior to the diaphragm, filling most of the right hypochondriac and epigastric regions. It is the body's largest gland, weighing about 1.4 kg (3 pounds). It has a tremendous variety of functions, but only one of them, the secretion of bile, contributes to digestion. Others are discussed in chapter 18, which provides a more thorough physiological basis for understanding nondigestive liver functions.

Gross Anatomy

The liver has four lobes—the right, left, quadrate, and caudate lobes. From an anterior view, we see only a large *right lobe* and smaller *left lobe.* They are separated by the *falciform*[11] *ligament,* a sheet of mesentery that suspends the liver from the diaphragm and anterior abdominal wall. The *round ligament,* also visible anteriorly, is a fibrous remnant of the umbilical vein, which carries blood from the umbilical cord to the liver of a fetus.

From the inferior view, we see a squarish *quadrate lobe* next to the gallbladder and a tail-like *caudate*[12] *lobe* posterior to that. The hepatic portal vein and hepatic artery enter the liver, and the bile passages leave, through an opening called the *hilum* on the inferior surface between these lobes. The gallbladder adheres to a depression between the right and quadrate lobes. The posterior aspect of the liver has a deep groove (sulcus) that accommodates the inferior vena cava. The liver is covered by a serosa except for its superior surface where it adheres to the diaphragm.

Microscopic Anatomy

The interior of the liver is filled with innumerable tiny cylinders called **hepatic lobules,** about 2 mm long by 1 mm in diameter. A lobule consists of a **central vein** passing down its core, surrounded by radiating sheets of cuboidal cells called **hepatocytes** (fig. 17.12d). Imagine spreading a book wide open until its front and back covers touch. The pages of the book would fan out around the spine somewhat like the plates of hepatocytes fan out from the central vein of a liver lobule.

Each plate of hepatocytes is an epithelium one or two cells thick. The spaces between the plates are blood-filled channels called **hepatic sinusoids.** These are bounded by a leaky fenestrated endothelium that freely allows blood plasma into the space around the hepatocytes, where it contacts their brush border of microvilli. Blood filtering through the sinusoids comes directly from the stomach and intestines. After a meal, the hepatocytes absorb glucose, amino acids, iron, vitamins, and other nutrients from it for metabolism or storage. They also remove and degrade hormones, toxins, bile pigments, and drugs. At the same time, they secrete albumin, lipoproteins, clotting factors, angiotensinogen, and other products into the blood. Between meals, they break down stored glycogen and release glucose into the circulation. The sinusoids also contain phagocytic cells called **hepatic macrophages,** which remove bacteria and debris from the blood.

After filtering through the sinusoids, blood flows into the central vein and from there to vessels that lead to the right and left hepatic veins. These leave the liver at its superior surface and drain immediately into the inferior vena cava.

The liver secretes about 500 to 1,000 mL of bile per day into narrow channels between sheets of hepatocytes. Bile passes to the *common hepatic duct.* A

---

[11] *falci* = sickle; *form* = shape

[12] *caud* = tail

short distance farther on, this is joined by the *cystic duct* coming from the gallbladder (fig. 17.13). Their union forms the **bile duct,** which descends through the lesser omentum toward the duodenum. Near the duodenum, the bile duct joins the duct of the pancreas and forms an expanded chamber called the **hepatopancreatic ampulla.** The ampulla terminates at a fold of tissue on the duodenal wall, the *major duodenal papilla.* Bile and pancreatic juices flow into the duodenum through an opening in the papilla.

The Gallbladder and Bile

The **gallbladder** is a pear-shaped sac on the underside of the liver that stores and concentrates bile. It is about 10 cm long and internally lined by a highly folded mucosa with a simple columnar epithelium. **Bile** is a yellow-green fluid containing minerals, cholesterol, phospholipids, bile pigments, and bile acids. The principal pigment is **bilirubin,** derived from the decomposition of hemoglobin. Further metabolic derivatives of bilirubin are responsible for the brown color of feces and yellow color of urine; without bile, human feces have a chalky gray color. **Bile acids (bile salts)** are steroids synthesized from cholesterol. They aid in fat digestion and absorption. All other components of the bile are wastes destined for excretion in the feces. When these waste products become excessively concentrated, they may form gallstones.

The Pancreas

The **pancreas** (fig. 17.13) is a spongy, flattened, retroperitoneal gland pressed between the posterior body wall and lower margin of the stomach. It measures 12 to 15 cm long and about 2.5 cm thick. The pancreas is both an endocrine and exocrine gland. Its endocrine part is the *pancreatic islets,* the source of insulin and glucagon,

Figure 17.13 Gross Anatomy of the Gallbladder, Pancreas, and Bile Passages. The liver is omitted to show more clearly the gallbladder, which adheres to its inferior surface, and the hepatic ducts, which emerge from the liver tissue. AP|R

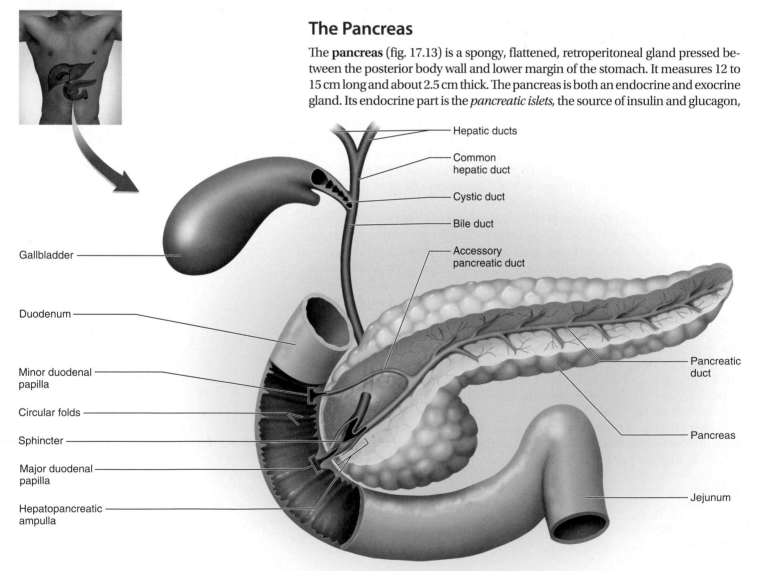

described in chapter 11. About 99% of the pancreas is exocrine tissue with a digestive function.

The exocrine pancreas secretes about 1,500 mL of **pancreatic juice** per day. This fluid flows to the duodenum by way of the **pancreatic duct,** which runs lengthwise through the gland and joins the bile duct at the hepatopancreatic ampulla. A sphincter in this ampulla regulates the release of pancreatic juice and bile simultaneously. Often, a second, smaller duct branches off the main one and empties separately into the duodenum, bypassing the sphincter.

Pancreatic juice is an alkaline solution of mainly enzymes and sodium bicarbonate. The bicarbonate buffers HCl arriving from the stomach. Two of the enzymes, **trypsin** and **chymotrypsin** (KY-mo-TRIP-sin), digest dietary protein; **pancreatic amylase** digests starch; **pancreatic lipase** digests fat; and **ribonuclease and deoxyribonuclease** digest RNA and DNA, respectively. Some of these enzymes are secreted in inactive forms called *proenzymes* and activated by other enzymes in the duodenum. For example, the proenzyme *trypsinogen* is converted to trypsin by an enzyme on the duodenal epithelial surface. We will later examine the mode of action of these enzymes on the nutrients.

Pancreatic juice and bile are not simply secreted continually, but only when needed. The following signal molecules activate their secretion:

1. **Acetylcholine (ACh)** is secreted by the vagus nerves and enteric plexus even before food is swallowed. It stimulates pancreatic secretion, but the fluid is retained in the pancreatic duct until chyme enters the duodenum.

2. **Cholecystokinin**[13] **(CCK)** is a hormone secreted by the mucosa of the duodenum in response to fat in the chyme. It stimulates the pancreas to secrete enzymes, contracts the gallbladder, and relaxes the sphincter of the ampulla so these secretions can enter the duodenum and mix with the fatty chyme. It is named for its stimulatory effect on the gallbladder.

3. **Secretin** is a hormone produced by the duodenum in response to the acidity of the chyme. It stimulates both the liver and pancreas to secrete sodium bicarbonate, which buffers the stomach acid and protects the duodenum from its erosive effect. The rise in duodenal pH due to this bicarbonate also activates digestive enzymes of the pancreatic juice and intestinal mucosa.

- -

Before You Go On

Answer these questions from memory. Reread the preceding section if there are too many you don't know.

15. What does the liver contribute to digestion?

16. Where is bile secreted? Where is it stored? Where does it perform its action?

17. List three enzymes secreted by the pancreas and state the function of each.

18. What stimulates cholecystokinin (CCK) secretion, and how does CCK affect other parts of the digestive system?

- -

[13]*chole* = bile; *cysto* = bladder (gallbladder); *kin* = action

17.5 The Small Intestine

Expected Learning Outcomes

When you have completed this section, you should be able to:

a. describe the gross and microscopic anatomy of the small intestine, including the differences in its three regions;

b. state how and for what functional reason the mucosa of the small intestine differs from that of the stomach;

c. give two reasons why the small intestine needs such a large surface area, and four structural features that meet that need; and

d. describe the types of movement that occur in the small intestine.

Nearly all chemical digestion and nutrient absorption occur in the small intestine. It efficiently carries out these tasks partly because it is the longest part of the digestive tract—about 2.7 to 4.5 m long in a living person. In the cadaver, where there is no muscle tone, it is 4 to 8 m long. The term *small* intestine refers not to its length but to its diameter—about 2.5 cm (1 in.).

Anatomy

The small intestine is a coiled mass filling most of the abdominal cavity inferior to the stomach and liver. It is divided into three regions (fig. 17.14): the duodenum, jejunum, and ileum.

The **duodenum** (dew-ODD-eh-num, DEW-oh-DEE-num) constitutes the first 25 cm (10 in.). It begins at the pyloric valve, arcs around the head of the pancreas, and curves to the left. Its name refers to its length, about equal to the width of 12 fingers.[14] Along with the pancreas, most of the duodenum is retroperitoneal. It has prominent *duodenal glands* in the submucosa. They secrete an abundance of bicarbonate-rich mucus, which neutralizes stomach acid and shields the mucosa from it. The duodenum receives stomach contents, pancreatic juice, and bile. Stomach acid is neutralized here, fats are physically broken up (emulsified) by the bile acids, pepsin is inactivated by the elevated pH, and pancreatic enzymes take over the job of chemical digestion.

The **jejunum** (jeh-JOO-num) is the first 40% of the small intestine beyond the duodenum—roughly 1.0 to 1.8 m in a living person. Its name refers to the fact that early anatomists typically found it to be empty.[15] The jejunum is located largely in the upper left quadrant of the abdomen. Its wall is relatively thick and muscular, and it has an especially rich blood supply. Most digestion and nutrient absorption occur here.

The **ileum**[16] forms the last 60% of the small intestine (about 1.6 to 2.7 m). It lies mostly in the lower right quadrant of the abdomen. Compared with the jejunum, its wall is thinner, less muscular, less vascular, and has a paler pink col-

---

[14] *duoden* = 12
[15] *jejun* = empty, dry
[16] from *eilos* = twisted

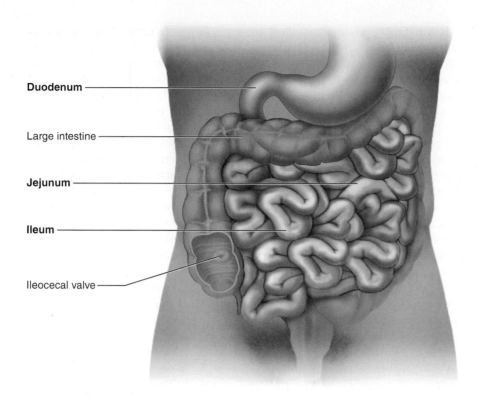

Duodenum

Large intestine

Jejunum

Ileum

Ileocecal valve

Figure 17.14 Gross Anatomy of the Small Intestine. Note the spatial relationships of its three main regions: duodenum, the first 25 cm past the stomach; jejunum, mostly in the upper left abdominal quadrant; and ileum, mostly in the lower right abdominal quadrant, ending at the ileocecal valve. **AP|R**

or. It has prominent lymphatic nodules in clusters called **Peyer**[17] **patches,** which are readily visible to the naked eye and become progressively larger approaching the large intestine.

The end of the small intestine is the **ileocecal** (ILL-ee-oh-SEE-cul) **junction,** where the ileum joins the *cecum* of the large intestine. The muscularis of the ileum is thickened at this point to form a sphincter, the **ileocecal valve,** which protrudes into the cecum and regulates the passage of food residue into the large intestine. Both the jejunum and ileum are intraperitoneal and thus covered with a serosa, which is continuous with the complex, folded mesentery that suspends the small intestine from the posterior abdominal wall.

Effective digestion and absorption require that the small intestine have a large internal surface area. This is provided by its relatively great length and by three kinds of internal folds or projections: the *circular folds, villi,* and *microvilli.* If the mucosa were smooth, like the inside of a hose, it would have a surface area of about 0.3 to 0.5 m², but with these surface elaborations, its actual surface area is about 200 m²—clearly a great advantage for nutrient absorption.

Circular folds, the largest elaborations of the intestinal wall, are transverse ridges up to 10 mm high that project into the lumen (fig. 17.13). They occur from the duodenum to the middle of the ileum, where they cause the chyme to flow on a spiral path along the intestine. This slows its progress, causes more contact with the mucosa, and promotes more thorough mixing and nutrient absorption.

[17] Johann K. Peyer (1653–1712), Swiss anatomist

Villi (VIL-eye; singular, *villus*) are small finger- to tongue-shaped projections about 0.5 to 1.0 mm high (fig. 17.15). They give the mucosa a fuzzy texture like a terrycloth towel. Villi are covered with simple columnar epithelium composed of two kinds of cells: columnar **absorptive cells (enterocytes)** and mucus-secreting **goblet cells.** The core of a villus is filled with loose connective tissue. Embedded in this are an arteriole, a capillary network, a venule, and a lymphatic capillary called a **lacteal** (LAC-tee-ul). Blood capillaries of the villus absorb most nutrients, but the lacteal absorbs most lipids. The lipids give its contents a milky appearance for which the lacteal is named.[18] The lacteal connects with larger lymphatic vessels that ultimately carry the lipids to the bloodstream (see chapter 14).

Microvilli form a fuzzy brush border on the surface of each absorptive cell. Being only about 1 μm high, microvilli can be resolved only with the electron microscope. They greatly increase the absorptive surface area of the small intestine and contain **brush border enzymes**—components of the plasma membrane that carry out some of the final stages of chemical digestion.

On the floor of the small intestine, between the bases of the villi, there are numerous pores that open into tubular glands called **intestinal crypts** (**crypts of Lieberkühn;**[19] LEE-ber-koohn). These crypts, similar to the gastric glands, extend as far as the muscularis mucosae. In the upper half, they consist of enterocytes and goblet cells like those of the villi. The lower half is dominated by dividing epithelial cells. In its life span of 3 to 6 days, an epithelial cell migrates up the crypt to the tip of the villus, where it is sloughed off and digested.

---

[18] *lact* = milk

[19] Johann N. Lieberkühn (1711–56), German anatomist

(a)

Figure 17.15 Intestinal Villi. (a) Villi. Each villus is about 1 mm high. (b) Structure of a villus. AP|R

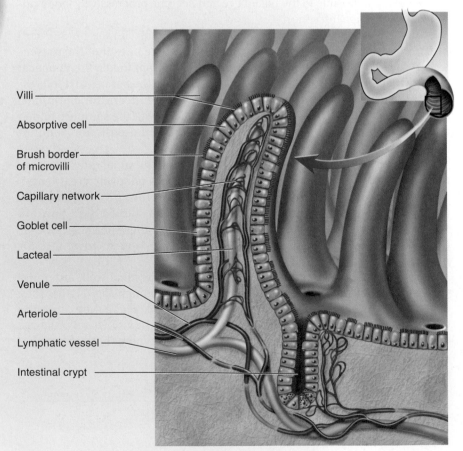

Villi

Absorptive cell

Brush border of microvilli

Capillary network

Goblet cell

Lacteal

Venule

Arteriole

Lymphatic vessel

Intestinal crypt

(b)

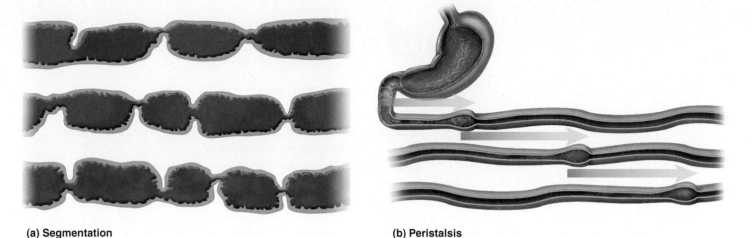

(a) Segmentation   (b) Peristalsis

Apply What You Know

Both the small intestine and the proximal convoluted tubule of the kidney (p. 543) are tubular structures specialized for absorption. Identify at least two features that they have in common to serve this purpose.

Intestinal Secretion and Motility

The small intestine secretes 1 to 2 L of alkaline *intestinal juice* per day, which helps to mix nutrients with digestive enzymes and protect the mucosa from stomach acid.

It exhibits two kinds of contractions—*segmentation* and *peristalsis* (fig. 17.16). These serve (1) to mix chyme with intestinal juice, bile, and pancreatic juice, allowing them to neutralize acid and digest nutrients more effectively; (2) to churn chyme and bring it into contact with the mucosa for digestion and absorption; and (3) to move residue toward the large intestine.

Segmentation is a movement in which ringlike constrictions appear at several places at once along the intestine and then relax as new constrictions form elsewhere. This is the most common movement of the small intestine. Its effect is to knead or churn the contents. When most nutrients have been absorbed and little remains but undigested residue, segmentation declines and peristalsis begins. These waves of contraction move chyme toward the colon over a period of about 2 hours.

Figure 17.16 Contractions of the Small Intestine. (a) Segmentation, in which circular constrictions of the intestine cut into the contents, churning and mixing them. (b) Peristalsis, in which successive waves of contraction overlap each other. Each wave travels partway down the intestine and milks the contents toward the colon.

Before You Go On

Answer these questions from memory. Reread the preceding section if there are too many you don't know.

19. What four aspects of its structure give the small intestine such a large surface area?

20. Why are mucous glands more important in the duodenum and Peyer patches more important in the ileum than either one is to other regions of the small intestine?

21. Sketch or describe a villus and label its epithelium, brush border, lamina propria, blood capillaries, and lacteal. How does the function of the blood capillaries differ from that of the lacteal?

22. Distinguish between segmentation and peristalsis of the small intestine. How do these differ in function?

17.6 Chemical Digestion and Absorption

Expected Learning Outcomes

When you have completed this section, you should be able to:

a. describe how each major class of nutrients is chemically digested and name the enzymes involved; and

b. describe how each type of nutrient is absorbed by the small intestine.

As food material passes from the mouth through the small intestine, nutrients are chemically degraded and absorbed. In this section, we focus on how this occurs, especially for the major classes of nutrients—carbohydrates, proteins, and lipids. These are summarized and compared in figure 17.17.

Carbohydrates

Most digestible dietary carbohydrate is starch. Cellulose is indigestible and is not considered here, although its importance as dietary fiber is discussed in the next chapter.

Starch (fig. 17.17a) is digested to the disaccharide maltose, and finally to glucose, which is absorbed by the small intestine. The process begins in the mouth, where salivary amylase begins to break down the starch into maltose and other small carbohydrate chains *(oligosaccharides[20])*. Salivary amylase functions best at pH 6.8 to 7.0, typical of the oral cavity. It is quickly denatured upon contact with stomach acid, although can continue digesting starch for up to 2 hours in the stomach if it is in the middle of a starchy food mass such as a pancake breakfast, where it can escape contact with the acid for a while. The stomach itself, however, produces no carbohydrate-digesting enzymes. About 50% of dietary starch is digested to maltose before it reaches the duodenum. Other dietary sugars include sucrose (cane or table sugar) and lactose (milk sugar), but these are not acted upon until they reach the small intestine.

Starch digestion resumes in the small intestine when the chyme mixes with *pancreatic amylase*. This breaks the remaining starch and oligosaccharides down to the maltose. Carbohydrate digestion is completed when the disaccharides contact the brush border of the epithelial cells. Here, they are acted upon by three brush border enzymes—*maltase,* which splits the maltose into glucose molecules; *sucrase,* which splits sucrose into glucose and fructose; and *lactase,* which digests lactose to glucose and galactose. (In most populations, however, lactase activity ceases or declines to a low level after age 4 and lactose becomes indigestible; see Clinical Application 17.2).

The monosaccharide products of carbohydrate digestion—glucose, fructose, and galactose—are immediately absorbed by carrier-mediated transport across the absorptive cell surface. Within the cell, most of the fructose is converted to glucose. Glucose, galactose, and any remaining fructose then pass out the base of the epithelial cells by facilitated diffusion, pass into the blood capillaries of the villi, and are carried off by the hepatic portal system to the liver for metabolism or storage.

---

[20] *oligo* = a few; *sacchar* = sugar

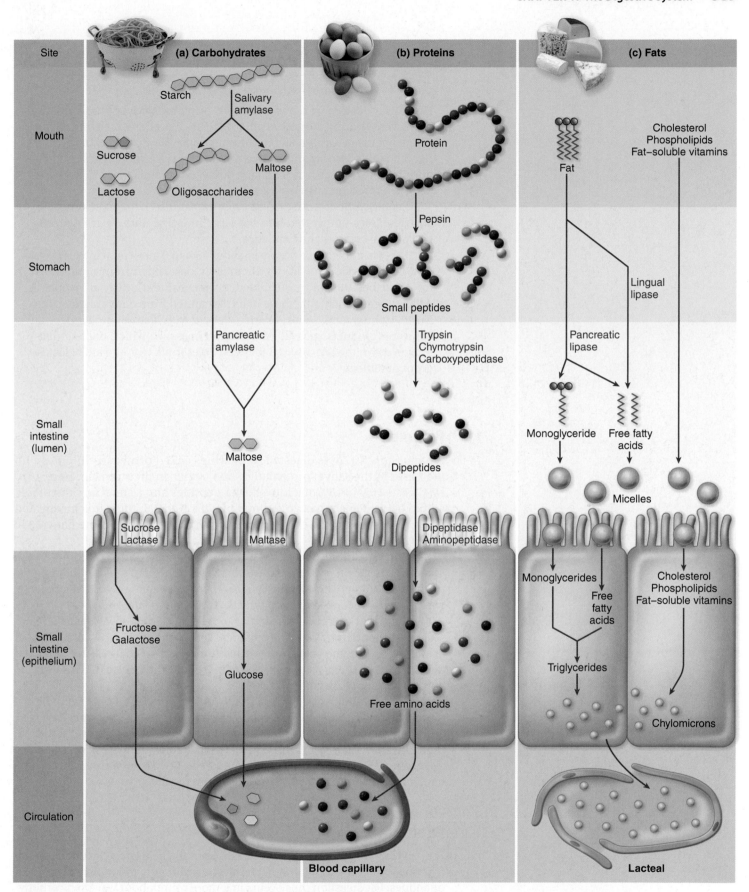

Figure 17.17 Digestion and Absorption of the Three Major Nutrient Classes. Enzyme names are indicated in red.

- *As they leave the small intestine, sugars and amino acids arrive at the liver before any other organ, but lipids do not. Why not? See chapter 13 and trace the route that a dietary fat would have to take to reach the liver.*

Clinical Application 17.2

LACTOSE INTOLERANCE

Humans, unlike most other mammals, drink milk beyond infancy, and moreover, we drink the milk of other species! This odd habit is largely limited to people of western and northern Europe, a few pastoral tribes of Africa, and their descendants in the Americas and elsewhere. These populations have a long history of milking domestic animals that has led to the persistent production of lactase into adulthood.

People without lactase have *lactose intolerance*. If they consume milk, lactose passes undigested into the large intestine and causes diarrhea, gas, painful cramps, and flatulence.

Lactose intolerance occurs in about 15% of American whites; 90% of American blacks, who are predominantly descended from nonpastoral African tribes; 70% or more of Mediterraneans; and nearly all people of Asian descent, including those of us descended from early migrants into North, Central, and South America. People with lactose intolerance can consume products such as yogurt and cheese, in which bacteria have broken down the lactose, and they can digest milk with the aid of lactase drops or tablets.

Proteins

Protein (fig. 17.17b) is digested by enzymes called **proteases**. There are no proteases in the saliva; protein digestion begins in the stomach. Here, *pepsin* digests 10% to 15% of it into shorter peptides and a small amount of free amino acids. Pepsin has an optimal pH of 1.5 to 3.5; thus, it is inactivated when it passes into the duodenum and mixes with the alkaline pancreatic juice (pH 8).

In the small intestine, three proteases from the pancreas continue the job—*trypsin, chymotrypsin,* and *carboxypeptidase.* Finally, the remaining dipeptides and other short chains are taken apart, one amino acid at a time, by two brush border enzymes—*dipeptidase* and *aminopeptidase.*

Amino acid absorption is similar to that of monosaccharides. There are several transport proteins in the epithelial cell membranes for different classes of amino acids. At the basal surfaces of the cells, amino acids exit by facilitated diffusion, enter the blood capillaries of the villus, and are carried away in the hepatic portal circulation.

Amino acids absorbed by the small intestine come not only from dietary protein but also from sloughed epithelial cells digested by the same enzymes, and from the enzymes digesting each other. Thus, the body recycles its own proteins rather than losing the valuable amino acids from these cells and enzymes in the feces. The amino acids from the last two sources total about 30 g/day, compared with about 44 to 60 g/day from the diet.

Fats

Fat (triglyceride) digestion (fig. 17.17c) is complicated by the fact that fats are hydrophobic and cannot enter the intestinal blood capillaries in significant quantities. Fat digestion thus occurs in a more roundabout way than carbohydrate and protein digestion.

Saliva contains an enzyme called *lingual lipase,* which is activated by stomach acid and digests a small amount of fat there. Most fat, however, reaches the small intestine in undigested globules. It is broken up by churning action (segmentation contractions) of the intestine into small **emulsification droplets,** which are coated with bile acids to prevent them from recombining, much like the action of scrubbing and dish soap on the grease in a frying pan. This *emulsification* process exposes far more of the fat's surface to enzyme action.

Lingual lipase no longer works at the pH of the small intestine, but *pancreatic lipase* takes over. It splits two of the fatty acids from each triglyceride molecule, leaving free fatty acids and a *monoglyceride.* These are taken up along with dietary cholesterol, phospholipids, and fat-soluble vitamins into small droplets called **micelles**[21] (my-SELLS), a component of the bile produced by the liver. When micelles contact the brush border, the lipids leave them and diffuse into the absorptive epithelial cells. Strangely, the absorptive cells reassemble the monoglycerides and free fatty acids into triglycerides. They next package all of these lipids into another type of droplet called **chylomicrons**[22] (KY-lo-MY-crons), coated with phospholipids and protein. These leave the base of the cells and enter the lacteal within the villus; they are too large to penetrate into the blood capillaries. From the lacteals, the white, fatty lymph *(chyle)* flows through larger and larger lymphatic vessels, eventually entering the bloodstream at the left subclavian vein (see chapter 14). The further fate of dietary fat is described in chapter 18.

Vitamins and Minerals

Fat-soluble vitamins A, D, E, and K are absorbed with other lipids as just described. If they are ingested without fat-containing food, they are not absorbed at all but are passed in the feces and wasted. Water-soluble vitamins (the B complex and vitamin C) are absorbed by simple diffusion, with the exception of vitamin B_{12}. This is an unusually large molecule that can only be absorbed if it binds to intrinsic factor from the stomach. Minerals (electrolytes) are absorbed along the entire length of the small intestine.

Iron and calcium are unusual in that they are absorbed in proportion to the body's need, whereas other minerals are absorbed at fairly constant rates regardless of need. Iron absorption is stimulated by the liver hormone *hepcidin.* Calcium is absorbed mostly by diffusion through the gaps between epithelial cells in the jejunum and ileum, but in the duodenum its absorption depends on membrane transport proteins and is regulated by parathyroid hormone and calcitriol (vitamin D).

Water

The digestive system is one of several systems involved in water balance. The digestive tract receives about 9 L of water per day. Most of this, about 6.7 L, consists of the digestive system's own secretions—saliva, gastric juice, bile, pancreatic juice, and intestinal juice. The rest (2.3 L) is from food and drink. The small intestine absorbs about 8 L of this, the large intestine absorbs about 0.8 L, and the remaining 0.2 L is voided in the daily fecal output. Water is absorbed by osmosis, following the absorption of salts and organic nutrients that create an osmotic gradient from the intestinal lumen to the ECF.

---

[21] *mic* = grain, crumb; *elle* = little

[22] *chyl* = juice; *micr* = small

23. Name the enzymes that digest starch, fat, and protein. Name the end products of their digestion.

24. Describe how digested sugars and amino acids get from the intestinal lumen into the blood.

25. Describe how lipids get from the intestinal lumen into the lymph and ultimately the blood.

26. What is the difference between fat-soluble and water-soluble vitamins in terms of absorption?

27. What are all the sources of water in the digestive tract? How much of it is lost by defecation, and what happens to the rest of it?

17.7 The Large Intestine

Expected Learning Outcomes

When you have completed this section, you should be able to:

a. describe the gross anatomy and mucosa of the large intestine;

b. state the physiological significance of intestinal bacteria;

c. discuss the types of contractions that occur in the colon; and

d. explain the neurological control of defecation.

The large intestine (fig. 17.18) receives about 500 mL of indigestible food residue per day, reduces it to about 150 mL of feces by absorbing water and salts, and eliminates the feces by defecation.

Anatomy

The **large intestine** measures about 1.5 m (5 feet) long and 6.5 cm (2.5 in.) in diameter. It begins with the **cecum,**[23] a blind pouch in the lower right abdominal quadrant inferior to the ileocecal valve. Attached to the lower end of the cecum is the **appendix,** a blind tube 2 to 7 cm long. The appendix is densely populated with lymphocytes and is a significant source of immune cells.

The **colon** is that portion of the large intestine between the ileocecal valve and the rectum (not including the cecum, rectum, or anal canal). It is divided into the ascending, transverse, descending, and sigmoid regions. The **ascending colon** begins at the ileocecal valve and passes up the right side of the abdominal cavity. It makes a 90° turn near the right lobe of the liver and becomes the **transverse colon.** This passes horizontally across the upper abdominal cavity and turns 90° downward near the spleen. Here it becomes the **descending colon,** which passes down the left side of the

---

[23] *cec* = blind

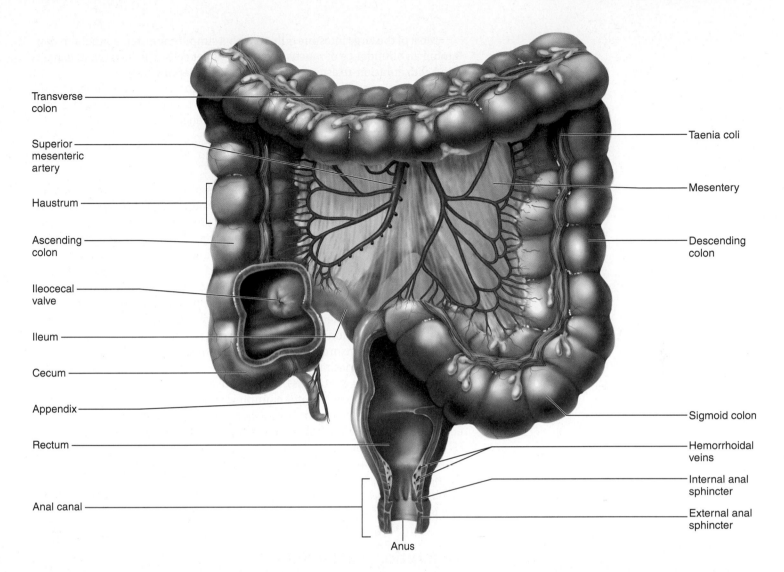

Transverse colon

Superior mesenteric artery

Haustrum

Ascending colon

Ileocecal valve

Ileum

Cecum

Appendix

Rectum

Anal canal

Taenia coli

Mesentery

Descending colon

Sigmoid colon

Hemorrhoidal veins

Internal anal sphincter

External anal sphincter

Anus

Figure 17.18 The Large Intestine. [AP|R]

- *Which anal sphincter is controlled by the autonomic nervous system? Which is controlled by the somatic nervous system? Explain the basis for your answers.*

abdominal cavity. Ascending, transverse, and descending colons thus form a squarish, three-sided frame around the small intestine.

The muscularis externa of the colon has longitudinal muscle cells arranged in three thickened, ribbonlike strips. Each strip is called a **taenia coli** (TEE-nee-ah CO-lye). The muscle tone of the taeniae coli (plural) contracts the colon lengthwise and causes its wall to bulge, forming pouches called **haustra**[24] (HAW-stra; singular, *haustrum*).

The pelvic cavity is narrower than the abdominal cavity, so the colon turns medially along the hip bone, then downward, forming a roughly S-shaped portion called the **sigmoid**[25] **colon.** In the pelvic cavity, the large intestine continues as the **rectum,**[26] about 15 cm long.

The final 3 cm of the large intestine is the **anal canal,** which passes through the pelvic floor and terminates at the anus. Prominent *hemorrhoidal veins* form superficial plexuses in the walls of the anus. Unlike veins in the limbs, they lack valves and are particularly subject to distension and venous pooling. Hemorrhoids are permanently distended veins that protrude into the anal canal or form bulges external to the anus.

---

[24] *haustr* = to draw

[25] *sigm* = sigma or S; *oid* = resembling

[26] *rect* = straight

Most of the large intestine is lined with a simple columnar epithelium containing an abundance of mucus-secreting goblet cells. The lower anal canal is lined with abrasion-resistant stratified squamous epithelium.

The anus is regulated by two sphincters: an **internal anal sphincter** composed of smooth muscle of the muscularis externa and an **external anal sphincter** composed of skeletal muscle of the pelvic floor.

Intestinal Bacteria and Gas

The large intestine harbors about 800 species of bacteria, most of which are harmless or beneficial. We provide them with room and board, while they provide us with nutrients from our food that we are not equipped to gain on our own. For example, they digest cellulose, pectin, and other plant polysaccharides for which we have no digestive enzymes, and we absorb the resulting sugars. Thus, we get more nutrition from our food because of these bacteria than we would get without them. Indeed, one person may get more calories than another from the same amount of food because of differences in their bacterial populations. Some bacteria also synthesize B vitamins and vitamin K. Vitamin K absorbed by the colon is important because the diet alone usually does not provide enough to ensure adequate blood clotting.

One of the less desirable and sometimes embarrassing products of these bacteria is intestinal gas **(flatus).** This is a mixture of bacterial gases and nitrogen from swallowed air. Painful cramping can result when undigested nutrients pass into the colon and furnish an abnormal substrate for bacterial action—for example, in lactose intolerance. Flatus is composed of nitrogen (N_2), carbon dioxide (CO_2), hydrogen (H_2), methane (CH_4), hydrogen sulfide (H_2S), and two amines—indole and skatole. Indole, skatole, and H_2S produce the odor of flatus and feces, whereas the others are odorless. The hydrogen is combustible and has sometimes ignited during the use of electrical cauterization in surgery.

Absorption and Motility

The large intestine takes about 12 to 24 hours to reduce the residue of a meal to feces. It reabsorbs water and electrolytes (especially NaCl). The feces consist of about 75% water and 25% solids. The solids are about 30% bacteria, 30% undigested dietary fiber, 10% to 20% fat, and smaller percentages of protein, epithelial cells, mucus, and salts. The fat is not from the diet but from broken-down epithelial cells and bacteria.

Colonic motility is predominantly a type of segmentation called **haustral contractions,** which occur about every 30 minutes. Distension of a haustrum with feces stimulates it to contract. This churns and mixes the residue, promotes water and salt absorption, and passes the residue distally to another haustrum. Stronger contractions called **mass movements** occur one to three times a day, last about 15 minutes, and move residue for several centimeters at a time. Mass movements occur especially in the transverse to sigmoid colon, and often within an hour after breakfast.

Defecation

Like urination (p. 550), the elimination of feces is not a continual event but is episodic—it is controlled by neurological reflexes and sphincters so it ordinarily occurs only in appropriate circumstances and times. Filling of the stomach with a new meal triggers a *gastrocolic reflex* through the enteric nervous system, inducing mass movements that propel feces from the upper colon toward the rectum. Stretching of the rectum then sets off both enteric and spinal reflexes called **defecation reflexes,** which account for the urge to defecate. Such responses are especially pronounced in the morning because of the overnight accumulation of feces in the upper colon.

One of the defecation reflexes, called the *intrinsic defecation reflex,* is controlled through the enteric plexus of the colonic wall. Stretch signals from the rectum are conducted through the plexus (1) to the descending and sigmoid colon, further activating a peristaltic wave that drives feces downward, and (2) to the internal anal sphincter, causing it to relax. In an infant, this alone results in defecation. The other reflex, called the *parasympathetic defecation reflex,* is spinal. Stretch signals are conducted to the spinal cord and return to the rectum through a visceral reflex arc via parasympathetic nerve fibers in the pelvic nerves. These signals likewise intensify peristalsis and relax the internal anal sphincter.

Obviously and fortunately, once we have acquired bowel control in childhood, we are not at the mercy of these two involuntary reflexes. The external anal sphincter is skeletal muscle and therefore under voluntary control. Feces usually are voided only if we voluntarily relax that sphincter in addition to the foregoing involuntary reflexes. The external anal and urinary sphincters are controlled together by inhibitory signals from the brainstem. This is why we find it hard to defecate without also urinating. Some spinal cord injuries, however, disconnect the brainstem from the lower spinal cord and abolish this voluntary control.

Defecation is also aided by the voluntary Valsalva maneuver, in which a breath hold and contraction of the abdominal muscles increase abdominal pressure, compress the rectum, and squeeze the feces from it.

Apply What You Know

One way to "potty train" a child or housebreak a puppy is to put the child on the toilet or the puppy outdoors within half an hour after a meal. Explain why this strategy would help, in view of what you now know about the neurological control of the colon.

Before You Go On

Answer these questions from memory. Reread the preceding section if there are too many you don't know.

28. List the parts of the large intestine in order from the ileocecal valve to the external anal sphincter.

29. Compare and contrast the mucosa and muscularis externa of the small and large intestines.

30. Discuss advantages and disadvantages of the abundant bacterial load in the large intestine.

31. Compare and contrast the defecation and urination (micturition) reflexes.

Aging of the Digestive System

Restaurants commonly offer a "seniors' menu" and reduced prices to customers over 60 or 65 years old, knowing that they tend to eat less. There are multiple reasons for this reduced appetite. Older people have lower metabolic rates and tend to be less active than younger people and, hence, need fewer calories. The stomach atrophies with age, and it takes less to fill it up. For many, food has less aesthetic appeal in old age because of losses in the senses of smell, taste, and even vision. In addition, older people secrete less saliva, making food less flavorful and swallowing more difficult. Reduced salivation also makes the teeth more prone to caries. Dentures are an unpleasant fact of life for many people over 65 who have lost their teeth to caries and periodontitis. Atrophy of the epithelium of the oral cavity and esophagus makes these surfaces more subject to abrasion and may further detract from the ease of chewing and swallowing.

The reduced mobility of old age makes shopping and meal preparation more troublesome, and with food losing its sensory appeal, some decide that it simply isn't worth the trouble. However, one's protein, vitamin, and mineral requirements remain essentially unchanged, so vitamin and mineral supplements may be needed to compensate for reduced food intake and poorer intestinal absorption. Malnutrition is common among older people and is an important factor in anemia and reduced immunity.

As the gastric mucosa atrophies, it secretes less acid and intrinsic factor. Acid deficiency reduces the absorption of calcium, iron, zinc, and folic acid. Heartburn becomes more common as the weakening lower esophageal sphincter fails to prevent the reflux of stomach contents into the esophagus. The most common digestive complaint of older people is constipation, which results from reduced muscle tone and weaker peristalsis of the colon. This seems to stem from a combination of factors: atrophy of the muscularis externa, reduced sensitivity to neurotransmitters that promote motility, less fiber and water in the diet, and less exercise. The liver, gallbladder, and pancreas show only slightly reduced function, but the drop in liver function reduces the rate of drug deactivation and can contribute to overmedication.

CAREER SPOTLIGHT

Dental Hygienist

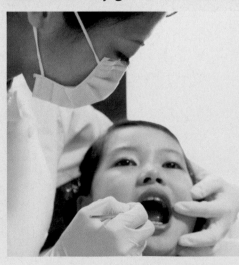

Dental hygienists are licensed health-care professionals who promote oral health and wellness. Their job includes assessment of the health of the oral cavity and surrounding structures of the head and neck to inspect for abnormalities or diseases such as oral cancer. They review health and dental history. They clean teeth by removing calculus and plaque. An important aspect of their care is teaching patients oral-hygiene techniques. They also apply cavity-preventive agents such as sealants and fluoride. In addition, they expose and develop dental radiographs (X-rays) for the dentist to interpret.

The majority of dental hygienists are employed by dentists and work in general dental offices. However, they may also serve in hospitals, nursing homes, public-health clinics, and schools. Dental hygienists must complete an accredited degree program and then pass both written and clinical examinations in order to be licensed by the state in which they practice. See appendix B for further career information in dental hygiene.

CONNECTIVE ISSUES

Ways in Which the Digestive System Affects Other Organ Systems

All Systems
Provides nutrients for all tissues.

Integumentary System
Vitamin C absorbed in the small intestine is necessary for collagen production and skin integrity.

Skeletal System
Small intestine adjusts calcium absorption in proportion to the needs of the skeletal system; dental decay or periodontitis can lead to bone infections and deterioration of mandibular and maxillary bone.

Muscular System
Liver disposes of lactic acid generated by muscles, thereby promoting recovery from muscular fatigue.

Nervous System
Some hormones secreted by the stomach and small intestine influence brain centers that regulate appetite.

Endocrine System
Stomach and small intestines produce some hormones and thus play an endocrine role; feeding stimulates the secretion of growth hormone and insulin; liver degrades hormones and influences rate of hormone clearance from the blood.

Circulatory System
Digestive tract absorbs fluid needed to maintain blood volume; liver degrades heme from dead erythrocytes and secretes clotting factors, blood albumin, and other plasma proteins except antibodies; liver regulates blood glucose level; intestinal epithelium stores iron and releases it as needed for hemoglobin synthesis; dental decay or periodontitis can lead to heart disease (infectious endocarditis).

Lymphatic and Immune Systems
GI mucosa is a site of lymphocyte production; infant intestine absorbs maternal IgA to confer passive immunity on the infant; acid, lysozyme, mucus, and other digestive secretions provide nonspecific defense against pathogens.

Respiratory System
Contraction of abdominal muscles aids in expiration by pushing digestive organs upward against underside of diaphragm.

Urinary System
Intestines complement the kidneys in water and electrolyte reabsorption; liver and kidneys collaborate in calcitriol synthesis; liver synthesizes urea, and kidneys excrete it.

Reproductive System
Provides nutrients for fetal growth and lactation.

Study Guide

Assess Your Learning Outcomes

To test your knowledge, discuss the following topics with a study partner or in writing, ideally from memory.

17.1 Overview of the Digestive System (p. 564)

1. Definition of *gastroenterology*
2. Five stages through which the digestive system processes food
3. Distinction between mechanical and chemical digestion and general purposes served by each
4. The two anatomical subdivisions of the digestive system
5. Layers of the digestive tract
6. The enteric nervous system and its function
7. Anatomical relationships among the parietal peritoneum, mesentery, greater omentum, and lesser omentum
8. Definitions of *intraperitoneal* and *retroperitoneal*

17.2 The Mouth Through Esophagus (p. 568)

1. Seven functions of the mouth
2. Anatomy and histology of the lips, cheeks, tongue, and palate
3. Tissues of the teeth and the types and numbers of teeth from midline to rear of the jaw
4. The three major pairs of major salivary glands; components of saliva and their functions
5. Anatomy of the pharynx and the digestive role of its sphincters
6. Anatomy of the esophagus, histology of its wall, and importance of its lower sphincter
7. The phases of swallowing

17.3 The Stomach (p. 572)

1. Definition of *chyme*
2. Role of the stomach in both mechanical and chemical digestion
3. Shape of the stomach, its four regions, and its pyloric sphincter
4. Histology of the stomach wall
5. Cells found in the gastric glands and their functions
6. Functions of hydrochloric acid
7. Two food macromolecules that are chemically digested in the stomach; the enzymes involved

8. Secretions of parietal and chief cells and the purposes they serve
9. The importance of intrinsic factor
10. Contractions of the stomach and the ejection of chyme into the duodenum
11. Vomiting, including its stimuli and risks
12. Events of the cephalic, gastric, and intestinal phases of gastric activity, and their neural and hormonal regulation
13. Three ways the stomach protects itself from self-digestion

17.4 The Liver, Gallbladder, and Pancreas (p. 579)

1. How liver function contributes to digestion
2. Gross anatomy of the liver
3. The structure of a hepatic lobule
4. The name and functions of the cuboidal epithelial cells of the liver
5. Structure of the gallbladder and bile passages
6. The composition and digestive role of bile
7. The location and gross anatomy of the pancreas
8. Function of the bicarbonate in pancreatic juice
9. Pancreatic enzymes and their individual functions
10. Hormonal regulation of pancreatic function

17.5 The Small Intestine (p. 584)

1. Length and diameter of the small intestine
2. The three regions of the small intestine, how they differ histologically, and what marks the beginning and end of each
3. Names of the internal folds and projections that increase the surface area of the small intestine
4. The structure of a villus, including its types of surface epithelial cells and the contents of its core
5. Differences between intestinal segmentation and peristalsis, and the function of each

17.6 Chemical Digestion and Absorption (p. 588)

1. Chemical digestion of starch in the mouth; the role and sources of amylase; how pH affects amylase
2. The actions of maltase, sucrase, and lactase and site where they perform chemical digestion
3. The absorption and fate of dietary monosaccharides
4. Sources of amino acids absorbed by the small intestine
5. The definition of *protease,* the names of proteases involved

in digestion, and the anatomical sites where different proteases work

6. How amino acids are absorbed by the small intestine
7. The role of bile and emulsification in fat digestion
8. The role of lipase and chylomicrons in fat digestion
9. The absorption of fat-soluble vitamins, water-soluble vitamins, and minerals
10. Varying modes of vitamin absorption in the small intestine
11. The hormonal regulation of calcium and iron absorption
12. The sources, absorption, and elimination of water in the digestive tract

17.7 The Large Intestine (p. 592)

1. Segments of the large intestine; the distinction between the large intestine and colon
2. Comparison of the wall of the small and large intestines and functional explanation of the anatomical differences
3. Reasons for the histological difference between the colon and anal canal
4. Beneficial and undesirable actions of intestinal bacteria; the composition of intestinal gas
5. Absorptive functions of the large intestine
6. Modes of contraction and neural control of the colon
7. The mechanism of defecation

Testing Your Recall

1. Which of the following enzymes acts in the stomach?
   a. chymotrypsin
   b. pepsin
   c. lactase
   d. lipase
   e. sucrase

2. Which of the following digests proteins?
   a. trypsin
   b. lipase
   c. maltase
   d. sucrase
   e. amylase

3. The lacteals absorb
   a. chylomicrons.
   b. micelles.
   c. emulsification droplets.
   d. amino acids.
   e. monosaccharides.

4. The infoldings visible on the inside of an empty stomach are
   a. circular folds.
   b. gastric rugae.
   c. haustra.
   d. microvilli.
   e. villi.

5. Which answer correctly lists the order of tissue layers of the small intestine, beginning with the innermost layer?
   a. mucosa, submucosa, serosa, muscularis externa
   b. submucosa, mucosa, muscularis externa, serosa
   c. serosa, muscularis externa, submucosa, mucosa
   d. mucosa, submucosa, muscularis externa, serosa
   e. mucosa, muscularis externa, submucosa, serosa

6. The gallbladder
   a. stores and concentrates bile.
   b. synthesizes cholesterol.
   c. synthesizes bile.
   d. secretes lipase.
   e. breaks down bilirubin.

7. Which structure is *not* lined with simple columnar epithelium?
   a. ascending colon
   b. duodenum
   c. stomach
   d. rectum
   e. anal canal

8. The pyloric region is part of what structure?
   a. the liver
   b. the stomach
   c. the duodenum
   d. the large intestine
   e. the pancreas

9. Which of the following secrete hydrochloric acid?
   a. hepatocytes
   b. parietal cells
   c. chief cells
   d. pancreatic cells
   e. enteroendocrine cells

10. Which of these is the voluntary phase of swallowing?
    a. oral phase
    b. cephalic phase
    c. gastric phase
    d. secretory phase
    e. pharyngeal phase

11. Most of the bulk of a tooth is composed of a yellowish tissue called _____.

12. The three regions of the small intestine from proximal to distal are the _____, _____, and _____.

13. The _____ suspends the liver from the abdominal wall and separates its right and left lobes.

14. Pepsinogen is produced by _____ cells.

15. Haustra are pouchlike outfoldings found on the _____.

16. The digestive tract has its own independent nervous plexus called the _____ nervous system.

17. _____, an enzyme that breaks down starch, is present in both pancreatic juice and _____.

18. The three pairs of salivary glands are the _____, _____, and _____.

19. What structure has hard and soft regions and makes it possible to continue breathing while chewing food?

20. _____ is a wave of muscular contraction produced by the muscularis externa that passes along material in the gastrointestinal lumen.

Answers in appendix A

True or False

Determine which five of the following statements are false, and briefly explain why.

1. Micelles are absorbed into lacteals.

2. Taeniae coli are found in the colon.

3. The ileocecal valve regulates the passage of chyme between the stomach and the duodenum.

4. Vitamin D is necessary for the intestinal absorption of iron.

5. Most monosaccharide molecules absorbed in the small intestine are glucose.

6. Trypsin is produced in the pancreas.

7. Swallowing is coordinated by a control center in the brainstem.

8. Most of the water in the digestive tract is reabsorbed in the large intestine.

9. Cholecystokinin stimulates the release of bile into the duodenum.

10. The jejunum is part of the large intestine.

Answers in appendix A

Testing Your Comprehension

1. Which do you think would have the most severe effect on digestion: surgical removal of the stomach, gallbladder, or pancreas? Explain.

2. Explain why most lipids must be absorbed by the lacteals rather than by the blood capillaries of a villus.

3. Follow a peanut butter sandwich through the digestive tract, beginning with the mouth. What major nutrient classes are in the sandwich? Where do mechanical and chemical digestion occur? Discuss at least three specific enzymes in your answer.

Nutrition and Metabolism

Mitochondria are the cell's "powerhouses," the organelles that make most of the ATP needed to sustain life. If all the body's mitochondria suddenly ceased to function, life itself would cease within one minute.

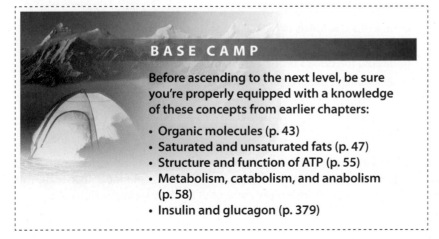

BASE CAMP

Before ascending to the next level, be sure you're properly equipped with a knowledge of these concepts from earlier chapters:

- Organic molecules (p. 43)
- Saturated and unsaturated fats (p. 47)
- Structure and function of ATP (p. 55)
- Metabolism, catabolism, and anabolism (p. 58)
- Insulin and glucagon (p. 379)

Chapter Outline

Anatomy & Physiology **REVEALED**
aprevealed.com

Module 12: Digestive System

Food is necessary for survival, fundamental to good health, and central to social interactions and cultural identity. It is the source of energy for cellular function and the raw material for building tissues and organs. Understanding food and nutrition is far from simple, however. We are surrounded by conflicting advice and opinions about what constitutes optimal nutrition. Through most of human history, people struggled to obtain enough food; currently, humans face the opposite problem—access to too much food, leading to obesity and associated health problems such as diabetes and cardiovascular disease. In chapter 17, we saw how the digestive system breaks nutrients down into usable forms and absorbs them into the blood and lymph. We now consider these nutrients in more depth, focusing on metabolism, the chemical changes that transform the food to energy for cellular function. In addition, we will explore regulation of appetite, body weight, and energy balance, and the related topic of body heat.

18.1 Nutrition

Expected Learning Outcomes
When you have completed this section, you should be able to:

a. define *calorie* and identify the three primary sources of dietary calories;
b. define *nutrient,* identify the six major categories of nutrients, and distinguish between macronutrients and micronutrients;
c. identify the major forms and sources of dietary carbohydrates;
d. name some forms of dietary fiber, explain why fiber is important in the diet yet is not a nutrient, and distinguish water-soluble from water-insoluble fiber;
e. discuss the forms of dietary lipid and explain why lipids are a more efficient means of energy storage than carbohydrates are;
f. describe the forms of blood lipoproteins and explain their respective functions;
g. discuss the body's protein and amino acid requirements; and
h. describe the minerals and vitamins required of the diet.

Good nutrition is part of a healthy lifestyle. Nutrients acquired from food support growth and maintenance of tissues, and fuel cellular activities. In this section, we will discuss the energy content of food and specific nutrients.

Calories

The subject of calories is familiar to those who are concerned about control of body weight. The calories listed on the back of a package of food indicate how much energy is in a serving. Energy content is directly related to capacity to do biological work. One calorie is the amount of heat that will raise the temperature of 1 g of water 1°C. One thousand calories is called a Calorie (capital *C*) when referring to diet and a **kilocalorie** (kcal) in biochemistry. (Confusingly, however, food labels often use both lowercase and capital c in referring to kilocalories.)

Nearly all dietary calories come from carbohydrates, proteins, and fats. Carbohydrates and proteins yield about 4 kcal/g when they are completely oxidized, and fats yield about 9 kcal/g.

When a chemical is described as **fuel** in this chapter, we mean it is oxidized solely or primarily to extract energy from it. The extracted energy is usually used to make adenosine triphosphate (ATP), the energy carrier necessary for all cellular work.

Nutrients

A **nutrient** is any ingested chemical that is used for fuel, growth, repair, or maintenance of the body. Nutrients fall into six major classes: water, carbohydrates, lipids, proteins, minerals, and vitamins (table 18.1). Water, carbohydrates, lipids, and proteins are considered **macronutrients** because they must be consumed in relatively large quantities. Minerals and vitamins are called **micronutrients** because only small quantities are required.

Table 18.1 Nutrient Classes and Their Principal Functions

Nutrient	Daily Requirement	Representative Functions
Water	2.5 L	Solvent; coolant; dilutes and eliminates metabolic wastes; supports blood volume and pressure
Carbohydrates	125–175 g	Fuel; component of glycoproteins, glycolipids, nucleic acids, and ATP; some structural functions
Lipids	80–100 g	Fuel; plasma membrane structure; myelin sheaths of nerve fibers; hormones; insulation; protective padding around organs
Proteins	44–60 g	Muscle contraction; structure of cellular membranes and extracellular material; major component of connective tissues; enzymes; some hormones; antibodies; transport of plasma lipids; oxygen binding and transport; blood clotting; blood viscosity and osmolarity; emergency fuel
Minerals	0.05–3,300 mg	Structure of bones and teeth; cofactors for many enzymes; electrolytes
Vitamins	0.002–60 mg	Coenzymes for many metabolic pathways; antioxidants

Recommended daily allowances (RDAs) of nutrients were first developed in 1943 by the National Research Council and National Academy of Sciences; they have been revised several times since. An RDA is a liberal but safe estimate of the daily intake that would meet the nutritional needs of most healthy people. Consuming less than the RDA of a nutrient does not necessarily mean you will be malnourished, but the probability of malnutrition increases in proportion to the amount of the deficit and how long it lasts.

Many nutrients can be synthesized by the body when they are unavailable from the diet. The body is incapable, however, of synthesizing minerals, most vitamins, eight of the amino acids, and one to three of the fatty acids. These are called **essential nutrients** because it is essential that they be included in the diet.

Carbohydrates

Carbohydrates—sugars and starch—serve primarily as fuel for cellular work and are required in greater amounts than any other nutrient. The brain alone consumes about 120 g of glucose per day. Carbohydrates also serve as a structural component of other molecules (table 18.1).

Dietary carbohydrates are derived primarily from plants in the form of polysaccharides (complex carbohydrates), disaccharides, and monosaccharides. The only significant polysaccharide nutrient is starch, obtained principally from grains and root vegetables such as potatoes. Cellulose is also an important dietary polysaccharide, but cannot be considered a nutrient (see the following section on fiber). Dietary disaccharides include sucrose, refined from sugarcane and sugar beets, and lactose, found in milk. A third disaccharide, maltose, comes mostly from the digestion of starch.

Digestion of starch and disaccharides ultimately results in the production of three monosaccharides: glucose, fructose, and galactose. Glucose is the only monosaccharide present in the blood in significant quantity and is therefore known as *blood sugar*. Its concentration is normally maintained at 70 to 110 mg/dL in peripheral venous blood. This level is carefully regulated. Most body cells rely on a combination of carbohydrates and fats to meet energy needs, but some cells such as neurons and erythrocytes depend almost exclusively on blood glucose. Even a brief period of **hypoglycemia** (glucose deficiency) causes nervous system disturbances felt as weakness or dizziness.

> *Apply What You Know*
>
> Glucose concentration is about 15 to 30 mg/dL higher in arterial blood than in most venous blood. Explain why.

Ideally, most carbohydrate intake should be in the form of starch. This is partly because foods that provide starch also usually provide other nutrients. Simple sugars not only provide empty calories but also promote tooth decay. In a typical American diet, only 50% of the carbohydrates come from starch and the other 50% from sucrose and corn syrup. Carbohydrate consumption in the United States has become excessive over the past century due to increased use of sugar in processed foods. A century ago, Americans consumed an average of 1.8 kg (4 pounds) of sugar per person per year. Now, with sucrose and high-fructose corn syrup so widely used in foods and beverages, the average American ingests over 28 kg (more than 60 pounds) of added sugar per year. A single nondiet soft drink contains 38 to 43 g (about 8 teaspoons) of sugar per 355 mL (12-ounce) serving.

Fiber

Dietary fiber refers to all fibrous materials of plant and animal origin that resist digestion. Most is plant matter—the carbohydrates cellulose and pectin and such noncarbohydrates as gums and lignin. The recommended daily allowance is about 30 g, but average intake varies greatly from country to country—from 40 to 150 g/day in India and Africa to only 12 g/day in the United States.

Fiber is not considered a nutrient because it is never absorbed into the human tissues, but it is an essential component of the diet. **Water-soluble fiber** includes pectin and certain other carbohydrates found in oats, beans, peas, carrots, brown rice, and fruits. It reduces blood cholesterol levels. **Water-insoluble** fiber includes cellulose and lignin. It apparently has no effect on cholesterol levels, but it absorbs water and swells, thereby softening the stool and increasing its bulk by 40% to 100%. This stretches the colon, stimulates peristalsis, and quickens the passage of feces. In doing so, water-insoluble fiber reduces the risk of constipation and diverticulitis. Contrary to previous popular and scientific belief, dietary fiber has no clear effect on the incidence of colorectal cancer.

Lipids

Fats and other lipids are reservoirs of energy and building blocks for plasma membranes, as well as support for organs. The fat stored in adipose tissue is composed of triglycerides, organic molecules that consist of fatty acids bound to a glycerol molecule (see chapter 2). In addition to the fat in adipose tissue, the body contains lesser amounts of other lipids such as phospholipids and cholesterol. These are major structural components of plasma membranes and myelin. Cholesterol is also important as a precursor of steroid hormones, bile acids, and vitamin D.

Fat and Energy Storage

Fat constitutes about 15% of the body mass of a healthy man in his early 20s, and about 25% of the body mass of a woman. Obesity, an excess of body fat, is discussed in Perspectives on Health (p. 610). Fat is an efficient way to store energy. It contains over twice as much energy as carbohydrates (9 kcal/g of fat compared with 4 kcal/g of carbohydrate). Furthermore, fat is hydrophobic, contains almost no water, and is therefore a more compact energy-storage substance than carbohydrates. Carbohydrates are hydrophilic, absorb water, and thus expand and occupy more space in the tissues. Typical fat reserves in a healthy young man contain enough energy for 119 hours of running, whereas his carbohydrate stores would suffice for only 1.6 hours.

Dietary Fat and Cholesterol

Dietary fat may be saturated or unsaturated. Saturated fats are predominantly of animal origin. They occur in meat, egg yolks, and dairy products but also in some plant products such as coconut and palm oils. Hydrogenated oils and vegetable shortening, found in many processed foods, are also high in saturated fat. Unsaturated fats predominate in nuts, seeds, and most vegetable oils.

A certain amount of dietary fat is necessary for health. Whereas the body can synthesize most fatty acids, **essential fatty acids** are those we cannot manufacture and must obtain from food. The *omega-3 fatty acids* may protect against cardiovascular disease. They are found in nuts, oily fish such as salmon, and some vegetable oils such as canola and sunflower oils.

Cholesterol is necessary for cell membrane stability and steroid hormone production. However, it is desirable to maintain a total plasma concentration of 200 mg/dL or less. Higher levels of blood cholesterol increase the risk of cardiovascular disease.

The richest dietary source of cholesterol is egg yolks, but it is also prevalent in dairy products; shellfish; organ meats such as liver and brains; and other mammalian meat. Most of the body's cholesterol, however, does not come from dietary sources but is synthesized internally. It may surprise you that lowering dietary cholesterol reduces the serum cholesterol level by no more than 5%. More important is the fact that certain saturated fatty acids (SFAs) raise serum cholesterol level. Some food advertising is deceptive on this point. It may truthfully advertise a food as being cholesterol-free, but neglect to mention that it contains SFAs that can raise the consumer's cholesterol level anyway. A moderate reduction of saturated fatty acid intake can lower blood cholesterol by 15% to 20%—considerably more effect than reducing dietary cholesterol alone.

Fat should account for no more than 30% of your daily caloric intake; no more than 10% of your fat intake should be saturated fat; and average cholesterol intake should not exceed 300 mg/day (one egg yolk contains about 240 mg). A typical American obtains 40% to 50% of his or her calories from fat, and ingests twice as much cholesterol as the recommended limit.

Blood Lipoproteins

Lipids must be transported to all cells of the body, yet they are hydrophobic and do not dissolve in the aqueous blood plasma. To overcome this, they are transported in the form of complexes called **lipoproteins**—tiny droplets with a core of cholesterol and triglycerides and a coating of proteins and phospholipids.

Lipoproteins are classified into categories by their density: **high-density lipoproteins (HDLs), low-density lipoproteins (LDLs),** and **very-low-density lipoproteins (VLDLs).** The higher the ratio of protein to lipid, the higher the density. VLDLs, produced in the liver, transport triglycerides to adipose tissue for storage; when triglycerides are removed, the VLDLs become LDLs, which contain mostly cholesterol (fig. 18.1a). Cells that need cholesterol, such as those that synthesize steroid hormones, absorb LDLs from the blood by receptor-mediated endocytosis.

HDLs start as protein shells produced in the liver. They travel in the blood and pick up excess cholesterol and phospholipids. The next time the HDL circulates through the liver, cholesterol is removed and eliminated in the bile (fig. 18.1b). Therefore, HDLs are a vehicle for removing excess cholesterol from the body. A high concentration of HDL protects against atherosclerosis, whereas a high LDL concentration is a warning sign because it signifies a high rate of cholesterol deposition in the arteries. Elevation of HDLs is promoted by a diet low in calories and saturated fats and by regular aerobic exercise. LDL levels rise in response to saturated fat consumption, cigarette smoking, and stress. In the popular press, HDLs and LDLs are often nicknamed "good cholesterol" and "bad cholesterol," respectively, but as we have seen, they are not cholesterol alone but complexes of proteins and diverse lipids.

Proteins

Proteins play an enormous diversity of roles in cellular processes and cell and tissue structure. They serve as signaling molecules (hormones and growth

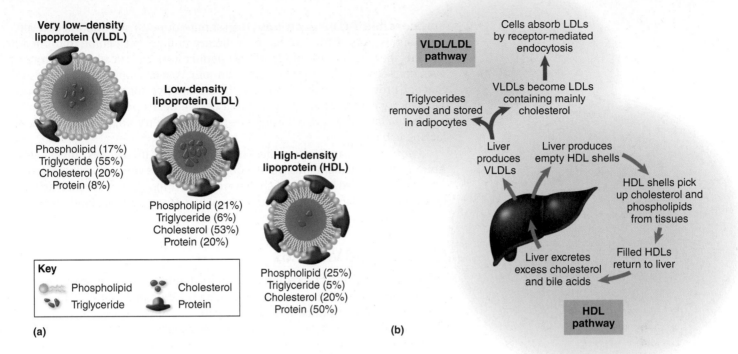

Very low–density lipoprotein (VLDL)

Phospholipid (17%)
Triglyceride (55%)
Cholesterol (20%)
Protein (8%)

Low-density lipoprotein (LDL)

Phospholipid (21%)
Triglyceride (6%)
Cholesterol (53%)
Protein (20%)

High-density lipoprotein (HDL)

Phospholipid (25%)
Triglyceride (5%)
Cholesterol (20%)
Protein (50%)

Key

Phospholipid
Triglyceride
Cholesterol
Protein

(a)

VLDL/LDL pathway

Cells absorb LDLs by receptor-mediated endocytosis

VLDLs become LDLs containing mainly cholesterol

Triglycerides removed and stored in adipocytes

Liver produces VLDLs

Liver produces empty HDL shells

HDL shells pick up cholesterol and phospholipids from tissues

Liver excretes excess cholesterol and bile acids

Filled HDLs return to liver

HDL pathway

(b)

Figure 18.1 Lipoprotein Processing. (a) Three types of serum lipoproteins. (b) Pathways of lipoprotein processing.

- *Why is a high HDL:LDL ratio healthier than a high LDL:HDL ratio?*

factors); protective antibodies; and membrane receptors, pumps, and ion channels. They function in oxygen transport (hemoglobin), muscle contraction (myosin and actin), and as enzymes that catalyze nearly every major reaction of cellular metabolism. Fibrous proteins such as collagen, elastin, and keratin make up much of the structure of skin, hair, nails, bones, cartilages, tendons, and ligaments. No other class of biomolecules has such a broad variety of functions.

Protein constitutes about 12% to 15% of the body's mass; 65% of it is in the skeletal muscles. For persons of average weight, the RDA of protein is 44 to 60 g, depending on age and sex. Multiplying your weight in pounds by 0.37 or your weight in kilograms by 0.8 gives an estimate of your RDA of protein in grams. A higher intake is recommended, however, under conditions of stress, infection, injury, and pregnancy. Infants and children require more protein than adults relative to body weight because proteins are necessary for building tissues during growth.

Total protein intake is not the only measure of dietary adequacy. The nutritional value of a protein depends on whether it supplies the right amino acids in the proportions needed to make human proteins. Adults can synthesize 12 of the 20 amino acids from other organic compounds when they are not available from the diet, but there are 8 **essential amino acids** that we cannot synthesize. These must be obtained by including the right proteins in the diet.

Cells do not store surplus amino acids for later use. When a protein is to be synthesized, all of the amino acids necessary must be present at once; if even one is missing, the protein cannot be made. High-quality **complete proteins** are those that provide all of the essential amino acids in the necessary proportions for human tissue growth and maintenance. Lower-quality **incomplete proteins** lack one or more essential amino acids. For example, cereals are low in lysine, and legumes are low in methionine.

Animal proteins found in meat, eggs, and dairy products closely match human proteins in amino acid composition. Thus, animal products provide high-quality complete protein, whereas plant proteins are incomplete. We typically use 70% to 90% of animal protein but only 40% to 70% of plant protein. It therefore takes a larger serving of plant protein than animal protein to meet our needs—for example, we need 400 g (about 14 ounces) of rice and beans to provide as much usable protein as 115 g (about 4 ounces) of hamburger. Nevertheless, this does not mean that your dietary protein must come from meat; indeed, about two-thirds of the world's population receives adequate protein from diets containing very little animal protein. We can combine plant foods so that one provides what another lacks—beans and rice, for example, are a complementary combination of legume and cereal. Beans provide the isoleucine and lysine lacking in grains, while rice provides the tryptophan and cysteine lacking in beans.

Reducing meat intake while increasing plant intake has advantages. Among other considerations, plant foods provide more vitamins, minerals, and fiber; less saturated fat; and no cholesterol. Also important in our increasingly crowded world is the fact that producing meat requires far more land than producing food crops.

Minerals and Vitamins

Minerals and vitamins are required in small quantities but have very potent effects on physiology. Minerals are inorganic elements that plants extract from soil or water and introduce into the food web. Vitamins are small organic compounds that are not used as fuel, but serve as coenzymes or cofactors that enable enzymes to function. Deficiencies or excessive amounts lead to disease states and are potentially lethal. Table 18.2 summarizes requirements and dietary sources of some important minerals and vitamins.

Minerals constitute about 4% of the body mass, with three-quarters of this being the calcium and phosphorus in the bones and teeth. Phosphorus is also a key structural component of phospholipids and ATP. Calcium, iron, magnesium, and manganese function as cofactors for enzymes. Iron is essential to the oxygen-carrying capacity of hemoglobin. Many mineral salts function as electrolytes. For example, sodium chloride (table salt) is required for nerve and muscle cell function and maintenance of blood volume. The recommended sodium intake is 1.1 g/day, but a typical U.S. diet contains about 3 to 7 g/day. This is due not just to the use of table salt but more significantly to the large amounts of salt in processed foods, much of it "disguised" in soy sauce, MSG (monosodium glutamate), baking soda, and baking powder. Broadly speaking, the best sources of minerals are vegetables, legumes, milk, eggs, fish, shellfish, and some other meats. Cereal grains are a relatively poor source, but processed cereals may be mineral-fortified.

Vitamins were originally named with letters in the order of their discovery, but they also have chemically descriptive names such as ascorbic acid (vitamin C) and riboflavin (vitamin B_2). Most vitamins must be obtained from the diet (table 18.2), but the body synthesizes some of them from precursors called *provitamins*—vitamin D from cholesterol, and vitamin A from carotene, abundantly present in carrots, squash, and other yellow vegetables and fruits. Vitamin K and folic acid are produced by the bacteria of the large intestine.

Vitamins are classified as water-soluble or fat-soluble. **Water-soluble vitamins** are absorbed with water from the small intestine, dissolve freely in the body fluids, and are quickly excreted by the kidneys. They cannot be stored in the body and therefore seldom accumulate to excess. The water-soluble vitamins are ascorbic acid (vitamin C) and the B vitamins. Ascorbic acid promotes hemoglobin synthesis, collagen synthesis, and sound connective tissue structure; it is an antioxidant that scavenges free radicals and possibly reduces the risk of cancer. The B vitamins function as **coenzymes**—small organic molecules that assist enzymes by transferring electrons from one metabolic reaction to another, making it possible for enzymes to catalyze these reactions. Some of their functions are discussed later in this chapter as we consider carbohydrate metabolism.

Table 18.2 Mineral and Vitamin Requirements and Some Dietary Sources

Mineral	RDA (mg)	Some Dietary Sources*
Major minerals		
Calcium	1,200	Milk, fish, shellfish, greens, tofu, orange juice
Phosphorus	1,200	Red meat, poultry, fish, eggs, milk, legumes, whole grains, nuts
Sodium	1,500	Table salt, processed foods; usually present in excess
Chloride	2,300	Table salt, some vegetables; usually present in excess
Magnesium	280–350	Milk, greens, whole grains, nuts, legumes, chocolate
Potassium	4,700	Red meat, poultry, fish, cereals, spinach, squash, bananas, apricots
Trace minerals		
Zinc	12–15	Red meat, seafood, cereals, wheat germ, legumes, nuts, yeast
Iron	10–15	Red meat, liver, shellfish, eggs, dried fruits, legumes, nuts, molasses
Fluoride	1.5–4.0	Fluoridated water and toothpaste, tea, seafood, seaweed
Iodine	0.15	Marine fish, fish oils, shellfish, iodized salt
Vitamin	**RDA (mg)**	**Some Dietary Sources**
Water-soluble vitamins		
Ascorbic acid (C)	60	Citrus fruits, strawberries, tomatoes, greens, cabbage, cauliflower, broccoli, brussels sprouts
B complex		
Thiamine (B_1)	1.5	Red meat, liver, other organ meats, eggs, greens, asparagus, legumes, whole grains, seeds, yeast
Riboflavin (B_2)	1.7	Widely distributed, and deficiencies are rare; all types of meat, milk, eggs, greens, whole grains, apricots, legumes, mushrooms, yeast
Niacin (nicotinic acid)	19	Readily synthesized from tryptophan, which is present in any diet with adequate protein; red meat, liver, other organ meats, poultry, fish, apricots, legumes, whole grains, mushrooms
Folic acid (folacin)	0.2	Eggs, liver, greens, citrus fruits, legumes, whole grains, seeds
Fat-soluble vitamins		
Retinol (A)	1.0	Fish oils, eggs, cheese, milk, greens, other green and yellow vegetables and fruits, margarine
Calcitriol (D)	0.01	Formed by exposure of skin to sunlight; fish, fish oils, milk
Tocopherol (E)	10	Fish oils, greens, seeds, wheat germ, vegetable oils, margarine, nuts

*"Red meat" refers to mammalian muscle such as beef and pork. Liver is specified separately and refers to beef, pork, and chicken livers, which are similar for most nutrients. "Other organ meats" refers to brain, pancreas, heart, kidney, and so on.

Fat-soluble vitamins are absorbed with dietary lipids. They are more varied in function than water-soluble vitamins. Vitamin A is a component of the visual pigments and promotes epithelial tissue maintenance. Vitamin D promotes calcium absorption and bone mineralization. Vitamin K is essential to blood clotting. Vitamins A and E are antioxidants, like ascorbic acid.

It is common knowledge that various diseases result from vitamin deficiencies, but it is less well known that vitamin excess also causes disease. A *deficiency* of vitamin A, for example, can result in night blindness, dry skin and hair, a dry conjunctiva and cloudy cornea, and increased incidence of urinary, digestive, and respiratory infections. This is the world's most common vitamin deficiency. An *excess* of vitamin A, however, may cause anorexia, nausea and vomiting, headache, pain and fragility of the bones, hair loss, an enlarged liver and spleen, and birth defects.

Some people take *megavitamins*—doses 10 to 1,000 times the RDA—thinking that they will improve athletic performance. Since vitamins are not burned as fuel, and small amounts fully meet the body's metabolic needs, there is no evidence that vitamin supplements improve performance except when used to correct a dietary deficiency. Megadoses of fat-soluble vitamins can be especially harmful.

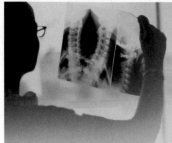

PERSPECTIVES ON HEALTH

Obesity

Obesity is clinically defined as a weight more than 20% above the recommended norm for one's age, sex, and height. In the United States, about 30% of the population are obese and another 35% are overweight; there has recently been an alarming increase in the number of children who are morbidly obese by the age of 10. *Body mass index (BMI)* is a useful tool for assessing obesity. BMI = W/H^2, where W is weight in kilograms and H is height in meters. (If using weight in pounds and height in inches, BMI = $703 \, W/H^2$.) A BMI of 20 to 25 is considered to be optimal for most people. A BMI over 27 is defined as overweight, and above 30 is considered obese.

The obesity epidemic is a serious health concern. Excess fat increases a person's risk of atherosclerosis, hypertension, and joint degeneration; cancer of the breast, uterus, and liver in women; and cancer of the colon, rectum, and prostate gland in men. Excess thoracic fat impairs breathing and results in increased blood P_{CO_2}, sleepiness, and reduced vitality. Obesity is also a significant impediment to successful surgery.

The rise in obesity has paralleled an increased incidence of type 2 diabetes mellitus. High levels of abdominal fat are associated with insulin insensitivity and lead to high blood sugar and all of the resulting complications. Until recently, type 2 diabetes almost exclusively developed in people who were age 50 and older, but the incidence of type 2 diabetes is rising in young people as more children become obese.

Heredity plays a significant role in obesity. However, a predisposition to obesity is often greatly worsened by overfeeding in infancy and childhood. Consumption of excess calories in childhood causes adipocytes to increase in size and number. In adulthood, adipocytes do not multiply except in some cases of extreme weight gain; their number remains constant. Weight gains and losses result from changes in cell size (cellular hypertrophy).

As so many dieters learn, it is very difficult to substantially reduce one's adult weight. Most diets are unsuccessful over the long run as dieters lose and regain the same weight over and over. From an evolutionary standpoint, this is not surprising. The body's appetite- and weight-regulating mechanisms are geared toward stimulation of appetite and eating rather than restraint. In our evolutionary past, scarcity of food was a more common problem than a food surplus, and ability to store fat was a key survival adaptation. Recent changes in human diet and activity have been too rapid for the gene pool to respond, with the result that billions of people around the world are overfat. Former survival mechanisms have now become mechanisms of pathology.

Understandably, pharmaceutical companies are keenly interested in developing effective weight-control drugs. There could be an enormous profit, for example, in drugs that affect appetite-regulating hormones—for example, by inhibiting the appetite-stimulating hormone *ghrelin* or enhancing or mimicking the appetite-suppressing hormone *leptin* (both discussed on p. 622). Such efforts have so far met with little success. An integrative approach to obesity requires better understanding of appetite and body weight regulation, as well as emphasis on increased exercise and consumption of varied, plant-based diets.

Before You Go On

Answer these questions from memory. Reread the preceding section if there are too many you don't know.

1. What are the four classes of macronutrients?
2. Why is fiber important in the diet, and why isn't it considered a nutrient?
3. Why is fat more efficient than carbohydrates for energy storage?
4. Contrast the functions of VLDLs, LDLs, and HDLs. Explain why a high level of blood HDLs is desirable, but a high level of LDLs is undesirable.
5. Why do some proteins have more nutritional value than others?
6. Define the terms mineral *and* vitamin. *Give several important examples of each.*

18.2 Carbohydrate Metabolism

Expected Learning Outcomes

When you have completed this section, you should be able to:

a. Define *metabolism, catabolism,* and *anabolism;*

b. describe the basic purposes and products of glycolysis, anaerobic fermentation, and aerobic respiration;

c. be able to trace the energy content of glucose through all these stages of metabolism, stating where the energy is contained by the end of each stage; and

d. describe the body's modes of glycogen synthesis and breakdown, and its synthesis of glucose from other noncarbohydrates.

As we studied in chapter 2, metabolism is the sum of all chemical changes in the body, and consists of two classes of reactions: energy-requiring synthesis reactions called *anabolism,* and energy-releasing breakdown reactions called *catabolism.* In this section, we focus mainly on the energy-releasing catabolism of glucose and the use of that energy to make ATP.

Glucose Catabolism

Most dietary carbohydrate is burned as fuel within a few hours of absorption. Glucose is the major product of carbohydrate digestion, and therefore the breakdown of glucose (glucose catabolism) provides energy from carbohydrates. The overall reaction for this is

$$C_6H_{12}O_6 + 6\ O_2 \rightarrow 6\ CO_2 + 6\ H_2O$$

The function of this reaction is not to produce carbon dioxide and water but to transfer energy from glucose to ATP.

If the preceding reaction were carried out in a single step, it would generate a short, intense burst of heat—like the burning of paper, which has the same chemical equation. Not only would this be useless to the body's metabolism, but it would kill the cells. In the body, however, the process is carried out in a series of small steps, each controlled by a separate enzyme. Energy is released in small manageable amounts, and as much of it as possible is transferred to ATP. The rest is released as heat.

Major pathways of glucose catabolism (fig. 18.2) follow.

1. **Glycolysis,** which splits a glucose molecule into two molecules of pyruvic acid, takes place in the cytoplasm.

2. **Anaerobic fermentation** processes pyruvic acid in a way that allows glycolysis to continue functioning in a state of oxygen deficiency.

3. **Aerobic respiration** occurs in the presence of oxygen and oxidizes pyruvic acid to carbon dioxide and water. The steps associated with aerobic respiration take place in the mitochondria.

Coenzymes are vitally important to these reactions. Enzymes remove electrons (as hydrogen atoms) from the intermediate compounds of these pathways and transfer them to coenzymes. Coenzymes become the temporary carriers of the energy extracted from glucose metabolites. Thus, the enzymes of glucose catabolism cannot function without their coenzymes.

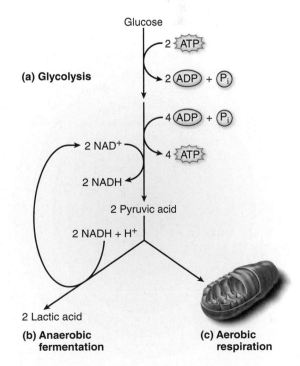

(a) Glycolysis

Glucose

2 ATP

2 ADP + Pi

4 ADP + Pi

2 NAD+

4 ATP

2 NADH

2 Pyruvic acid

2 NADH + H+

2 Lactic acid

(b) Anaerobic fermentation

(c) Aerobic respiration

Figure 18.2 Pathways of Glucose Catabolism. (a) Glycolysis. (b) Anaerobic fermentation. (c) Aerobic respiration.

Two coenzymes of special importance are **NAD⁺** (nicotinamide adenine dinucleotide) and **FAD** (flavin adenine dinucleotide). Both are derived from B vitamins: NAD^+ from niacin and FAD from riboflavin.

Glycolysis

Glycolysis[1] consists of a series of enzymatic steps that convert glucose, a six-carbon sugar (C_6), into two three-carbon (C_3) molecules of **pyruvic acid** (fig. 18.2a). Glycolysis requires an "initial investment" of 2 ATP, but pays it back "with interest"—generating 4 ATP, for a net yield of 2 ATP. It also requires 2 molecules of the coenzyme NAD^+. As glycolysis extracts energy from glucose, it transfers it to NAD^+, creating a higher-energy form, NADH. At the end of glycolysis, some of the energy that was in the glucose has escaped as heat, but energy still useful to the body remains in the 2 ATP, 2 NADH, and mostly, the 2 pyruvic acid.

Anaerobic Fermentation

The fate of pyruvic acid depends on whether or not oxygen is available. In an exercising muscle, the demand for ATP may exceed the supply of oxygen. The only ATP the cells can make under these circumstances is the 2 ATP produced by glycolysis. Cells without mitochondria, such as erythrocytes, are also restricted to making ATP by this method.

In the absence of oxygen, a cell resorts to a one-step reaction called *anaerobic fermentation.* (This is often inaccurately called *anaerobic respiration,* but strictly speaking, human cells do not carry out anaerobic respiration; that is a process found only in certain bacteria.) In this pathway, pyruvic acid is reduced to **lactic acid** (fig. 18.2b). This step does not liberate any more of the original glucose energy; its purpose is to regenerate the NAD^+ necessary to keep glycolysis running.

Skeletal muscle is relatively tolerant of anaerobic fermentation, and cardiac muscle less so. The brain employs almost no anaerobic fermentation. Most cells can survive brief periods of oxygen deficiency by generating ATP indirectly in this way. However, oxygen is required to produce enough ATP to sustain the body for more than a few minutes.

Aerobic Respiration

In the presence of oxygen, pyruvic acid enters the mitochondria and the pathway of *aerobic respiration* (fig. 18.2c). Most of our ATP is produced through oxidation of pyruvic acid in two principal steps:

- **Matrix reactions**, so named because their controlling enzymes are in the fluid of the mitochondrial matrix
- **Membrane reactions**, so named because their controlling enzymes are bound to the membranes of the mitochondrial cristae

The Matrix Reactions

Most of the matrix reactions (fig. 18.3) constitute a series of enzymatic steps called the **citric acid (Krebs**[2]**) cycle.** For every glucose molecule that enters glycolysis, all of the matrix reactions occur twice (once for each pyruvic acid). As the products of pyruvic acid pass through the cycle, carbon atoms are stripped and carried away as CO_2, to be exhaled as metabolic waste. The energy extracted from pyruvic acid is transferred to ATP, more NADH, and the reduced (high-energy) form of the other coenzyme, $FADH_2$. At the end of the cycle, there

---

[1] *glyco* = sugar; *lysis* = splitting
[2] Sir Hans Krebs (1900–1981), German biochemist

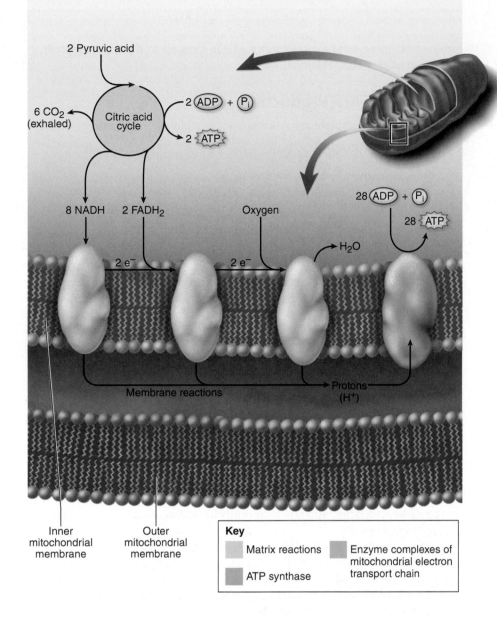

Figure 18.3 Aerobic Respiration. Of the 8 NADH generated by the matrix reactions, 2 are produced where pyruvic acid enters the citric acid cycle and 6 are generated by the citric acid cycle itself.

- *What two coenzymes import energy into this reaction chain, supplying the energy that becomes stored in ATP?*

is nothing left of the organic matter of the glucose; its carbon atoms have all been carried away. What remains of value at the end of the matrix reactions is the energy content of another 2 ATP, 8 NADH, and 2 FADH$_2$. These reduced coenzymes provide energy for the relatively massive production of ATP in the membrane reactions that follow.

The Membrane Reactions

The membrane reactions further oxidize NADH and FADH$_2$ and transfer their energy to ATP. This is accomplished by passing their electrons along a series of *enzyme complexes* called the **mitochondrial electron-transport chain**, bound to the inner mitochondrial membrane (fig. 18.3). The members of the chain are arranged in a precise order that enables each one to receive a pair of electrons and pass them to the next—like a row of people passing along a hot potato. By the time the "potato" reaches the last member in the chain, it is relatively "cool"—its energy has been used to make ATP. The final electron acceptor in the chain is oxygen. Each oxygen atom (O) receives two electrons (2 e$^-$) from the transport chain and two protons (2 H$^+$) from the matrix reactions, generating water (H$_2$O) as an end product. An enzyme called *ATP synthase* uses energy from this process to generate more ATP. This is a relatively prolific mode of ATP production, generating up to 28 more ATP in contrast to the 2 from glycolysis and 2 from the matrix reactions.

Overview of ATP Production

In summary, the complete aerobic respiration of glucose can be represented

$$C_6H_{12}O_6 + 6\ O_2 + 32\ ADP + 32\ P_i \rightarrow 6\ CO_2 + 6\ H_2O + 32\ ATP$$

and the sources of ATP can be summarized

$$\begin{array}{l} 2\ \text{ATP net yield of glycolysis} \\ 2\ \text{ATP from matrix reactions} \\ \underline{+\ 28\ \text{ATP from membrane reactions}} \\ \quad 32\ \text{ATP per glucose} \end{array}$$

It is clear that oxygen is required for the most efficient ATP production, and that most of the ATP in our bodies is generated by the membrane reactions of the mitochondria.

Glycogen Metabolism

ATP is quickly used after it is synthesized—it is an *energy-transfer* molecule, not an *energy-storage* molecule. Therefore, if the body has an ample amount of ATP and there is still more glucose in the blood, it converts the glucose to other compounds better suited for energy storage—namely, glycogen and fat. Here we consider the synthesis and use of glycogen (see fig. 11.11, p. 380).

Glycogenesis, the synthesis of glycogen, is stimulated by insulin. It consists of polymerization reactions that link thousands of glucose molecules into branched chains of glycogen for storage. The average adult body contains about 400 to 450 g of glycogen: nearly one-quarter of it in the liver, three-quarters of it in the skeletal muscles, and small amounts in cardiac muscle, the uterine lining, and other tissues.

Glycogenolysis, the hydrolysis of glycogen, releases glucose when new glucose is not being ingested. The process is stimulated by glucagon and epinephrine. The liver releases the resulting glucose into the blood, maintaining blood sugar levels between meals. The other glycogen-storing tissues use the free glucose only for their own needs.

When the body contains insufficient glycogen and free glucose to meet its needs, the liver carries out **gluconeogenesis**,[3] the synthesis of glucose from noncarbohydrates such as glycerol and amino acids. This is our means of weight loss when we control calorie intake and force the body to convert fat to glycogen. The distinctions among these similar terms are summarized in table 18.3.

---

[3] *gluco* = sugar, glucose; *neo* = new; *genesis* = production of

Table 18.3 Terminology Related to Glucose and Glycogen Metabolism

Anabolic (synthesis) reactions	
Glycogenesis	The synthesis of glycogen by polymerizing glucose
Gluconeogenesis	The synthesis of glucose from noncarbohydrates such as glycerol and amino acids
Catabolic (breakdown) reactions	
Glycolysis	The splitting of glucose into two molecules of pyruvic acid in preparation for anaerobic fermentation or aerobic respiration
Glycogenolysis	The hydrolysis of glycogen to release glucose

Before You Go On

Answer these questions from memory. Reread the preceding section if there are too many you don't know.

7. In the laboratory, glucose can be oxidized in a single step to CO_2 and H_2O. Why is it done in so many little steps in cells?

8. What are the end products of glycolysis? By the end of glycolysis, where is the energy that used to be in the glucose?

9. What is the main advantage of aerobic respiration over anaerobic fermentation?

10 What is the total yield of ATP for all steps from glucose through the end of aerobic respiration? State what quantities of ATP are produced at each major stage.

11. What conditions promote glycogenesis, glycogenolysis, and gluconeogenesis?

18.3 Lipid and Protein Metabolism

Expected Learning Outcomes

When you have completed this section, you should be able to:

a. describe the process that uses the energy of triglycerides for synthesizing ATP, including the fates of the glycerol and fatty acid components;

b. describe the process of protein catabolism and why it generates nitrogenous wastes; and

c. identify several nondigestive functions of the liver.

In the foregoing discussion, glycolysis and the mitochondrial reactions were treated from the standpoint of carbohydrate oxidation. When blood glucose concentration is low, most cells can break down other nutrient molecules to produce ATP. Here, we examine metabolic pathways associated with the oxidation of lipids and proteins as fuel.

Lipids

The total amount of triglycerides stored in the body's adipocytes usually remains quite constant, but there is continual turnover as lipids are released, transported in the blood, and either oxidized for energy or redeposited in other adipocytes. As we all know, a diet high in sugars causes us to gain weight, mostly in the form of fat. When sugars and amino acids exceed the body's immediate needs, they can be used to synthesize triglyceride precursors, glycerol and fatty acids. These can then be condensed to form a triglyceride, which is stored in the adipose tissue.

In contrast, if glucose supplies are low, fat is used for fuel. Breaking down fat begins when a triglyceride is split into glycerol and fatty acids—a process stimulated by epinephrine, norepinephrine, glucocorticoids, thyroid hormone, and growth hormone. The glycerol and fatty acids are further oxidized by separate pathways (fig. 18.4). Glycerol enters the pathway of glycolysis. It generates only half as much ATP as glucose, however, because it is a C_3 compound compared with glucose (C_6); thus, it leads to the production of only half as much pyruvic acid.

The fatty acids are oxidized in the mitochondria into 2-carbon fragments called *acetyl groups* and fed into the matrix reactions. A typical fatty acid of 16 carbon atoms can yield 129 molecules of ATP—obviously a much richer source of energy than a glucose molecule.

Excessively rapid fat breakdown, as in diabetes mellitus, can produce an overload of acetyl groups, which the liver converts to *ketone bodies*. Accumulation of ketone bodies can cause a pH imbalance called *ketoacidosis*. In untreated diabetes, cells cannot take up glucose and therefore oxidize fats. This leads to a telltale sweet odor on the breath from exhaled ketones, a diagnostic presence of ketones in the urine, and often a ketoacidosis-induced coma and death.

> **Apply What You Know**
>
> Explain why elevation of blood ketones and even ketoacidosis may result from extreme low-carbohydrate diets.

Proteins

Amino acids are primarily used for protein synthesis—the process of building the thousands of different proteins necessary for cellular function and structure. Each protein requires a specific combination of amino acids, assembled in proper order according to instructions contained in the DNA. Protein manufacture, described on page 84, is stimulated by various hormones such as growth hormone and thyroid hormone.

Amino acids can also be converted to glucose or fat or used directly as fuel. Although they are not the preferred fuel of the body, amino acids may be converted to pyruvic acid or one of the acids of the citric acid cycle (fig. 18.4). Before they enter the citric acid cycle, amino acids are **deaminated**—the amino group ($-NH_2$) is removed and becomes ammonia (NH_3), which is extremely toxic to cells. The liver quickly combines ammonia with carbon dioxide to produce a less toxic waste, urea, which is then excreted in the urine. Urea is the body's main nitrogenous waste. When a diseased liver cannot produce urea, ammonia accumulates in the blood and death from *hepatic coma* may ensue within a few days.

Liver Functions in Metabolism

The liver plays a wide variety of roles in the processes discussed in this chapter—especially carbohydrate, lipid, and protein metabolism. Although it is connected to the digestive tract and generally regarded as a digestive gland, the overwhelming majority of its functions are nondigestive (table 18.4). All of these are performed by the cuboidal hepatocytes described in chapter 17. Such functional diversity is remarkable in light of the uniform structure of these cells. Because of the numerous critical functions performed by the liver, degenerative liver diseases such as hepatitis, cirrhosis, and liver cancer are especially life-threatening (see Clinical Application 18.1).

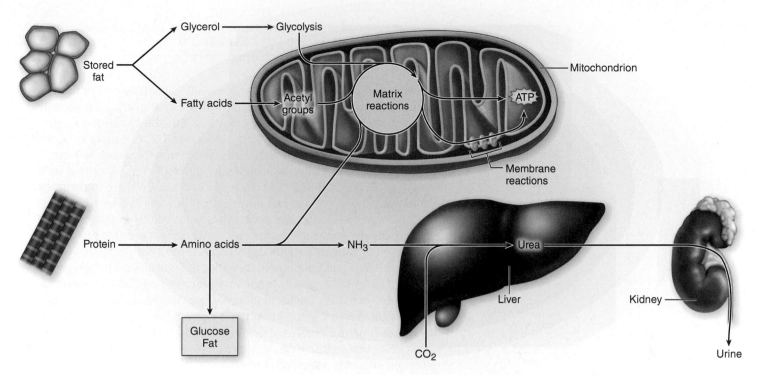

Figure 18.4 Fat and Protein Catabolism. Pathways for burning fat and protein as fuel to generate ATP.

Table 18.4 Functions of the Liver

Carbohydrate metabolism

Converts dietary fructose and galactose to glucose. Stabilizes blood glucose concentration by storing excess glucose as glycogen (glycogenesis); releasing glucose from glycogen when needed (glycogenolysis); and synthesizing glucose from fats and amino acids (gluconeogenesis) when glucose demand exceeds glycogen reserves.

Lipid metabolism

Carries out most of the body's fat synthesis and synthesizes cholesterol and phospholipids; produces VLDLs to transport lipids to adipose tissue and other tissues for storage or use; and stores fat in its own cells. Carries out most oxidation of fatty acids; produces ketone bodies. Produces HDL shells, which pick up excess cholesterol from other tissues and return it to the liver; excretes the excess cholesterol in bile.

Protein and amino acid metabolism

Carries out most deamination of amino acids. Converts ammonia to urea. Synthesizes nonessential amino acids.

Synthesis of plasma proteins

Synthesizes nearly all the proteins of blood plasma, including albumin, globulins, fibrinogen, and several clotting factors.

Vitamin and mineral metabolism

Converts vitamin D_3 to calcidiol, a step in the synthesis of calcitriol; stores a 3- to 4-month supply of vitamin D. Stores a 10-month supply of vitamin A and enough vitamin B_{12} to last for several years. Stores and releases iron as needed. Excretes excess calcium by way of the bile.

Digestion

Synthesizes bile acids and lecithin, which emulsify fat and promote its digestion.

Disposal of drugs, toxins, and hormones

Detoxifies alcohol, antibiotics, and many other drugs. Degrades hormones, removing them from circulation.

Clinical Application 18.1

HEPATITIS

Hepatitis, inflammation of the liver, is usually caused by one of the five strains of hepatitis viruses. They differ in mode of transmission, severity of the resulting illness, affected age groups, and the best strategies for prevention. Hepatitis A is common and mild. Over 45% of people in urban areas of the United States have had it. It spreads rapidly in such settings as day-care centers, and it can be acquired by eating uncooked seafood such as oysters. Hepatitis B and C are far more serious. Both are transmitted sexually and through blood and other body fluids; the incidence of hepatitis C has surpassed AIDS as a sexually transmitted disease. The diseases may be passed among injecting drug users who share needles and may be transmitted by using nonsterile instruments during body piercings or tattoos. Initial signs and symptoms of hepatitis include fatigue, nausea, vomiting, and weight loss. The liver becomes enlarged and tender. Jaundice, or yellowing of the skin, tends to follow as hepatocytes are destroyed, bile passages are blocked, and bile pigments accumulate in the blood. Hepatitis A causes up to 6 months of illness, but most people recover and then have permanent immunity to it. Hepatitis B and C, however, often lead to chronic hepatitis, which can progress to cirrhosis or liver cancer. More liver transplants are necessitated by hepatitis C than by any other cause.

Before You Go On

Answer these questions from memory. Reread the preceding section if there are too many you don't know.

12. What two products result from the hydrolysis of triglycerides? How is each of them used to make ATP, and how does the ATP yield from this compare to the yield from glucose?
13. Why can an abnormally rapid rate of fat oxidation result in a pH imbalance in the body?
14. What must be done to an amino acid before it can enter the citric acid cycle? What happens to the by-product of this preparatory reaction?
15. List at least six nondigestive functions of the liver.

18.4 Metabolic States and Metabolic Rate

Expected Learning Outcomes

When you have completed this section, you should be able to:

a. distinguish the absorptive (fed) state from the postabsorptive (fasting) state;
b. explain what happens to carbohydrates, fats, proteins, and amino acids in each of these states;
c. describe the hormonal and nervous regulation of each state;
d. define *metabolic rate* and *basal metabolic rate*; and
e. describe some factors that alter the metabolic rate.

Your metabolism changes from hour to hour depending on how long it has been since your last meal. For about 4 hours during and after a meal, you are in the **absorptive (fed) state**. This is a time in which nutrients are being absorbed and may be used immediately to meet energy and other needs. The **postabsorptive (fasting) state** prevails in the late morning, late afternoon, and overnight. During this time, the stomach and small intestine are empty and the body's energy needs are met from stored fuels.

The Absorptive State

In the absorptive (fed) state, blood glucose is readily available for ATP synthesis. It serves as the primary fuel and spares the body from having to draw on stored fuels. The status of major nutrient classes during this phase is as follows.

- **Carbohydrates**. Absorbed sugars are transported by the hepatic portal system to the liver. Most glucose passes through the liver and becomes available to cells everywhere in the body. Glucose in excess of immediate need, however, is absorbed by the liver and may be converted to glycogen or fat. Most fat synthesized in the liver is released into the circulation.
- **Fats**. Dietary fats enter the lymph before entering the bloodstream. Unlike other nutrients, they initially bypass the liver. Fats are taken up by tissues, especially adipose and muscular tissue, for which they are a major energy source.
- **Amino acids**. Amino acids, like sugars, circulate first to the liver. Most pass through and become available to other cells for protein synthesis. Some, however, are removed by the liver and are (1) used for protein synthesis, (2) deaminated and used as fuel for ATP synthesis, or (3) deaminated and used for fatty acid synthesis.

Regulation of the Absorptive State

The absorptive state is regulated largely by insulin, which is secreted in response to elevated blood glucose and amino acid levels and to the intestinal hormones gastrin, secretin, and cholecystokinin. Insulin has the following effects on its target cells:

- Within minutes, it increases the cellular uptake of glucose by as much as 20-fold. As cells absorb glucose, the blood glucose concentration falls.
- It stimulates glucose oxidation, glycogenesis, and fat synthesis.
- It stimulates the active transport of amino acids into cells and promotes protein synthesis.

The Postabsorptive State

The essence of the postabsorptive (fasting) state is to prevent blood glucose concentration from falling below a range of about 90 to 100 mg/dL. This is especially critical to the brain, which does not normally use alternative energy substrates. The postabsorptive status of major nutrients is as follows:

- **Carbohydrates**. Glucose is drawn from the body's glycogen reserves (glycogenolysis) or synthesized from other compounds (gluconeogenesis). The liver usually stores enough glycogen after a meal to support 4 hours of postabsorptive metabolism before significant gluconeogenesis occurs.

- **Fats**. Adipocytes and hepatocytes hydrolyze fats and convert the glycerol to glucose. Free fatty acids (FFAs) can be used as a source of energy by most body cells.
- **Proteins**. If glycogen and fat reserves are depleted, the body begins to use proteins as fuel.

Regulation of the Postabsorptive State

Postabsorptive metabolism is more complex than the absorptive state. It is regulated mainly by the sympathetic nervous system and glucagon. As blood glucose levels drop, insulin secretion declines and pancreatic alpha cells secrete glucagon. Glucagon promotes glycogenolysis and gluconeogenesis, which raise the blood glucose level, and it promotes fat breakdown and a rise in FFA levels. Thus, it makes both glucose and lipids available for fuel.

The sympathetic nervous system also promotes glycogenolysis and fat breakdown, especially under conditions of injury, fear, anger, and other forms of stress. Adipose tissue is richly innervated by the sympathetic nervous system. Adipocytes, hepatocytes, and muscle cells also respond to epinephrine from the adrenal medulla. In circumstances where there is likely to be injury, the sympathetic system mobilizes stored energy reserves and makes them available to meet the demands of tissue repair. Stress also stimulates the release of cortisol, which promotes fat and protein catabolism and gluconeogenesis (see p. 385).

Metabolic Rate

Metabolic rate refers to the amount of energy liberated in the body per unit of time, expressed in such terms as kcal/h or kcal/day. Metabolic rate depends on physical activity, mental state, absorptive or postabsorptive status, thyroid hormone and other hormones, and other factors. The **basal metabolic rate (BMR)** is a baseline when one is awake but relaxed, in a postabsorptive state 12 to 14 hours since the last meal. It is not the minimum metabolic rate needed to sustain life, however. When one is asleep, the metabolic rate is slightly lower than the BMR. **Total metabolic rate** is the sum of BMR and energy expenditure for voluntary activities, especially muscular contractions.

The BMR of an average adult is about 2,000 kcal/day for a male and slightly less for a female. Roughly speaking, one must therefore consume at least 2,000 kcal/day to fuel one's essential metabolic tasks—active transport, muscle tone, brain activity, cardiac and respiratory rhythms, renal function, and other essential processes. Even a relatively sedentary lifestyle requires another 500 kcal/day to support a low level of physical activity, and someone who does hard physical labor may require as much as 5,000 kcal/day.

Aside from physical activity, some factors that raise the total metabolic rate (TMR) and caloric requirements include pregnancy, anxiety (which stimulates epinephrine release and muscle tension), fever (TMR rises about 14% for each 1°C of body temperature), and thyroid hormone. TMR is relatively high in children and declines with age. Therefore, as we reach middle age we often find ourselves gaining weight with no apparent change in food intake.

Some factors that lower TMR include apathy, depression, and prolonged starvation. In weight-loss diets, loss is often rapid at first and then goes more slowly. This is partly because the initial loss is largely water and partly because the TMR drops over time, fewer dietary calories are "burned off," and there is more fat synthesis even with the same caloric intake. As one reduces food intake, the body reduces its metabolic rate to conserve body mass—thus making weight loss all the more difficult.

Before You Go On

Answer these questions from memory. Reread the preceding section if there are too many you don't know.

16. *Define* absorptive state *and* postabsorptive state. *In which state is the body storing excess fuel? In which state is it drawing from these stored fuel reserves?*

17. *What hormone primarily regulates the absorptive state, and what are the major effects of this hormone?*

18. *What is the main priority of postabsorptive metabolism? What are the principal nervous and hormonal means of regulating it?*

19. *Define* basal metabolic rate. *Explain why this is not the minimal metabolic rate to sustain life.*

20. *List a variety of factors and conditions that raise a person's total metabolic rate above basal metabolic rate.*

18.5 Energy Balance and Appetite Regulation

Expected Learning Outcomes

When you have completed this section, you should be able to:

a. explain the concept of energy balance;

b. name the hormones that regulate short- and long-term appetite and describe the effects of each;

c. name the appetite-stimulating and -inhibiting signals secreted in the hypothalamus, and describe their effects.

The subjects of nutrition and metabolism quickly bring to mind the subject of body weight and the popular desire to control it. Weight is determined by the body's **energy balance**—if energy intake and output are equal, body weight is stable. We gain weight if energy intake exceeds output and lose weight if output exceeds intake.

Despite variation in daily food intake and energy expenditure, energy balance is closely regulated, and most people maintain a fairly constant body weight. However, humans, unlike other animals, often ignore feelings of satiety to eat more than is comfortable. In modern times, access to abundant, appealing, high-calorie food is coupled with a relatively inactive lifestyle. Consequently, many people have excess amounts of body fat and obesity has become common. Evidence from twin and adoption studies and family pedigrees indicates that about 30% to 50% of the variation in human weight is due to heredity; the rest is due to environmental factors such as exercise and eating habits.

Appetite

The struggle for weight control often seems to be a struggle against the appetite. Since the early 1990s, physiologists have discovered a still-growing list of molecules that control short- and long-term appetite and body weight. Only a few of them will be described here.

The following three peptides work over the short term, minutes to hours, first making one feel hungry and motivated to eat, then causing a feeling of satiety that signals an end to the meal.

- **Ghrelin**[4] is secreted by parietal cells in the stomach. It has been called the "hunger hormone" because it is a powerful appetite stimulator. Shortly after eating, ghrelin concentration in the blood drops.
- **Peptide YY (PYY)** is secreted by cells in the small intestine (ileum) and colon that sense that food has arrived even as it enters the stomach. The primary effect of PYY is to signal satiety and terminate eating. Thus, ghrelin is one of the signals that begins a meal, and PYY is one of the signals that ends it. PYY remains elevated well after a meal. It acts as a kind of brake that prevents the stomach from emptying too quickly; thus, it prolongs the sense of satiety.
- **Cholecystokinin (CCK)** is secreted by cells in the duodenum and jejunum. It stimulates the brain and produces an appetite-suppressing effect. Thus, it joins PYY as a signal to stop eating. (Earlier, we saw that CCK also regulates activity of the gallbladder and pancreas; p. 583.)

Other peptides regulate appetite, metabolic rate, and body weight over the longer term, thus governing one's average rate of caloric intake and energy expenditure over periods of weeks to years. The following two members of this group work as "adiposity signals," informing the brain about amounts of adipose tissue and activating mechanisms for adding or reducing fat. Adipose tissue is increasingly seen as an important source of multiple hormones that influence the body's energy balance.

- **Leptin**[5] is secreted mainly by adipocytes throughout the body. Its level is proportional to one's fat stores, so this is the brain's primary way of knowing how much body fat we have. Leptin *inhibits* fat deposition. Animals with a leptin deficiency or a defect in leptin receptors overeat and become extremely obese. With few exceptions, however, obese humans are not leptin-deficient and are not aided by leptin injections. More commonly, it seems that human obesity is linked to unresponsiveness to leptin—a receptor defect rather than a hormone deficiency.
- **Insulin** is secreted by the pancreatic beta cells in response to eating. As we have seen, it lowers blood glucose levels by causing it to be transported into cells. In addition, like leptin, it signals the status of the body's fat stores, but in contrast to leptin, it promotes fat deposition. When average daily insulin secretion rises, food intake and body weight increase.

An important brain center for appetite regulation is the hypothalamus. All five of the aforementioned peptides have receptors in the hypothalamus, which in turn, secretes signals according to information it receives (fig. 18.5). It secretes **neuropeptide Y** (NPY), a potent appetite stimulant. Ghrelin stimulates NPY secretion, whereas insulin, PYY, and leptin inhibit it. The hypothalamus also secretes **melanocortin**, which inhibits eating. Leptin stimulates melanocortin secretion and inhibits the secretion of appetite stimulants called *endocannabinoids*, named for their resemblance to the tetrahydrocannabinol (THC) of marijuana.

Hunger is stimulated not only by chemical signals but also by movements of the stomach and intestinal walls. Mild **hunger contractions** begin soon after the stomach is emptied and increase in intensity over a period of hours. They

---

[4] Named partly from *ghre* = growth, and partly as an acronym derived from *growth* h*ormone* rel*easing*

[5] *lept* = thin

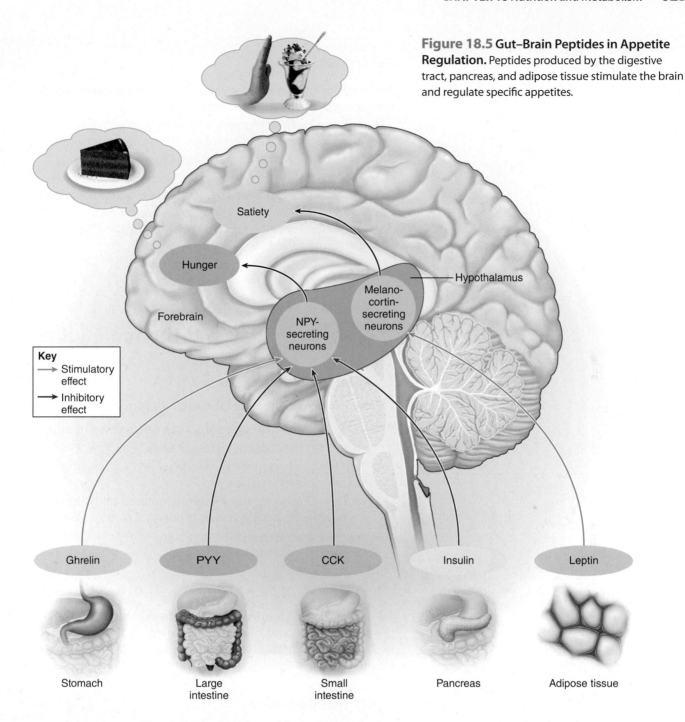

Figure 18.5 Gut–Brain Peptides in Appetite Regulation. Peptides produced by the digestive tract, pancreas, and adipose tissue stimulate the brain and regulate specific appetites.

Key
→ Stimulatory effect
→ Inhibitory effect

Satiety

Hunger

Hypothalamus

Melano-cortin-secreting neurons

Forebrain

NPY-secreting neurons

Ghrelin · PYY · CCK · Insulin · Leptin

Stomach · Large intestine · Small intestine · Pancreas · Adipose tissue

can become quite a painful and powerful incentive to eat, yet they do not affect the amount of food consumed. Food intake is terminated not just by chemical signals. Merely chewing and swallowing food briefly satisfy the appetite. Inflating the stomach with a balloon inhibits hunger even in an animal that has not actually swallowed any food. Satiety produced by these mechanisms, however, is short-lived. Lasting satiety depends on the absorption of nutrients into the blood.

Appetite is a question of not merely *how much* but also *what kind* of food is consumed. Humans are adapted to eat a wide variety of foods including plants and meat. Some foods provide nutrients that others do not. Different neurotransmitters seem to govern the appetite for different classes of nutrients. For example, *norepinephrine* stimulates the appetite for carbohydrates, *galanin* for fatty foods, and *endorphins* for protein.

Before You Go On

Answer these questions from memory. Reread the preceding section if there are too many you don't know.

21. How does the effect of ghrelin differ from the effects of peptide YY and cholecystokinin? Where is each of these hormones secreted?
22. Contrast the effects of leptin and insulin on long-term energy balance.
23. Where are neuropeptide Y and melanocortin secreted? What are their effects?
24. Identify some neurotransmitters that stimulate the appetite for specific categories of nutrients.

18.6 Body Heat and Thermoregulation

Expected Learning Outcomes

When you have completed this section, you should be able to:

a. state the normal ranges of human core and shell temperature;
b. identify the principal sources of body heat;
c. define and contrast three modes of heat loss and their relative contributions to total heat loss;
d. describe how the hypothalamus monitors body temperature and how the heat-losing and heat-promoting centers work; and
e. name and describe the pathological states of excessively high and low body temperatures.

The enzymes that control our metabolism depend on an optimal, stable temperature. In order to maintain this, heat loss from the body must be matched by heat generation. **Thermoregulation**, the balance between heat production and loss, is therefore a critically important aspect of homeostasis.

Body Temperature

Body temperature fluctuates about 1°C (1.8°F) in a 24-hour cycle. It tends to be lowest in the early morning and highest in the late afternoon. Temperature also varies from one place in the body to another, and from person to person. Therefore, "normal" body temperature is a relative term.

The most important body temperature is the **core temperature**—the temperature of organs in the cranial, thoracic, and abdominal cavities. Rectal temperature is relatively easy to measure and gives an estimate of core temperature: usually 37.2° to 37.6°C (99.0°–99.7°F). It may be as high as 38.5°C (101°F) in active children and adults.

Shell temperature is the cooler temperature closer to the surface, especially skin and oral temperature. Adult oral temperature is typically 36.6° to 37.0°C (97.9°–98.6°F) but may be as high as 40°C (104°F) during hard exercise. Shell temperature fluctuates as a result of processes that serve to maintain a stable core temperature.

Heat Production and Loss

Most body heat comes from metabolic reactions related to nutrient oxidation and ATP use. At rest, most heat is generated by the brain, heart, liver, and endocrine glands. Skeletal muscles contribute about 20% to 30% of the total resting heat. However, during vigorous exercise, they produce 30 to 40 times as much heat as the rest of the body.

The body loses heat in three ways: radiation, conduction, and evaporation.

1. **Radiation** is the loss of thermal energy through infrared (IR) rays emitted by all moving molecules. We noticeably absorb radiant heat from heat lamps or the sun, but indeed from every object around us. Since we are usually warmer than the objects around us, however, we usually lose more heat this way than we gain.

2. **Conduction** is the transfer of kinetic energy from molecule to molecule as they collide with one another. Heat from your body warms the clothes you wear, the chair you sit in, and the air around you through conduction. Conduction is aided by *convection*, the movement of warm air away from the body and its replacement by a current of cooler air (or water, if one is immersed in a cool lake or bath).

3. **Evaporation** is the loss of water from the body surface through its change to a gaseous state. When water evaporates, it carries a substantial amount of heat with it. This is the significance of perspiration. Sweat wets the skin and its evaporation carries heat away. In extreme conditions, the body can lose 2 L or more of sweat and with it, up to 600 kcal of heat, per hour. Evaporative heat loss is increased by moving air, so a breeze or a fan enhances heat loss.

The relative amounts of heat lost by different methods depend on prevailing conditions. A nude body at an air temperature of 21°C (70°F) loses about 60% of its heat by radiation, 18% by conduction, and 22% by evaporation. If air temperature is higher than skin temperature, evaporation becomes the only means of heat loss because radiation and conduction add more heat to the body than they remove. Hot, humid weather hinders even evaporative cooling because there is less of a humidity gradient from skin to air.

Thermoregulation

Thermoregulation is achieved through several negative feedback loops. The hypothalamus monitors the temperature of the blood and also receives signals from nerve endings (thermoreceptors) in the skin. In turn, it sends appropriate signals to other hypothalamic nuclei called the *heat-losing* and *heat-promoting centers*.

The **heat-losing center** promotes cutaneous vasodilation, which increases blood flow close to the body surface and thus promotes heat loss. If this fails to restore normal temperature, the heat-losing center triggers sweating.

The **heat-promoting center** activates mechanisms to conserve body heat or generate more. By way of the sympathetic nervous system, it causes cutaneous vasoconstriction. This retains warm blood deeper in the body and less heat is lost through the skin. If this fails to restore or maintain normal core temperature, the body resorts to shivering. On a cold day, you may notice that your muscles become tense and you begin to shiver. Muscle contraction releases heat from ATP, and shivering can increase the body's heat production as much as fourfold.

A more long-term mechanism for generating heat is to raise metabolic rate. After several weeks of cold weather, for example, metabolic rate increases as much as 30%. Activation of the sympathetic nervous system raises metabolic rate, with the result that more nutrients are burned as fuel. We consume more

Clinical Application 18.2

HYPOTHERMIA AND FROSTBITE

Hypothermia occurs when the body cannot generate heat as quickly as it is lost to the environment. All of us have probably experienced a drop in body temperature in response to cold weather; a decrease in temperature of 1 to 2°C (1.8 to 3.6°F) leads to shivering, goose bumps, and numb hands. Shivering and vasoconstriction of surface blood vessels serve to maintain a high temperature in the core of the body and the brain, and behavioral responses such as moving indoors or adding more clothing prevent a spiral to more severe heat loss. In certain conditions of extreme cold, however, the body temperature drops to a level where the hypothalamus can no longer produce an adequate response, and cellular metabolic processes slow to the point of disrupting normal body function. Mountain climbers exposed to low temperatures and windy conditions must be careful to notice the symptoms of hypothermia in themselves and their companions— poor muscle coordination, difficulty speaking, blue fingertips and lips, and impaired cognitive ability. Drinking warm liquids, getting to shelter, and warming with blankets or by sharing body heat are essential to prevent a further slide toward nervous system depression and possible death. Another concern is frostbite, which results from exposure of body surfaces to temperatures low enough to freeze tissues. It occurs most often in fingers and toes, earlobes, and the tips of the nose. If ice crystals form in cells, tissues are damaged and blood circulation is impaired. In severe cases, gangrene follows thawing and the affected areas must be surgically removed.

calories to "stoke the furnace" and, consequently, have greater appetites in the winter than in the summer.

In addition to these physiological mechanisms, humans and other animals practice **behavioral thermoregulation**. For example, just getting out of the sun greatly cuts down heat gain by radiation; shedding heavy clothing helps to cool the body; or putting on a sweater or adding another blanket to the bed conserves heat.

In summary, you can see that thermoregulation is a function of multiple organs: the brain, autonomic nerves, skin, blood vessels, and skeletal muscles.

Disturbances of Thermoregulation

Either excessively low body temperature, called **hypothermia**, or excessively high body temperature, called **hyperthermia**, can disrupt metabolism to the point of death. Hypothermia can result from exposure to cold weather or immersion in icy water (see Clinical Application 18.2). If the core temperature falls below 33°C (91°F), the metabolic rate drops so low that heat production cannot keep pace with heat loss, and the temperature falls even more. Death from cardiac fibrillation may occur below 32°C (90°F), but some people survive body temperatures as low as 29°C (84°F) in a state of suspended animation. A body temperature below 24°C (75°F) is usually fatal.

Hyperthermia is characterized by heat cramps, heat exhaustion, and heatstroke. **Heat cramps** are painful muscle spasms that result from excessive electrolyte loss in the sweat. They occur especially when a person begins to relax after strenuous exertion and heavy sweating. **Heat exhaustion** results from more severe water and electrolyte loss and is characterized by low blood pressure, dizziness, vomiting, and sometimes fainting. **Heatstroke (sunstroke)** occurs when the core body temperature is over 40°C (104°F); the skin is hot and dry; and the subject exhibits nervous system dysfunctions such as delirium, convulsions, or coma. It is also accompanied by tachycardia, hyperventilation, inflammation, and multiorgan dysfunction; it is often fatal.

Chapter 14 described the mechanism of fever and its importance in combating infection. Fever is a normal protective mechanism that should be allowed to run its course if temperature is not extraordinarily high. A body temperature above 42° to 43°C (108°–110°F), however, is dangerous. At a core temperature of 44° to 45°C (111°–113°F), metabolic dysfunction and neurological damage can be fatal.

Before You Go On

Answer these questions from memory. Reread the preceding section if there are too many you don't know.

25. *Define* core body temperature. *How much lower is the typical shell temperature?*

26. *What is the primary source of body heat? What are some lesser sources? How does the proportion of heat produced by skeletal muscles change during exercise compared to rest?*

27. *Distinguish between radiation, conduction, and evaporation as modes of heat loss. Which of these is aided by convection? Consider a person outdoors on a day when the air temperature is 38°C (101°F); what would be the person's sole mode of body heat loss?*

28. *Explain how the hypothalamus acts to return body temperature to normal if it senses that it is too warm or too cool.*

29. *Describe the effects of hyperthermia and hypothermia.*

30. *Distinguish between heat cramps, heat exhaustion, and heatstroke.*

CAREER SPOTLIGHT

Dietitian

A registered dietitian (RD) is a food and nutrition expert who has completed a bachelor's degree. In addition, an RD must complete an internship approved by the American Dietetic Association and pass a national examination. Registered dietitians focus on prevention and treatment of disease using a nutritional perspective. They work in a variety of settings, often as members of a medical team in hospitals, clinics, or other care facilities such as nursing homes. They offer nutritional counseling to schools, child-care centers, and correctional facilities. They may also provide nutritional support to both professional and recreational athletes in sports facilities. Dietitians serve in the food and nutrition industry and are also active in nutrition research.

A dietetic technician, registered (DTR), is a nutrition practitioner who has completed an associate degree at an accredited institution, an approved internship, and a national DTR examination. The majority of DTRs work with RDs assisting with nutrition support in hospitals, clinics, or other health-care facilities, or public health settings such as schools, child-care centers, or weight-management centers as nutrition counselors. In an era when nutrition is a focus of health because of the obesity epidemic and rising incidence of diabetes mellitus, these professions are recognized as being critical for disease prevention and management.

See appendix B for further information on careers in dietetics.

Study Guide

Assess Your Learning Outcomes

To test your knowledge, discuss the following topics with a study partner or in writing, ideally from memory.

18.1 Nutrition (p. 602)

1. Definitions of *calorie* (with a lowercase *c*) and *kilocalorie* (kcal)
2. The definition of *nutrient,* and why some substances can be important in the diet yet not be considered nutrients
3. Energy in carbohydrates and proteins compared with that of fat
4. The meaning of *macronutrients* and *micronutrients,* with examples of each
5. The essential fatty acids and amino acids, and why these are called "essential" in contrast to other nutrients
6. The chief dietary mono-, di-, and polysaccharides and the sources of each
7. Examples of dietary fiber and why fiber is important
8. Reasons why fat is superior to carbohydrate for long-term energy storage
9. Dietary sources of saturated and unsaturated fats and the difference between them
10. Why cholesterol is important to normal function, and some foods high in cholesterol
11. The relative effect of dietary cholesterol versus other factors on one's blood cholesterol level
12. How blood lipoproteins enable lipids to be transported in spite of their hydrophobic nature
13. The three kinds of blood lipoproteins and the differences between their composition and function
14. The true meaning of "good cholesterol" versus "bad cholesterol" and why these blood lipotroteins have acquired those colloquial expressions
15. Several functions of dietary proteins
16. How proteins of plant and animal origin differ in dietary sufficiency, and how one can obtain all necessary amino acids from a diet free of animal protein
17. Definition of *essential amino acid* and how essential amino acids are obtained
18. Environmental source and best dietary source of minerals; important minerals in the body and their general functions

19. General functions of vitamins and how they differ from other small organic compounds
20. The difference between water-soluble and fat-soluble vitamins and examples of each
21. The hazards of excessive vitamin intake

18.2 Carbohydrate Metabolism (p. 611)

1. The definition of *metabolism* and its two subdivisions
2. The equation for the complete oxidation of glucose
3. Why the human body can't simply break down glucose in one reaction step
4. The three major pathways of glucose metabolism
5. The two coenzymes involved in glucose oxidation and the general role that they have in common
6. The end products of glycolysis
7. The purpose of anaerobic fermentation, and its main reactant and end product
8. Where aerobic respiration occurs in the cell; differences between the matrix and membrane reactions; and the relative ATP yield of each
9. Distinctions between glycolysis, glycogenesis, glycogenolysis, and gluconeogenesis, including the purpose, input, and output (products) of each

18.3 Lipid and Protein Metabolism (p. 615)

1. The breakdown products of triglyceride hydrolysis and how each of these can be used to make ATP
2. The relative ATP yield of fat versus glucose oxidation
3. The cause and pathological effects of ketoacidosis
4. The process of catabolizing proteins for fuel and how it gives rise to ammonia and urea
5. Five (or more) nondigestive functions of the liver

18.4 Metabolic States and Metabolic Rate (p. 618)

1. The distinction between the absorptive and postabsorptive metabolic states
2. How carbohydrates, fats, and proteins (or amino acids) are processed during each of those states
3. The hormones and nervous mechanisms for regulating these two states
4. Definitions of *basal* and *total metabolic rate,* and some typical values of each
5. Factors that raise and lower metabolic rate

18.5 Energy Balance and Appetite Regulation (p. 621)

1. The concept of energy balance
2. Roles of ghrelin, peptide YY, and cholecystokinin as short-term regulators of appetite
3. The two long-term adiposity signals to the brain and their effects on body fat stores
4. The role of the hypothalamus in appetite regulation; two peptides produced by the hypothalamus that stimulate or inhibit appetite

18.6 Body Heat and Thermoregulation (p. 624)

1. The difference between core temperature and shell temperature, and where one would place a thermometer to measure each
2. The source of most body heat when one is at rest
3. The three ways in which the body loses heat
4. How the hypothalamus monitors and regulates body temperature
5. The terms for abnormally high or low body temperatures and causes and effects of each

Testing Your Recall

1. ___ are not used as fuel and are required in relatively small quantities.
   a. Micronutrients
   b. Macronutrients
   c. Essential nutrients
   d. Proteins
   e. Lipids

2. The only significant digestible polysaccharide in the diet is
   a. glycogen.
   b. cellulose.
   c. starch.
   d. maltose.
   e. fiber.

3. Which of the following store(s) the greatest amount of energy for the smallest amount of space in the body?
   a. glucose
   b. triglycerides
   c. glycogen
   d. proteins
   e. vitamins

4. A/an ___ is a nutrient that cannot be synthesized by the body and therefore must come from the diet.

   a. omega-3 fatty acid
   b. saturated fat
   c. high-density lipoprotein
   d. extrinsic fatty acid
   e. essential fatty acid

5. Which of the following is most likely to make you hungry?
   a. leptin
   b. ghrelin
   c. cholecystokinin
   d. peptide YY
   e. melanocortin

6. Which of the following normally releases glucose from the liver between meals?
   a. insulin
   b. glycogenesis
   c. glycolysis
   d. glycogenolysis
   e. gluconeogenesis

7. What function is served by oxygen in the process of aerobic respiration?
   a. It provides the energy necessary to get glycolysis started.
   b. It oxidizes glucose to pyruvic acid.
   c. It converts pyruvic acid to lactic acid.
   d. It transfers high-energy electrons between mitochondrial enzymes.
   e. It serves as the final electron acceptor in the electron-transport chain.

8. Glycolysis has a net yield of ____ molecules of ATP for each molecule of glucose.
   a. 2
   b. 4
   c. 8
   d. 32
   e. 38

9. Which of these occurs in the mitochondrial matrix?
   a. glycolysis
   b. glycogenesis
   c. the electron-transport chain
   d. the citric acid cycle
   e. anaerobic fermentation

10. Absorbing heat from the sun is an example of
    a. conduction.
    b. convection.
    c. hyperthermia.
    d. radiation.
    e. evaporation.

11. A/an _____ protein lacks one or more essential amino acids.

12. In the postabsorptive state, glycogen is hydrolyzed to liberate glucose. This process is called _____.

13. Synthesis of glucose from amino acids or triglycerides is called _____.

14. The major nitrogenous waste resulting from protein catabolism is _____.

15. The organ that synthesizes the nitrogenous waste in question 14 is the _____.

16. The hormone _____ is the main regulator of the absorptive state.

17. The temperature of organs in the body cavities is called _____.

18. The appetite hormones leptin and ghrelin act on the _____, an important brain center for appetite regulation.

19. In the process of anaerobic fermentation, pyruvic acid is reduced to _____.

20. A biochemist or physiologist would generally use the term _____ and the abbreviation _____ for what a dietitian would call one Calorie or 1,000 calories.

Answers in appendix A

True or False

Determine which five of the following statements are false, and briefly explain why.

1. Ghrelin and leptin are two hormones that stimulate the appetite.

2. Water is a nutrient, but oxygen and cellulose are not.

3. An extremely low-fat diet can cause vitamin-deficiency diseases.

4. Most of the body's cholesterol comes from the diet.

5. The majority of the liver's functions are digestive.

6. Aerobic respiration produces more ATP than anaerobic fermentation does.

7 Reactions occurring on the mitochondrial inner membrane produce more ATP than glycolysis and the matrix reactions combined.

8. Gluconeogenesis occurs especially in the absorptive state during and shortly after a meal.

9. Hepatitis B and C usually lead to more serious complications—such as liver cancer or cirrhosis—than hepatitis A.

10. At a comfortable air temperature, the body loses more heat by conduction than by any other means.

Answers in appendix A

Testing Your Comprehension

1. Cyanide blocks the transfer of electrons from the final cytochrome to oxygen in the electron-transport chain. In light of this, explain why it causes sudden death. Also explain whether cyanide poisoning could be treated by giving a patient supplemental oxygen, and justify your answer.

2. People with type 1 diabetes mellitus inject insulin to help lower blood sugar. Explain why they may sometimes become hypoglycemic (experience low blood sugar). What hormone would be secreted to counteract the effects of hypoglycemia?

3. Your friend is trying to reduce his cholesterol and is discouraged. He says to you, "I only eat foods that say they are cholesterol-free on the labels! How could I still have high cholesterol?" Explain to your friend why he still may have high cholesterol even if he eats foods that are advertised as cholesterol-free.

Chapter

19

The Reproductive System

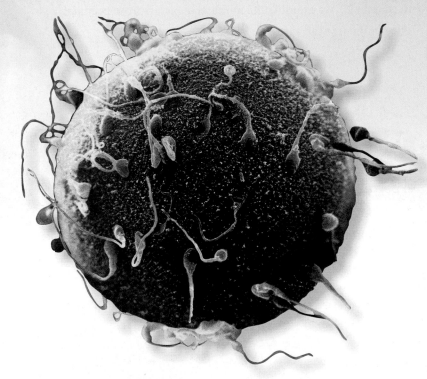

Only one sperm will enter this egg; the egg has mechanisms for blocking all other sperms from entering and giving it a lethal gene overdose.

Chapter Outline

BASE CAMP

Before ascending to the next level, be sure you're properly equipped with a knowledge of these concepts from earlier chapters:

- Mitosis (p. 89)
- Gonadotropins of the pituitary (p. 373)

Module 14: Reproductive System

From all we have learned of the structure and function of the human body, it seems a wonder that it works at all! The fact is, however, that even with modern medicine we cannot keep it working forever. The body suffers various degenerative changes as we age, and eventually our time is up and we must say good-bye. Yet our genes live on in new containers—our offspring. The production of offspring is the subject of these last two chapters. In this chapter, we examine the male and female roles in producing a new life, and in chapter 20, we examine the human life history from prenatal development to old age.

19.1 Essentials of Sexual Reproduction

Expected Learning Outcomes

When you have completed this section, you should be able to:

a. explain why sexual reproduction in humans requires two different types of gametes;

b. enumerate the functions of the male and female reproductive systems; and

c. distinguish between the gonads of the two sexes, and between the internal and external genitalia.

The essence of sexual reproduction is that it is biparental—the offspring receive genes from two parents and therefore are not genetically identical to either one. To achieve this, the parents must produce **gametes**[1] (sex cells) that meet and combine their genes in a **zygote**[2] (fertilized egg). The gametes require two properties to succeed at this: motility so they can establish contact, and nutrients for the embryo. A single cell cannot perform both roles well, because to carry ample nutrients means to be relatively heavy, which is inconsistent with mobility. Therefore, these tasks are divided between two kinds of gametes. The small motile one—little more than DNA with a propeller—is the **sperm**, and the large nutrient-laden one is the **egg (ovum)**. The **reproductive systems** of the two sexes jointly serve to produce and unite these gametes, harbor the fetus and give birth, and nourish the infant.

---

[1] *gam* = marriage, union
[2] *zygo* = yoke, union

The organs that produce the gametes—the *testes* of the male and *ovaries* of the female—are collectively called the **gonads**[3] or *primary sex organs*. In addition to producing gametes, they produce sex hormones with widespread effects throughout the body, such as bone growth and maintenance, muscle growth, hair growth, and fat deposition. The other organs necessary to reproduction are called *secondary sex organs*, such as the penis, uterus, mammary glands, and others to be considered in this chapter.

According to location, reproductive organs are also classified as **external** and **internal genitalia**. The external genitalia are located in the perineum between the thighs. Most of them are externally visible or located superficially beneath the skin. The internal genitalia are located mainly in the pelvic cavity, except for the male testes and some associated ducts in the scrotum.

Before You Go On

Answer these questions from memory. Reread the preceding section if there are too many you don't know.

1. Define gonad *in a way that includes the word* gamete(s).
2. Distinguish between a gamete and a zygote.
3. Distinguish between internal and external genitalia.

19.2 The Male Reproductive System

Expected Learning Outcomes

When you have completed this section, you should be able to:

a. describe the anatomy of the male reproductive tract;

b. trace the pathway taken by a sperm cell from its formation to ejaculation, naming all the passages that it travels;

c. describe the structure, locations, and functions of the male accessory glands, scrotum, and penis; and

d. discuss male sexual development from puberty through andropause.

Anatomy AP|R

This section surveys the male reproductive system (fig. 19.1), which is concerned with the production and delivery of sperm. We will proceed roughly in the order of sperm formation, transport, and emission, thus beginning with the testes, continuing through the duct system and accessory organs, and ending with the penis.

---

[3] *gon* = seed

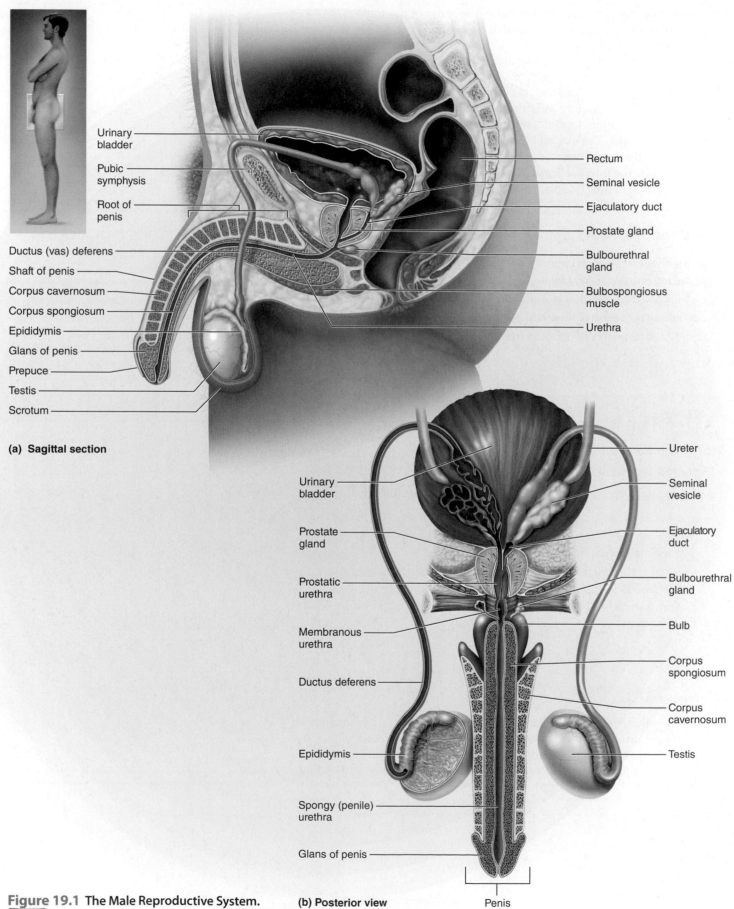

Urinary bladder

Pubic symphysis

Root of penis

Ductus (vas) deferens

Shaft of penis

Corpus cavernosum

Corpus spongiosum

Epididymis

Glans of penis

Prepuce

Testis

Scrotum

Rectum

Seminal vesicle

Ejaculatory duct

Prostate gland

Bulbourethral gland

Bulbospongiosus muscle

Urethra

(a) Sagittal section

Urinary bladder

Prostate gland

Prostatic urethra

Membranous urethra

Ductus deferens

Epididymis

Spongy (penile) urethra

Glans of penis

Ureter

Seminal vesicle

Ejaculatory duct

Bulbourethral gland

Bulb

Corpus spongiosum

Corpus cavernosum

Testis

Penis

(b) Posterior view

Figure 19.1 The Male Reproductive System.

AP|R

The Testes

The **testes (testicles)** produce both sex hormones and sperm. They are oval and slightly flattened, about 4 cm long and 2.5 cm wide (fig. 19.2). The testis has a white fibrous capsule with extensions *(septa)* that divide the interior tissue into about 300 wedge-shaped *lobules.* Each lobule contains one to three **seminiferous**[4] (SEM-ih-NIF-er-us) **tubules**—slender ducts up to 70 cm

[4] *semin* = seed, sperm; *fer* = to carry

Figure 19.2 The Testis and Associated Structures. (a) The scrotum is opened and folded downward to reveal the testis and associated organs. (b) Anatomy of the testis, epididymis, and spermatic cord. (c) Electron micrograph of a seminiferous tubule. (d) Light micrograph.

[Part (c) © Dr. Richard Kessel & Dr. Kardon/Tissues and Organs/Visuals Unlimited, Inc.]

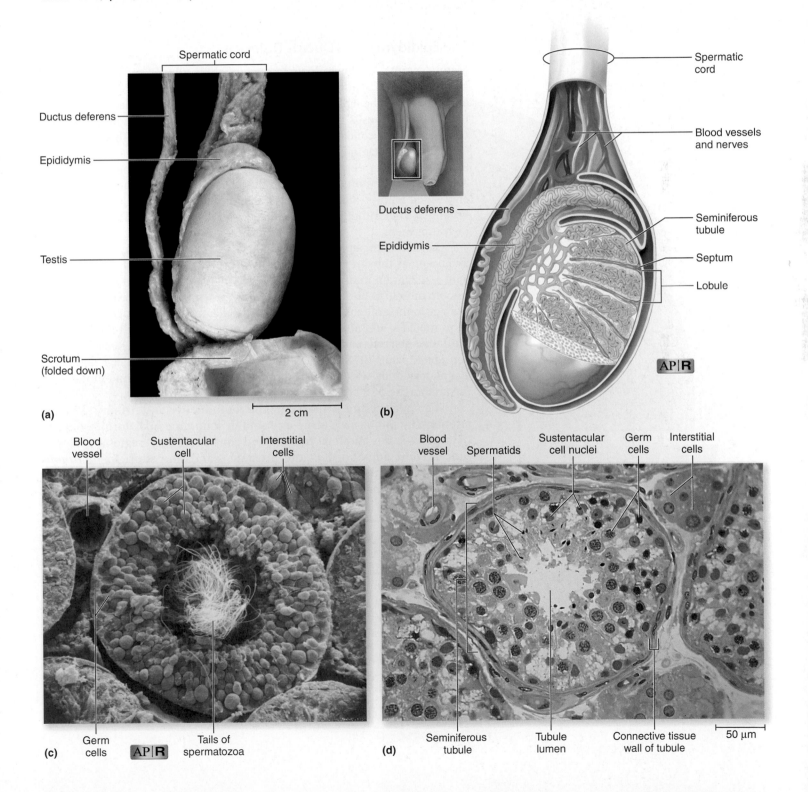

(a)

Spermatic cord

Ductus deferens

Epididymis

Testis

Scrotum (folded down)

2 cm

(b)

Spermatic cord

Blood vessels and nerves

Ductus deferens

Epididymis

Seminiferous tubule

Septum

Lobule

AP|R

(c)

Blood vessel

Sustentacular cell

Interstitial cells

Germ cells   AP|R

Tails of spermatozoa

(d)

Blood vessel

Spermatids

Sustentacular cell nuclei

Germ cells

Interstitial cells

Seminiferous tubule

Tubule lumen

Connective tissue wall of tubule

50 μm

long each, in which the sperm are produced. Between the tubules are clusters of endocrine **interstitial cells**, the source of testosterone.

A seminiferous tubule has a narrow lumen lined by a thick epithelium (fig. 19.2c, d). The epithelium consists of several layers of **germ cells** in the process of becoming sperm, and a much smaller number of tall **sustentacular**[5] **cells**, which support and protect the germ cells and promote their development. The germ cells depend on the sustentacular cells for nutrients, waste removal, growth factors, and other needs. The sustentacular cells also secrete a hormone, *inhibin*, that regulates the rate of sperm production, as we will see later.

The Epididymis and Ductus Deferens

Mature sperm drift slowly down the seminiferous tubules and leave the testis by way of about 12 short ducts on its posterior side (fig. 19.2a, b). These lead to an organ called the **epididymis**[6] (EP-ih-DID-ih-miss) that adheres to the posterior side of the testis. The epididymis, a site of sperm maturation and storage, contains a single coiled duct embedded in connective tissue. This duct, about 6 m (almost 20 feet) long, reabsorbs about 90% of the fluid secreted by the testis. Sperm mature as they travel down the epididymis over a period of about 20 days. They are stored here and in the adjacent portion of the ductus deferens, which follows. Stored sperm remain fertile for 40 to 60 days, but if they become too old without being ejaculated, they disintegrate and the epididymis reabsorbs them.

The duct of the epididymis straightens out at the inferior end, turns 180°, and becomes the **ductus (vas) deferens.** This is a muscular tube about 45 cm long and 2.5 mm in diameter. It has a very narrow lumen and a thick wall of smooth muscle important in ejaculation. The ductus travels upward through a passage in the groin, the **inguinal canal,** and enters the pelvic cavity. There, it turns medially and approaches the urinary bladder. Posterior to the bladder, the ductus deferens ends by uniting with the duct of the seminal vesicle, a gland considered later. The two ducts unite into a short (2 cm) passage called the **ejaculatory duct**, which passes through the prostate gland and empties into the urethra (fig. 19.1).

The male urethra is shared by the reproductive and urinary systems, although it cannot pass urine and semen simultaneously for reasons explained later. It is about 20 cm long and consists of three regions: the *prostatic urethra* embedded in the prostate gland, the short *membranous urethra* passing through the pelvic floor, and the long *spongy (penile) urethra* passing through the penis (fig. 19.1).

The Scrotum

The testes are contained in the **scrotum,**[7] a pendulous pouch of skin, muscle, and fibrous connective tissue (fig. 19.3). The left testis is usually suspended lower than the right so the two are not compressed against each other between the thighs. The scrotum is divided into right and left compartments by a median septum.

The scrotum also contains the epididymis and spermatic cord associated with each testis. The **spermatic cord** is a fibrous bundle that contains the

---

[5] *sustentacul* = support
[6] *epi* = upon; *didym* = twins, testes
[7] *scrotum* = bag

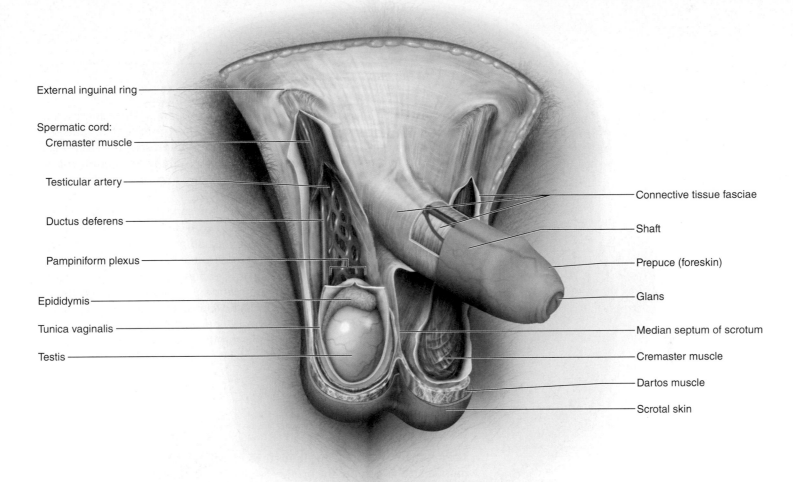

External inguinal ring

Spermatic cord:
  Cremaster muscle

Testicular artery

Ductus deferens

Pampiniform plexus

Epididymis

Tunica vaginalis

Testis

Connective tissue fasciae

Shaft

Prepuce (foreskin)

Glans

Median septum of scrotum

Cremaster muscle

Dartos muscle

Scrotal skin

Figure 19.3 The Scrotum and Spermatic Cord. The cremaster muscle, pampiniform plexus, and dartos muscle serve in various mechanisms for regulating the temperature of the testis. **AP|R**

ductus deferens, blood and lymphatic vessels, and nerves. It passes upward behind and superior to the testis, then anterior to the pubis and into the inguinal canal.

Sperm production requires a temperature of about 35°C—or 2°C cooler than the temperature in the pelvic cavity—and the scrotum has three mechanisms for controlling testicular temperature. In response to cold, the **cremaster**[8] **muscle** of the spermatic cord contracts and draws the testes closer to the body, while the **dartos muscle**, a layer of smooth muscle beneath the skin of the scrotum, contracts and tautens the scrotum. When warmer, both muscles relax and the testes are suspended farther away from the pelvic floor. In addition, the spermatic cord has a network of veins, the **pampiniform plexus**, that carries heat away from the arterial blood on its way to the testis, preventing the warm blood from overheating the testis.

The Seminal Vesicles

A pair of glands called **seminal vesicles** occurs on the posterior side of the urinary bladder (see fig. 19.1). Each has a connective tissue capsule, smooth muscle, and a single, highly convoluted and branched duct that exits the gland and joins the ductus deferens. The vesicles produce a yellowish secretion that forms about 60% of the semen. Its composition and functions are discussed later.

[8] *cremaster* = suspender

Clinical Application 19.1

PROSTATE DISEASES

The prostate gland remains at a stable size from age 20 to 45 or so, then begins to grow slowly. By age 70, over 90% of men show some degree of noncancerous enlargement called *benign prostatic hyperplasia (BPH)*. The main concern in BPH is that it compresses the urethra, makes it harder to empty the bladder, requires more frequent urination since the bladder refills more quickly, and may promote bladder and kidney infections.

Prostate cancer is the second most common cancer in men (after lung cancer); it affects about 9% of men over the age of 50. Tumors tend to form near the periphery of the gland, where they do not obstruct urine flow, and go unnoticed until they cause pain. Prostate cancer often metastasizes to nearby lymph nodes and then to the lungs and other organs. Once it metastasizes, it has a 50% to 90% mortality rate, but is highly survivable if detected and treated before then. It can be detected by palpation through the rectal wall, called *digital rectal examination (DRE)*, or by blood tests for certain prostatic enzymes. One of these enzymes is known as *prostate-specific antigen*, so the blood assay is called a PSA test.

The Prostate Gland

The **prostate**[9] (PROSS-tate) **gland** surrounds the urethra and ejaculatory duct immediately inferior to the urinary bladder (see Clinical Application 19.1). It measures about 2 × 4 × 3 cm and opens into the prostatic urethra by numerous pores in the urethral wall. The thin, milky secretion of the prostate constitutes about 30% of the semen. Its functions, too, are considered later.

The Bulbourethral Glands

The **bulbourethral glands** are brownish, spherical glands about 1 cm in diameter, with a short duct to the uretha at the root of the penis (see fig. 19.1). During sexual arousal, they produce a clear slippery fluid that provides some of the lubrication for intercourse, but more importantly, neutralizes the acidity of residual urine in the urethra (thus protecting the sperm).

The Penis

The **penis**[10] serves to deposit semen in the vagina. The externally visible portion of it, composed of the **shaft** and **glans**[11] (fig. 19.4), is half its total length. In the flaccid (nonerect) state, the external portion measures about 8 to 10 cm (3–4 in.) long and 3 cm in diameter; the typical dimensions of an erect penis are 13 to 18 cm (5–7 in.) long and 4 cm in diameter. The glans is the expanded head at the distal end of the penis with the external urethral orifice at its tip. The skin is loosely attached to the shaft, allowing for expan-

---

[9] *pro* = before; *stat* = to stand; commonly misspelled and mispronounced *prostrate*

[10] *penis* = tail

[11] *glans* = acorn

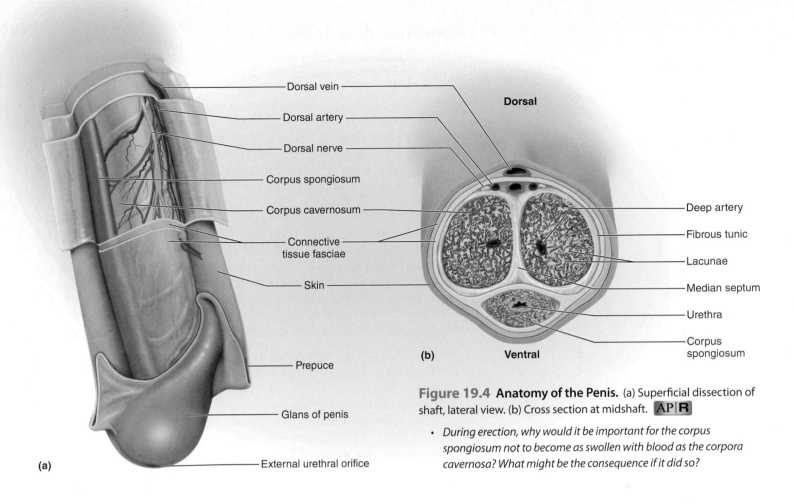

Figure 19.4 Anatomy of the Penis. (a) Superficial dissection of shaft, lateral view. (b) Cross section at midshaft. AP|R

- *During erection, why would it be important for the corpus spongiosum not to become as swollen with blood as the corpora cavernosa? What might be the consequence if it did so?*

sion during erection. It continues over the glans as the **prepuce**, or foreskin, which is sometimes removed by circumcision.

The penis consists mainly of three cylindrical bodies called **erectile tissues** (fig. 19.4b), which fill with blood during sexual arousal and account for its enlargement and erection. A single erectile body, the **corpus spongiosum**, passes along the ventral side of the penis and encloses the urethra. It expands at the distal end to fill the entire glans. The dorsal side of the penis has a **corpus cavernosum** (plural, *corpora cavernosa*) on each side, but this extends only as far as the margin of the glans. The corpora cavernosa are separated by a median septum and sheathed in a tight fibrous tunic, which plays a role in erection explained later.

All three cylinders of erectile tissue are spongy in appearance and contain tiny blood sinuses called **lacunae**. The partitions between lacunae, called **trabeculae**, are composed of connective tissue and smooth muscle. In the flaccid penis, muscle tone collapses the lacunae, which appear as tiny slits in the tissue.

At the body surface, the penis turns 90° posteriorly and continues inward as the **root**. The corpus spongiosum ends internally as a dilated **bulb**, which is ensheathed in the *bulbospongiosus muscle*. The corpora cavernosa diverge like the arms of a Y and attach the penis to the pubic arch of the pelvic girdle.

From Puberty to Andropause

Unlike any other organ system, the reproductive system remains dormant for several years after birth. Around age 10 to 12 in most boys and 8 to 10 in girls, a surge of pituitary gonadotropins awakens the reproductive system and begins preparing it for adult reproductive function. This is the onset of puberty.

Adolescence[12] is the period from the onset of gonadotropin secretion and reproductive development until a person attains full adult height, usually in the late teens to early twenties. **Puberty**[13] is the first few years of adolescence, until the first menstrual period in girls and the first ejaculation of viable sperm in boys. In North America, this is typically attained around age 12 in girls and 13 in boys.

As the brain matures, the hypothalamus begins to secrete **gonadotropin-releasing hormone (GnRH)**. GnRH stimulates the anterior pituitary gland to secrete two *gonadotropins*—**follicle-stimulating hormone (FSH)** and **luteinizing hormone (LH)**. Although named for their functions in females, LH also stimulates the interstitial cells of the testis to secrete testosterone, and FSH renders the seminiferous tubules more sensitive to testosterone. Without FSH, the tubules don't respond to testosterone at all. Testosterone produces many conspicuous effects on the adolescent male:

- Growth of the sex organs, beginning with the testes and scrotum, then the penis and internal ducts and glands. This is accompanied by sperm production.

- A burst of generalized body growth, due especially to elongation of the limb bones and increasing muscle mass under the influence of both testosterone and growth hormone. This is accompanied by enlargement of the larynx, deepening of the voice, and an elevated metabolic rate and appetite.

- Appearance of the pubic hair, axillary hair, and later the facial hair, along with the apocrine sweat glands in these areas.

- Awakening of the sex drive, or **libido**, accompanied by frequent erection and spontaneous ejaculation, especially in the sleep (*nocturnal emission* or "wet dreams").

Throughout adulthood, testosterone sustains the male reproductive tract, sperm production, and libido. The sustentacular cells also secrete a hormone called **inhibin**. By inhibiting FSH but not LH secretion, this hormone can slow down sperm production without reducing the body's other responses to testosterone.

Testosterone secretion peaks at about 7 mg/day at age 20, then declines steadily to as little as one-fifth of this level by age 80. Since testicular hormones normally inhibit pituitary gonadotropin secretion, the declining level of testosterone is accompanied by a rising level of FSH and LH after age 50. These hormonal changes produce effects called **andropause (male climacteric)**. Andropause has little or no noticeable effect in most men, but in some cases there are mood changes, hot flashes, or illusions of suffocation—symptoms similar to those of menopause in women. Despite joking references to "male menopause," the term *menopause* refers to the cessation of menstruation and therefore makes no sense in the context of male physiology. Men normally produce sperm to the end of life, although in reduced numbers.

---

[12] *adolesc* = to grow up

[13] *puber* = grown up

We will soon explore the process of sperm production and physiology of intercourse for the male, but first we will survey the female reproductive system. We can then take a more integrative view of what sperm and egg production, and the responses of intercourse, have in common for both sexes.

Before You Go On

Answer these questions from memory. Reread the preceding section if there are too many you don't know.

4. What are seminiferous tubules? What types of cells compose them? What is their relationship to the interstitial cells?

5. Describe the pathway taken by the sperm from the time they leave the testis until they are ejaculated. What point in this pathway is the main storage place for sperm awaiting ejaculation?

6. Describe three ways in which the body regulates the temperature of the testes, and explain why this is important.

7. Name three kinds of accessory glands of the male reproductive system, describe their locations, and state their functions.

8. What are the two kinds of erectile tissues in the penis, and where are they located?

9. What pituitary hormones regulate the secretion and action of testosterone? Describe several effects of testosterone.

19.3 The Female Reproductive System

Expected Learning Outcomes

When you have completed this section, you should be able to:

a. describe the anatomy and histology of the ovaries;

b. describe the gross anatomy and histology of the female reproductive tract;

c. describe the structure and function of the glands and other accessory organs of the female reproductive system; and

d. discuss female sexual development from puberty through menopause.

The female reproductive system is more complex than the male's because it serves more purposes. Whereas the male needs only to produce and deliver gametes, the female must do these as well as provide nutrition and safe harbor for fetal development, then give birth and nourish the infant. Furthermore, female reproductive physiology is more conspicuously cyclic, and female hormones are secreted in a more complex sequence compared with the relatively steady, simultaneous secretion of regulatory hormones in the male. AP|R

The Ovaries and Reproductive Tract

The female *reproductive tract* consists of a pair of uterine (fallopian) tubes, the uterus, and the vagina. The ovaries, strictly speaking, are not part of the reproductive tract, but are most conveniently considered here as a starting point for our description.

The Ovaries

The **ovaries**[14] (fig. 19.5) are the female gonads; they produce egg cells (ova) and sex hormones. Each ovary lies in a shallow depression of the posterior pelvic wall and is held in place by several connective tissue ligaments (fig. 19.7). The ovary is almond-shaped, about 3 cm long, 1.5 cm wide, and 1 cm thick. It is enclosed in a white fibrous capsule, and internally it is loosely divided into a central core *(medulla)* of fibrous connective tissue and blood vessels, and an outer zone *(cortex)* where the eggs develop.

Unlike the testis, the ovary has no ducts. Instead, each egg develops in its own bubblelike **follicle** and is released by *ovulation,* the bursting of the follicle. A child's ovaries are smooth-surfaced, but during the reproductive years, they become corrugated with bulges created by growing follicles. Figure 19.5 shows several types of follicles that coincide with different stages of egg maturation, as discussed later. We will later examine these follicles and their cyclic development in more detail.

Figure 19.5 Structure of the Ovary. Arrows indicate the developmental sequence of the ovarian follicles; the follicles do not migrate around the ovary, and not all the follicle types illustrated here are present simultaneously. AP|R

[14] *ov* = egg; *ary* = place for

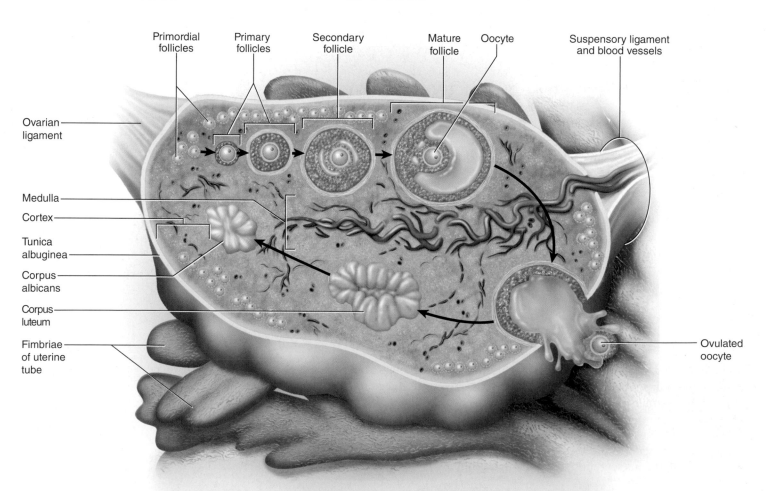

The Uterine Tubes

The **uterine (fallopian**[15]**) tube** is a muscular canal about 10 cm long that extends from the ovary to the uterus (fig. 19.6). At the distal (ovarian) end, it flares into a trumpetlike **infundibulum**[16] with feathery projections called **fimbriae**[17] (FIM-bree-ee). The mucosa of the uterine tube is folded into longitudinal ridges and has an epithelium of ciliated cells and a smaller number of secretory cells (fig. 19.7a). The cilia beat toward the uterus and, with the help of muscular contractions of the tube, convey the egg or developing embryo in that direction.

The Uterus

The **uterus**[18] is a muscular chamber that opens into the roof of the vagina and usually tilts forward over the urinary bladder (fig. 19.5). It harbors the fetus, provides nutrition, and expels the fetus at the end of its development. It is somewhat pear-shaped, with a broad superior curvature called the **fundus**, a midportion called the **body**, and a cylindrical inferior end called the **cervix**. The uterus measures about 7 cm from cervix to fundus, 4 cm wide at its broadest point on the fundus, and 2.5 cm thick, but it is somewhat larger in women who have been pregnant.

---

[15] Gabriele Fallopio (1523–62), Italian anatomist and physician

[16] *infundibulum* = funnel

[17] *fimbria* = fringe

[18] *uterus* = womb

Figure 19.6 The Female Pelvic Cavity and Reproductive Organs (Sagittal Section). AP|R

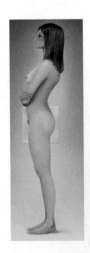

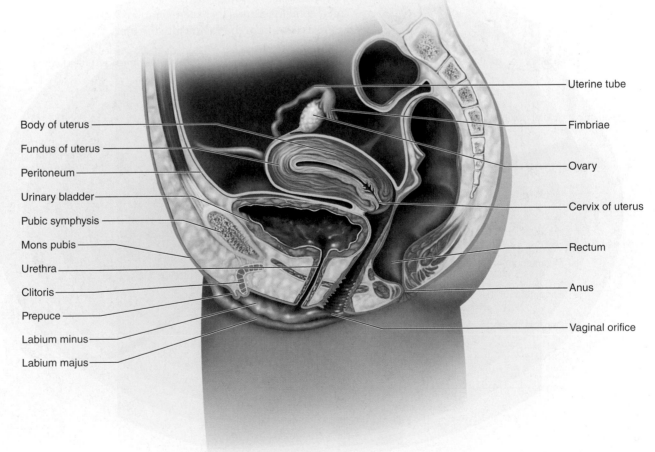

Body of uterus

Fundus of uterus

Peritoneum

Urinary bladder

Pubic symphysis

Mons pubis

Urethra

Clitoris

Prepuce

Labium minus

Labium majus

Uterine tube

Fimbriae

Ovary

Cervix of uterus

Rectum

Anus

Vaginal orifice

Clinical Application 19.2

CERVICAL CANCER

Cervical cancer occurs especially among women from ages 30 to 50, especially those who smoke, began sexual activity at an early age, or have histories of frequent sexually transmitted diseases or cervical inflammation. It is often caused by the human papillomavirus (HPV) (see p. 663). Cervical cancer usually begins in the lower cervix, develops slowly, and remains a local, easily removed lesion for several years. If the cancerous cells spread to the subepithelial connective tissue, however, the cancer is said to be *invasive* and is much more dangerous.

The best protection against cervical cancer is early detection by means of a *Pap*[19] *smear.* Loose cells are scraped from the cervix and microscopically examined for signs of *dysplasia* (abnormal development) or *carcinoma* (cancer) (fig. 19.8). Different grades of abnormality call for measures ranging from a repeat smear in a few months, to a biopsy. Confirmed cases of cervical cancer can be treated with radiation therapy, electrosurgical excision, or hysterectomy.[20] An average woman is typically advised to have annual Pap smears for 3 years and may then have them less often at the discretion of her physician. Women with any of the foregoing risk factors may be advised to have more frequent examinations.

Figure 19.7 **The Female Reproductive Tract.**
(a) Electron micrograph of the mucosa of the uterine tube, showing ciliated cells (yellow) and mucous cells (red). (b) Posterior view of the reproductive tract and associated organs. (c) Relationship of the ovary to ligaments of the uterus and uterine tube. AP|R

The lumen is roughly triangular, with its upper corners opening into the uterine tubes and the lower end communicating with the vagina by way of a narrow passage through the cervix called the **cervical canal** (fig. 19.7b). The canal contains glands that secrete mucus, thought to inhibit the spread of microorganisms from the vagina into the uterus.

[19] George N. Papanicolaou (1883–1962), Greek–American physician and cytologist
[20] *hyster* = uterus; *ectomy* = cutting out

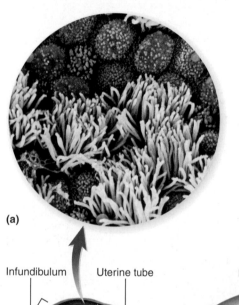

(a)

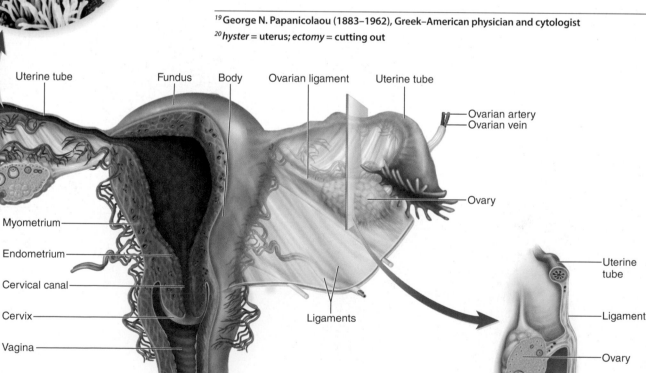

(b)

(c)

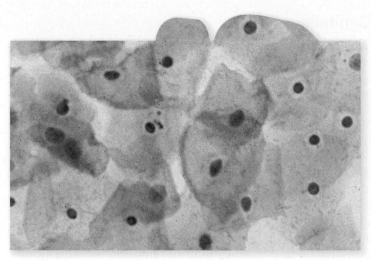

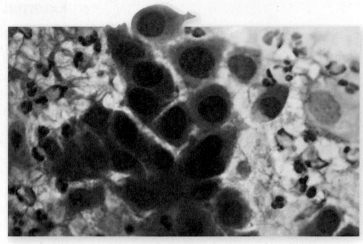

(a) Normal cells

(b) Malignant cells of cervical cancer

Figure 19.8 Pap Smears. These are smears of epithelial cells from the cervix. In the malignant (cancerous) cells, note the loss of cell volume and the greatly enlarged nuclei.

The uterine wall consists of a thin external serosa called the *perimetrium,* a thick middle layer called the *myometrium,* and an inner mucosa called the *endometrium.* The **myometrium**[21] consists mainly of smooth muscle and is responsible for the labor contractions that help to expel the fetus.

The **endometrium**[22] has a simple columnar epithelium and deep glandular pits with coiled *spiral arteries* between them (see fig. 19.16). The superficial half to two-thirds of the endometrium, called the **functional layer,** is shed in each menstrual period. The deeper layer, called the **basal layer,** stays behind and regenerates a new functional layer in the next cycle. When pregnancy occurs, the endometrium is the site of attachment of the embryo and forms part of the *placenta* from which the fetus is nourished.

The Vagina

The **vagina**[23] (see figs. 19.6, 19.7) is a tube 8 to 10 cm long that allows for the discharge of menstrual fluid, receipt of the penis and semen, and birth of a baby. The vaginal wall is thin but very distensible. It has no glands, but it is lubricated by serous fluid seeping through its wall and by mucus draining from the cervix. The vaginal epithelium is nonkeratinized stratified squamous—the form of tissue best adapted for lubrication and abrasion resistance. Its cells are rich in glycogen. Bacteria ferment this to lactic acid, which produces a low vaginal pH (about 3.5–4.0) that inhibits the growth of pathogens.

At the vaginal orifice, the mucosa folds inward and forms a membrane, the **hymen,** which stretches across the opening. The hymen has one or more openings to allow menstrual fluid to pass through, but it usually ruptures during or before the first intercourse, sometimes in the course of medical examinations, tampon use, or exercise.

---

[21] *myo* = muscle; *metr* = uterus

[22] *endo* = inside; *metr* = uterus

[23] *vagina* = sheath

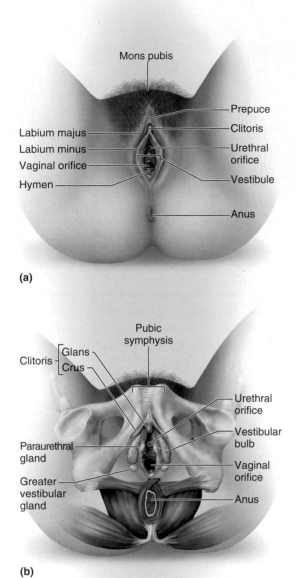

Mons pubis

Prepuce

Clitoris

Labium majus

Labium minus

Urethral orifice

Vaginal orifice

Vestibule

Hymen

Anus

(a)

Pubic symphysis

Clitoris { Glans
Crus

Urethral orifice

Vestibular bulb

Paraurethral gland

Vaginal orifice

Greater vestibular gland

Anus

(b)

Figure 19.9 The Female Perineum. (a) Surface anatomy. (b) Subcutaneous structures. AP|R

External Genitalia

The external genitalia of the female are collectively called the **vulva;**[24] they include the mons pubis, labia majora and minora, clitoris, vaginal orifice, and accessory glands and erectile tissues (fig. 19.9).

The **mons**[25] **pubis** is an anterior mound of adipose tissue and skin overlying the pubic symphysis, bearing most of the pubic hair. The **labia majora**[26] (singular, *labium majus*) are a pair of thick folds of skin and adipose tissue forming the lateral borders of the vulva; they bear pubic hair only on their lateral surfaces. Medial to these are the hairless **labia minora**[27] (singular, *labium minus*), much thinner because they are not adipose. The area enclosed by them, called the **vestibule**, contains the urinary and vaginal orifices. At the anterior margin of the vestibule, the labia minora meet and form a hoodlike **prepuce** over the clitoris.

The **clitoris** consists of a pair of corpora cavernosa, similar to the dorsal half of the penis; it has no corpus spongiosum and it does not enclose the urethra. Only its head, the **glans,** is external, protruding slightly from the prepuce; a leglike *crus* anchors it to the pelvic girdle on each side. The function of the clitoris is entirely sensory, serving as the primary center of sexual stimulation.

Just deep to the labia majora, a pair of subcutaneous erectile tissues called the **vestibular bulbs** brackets the vagina like parentheses. They become congested with blood during sexual excitement and cause the vagina to tighten somewhat around the penis, enhancing sexual stimulation for both partners. On each side of the vagina is a pea-size **greater vestibular (Bartholin**[28]**) gland** with a short duct opening into the vestibule or lower vagina. They moisten the vulva, and during sexual excitement they provide most of the lubrication for intercourse. The vestibule is also lubricated by a number of **lesser vestibular glands**. A pair of mucous **paraurethral (Skene**[29]**) glands** opens into the vestibule near the external urethral orifice. They eject fluid, sometimes abundantly, during orgasm. They arise from the same embryonic structure as the male's prostate gland, and their fluid is similar to the prostatic secretion.

Breasts

The **breast** (fig. 19.10) is a mound of tissue overlying the pectoralis major in both the male and female. The female breast markedly enlarges at puberty and remains so for life, but usually contains very little mammary gland. The **mammary gland** develops within the breast during pregnancy, remains active in the lactating breast, and atrophies when a woman ceases to nurse.

The breast has two principal regions: the conical to pendulous **body**, with the nipple at its apex, and an extension toward the armpit called the **axillary tail**. Lymphatics of the axillary tail are especially important as a route of breast cancer metastasis, making this region particularly important to include in breast self-examination.

The nipple is surrounded by a circular colored zone, the **areola**. Dermal blood capillaries and nerves come closer to the surface here than in the surrounding skin and make the areola more sensitive and deeper in color. The areola has sparse hairs and **areolar glands**, visible as small bumps on the surface. The areolar glands and sebaceous glands of the region oil the skin and help to prevent chapping in a nursing mother.

[24] *vulva* = covering

[25] *mons* = mountain

[26] *labi* = lip; *major* = larger, greater

[27] *minor* = smaller, lesser

[28] Caspar Bartholin (1655–1738), Danish anatomist

[29] Alexander J. C. Skene (1838–1900), American gynecologist

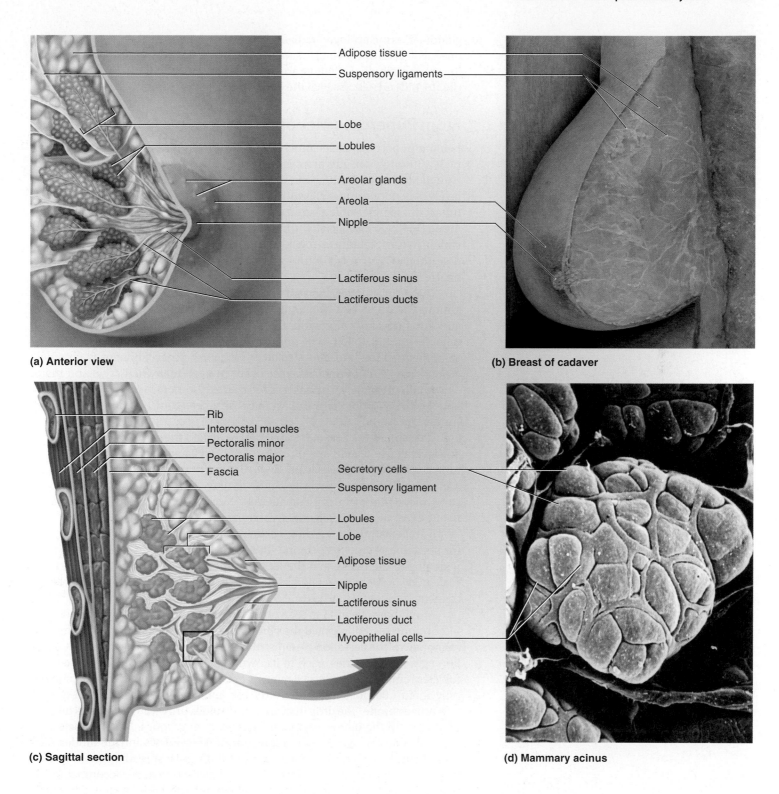

Adipose tissue
Suspensory ligaments
Lobe
Lobules
Areolar glands
Areola
Nipple
Lactiferous sinus
Lactiferous ducts

(a) Anterior view

(b) Breast of cadaver

Rib
Intercostal muscles
Pectoralis minor
Pectoralis major
Fascia
Secretory cells
Suspensory ligament
Lobules
Lobe
Adipose tissue
Nipple
Lactiferous sinus
Lactiferous duct
Myoepithelial cells

(c) Sagittal section

(d) Mammary acinus

Internally, the nonlactating breast is mostly adipose and collagenous tissue. Breast size is determined by the adipose tissue and is unrelated to the amount of milk the mammary gland can produce. Even in the nonlactating breast, there is a system of **lactiferous**[30] **ducts** radiating inwardly from the nipple. During pregnancy, these branch and develop into 15 to 20 lobes of

[30] *lact* = milk; *fer* = to carry

Figure 19.10 The Breast. Parts (a), (c), and (d) depict the breast in a lactating state. Some of the features in (a) and (c) are absent from the nonlactating breast in part (b). The cluster of lobules boxed in (c) would contain numerous microscopic acini like the one in (d). [Part (d) is not to be construed as an enlargement of the entire boxed area.] AP|R

glandular tissue arranged radially around the nipple. Each lobe is drained by one lactiferous duct, which dilates to form a *lactiferous sinus* just before opening on the nipple. Lactation and functional details of the mammary acinus (fig. 19.10d) are discussed at the end of this chapter.

From Puberty to Menopause

Female puberty is under much the same hormonal control as that of the male, although it begins at a slightly younger age—typically age 9 or 10 in the United States and Europe, but as young as age 3 in 1% to 3% of girls. Rising levels of gonadotropin-releasing hormone (GnRH) stimulate the anterior pituitary to secrete follicle-stimulating hormone (FSH) and luteinizing hormone (LH). FSH, especially, stimulates development of the ovarian follicles, which, in turn, secrete estrogens, progesterone, inhibin, and a small amount of androgen. **Estrogens**[31] are feminizing hormones with widespread effects on the body.

The earliest noticeable sign of female puberty is breast development, beginning with the *breast bud,* a small conical elevation of the areola. The lobules and ducts develop early under the influence of estrogens, progesterone, and other hormones, followed by the growth of adipose and fibrous tissue and development of the bud into a fuller breast. Breast development is complete around age 20, but minor changes occur in each menstrual cycle and major changes in pregnancy. The onset of breast development is soon followed by the appearance of pubic and axillary hair, sebaceous (oil) glands, and axillary (underarm) glands.

Next comes the first menstrual period, called **menarche**[32] (men-AR-kee). This occurs when a girl attains about 17% body fat, and therefore depends on nutrition. The average age at menarche in Europe and America is now about 12. Menarche does not necessarily signify fertility. A girl's first few cycles are typically *anovulatory* (no egg is ovulated). Most girls begin ovulating regularly about a year after they begin menstruating.

Estrogen stimulates growth of the ovaries and secondary sex organs, and the secretion of growth hormone. This brings about a rapid increase in height, widening of the pelvis, and the fat deposition that contours the maturing female body. Inhibin is secreted in girls, as it is in boys, and regulates the development of eggs and follicles.

Around the age of 45 to 55, women, like men, go through a midlife change in hormone secretion called the **climacteric**. In women, it is accompanied by **menopause**, the cessation of menstruation. A female is born with about 2 million eggs in her ovaries, each in its own follicle. The older she gets, the fewer follicles remain. Climacteric begins not at any specific age, but when she has only about 1,000 follicles left. Even the remaining follicles secrete less estrogen and progesterone. The drop in ovarian hormones brings about various degrees of atrophy in the uterus, vagina, breasts, skin, and bones. Blood vessels constrict and dilate in response to shifting hormone balances, and the sudden dilation of cutaneous arteries may cause **hot flashes**—a spreading sense of heat from the abdomen to the thorax, neck, and face, sometimes accompanied by headaches resulting from the sudden dilation of arteries in the head. It is difficult to precisely establish the time of menopause because the menstrual periods can stop for several months and then begin again. Menopause is generally considered to have occurred when there has been no menstruation for a year or more.

---

[31] *estro* = desire, frenzy; *gen* = to produce
[32] *men* = monthly ; *arche* = beginning

Before You Go On

Answer these questions from memory. Reread the preceding section if there are too many you don't know.

10. What structure in the ovary serves the same purpose as seminiferous tubules in the testis? In what ways does it differ from a seminiferous tubule?

11. What structure transports an egg from the ovary to the uterus? What feature within this structure moves the egg along this route?

12. Name the upper, middle, and lower parts of the uterus and the three layers of the uterine wall.

13. Identify all structures enclosed within the labia majora, including the subcutaneous glands, and describe their relative positions.

14. What is the difference between the breast and mammary gland? What traces of mammary gland can be found in the nonlactating breast?

15. What hormone from the brain stimulates the onset of puberty? How does that hormone lead to the production of estrogen?

19.4 The Production and Union of Sex Cells

Expected Learning Outcomes

When you have completed this section, you should be able to:

a. explain the relevance of meiosis to sexual reproduction, state the stages of meiosis, and describe how it differs from mitosis;

b. describe the stages in the production of sperm and eggs, how these stages relate to meiosis, and the major differences between sperm and egg production;

c. give a functional description of the major components of semen and sperm cells;

d. relate the process of egg production to the cyclic changes in the ovary and uterus;

e. describe the production of eggs and how it is correlated with cyclic changes in the ovaries and uterus; and

f. describe the physiological processes that occur in the male and female during sexual intercourse.

We've examined the equipment; now let's examine its purpose—to produce the sex cells and get them together.

Meiosis and Gametogenesis

Sexual reproduction poses a problem, given that human cells normally have 46 chromosomes and two humans intent on producing a child have to get two of those cells together. What's to prevent the child from having 92 chromosomes per cell, and the generation after that having 184, and so on? Sexual reproduction has a great advantage for species survival (genetic diversity), but it has to overcome this matter of doubling the chromosome number in each new generation. This is where **meiosis**[33] comes in.

To understand meiosis, it is important to know that the chromosomes of most cells occur in 23 pairs—designated *chromosomes 1–22* and one pair of *sex chromosomes,* designated *X* and *Y* in the male and two *X*s in the female. In all 23 pairs, one chromosome was contributed by the individual's mother and the other by the father. Within each pair except the male's XY, the two chromosomes look identical and are called **homologous chromosomes**. Any cell that has 23 chromosome pairs (46 chromosomes in all) is called a **diploid** cell, symbolized **2*n***. In contrast, the germ cells (sperm, eggs, and some of the developmental stages leading to them) have 23 unpaired chromosomes, and are called **haploid (*n*)**. One of the functions of meiosis is to reduce diploid stem cells to haploid germ cells; this is why meiosis is also called *reduction division.*

Meiosis has many similarities to mitosis (see p. 89), but several important differences:

1. Mitosis serves a range of functions including fetal development, childhood growth, and tissue maintenance and repair, whereas meiosis is used only for the production of gametes.

2. Mitosis maintains a constant number of chromosomes from parent cell to daughter cells, whereas meiosis reduces the chromosome number from diploid to haploid.

3. In mitosis, chromosomes do not change in genetic makeup. In an early stage of meiosis, however, the chromosomes of each pair join and exchange portions of their DNA. This creates new combinations of genes—some from one's mother and some from one's father on each chromosome—so the chromosomes we pass to our offspring are not the same ones that we inherited from our parents.

4. In mitosis, each parent cell produces only two daughter cells. In meiosis, it produces four. In the male, four equal-sized sperm develop from each original germ cell. In the female, though, three of the four daughter cells are tiny cells that soon die, and only one large, mature egg is produced.

Meiosis consists of two cell divisions in succession (fig. 19.11). The first division produces two haploid daughter cells, and the second division divides each of these into two more, so that the ultimate outcome is four haploid cells. In the first division *(meiosis I),* after two homologous chromosomes exchange segments of their DNA, they migrate to separate daughter cells. The daughter cells thus have only 23 chromosomes, but each chromosome retains its double-stranded form, composed of two chromatids joined at the centromere (see fig. 3.16, p. 91). In the second division *(meiosis II),* each centromere divides and each daughter cell now receives 23 single-stranded chromosomes.

Spermatogenesis, Sperm, and Semen

Both sperm and egg production rely on mitosis *and* meiosis. Sperm production is called **spermatogenesis**. In the average young man, it occurs at the remarkable rate of about 300,000 sperm per minute, or 400 million per day. Yet these sperm constitute only about 10% of the fluid that is ejaculated, called the *semen.*

---

[33] *meio* = less, fewer

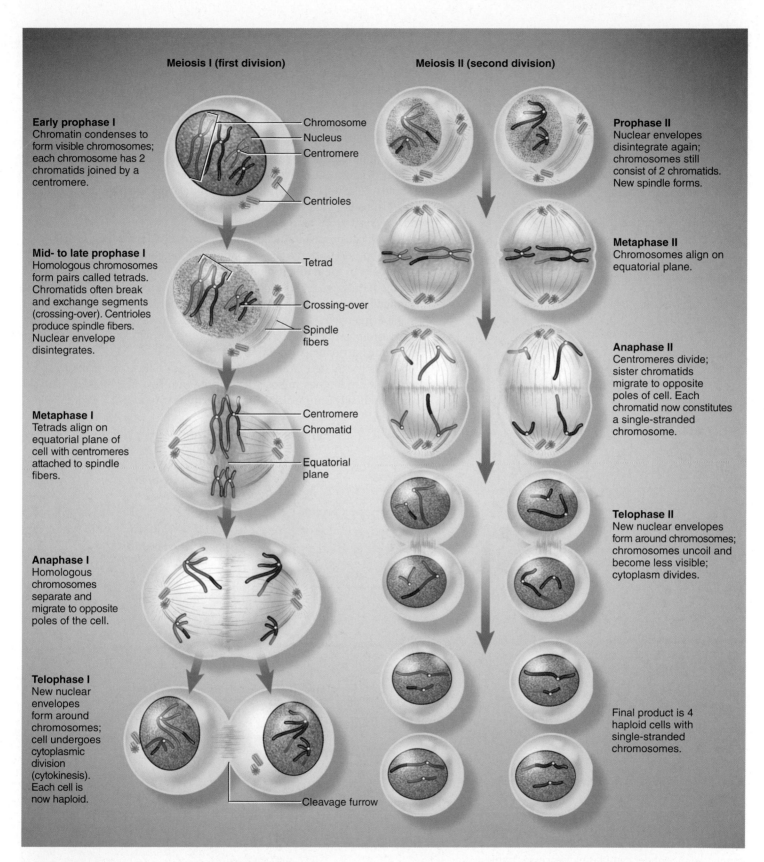

Meiosis I (first division)

Meiosis II (second division)

Early prophase I
Chromatin condenses to form visible chromosomes; each chromosome has 2 chromatids joined by a centromere.

Chromosome
Nucleus
Centromere
Centrioles

Prophase II
Nuclear envelopes disintegrate again; chromosomes still consist of 2 chromatids. New spindle forms.

Mid- to late prophase I
Homologous chromosomes form pairs called tetrads. Chromatids often break and exchange segments (crossing-over). Centrioles produce spindle fibers. Nuclear envelope disintegrates.

Tetrad
Crossing-over
Spindle fibers

Metaphase II
Chromosomes align on equatorial plane.

Metaphase I
Tetrads align on equatorial plane of cell with centromeres attached to spindle fibers.

Centromere
Chromatid
Equatorial plane

Anaphase II
Centromeres divide; sister chromatids migrate to opposite poles of cell. Each chromatid now constitutes a single-stranded chromosome.

Anaphase I
Homologous chromosomes separate and migrate to opposite poles of the cell.

Telophase II
New nuclear envelopes form around chromosomes; chromosomes uncoil and become less visible; cytoplasm divides.

Telophase I
New nuclear envelopes form around chromosomes; cell undergoes cytoplasmic division (cytokinesis). Each cell is now haploid.

Cleavage furrow

Final product is 4 haploid cells with single-stranded chromosomes.

Figure 19.11 Meiosis. For simplicity, the cell is shown with only two pairs of homologous chromosomes. Human cells begin meiosis with 23 pairs. AP|R

• *Although we pass the same genes to our offspring as we inherit from our parents, we do not pass on the same chromosomes. What event in this figure accounts for the latter fact?*

Spermatogenesis

Figure 19.12 relates meiosis to sperm production. At birth, the testes contain stem cells called *spermatogonia,* which lie dormant for years. At puberty, testosterone reactivates them and initiates spermatogenesis. The essential steps of spermatogenesis are as follows, numbered to match the figure. Until the end of the process when fully formed sperm are released, all these cells remain enfolded in the membranes of the sustentacular cells of the testis.

① **Spermatogonia** divide by mitosis, thereby keeping up a man's lifetime supply of sperm. One daughter cell from each division remains near the tubule wall as a stem cell, thus maintaining a lifelong population of stem cells. The other daughter cell migrates slightly away from the tubule wall on its way to producing sperm.

② The latter cell enlarges and becomes a **primary spermatocyte**. Since this cell is about to undergo meiosis and become genetically different from other cells of the body, it must be protected from the immune system. Ahead of the primary spermatocyte, the tight junction between two sustentacular cells is dismantled, while a new tight junction forms behind the spermatocyte. The spermatocyte moves toward the lumen, like an astronaut passing through a double-doored airlock, and is now protected by the barrier closing behind it.

③ Now safely isolated from blood-borne antibodies, the primary spermatocyte undergoes meiosis I, dividing into two equal-sized, haploid and genetically unique **secondary spermatocytes**.

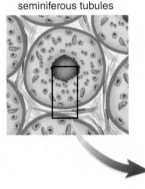

Cross section of
seminiferous tubules

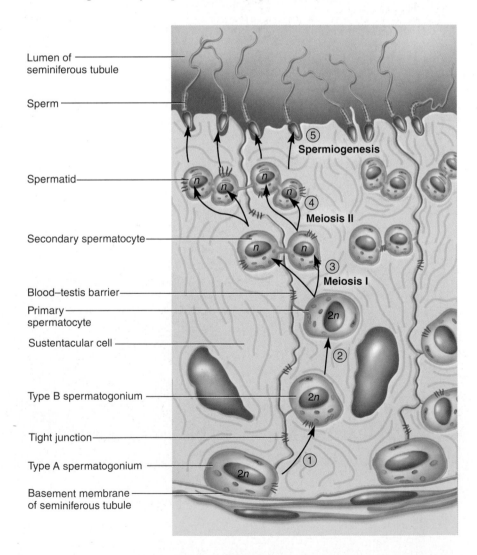

Lumen of
seminiferous tubule

Sperm

⑤ **Spermiogenesis**

Spermatid

④ **Meiosis II**

Secondary spermatocyte

③ **Meiosis I**

Blood–testis barrier

Primary
spermatocyte

Sustentacular cell

②

Type B spermatogonium

Tight junction

①

Type A spermatogonium

Basement membrane
of seminiferous tubule

Figure 19.12 Spermatogenesis. The process proceeds from the bottom of the figure (periphery of the seminiferous tubule) to the top (lumen of the tubule). The daughter cells from secondary spermatocytes through spermatids remain connected by slender cytoplasmic processes until spermiogenesis is complete and individual spermatozoa are released. See text for explanation of steps 1 to 5. ($2n$ = diploid cells, n = haploid cells) **AP|R**

④ Each secondary spermatocyte undergoes meiosis II, dividing into two **spermatids**—a total of four for each spermatogonium.

⑤ A spermatid divides no further, but transforms into a single sperm cell. This transformation, called **spermiogenesis**, consists essentially of sprouting a tail (flagellum) and shedding excess cytoplasm, making the sperm a lightweight, mobile, self-propelled cell. It will not move under its own power, however, until ejaculation.

When fully formed, the sperm depart from their supportive sustentacular cells and are washed down the seminiferous tubule by a slow flow of fluid. After division of the spermatogonium, it takes about 74 days for the one daughter cell to become mature sperm.

Sperm

The **spermatozoon (sperm cell)** has two parts: a pear-shaped head and a long flagellum, or tail (fig. 19.13). The **head** is filled mostly by two structures: a nucleus and acrosome. The most important of these is the nucleus, which fills most of the head and contains a haploid set of condensed, genetically inactive chromosomes. The **acrosome**[34] is a lysosome in the form of a thin cap covering the apical half of the nucleus. It contains enzymes that a successful sperm later uses to penetrate the egg.

The **tail** is divided into three regions called the midpiece, principal piece, and endpiece. The **midpiece** is the thickest part. It contains large mitochondria coiled around the core (axoneme) of the flagellum. They produce the ATP needed for sperm movement. The **principal piece** constitutes most of the tail and consists of the axoneme surrounded by a sheath of supportive fibers, which stiffen the tail and enhance its propulsive power. The principal piece provides most propulsion of the sperm. It tapers to a short **endpiece** composed of axoneme only.

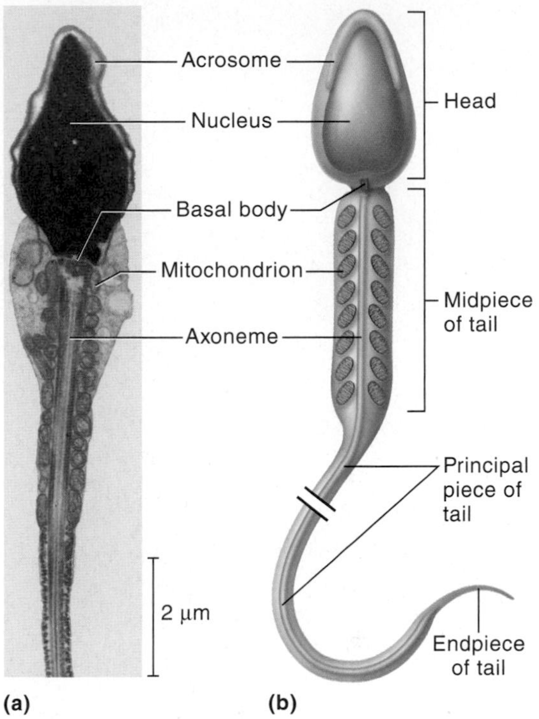

Semen

The fluid expelled in orgasm is called **semen**[35] (**seminal fluid**). A typical ejaculation is 2 to 5 mL of semen, composed mainly of sperm (10%), prostatic fluid (30%), and seminal vesicle fluid (60%), with a trace of other secretions. Most sperm emerge in the first one or two jets of semen. The semen usually has a **sperm count** of 50 to 120 million sperm/mL. A sperm count lower than 20 to 25 million sperm/mL is usually associated with **infertility (sterility)**, the inability to fertilize an egg.

The prostate gland contributes a thin, milky white fluid containing calcium and other ions; a clotting enzyme; and a protein-hydrolyzing enzyme called *serine protease* (also known as prostate-specific antigen, PSA; see Clinical Application 19.1). The seminal vesicles contribute a viscous yellowish fluid. This is the last component of the semen to emerge, and it flushes remaining sperm from the urethra. It contains fructose and other carbohydrates, prostaglandins (discovered in and named for the prostate, but more abundant in the seminal vesicle fluid), and a protein called *prosemenogelin*.

A well-known property of semen is its stickiness. This arises when the clotting enzyme from the prostate converts prosemenogelin into a sticky fibrinlike protein, semenogelin. **Semenogelin** entangles the sperm, sticks to the walls of the vagina and cervix, and ensures that the semen doesn't simply drain back out of the vagina. It may also promote the uptake of sperm-laden clots of semen into the uterus.

About 20 to 30 minutes after ejaculation, the serine protease of the prostatic fluid breaks down semenogelin and liquifies the semen. The sperm, which lay still until then, now become very active, thrashing with their tails and

(a)   **(b)**

Figure 19.13 The Mature Spermatozoon.
(a) Head and part of the tail of a spermatozoon (electron micrograph). (b) Sperm structure.

---

[34] *acro* = tip, peak; *some* = body

[35] *semen* = seed

crawling up the mucosa of the vagina and uterus. Their motility depends on ATP, which is made in the midpiece mitochondria by oxidizing the fructose in the semen. You could think of the sperm midpiece as the "motor," the tail as the "propeller," and the head or its nucleus as the "cargo."

Oogenesis and the Ovarian–Menstrual Cycle

In the absence of pregnancy, women have two interrelated monthly cycles controlled by shifting patterns of hormone secretion: the **ovarian cycle**, consisting of events in the ovaries, and the **menstrual cycle**, consisting of parallel changes in the uterus. The ovarian cycle is concerned with *oogenesis,* the production of eggs, and *folliculogenesis,* the parallel developments in the follicles that enclose them.

Oogenesis

Oogenesis[36] (OH-oh-JEN-eh-sis), like spermatogenesis, produces a haploid gamete by means of meiosis (fig. 19.14). A major difference, however, is that whereas males produce sperm continually, oogenesis is conspicuously cyclic and usually produces only one mature egg per month. It is accompanied by cyclic changes in hormone secretion and in the histological structure of the ovaries and uterus; the uterine changes result in the monthly menstrual flow.

Stem cells called **oogonia** (OH-oh-GO-nee-uh) multiply by mitosis in the ovaries of a female fetus. Most of them degenerate and die before she is born, but about 2 million remain at birth. By the age of 6 months, all of these transform into **primary oocytes** and proceed as far as early meiosis I, but then go into developmental arrest until puberty. The term *egg,* or *ovum,* applies loosely to any stage from the primary oocyte to the time of fertilization. Most of a girl's primary oocytes degenerate and die during childhood, until only about 400,000 remain at the onset of puberty. This is her lifetime supply of eggs, but it is ample. Even if a female ovulated every 28 days from age 14 to 50, she would ovulate only 480 times. All of the other eggs degenerate without ovulating between puberty and menopause.

Oogenesis resumes in adolescence, when FSH stimulates monthly cohorts of about two dozen oocytes to complete meiosis I. Each primary oocyte divides into two haploid daughter cells of unequal size and different destinies. It is important to produce an egg with as much cytoplasm as possible, because if fertilized, it must divide repeatedly and produce numerous daughter cells. Splitting each oocyte into four equal but small parts would run counter to this purpose. Therefore, meiosis I produces a large daughter cell called the **secondary oocyte** and a much smaller one called the **first polar body**. The polar body disintegrates; it is merely a means of discarding the extra set of chromosomes.

The secondary oocyte proceeds as far as metaphase II, then arrests until after ovulation. If it is not fertilized, it dies and never finishes meiosis. If fertilized, it completes meiosis II and casts off a **second polar body**, which disposes of one chromatid from each chromosome. The chromosomes of the large remaining egg unite with those of the sperm, and a zygote has been achieved. Further development of the zygote is discussed in chapter 20.

Folliculogenesis

As an egg undergoes oogenesis, the follicle around it undergoes as many as five stages of growth, collectively called **folliculogenesis**.

1. **Primordial follicles**. A primordial follicle consists of a primary oocyte surrounded by a single layer of squamous *follicular cells* (fig. 19.14).

---

[36] *oo* = egg; *genesis* = production

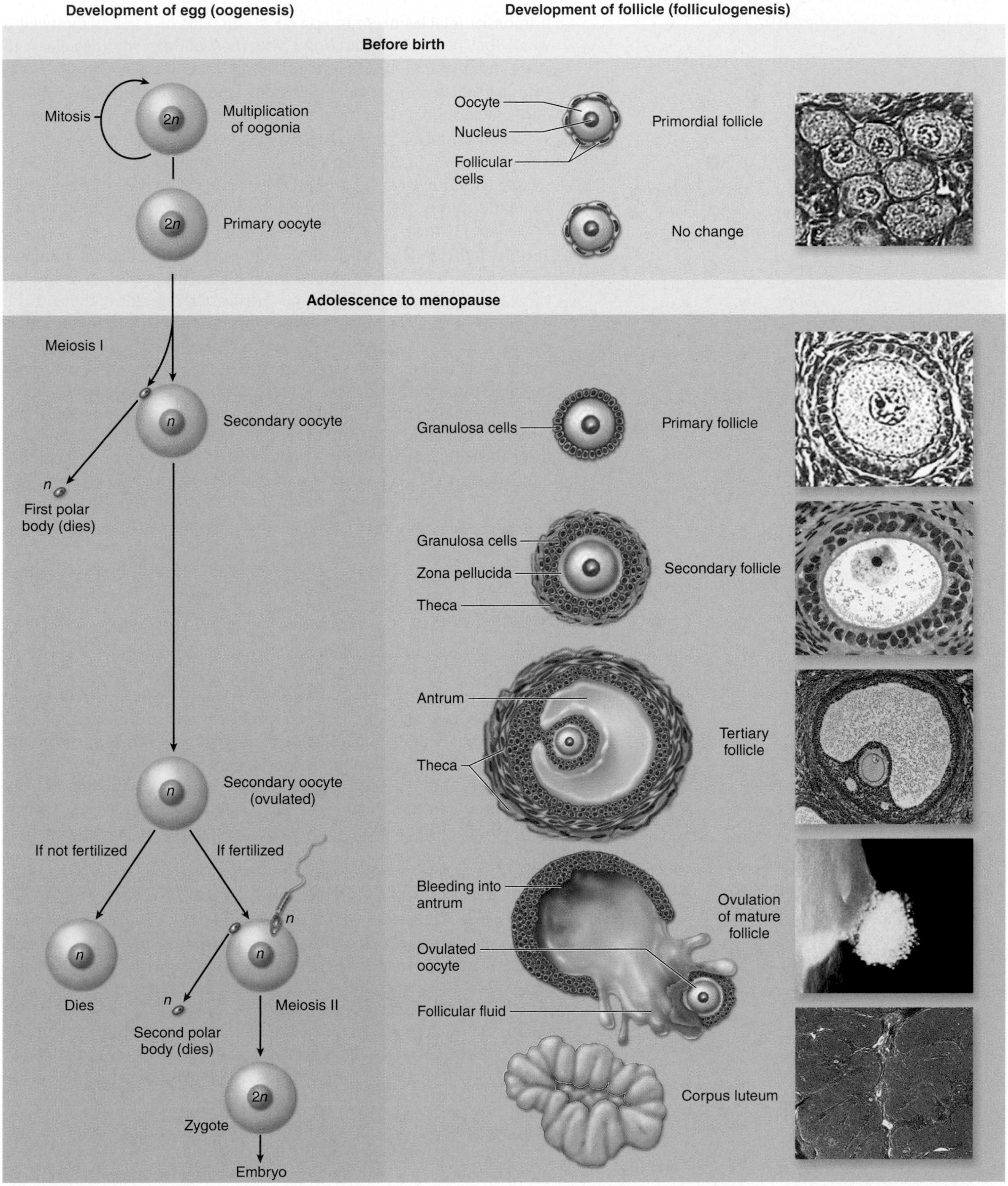

Development of egg (oogenesis)

Development of follicle (folliculogenesis)

Before birth

Mitosis — Multiplication of oogonia (2n)

Primary oocyte (2n)

Oocyte
Nucleus
Follicular cells

Primordial follicle

No change

Adolescence to menopause

Meiosis I

Secondary oocyte (n)

n First polar body (dies)

Granulosa cells — Primary follicle

Granulosa cells
Zona pellucida
Theca

Secondary follicle

Antrum
Theca

Tertiary follicle

Secondary oocyte (ovulated) (n)

If not fertilized / If fertilized

Dies (n)

n Second polar body (dies)

n

Meiosis II

Bleeding into antrum
Ovulated oocyte
Follicular fluid

Ovulation of mature follicle

Corpus luteum

Zygote (2n)

Embryo

Figure 19.14 Oogenesis (Left) and Corresponding Development of the Follicle (Right). AP|R

- *How many eggs are produced per primary oocyte? How does this compare to the number of sperm per primary spermatocyte? What is the functional importance of this difference?*

Primordial follicles begin to appear in the fetus as early as the twelfth week. They persist into adulthood, with most of them waiting at least 13 years and some as long as 50 years before they develop any further.

2. **Primary follicles.** The follicular cells become cuboidal but still form just one layer around the now larger oocyte.

3. **Secondary follicles.** The follicular cells divide and pile on top of each other in layers; the follicular cells are now called *granulosa cells.* They secrete a layer of clear gel around the oocyte called the *zona pellucida,* while the connective tissue around the follicle condenses to form a fibrous husk, or *theca.*

4. **Tertiary follicles.** Next, the follicle cells secrete a fluid that accumulates in little pools in the follicle wall. As they enlarge, the pools merge and become a single fluid-filled cavity, the **antrum.** On one side of the antrum, a mound of granulosa cells covers the oocyte and secures it to the follicle wall. The innermost layer of cells around the egg creates a barrier that ensures that nothing from the bloodstream can get to the oocyte except by going through (not between) these cells. The granulosa cells selectively allow nutrients and hormones to pass through to the egg, while screening out antibodies and other potentially harmful chemicals. The theca develops a rich supply of blood vessels and collaborates with the follicle cells to secrete sex hormones, especially estrogen.

5. **Mature (graafian[37]) follicles.** Normally only one follicle in each monthly cohort becomes a mature follicle, destined to ovulate while the rest degenerate.

The Ovarian Cycle

We can now relate oogenesis and folliculogenesis to the timetable of events in the monthly *sexual cycle* (fig. 19.15). The cycle varies from 20 to 45 days in length, differing from person to person and from month to month in the same person, but it averages 28 days, so we will use this as a basis for discussion. As you study this cycle, bear in mind that hormones of the hypothalamus regulate the pituitary gland; pituitary hormones regulate the ovaries; and the ovaries, in turn, secrete hormones that regulate the uterus. That is, the basic hierarchy of control can be represented: hypothalamus → pituitary → ovaries → uterus. However, the ovaries also exert feedback control over the hypothalamus and pituitary.

We begin with a brief preview of the cycle. It starts with a 2-week *follicular phase.* The first 3 to 5 days are marked by menstruation, in which blood and endometrial tissue are discharged vaginally. While this is going on, a cohort of ovarian follicles grows until one of them ovulates around day 14. Endometrial tissue is regenerated and thickens between the end of menstruation and ovulation. After ovulation, the remainder of that follicle becomes a body called the *corpus luteum.* Over the next 2 weeks, called the *luteal phase,* the corpus luteum stimulates endometrial secretion, making the endometrium thicken still more. If pregnancy does not occur, the endometrium breaks down again in the last 2 days. As loose tissue and blood accumulate, menstruation begins and the cycle starts over. Now the details.

1. **Follicular phase.** This phase is marked by the growth of a cohort of follicles stimulated by follicle-stimulating hormone (FSH). As they grow, these follicles secrete estrogen. By unknown means, one of them has already been selected during the previous month's cycle to become the

---

[37] Reijnier de Graaf (1641–73), Dutch physiologist and histologist

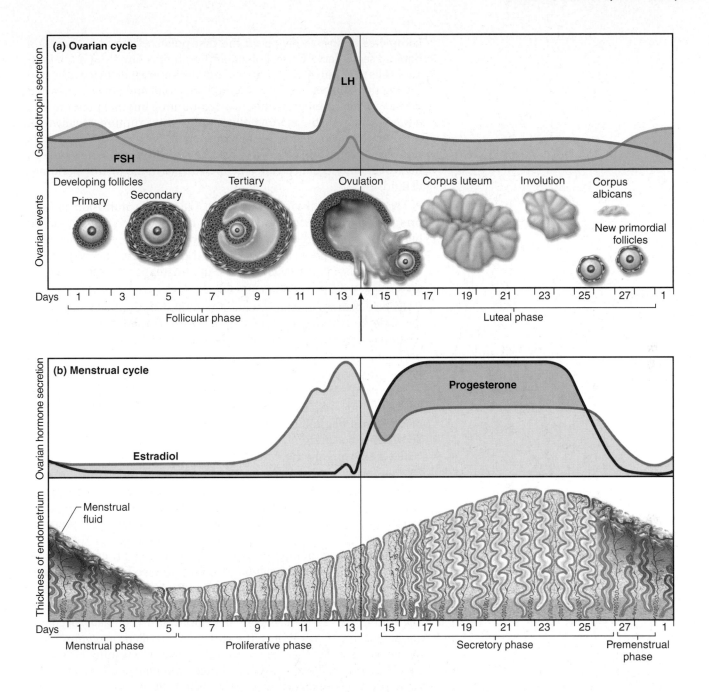

dominant follicle—the one destined to ovulate in this cycle. The dominant follicle is the most sensitive of all of them to FSH, luteinizing hormone (LH), and estrogen, so it races ahead of the others toward the day of ovulation. At the same time, estrogen from the ovaries indirectly inhibits the pituitary from secreting FSH. All of the less sensitive follicles therefore degenerate, while the highly sensitive dominant follicle continues to grow. It reaches a diameter of about 20 mm and bulges from the ovarian surface like a great blister, ready to ovulate. It is now considered a mature follicle.

2. **Ovulation**. Ovulation is the rupture of the mature follicle and the release of its egg and some of the surrounding cells. Dramatic changes over the preceding day signify its imminence. Luteinizing hormone (LH) output from the pituitary rises during the preceding events, but on the day just before ovulation, there is a sharp spike in LH output (fig. 19.15a). This induces several momentous events. The primary oocyte

Figure 19.15 The Female Sexual Cycle. (a) The ovarian cycle (events in the ovary). (b) The menstrual cycle (events in the uterus). **AP|R**

completes meiosis I, giving off the first polar body and becoming a haploid secondary oocyte. Follicular fluid builds rapidly and the follicle swells to as much as 25 mm in diameter—more than twice the usual thickness of the whole ovary. The follicular wall and adjacent ovarian tissue weaken by inflammation. With mounting internal pressure and a weakening wall, the mature follicle approaches rupture. Meanwhile, the uterine tube prepares to catch the oocyte. It swells with edema; its fimbriae envelop and caress the ovary in synchrony with the woman's heartbeat; and its cilia create a gentle current in the nearby peritoneal fluid.

Ovulation itself takes only 2 or 3 min. A nipplelike protrusion appears over the follicle; it seeps fluid for 1 or 2 minutes; and then the follicle bursts. The remaining fluid oozes out, carrying the oocyte and its companion cells (see photo in figure 19.14). These are normally swept up by the ciliary current and taken into the uterine tube, although many oocytes fall into the pelvic cavity and die.

Follicle maturation and ovulation normally occur in only one ovary per cycle, and the ovaries usually alternate from month to month. Some women can feel the moment of ovulation as a slight twinge of abdominal pain.

3. **Luteal phase.** Days 15 to 28, from just after ovulation to the onset of menstruation, are called the **luteal (postovulatory) phase**. When the follicle ruptures, it collapses and bleeds into the antrum. As the clotted blood is absorbed, follicular and theca cells multiply and fill the antrum, and a dense bed of blood capillaries grows amid them. The follicle is now called the **corpus luteum**,[38] named for a yellow lipid that accumulates in its cells.

LH stimulates this transformation from ruptured follicle to corpus luteum, as well as its continued growth. The corpus luteum secretes rising levels of estrogen and especially progesterone. Progesterone prepares the uterus for the possibility of pregnancy (and is named for this fact[39]). At the same time, hormones from the corpus luteum exert negative feedback inhibition on the pituitary and suppress its output of FSH and LH. (This is the basis for hormonal birth control pills.)

In the absence of pregnancy, the corpus luteum begins shrinking around day 22 (8 days after ovulation) and eventually becomes an inactive bit of scar tissue. The shrinkage is called *involution*[40] and the resulting scar is the *corpus albicans*.[41] With the waning of ovarian steroid secretion, the pituitary is no longer inhibited and FSH levels begin to rise again, ripening a new cohort of follicles.

The most variable part of the foregoing cycle is the follicular phase, from day 1 of menstruation until ovulation. This is unfortunate, because it makes it seldom possible to reliably predict the date of ovulation for the purposes of family planning or pregnancy avoidance.

The Menstrual Cycle

The menstrual cycle consists of uterine changes in parallel with the ovarian events—a buildup of endometrium through most of the cycle, followed by its breakdown and vaginal discharge. The menstrual cycle is divided into a

---

[38] *corpus* = body; *lute* = yellow

[39] *pro* = favoring, promoting; *gest* = pregnancy; *sterone* = steroid hormone

[40] *in* = inward; *volution* = turning

[41] *corpus* = body; *albicans* = white

proliferative phase, secretory phase, premenstrual phase, and *menstrual phase* (fig. 19.16). The menstrual phase averages 5 days long, and the first day of noticeable vaginal discharge is defined as day 1 of the sexual cycle. But even though it begins our artificial timetable for the cycle, the reason for menstruation is best understood after you become acquainted with the buildup of endometrial tissue that precedes it. Thus, we begin our survey of the cycle with the proliferative phase.

1. **Proliferative phase**. In this phase, the uterus rebuilds the layer of tissue lost in the last menstruation. At day 5 (or the end of menstruation), the endometrium is about 0.5 mm thick and consists only of the permanent *basal layer* of tissue that is never shed in menstruation. But as a new cohort of follicles develops, they secrete more and more estrogen. Estrogen stimulates mitosis in the basal layer as well as a prolific regrowth of blood vessels, thus regenerating the *functional layer* that was lost at menstruation (fig. 19.16a). By day 14, the endometrium is 2 to 3 mm thick. Estrogen also stimulates endometrial cells to produce progesterone receptors, priming them for the progesterone-dominated phase to follow.

2. **Secretory phase**. In this phase, the endometrium thickens still more, but because of secretion and fluid accumulation rather than mitosis. This phase extends from day 15 (after ovulation) to day 26 of a typical cycle. After ovulation, the corpus luteum secretes mainly progesterone, which stimulates the endometrial glands to secrete glycogen. The glands grow wider, longer, and more coiled, and the endometrium swells with tissue fluid (fig. 19.16b). By the end of this phase, the endometrium is 5 to 6 mm thick—a soft nutritious bed available for embryonic development in the event of pregnancy.

3. **Premenstrual phase**. If there is no pregnancy, then the last 2 days of the cycle are marked by endometrial degeneration. This comes about because the dying corpus luteum secretes less and less progesterone. The

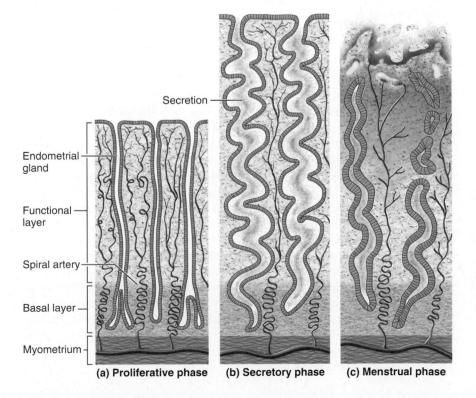

Secretion

Endometrial gland

Functional layer

Spiral artery

Basal layer

Myometrium

(a) **Proliferative phase**   (b) **Secretory phase**   (c) **Menstrual phase**

Figure 19.16 Endometrial Changes Through the Menstrual Cycle. (a) Late proliferative phase. The endometrium is 2 to 3 mm thick and has relatively straight, narrow endometrial glands. Spiral arteries penetrate upward between the endometrial glands. (b) Secretory phase. The endometrium has thickened to 5 to 6 mm by accumulating glycogen and mucus. The endometrial glands are much wider and more distinctly coiled, showing a zigzag or "sawtooth" appearance in histological sections. (c) Menstrual phase. Dying tissue has begun to fall away from the uterine wall, with bleeding from broken blood vessels and pooling of blood in the tissue and uterine lumen.

loss of progesterone triggers spasmodic contractions of the spiral arteries of the endometrium. The interrupted blood flow brings on tissue necrosis and sometimes menstrual cramps. As the endometrial glands, stroma, and blood vessels degenerate, pools of blood accumulate in the tissue. Pieces of dying endometrium fall away from the uterine wall, mix with blood and serous fluid in the lumen, and form the **menstrual fluid** (fig. 19.16c). Eventually, all of the functional layer is lost in the menstrual fluid and only the basal layer remains.

4. **Menstrual phase (menses).** This phase begins when enough fluid has accumulated in the uterus to start being discharged from the vagina. The first day of discharge marks day 1 of the menstrual cycle. The average woman expels about 40 mL of blood and 35 mL of serous fluid over a 5-day period.

In summary, the ovaries go through a follicular phase characterized by growing follicles; then ovulation; and then a postovulatory (mostly luteal) phase dominated by the corpus luteum. The uterus, in the meantime, goes through a menstrual phase in which it discharges its functional layer; then a proliferative phase in which it replaces that tissue by mitosis; then a secretory phase in which the endometrium thickens by the accumulation of secretions; and finally, a premenstrual phase in which the functional layer breaks down again. The first half of the cycle is governed largely by follicle-stimulating hormone (FSH) from the pituitary gland and estrogen from the ovaries. Ovulation is triggered by luteinizing hormone (LH) from the pituitary, and the second half of the cycle is governed mainly by LH from the pituitary and progesterone from the ovaries.

Uniting the Gametes AP|R

We have just seen how the sperm and eggs are produced—now to get them together. The physiology of sexual intercourse was a taboo subject for many decades until, in the 1950s, William Masters and Virginia Johnson daringly launched the first physiological studies of sexual response in the laboratory. Thanks to their pioneering work and those who have followed, sexual response is now studied on a level as matter-of-fact as the physiology of digestion or respiration.

Anatomical Foundations

To understand sexual function, we must give closer attention to the blood circulation and nerve supply to the genitalia, especially the penis. Each internal iliac artery gives off a **penile artery** that enters the root of the penis and divides in two. One branch, the **dorsal artery**, travels dorsally along the penis not far beneath the skin (see fig. 19.4), supplying blood to the skin, fascia, and corpus spongiosum. The other branch, the **deep artery**, travels through the core of the corpus cavernosum and gives off smaller arteries that penetrate the trabeculae and empty into the lacunae. When the deep artery dilates, the lacunae fill with blood and the penis becomes erect. Blood drains from the penis through a median *deep dorsal vein,* which runs between the two dorsal arteries near the penile surface. The blood supply to the female's clitoris is essentially the same except that there is no corpus spongiosum.

The penis and clitoris are richly innervated by sensory and motor nerve fibers. The glans of each has an abundance of tactile, pressure, and temperature receptors. Signals travel out of the penis or clitoris by way of a pair of prominent *dorsal nerves* and ultimately to the sacral spinal cord. Sensory fibers of the penile shaft and scrotum of the male and the labia and vestibule of the female are also highly important to sexual stimulation.

Both autonomic and somatic motor fibers carry impulses from the spinal cord to the genitalia. Sympathetic fibers from spinal cord levels T12 to L2 innervate the arteries of the penis and clitoris, the trabecular muscle, the accessory glands, and the male spermatic ducts. They dilate the arteries and can induce erection even when the sacral region of the spinal cord is damaged. The arteries also receive parasympathetic fibers from spinal cord segments S2 to S4 of the spinal cord. These nerves are involved in an autonomic reflex arc that causes erection in response to direct stimulation of the genitalia and perineal region.

Sexual Response

At the onset of sexual arousal, the genitalia become swollen with blood, muscle tension rises throughout the body, and the heart rate, blood pressure, and respiratory rate rise. The bulbourethral glands of the male and greater vestibular glands of the female secrete their fluids during this phase, with the female secretion, especially, providing lubrication for intercourse. The vagina contains no glands but becomes moistened by seepage of serous fluid from its wall.

The most obvious manifestation of male sexual excitement is **erection** of the penis, caused by the rapid inflow of blood through the dilated deep arteries. This is an autonomic reflex mediated mainly by parasympathetic nerve fibers from the spinal cord. As lacunae near the deep arteries fill with blood, they compress lacunae closer to the periphery of the erectile tissue. This is where blood leaves the erectile tissues, so compression blocks the outflow of blood. Each corpus cavernosum is wrapped in a tight fibrous tunic or sleeve. This tunic cannot expand very much laterally, so it contributes to the tension and firmness of the corpus cavernosum and the penis expands more in length than in width. Once entry into the vagina is achieved, tactile and pressure sensations produced by vaginal massaging of the penis further accentuate the erection reflex. In the female, the clitoris becomes similarly engorged and erect, swelling to two or three times its usual size. But since the clitoris cannot swing upward away from the body like the penis, it tends to withdraw beneath the prepuce.

The male corpus spongiosum has neither a central artery nor a fibrous tunic. It swells and becomes more visible as a cordlike ridge along the ventral surface of the penis, but it does not become nearly as engorged and hardened as the corpora cavernosa.

> ### *Apply What You Know*
> Why is it important that the corpus spongiosum not become as engorged as the corpora cavernosa? What might be the consequence if it did so?

Orgasm[42] is a short but intense reaction that lasts from 3 to 15 seconds; it is usually marked by the discharge of semen in males and by vaginal and uterine contractions in females. From the standpoint of producing offspring, the most significant aspect of male orgasm is the **ejaculation**[43] of semen into the vagina. It occurs in two stages called *emission* and *expulsion*. In **emission**, the sympathetic nervous system induces peristalsis in the ductus deferens, which propels sperm from the epididymis to the prostatic urethra. The prostate gland expels its fluid into the urethra, followed quickly by the seminal vesicles. The contractions and fluid flow in this phase create an urgent sensation that ejaculation is inevitable.

Semen in the urethra then activates somatic and sympathetic reflexes that result in its **expulsion**. Sensory signals travel to an integrating center in the upper lumbar region of the spinal cord. Sympathetic signals return to the prostate

---

[42] *orgasm* = swelling

[43] *e* = *ex* = out; *jacul* = to throw

gland and seminal vesicles, causing them to expel more fluid into the urethra. The internal urethral sphincter constricts so urine cannot enter the urethra and semen cannot enter the bladder. Somatic motor signals from the spinal cord stimulate the **bulbospongiosus muscle**, which ensheaths the root of the penis. In five or six spasmodic contractions, this muscle compresses the urethra and forcibly expels the semen. Most sperm emerge in the first milliliter of semen, mixed primarily with prostatic fluid. The seminal vesicle secretion follows and flushes most remaining sperm from the urethra.

In female orgasm, the lower end of the vagina gives three to five strong contractions about 0.8 seconds apart, while the cervix plunges spasmodically into the vagina and into the pool of semen, should this be present. The uterus exhibits peristaltic waves of contraction, which might help to draw semen from the vagina.

Following orgasm, sympathetic stimulation constricts the arteries and reduces the flow of blood into the penis and clitoris. It also causes contraction of the trabecular muscles, which squeeze blood from the lacunae of the erectile tissues. The penis may remain semierect long enough to continue intercourse, but gradually it becomes soft and flaccid again. In men, this is usually followed by a *refractory period* of anywhere from 10 minutes to a few hours, in which it is usually impossible to attain another erection and orgasm. Women do not have a refractory period and may quickly experience additional orgasms.

In the decades before effective contraceptives were widely available, it was thought that douching (flushing the vagina) shortly after intercourse could prevent pregnancy. However, sperm can make it well into the uterine tubes and well beyond the reach of a douche as quickly as 5 minutes after ejaculation. For reasons explained in chapter 20, however, it is about 10 hours before they become capable of penetrating an egg and initiating a pregnancy.

Before You Go On

Answer these questions from memory. Reread the preceding section if there are too many you don't know.

16. Why is meiosis necessary in sexual reproduction? At what stages in meiosis do we see a change in the number of chromosomes per cell and the number of chromatids per chromosome?

17. List the stages of spermatogenesis; identify which stages involve mitosis, meiosis I, and meiosis II; and state the chromosome number at each stage. Then do the same for oogenesis.

18. Identify the main parts of a mature spermatozoon and the function of each part.

19. Spermatogenesis produces four functional gametes per spermatogonium and oogenesis produces only one gamete per oogonium. Why?

20. Name the five kinds of ovarian follicles and identify the essential features that distinguish one from another.

21. Describe the changes occurring in the ovaries and uterus in the 2 weeks before ovulation and in the 2 weeks after.

22. What is a corpus luteum? Where does it come from and what purpose does it serve?

23. Describe some similarities and differences in the physiology of orgasm between the male and female.

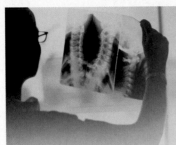

PERSPECTIVES ON HEALTH

Sexually Transmitted Diseases

Sexually transmitted diseases (STDs) have been well known since the writings of Hippocrates. Here we discuss three bacterial STDs—gonorrhea, chlamydia, and syphilis—and three viral STDs—genital herpes, genital warts, and hepatitis. AIDS is discussed in chapter 14.

All of these STDs have an *incubation period* in which the pathogen multiplies without symptoms, and a *communicable period* in which one can transmit the disease to others even in the absence of symptoms. Aside from recent advances with human papillomavirus (HPV), there are few vaccines for STDs, because the bacteria and viruses live inside host cells and go undetected by the immune system. STDs often cause fetal deformity, stillbirth, and neonatal death (see p. 698).

Gonorrhea (GON-oh-REE-uh) is caused by the bacterium *Neisseria gonorrhoeae.* Galen, thinking the pus discharged from the penis was semen, named the disease *gonorrhea* ("flow of seed"). Gonorrhea causes abdominal discomfort, genital pain and discharge, painful urination, and abnormal uterine bleeding, but most infected women are asymptomatic. It can cause scarring of the uterine tubes, resulting in infertility. Gonorrhea is treated with antibiotics.

Nongonococcal urethritis (NGU) is any urethral inflammation caused by agents other than the gonorrhea bacterium. NGU often produces pain or discomfort on urination. The most common bacterial NGU is infection by *Chlamydia trachomatis.* Most chlamydia infections are asymptomatic, but they may cause urethral discharge and pain in the testes or pelvic region. Gonorrhea and chlamydia frequently occur together.

Pelvic inflammatory disease (PID) is bacterial infection of the female pelvic organs, usually with *Chlamydia* or *Neisseria.* It often results in sterility and may require surgical removal of infected uterine tubes or other organs. The incidence of PID in the United States has increased from 17,800 cases in 1970 to more than a million cases per year currently. PID is responsible for many cases of ectopic pregnancy (see Clinical Application 20.1, p. 682).

Syphilis is caused by a corkscrew-shaped bacterium named *Treponema pallidum.* After an incubation period of 2 to 6 weeks, a small, hard lesion called a *chancre* (SHAN-kur) appears at the site of infection—usually on the penis of a male but sometimes out of sight in the vagina of a female. It disappears in 4 to 6 weeks, ending the first stage of syphilis and often creating an illusion of recovery. A second stage ensues, however, with a widespread pink rash, other skin eruptions, fever, joint pain, and hair loss. This subsides in 3 to 12 weeks, but symptoms can come and go for up to 5 years. A person is contagious even when symptoms

are not present. The disease may progress to a third stage, *tertiary syphilis (neurosyphilis),* with cardiovascular damage and brain lesions that can cause paralysis and dementia. Syphilis is treatable with antibiotics.

Genital herpes is the most common STD in the United States, with 20 to 40 million infected people at a given time. It is usually caused by the *herpes simplex virus type 2 (HSV-2).* After an incubation period of 4 to 10 days, the virus causes blisters on the penis of the male; on the labia, vagina, or cervix of the female; and sometimes on the thighs and buttocks of either sex. Over 2 to 10 days, these blisters rupture, seep fluid, and begin to form scabs. The initial infection may be painless or it may cause intense pain, urethritis, and watery discharge from the penis or vagina. The lesions heal in 2 to 3 weeks and leave no scars.

During this time, however, HSV-2 colonizes sensory nerves and ganglia. Here the virus can lie dormant for years, later migrating along the nerves and causing epithelial lesions anywhere on the body. The movement from place to place is the basis of the name *herpes.*[44] Most patients have five to seven recurrences, ranging from several years apart to several times a year. An infected person is contagious to a sexual partner when the lesions are present and sometimes even when they are not. HSV-2 may increase the risk of cervical cancer and AIDS.

Genital warts (condylomas) are one of the most rapidly increasing STDs today, with about 6.2 million new cases per year. They are caused by various *human papillomavirus (HPV)* strains. In the male, lesions usually appear on the penis, perineum, or anus; in the female, they are usually on the cervix, vaginal wall, perineum, or anus. Lesions are sometimes small and almost invisible. HPV has been implicated in cancer of the penis, vagina, cervix, and anus; it is found in about 90% of cervical cancers. About 90% of genital warts, however, involve strains that have not been linked to cancer. Genital warts are sometimes treated with cryosurgery (freezing and excision), laser surgery, or interferon.

Hepatitis B and *C,* introduced in Clinical Application 18.1 (p. 618), are inflammatory liver diseases caused by the hepatitis B and C viruses (HBV, HCV). Although they can be transmitted by means other than sex, they are becoming increasingly common as STDs. Hepatitis C threatens to become a major epidemic of the twenty-first century. It already far surpasses the prevalence of AIDS and is the leading reason for liver transplants in the United States.

---

[44] *herp* = to creep

19.5 Pregnancy, Childbirth, and Lactation

Expected Learning Outcomes

When you have completed this section, you should be able to:

a. itemize the major hormones of pregnancy and describe their effects;

b. describe the effects of pregnancy on a woman's body;

c. explain what happens in each stage of childbirth;

d. discuss the hormonal control of lactation; and

e. discuss the composition of colostrum and breast milk.

This section treats pregnancy from the maternal standpoint—adjustments of the woman's body to pregnancy and the mechanism of childbirth—followed by the physiology of lactation. Development of the fetus is described in chapter 20.

Pregnancy

Pregnancy (gestation) lasts an average of 266 days from conception (fertilization of the egg) to childbirth, but the gestational calendar is usually measured from the first day of the woman's last menstrual period (LMP). Thus, the birth is predicted to occur 280 days (40 weeks) from the LMP. The duration of pregnancy, called its *term,* is commonly described in 3-month intervals called **trimesters**.

Pregnancy places considerable stress on a woman's body and requires adjustments in nearly all the organ systems. Some of these effects are as follows:

Integumentary System. The skin grows to accommodate expansion of the abdomen and breasts and the added fat deposition in the hips and thighs. Stretching of the dermis often tears the connective tissue and causes *stretch marks (striae).* Increased melanin synthesis commonly darkens the areola of the breast, making the nipple a more easily seen target for a nursing infant. A dark line *(linea nigra)* may appear from the umbilicus to the pubic region, and some women develop a blotchy facial "mask of pregnancy" *(chloasma),* but these lighten or disappear when the pregnancy is over.

Circulatory System. A woman gains 1 to 2 liters of blood over the course of a pregnancy, with about 625 mL/min. going to supply the placenta. Another effect of pregnancy is that the growing uterus puts pressure on the large pelvic blood vessels, thus interfering with venous return from the lower body. This can result in hemorrhoids, varicose veins, and edema of the feet.

Respiratory System. Breathing also becomes more labored as the growing uterus puts pressure on the diaphragm from below. In the last month, however, the fetus drops lower in the abdominopelvic cavity, taking some pressure off the diaphragm and allowing the woman to breathe more easily.

Urinary System. A pregnant woman's urine output is slightly elevated, enabling her to dispose of both her own and the fetus's metabolic wastes. As the pregnant uterus compresses the bladder and reduces its capacity, urination becomes more frequent and some women experience uncontrollable leakage of urine (incontinence).

Digestive System. For many women, one of the first signs of pregnancy is **morning sickness**—nausea, especially after rising from bed, in the first few months of gestation. This sometimes progresses to vomiting, occasionally severe enough to require hospitalization to stabilize a woman's fluid and electrolyte balance.

Metabolism and Nutritional Needs. Pregnancy raises a woman's metabolic rate and often makes her feel overheated. It stimulates the appetite, but all too many pregnant women overeat as a result. One needs only an extra 300 kcal/day even in the last trimester, and a healthy average weight gain is 11 kg (24 pounds). Some women, however, gain as much as 34 kg (75 pounds). About 3 kg (7 pounds) of the healthy weight gain is the fetus; the rest is the placenta, fetal membranes, amniotic fluid, uterus, breasts, maternal body fluids, and fat. Pregnancy warrants taking supplemental iron, vitamin D, and vitamin K. Extra folic acid can prevent the risk of neurological disorders in the fetus such as spina bifida, but it has to be taken even before one is pregnant. By the time one is aware of being pregnant, it is too late for extra folic acid to be effective (see Clinical Application 8.2, p. 277).

Childbirth

The uterus weighs about 50 g when a woman is not pregnant and about 900 g (2 pounds) by full term. Most of this increase is due to muscle growth in preparation for pregnancy. Ancient authorities thought that the fetus kicked against the uterus and pushed itself out head first. The fetus, however, is a rather passive player in its own birth; its expulsion is achieved only by the contractions of the mother's uterine and abdominal muscles.

During gestation, the uterus exhibits relatively weak **Braxton Hicks**[45] **contractions**. These become stronger in late pregnancy and often send women rushing to the hospital with "false labor." At full term, however, these contractions transform suddenly into the more powerful **labor contractions**. True labor contractions mark the onset of **parturition** (PAR-too-RISH-un), the process of giving birth.

The increasing contractility of the uterus may stem from the rising ratio of estrogen to progesterone toward the end of pregnancy. Also, as the pregnancy nears full term, the posterior lobe of the pituitary gland releases increasing amounts of **oxytocin (OT)**, while progesterone stimulates the uterine muscle to develop more OT receptors. Oxytocin directly stimulates the uterine muscle to contract. Stretching of the uterus by the growing fetus also plays a role in initiating labor contractions. This is probably why twins are born an average of 19 days earlier than solitary infants.

Labor contractions typically begin about 30 minutes apart, then become more frequent and intense, eventually occuring every 1 to 3 minutes. It is important that they be intermittent rather than one long, continual contraction. Each contraction compresses blood vessels and reduces blood flow to the placenta, so the uterus must periodically relax to restore flow and oxygen delivery to the fetus. Contractions are strongest in the fundus and body of the uterus and weaker near the cervix, thus pushing the fetus downward.

According to the *positive feedback theory of labor,* contractions are induced by stretching of the cervix. This triggers a reflex contraction of the uterine body that pushes the fetus downward and stretches the cervix still more. Thus, there is a self-amplifying cycle of stretch and contraction. In addition,

---

[45] John Braxton Hicks (1823–97), British gynecologist

stretching triggers a reflex through the spinal cord, hypothalamus, and posterior pituitary. The pituitary releases oxytocin, which is carried in the blood and stimulates the uterine muscle. This, too, is a positive feedback cycle: cervical stretching → oxytocin secretion → uterine contraction → cervical stretching (see fig. 1.5, p. 12).

As labor progresses, a woman feels a growing urge to "bear down." A somatic reflex arc extends from the uterus to the spinal cord and back to the skeletal muscles of the abdomen. Contraction of these muscles—partly reflexive and partly voluntary—aids in expelling the fetus, especially when combined with breath holding (the Valsalva maneuver) for increasing intra-abdominal pressure.

Labor occurs in three stages (fig. 19.17), each lasting longer in a woman giving birth for the first time (a *primipara*) than in a woman who has given birth before *(multipara).*

1. **Dilation stage**. This stage is marked by the **dilation** (widening) of the cervical canal and **effacement** (thinning) of the cervix. The cervix dilates to about 10 cm (the diameter of the baby's head), and usually the fetal membranes rupture and discharge the *amniotic fluid* (the "breaking of the waters"). This stage typically lasts 8 to 24 hours in a primipara but as little as a few minutes in a multipara.
2. **Expulsion (second) stage**. This stage typically lasts 30 to 60 minutes in a primipara and as little as 1 minute in a multipara. It begins when the baby's head enters the vagina and lasts until the baby is entirely expelled. Delivery of the head is the most difficult part, with the rest of the body following much more easily.
3. **Placental (third) stage**. The uterus continues to contract after expulsion of the baby. The placenta, however, is a nonmuscular organ that cannot contract, so it buckles away from the uterine wall. The placenta and fetal membranes *(afterbirth)* are then expelled by uterine contractions. The membranes must be carefully inspected to be sure everything has been expelled. If any of these structures remain in the uterus, they can cause postpartum hemorrhaging. About 350 mL of blood is typically lost in the placental stage, but uterine contractions compress the blood vessels and prevent more extensive bleeding.

Postpartum Changes

Over the first 6 weeks **postpartum** (after birth), the mother's anatomy and physiology stabilize and her reproductive organs return nearly to their condition prior to pregnancy. The shrinkage of the uterus during this period is called **involution** (the same word as for the shrinkage of the corpus luteum described earlier). It is achieved through the autolysis (self-digestion) of uterine cells by their own enzymes. For about 10 days, this produces a vaginal discharge called **lochia**[46] (LO-kee-ah) which is bloody at first and then turns clear and serous. Breast-feeding hastens involution because (1) it suppresses estrogen secretion, which would otherwise cause the uterus to remain more flaccid; and (2) it stimulates oxytocin secretion, which causes the myometrium to contract and firm up the uterus sooner.

Lactation

Lactation is the synthesis and ejection of milk from the mammary glands. It lasts for as little as a week in women who do not breast-feed their infants, but it can continue for many years as long as the breast is stimulated by a nursing

[46] *lochos* = childbirth

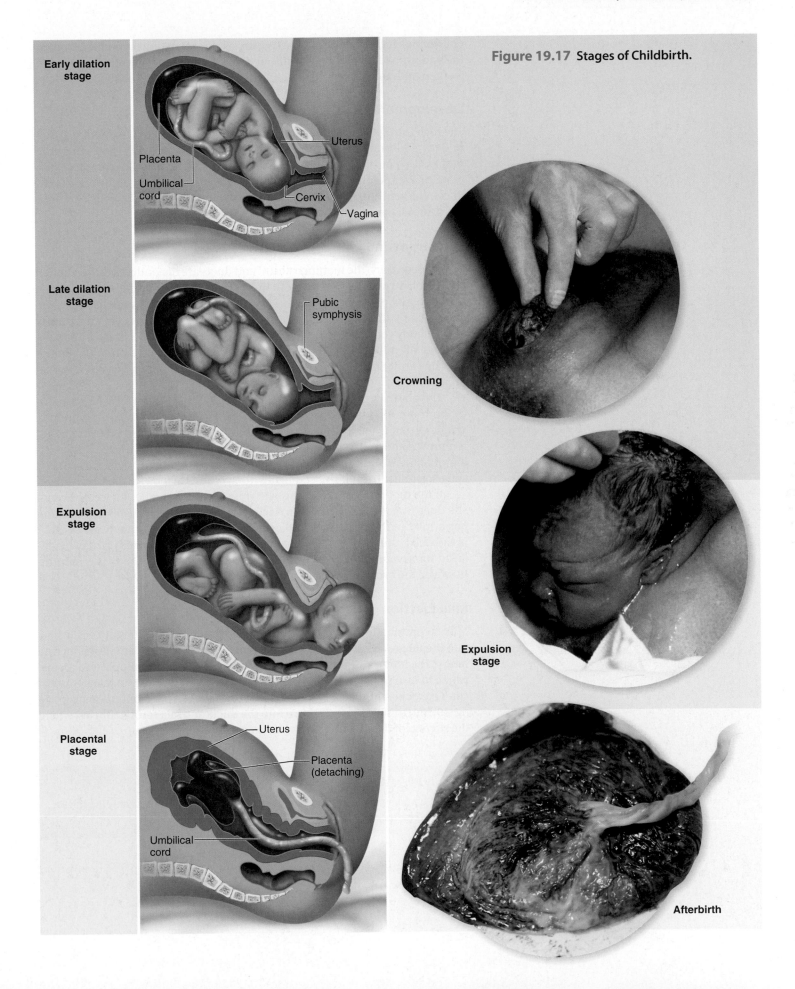

Figure 19.17 Stages of Childbirth.

Early dilation stage

Placenta

Umbilical cord

Uterus

Cervix

Vagina

Late dilation stage

Pubic symphysis

Crowning

Expulsion stage

Expulsion stage

Placental stage

Uterus

Placenta (detaching)

Umbilical cord

Afterbirth

child or mechanical device (breast pump). Numerous studies conducted before the widespread marketing of artificial infant formulas suggest that worldwide, women traditionally nursed their infants until a median age of about 2.8 years.

Development of the Mammary Glands in Pregnancy

The high estrogen level in pregnancy causes the ducts of the mammary glands to grow and branch extensively. Growth hormone, prolactin, and other hormones also contribute to this development. Once the ducts are complete, progesterone stimulates the budding and development of acini at the ends of the ducts. The acini are organized into grapelike clusters (lobules) within each lobe of the breast (see fig. 19.10).

Colostrum and Milk Synthesis

In late pregnancy, the mammary acini and ducts are distended with a secretion called **colostrum**. This is similar to breast milk in protein and lactose content but has about one-third less fat. It is the infant's only natural source of nutrition for the first 1 to 3 days postpartum. Colostrum has a thin watery consistency and a cloudy yellowish color. The daily volume of colostrum is at most 1% of the volume of milk secreted later. A major benefit of colostrum is that it contains antibodies that protect the infant from infection and gastrointestinal inflammation.

The pituitary gland secretes prolactin beginning 5 weeks into the pregnancy, but no milk is synthesized until after birth. Placental steroids inhibit milk synthesis, but when the placenta is discharged in the afterbirth, this inhibition is removed and prolactin begins to exert its effect. Milk is synthesized in increasing quantity over the following week. Each time the infant nurses, there is a spike in prolactin level. This stimulates the synthesis of milk for the next feeding.

If the mother does not nurse, she stops lactating in about a week. Even if she does nurse, milk production declines after 7 to 9 months. Breast-feeding has a mild (although unreliable) contraceptive effect. In women who do not breast-feed, the ovarian cycle resumes in a few weeks, whereas it may be delayed for several months in women who breast-feed. For the first 6 months, however, the cycles are usually anovulatory (no egg is released).

Milk Ejection

Milk is continually secreted into the mammary acini, but it does not easily flow into the ducts. Its flow, called **milk ejection**, is controlled by a neuroendocrine reflex. The infant's suckling stimulates nerve endings of the nipple and areola, which in turn signal the hypothalamus and posterior pituitary to release oxytocin. Oxytocin stimulates contractile *myoepithelial cells* around the acini (see fig. 19.10d) to squeeze milk down into the ducts and lactiferous sinuses, where the suction of the infant's mouth can draw it out.

Apply What You Know

When a woman is nursing her baby at one breast, would you expect only that breast, or both breasts, to eject milk? Explain why.

Breast Milk

Breast milk changes composition over the first 2 weeks, varies from one time of day to another, and changes even during the course of a single feeding. For example, at the end of a feeding there is less lactose and protein in the milk, but six times as much fat, as there is at the beginning. Therefore, a baby may not receive complete nourishment if its feeding is interrupted. A woman nursing one baby eventually produces about 1.5 L of milk per day; women with twins produce more. Lactation places a great demand on the mother for calcium, phosphate, protein, fat, and other nutrients, and requires appropriate dietary adjustments so that mineralization of the infant's skeleton does not come at the expense of the mother's bone tissue.

It used to be common to feed infants with cow's milk, but this is a poor substitute for colostrum and breast milk. Table 19.1 compares their composition. Cow's milk is more difficult for an infant to digest because of its excessive protein content, it causes more diaper rash, and it fails to provide the immunity and other benefits of breast-feeding.

Before You Go On

Answer these questions from memory. Reread the preceding section if there are too many you don't know.

24. What are the roles of oxytocin, prostaglandins, and uterine stretching in childbirth?

25. In childbirth, what are dilation, effacement, and crowning? In what stage of childbirth does each one occur?

26. How does colostrum differ from breast milk? What purposes does it serve?

27. Contrast the roles of prolactin and oxytocin in breast-feeding.

Table 19.1 A Comparison of Colostrum, Human Milk, and Cow's Milk

Constituents	Human Colostrum	Human Milk	Cow's Milk
Total protein (g/L)	22.9	10.6	30.9
Lactalbumin (g/L)	—	3.7	25.0
Casein (g/L)	—	3.6	2.3
Immunoglobulins (g/L)	19.4	0.09	0.8
Fat (g/L)	29.5	45.4	38.0
Lactose (g/L)	57	71	47
Calcium (mg/L)	481	344	1370
Phosphorus (mg/L)	157	141	910

Aging of the Reproductive System

Male reproductive function changes relatively gradually after middle age, with declining testosterone secretion, sperm count, and libido. By age 65, sperm count is about one-third of what it was in a man's 20s. Men remain fertile (able to father children) well into old age. However, erectile dysfunction (ED)—the inability to maintain an erection long enough to complete intercourse—becomes more common for a variety of reasons, including atherosclerosis, medications, and psychological factors. About 20% of men in their 60s and 50% of men in their 80s experience some degree of ED. Nearly all men with ED, however, remain able to ejaculate.

In women, the changes in reproductive function are more abrupt and pronounced. Over the course of menopause, gametogenesis ceases, active follicles are depleted, and the ovaries stop secreting sex hormones. This can result in vaginal dryness, genital atrophy, and reduced libido. With the loss of ovarian steroids, postmenopausal women have elevated risks of atherosclerosis and osteoporosis. Women who become pregnant in their 30s and 40s or beyond have an increasing risk of having infants with chromosomal birth defects such as Down syndrome, for a reason explained in chapter 20 (p. 697).

CAREER SPOTLIGHT

Midwife

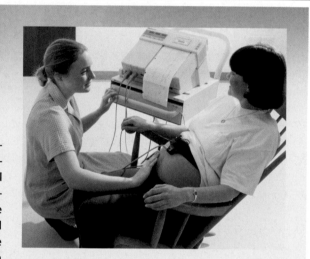

A midwife is a person who provides prenatal counseling and care to expectant mothers, attends the birth, and provides postpartum care to the mother and infant. Midwives encourage natural birth with minimal technological intervention. They work with relatively healthy women and low-risk pregnancies, but are trained to recognize complications beyond their expertise and to refer these to obstetricians for advanced care. Midwives often attend births in the home or at a birthing center, but in some hospitals today, the majority of babies are delivered by midwives. The term *midwife* derives from Old English "with (*mid*) woman (*wif*)," referring to the pregnant woman and not the practitioner; both women and men work as midwives.

The best career opportunities and salaries go to certified nurse-midwives (CNMs), who must earn a nursing degree and then pursue about 2 years of specialized training in midwifery, often leading to a master's degree. CNMs must pass a national board exam, and can then be licensed in any of the 50 states. Most CNM programs are given at universities, and most CNMs practice in hospitals.

There also are *direct-entry midwives* who do not hold nursing degrees but have more limited employment opportunities. These range from lay midwives (LMs), educated through self-study and apprenticeship, to certified professional midwives (CPMs), who complete formal programs of study. Prerequisites for such programs vary by state but may include general biology, microbiology, human anatomy and physiology, mathematics, English, and social science. Many states do not recognize or license direct-entry midwives. Where licensed, these midwives usually practice outside of hospitals and sometimes establish their own birthing centers.

See appendix B for further information on careers in midwifery.

CONNECTIVE ISSUES

Ways in Which the Reproductive System Affects Other Organ Systems

Integumentary System

Androgens of puberty stimulate growth of body hair and apocrine glands and increased secretion by sebaceous glands; estrogens stimulate fat deposition and breast development in females; pregnancy may cause stretch marks and pigmentation changes in skin.

Skeletal System

Androgens and estrogens stimulate adolescent skeletal growth and maintain bone density in adulthood; pregnancy and lactation can draw calcium and phosphorus from bones if dietary intake is inadequate to meet these needs.

Muscular System

Androgens and estrogens stimulate muscle growth.

Nervous System

Androgens stimulate libido; hormones from gonads and placenta exert negative feedback control of hypothalamus.

Endocrine System

Gonads and placenta are part of the endocrine system and exert negative feedback control of anterior pituitary gland; pregnancy reduces a woman's insulin sensitivity and sometimes causes gestational diabetes; the onset of labor stimulates oxytocin secretion by the posterior pituitary; nursing an infant stimulates prolactin secretion by the anterior pituitary.

Circulatory System

Androgens stimulate erythropoiesis; estrogens may inhibit atherosclerosis in females; pregnancy increases blood volume and cardiac output and sometimes causes varicose veins.

Lymphatic and Immune Systems

Blood–testis barrier in testes and zona pellucida in ovaries isolate sperm and eggs from the immune system. Sexually transmitted diseases present many challenges to the immune system, and AIDS undermines immunity.

Respiratory System

Pulmonary ventilation increases in response to sexual arousal and in pregnancy.

Urinary System

Sexual arousal constricts the internal urethral sphincter; prostatic hyperplasia may impede urine flow; pregnancy crowds the urinary bladder and often causes incontinence.

Digestive System

Fetus crowds digestive organs, contributing to heartburn and constipation; pregnancy increases nutritional requirements.

671

Study Guide

Assess Your Learning Outcomes

To test your knowledge, discuss the following topics with a study partner or in writing, ideally from memory.

19.1 Essentials of Sexual Reproduction (p. 632)

1. The meaning of *gamete* and *zygote*, and why sexual reproduction employs two different kinds of gametes
2. The meaning of *gonads, primary* and *secondary sex organs,* and *external* and *internal genitalia*

19.2 The Male Reproductive System (p. 633)

1. Gross anatomy of the testis and the internal organization of its tissues
2. Structure of the seminiferous tubules and the functions of its germ cells and sustentacular cells
3. Anatomy and functions of the epididymis and ductus deferens
4. Structure of the scrotum and especially the components of the spermatic cord
5. The mechanisms by which the scrotum and spermatic cord regulate the temperature of the testes, and the reason why this temperature regulation is so important
6. The locations and structure of the seminal vesicles and prostate gland and their contributions to the semen
7. The location and function of the bulbourethral glands
8. Gross anatomy of the penis and the microscopic anatomy of its three erectile tissues
9. Hormonal events that bring on puberty, and the bodily changes that these hormones produce
10. The hormonal events of andropause and the effects of these changes experienced by some men

19.3 The Female Reproductive System (p. 641)

1. Location, gross anatomy, and internal structure of the ovaries
2. The regions and internal histology of the uterine tubes
3. The regions and shape of the uterus
4. The three layers of the uterine wall and two subdivisions of the endometrium
5. The gross anatomy and histology of the vagina
6. Anatomy of the vulva, including the locations and structures of the mons pubis, labia majora and minora, clitoris, prepuce, and vaginal orifice
7. The locations and functions of the vestibular bulbs, greater and lesser vestibular glands, and paraurethral glands
8. External regions and internal structure of the nonlactating breast
9. Hormonal events that bring on puberty, and the bodily changes that these hormones produce
10. The hormonal events of female climacteric and menopause and the effects of these changes experienced by some women

19.4 The Production and Union of Sex Cells (p. 649)

1. The homologous pairs of human chromosomes, the sex chromosomes, and the distinction between diploid and haploid cells
2. The reason that meiosis is integral to sexual reproduction, how it converts diploid cells to haploid cells, and how it contributes to genetic diversity
3. Similarities and differences between mitosis and meiosis
4. The stages of meiosis and principal events that characterize each stage
5. The stages of spermatogenesis, the points at which mitosis and meiosis are involved, and the point at which the chromosome number changes from diploid to haploid
6. Structure of the mature spermatozoon and the functions of its parts
7. Constituents of semen, their sources, and their functions
8. The stages of oogenesis, and the similarities and differences between oogenesis and spermatogenesis
9. The stages of folliculogenesis, the features that distinguish ovarian follicles of each type, the correlation between folliculogenesis and oogenesis, and the structure of a mature follicle
10. The phases and events of the ovarian cycle, how they relate to changes in the follicles, and the hormones that dominate each phase
11. The phases of the menstrual cycle, the endometrial changes that occur in each, and how these relate to events in the ovaries and their hormonal output

12. Circulatory anatomy, erectile tissues, and innervation of the penis and clitoris and how these relate to erectile function

13. The mechanism of erection, physiology of orgasm and ejaculation, and processes that follow orgasm in both sexes

19.5 Pregnancy, Childbirth, and Lactation (p. 664)

1. The timetable of pregnancy

2. Adaptations to pregnancy seen in a woman's integumentary, circulatory, respiratory, urinary, and digestive systems, and in her metabolism and nutritional needs

3. Factors responsible for the increase in uterine contractility toward the end of pregnancy, for the onset of true labor, and for its progression

4. The stages of labor and the defining events of each stage

5. Changes in the maternal body over the first 6 weeks after childbirth

6. Development of the mammary glands during pregnancy and the hormones that regulate this

7. The nature of colostrum, how it differs from milk, and what purposes it serves

8. The influence of breast-feeding on ovarian cycling and fertility

9. The roles of prolactin and oxytocin in lactation, and the mechanism by which oxytocin stimulates milk ejection

10. The composition of breast milk and reasons why it is superior to cow's milk for infant nutrition

Testing Your Recall

1. Thinning of the cervical tissue in childbirth is called
   a. chloasma.
   b. effacement.
   c. lochia.
   d. involution.
   e. dilation.

2. Prior to ejaculation, sperm are stored mainly in
   a. the seminiferous tubules.
   b. the seminal vesicles.
   c. the ejaculatory ducts.
   d. the epididymis.
   e. the prostate gland.

3. The hormone that triggers the onset of puberty in both sexes is
   a. gonadotropin-releasing hormone.
   b. human chorionic gonadotropin.
   c. follicle-stimulating hormone.
   d. testosterone.
   e. estrogen.

4. The 46 chromosomes of a diploid set separate into two haploid sets at
   a. prophase I.
   b. prophase II.
   c. anaphase I.
   d. anaphase II.
   e. telophase II.

5. Sperm motility requires ATP, which is produced in the sperm's
   a. acrosome.
   b. nucleus.
   c. midpiece.
   d. polar body.
   e. zona pellucida.

6. Menstruation is brought on by a drop in the level of
   a. progesterone.
   b. estrogen.
   c. oxytocin.
   d. FSH.
   e. LH.

7. The ovaries secrete all of the following except
   a. estrogen.
   b. progesterone.
   c. androgen.
   d. follicle-stimulating hormone.
   e. inhibin.

8. The greatest fraction of the volume of the semen comes from
   a. the testes.
   b. the epididymis.
   c. the prostate gland.
   d. the corpus cavernosum.
   e. the seminal vesicles.

9. The tissue lost in menstruation is
   a. the functional layer of the endometrium.
   b. the basal layer of the endometrium.
   c. the myometrium.
   d. the infundibulum.
   e. the corpus luteum.

10. Testosterone is secreted by
    a. the hypothalamus.
    b. the anterior pituitary.
    c. the posterior pituitary.
    d. interstitial cells of the testis.
    e. germ cells of the testis.

11. A fertilized egg is called a/an _____.

12. Meiosis I produces two secondary spermatocytes in a male, but produces a secondary oocyte and _____ in the female.

13. A tertiary follicle is defined by the appearance of a cavity called the _____.

14. Days 15 to 28 of the ovarian cycle are called the _____ phase.

15. The male ejaculatory duct is formed by the convergence of a ductus deferens and the duct from a _____.

16. The hormone that stimulates both labor contractions and milk ejection is _____.

17. Sperm develop in microscopic ducts called _____ in the testes.

18. In the first 1 to 3 days postpartum, the mammary glands secrete _____ rather than milk.

19. Ovulation is triggered by a spike in the secretion of _____ around day 13 of a typical ovarian cycle.

20. Erection of the penis and clitoris results from the accumulation of blood in little sinuses called _____ in the erectile tissues.

Answers in appendix A

True or False

Determine which five of the following statements are false, and briefly explain why.

1. Testosterone is secreted by the sustentacular cells of the testes.

2. A sperm digests its way into an egg by means of enzymes in its acrosome.

3. More than half of the volume of the semen comes from the seminal vesicles.

4. Like the penis, the clitoris has three erectile tissues—a corpus spongiosum and a pair of corpora cavernosa.

5. Androgens play a significant role in both male and female puberty.

6. The vagina is lubricated by an abundance of mucous glands in its wall.

7. Usually, there are no lactiferous ducts in the breast; these begin to develop only in pregnancy.

8. Unlike women, men remain fertile after climacteric.

9. Spermiogenesis is a stage in which the male sex cells change form, but do not divide any further.

10. *Climacteric* refers to the phase of sexual response in which both men and women experience orgasm.

Answers in appendix A

Testing Your Comprehension

1. A breast-feeding mother leaves her baby at home and goes grocery shopping. She hears someone else's baby crying in another aisle, and notices her blouse becoming wet with a little exuded milk. Explain the physiological connection between hearing that sound and the ejection of milk.

2. Explain why sperm and eggs cannot be produced by mitosis alone. Why is meiosis also necessary?

3. What cells in the ovary have the closest functional similarity to the interstitial cells of the testis? Explain your answer.

Human Development and Aging

A new life in the making

BASE CAMP

Before ascending to the next level, be sure you're properly equipped with a knowledge of these concepts from earlier chapters:

- Anatomy and histology of the uterus (p. 643)
- Structure of the spermatozoon (p. 653)
- Stages of oogenesis (p. 654)

Anatomy & Physiology | REVEALED®
aprevealed.com

Module 14: Reproductive System

Perhaps the most dramatic, miraculous aspect of human life is the transformation of a one-celled fertilized egg into an independent, fully developed individual. From the beginning of recorded thought, people have pondered how a baby forms in the mother's body and how two parents can produce another human being who, although unique, possesses characteristics of each. In his quest to understand prenatal development, Aristotle speculated that the hereditary traits of a child resulted from the mixing of the male's semen with the female's menstrual blood. In the seventeenth century, some scientists thought that the head of the sperm had a miniature human curled up in it, while others thought that the miniature person existed in the egg and the sperm were parasites in the semen. The modern science of **embryology**—the study of prenatal development—was not born until the nineteenth century when advances in microscope technology allowed for better study of the early stages. Late-twentieth-century studies of regulatory genes that control prenatal development and evolution have tremendously deepened our insight into human development.

Embryology can be viewed as one aspect of **developmental biology,** a broader science that embraces changes in form and function from fertilized egg through old age. Human development from conception to death is the scope of this chapter.

20.1 Fertilization and Preembryonic Development

Expected Learning Outcomes

When you have completed this section, you should be able to:

a. describe the processes of sperm migration and fertilization;

b. explain how an egg prevents fertilization by more than one sperm;

c. outline the timetable of prenatal development from the two perspectives of clinical trimesters and developmental stages;

d. describe the major events that transform a fertilized egg into an embryo; and

e. describe the implantation of the preembryo in the uterine wall.

Authorities attach different meanings to the word *embryo.* Some use it to denote stages beginning with the fertilized egg or at least with the two-celled stage produced by its first division. Others first apply the word to an individual 16 days old, when it consists of three **primary germ layers** called the *ectoderm, mesoderm,* and *endoderm.* The events leading up to that stage are called *embryogenesis,* and the first 16 days after fertilization are thus called the *preembryonic stage.* This is the sense in which we will use such terms in this book. We begin with the process in which a sperm locates and fertilizes the egg.

Sperm Migration and Capacitation

A human egg survives only 12 to 24 hours after it is ovulated if it is not fertilized; yet it takes about 72 hours to reach the uterus. Therefore, sperm must travel well up the uterine tube to meet the egg before it dies. The vast majority of them never make it that far—they are destroyed by vaginal acidity, fail to get through the cervical canal, succumb to attack by leukocytes in the uterus, or travel up the wrong uterine tube. The average ejaculation discharges about 200 million sperm into the vagina, yet only 200 of these—1 in a million—make it to the vicinity of an egg.

Sperm can reach the distal region of the uterine tube in half an hour or less, but they cannot fertilize an egg for about 10 hours. First they must undergo a process called **capacitation** that makes it possible to penetrate an egg. This happens as fluids of the female reproductive tract dilute inhibitory factors in the semen and weaken the membrane covering the head of the sperm.

Sperm survive for up to 6 days after ejaculation, so there is little chance of pregnancy from intercourse occurring more than a week before ovulation. Fertilization also is unlikely if intercourse takes place more than 14 hours after ovulation, because the egg would no longer be viable by the time the sperm became capacitated. For those wishing to conceive a child, the optimal window of opportunity is therefore from a few days before ovulation to 14 hours after. Those wishing to avoid pregnancy, however, should allow a wider margin of safety for variations in sperm and egg longevity, capacitation time, and time of ovulation.

Fertilization and Meiosis II

When the sperm encounters an egg, its acrosome undergoes exocytosis, releasing enzymes needed to penetrate the egg. However, the first sperm to reach an egg is not the one to fertilize it. It requires many sperm to clear a path through the barriers around the egg for the one that fertilizes it (fig. 20.1). That sperm binds to the zona pellucida and releases enzymes that digest a path to the egg itself.

The sperm head enters the egg and adds its haploid (n) set of sperm chromosomes to the haploid set of egg chromosomes, producing a diploid $(2n)$ set. Fertilization by two or more sperm, called **polyspermy,** would produce a triploid $(3n)$ or larger set of chromosomes and the egg would die of a "gene overdose." Thus, it is important for the egg to prevent this, and it has two mechanisms for doing so. First, the binding of a sperm to the egg triggers a rapid inflow of Na^+ that depolarizes the egg membrane; this inhibits the attachment of any more sperm. Second, secretory vesicles just beneath the egg membrane undergo exocytosis, releasing a secretion that swells with water, pushes any

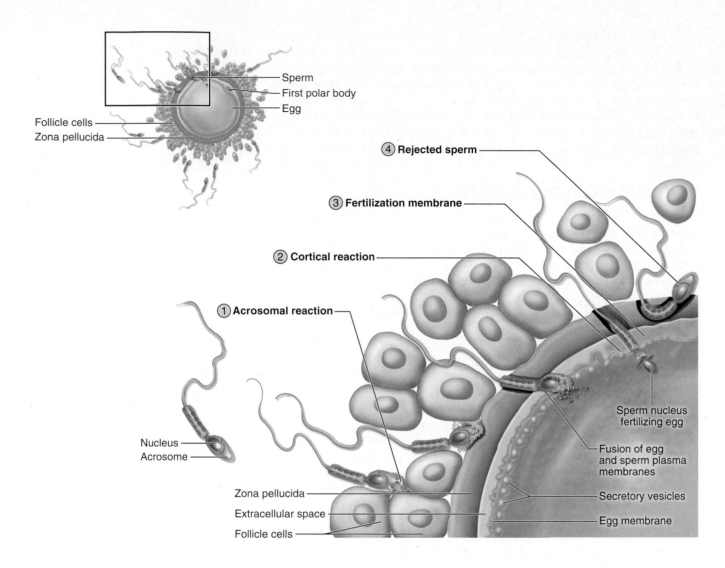

Sperm
First polar body
Egg
Follicle cells
Zona pellucida

④ **Rejected sperm**

③ **Fertilization membrane**

② **Cortical reaction**

① **Acrosomal reaction**

Nucleus
Acrosome

Zona pellucida
Extracellular space
Follicle cells

Sperm nucleus fertilizing egg

Fusion of egg and sperm plasma membranes

Secretory vesicles

Egg membrane

Figure 20.1 Fertilization and Prevention of Polyspermy. AP|R

remaining sperm away from the egg, and creates an impenetrable **fertilization membrane** between the egg and zona pellucida.

A secondary oocyte begins meiosis II before ovulation and completes it only if fertilized. Through the formation of a second polar body, the fertilized egg discards one chromatid from each chromosome. The sperm and egg nuclei then swell and rupture, spilling their chromosomes into a single diploid set. The fertilized egg, now called a **zygote,** is ready for its first mitotic division.

We will use the term *conceptus* for everything that arises from this zygote—not only the future fetus but also the placenta, umbilical cord, and membranes associated with it.

Major Stages of Prenatal Development

Clinically, the course of a pregnancy is divided into 3-month intervals called **trimesters:**

1. The **first trimester** extends from fertilization through 12 weeks. This is a precarious stage in which more than half of all embryos die. The conceptus is most vulnerable to stress, drugs, and nutritional deficiencies during this time.

Table 20.1 The Stages of Prenatal Development

Stage	Age*	Major Developments and Defining Characteristics
Preembryonic stage		
Zygote	0–30 hours	A single diploid cell formed by the union of egg and sperm
Cleavage	30–72 hours	Mitotic division of the zygote into smaller, identical cells called blastomeres
Morula	3–4 days	A spheroidal stage consisting of 16 or more blastomeres
Blastocyst	4–16 days	A fluid-filled, spheroidal stage with an outer mass of trophoblast cells and inner mass of embryoblast cells; becomes implanted in the endometrium; inner cell mass forms an embryonic disc and differentiates into the three primary germ layers
Embryonic stage	16 days–8 weeks	A stage in which the primary germ layers differentiate into organs and organ systems; ends when all organ systems are present
Fetal stage	8–38 weeks	A stage in which organs grow and mature at a cellular level to the point of being capable of supporting life independently of the mother

* From the time of fertilization

2. The **second trimester** (weeks 13 through 24) is a period in which the organs complete most of their development. It becomes possible with sonography to see good anatomical detail in the fetus. By the end of this trimester, the fetus looks distinctly human and, with intensive care, infants born at the end of the second trimester have a chance of survival.

3. In the **third trimester** (week 25 to birth), the fetus grows rapidly and the organs develop enough to support life outside the womb. At 35 weeks from fertilization, the fetus typically weighs about 2.5 kg (5.5 pounds). It is considered mature at this weight, and usually survives if born early. Most twins are born at about 35 weeks and solitary infants around 38 weeks.

From a more biological than clinical standpoint, human development is divided into the *preembryonic, embryonic,* and *fetal stages.* The timetable and landmark events that distinguish these are outlined in table 20.1 and described in the following pages.

The Preembryonic Stage

The **preembryonic stage** comprises the first 16 days of development, leading to the existence of an embryo. It involves three major processes: cleavage, implantation, and embryogenesis.

Cleavage

Cleavage refers to mitotic divisions that occur in the first 3 days while the conceptus migrates down the uterine tube (fig. 20.2). The first cleavage occurs about 30 hours after fertilization and produces the first two daughter cells, or **blastomeres.**[1] These divide simultaneously at shorter and shorter time intervals, doubling the number of blastomeres each time. By the time the conceptus arrives in the uterus, about 72 hours after ovulation, it consists of 16 or more cells and somewhat resembles a mulberry—hence, it is called a **morula.**[2] The morula is no larger than the zygote; cleavage merely produces smaller and smaller blastomeres. This increases the ratio of cell surface area to volume, which favors rapid nutrient uptake and waste removal, and it produces a larger number of cells from which to form different embryonic tissues.

The morula lies free in the uterine cavity for 4 to 5 days and divides into 100 cells or so. Meanwhile, the zona pellucida disintegrates and releases the conceptus, which is now at a stage called the **blastocyst**—a hollow sphere with an outer layer of squamous cells called the **trophoblast**[3] and an inner cell mass called the **embryoblast** (fig. 20.3a). The trophoblast is destined to form part of the placenta and play an important role in nourishment of the embryo, whereas the embryoblast is destined to become the embryo itself.

In some cases, the embryoblast splits into two cell masses, resulting in *monozygotic* (identical) *twins,* which usually share the same placenta. *Dizygotic* (nonidentical) *twins* and most other multiple births occur when two or more eggs are ovulated in the same month and fertilized by separate sperm.

---

[1] *blast* = bud, precursor; *mer* = segment, part

[2] *mor* = mulberry; *ula* = little

[3] *troph* = food, nourishment

Figure 20.2 Migration of the Conceptus. The egg is fertilized in the distal end of the uterine tube, and the preembryo begins cleavage as it migrates to the uterus.

• *Why can't the egg be fertilized in the uterus?*

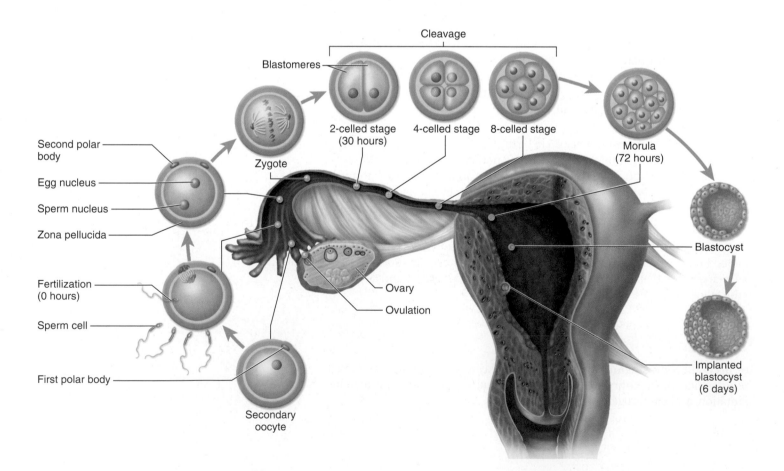

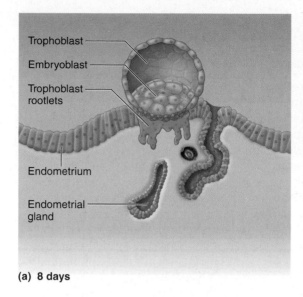

(a) 8 days

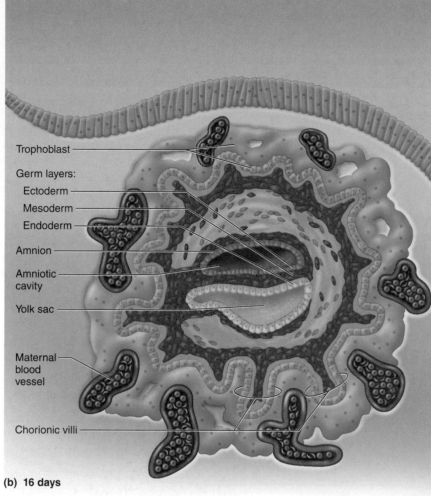

Figure 20.3 Implantation. (a) Early implantation about 8 days after ovulation. The trophoblast has begun growing rootlets, which penetrate the endometrium. (b) By 16 days, the conceptus is completely covered by endometrial tissue. The embryo is now flanked by a yolk sac and amnion and is composed of three primary germ layers.

(b) 16 days

They are no more or less genetically similar than other siblings. They attach separately to the uterine wall and each forms its own placenta.

> **Apply What You Know**
>
> Your friend John says that he and his sister Beth are identical twins. Even though they do look strikingly similar, you are skeptical of his claim. Other than tactful silence, how might you respond?

Implantation

About 6 days after ovulation, the blastocyst adheres to the endometrium and becomes embedded in it—a process called **implantation.** The trophoblast cells on the attached side proliferate and grow into the uterus like little roots, digesting endometrial cells along the way. The endometrium reacts to this injury by growing over the blastocyst and eventually covering it (fig. 20.3). Implantation takes about a week and is completed about the time the next menstrual period would have occurred if the woman had not become pregnant.

Another role of the trophoblast is to secrete the hormone **human chorionic gonadotropin (HCG)** (CORE-ee-ON-ic). HCG stimulates the corpus luteum to secrete estrogen and progesterone, and progesterone prevents menstruation. Pregnancy tests are based on detecting HCG in a woman's urine. The level of HCG in the mother's blood rises until the end of the second month. During this

Clinical Application 20.1

ECTOPIC PREGNANCY

In about 1 out of 300 pregnancies, the blastocyst implants somewhere other than the uterus, producing an *ectopic*[4] *pregnancy.* Most cases are *tubal pregnancies,* implantation in the uterine tube. This usually occurs because the conceptus encounters an obstruction such as a constriction resulting from earlier pelvic inflammatory disease, tubal surgery, previous ectopic pregnancies, or other causes. The uterine tube cannot expand enough to accommodate the growing conceptus for long; unless detected and treated early, the tube usually ruptures within 12 weeks, with potentially fatal hemorrhaging. Occasionally, a conceptus implants in the abdominopelvic cavity, producing an *abdominal pregnancy.* It can grow anywhere it finds an adequate blood supply—for example, on the outside of the uterus, colon, or bladder. About 1 pregnancy in 7,000 is abdominal. This is a serious threat and usually requires aborting the pregnancy to save the woman's life, but about 9% of abdominal pregnancies result in live birth by cesarean section.

time, the trophoblast develops into a membrane called the *chorion,* which takes over the role of the corpus luteum and makes HCG unnecessary. The ovaries then become inactive for the rest of the pregnancy, but estrogen and progesterone levels rise dramatically as they are secreted by the ever-growing chorion.

Embryogenesis

During implantation, the embryoblast undergoes **embryogenesis**—arrangement of the blastomeres into the three primary germ layers: *ectoderm, mesoderm,* and *endoderm.* At the beginning of this phase, the embryoblast separates slightly from the trophoblast, creating a narrow space between them called the **amniotic cavity** lined by a membrane, the *amnion.* The embryoblast flattens into an **embryonic disc** composed of two cell layers—a deep layer facing the amniotic cavity and a superficial layer facing away. The superficial layer produces a fluid-filled *yolk sac.* Now the embryonic disc is flanked by two spaces: the amniotic cavity on one side and the yolk sac on the other.

Meanwhile, the disc elongates and a thickened ridge forms along the midline, with a **primitive groove** down its middle. These events make the embryo bilaterally symmetric and define its future right and left sides, dorsal and ventral surfaces, and cephalic (head) and caudal (tail) ends.

The next step is **gastrulation**—multiplying cells migrate medially toward the primitive groove and down into it (fig. 20.4). They become a layer called **endoderm,** which will become the inner lining of the digestive tract among other things. A day later, migrating cells form a third layer between the first two, called **mesoderm.** Once this is formed, the upper surface is called **ectoderm.** The ectoderm and endoderm are epithelia composed of tightly joined cells, but the mesoderm is a more loosely organized tissue. It later becomes a loose, gelatinous, fetal connective tissue called **mesenchyme,** which gives rise to such tissues as muscle, bone, and blood. Precise control of these embryonic cell migrations is crucial, because any misplacement of cells at this stage results in gross deformities of the embryo and a likelihood of prenatal death.

[4] *ec* = outside; *top* = place

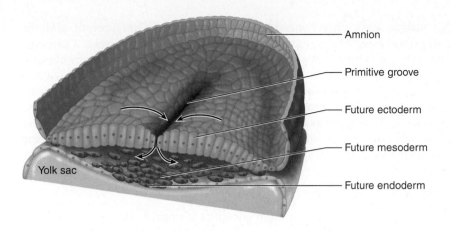

Amnion

Primitive groove

Future ectoderm

Future mesoderm

Yolk sac

Future endoderm

Figure 20.4 Formation of the Primary Germ Layers (Gastrulation). Composite view of the embryonic disc at 16 days, corresponding to the center of figure 20.3b. Surface cells migrate into the primitive groove, first forming a layer of endoderm beneath, then filling the space between these first two layers with mesoderm. Upon completion of this process, the uppermost layer is considered ectoderm.

- *Name any three adult organs that develop from the red cells in the middle.*

Once the three primary germ layers are formed, embryogenesis is complete and the individual is considered an embryo. It is about 2 mm long and 16 days old at this point.

Before You Go On

Answer these questions from memory. Reread the preceding section if there are too many you don't know.

1. *How soon can a sperm reach an egg after ejaculation? How soon can it fertilize an egg? What accounts for the difference?*
2. *Describe two ways a fertilized egg prevents the entry of excess sperm.*
3. *In the blastocyst, what are the cells called that eventually give rise to the embryo? What are the cells that carry out implantation?*
4. *In which clinical trimester does a conceptus reach the embryonic stage? The fetal stage?*
5. *What defines a true embryo and at what age is this stage reached?*

20.2 The Embryonic and Fetal Stages

Expected Learning Outcomes

When you have completed this section, you should be able to:

a. identify the major tissues derived from the primary germ layers;
b. explain how the conceptus is nourished before the placenta takes over this function;
c. describe the formation of the placenta and list its multiple functions;
d. describe the four embryonic membranes and their functions;
e. describe the conversion of a flat embryonic disc into a cylindrical form enclosing a primitive gut; and
f. describe how the fetal circulatory system differs from that of a neonate.

Sixteen days after conception, the germ layers are present and the **embryonic stage** of development begins. Over the next 6 weeks, a placenta forms on the uterine wall and becomes the embryo's primary means of nutrition, while the germ layers differentiate into organs and organ systems—a process called **organogenesis** (table 20.2). Although these organs are still far from functional, it is their presence at 8 weeks that marks the transition from the embryonic stage to the fetal stage. Here we will see how the embryo becomes a fetus, how the membranes collectively known as the "afterbirth" develop around the fetus, and how the conceptus is nourished throughout its gestation.

Prenatal Nutrition

Over the course of gestation, the conceptus is nourished in three different, overlapping ways. As it travels down the uterine tube and lies free in the uterine cavity before implantation, it is nourished by a glycogen-rich secretion of the uterine tubes and endometrial glands called *uterine milk.*

As it implants, the conceptus shifts to **trophoblastic nutrition,** in which it consumes so-called *decidual[5] cells* of the endometrium. Progesterone from the corpus luteum stimulates these cells to proliferate and accumulate a store of glycogen, protein, and lipid. As the conceptus burrows into the endometrium, the trophoblast digests them and relays the nutrients to the embryoblast. This is the only mode of nutrition for the first week after implantation. It remains the dominant source of nutrients through the end of week 8; the period from implantation through week 8 is therefore called the **trophoblastic phase** of the pregnancy. Trophoblastic nutrition wanes as placental nutrition takes over, and ceases entirely by the end of week 12.

In **placental nutrition,** nutrients diffuse from the mother's blood through the placental wall into the fetal blood. The **placenta**[6] is a disc-shaped organ attached to the uterus on one side and, on the other side, attached by way of an **umbilical cord** to the fetus (fig. 20.5). The uterine surface of the placenta bears shaggy **chorionic villi** (CORE-ee-ON-ic VILL-eye) embedded in the endometrium. When fully developed, the placenta is about 20 cm in diameter, 3 cm thick, and weighs about one-sixth as much as the newborn infant.

---

[5] *decid* = falling off

[6] *placenta* = flat cake

Table 20.2 Derivatives of the Three Primary Germ Layers

Germ Layer	Major Derivatives
Ectoderm	Epidermis; hair follicles and piloerector muscles; cutaneous glands; nervous system; adrenal medulla; pineal and pituitary glands; lens, cornea, and intrinsic muscles of the eye; internal and external ear; salivary glands; epithelia of the nasal cavity, oral cavity, and anal canal
Mesoderm	Skeleton; skeletal, cardiac, and most smooth muscle; cartilage; adrenal cortex; middle ear; dermis; blood; blood and lymphatic vessels; bone marrow; lymphoid tissue; epithelium of kidneys, ureters, gonads, and genital ducts; mesothelium of the abdominal and thoracic cavities
Endoderm	Most of the mucosal epithelium of the digestive and respiratory tracts; mucosal epithelium of urinary bladder and parts of urethra; epithelial components of accessory reproductive and digestive glands (except salivary glands); thyroid and parathyroid glands; thymus

Figure 20.5 The Placenta and Embryonic Membranes.
(a) A 12-week fetus. The placenta is completely formed. (b) A portion of the mature placenta and umbilical cord, showing the relationship between fetal and maternal circulation. (c) The fetal side of the placenta, showing blood vessels, the umbilical cord, and some of the amniotic sac attached to the lower left margin. (d) The maternal (uterine) side, where chorionic villi give the placenta a rougher texture.

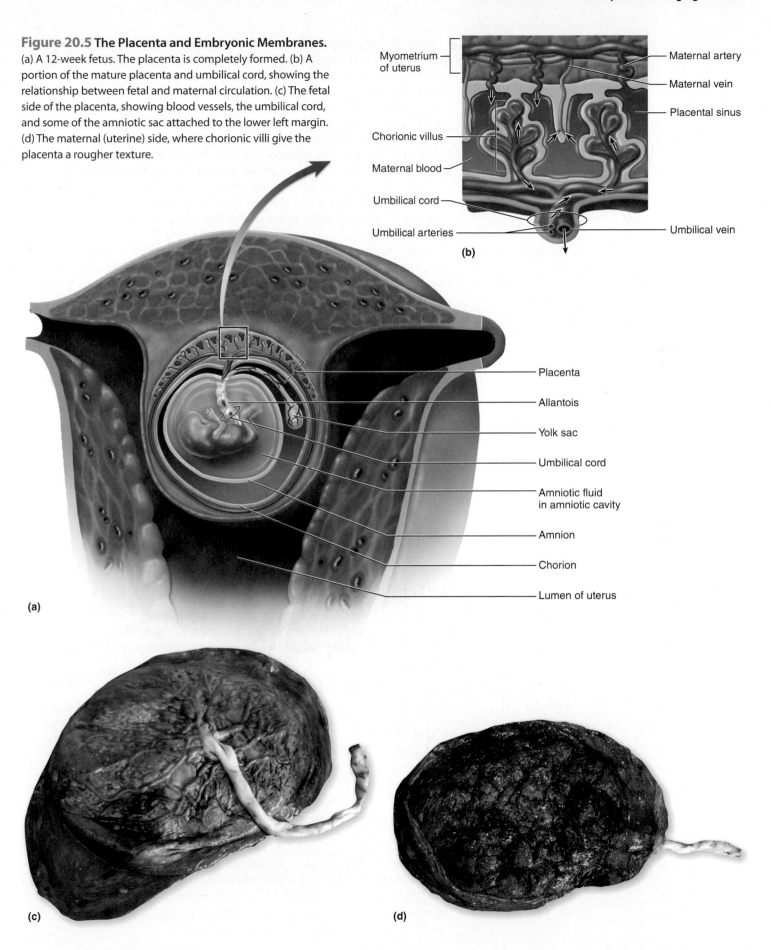

(b)

Myometrium of uterus

Chorionic villus

Maternal blood

Umbilical cord

Umbilical arteries

Maternal artery

Maternal vein

Placental sinus

Umbilical vein

Placenta

Allantois

Yolk sac

Umbilical cord

Amniotic fluid in amniotic cavity

Amnion

Chorion

Lumen of uterus

(a)

(c)

(d)

The placenta begins to develop about 11 days after conception, becomes the dominant mode of nutrition around the beginning of week 9, and is the sole mode of nutrition from the end of week 12 until birth. The period from week 9 until birth is called the **placental phase** of the pregnancy.

The development of the placenta begins during implantation, as extensions of the trophoblast penetrate more and more deeply into the endometrium, like the roots of a tree penetrating into the nourishing soil of the uterus. These roots are the early chorionic villi. As they invade uterine blood vessels, they become surrounded by pools of maternal blood. These pools eventually merge to form a single blood-filled sinus. Maternal blood stimulates rapid growth of the villi, which become branched and treelike. Mesoderm grows into the villi and gives rise to the blood vessels that connect to the embryo by way of the umbilical cord.

The umbilical cord contains two **umbilical arteries** and one **umbilical vein.** Pumped by the fetal heart, blood flows to the placenta by way of the umbilical arteries and then returns to the fetus by way of the umbilical vein. The chorionic villi are thus *filled with* fetal blood and *surrounded by* maternal blood (fig. 20.5b); the two bloodstreams do not mix unless there is damage to the placental barrier. The barrier, however, is only 3.5 μm thick—half the diameter of one red blood cell. As development progresses, this barrier becomes more and more permeable to chemicals. Oxygen and nutrients pass from the maternal blood, through the barrier, into the fetal blood, while fetal wastes pass the other way to be eliminated by the mother. Unfortunately, the placenta is also permeable to nicotine, alcohol, and most other drugs that may be present in the maternal bloodstream and harmful to the embryo.

Table 20.3 describes the many functions of the placenta.

Apply What You Know

What basic law of diffusion (see chapter 3) determines that at the placenta, fetal wastes pass from the fetal blood into the maternal blood, and oxygen passes from the maternal blood into the fetal blood, rather than in the opposite directions?

Embryonic Membranes

The placenta is only one of several accessory organs that develop alongside the embryo; there are also four embryonic membranes called the *amnion, yolk sac, allantois,* and *chorion* (figs. 20.5a and 20.6). The **amnion** is a transparent sac

Table 20.3 Functions of the Placenta

Nutritional roles	Permits nutrients such as glucose, amino acids, fatty acids, minerals, and vitamins to diffuse from the maternal blood to the fetal blood; stores nutrients such as carbohydrates, protein, iron, and calcium in early pregnancy and releases them to the fetus later, when fetal demand is greater than the mother can absorb from the diet
Excretory roles	Permits nitrogenous wastes such as ammonia, urea, uric acid, and creatinine to diffuse from the fetal blood to the maternal blood
Respiratory roles	Permits O_2 to diffuse from mother to fetus and CO_2 from fetus to mother
Endocrine roles	Secretes estrogens, progesterone, human chorionic gonadotropin, and other hormones; allows hormones synthesized by the conceptus to pass into the mother's blood and maternal hormones to pass into the fetal blood
Immune roles	Transfers maternal antibodies (especially IgG) into fetal blood to confer passive immunity on fetus

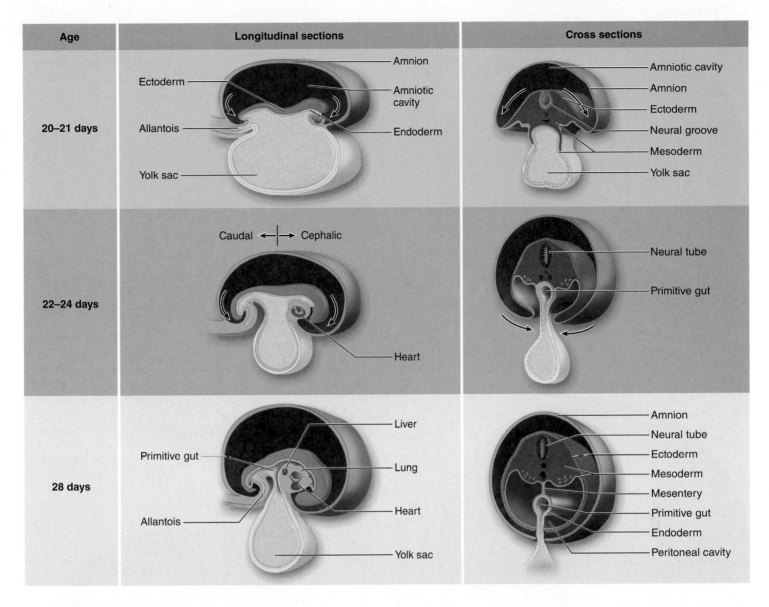

Age	Longitudinal sections	Cross sections
20–21 days	Amnion Ectoderm Amniotic cavity Allantois Endoderm Yolk sac	Amniotic cavity Amnion Ectoderm Neural groove Mesoderm Yolk sac
22–24 days	Caudal ⟵|⟶ Cephalic Heart	Neural tube Primitive gut
28 days	Liver Primitive gut Lung Heart Allantois Yolk sac	Amnion Neural tube Ectoderm Mesoderm Mesentery Primitive gut Endoderm Peritoneal cavity

that encloses the embryo and is penetrated only by the umbilical cord. It is filled with **amniotic fluid** (fig. 20.7d, e), which protects the embryo from trauma, infection, and temperature fluctuations; allows the freedom of movement important to muscle development; enables the embryo to develop symmetrically; prevents body parts from adhering to each other, such as an arm to the trunk; and stimulates lung development as the fetus "breathes" the fluid. The fetus also repeatedly swallows amniotic fluid and urinates it back into the amniotic sac. At term, the amnion contains 700 to 1,000 mL of fluid, released vaginally when the amnion ruptures during labor (the "breaking of the waters").

The **yolk sac,** named for its role in egg-laying vertebrates, is a small bag suspended from the underside of the embryo. It contributes to the formation of the digestive tract and produces the first blood cells and future egg or sperm cells.

The **allantois** (ah-LON-toe-iss) begins as an outgrowth of the caudal end of the gut. It forms the foundation for the umbilical cord and becomes part of the urinary bladder.

The **chorion** is the outermost membrane, enclosing all the others and the embryo. Initially, it has shaggy chorionic villi around its entire surface (see fig. 20.3b), but as the pregnancy advances, the villi of the placental region grow and branch while the rest of them degenerate. The chorion thus

Figure 20.6 Embryonic Folding. The right-hand figures are cross sections cut about midway along the figures on the left. From 22 to 28 days, the two ends of the embryo curl toward each other (left-hand figures) until the embryo assumes a C shape, and the flanks of the embryo fold laterally (right-hand figures), converting the flat embryonic disc into a more cylindrical body and eventually enclosing a body cavity.

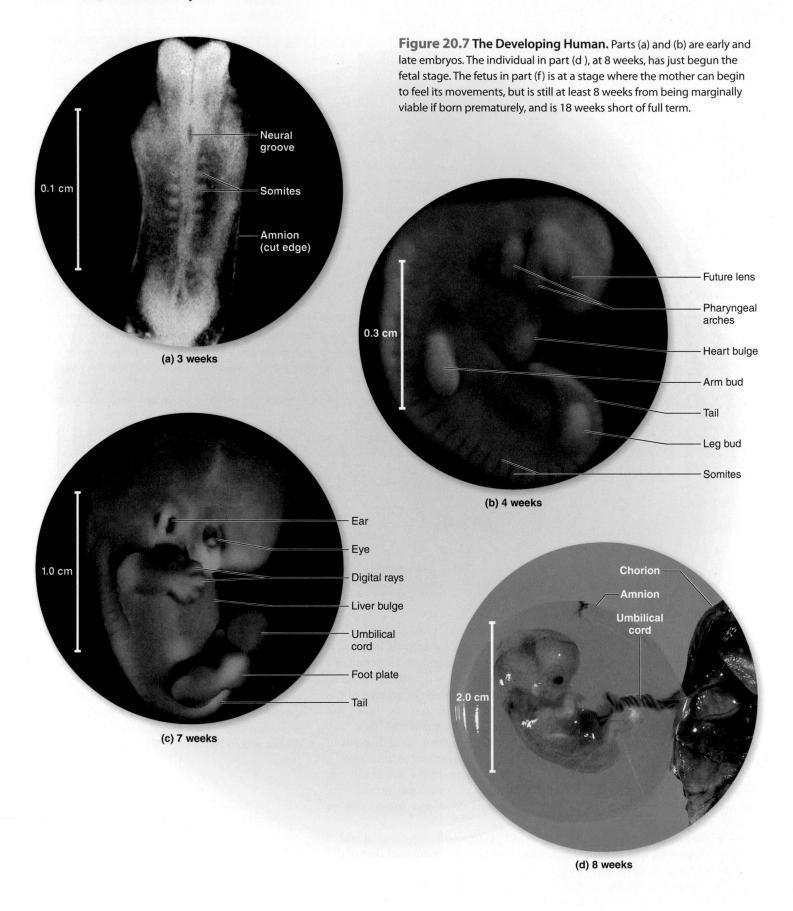

Figure 20.7 The Developing Human. Parts (a) and (b) are early and late embryos. The individual in part (d), at 8 weeks, has just begun the fetal stage. The fetus in part (f) is at a stage where the mother can begin to feel its movements, but is still at least 8 weeks from being marginally viable if born prematurely, and is 18 weeks short of full term.

(a) 3 weeks

Neural groove

Somites

Amnion (cut edge)

0.1 cm

(b) 4 weeks

Future lens

Pharyngeal arches

Heart bulge

Arm bud

Tail

Leg bud

Somites

0.3 cm

(c) 7 weeks

Ear

Eye

Digital rays

Liver bulge

Umbilical cord

Foot plate

Tail

1.0 cm

(d) 8 weeks

Chorion

Amnion

Umbilical cord

2.0 cm

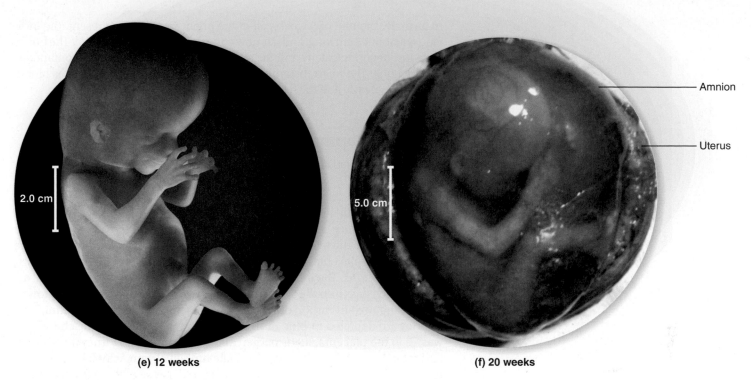

(e) 12 weeks   (f) 20 weeks

Figure 20.7 *(continued)*

becomes smooth surfaced away from the placental region (fig. 20.5a). The chorion secretes pregnancy-regulating hormones and, through its role in the placenta, provides an avenue of fetal nutrition.

Embryonic Folding and Organ Development

One of the major transformations to occur in the embryonic stage is conversion of the flat embryonic disc into a somewhat cylindrical form. This occurs during week 4 as the embryo rapidly grows and folds around the yolk sac (fig. 20.6). As the cephalic and caudal ends curve around the ends of the yolk sac, the embryo becomes C-shaped, with the head and tail almost touching. At the same time, the lateral margins of the disc fold around the sides of the yolk sac to form the ventral surface of the embryo. This lateral folding encloses a longitudinal channel, the *primitive gut,* which later becomes the digestive tract.

As a result of embryonic folding, the entire surface becomes covered with ectoderm, which later produces the epidermis of the skin. In the meantime, the mesoderm splits into two layers. One of them adheres to the ectoderm and the other to the endoderm, thus opening a body cavity between them. This cavity is soon subdivided by a diaphragm, giving rise to the thoracic cavity and peritoneal cavity. By the end of week 5, the thoracic cavity further divides into pleural and pericardial cavities.

Two more significant events in organogenesis are the appearance of a **neural tube,** which will later become the brain and spinal cord, and segmentation of the mesoderm into blocks of tissue called **somites** (fig. 20.7a, b), which will give rise to the vertebral column, trunk muscles, and dermis of the skin.

We cannot delve at greater length into development of all the organ systems here, but this description is at least enough to see how some of them begin to form. By the end of 8 weeks, all of the organ systems are present, the individual is about 3 cm long, and it is now considered a **fetus** (fig. 20.7c). The bones have just begun to calcify and the skeletal muscles exhibit weak spontaneous contractions. The heart, beating since the fourth week, now circulates blood. The head is nearly half the total body length.

Fetal Development

The fetus is the final stage of prenatal development, from the start of the ninth week until birth. The organs that formed during the embryonic stage now undergo growth and cellular differentiation, acquiring the functional capability to support life outside the mother.

The circulatory system shows the most conspicuous changes from the prenatal state to the neonatal state. The unique aspects of fetal circulation are the umbilical–placental circuit and the presence of three circulatory shortcuts called *shunts*. The internal iliac arteries give rise to the two umbilical arteries, which pass into the umbilical cord. The blood in these arteries is low in oxygen and high in carbon dioxide and other fetal wastes; thus, the umbilical arteries are depicted in blue in figure 20.8. The arterial blood discharges its wastes in the placenta, loads oxygen and nutrients coming from the mother, and returns to the fetus by way of a single umbilical vein, which leads toward the liver. The umbilical vein is depicted in red because of its well-oxygenated blood. A little of this venous blood filters through the liver to nourish it, but most of it bypasses the liver by way of a shunt called the **ductus venosus,** which leads to the inferior vena cava.

In the inferior vena cava, placental blood mixes with venous blood from the fetus's body and flows to the right atrium of the heart. After birth, the right ventricle pumps all of its blood into the lungs, but there is little need for this in the fetus because the lungs are not yet functional. Therefore, most fetal blood bypasses the pulmonary circuit. Some goes directly from the right to left atrium through the **foramen ovale,** a hole in the interatrial septum. Some also goes into the right ventricle and is pumped into the pulmonary trunk, but most of this is shunted directly into the aorta by way of a short passage called the **ductus arteriosus.** The lungs receive only a trickle of blood, sufficient to meet their metabolic needs during development. Blood leaving the left ventricle enters the general systemic circulation. This circulatory pattern changes dramatically at birth, when the neonate is cut off from the placenta and the lungs expand with air. Those changes will be described later.

Fetal growth is charted by weight and body length (table 20.4). Body length is customarily measured from the crown of the head to the curve of the buttocks in a sitting position *(crown-to-rump length, CRL),* thus excluding the lower limbs. Full-term fetuses have an average CRL of about 36 cm (14 in.) and average weight of about 3.0 to 3.4 kg (6.6–7.5 pounds). The fetus gains about 50% of its birth weight in the last 10 weeks.

Additional aspects of embryonic and fetal development are listed in table 20.4 and depicted in figure 20.7.

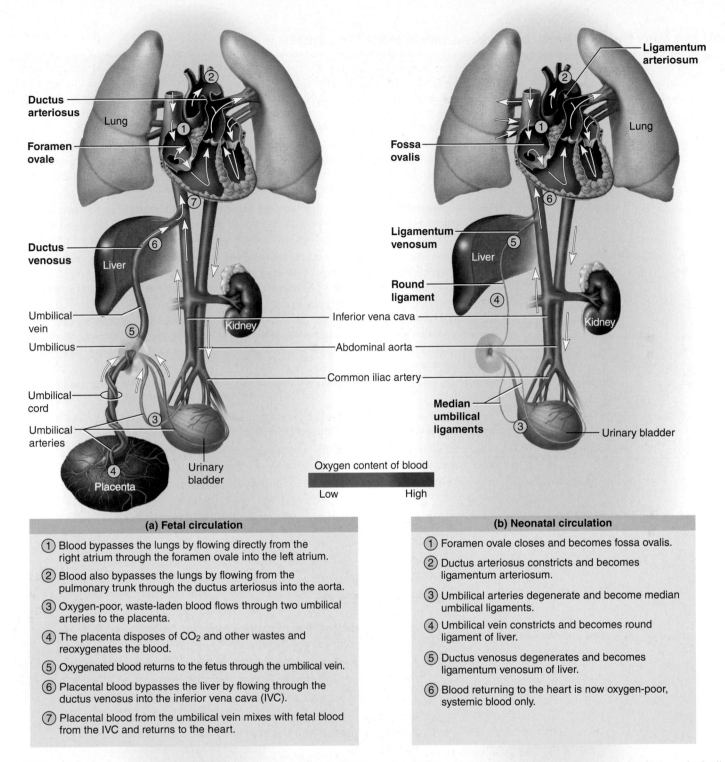

Oxygen content of blood

Low   High

(a) Fetal circulation

① Blood bypasses the lungs by flowing directly from the right atrium through the foramen ovale into the left atrium.

② Blood also bypasses the lungs by flowing from the pulmonary trunk through the ductus arteriosus into the aorta.

③ Oxygen-poor, waste-laden blood flows through two umbilical arteries to the placenta.

④ The placenta disposes of CO_2 and other wastes and reoxygenates the blood.

⑤ Oxygenated blood returns to the fetus through the umbilical vein.

⑥ Placental blood bypasses the liver by flowing through the ductus venosus into the inferior vena cava (IVC).

⑦ Placental blood from the umbilical vein mixes with fetal blood from the IVC and returns to the heart.

(b) Neonatal circulation

① Foramen ovale closes and becomes fossa ovalis.

② Ductus arteriosus constricts and becomes ligamentum arteriosum.

③ Umbilical arteries degenerate and become median umbilical ligaments.

④ Umbilical vein constricts and becomes round ligament of liver.

⑤ Ductus venosus degenerates and becomes ligamentum venosum of liver.

⑥ Blood returning to the heart is now oxygen-poor, systemic blood only.

Figure 20.8 Blood Circulation in the Fetus and Newborn. Boldfaced terms in (a) indicate the three shunts in the fetal circulation, which allow most blood to bypass the liver and lungs. Boldfaced terms in (b) indicate the postpartum vestiges of fetal structures.

Table 20.4 Major Events of Prenatal Development, with Emphasis on the Fetal Stage

End of Week	Typical Crown-to-Rump Length (CRL) and Weight	Developmental Events
4	0.6 cm; <1 g	Vertebral column and central nervous system begin to form; limbs represented by small limb buds; heart begins beating around day 22; no visible eyes, nose, or ears
8	3 cm; 1 g	Eyes form, eyelids fused shut; nose flat, nostrils evident but plugged with mucus; head nearly as large as the rest of the body; brain waves detectable; bone calcification begins; limb buds form paddlelike hands and feet with ridges called *digital rays,* which then separate into distinct fingers and toes; blood cells and major blood vessels form; genitals present but sexes not yet distinguishable
12	9 cm; 45 g	Eyes well developed, facing laterally; eyelids still fused; nose develops bridge; external ears present; limbs well formed, digits exhibit nails; fetus swallows amniotic fluid and produces urine; fetus moves, but too weakly for mother to feel it; liver is prominent and produces bile; palate is fusing; sexes can be distinguished
16	14 cm; 200 g	Eyes face anteriorly, external ears stand out from head, face looks more human; body larger in proportion to head; skin is bright pink, scalp has hair; joints forming; lips exhibit sucking movements; kidneys well formed; digestive glands forming and *meconium*[7] (fetal feces) accumulating in intestine; heartbeat can be heard with a stethoscope
20	19 cm; 460 g	Body covered with fine hair called *lanugo*[8] and cheeselike sebaceous secretion called *vernix caseosa,*[9] which protects it from amniotic fluid; skin bright pink; brown fat forms and will be used for postpartum heat production; fetus is now bent forward into "fetal position" because of crowding; *quickening* occurs—mother can feel fetal movements
24	23 cm; 820 g	Eyes partially open; skin wrinkled, pink, and translucent; lungs begin producing surfactant; rapid weight gain
28	27 cm; 1,300 g	Eyes fully open; skin wrinkled and red; full head of hair present; eyelashes formed; fetus turns into upside-down *vertex position;* testes begin to descend into scrotum; marginally viable if born at 28 weeks
32	30 cm; 2,100 g	Subcutaneous fat deposition gives fetus a more plump, babyish appearance, with lighter, less wrinkled skin; testes descending; twins usually born at this stage
36	34 cm; 2,900 g	More subcutaneous fat deposited, body plump; lanugo is shed; nails extend to fingertips; limbs flexed; firm hand grip
38	36 cm; 3,400 g	Prominent chest, protruding breasts; testes in inguinal canal or scrotum; fingernails extend beyond fingertips

---

Before You Go On

Answer these questions from memory. Reread the preceding section if there are too many you don't know.

6. Name three mature structures that develop from the ectoderm, three from the mesoderm, and three from the endoderm.

7. How is trophoblastic nutrition different from placental nutrition?

8. Identify the two sources of blood to the placenta. Where do these two bloodstreams come closest to each other? What keeps them separated?

9. State the functions of the placenta, amnion, chorion, yolk sac, and allantois.

10. What developmental characteristic distinguishes a fetus from an embryo? At what gestational age is this attained?

11. Identify the three circulatory shunts of the fetus. Why does the blood take these "shortcuts" before birth?

---

[7] *mecon* = poppy juice, opium; refers to an appearance similar to black tar opium

[8] *lan* = down, wool

[9] *vernix* = varnish; *caseo* = cheese

PERSPECTIVES ON HEALTH

Methods of Contraception

Contraception means any procedure or device intended to prevent pregnancy (the presence of an implanted conceptus in the uterus). This essay summarizes the most popular methods and some issues involved in choosing among them.

Behavioral Methods

Abstinence (refraining from intercourse) is, obviously, a completely reliable method if used consistently. The rhythm method (periodic abstinence) is based on avoiding intercourse near the time of expected ovulation. Among typical users, it has a 25% failure rate, partly due to lack of restraint and partly because it is difficult to predict the exact date of ovulation. Intercourse must be avoided for at least 7 days before ovulation so there will be no surviving sperm in the reproductive tract when the egg is ovulated, and for at least 2 days after ovulation so there will be no fertile egg present when sperm are introduced.

Withdrawal (coitus interruptus) requires the male to withdraw the penis before ejaculation. This often fails because of lack of willpower, because some sperm are present in the preejaculatory fluid, and because sperm ejaculated anywhere in the vulva can potentially get into the reproductive tract.

Barrier and Spermicidal Methods

Barrier methods are designed to prevent sperm from getting into or beyond the vagina. They are most effective when used with chemical spermicides, available as nonprescription foams, creams, and jellies. Second only to birth-control pills in popularity is the male *condom*, a sheath usually made of latex, worn over the penis. Condoms are the only contraceptive methods that also protect against disease transmission, although animal-membrane condoms do not protect against HIV or hepatitis B viruses. Condoms have the advantages of being inexpensive and requiring no medical examination or prescription.

The *diaphragm* is a latex dome worn over the cervix to block sperm migration. It requires a physical examination and prescription to ensure proper fit, but is otherwise comparable to the condom in convenience and reliability, provided it is used with a spermicide. Without a spermicide, it is not very effective.

The *sponge* is a concave foam disc inserted before intercourse to cover the cervix. It is impregnated with spermicidal gel and acts by absorbing semen and killing the sperm. It requires no prescription or fitting. The sponge provides protection for up to 12 hours, and must be left in place for 6 hours after intercourse.

Hormonal Methods

Most hormonal methods of contraception are aimed at preventing ovulation. They mimic the negative feedback effect of ovarian hormones on the pituitary, inhibiting FSH and LH secretion so follicles do not mature. For most women, they are highly effective and present minimal complications.

The oldest and still the most widely used hormonal method in the United States is the *combined oral contraceptive (birth-control pill)*. It is composed of estrogen and progestin, a synthetic progesterone. It must be taken daily, at the same time of day, for 21 days each cycle. The 7-day withdrawal allows for menstruation. Side effects include an elevated risk of heart attack or stroke in smokers and in women with a history of diabetes, hypertension, or clotting disorders.

Other hormonal methods avoid the need to remember a daily pill. One option is a skin patch that releases estrogen and progestin transdermally. It is changed at 7-day intervals (three patches per month and one week without). The NuvaRing is a soft flexible vaginal ring that releases estrogen and progestin for absorption through the vaginal mucosa. It must be worn continually for 3 weeks and removed for the fourth week of each cycle. Medroxyprogesterone (trade name Depo-Provera) is a progestin administered by injection two to four times per year. It provides highly reliable, long-term contraception, although in some women it causes headaches, nausea, or weight gain.

Some drugs can be taken orally after intercourse to prevent implantation of a conceptus. These are called emergency contraceptive pills (ECPs), or "morning-after pills" (trade names Plan B, Levonelle). An ECP is a high dose of estrogen and progestin or a progestin alone. It can be taken within 72 hours after intercourse and induces menstruation within 2 weeks. ECPs inhibit ovulation, inhibit sperm or egg transport in the uterine tube, and prevent implantation. They do not work if a blastocyst is already implanted.

Intrauterine Devices

Intrauterine devices (IUDs) are springy, often T-shaped devices inserted through the cervical canal into the uterus. Some IUDs act by releasing a synthetic progesterone, but most have a copper wire wrapping or copper sleeve. IUDs irritate the uterine lining and interfere with blastocyst implantation, and copper IUDs also inhibit sperm motility. An IUD can be left in place for 5 to 12 years.

(continued on next page)

(continued from previous page)

Surgical Methods

People who are confident that they do not want more children (or any) often elect to be surgically sterilized. This entails the cutting and tying or clamping of the genital ducts, thus blocking the passage of sperm or eggs. Surgical sterilization has the advantage of convenience, since it requires no further attention. Its initial cost is higher, however, and for people who later change their minds, surgical reversal is much more expensive than the original procedure and is often unsuccessful. *Vasectomy* is the severing of the male's ductus (vas) deferens, done through a small incision in the back of the scrotum. In *tubal ligation,* the uterine tubes are cut. This can be done through small abdominal incisions to admit a cutting instrument and laparoscope (viewing device).

Issues in Choosing a Contraceptive

We have not considered all the currently available methods of contraception or all the issues important to the choice of a contraceptive. No one method can be recommended as best for all people. Many issues enter into one's choice, including personal preference, pattern of sexual activity, medical history, religious views, convenience, initial and ongoing costs, and disease prevention. For most people, however, the two primary issues are safety and reliability. Further information necessary to a sound choice and proper use of contraceptives should be sought from a health department, college health service, physician, reliable Internet sites, or other such sources.

20.3 The Neonate

Expected Learning Outcomes

When you have completed this section, you should be able to:

a. describe how the respiratory and circulatory systems change and affect each other at birth;

b. explain why the neonate faces unique challenges in immunity, thermoregulation, and fluid balance;

c. describe the most critical physiological problems of a premature infant; and

d. discuss and classify several common causes of birth defects.

Development is by no means complete at birth. For example, the liver and kidneys still are not fully functional, most joints are not yet ossified, and myelination of the nervous system is not completed until adolescence. Indeed, humans are born in a very immature state compared with other mammals—a fact necessitated by the narrow outlet of the female pelvis.

Adapting to Life Outside the Womb

The first 6 weeks of postpartum life constitute the **neonatal period.** This period is a crisis in which the neonate must adapt to life outside the mother's body. Here we'll consider some of the major adjustments of the neonate to this new way of life.

Respiration

The most obvious adaptation to external life is that the infant must breathe on its own. It is an old misconception that a neonate must be spanked to stimulate it to breathe. During birth, CO_2 accumulates in the baby's blood and strongly stimulates the respiratory chemoreceptors. Unless the infant is depressed by oversedation of the mother, it normally begins breathing spontaneously. It requires a great effort, however, to take the first few breaths and inflate the collapsed alveoli. For the first 2 weeks, a baby takes about 45 breaths per minute, but subsequently stabilizes at about 12 breaths per minute.

Circulation

When the lungs expand with air, resistance and blood pressure in the pulmonary circulation drop rapidly and pressure in the right side of the heart falls below that in the left. Blood flows briefly from the left atrium to the right through the foramen ovale, pushing two flaps of tissue into place to close this shunt. Usually these flaps fuse and permanently seal the foramen during the first year, leaving a depression, the *fossa ovalis,* in the interatrial septum. Pressure changes in the pulmonary trunk and aorta also cause the ductus arteriosus to collapse. It closes permanently around 3 months of age, leaving a permanent cord, the *ligamentum arteriosum,* between the two vessels.

After the umbilical cord is clamped and cut, the umbilical arteries and vein collapse and become fibrotic. The proximal part of each umbilical artery becomes the *superior vesical artery,* which supplies the bladder. Other obliterated vessels become fibrous ligaments of the liver and abdominal wall (fig. 20.8b).

Immunity

Cellular immunity begins early in fetal development, but the immune responses of the neonate are still weak. Fortunately, an infant is born with a near-adult level of antibodies in the IgG class acquired from the mother through the placenta. Maternal IgG breaks down rapidly after birth, but its level remains high enough for 6 months to protect the infant from measles, diphtheria, polio, and most other infectious diseases (but not whooping cough). By 6 months, the infant's own IgG reaches about half the typical adult level. The lowest total (maternal + infant) level of IgG exists around 5 to 6 months of age, and respiratory infections are especially common at that age. A breast-fed neonate also acquires protection from the IgA present in the colostrum.

Thermoregulation

Thermoregulation is another critical aspect of neonatal physiology. An infant has a large ratio of surface area to volume, so it loses heat easily. One of its defenses against hypothermia is brown fat, a special adipose tissue deposited from weeks 7 to 20 of fetal development. The mitochondria of brown fat release all the energy of pyruvic acid as heat rather than using it to make ATP; thus, this is a heat-generating tissue. Nevertheless, body temperature is more variable in infants and children than in adults.

Fluid Balance

The kidneys are not fully developed at birth and cannot concentrate the urine as much as a mature kidney can. Consequently, infants have a relatively high rate of water loss and require more fluid intake, relative to body weight, than adults do.

Premature Infants

Neonates weighing under 2.5 kg (5.5 pounds) are generally considered **premature.** They have multiple difficulties with respiration, thermoregulation, excretion, digestion, and liver function. Most neonates weighing 1.5 to 2.5 kg are viable, but with difficulty. Neonates weighing under 500 g rarely survive.

Infants born before 7 months commonly suffer **infant respiratory distress syndrome (IRDS)**—the most common cause of death in premature babies. It occurs because the lungs have not produced enough pulmonary surfactant (see p. 512). Consequently, when the infant exhales, the alveoli do not remain open, but collapse like sheets of wet paper clinging together. A great deal of effort is needed to reinflate them. The infant becomes very fatigued by the high energy demand of breathing. IRDS may be treated by ventilating the lungs with oxygen-enriched air at a positive pressure to keep the lungs inflated between breaths, and by administering surfactant as an inhalant.

A premature infant has an incompletely developed hypothalamus and therefore cannot thermoregulate effectively. Body temperature must be controlled by placing the infant in a warmer.

It is difficult for premature infants to ingest milk because of their small stomach volume and undeveloped sucking and swallowing reflexes. Some must be fed by nasogastric or nasoduodenal tubes.

The liver is also poorly developed, and bearing in mind its very diverse functions (see table 18.4 p. 617), you can probably understand why this would have several serious consequences. The liver synthesizes insufficient albumin, so the baby suffers hypoproteinemia. This upsets the osmotic balance between capillary filtration and reabsorption, leading to edema. The infant bleeds easily because of a deficiency of the clotting factors synthesized by the liver. Jaundice is common in neonates, especially premature babies, because the liver cannot dispose of bile pigments efficiently.

Birth Defects

A birth defect, or **congenital anomaly,**[10] is the abnormal structure or position of an organ at birth, resulting from a defect in prenatal development. The study of birth defects is called **teratology.**[11] Birth defects are the single most common cause of infant mortality in North America. Some congenital defects are not detected until months to years after birth. Thus, by the age of 2 years, 6% of children are diagnosed with congenital anomalies, and by age 5 the incidence is 8%. The following sections discuss some known causes of congenital anomalies, but in many cases, the cause is unknown.

Mutations and Genetic Anomalies

Genetic anomalies are the most common cause of birth defects, accounting for one-third of all cases and 85% of those with identifiable causes. Some defects stem from **mutations,** or changes in DNA structure—for example, achondroplastic dwarfism, microcephaly (abnormal smallness of the head), stillbirth, and some childhood cancer. Mutations can occur through errors in DNA replication during the cell cycle or under the influence of environmental agents called **mutagens,** including some chemicals, viruses, and radiation.

Some of the most common genetic disorders result, however, from the failure of homologous chromosomes to separate during meiosis. Homologous chromosomes pair up during prophase I and normally separate from each other at anaphase I (see p. 651). This separation, called *disjunction,* produces daughter cells with 23 chromosomes each. In **nondisjunction,** a pair of chromosomes fails to separate. Both chromosomes of that pair then go to the same daughter cell, which receives 24 chromosomes while the other daughter cell receives 22. **Aneuploidy**[12] (AN-you-PLOY-dee), the presence of an extra chromosome or lack of one, accounts for about 50% of spontaneous abortions. Aneuploidy can be detected prior to birth by *amniocentesis*—the examination of cells in a sample of amniotic fluid—or by *chorionic villus sampling (CVS),* the biopsy of cells from the chorion.

---

[10] *con* = with; *gen* = born; *a* = without; *nomaly* = evenness, symmetry

[11] *terato* = monster; *logy* = study of

[12] *an* = not, without; *eu* = true, normal; *ploid, from diplo* = double, paired

Figure 20.9 Down Syndrome. (a) A child with Down syndrome (right) and her sister. (b) The chromosomes of a person with Down syndrome, showing the trisomy of chromosome 21. (c) Characteristics of the hand in Down syndrome. (d) The epicanthal fold over the medial commissure (canthus) of the left eye.

- *What was the sex of the person from whom the karyotype in part (b) was obtained?*

(a)

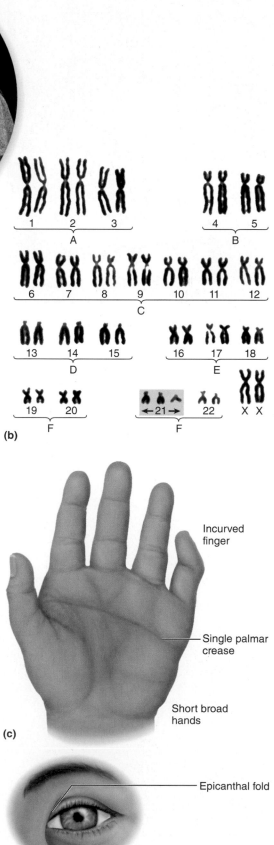

(b)

One example of aneuploidy involving the sex chromosomes is **Klinefelter**[13] **syndrome.** This occurs when the two X chromosomes fail to separate in oogenesis, and the resulting XX egg is fertilized by a Y-bearing sperm. The resulting XXY zygote develops into a sterile male, usually of average intelligence, but with undeveloped testes, sparse body hair, unusually long arms and legs, and enlarged breasts *(gynecomastia)*. The aneuploidy often goes undiagnosed until puberty, when failure to develop the secondary sex characteristics may result in referral to a genetic counselor or medical geneticist.

The other 22 pairs of chromosomes (the *autosomes*) are also subject to nondisjunction. An individual with three copies of an autosome is said to exhibit **trisomy.** Only three trisomies are survivable: those involving chromosomes 13, 18, and 21. The reason is that these three chromosomes are relatively gene-poor. In all other trisomies, the embryo gets a lethal "gene overdose." Nearly all fetuses with trisomy-13 *(Patau syndrome)* and trisomy-18 *(Edward syndrome)* die before birth. Infants born with these trisomies are severely deformed, and fewer than 5% survive for 1 year.

Trisomy-21 (Down[14] **syndrome)** is the most survivable of the three, and therefore the most common among children and adults. Its signs include impaired physical development; short stature; a relatively flat face with a flat nasal bridge; low-set ears; *epicanthal folds* at the medial corners of the eyes; an enlarged, protruding tongue; stubby fingers; and a short broad hand with only one palmar crease (fig. 20.9). People with Down syndrome tend to have outgoing, affectionate personalities. Mental retardation is common and sometimes severe, but is not inevitable. About 75% of fetuses with trisomy-21 die before birth, but it occurs in about 1 out of 700 to 800 live births in the United States. Many persons born with it die by age 10 from such causes as immune deficiency and abnormalities of the heart or kidneys, but for those who survive beyond that age, modern medical care has extended life expectancy to about 60 years. After age 40, however, many of these develop early-onset Alzheimer disease, linked to a gene on chromosome 21.

Aneuploidy is far more common in humans than in any other species, and 90% of cases are of maternal rather than paternal origin. These facts seem to result from the extraordinarily long time it takes for human oocytes to complete meiosis—as long as 50 years (see p. 656). For various reasons including defects in

Incurved finger

Single palmar crease

Short broad hands

(c)

Epicanthal fold

(d)

---

[13] Harry F. Klinefelter Jr. (1912–90), American physician
[14] John Langdon H. Down (1828–96), British physician

Figure 20.10 Schoolboy Showing the Effect of Thalidomide on Upper Limb Development.

Clinical Application 20.2

THE THALIDOMIDE TRAGEDY

Perhaps the most notorious teratogenic drug is thalidomide, a sedative first marketed in West Germany in 1957. Thalidomide was taken by many women in early pregnancy to relieve morning sickness, and by others as a sleeping aid even before they knew they were pregnant. By the time it was removed from the world market in 1961, it had affected an estimated 10,000 to 20,000 babies worldwide, many of them born with unformed arms or legs and often with defects of the ears, heart, and intestines. The U.S. Food and Drug Administration never approved thalidomide for market, but many American women obtained it by participation in clinical drug trials or from foreign sources. Thalidomide has recently been reintroduced and used under tightly controlled conditions for more limited purposes such as leprosy. In some Third World countries, however, people still take thalidomide in a misguided attempt to treat AIDS and other diseases, resulting in an upswing in severe birth defects (fig. 20.10). A general lesson to be learned from the thalidomide tragedy and other cases is that pregnant women should avoid all sedatives, barbiturates, and opiates. Even the acne medicine isotretinoin (Accutane) has caused severe birth defects.

the mitotic spindle and in chromosomal crossing over, aging eggs become less able to separate their chromosomes into two identical sets. This is evident in the statistics of Down syndrome: The chance of having a child with Down syndrome is about 1 in 3,000 for a woman under 30, 1 in 365 by age 35, and 1 in 9 by age 48.

Teratogens

Environmental agents that cause birth defects are called **teratogens.**[15] They fall into three major categories: radiation, infectious diseases, and chemicals. For this reason, diagnostic X-rays should be avoided by pregnant women. Infectious diseases are largely beyond the scope of this book, but several microorganisms can cross the placenta and cause serious congenital anomalies, stillbirths, or neonatal death. Teratogenic viruses include herpes simplex, rubella, cytomegalovirus, and human immunodeficiency virus. Congenital bacterial infections include gonorrhea and syphilis. *Toxoplasma,* a protozoan contracted from meat, unpasteurized milk, and housecats, is another common cause of fetal deformity. Fetuses are more vulnerable than adults to these agents because of their immature immune systems.

Alcohol causes more birth defects than any other drug. Even one drink a day can have noticeable effects on fetal and childhood development, some of which are not noticed until a child begins school. Alcohol abuse during pregnancy can cause **fetal alcohol syndrome (FAS),** characterized by a small head, malformed facial features, cardiac and central nervous system defects, stunted growth, and behavioral symptoms such as hyperactivity, nervousness, and poor attention span. Cigarette smoke also contains potent teratogens that contribute to fetal and infant mortality, ectopic pregnancy, anencephaly (failure of the cerebrum to develop), cleft palate and lip, and cardiac abnormalities. Some other chemical teratogens are discussed in Clinical Application 20.2.

[15] *terato* = monster; *gen* = producing

The effect of a teratogen depends on the genetic susceptibility of the embryo, the dosage of the teratogen, and the time of exposure. Teratogen exposure during the first 2 weeks usually does not cause birth defects, but may cause spontaneous abortion. The period of greatest vulnerability is 3 to 8 weeks. Different organs have different critical periods. For example, limb abnormalities are most likely to result from teratogen exposure at 24 to 36 days, and brain abnormalities from exposure at 3 to 16 weeks.

Before You Go On

Answer these questions from memory. Reread the preceding section if there are too many you don't know.

12. How does inflation of the lungs at birth affect the route of blood flow through the heart?

13. How is an infant protected from infection up to the time when it produces enough antibodies of its own?

14. Why is the regulation of body temperature so challenging for an infant, and even more so for a premature infant?

15. Why is respiratory distress syndrome common in premature infants?

16. Why does immaturity of the liver of a premature infant lead to such a broad spectrum of problems?

17. Genetic birth defects can result from either mutation or nondisjunction. Explain the difference and give examples of each.

18. What is a teratogen? What are the three classes of teratogens? Give examples of each.

20.4 Aging, Senescence, and Death

Expected Learning Outcomes

When you have completed this section, you should be able to:

a. distinguish between aging and senescence;

b. discuss the relationship of exercise to senescence;

c. summarize some current theories of senescence;

d. explain why evolution is unlikely to eliminate genes for the diseases of old age; and

e. distinguish between life expectancy and life span.

Like Ponce de León searching for the legendary fountain of youth in Florida, people yearn for a way to preserve their youthful appearance and function. Our real concern, however, is not aging but senescence. The term **aging** is used in various ways, but is taken here to mean all changes that occur in the body with the passage of time—including the growth, development, and increasing functional efficiency that occur from childhood to adulthood, as well as the degenerative changes that occur later in life. **Senescence** is the degeneration that occurs in an organ system after the age of peak functional efficiency. It includes a gradual loss of reserve capacities, reduced ability to repair damage and compensate for stress, and increased susceptibility to disease.

Senescence is an issue of paramount importance for health-care providers. One in nine Americans is 65 or older. As the average age of the population rises, health-care professionals find themselves increasingly occupied by the prevention and treatment of the diseases of age. The leading causes of death change markedly with age. Accidents, homicide, suicide, and AIDS figure prominently in the deaths of people 18 to 34 years old, whereas the major causes of death after age 55 are clearly related to senescence of the organ systems: heart disease, cancer, stroke, diabetes, and lung disease. The causes of senescence, however, remain as much a scientific mystery today as cancer was 50 years ago and heredity was 100 years ago. The scientific and clinical study of aging is called **gerontology**[16] and the clinical diagnosis and management of the medical conditions of old age is called **geriatrics.**[17]

This subject also raises many points relevant to personal health and fitness choices that can lessen the rate and effects of senescence. In addition, the study of senescence calls renewed attention to the multiple interactions among organ systems. As we have already seen in earlier chapters, the senescence of one organ system typically contributes to the senescence of others.

Exercise and Rates of Senescence

The organ systems do not all degenerate at the same rate or to the same extent. For example, from ages 30 to 80, the speed of nerve conduction declines only 10% to 15%, but the number of functional glomeruli in the kidneys declines about 60%. Some physiological functions show only moderate changes at rest but more pronounced differences when tested under exercise conditions. The organ systems also vary widely in the age at which senescence becomes noticeable. There are forerunners of atherosclerosis, for example, even in infants, and visual and auditory sensitivity begin to decline soon after puberty. By contrast, the female reproductive system shows no significant senescence until menopause, and then declines relatively abruptly. Aside from these unusual examples, most physiological measures of performance peak between the late teens and age 30 and then decline at a rate influenced by the level of use of the organs.

The rate of senescence also depends on one's habits of use or abuse of an organ system. Other than the mere passage of time, senescence results from obesity and insufficient exercise more than from any other causes. Conversely, good nutrition and exercise are the best ways to slow its progress. There is no clear evidence that exercise will prolong your life, but there is little doubt that it improves the quality of life in old age. It maintains endurance, strength, and joint mobility; it also reduces the incidence and severity of hypertension, osteoporosis, musculoskeletal pain, obesity, and diabetes mellitus. This is especially true if you begin a program of regular physical exercise early in life and make a lasting habit of it. If you stop exercising regularly, the body rapidly becomes deconditioned; appreciable reconditioning can be achieved, however, even when an exercise program is begun late in life. A person in his or her 90s can increase muscle strength two- or threefold in 6 months with as little as 40 minutes of isometric exercise a week. The improvement results from a combination of muscle hypertrophy and neural efficiency.

Resistance training may be the most effective way of reducing accidental injuries such as bone fractures, whereas endurance training reduces body fat

---

[16] *geronto* = old age; *logy* = study of

[17] *ger* = old age; *iatr* = treatment of

and increases cardiac output, maximum oxygen uptake, and insulin sensitivity. A general guideline for ideal endurance training is to have three to five periods of aerobic exercise per week, each 20 to 60 minutes long and vigorous enough to reach 60% to 90% of your maximum heart rate. The maximum is best determined by a stress test but averages about 220 beats per minute minus your age in years.

An exercise program should ideally be preceded by a complete physical examination and stress test. Warm-up and cool-down periods are especially important in avoiding soft tissue injuries. Because of their lower capacity for thermoregulation, older people must be careful not to overdo exercise, especially in hot weather. At the outset of a new exercise program, it is best to "start low and go slow."

Theories of Senescence

Why do our organs wear out? Why must we die? There still is no general theory on this. The question actually comes down to two issues: (1) What are the mechanisms that cause the organs to deteriorate with age? (2) Why has natural selection not eliminated these and produced bodies capable of longer life?

Mechanisms of Senescence

Numerous theories have been proposed and discarded to explain why organ function degenerates with age. Some authorities maintain that senescence is governed by inevitable or even preprogrammed changes in cellular function. Others attribute senescence to environmental factors that progressively damage our cells over the course of a lifetime.

There is good evidence of a hereditary component to life expectancy. Unusually long and short lives tend to run in families. Monozygotic twins are more likely than dizygotic twins to die at a similar age. Some genetic diseases such as *progeria*[18] (fig. 20.11) are associated with rapid, early senescence and premature death. Such syndromes differ from normal senescence, but nevertheless demonstrate that many changes associated with old age can be brought on by a genetic anomaly.

Knowing that senescence is partially hereditary does not answer the question about why tissues degenerate. Quite likely, no one theory explains all forms of senescence, but we can briefly examine some of them.

Replicative Senescence. Normal organ function depends on a rate of cell renewal that keeps pace with cell death. There is a limit, however, to how many times cells can divide. Certain cultured human cells divide 80 to 90 times if taken from a fetus, but only 20 to 30 times if taken from older people. After their maximum number of divisions, the cells degenerate and die. This decline in mitotic potential with age is called **replicative senescence.**

The reason for this may lie in the **telomere,**[19] a "cap" on the end of each chromosome analogous to the plastic tip of a shoelace. In humans, it consists of a noncoding nucleotide sequence CCCTAA repeated 1,000 times or more. During DNA replication, DNA polymerase cannot reproduce the very ends of the molecule (see p. 88). If there were functional genes at the end, they would not get duplicated. The telomere may therefore provide a bit of "disposable" DNA at the end, so that DNA doesn't fail to replicate genes that would otherwise be there. Every time DNA is replicated, 50 to 100 bases are lost from the telomere. Once the telomere is used up, the polymerase may fail to replicate some of the terminal genes, making old chromosomes more vulnerable to damage, replication errors, or both, and causing old cells to be increasingly dysfunctional.

[18]*pro* = before; *ger* = old age

[19]*telo* = end; *mer* = piece

Figure 20.11 Progeria. This is a genetic disorder in which senescence appears to be greatly accelerated. The individuals here, from left to right, are 15, 12, and 26 years old. Few people with progeria live as long as the woman on the right.

The telomere theory is clearly not the entire answer to why organs degenerate, however. Skeletal muscles and the brain exhibit extreme senescence, yet muscle fibers and neurons are nonmitotic. Their senescence obviously is not a result of repeated mitosis and cumulative telomere damage.

Protein Abnormalities. Collagen and many other proteins exhibit increasingly abnormal structure in older tissues and cells. The changes are not in amino acid sequence—therefore not attributable to DNA mutations—but lie in the way the proteins are folded, cross-linked with each other through disulfide bridges, or bound to components such as carbohydrates. This is another reason that cells accumulate more dysfunctional proteins as they age. Collagen cross-linking makes proteins less soluble and more stiff, and is thought to be a factor in several of the most noticeable changes of the aging body such as stiffening of the joints, lenses, and arteries.

Free Radical Theory. Free radicals have very destructive effects on macromolecules (see p. 35). We have antioxidants to protect us from them, but it is believed that some of these antioxidants become less abundant with age and are eventually overwhelmed by free radicals, or that some of the molecules damaged by free radicals are long-lived and accumulate in cells. Free radical damage may therefore be a contributing factor in some of the other mechanisms of senescence discussed here.

Autoimmune Theory. Some of the altered macromolecules described previously may be recognized as foreign antigens, stimulating lymphocytes to mount an immune response against the body's own tissues. Autoimmune diseases such as rheumatoid arthritis do, in fact, become more common in old age.

Evolution and Senescence

If certain genes contribute to senescence, why doesn't natural selection eliminate them, much as it tends to eliminate other harmful genes from the population? Biologists once postulated that senescence and death were for the good of the species—a way for older, worn-out individuals to make way for younger, healthier ones. We can see the importance of death by imagining that science had put an end to senescence and people of all ages died at the same rate that 18-year-olds do now: 1 per 1,000 per year. If so, the median age of the population would be 163, and 13% of us would live to be 2,000 years old. The implications for world population and competition for resources would be staggering. Thus, it is easy to understand why scientists once interpreted death as a self-sacrificing phenomenon for the good of the species.

But this hypothesis has several weaknesses. One is the fact that natural selection works exclusively through the effects of genes on the reproductive rates of individuals. A species evolves only because some members reproduce more than others. A gene that doesn't affect reproductive rate can be neither eliminated nor favored by natural selection. Genes for many disorders of old age have little or no effect until a person is past reproductive age. Our prehistoric and even fairly recent ancestors usually died of accidents, predation, starvation, weather, and infectious diseases at an early age. Few people lived long enough to be affected by colon cancer, atherosclerosis, or Alzheimer disease. Natural selection is "blind" to such death-dealing genes, which escape the selection process and remain with us today.

Life Expectancy and Death

Life expectancy, the average length of life in a given population, has increased impressively in industrialized countries. People born in the United States at the beginning of the twentieth century had a life expectancy of only 45 to 50 years;

nearly half of them died of infectious disease. The average boy born today can expect to live 75 years and the average girl 80 years. This is due mostly to victories over infant and child mortality, not to advances at the other end of the life span. **Life span,** the maximum age attainable by humans, has not increased for many centuries and there seems to be little prospect that it ever will. There is no verifiable record of anyone living past the age of 122 years.

Ninety-nine percent of us will die before age 100, and there is little chance that this outlook will change within our lifetimes. We don't presently foresee any "cure for old age" or significant extension of the human life span. The real issue is to maintain the best possible quality of life, and when the time comes to die, to be able to do so in comfort and with dignity.

Before You Go On

Answer these questions from memory. Reread the preceding section if there are too many you don't know.

19. Define and distinguish between aging and senescence.

20. Explain why both endurance and resistance exercises are important in old age. Support your argument with specific examples.

21. Summarize five mechanisms that may be responsible for senescence.

22. Explain why natural selection tends to remove many harmful genes from a population, but is unlikely to affect genes for the diseases of old age.

23. Distinguish between life expectancy and life span, and explain why one of these has changed significantly over the past century while the other one has not.

CAREER SPOTLIGHT

Genetic Counselor

Genetic counselors provide information, support, and advocacy for families with members who have genetic disorders or birth defects and couples concerned about their probability of transmitting genetic disorders. They trace family histories of disease, identify risk factors, interpret information for their clients, advise them on testing and prevention options and coping strategies, and refer affected individuals to community support services. Among the numerous disorders studied by genetic counselors are cystic fibrosis, sickle-cell disease, breast cancer, Down syndrome, cleft palate, hemophilia, and mental retardation. Genetic counselors also advise women who are pregnant or planning to become so, and who are concerned about their age or exposure to risk factors such as certain drugs, infectious agents, or radiation. Genetic counselors come from backgrounds as diverse as biology, nursing, and social work, and work in hospitals, clinics for pediatric care and adult genetics, diagnostic laboratories and biotechnology companies, research institutions, private practice, and government agencies.

To become a certified genetic counselor, one must first earn a bachelor's degree in a relevant biological or social science, then gain a master's degree and pass a board examination. See appendix B for links to further information.

Study Guide

Assess Your Learning Outcomes

To test your knowledge, discuss the following topics with a study partner or in writing, ideally from memory.

20.1 Fertilization and Preembryonic Development (p. 676)

1. The distinction between a preembryo and embryo, and the age of transition from one to the next

2. Where the sperm and egg meet, why the sperm must travel most of the distance, how many sperm make it, and why so many of them do not

3. Why a freshly ejaculated sperm cannot fertilize an egg, and what must happen to make it capable of doing so

4. The time frame around ovulation and ejaculation when fertilization is possible

5. How a sperm penetrates the egg and how the egg usually prevents more than one sperm from doing so

6. Events that occur between the moment of sperm penetration and the existence of a true zygote

7. Distinguishing aspects of the three clinical trimesters of pregnancy

8. The three major processes that occur in the preembryonic stage of development

9. Distinctions between a zygote, blastomere, morula, and blastocyst; the two cell masses of the blastocyst; and the destiny of each of those cell masses

10. The two functions of the trophoblast

11. The two ways in which twins are produced and the distinction between monozygotic and dizygotic twins

12. The mechanism of implantation of a conceptus in the uterine wall

13. The defining processes of embryogenesis—how the three primary germ layers are created and how the embryoblast becomes an embryo

20.2 The Embryonic and Fetal Stages (p. 683)

1. The term for the overall process that transforms an embryo into a fetus; how we define when a fetus exists; and at what gestational age this is attained

2. The three modes of prenatal nutrition; what distinguishes the trophoblastic from the placental phases of nutrition; and the timetable on which these develop and overlap

3. How the placenta is produced, and its five functions in pregnancy

4. The anatomy of the placenta and umbilical cord, including the origin of the umbilical arteries and destination of the umbilical vein

5. The anatomy and functions of the amnion, yolk sac, allantois, and chorion, and the functions of amniotic fluid

6. How the flat embryonic disc becomes a cylindrical body; how its primitive gut, thoracic cavity, peritoneal cavity, neural tube, and somites come into being; and what mature tissues arise from the somites

7. How the route of blood flow in the fetus differs from its route after birth, including the names and anatomy of the three circulatory shunts

8. The meaning of the following features of the fetus—*digital rays, meconium, lanugo,* and *vernix caseosa*—and the meaning of *quickening* and the *vertex position* of the fetus

20.3 The Neonate (p. 683)

1. The time frame of the neonatal period

2. How the neonate's respiratory and circulatory systems change and influence each other at birth

3. What gives the neonate immunity to pathogens until it can form its own antibodies

4. Why a neonate is at risk of hypothermia and fluid loss; how its body is adapted to minimize heat loss; and the implications of these risks for neonatal care

5. The criterion for a premature infant and the reasons why premature infants have such problems with respiration, thermoregulation, feeding, and liver functions

6. The subject of teratology and why it is of such public-health importance

7. The importance of mutation as a cause of birth defects, and some examples of genetic birth defects

8. The meaning of *nondisjunction* and *aneuploidy;* the role of aneuploidy in producing birth defects; and methods for the prenatal diagnosis of aneuploidy

9. Examples of birth defects related to the nondisjunction of sex chromosomes and autosomes

10. The causation, characteristics, and prognosis for persons with Down syndrome

11. The three basic categories of teratogens; examples of each; and why the effects of a teratogen can vary according to the gestational age at exposure

20.4 Aging, Senescence, and Death (p. 699)

1. The distinction between aging and senescence; why senescence is so relevant to the health-care professional's work; and why it has such personal relevance
2. Examples of variation in the rate of senescence from one organ system to another
3. The extent to which regular exercise can have an impact on senescence and life expectancy
4. Evidence for a hereditary influence on longevity
5. The possible role of telomere damage in replicative senescence, and the limitations of the telomere theory
6. Other theories of senescence related to molecular cross-linking, other changes in protein structure, free radical damage, and autoimmunity
7. The theory of why evolution by natural selection has not eliminated genes for the diseases of old age, even though it does tend to eliminate other harmful genes from a population
8. The difference between life expectancy and life span, and why one of these has changed so much and the other has changed so little over the past century or more

Testing Your Recall

1. A 32-celled stage that has developed from a fertilized egg is best classified as
   a. a morula.
   b. a blastocyst.
   c. a zygote.
   d. an embryo.
   e. an embryoblast.

2. Which of these tissues does *not* exist prior to embryonic folding?
   a. the trophoblast
   b. the mesoderm
   c. the endoderm
   d. the epidermis
   e. the ectoderm

3. A newly ejaculated sperm cannot fertilize an egg because it has not yet undergone
   a. polyspermy.
   b. capacitation.
   c. the mobilization reaction.
   d. quickening.
   e. cleavage.

4. The fertilization membrane around a zygote is a product of
   a. the acrosome.
   b. the zona pellucida.
   c. exocytosis by the egg.
   d. gastrulation.
   e. embryogenesis.

5. All of these occur in the preembryonic stage *except*
   a. cleavage.
   b. implantation.
   c. trophoblastic nutrition.
   d. organogenesis.
   e. embryogenesis.

6. Early in pregnancy, the corpus luteum remains active under the influence of _____ from the conceptus.
   a. estrogen
   b. progesterone
   c. meconium
   d. lanugo
   e. human chorionic gonadotropin

7. The lanugo is
   a. fetal hair.
   b. a sperm enzyme used to penetrate the egg.
   c. contents of the fetal colon at full term.
   d. a form of neonatal respiratory distress.
   e. a circulatory shunt that bypasses the liver.

8. Fetal alcohol syndrome results from
   a. a mutagen.
   b. a teratogen.
   c. a carcinogen.
   d. nondisjunction.
   e. aneuploidy.

9. One theory of replicative senescence is that it results from depletion of the _____ at the ends of the chromosomes.
   a. telomeres
   b. sarcomeres
   c. centromeres
   d. somites
   e. acrosomes

10. Another theory of senescence is that we suffer more and more tissue damage from _____ as our antioxidant levels decline with age.
    a. autoimmune disease
    b. cross-linking
    c. polyspermy
    d. nondisjunction
    e. free radicals

11. Endoderm and mesoderm are formed by the migration of cells down into the _____ of the embryonic disc.

12. The process of migration described in question 11, resulting in three primary germ layers, is called _____.

13. Muscle, bone, and blood arise from the primary germ layer called _____.

14. The oldest age to which any human being has lived is called the human _____.

15. Prior to implantation, a blastocyst is nourished by a secretion from the _____ glands.

16. In trophoblastic nutrition, the conceptus digests and consumes _____ cells of the endometrium.

17. Treelike processes called _____ contain fetal blood and project into the placental sinus, where they are surrounded by maternal blood.

18. The "breaking of the waters" as a woman goes into labor is a release of the _____ fluid in which the fetus has been suspended for most of its gestation.

19. The vertebrae, trunk muscles, and dermis arise from blocks of tissue called _____ that lie in series alongside the median plane of the embryo.

20. Any chemical, infectious organism, or radiation capable of causing a birth defect is called a/an _____.

Answers in appendix A

6. The placenta serves not only for nutrition of the fetus but also as a source of hormones.

7. The ductus venosus allows most fetal blood to bypass the liver.

8. Gerontologists are optimistic that the human life span will increase significantly in the twenty-first century.

9. The ability of sperm to fertilize an egg declines rapidly in the first 4 to 5 hours after ejaculation.

10. The developing individual is already a fetus by the time it enters the second trimester of pregnancy.

Answers in appendix A

True or False

Determine which five of the following statements are false, and briefly explain why.

1. A conceptus is considered a preembryo in the first trimester, an embryo in the second trimester, and a fetus in the third trimester of pregnancy.

2. Even though cleavage divides a morula into more and more blastomeres, the morula does not get significantly larger.

3. Twins result from disjunction, the separation of cells of the morula into two masses that each produce an embryo.

4. By the time implantation is complete, the blastocyst is completely covered over by endometrium.

5. Trophoblastic nutrition of the embryo must end before placental nutrition can begin.

Testing Your Comprehension

1. Some health-supplement vendors market the enzyme superoxide dismutase (SOD) as an oral antioxidant to retard senescence. Explain why it would be a waste of your money to buy it.

2. In some children, the ductus arteriosus fails to close after birth—a condition that eventually requires surgery. Predict how this condition would affect (a) pulmonary blood pressure, (b) systemic diastolic pressure, and (c) the right ventricle of the heart.

3. Only one sperm is needed to fertilize an egg, yet a man who ejaculates fewer than 10 million sperm is usually infertile. Explain this apparent contradiction. Supposing 10 million sperm were ejaculated, predict how many would come within close range of the egg. How likely is it that any one of these sperm would fertilize it?

Appendix A

ANSWER KEYS

This appendix provides answers to the end-of-chapter Testing Your Recall, True or False, and figure legend questions.

Chapter 1

Testing Your Recall

1.	a	9.	e	16.	medial
2.	b	10.	c	17.	parietal, visceral
3.	b	11.	palpation	18.	negative feedback
4.	e	12.	computed		loop
5.	c		tomography	19.	anatomical position
6.	a	13.	homeostasis	20.	proximal
7.	d	14.	organ		
8.	b	15.	wrist, ankle		

True or False (These items are false for the reasons given; all others are true.)

2. The diaphragm is inferior to the lungs.
4. These cavities are lined with serous membranes.
5. Abnormal skin appearances are discerned by inspection, not by listening (ausculation).
8. Negative feedback has beneficial, homeostasis-restoring effects.
9. There are numerous organelles in every cell.

Answers to Figure Legend Questions

1.1 A PET scan requires an injection of radioisotopes, which is an invasion of the body surface. No invasive procedure is needed for an MRI.
1.2 Sonography involves no ionizing radiation that could harm a fetus.
1.4 Vasodilation allows warm blood to flow closer to the body surface and lose more heat to the environment.
1.5 Yes, childbirth could be regarded as a homeostatic response to pregnancy that culminates in ending the pregnancy and returning the body to its more usual state.

Chapter 2

Testing Your Recall

1.	a	8.	c	15.	dehydration
2.	c	9.	b		synthesis
3.	a	10.	d	16.	-ose, -ase
4.	c	11.	cation	17.	phospholipids
5.	a	12.	free radicals	18.	ATP
6.	e	13.	catalyst, enzymes	19.	catabolism
7.	b	14.	anabolism	20.	substrate

True or False (These items are false for the reasons given; all others are true.)

1. The monomers of a polysaccharide are monosaccharides.
3. ATP is a very short-lived molecule, not a long-term energy-storage molecule.
6. A saturated fat is one to which no more hydrogen can be added.
8. Humans cannot digest cellulose and therefore get no calories from it.
9. Two isomers do not have identical chemical properties. Ethanol and ether, for example, are isomers with very different properties.

Answers to Figure Legend Questions

2.6 Relatively little energy is required to make an N_2 molecule vaporize because it has no attraction to other molecules in the liquid state. Thus, N_2 boils at a low temperature.
2.7 Blood plasma is a solution of (for example) sodium and glucose and a colloid of albumin and other proteins. With the blood cells removed, it is not a suspension.

Chapter 3

Testing Your Recall

1.	e	9.	d	17.	nucleus,
2.	d	10.	b		mitochondria
3.	b	11.	micrometers	18.	smooth ER,
4.	b	12.	receptor		peroxisomes
5.	e	13.	gates	19.	cell-adhesion
6.	a	14.	glycocalyx		molecules
7.	d	15.	anaphase	20.	phagocytosis
8.	a	16.	phospholipids		

True or False (These items are false for the reasons given; all others are true.)

2. DNA replication occurs in the S phase, not during mitosis.
6. Movement of substances down a concentration gradient requires no ATP.
7. Osmosis is a passive process, not active transport.
9. Desmosomes are just spots of intercellular attachment; they have no channels for transferring material from cell to cell.
10. Organelles all lie outside the nucleus, in the cytoplasm.

Answers to Figure Legend Questions

3.2 The phospholipid heads are hydrophilic and thus attracted to the water inside and outside the cell.
3.4 The photo shows an intestinal absorptive cell, which has prominent microvilli for nutrient absorption and a glycocalyx for protection from acid and enzymes of the digestive tract.
3.6 Gap junctions
3.7 Simple diffusion
3.10 The nucleus, endoplasmic reticulum, and Golgi complex participate directly in protein synthesis; mitochondria participate indirectly since protein synthesis requires the ATP that they provide.
3.16 Chromatids are also composed of protein.

Chapter 4

Testing Your Recall

1.	a	8.	a	15.	collagen
2.	b	9.	b	16.	glial cells
3.	d	10.	c	17.	basement
4.	e	11.	necrosis		membrane
5.	c	12.	simple squamous	18.	matrix
6.	e	13.	lacunae	19.	holocrine
7.	b	14.	periosteum	20.	simple

True or False (These items are false for the reasons given; all others are true.)

1. The esophageal epithelium is nonkeratinized.
5. Adipose tissue is an exception to this rule.
6. Adipocytes also occur in areolar connective tissue.
9. All living cells are electrically excitable, but nerve and muscle cells have this developed to the highest degree.
10. Fibrocartilage and hyaline articular cartilages have no perichondrium.

Answers to Figure Legend Questions

4.8 Exfoliation of squamous cells also occurs in the epidermis, oral cavity, esophagus, anal canal, vagina, and a few other places.
4.29 Holocrine glands require a high rate of stem cell mitosis to replace the cells that continually disintegrate to become the secretion.
4.30 Mucous membranes are also found throughout the digestive, reproductive, and urinary tracts and other regions of the respiratory tract in addition to the trachea cited. Serous membranes are also found in the pericardium, pleurae, parietal peritoneum, mesenteries, and serosae of many other abdominal organs in addition to the small intestine cited.

Chapter 5

Testing Your Recall

1. d	8. a	14. cyanosis
2. b	9. a	15. dermal papillae
3. d	10. e	16. earwax
4. a	11. insensible	17. sebaceous glands
5. e	perspiration	18. keratin, collagen
6. b	12. piloerector	19. vitamin D
7. c	13. dermato-, cutane-	20. keratinocytes

True or False (These items are false for the reasons given; all others are true.)

3. The dermis is mostly collagen; keratin occurs in the epidermis.
4. Vitamin D is synthesized by collaboration of epidermal keratinocytes, the liver, and the kidneys.
7. The hypodermis is not part of the skin.
8. People of all races have about equal densities of melanocytes.
9. Pallor is a temporary paleness, not a genetic lack of pigment.

Answers to Figure Legend Questions

5.3 Keratinocytes
5.4 Shampoo only contacts the dead shafts of the hairs, which therefore cannot be nourished. The only nourishment a hair can receive is from the dermal papilla at the base of the hair root, from blood-borne nutrients.
5.7 Basal cell carcinoma is the more common of the two but melanoma is the more dangerous.

Chapter 6

Testing Your Recall

1. e	8. e	15. hypocalcemia
2. a	9. a	16. osteoblasts
3. a	10. c	17. kinesiology
4. d	11. temporal	18. menisci
5. b	12. canaliculi	19. sutures
6. a	13. appositional	20. anulus fibrosus
7. d	14. synovial fluid	

True or False (These items are false for the reasons given; all others are true.)

2. The growth zone of a child's bone is the metaphysis (or epiphyseal plate).
3. Ligaments connect one bone to another; tendons connect muscle to bone.

5. This action involves hyperextension of the shoulder.
7. The bursae contain synovial fluid but do not secrete it; the synovial membrane does.
8. The arm contains only one bone (the humerus), but the leg contains two (tibia and fibula).

Answers to Figure Legend Questions

6.1 Bone marrow
6.2 Osteoblasts engage actively in synthesizing the collagen of the bone matrix and require a lot of rough endoplasmic reticulum to carry this out. Osteogenic cells do not synthesize proteins for export from the cell.
6.3 Spongy bone has more surface area and is therefore more strongly affected by osteoclasts.
6.10 The appendicular skeleton includes the pectoral and pelvic girdles, which are within the axial region of the body.
6.13 The temporal bone articulates with the occipital, parietal, frontal, sphenoid, and zygomatic bones and the mandible.
6.20 Rupture of this ligament could allow the head and atlas to slip posteriorly and thus cause the dens of the axis to rip through the cervical spinal cord.
6.21 Transverse foramina and a forked spinous process
6.33 Interphalangeal joints are not subjected to very much compression.
6.46 Metacarpal bones and proximal phalanges; proximal phalanges and middle phalanges

Chapter 7

Testing Your Recall

1. a	8. d	15. calcium
2. b	9. c	16. involuntary
3. e	10. d	17. sphincter
4. c	11. myosin	18. quadriceps femoris
5. b	12. acetylcholine	19. latissimus dorsi
6. c	13. tonus	20. gastrocnemius
7. a	14. fascicle	

True or False (These items are false for the reasons given; all others are true.)

1. A motor neuron branches out to anywhere from 3 to 1,000 muscle fibers or so.
2. Calcium binds to troponin, not to myosin.
4. Myofilaments slide over each other; they do not shorten.
6. Its origin is on the clavicle and sternum.
8. It refers to events from calcium release by the sarcoplasmic reticulum to movement of tropomyosin away from the active sites of the actin thin filaments. Myosin–actin binding follows excitation–contraction coupling.

Answers to Figure Legend Questions

7.1 Skeletal muscle cells are long and threadlike; cardiac and smooth myocytes are not.
7.4 A bands are a region of thick and thin filament overlap, so there is a greater density of protein here than in the I bands, which have only thin filaments.
7.9 I bands would be absent because there is no region of the sarcomere occupied by thin filaments only.
7.13 For the peristaltic contractions of swallowing
7.15 The triceps brachii would be the prime mover whereas the biceps brachii and brachialis muscles would exert a braking action.
7.17 The pterygoid muscles approach the mandible obliquely and pull it from side to side, whereas the temporalis muscles approach it from directly above and can only pull the mandible upward.
7.20 The hip and the knee

Chapter 8

Testing Your Recall

1. a	8. b	15. node of Ranvier
2. c	9. c	16. bipolar
3. a	10. a	17. ependymal
4. e	11. dura mater	18. action potential
5. d	12. central	19. neurotransmitters
6. b	13. afferent	20. gray matter
7. e	14. brachial	

True or False (These items are false for the reasons given; all others are true.)

2. The arachnoid mater lies between the dura mater and pia mater.
4. There are no spinal nerve plexuses in the thoracic region.
5. Some somatic reflex arcs pass through the brainstem.
8. A nerve fiber is just the axon of a neuron, not a synonym for *neuron,* the entire cell.
9. The myelin sheath in the brain is produced by oligodendrocytes.

Answers to Figure Legend Questions

8.5 Schwann cells
8.10 Proximal to the action potential, the axolemma is in its refractory period and cannot immediately be stimulated to fire again.
8.19 Such a reflex in the upper limb would stimulate the biceps brachii and brachialis and inhibit the triceps brachii.

Chapter 9

Testing Your Recall

1. c	8. d	14. cholinergic
2. c	9. e	15. choroid plexus
3. a	10. e	16. precentral
4. e	11. corpus callosum	17. proprioceptors
5. a	12. ventricles,	18. association
6. b	cerebrospinal	19. occipital
7. d	13. arbor vitae	20. Broca area

True or False (These items are false for the reasons given; all others are true.)

1. The longitudinal fissure separates the cerebral (not cerebellar) hemispheres.
5. The sympathetic nervous system inhibits digestion.
6. Hearing is a function of the temporal lobe; the occipital lobe is visual.
8. The parasympathetic division has long cholinergic preganglionic fibers.
10. The optic nerve does not move the eye; the oculomotor, abducens, and trochlear nerves do.

Answers to Figure Legend Questions

9.3 The cranial dura mater is pressed against the surrounding bone, and even attached to it at places, with no intervening epidural space or fat; it has two layers, separated in some places by dural sinuses; and it folds inward to form partitions between parts of the brain.
9.8 The amygdala
9.9 One would lose somatosensory sensations such as touch and pain from the lower limb on the opposite side of the body.
9.17 Such an injury could abolish cardioacceleration in fear because it would disrupt the conduction of signals from the brain (fear centers) down the spinal cord to the origins of the cardiac nerves. It would not affect gastric secretion, because that is mediated through the vagus nerves, which originate above the point of injury.

Chapter 10

Testing Your Recall

1. b	8. a	15. saccule
2. e	9. d	16. accommodation
3. a	10. e	17. olfactory bulbs
4. c	11. lamellar corpuscle	18. hair cells
5. b	12. fovea	19. cone
6. d	13. optic chiasm	20. proprioceptors
7. b	14. lacrimal	

True or False (These items are false for the reasons given; all others are true.)

1. Pain has its own specialized receptors (nociceptors) and is not the result of overstimulation of touch receptors.
4. A hair receptor is a nerve fiber associated with a hair follicle, not to be confused with hair cells of the inner ear.
5. Most refraction occurs when light rays pass from air into the cornea of the eye.
8. The auditory tube lets air into the middle ear, not the inner ear.
10. Lingual papillae are visible bumps on the tongue. Most types contain taste buds, but they are not the same as the taste buds.

Answers to Figure Legend Questions

10.1 Small receptive fields; the lips are very sensitive to fine two-point discrimination.
10.6 Vestibulocochlear, CN VIII
10.11 Inner hair cells. The outer hair cells tune the cochlea to enable the brain to discriminate one pitch of sound from another.
10.12 Air pressure behind the eardrum would dampen its vibrations and interfere with transmission of sound through the ossicles to the inner ear.
10.14 The eyes would be more prone to watering because the tears could not drain away into the nasal cavity.
10.17 It is the subject's right eye. The optic disc is medial to the fovea (the optic nerves converge medially from the two eyes), so the subject's nose would be toward page right, making this the subject's right eye.
10.20 The inner segment, because this is where the mitochondria are concentrated.
10.22 By tracing fibers from the upper optic radiation in this figure to the retinas, it can be seen that one would lose vision from the lateral retina of the right eye and medial retina of the left eye. These two retinal areas cover the left peripheral vision, so the effect would be blindness in the left half of the visual field.

Chapter 11

Testing Your Recall

1. b	9. e	15. anterior
2. d	10. d	pituitary gland
3. a	11. cortisol	16. interstitial
4. c	12. hypothalamus,	17. ovaries, testes
5. c	posterior	18. pineal
6. b	pituitary gland	19. thymus
7. e	13. leptin	20. oxytocin
8. b	14. ghrelin	

True or False (These items are false for the reasons given; all others are true.)

2. Many hormones are secreted by organs and tissues other than endocrine glands, such as the heart, brain, and stomach.
4. The pineal gland and thymus shrink (involute) with age.
5. The adrenal cortex secretes steroids.
8. The infundibulum is not a duct and no hormones are secreted through ducts.
9. Thyroxine raises metabolic rate and heat production.

Answers to Figure Legend Questions

11.1 Liver, kidneys, stomach, heart, skin, small intestine, bones, placenta
11.4 Mechanism (b), since testosterone is a steroid
11.6 The posterior lobe
11.8 The C cells secrete calcitonin.
11.10 The exocrine cells secrete digestive enzymes.

Chapter 12

Testing Your Recall

1.	e	8.	d	15.	serum
2.	a	9.	d	16.	hemophilia
3.	b	10.	a	17.	leukopenia
4.	d	11.	hemopoiesis	18.	albumin
5.	e	12.	hematocrit	19.	polycythemia
6.	d	13.	macrophages	20.	erythropoiesis
7.	b	14.	hemostasis		

True or False (These items are false for the reasons given; all others are true.)

3. A low blood oxygen level is the result, not the cause, of anemia.
4. Neutrophils are the most important bacteria-fighting WBCs.
6. Neutrophils are the most abundant WBCs in the blood.
9. RBCs live longer than most granulocytes in spite of their lack of a nucleus.
10. Leukemia is not a WBC deficiency but an overproduction of immature WBCs.

Answers to Figure Legend Questions

12.1 Nuclei
12.3 The nucleus of the stem cell line was here; the cell caves in at the center when the nucleus is lost.
12.4 To the iron atom at the center of the heme group
12.12 Hemophilia affects the intrinsic pathway because it usually involves a lack of factor VIII, part of the intrinsic pathway.
12.13 Although superabundant, these cells are immature and unable to perform their defensive roles.

Chapter 13

Testing Your Recall

1.	d	9.	e	16.	hypertension
2.	d	10.	a	17.	baroreceptors
3.	b	11.	systole, diastole	18.	myocardial
4.	b	12.	sphygmomanometer		infarction
5.	d	13.	angiogenesis	19.	T wave
6.	d	14.	vagus	20.	left coronary artery
7.	b	15.	cerebral arterial		
8.	a		circle		

True or False (These items are false for the reasons given; all others are true.)

1. The coronary circulation is part of the systemic circuit; there is also a pulmonary circuit.
3. Aldosterone causes salt retention and tends to sustain blood pressure rather than lowering it.
5. Valves are found in medium veins, not in arteries.
6. A doubling of vessel radius increases blood flow 16-fold.
8. An ECG is a composite recording from all cardiomyocytes of the heart, and is not an individual action potential.

Answers to Figure Legend Questions

13.1 Both; they receive pulmonary blood from the pulmonary arteries and systemic blood from the bronchial arteries.
13.2 Most of it lies to the left of the median plane.
13.7 The right atrium receives the signal and begins contracting a little sooner than the left because the pacemaker is in the right atrium.

13.9 This is the point in time when the aortic valve opens and the left ventricle begins ejecting blood into the aorta.
13.10 Being closer to the heart, the arteries are subjected to greater stress from blood pressure surges and must be more elastic to withstand this.
13.13 Vasodilation lowers blood pressure.
13.20 Aorta, left common carotid artery, left external carotid artery

Chapter 14

Testing Your Recall

1.	b	9.	a	16.	antigen-
2.	c	10.	d		presenting cells
3.	e	11.	pathogen	17.	palatine tonsils
4.	c	12.	lacteals	18.	thymus
5.	b	13.	regulatory T (T_R)	19.	neutrophils
6.	c		cells	20.	histamine
7.	b	14.	interferons		
8.	e	15.	plasma		

True or False (These items are false for the reasons given; all others are true.)

1. B cells are involved only in adaptive immunity.
4. Helper T cells are involved in both humoral and cell-mediated immunity.
6. Both B and T cells reside in lymph nodes.
7. HIV infects primarily helper T cells.
9. Plasma cells secrete about 2,000 antibody molecules per second.

Answers to Figure Legend Questions

14.2 Lymphatic capillaries have larger gaps between their endothelial cells than blood capillaries do.
14.4 Axillary lymph nodes are the first place where cells metastasizing from a breast tumor would lodge.
14.5 Without lymphatic valves, there could be no steady one-way flow of lymph toward the subclavian veins.
14.17 Helper T cells are important in nonspecific defense, humoral immunity, and cellular immunity, so HIV impairs all three of those process by destroying helper T cells.
14.20 It synthesizes antibody molecules.

Chapter 15

Testing Your Recall

1.	b	9.	d	15.	epiglottis
2.	b	10.	c	16.	carbonic acid
3.	e	11.	spirometry	17.	chemoreceptors
4.	b	12.	ventral respiratory	18.	mucociliary
5.	e		group		escalator
6.	d	13.	thyroid	19.	hilum
7.	d	14.	alveolar	20.	asthma
8.	a		macrophages		

True or False (These items are false for the reasons given; all others are true.)

1. Tidal volume averages about 500 mL.
3. Most CO_2 transport is in the form of carbonic acid–bicarbonate ions.
6. The most numerous cells in the lung are alveolar macrophages.
7. P_{O_2} normally has very little effect on breathing rate and depth.
10. The lungs fill as the result of chest expansion, not as the cause.

Answers to Figure Legend Questions

15.2 The line should be drawn across the upper end of the trachea.
15.5 The right main bronchus is wider and more vertical than the left, so it is easier for aspirated objects to fall into the right one.
15.8 It has no cartilage in the wall; if it were a bronchus, it would have supportive cartilage.

Chapter 16

Testing Your Recall

1. a	8. d	14. macula densa
2. d	9. c	15. podocytes
3. b	10. a	16. nephron
4. b	11. micturition	17. antidiuretic
5. d	12. glomerular	18. internal urethral
6. b	filtration rate	19. sodium, potassium
7. e	13. ureters	20. fibrous capsule

True or False (These items are false for the reasons given; all others are true.)

2. Urea is the most abundant urine solute.
4. Sodium is reabsorbed in the PCT, nephron loop, and DCT.
5. Natriuretic peptides promote sodium excretion and lower blood volume.
8. Urine can be as dilute as 50 mOsm/L.
9. The PNS relaxes the internal urethral sphincter to permit emptying of the bladder.

Answers to Figure Legend Questions

16.5 The afferent arteriole is larger than the efferent arteriole. Because it has a large inlet and small outlet, the glomerulus has an unusually high blood pressure (BP) compared to other capillaries. This high BP drives the glomerular filtration process.

16.12 The shorter urethra of women makes it easier for bacteria to invade the bladder.

16.13 The urine volume would decrease.

Chapter 17

Testing Your Recall

1. b	9. b	16. enteric
2. a	10. a	17. amylase, saliva
3. a	11. dentine	18. parotid,
4. b	12. duodenum,	submandibular,
5. d	jejunum, ileum	sublingual
6. a	13. falciform ligament	19. palate
7. e	14. chief	20. peristalsis
8. b	15. colon	

True or False (These items are false for the reasons given; all others are true.)

1. Chylomicrons are absorbed into lacteals; micelles remain in the small intestine.
3. The pyloric valve regulates the passage of chyme from stomach to duodenum.
4. Vitamin D promotes absorption of calcium, not iron.
6. Trypsinogen is produced in the pancreas and converted to trypsin after secretion.
10. The jejunum is a segment of the small intestine.

Answers to Figure Legend Questions

17.5 The two premolars and the third molar of each side of the jaw are not yet present at age 3.

17.7 The root of the tongue blocks the oral cavity and the soft palate rises to block the nasal cavity so food cannot be forced into either one from the pharynx.

17.8 The muscularis externa has three layers in the stomach, only two in the esophagus. The esophagus lacks the oblique layer. The addition of a layer in the stomach is related to the intense churning contractions that occur in this organ.

17.17 Sugars and amino acids are absorbed into the blood, which goes from the small intestine directly to the liver via the hepatic portal system. Dietary fat is absorbed by the intestinal lacteals and travels the route: lacteal → lymphatic trunk → thoracic duct → left subclavian vein → left brachiocephalic vein → superior vena cava → right side of heart → lungs → left side of heart → aorta → celiac trunk → common hepatic artery → liver.

17.18 The internal anal sphincter is controlled by the autonomic nervous system because it is composed of smooth muscle. The external anal sphincter is composed of skeletal muscle and therefore controlled by the somatic nervous system.

Chapter 18

Testing Your Recall

1. a	8. a	15. liver
2. c	9. d	16. insulin
3. b	10. d	17. core temperature
4. e	11. incomplete	18. hypothalamus
5. b	12. glycogenolysis	19. lactic acid
6. d	13. gluconeogenesis	20. kilocalorie, kcal
7. e	14. urea	

True or False (These items are false for the reasons given; all others are true.)

1. Leptin inhibits the appetite.
4. Most of the body's cholesterol is made by the body itself.
5. Most liver functions are nondigestive; only the secretion of bile acids and lecithin aids digestion.
8. Gluconeogenesis occurs in the fasting state.
10. At a comfortable air temperature, the body loses most of its heat by radiation.

Answers to Figure Legend Questions

18.1 HDLs remove excess cholesterol from the tissues whereas LDLs deliver it to the tissues to be deposited. A high HDL level therefore indicates a high rate of clearance of excess cholesterol from the body.

18.3 FADH and NADH$_2$

Chapter 19

Testing Your Recall

1. b	9. a	16. oxytocin
2. d	10. d	17. seminiferous
3. a	11. zygote	tubules
4. c	12. first polar body	18. colostrum
5. c	13. antrum	19. luteinizing
6. a	14. luteal	hormone
7. d	(postovulatory)	20. lacunae
8. e	15. seminal vesicle	

True or False (These items are false for the reasons given; all others are true.)

1. Interstitial cells secrete testosterone.
4. The clitoris has no corpus spongiosum.
6. The vagina has no mucous glands.
7. Lactiferous ducts exist even in the nonlactating breast.
10. Climacteric is a midlife decline in male reproductive function associated with andropause, and decline and cessation of reproductive function in women associated with menopause.

Answers to Figure Legend Questions

19.4 Engorgement of the corpus spongiosum would compress the urethra, block the passage of semen, and make ejaculation difficult or impossible.

19.11 Crossing-over in prophase I

19.14 One egg per primary oocyte; four sperm per primary spermatocyte. It is important to produce a large egg with adequate nutrients to support early development and cytoplasm to be divided among multiple blastomeres, so the oocyte should not be divided into small cells.

Chapter 20

Testing Your Recall

1. a
2. d
3. b
4. c
5. d
6. e
7. a
8. b
9. a
10. e
11. primitive groove
12. gastrulation
13. mesoderm
14. life span
15. uterine
16. decidual
17. chorionic villi
18. amniotic
19. somites
20. teratogen

True or False (These items are false for the reasons given; all others are true.)

1. These stages do not correspond to trimesters; the preembryo extends only to day 16.

3. Disjunction is the normal separation of chromosomes in meiosis.
5. Trophoblastic and placental nutrition overlap.
8. Despite significant improvements in life expectancy, there is little expectation of increasing life span.
9. The sperm's ability to fertilize an egg increases in the hours after ejaculation because of capacitation.

Answers to Figure Legend Questions

20.2 An unfertilized oocyte would already be dead by the time it arrived in the uterus; it does not live long enough to survive the 3-day migration.

20.4 Answers may vary but could include bones, muscles, and blood vessels (any mesodermal derivative).

20.9 The chromosomes shown came from a female, as we can see from the presence of two X chromosomes.

Appendix B
HEALTH SCIENCE CAREERS

Further information is available from the following organizations and websites for each of the health-science careers featured in the Career Spotlight essays. These URLs were last confirmed as of press time for this book, and will be reconfirmed and updated as needed with each 3-year revision. However, if any of them should be found inactive, relevant information may be found by entering the organizational names and a key word such as *careers* or *jobs* in any search engine, or by going to the organization's home page and looking for a career or job link. On this book's website, www.mhhe.com/saladinessentials, you can click hyperlinks for all of the following that will take you directly to these career websites. Education and licensing requirements vary from state to state, so you may also wish to search by career title and your own state for specific information on what may be required of you to enter one of these careers.

This chapter-by-chapter list is followed by some general sources of health-career information and a list of many more career ideas than the ones addressed in the chapter Career Spotlight essays.

Chapter 1—Radiologic Technologist
American Society of Radiologic Technologists
www.asrt.org/main/careers/careers-in-radiologic-technology
U.S. Bureau of Labor Statistics, Occupational Outlook Handbook
www.bls.gov/ooh/Healthcare/Radiologic-technologists.htm

Chapter 2—Medical Technologist
American Medical Technologists
www.americanmedtech.org/Certification/MedicalTechnologist.aspx
American Society for Clinical Laboratory Science
http://ascls.site-ym.com/?page=Stud_Cent

Chapter 3—Cytotechnologist
Mississippi Hospital Association, Health Careers Center
www.mshealthcareers.com/careers/cytotechnologist.htm
ExploreHealthCareers.org
http://explorehealthcareers.org/en/Career/27/Cytotechnologist

Chapter 4—Histotechnician
National Society for Histotechnology
www.nsh.org/what-histotechnology
American Society for Clinical Pathology
www.ascp.org/functional-nav/career-center

Chapter 5—Dermatology Nurse
Dermatology Nurses' Association
www.dnanurse.org/education/certification
Johnson & Johnson Services, The Campaign for Nursing's Future
www.discovernursing.com/specialty/dermatology-nurse#.UCG-281ttZo

Chapter 6—Orthopedic Nurse
NurseConnect
www.nurseconnect.com/Careers/SpecialtyProfile.aspx?Id=173514
Johnson & Johnson Services, The Campaign for Nursing's Future
www.discovernursing.com/specialty/orthopaedic-nurse#.UCG_kM1ttZo

Chapter 7—Massage Therapist
American Massage Therapy Association
www.amtamassage.org/professional_development/starting.html
Massage Career Guides and Massage Professionals
www.massage-career-guides.com/

Chapter 8—Occupational Therapist
American Occupational Therapy Association
www.aota.org/Students/Prospective.aspx
Mississippi Hospital Association, Health Careers Center
www.mshealthcareers.com/careers/occupational.htm

Chapter 9—Electroneurodiagnostic Technologist
ExploreHealthCareers.org
http://explorehealthcareers.org/en/Career/137/Electroneurodiagnostic_Technologist/
Mississippi Hospital Association, Health Careers Center
www.mshealthcareers.com/careers/electroneurodiagnostictechnologist.htm

Chapter 10—Optician
Mississippi Hospital Association, Health Careers Center
www.mshealthcareers.com/careers/optician.htm
ExploreHealthCareers.org
http://explorehealthcareers.org/en/Career/13/Optician_Dispensing

Chapter 11—Diabetes Educator
American Association of Diabetes Educators
www.diabeteseducator.org/ProfessionalResources/cpcp/

Johnson & Johnson Services, The Campaign for Nursing's Future

> www.discovernursing.com/specialty/diabetes-nurse#.
> UCHAAc1ttZo

Chapter 12—Phlebotomist

Phlebotomist.net

> www.phlebotomist.net

American Medical Technologists

> www.americanmedtech.org/Certification/Phlebotomist.aspx

Mississippi Hospital Association, Health Careers Center

> www.mshealthcareers.com/careers/phlebotomist.htm

Chapter 13—Electrocardiographic Technician

Mississippi Hospital Association, Health Careers Center

> www.mshealthcareers.com/careers/
> electrocardiographtechnician.htm

Radiology-Schools.com

> www.radiology-schools.com/how-to-become-ekg-
> technician.html

Chapter 14—Public Health Nurse

ExploreHealthCareers.org

> http://explorehealthcareers.org/en/Career/149/
> Public_Health_Nurse

Johnson & Johnson Services, The Campaign for Nursing's Future

> www.discovernursing.com/specialty/public-health-nurse#.
> UCHASM1ttZo

Chapter 15—Respiratory Therapist

U.S. Bureau of Labor Statistics, Occupational Outlook Handbook

> www.bls.gov/ooh/healthcare/respiratory-therapists.htm

American Association for Respiratory Care

> www.aarc.org/career/be_an_rt

Mississippi Hospital Association, Health Careers Center

> www.mshealthcareers.com/careers/respiratory.htm

Chapter 16—Dialysis Technician

Dialysis Technicians

> www.dialysistechnicians.org/

National Association of Nephrology Technicians

> www.dialysistech.net/

Dialysis Technician Resources

> www.renalweb.com/topics/technical/technical.htm

Chapter 17—Dental Hygienist

American Dental Hygienist Association

> http://careers.adha.org/

American Dental Association

> www.ada.org/357.aspx

Chapter 18—Dietitian

Academy of Nutrition and Dietetics

> www.eatright.org/BecomeanRDorDTR

U.S. Bureau of Labor Statistics, Occupational Outlook Handbook

> www.bls.gov/ooh/Healthcare/Dietitians-
> and-nutritionists.htm

The British Dietetic Association

> www.bda.uk.com/

Chapter 19—Midwife

American College of Nurse Midwives

> www.midwife.org/Become-a-Midwife

ExploreHealthCareers.org

> http://explorehealthcareers.org/en/Career/71/Nurse_Midwife

Chapter 20—Genetic Counselor

The American Board of Genetic Counseling (ABGC)

> www.abgc.net/ABGC/AmericanBoardofGenetic
> Counselors.asp

ExploreHealthCareers.org

> http://explorehealthcareers.org/en/Career/53/Genetic_
> Counselor

Additional Careers

Numerous health-science and health-care professions are available, going far beyond the 20 that we have described in connection with individual chapters. Here we list some sources of general information that cover many or most of these professions, and a list of professions you may wish to explore. The requirements for entering these professions range from programs that extend only 1 to 2 years beyond high school, to associate degrees, bachelor's degrees, and graduate degrees. We have not listed doctoral-level careers (physician, dentist, etc.) or the many specialties within registered nursing, which generally require a B.S.N. degree followed by experience and specialized training. An Internet search of any of these career names plus a search term such as *careers, jobs,* or *programs* should yield useful information on training and entry requirements. In some cases, we combine two similar careers on one line with designations such as technologist/technician and assistant/aide. Your career research will show the differences between the two.

General Sources

American Medical Association, Health Care Careers Directory

> www.ama-assn.org/ama/pub/education-careers/
> careers-health-care/directory.page?

ExploreHealthCareers.org

> http://explorehealthcareers.org

Mississippi Hospital Association, Health Careers Center

> www.mshealthcareers.com

U.S. Bureau of Labor Statistics, Occupational Outlook Handbook, Healthcare Occupations

> www.bls.gov/ooh/Healthcare

Wikipedia has synopses of most careers in this appendix and has links to other websites on each.

> http://en.wikipedia.org

Careers

Acupuncturist
Allied dental educator
Anesthesiologist assistant
Athletic trainer
Biomedical equipment technician
Blood bank technology specialist
Cardiopulmonary rehabilitation specialist
Cardiovascular technologist/technician
Clinical assistant
Clinical laboratory technologist/technician
Community health worker
Cytogenetic technologist
Cytotechnologist
Dental assistant
Dental biller
Dental informatics specialist
Dental laboratory technician
Diagnostic medical sonographer
Diagnostic molecular scientist
Dietetic technician
Dosimetrist
Electrocardiograph technician
Emergency medical dispatcher
Emergency medical technician/paramedic
Environmental health specialist
Health-care administrator
Health-care documentation specialist
Health-care interpreter
Health counselor
Health educator
Health information technician
Health unit coordinator
Home care assistant/aide
Hospital biller
Kinesiotherapist
Licensed practical nurse
Licensed vocational nurse
Lymphedema therapist
Long-term care specialist
Mammographer
Medical animation specialist

Medical assistant
Medical biller
Medical coder
Medical dosimetrist
Medical illustrator
Medical insurance specialist
Medical laboratory technician
Medical librarian
Medical office administrator
Medical photographer
Medical transcriptionist
Music therapist
Nuclear medicine technologist
Nurses aide/assistant
Nursing informatics specialist
Nutritionist
Occupational health and safety specialist
Occupational therapy assistant/aide
Ophthalmic medical technician
Optical laboratory technician
Optometric technician
Orientation and mobility specialist
Orthopedic technician
Orthoptist
Orthotist/prosthetist
Pathologists' assistant
Perfusionist
Pharmacy technician
Physical therapist assistant
Physician assistant
Psychiatric technician/aide
Radiation therapist
Radiographer
Radiologic technologist
Recreational therapist
Rehabilitation counselor
Sleep technologist
Sonographer
Speech–language therapist
Surgical technologist
Vascular technologist
X-ray technologist

Appendix C
SYMBOLS, WEIGHTS, AND MEASURES

UNITS OF LENGTH

m	meter
cm	centimeter (10^{-2} m)
mm	millimeter (10^{-3} m)
μm	micrometer (10^{-6} m)
nm	nanometer (10^{-9} m)

UNITS OF MASS

g	gram
kg	kilogram (10^3 g)
mg	milligram (10^{-3} g)

UNITS OF VOLUME

L	liter
dL	deciliter (10^{-1} L) (= 100 mL)
mL	milliliter (10^{-3} L)
μL	microliter (10^{-6} L) (= 1 mm³)

GREEK LETTERS

α	alpha (as in α helix)
β	beta (as in β hemoglobin)
γ	gamma (as in γ globulin)
Δ	delta (uppercase)
δ	delta (lowercase) (as in a small positive charge, δ+)
μ	mu (as in micrometer, μm)

CONVERSION FACTORS

1 cm = 0.394 in	1 in = 2.54 cm
1 mL = 0.034 fl oz	1 fl oz = 29.6 mL
1 L = 1.057 qt	1 qt = 0.946 L
1 g = 0.0035 oz	1 oz = 28.38 g
1 kg = 2.2 lb	1 lb = 0.45 kg
°C = (5/9)(°F−32)	°F = (9/5)(°C) + 32

UNITS OF CONCENTRATION

Chemical concentrations—the amounts of solute in a given volume of solution—are expressed in different ways for different scientific or clinical purposes. Some of these are explained here, particularly those used in this book.

Weight per Volume

A simple way to express concentration is the weight of solute in a given volume of solution. For example, intravenous (I.V.) saline typically contains 8.5 grams of NaCl per liter of solution (8.5 g/L). For many biological purposes, however, we deal with smaller quantities such as milligrams per deciliter (mg/dL). For example, a typical serum cholesterol concentration may be 200 mg/dL, also expressed as 200 mg/100 mL or 200 milligram-percent (mg-%).

Percentages

Percentage concentration is also simple to compute, but it is necessary to specify whether the percentage refers to the weight or to the volume of solute in a given volume of solution. For example, if we begin with 5 g of dextrose (an isomer of glucose) and add enough water to make 100 mL of solution, the resulting concentration will be 5% weight per volume (w/v). A commonly used intravenous fluid is D5W, which stands for 5% w/v dextrose in distilled water.

If the solute is a liquid, such as ethanol, percentages refer to volume of solute per volume of solution. Thus, 70 mL of ethanol diluted with water to 100 mL of solution produces 70% volume per volume (70% v/v) ethanol.

Percentage concentrations are easy to prepare, but that unit of measurement is inadequate for many purposes.

Molarity

The physiological effect of a chemical depends on how many molecules of it are present in a given volume, not the weight of the chemical. Five percent glucose, for example, contains almost twice as many sugar molecules as the same volume of 5% sucrose. Each solution contains 50 g of sugar per liter, but glucose has a molecular weight (MW) of 180 and sucrose has a MW of 342. Since each molecule of glucose is lighter, 50 g of glucose contains more molecules than 50 g of sucrose.

To produce solutions with a known number of molecules per volume, we must factor in the molecular weight. If we know the MW and weigh out that many grams of the substance, we have a quantity known as its **gram molecular weight,** or 1 **mole.** One mole of glucose is 180 g and 1 mole of sucrose is 342 g. Each quantity contains the same number of molecules of the respective sugar—a number known as Avogadro's number, 6.023×10^{23} molecules per mole.

Molarity (M) is the number of moles of solute per liter of solution. A one-molar (1.0 M) solution of glucose contains 180 g/L, and 1.0 M solution of sucrose contains 342 g/L. Both have the same number of solute molecules in a given volume. Body fluids and laboratory solutions usually are less concentrated than 1 M, so biologists and clinicians more often work with **millimolar (mM, 10^{-3} M)** and **micromolar (μM, 10^{-6} M)** concentrations.

Osmolarity and Osmolality

The osmotic concentration of body fluids has a great effect on cellular function. If it is greater than the osmotic concentration within cells, cells lose water and shrivel; if it is less, cells absorb water and swell. Either of these can cause potentially fatal cellular breakdown or dysfunction. Osmotic concentration differences also influence such things as intestinal nutrient absorption, the sense of thirst, urinary water loss, total body water volume, and blood pressure. Thus, it is important to quantify osmotic concentrations in physiology and in clinical practice (as when giving I.V. fluid therapy).

One **osmole** is 1 mole of dissolved particles. If a solute does not ionize in water, then 1 mole of the solute yields 1 osmole (osm) of dissolved particles. A solution of 1 M glucose, for example, is also 1 osm/L. If a solute does ionize, it yields two or more dissolved particles in solution. A 1 M solution of NaCl, for example, contains 1 mole/L of sodium ions and 1 mole/L of chloride ions. Both ions equally affect osmosis and must be separately counted in a measure of osmotic concentration. Thus, 1 M NaCl = 2 osm/L. Calcium chloride ($CaCl_2$) would yield three ions if it dissociated completely (one Ca^{2+} and two Cl^-), so 1 M $CaCl_2$ = 3 osm/L.

Osmolality is the number of osmoles of solute per kilogram of water, and **osmolarity** is the number of osmoles per liter of solution. Most clinical calculations are based on osmolarity, since it is easier to measure the volume of a solution than the weight of water it contains. The difference between osmolality and osmolarity can be important in experimental work, but at the concentrations of human body fluids, there is less than 1% difference between the two, and the two are essentially interchangeable for clinical purposes.

All body fluids and many clinical solutions are mixtures of many chemicals. The osmolarity of such a solution is the total osmotic concentration of all of its dissolved particles.

A concentration of 1 osm/L is substantially higher than we find in most body fluids, so physiological concentrations are usually expressed in terms of **milliosmoles per liter (mOsm/L)** (1 mOsm/L = 10^{-3} osm/L). Blood plasma, tissue fluid, and intracellular fluid measure about 300 mOsm/L.

Milliequivalents per Liter

Electrolyte concentrations are typically expressed in **milliequivalents per liter (mEq/L).** Electrolytes are important for their chemical, physical (osmotic), and electrical effects on the body. Their electrical effects, which determine such things as nerve, heart, and muscle actions, depend not only on their concentration but also on their electrical charge. A calcium ion (Ca^{2+}) has twice the electrical effect of a sodium ion (Na^+), for example, because it carries twice the charge.

In measuring electrolyte concentrations, one must take the charges into account.

One **equivalent (Eq)** of an electrolyte is the amount that would electrically neutralize 1 mole of hydrogen ions (H^+) or hydroxide ions (OH^-). For example, 1 mole (58.4 g) of NaCl yields 1 mole, or 1 Eq, of Na^+ in solution. Thus, an NaCl solution of 58.4 g/L contains 1 equivalent of Na^+ per liter (1 Eq/L). One mole (98 g) of sulfuric acid (H_2SO_4) yields 2 moles of positive charges (H^+). Thus, 98 g of sulfuric acid per liter would be a solution of 2 Eq/L.

The electrolytes in human body fluids have concentrations less than 1 Eq/L, so we more often express their concentrations in milliequivalents per liter (mEq/L). If you know the millimolar concentration of an electrolyte, you can easily convert this to mEq/L by multiplying it by the charge on the ion:

$$1 \text{ mM Na}^+ = 1 \text{ mEq/L}$$
$$1 \text{ mM Ca}^{2+} = 2 \text{ mEq/L}$$
$$1 \text{ mM Fe}^{3+} = 3 \text{ mEq/L}$$

Acidity and Alkalinity (pH)

Acidity is expressed in terms of **pH,** a measure derived from the molarity of H^+. Molarity is represented by square brackets, so the molarity of H^+ is symbolized $[H^+]$. pH is the negative logarithm of hydrogen ion molarity; that is, pH = $-\log [H^+]$.

In pure water, 1 in 10 million molecules of H_2O ionizes into hydrogen and hydroxide ions: $H_2O \rightleftharpoons H^+ + OH^-$. Pure water has a neutral pH because it contains equal amounts of H^+ and OH^-. Since 1 in 10 million molecules ionize, the molarity of H^+ and the pH of water are

$$[H^+] = 0.0000001 \text{ molar} = 10^{-7} \text{ M}$$
$$\log [H^+] = -7$$
$$pH = -\log [H^+] = 7$$

The pH scale ranges from 0 to 14 and is logarithmic, so each integer up or down the scale represents a tenfold difference in $[H^+]$. That is, for three strongly acidic pH values:

If $[H^+]$ = 0.1 M	pH = $-\log 10^{-1}$ = 1.0
If $[H^+]$ = 0.01 M	pH = $-\log 10^{-2}$ = 2.0
If $[H^+]$ = 0.001 M	pH = $-\log 10^{-3}$ = 3.0

The less concentrated the H^+, the higher the pH. pH values below 7.0 are considered acidic. Above 7.0, they are basic or alkaline. At the basic end of the scale:

If $[H^+]$ = 0.000000000001 M	pH = $-\log 10^{-12}$ = 12.0
If $[H^+]$ = 0.0000000000001 M	pH = $-\log 10^{-13}$ = 13.0
If $[H^+]$ = 0.00000000000001 M	pH = $-\log 10^{-14}$ = 14.0

Appendix D
BIOMEDICAL WORD ROOTS, PREFIXES, AND SUFFIXES

a- no, not, without (atom, agranulocyte)
ab- away (abducens, abduction)
acetabulo- small cup (acetabulum)
acro- tip, extremity, peak (acromion, acromegaly)
ad- to, toward, near (adsorption, adrenal)
adeno- gland (lymphadenitis, adenohypophysis)
aero- air, oxygen (aerobic)
af- toward (afferent)
ag- together (agglutination)
-al pertaining to (parietal, pharyngeal)
ala- wing (ala nasi)
albi- white (albicans, linea alba, albino)
algi- pain (analgesic, myalgia)
aliment- nourishment (alimentary)
allo- other, different (allele)
amphi- both, either (amphipathic, amphiarthrosis)
an- without (anaerobic, anemic)
ana- **1.** up, buildup (anabolic, anaphylaxis). **2.** apart (anaphase, anatomy). **3.** back (anastomosis)
andro- male (androgen, andropause)
angi- vessel (angiogram, hemangioma)
ante- before, in front (antebrachium)
antero- forward (anterior, anterograde)
anti- against (antidiuretic, antibody, antagonist)
apo- from, off, away, above (apocrine, aponeurosis)
arbor- tree (arborization, arbor vitae)
artic- **1.** joint (articulation). **2.** speech (articulate)
-ary pertaining to (axillary, coronary)
-ase enzyme (polymerase, amylase)
ast-, astro- star (aster, astrocyte)
-ata plural of **-a** (carcinomata, chiasmata)
-ate possessing, like (hamate, cruciate)
athero- fat (atheroma, atherosclerosis)
atrio- entryway (atrium, atrioventricular)
auri- ear (auricle, binaural)
auto- self (autorhythmic, autoimmune)
axi- axis, straight line (axial, axoneme, axon)
baro- pressure (baroreceptor, hyperbaric)
bene- good, well (benign, beneficial)
bi- two (bipedal, biceps)
bili- bile (biliary, bilirubin)
bio- life, living (biology, biopsy, microbial)
blasto- precursor, producer (fibroblast, blastomere)
brachi- arm (brachium, brachial)
brady- slow (bradycardia, bradypnea)
bucco- cheek (buccal, buccinator)

burso- purse (bursa, bursitis)
calc- calcium, stone (calcaneus, hypocalcemia)
callo- thick (callus, callosum)
calori- heat (calorie, calorigenic)
calvari- bald, skull (calvaria)
calyx cup, chalice (glycocalyx, renal calyx)
capito- head (capitis, capitulum)
capni- smoke, CO_2 (hypocapnia)
carcino- cancer (carcinogen, carcinoma)
cardi- heart (cardiology, pericardium)
carot- **1.** carrot (carotene). **2.** stupor (carotid)
carpo- wrist (carpus, metacarpal)
case- cheese (caseosa, casein)
cata- down, break down (catabolism)
cauda- tail (cauda equina, caudate lobe)
-cel little (pedicel)
celi- belly, abdomen (celiac)
centri- middle (centromere, centriole)
cephalo- head (cephalic, encephalitis)
cervi- neck, narrow part (cervix, cervical)
chiasm- cross, X (optic chiasm)
choano- funnel (choana)
chole- bile (cholecystokinin)
chondro- **1.** grain (mitochondria). **2.** cartilage, gristle (chondrocyte)
chromo- color (chromatin, cytochrome)
chrono- time (chronotropic, chronic)
cili- eyelash (cilium, supraciliary)
circ- about, around (circumduction)
cis- cut (incision, incisor)
cisterna reservoir (terminal cisterna)
clast- break down, destroy (osteoclast)
clavi- hammer, key (clavicle, supraclavicular)
-cle little (tubercle, corpuscle)
cleido- clavicle (sternocleidomastoid)
cnemo- leg (gastrocnemius)
co- together (coenzyme, cotransport)
collo- **1.** hill (colliculus). **2.** glue (colloid, collagen)
contra- opposite (contralateral)
corni- horn (cornified, corniculate)
corono- crown (coronary, corona)
corpo- body (corpus luteum)
corti- bark, rind (cortex, cortical)
costa- rib (intercostal, subcostal)
coxa- hip (os coxae, coxal)
crani- helmet (cranium, epicranius)

cribri- sieve, strainer (cribriform plate)
crino- separate, secrete (holocrine, endocrine)
crista- crest (crista ampullaris, crista galli)
crito- to separate (hematocrit)
cruci- cross (cruciate ligament)
-cul small (canaliculus, trabecula)
cune- wedge (cuneiform, cuneatus)
cutane-, cuti- skin (subcutaneous, cuticle)
cysto- bladder (cystitis, polycystic)
cyto- cell (cytology, monocyte)
de- down (defecate, dehydration)
demi- half (demifacet, demilune)
den-, denti- tooth (dentition, dens)
dendro- tree, branch (dendrite, oligodendrocyte)
dermo- skin (dermatology, hypodermic)
desmo- band, bond, ligament (desmosome, syndesmosis)
dia-across, through, separate (diaphragm, dialysis)
dis- 1. apart (dissect, dissociate). 2. opposite, absence (disinfect, disability)
diure- pass through, urinate (diuretic)
dorsi- back (dorsal, latissimus dorsi)
duc- to carry (duct, adduction, abducens)
dys- bad, abnormal, painful (dyspnea, dystrophy)
e- out (ejaculate, eversion)
-eal pertaining to (hypophyseal, pineal)
ec-, ecto- out, external (ectopic, ectoderm, splenectomy)
ef- out of (efferent, effusion)
-el, -elle small (fontanel, organelle)
electro- electricity (electrocardiogram, electrolyte)
em- in, within (embolism, embedded)
emesi-, emeti- vomiting (emetic, hyperemesis)
-emia blood condition (anemia, hypoxemia)
en- in, into (enzyme, parenchyma)
encephalo- brain (encephalitis)
enchymo- poured in (mesenchyme, parenchyma)
endo- within, into, internal (endocrine, endocytosis)
entero- gut, intestine (mesentery, myenteric)
epi- upon, above (epidermis, epiphysis, epididymis)
ergo- work, energy, action (allergy, adrenergic)
eryth-, erythro- red (erythema, erythrocyte)
esthesio- sensation, feeling (anesthesia, somesthetic)
eu- good, true, normal, easy (eupnea, aneuploidy)
exo- out (exocytosis, exocrine)
facili- easy (facilitated)
fasci- band, bundle (fascia, fascicle)
fenestr- window (fenestrated)
fer- to carry (efferent, uriniferous)
ferri- iron (ferritin, transferrin)
fibro- fiber (fibroblast, fibrosis)
fili- thread (myofilament)
flagello- whip (flagellum)
foli- leaf (folic acid, folia)
-form shape (cuneiform, fusiform)
fove- pit, depression (fovea)
funiculo- little rope, cord (funiculus)

fusi- 1. spindle (fusiform). 2. pour out (perfusion)
gamo- marriage, union (gamete, monogamy)
gastro- belly, stomach (digastric, gastrointestinal)
-gen, -genic, -genesis producing, giving rise to (pathogen, carcinogenic, glycogenesis)
germi- 1. sprout, bud (germ cell). 2. microbe (germicide)
gero- old age (geriatrics, gerontology)
gesto- 1. to bear, carry (ingest). 2. pregnancy (gestation, progesterone)
glia- glue (neuroglia, glioma)
globu- ball, sphere (globulin, hemoglobin)
glom- ball (glomerulus)
glosso- tongue (hypoglossal, glossopharyngeal)
glyco- sugar (glycogen, hypoglycemia)
gono- 1. angle, corner (trigone). 2. seed, sex cell, generation (gonad, oogonium, gonorrhea)
gradi- walk, step (retrograde, gradient)
-gram recording of (sonogram, electrocardiogram)
-graph recording instrument (sonograph, electrocardiograph)
-graphy recording process (sonography, radiography)
gravi- severe, heavy (gravid, myasthenia gravis)
gyro- turn, twist (gyrus)
hallu- great toe (hallux, hallucis)
hemi- half (hemidesmosome, hemisphere)
-hemia blood condition (polycythemia)
hemo- blood (hemoglobin, hematology)
hetero- different, other, various (heterozygous, heterograft)
histo- tissue, web (histology, histone)
holo- whole, entire (holistic, holocrine)
homeo- constant, unchanging, uniform (homeostasis, homeothermic)
homo- same, alike (homologous, homozygous)
hyalo- clear, glassy (hyaline, hyaluronic acid)
hydro- water (dehydration, hydrolysis, hydrophobic)
hyper- above, above normal, excessive (hyperkalemia, hypertonic)
hypo- below, below normal, deficient (hypodermis, hyponatremia)
-ia condition (anemia, hypocalcemia)
-ic pertaining to (isotonic, antigenic)
-icle, -icul small (ossicle, canaliculus, reticular)
ilia- flank, loin (ilium, iliac)
-illa, -illus little (bacillus)
-in protein (trypsin, fibrin, globulin)
infra- below (infraspinous, infrared)
insulo- island (insula, insulin)
inter- between (intercellular, intervertebral)
intra- within (intracellular, intraocular)
iono- ion (ionotropic, cationic)
ischi- to hold back (ischium, ischemia)
-ism process, state, condition (metabolism, rheumatism)
iso- same, equal (isometric, isotonic, isomer)
-issimus most, greatest (latissimus, longissimus)
-ite little (dendrite, somite)

-itis inflammation (dermatitis, gingivitis)
jug- to join (conjugated, jugular)
juxta- next to (juxtamedullary, juxtaglomerular)
kali- potassium (hypokalemia)
karyo- seed, nucleus (megakaryocyte, karyotype)
kerato- horn (keratin, keratinocyte)
kine- motion, action (kinetic, kinase, cytokinesis)
labi- lip (labia majora, levator labii)
lacera- torn, cut (laceration)
lacrimo- tear, cry (lacrimal gland, nasolacrimal)
lacto- milk (lactose, lactation, prolactin)
lamina- layer (lamina propria)
latero- side (bilateral, ipsilateral)
lati- broad (fascia lata, latissimus dorsi)
-lemma husk (sarcolemma, neurilemma)
lenti- lens (lentiform)
-let small (platelet)
leuko- white (leukocyte, leukemia)
levato- to raise (levator labii, elevation)
ligo- to bind (ligand, ligament)
line- line (linea alba, linea aspera)
litho- stone (otolith, lithotripsy)
-logy study of (histology, hematology)
lucid- light, clear (stratum lucidum, zona pellucida)
lun- moon, crescent (lunate, lunule, semilunar)
lute- yellow (macula lutea, corpus luteum)
lyso-, lyto- split apart, break down (lysosome, hydrolysis, electrolyte, hemolytic)
macro- large (macromolecule, macrophage)
macula- spot (macula lutea, macula densa)
mali- bad (malignant, malformed)
malle- hammer (malleus, malleolus)
mammo- breast (mammary, mammillary)
mano- hand (manus, manipulate)
manubri- handle (manubrium)
masto- breast (mastoid, mastectomy)
medi- middle (medial, mediastinum)
medullo- marrow, pith (medulla)
mega- large (megakaryocyte, hepatomegaly)
melano- black (melanin, melanocyte, melancholy)
meno- month (menstruation, menopause)
mento- chin (mental, mentalis)
mero- part, segment (isomer, centromere)
meso- in the middle (mesoderm, mesentery)
meta- beyond, next in a series (metaphase, metacarpal)
metabolo- change (metabolism, metabolite)
-meter measuring device (spirometer, sphygmomanometer)
metri- 1. length, measure (isometric, emmetropic). 2. uterus (endometrium)
micro- small (microscopic, microglia)
mito- thread, filament (mitochondria, mitosis)
mono- one (monocyte, monogamy)
morpho- form, shape, structure (morphology, amorphous)
muta- change (mutagen, mutation)

myelo- 1. spinal cord (poliomyelitis, myelin). 2. bone marrow (myeloid, myelocytic)
myo-, mysi- muscle (myosin, epimysium)
natri- sodium (hyponatremia, natriuretic)
neo- new (neonatal, gluconeogenesis)
nephro- kidney (nephron)
neuro- nerve (aponeurosis, neurosoma)
nucleo- nucleus, kernel (nucleolus, nucleic acid)
ob- 1. life (aerobic, microbe). 2. against, toward, before (obstetrics, obturator, obstruction)
oculo- eye (oculi, oculomotor)
odonto- tooth (odontoblast, periodontal)
-oid like, resembling (colloid, sigmoid)
-ole small (arteriole, bronchiole)
oligo- few, a little, scanty (oligopeptide, oliguria)
-oma tumor, mass (carcinoma, hematoma)
omo- shoulder (omohyoid, acromion)
onycho- nail, claw (hyponychium)
oo- egg (oogenesis, oocyte)
op- vision (optics, myopia, photopic)
-opsy viewing, to see (biopsy, rhodopsin)
or- mouth (oral, orbicularis oris)
orbi- circle (orbicularis, orbit)
organo- tool, instrument (organ, organelle)
ortho- straight (orthodontics, orthopedics)
-ose 1. full of (adipose). 2. sugar (sucrose, glucose)
-osis 1. process (osmosis, exocytosis). 2. condition, disease (cyanosis, thrombosis). 3. increase (leukocytosis)
osmo- push (osmosis)
osse-, oste- bone (osseous, osteoporosis)
oto- ear (otolith, otitis, parotid)
-ous 1. full of (nitrogenous, edematous). 2. pertaining to (mucous, nervous). 3. like, characterized by (squamous, filamentous)
ovo- egg (ovum, ovary, ovulation)
oxy- 1. oxygen (oxyhemoglobin, hypoxia). 2. sharp, quick (oxytocin)
palli- pale (pallor, globus pallidus)
palpebro- eyelid (palpebrae)
pan- all (pancreas, panhysterectomy)
papillo- nipple (papilla, papillary)
par- birth (postpartum, parturition)
para- next to (parathyroid, parotid)
parieto- wall (parietal)
patho- 1. disease (pathology, pathogen). 2. feeling (sympathetic)
pecto- 1. chest (pectoralis). 2. comblike (pectineus)
pedi- 1. foot (bipedal, pedicle). 2. child (pediatrics)
pelvi- basin (pelvis, pelvic)
-penia deficiency (leukopenia, osteopenia)
penna- feather (unipennate, bipennate)
peri- around (periosteum, peritoneum, periodontal)
phago- eat (phagocytosis, macrophage)
philo- loving, attracted (hydrophilic)
phobo- fearing, repelled (hydrophobic)

phragm- partition (diaphragm)
phreno- diaphragm (phrenic nerve)
physio- nature, natural cause (physiology, physician)
-physis growth (diaphysis, hypophysis)
pilo- hair (piloerector)
pino- drink, imbibe (pinocytosis)
planto- sole of foot (plantaris, plantar wart)
plasi- growth (hyperplasia)
plasm- shaped, molded (cytoplasm, endoplasmic)
plasti- form (thromboplastin)
platy- flat (platysma)
pnea- breath, breathing (eupnea, dyspnea)
pneumo- air, breath, lung (pneumonia, pneumothorax)
podo- foot (pseudopod, podocyte)
poies- forming (hemopoiesis, erythropoietin)
poly- many, much, excessive (polypeptide, polyuria)
primi- first (primary, primipara)
pro- 1. before, in front, first (prokaryote, prophase, prostate).
  2. promote, favor (progesterone, prolactin)
pseudo- false (pseudopod)
ptero-, pterygo- wing (pterygoid)
-ptosis dropping, falling, sagging (apoptosis, nephroptosis)
puncto- point (puncta)
pyro- fire (pyrogen, antipyretic)
quadri- four (quadriceps, quadratus)
quater- fourth (quaternary)
radiat- radiating (corona radiata)
rami- branch (ramus)
recto- straight (rectus abdominis, rectum)
reno- kidney (renal, renin)
reti- network (reticular, rete testis)
retinac- retainer, bracelet (retinaculum)
retro- behind, backward (retroperitoneal, retrovirus)
rhombo- rhombus (rhomboideus)
rubo-, rubro- red (bilirubin)
rugo- fold, wrinkle (ruga, corrugator)
sacculo- little sac (saccule)
sarco- flesh, muscle (sarcoplasm, sarcomere)
scala- staircase (scala tympani, scalene)
sclero- hard, tough (sclera, sclerosis)
scopo- see (microscope, endoscopy)
secto- cut (section, dissection)
semi- half (semilunar, semimembranosus)
sepsi- infection (asepsis, septicemia)
-sis process (diapedesis, amniocentesis)
sole- sandal, sole of foot (sole, soleus)
soma-, somato- body (somatic, somatotropin)
spheno- wedge (sphenoid)
spiro- breathing (inspiration, spirometry)
spleno- 1. bandage (splenius capitis). **2.** spleen (splenic artery)
squamo- scale, flat (squamous, desquamation)
stasi-, stati- put, remain, stay the same (hemostasis, homeostatic)

steno- narrow (stenosis)
ster-, stereo- solid, three-dimensional (steroid, stereoscopic)
sterno- breast, chest (sternum, sternocleidomastoid)
stria- stripe (striated, corpus striatum)
sub- below (subcutaneous, subclavicular)
sulc- furrow, groove (sulcus)
supra- above (supraspinous, supraclavicular)
sura- calf of leg (triceps surae)
sym- together (sympathetic, symphysis)
syn- together (synostosis, synovial)
tachy- fast (tachycardia, tachypnea)
tarsi- ankle (tarsus, metatarsal)
tecto- roof, cover (tectorial membrane)
telo- last, end (telophase, telencephalon)
tempo- time (temporal)
terti- third (tertiary)
theli- nipple, female, tender (epithelium, thelarche)
thermo- heat (thermogenesis, hypothermia)
thrombo- blood clot (thrombosis, thrombin)
thyro- shield (thyroid, thyrotropin)
-tion process (circulation, pronation)
toci- birth (oxytocin)
tomo- 1. cut (tomography, atom, anatomy). **2.** segment (dermatome, myotome, sclerotome)
tono- force, tension (isotonic, tonus, myotonia)
topo- place, position (isotope, ectopic)
trabo- plate (trabecula)
trans- across (transpiration, transdermal)
trapezi- 1. table, grinding surface (trapezium). **2.** trapezoid (trapezius)
tri- three (triceps, triglyceride)
tricho- hair (trichosiderin, hypertrichosis)
trocho- wheel, pulley (trochlea)
troph- 1. food, nourishment (trophic, trophoblast). **2.** growth (dystrophy, hypertrophy)
tropo- to turn, change (metabotropic, gonadotropin)
tunica coat (tunica intima, tunica vaginalis)
tympano- drum, eardrum (tympanic, tensor tympani)
-ul small (tubule, capitulum, glomerulus)
-uncle, -unculus small (homunculus, caruncle)
uni- one (unipennate, unipolar)
uri- urine (glycosuria, urinalysis, diuretic)
utriculo- little bag (utriculus)
vagino- sheath (tunica vaginalis)
vago- wander (vagus)
vaso- vessel (vascular, vas deferens)
ventro- belly, lower part (ventral, ventricle)
vermi- worm (vermis)
vertebro- spine (vertebrae, intervertebral)
vesico- bladder, blister (vesical, vesicular)
villo- hair, hairy (microvillus)
vitre- glass (in vitro, vitreous humor)
zygo- union, join, mate (zygomatic, zygote, azygos)

Terms defined here are those that are used most often in this book and not redefined each time they arise. The index indicates where you can find definitions or explanations of additional terms. Terms are defined only in the sense that they are used in this book; some have broader meanings, even within biology and medicine, that are beyond its scope. Commonly abbreviated terms such as *ATP* are defined under the full spelling.

A

abdominal cavity The body cavity between the diaphragm and pelvic brim. (fig. 1.12)

abduction (ab-DUC-shun) Movement of a body part away from the median plane, as in raising an arm away from the side of the body. (fig. 6.36)

absorption 1. Process in which a chemical passes through a membrane or tissue surface and becomes incorporated into a body fluid or tissue. 2. Any process in which one substance passes into another and becomes a part of it.

acetylcholine (ACh) (ASS-eh-till-CO-leen) A neurotransmitter released by somatic motor fibers, parasympathetic fibers, and some other neurons, composed of choline and an acetyl group.

acetylcholinesterase (AChE) (ASS-eh-till-CO-lin-ESS-ter-ase) An enzyme that breaks down acetylcholine, thus halting signal transmission at a synapse.

acid A proton (H⁺) donor; a chemical that releases protons into solution.

acidosis An acid–base imbalance in which the blood pH is lower than 7.35.

acinus (ASS-ih-nus) A sac of secretory cells at the inner end of a gland duct. (fig. 4.28)

actin A filamentous intracellular protein that provides cytoskeletal support and interacts with other proteins, especially myosin, to cause cellular movement; important in muscle contraction and membrane actions such as phagocytosis, ameboid movement, and cytokinesis.

action The movement produced by the contraction of a particular muscle.

action potential A rapid voltage change in which a plasma membrane briefly reverses electrical polarity; has a self-propagating

effect that produces a traveling wave of excitation in nerve and muscle cells.

active site The region of a protein that binds to another molecule, such as the substrate-binding site of an enzyme or the hormone-binding site of a receptor.

active transport Transport of particles through a selectively permeable membrane, up their concentration gradient, with the aid of a carrier that consumes ATP.

adduction (ah-DUC-shun) Movement of a body part toward the median plane, such as bringing the feet together from a spread-legged position. (fig. 6.36)

adenosine triphosphate (ATP) (ah-DEN-oh-seen tri-FOSS-fate) A molecule composed of adenine, ribose, and three phosphate groups that functions as a universal energy-transfer molecule; yields adenosine diphosphate (ADP) and an inorganic phosphate group (Pi) upon hydrolysis. (fig. 2.20)

adipocyte (AD-ih-po-site) A fat cell.

adipose tissue A connective tissue composed predominantly of adipocytes; fat.

aerobic respiration Oxidation of organic compounds in a reaction series that requires oxygen and produces ATP.

afferent (AFF-uh-rent) Carrying toward, as in *afferent neurons,* which carry signals toward the central nervous system, and *afferent arterioles,* which carry blood toward a tissue.

agglutination (ah-GLUE-tih-NAY-shun) Clumping of cells or molecules by antibodies. (fig. 14.22)

albumin (al-BYU-min) A class of small proteins constituting about 60% of the protein fraction of the blood plasma; plays roles in blood viscosity, osmosis, and solute transport.

aldosterone (AL-doe-steh-RONE, al-DOSS-teh-rone) A steroid hormone secreted by the adrenal cortex that acts on the kidneys to promote sodium retention and potassium excretion; indirectly promotes water retention and thereby opposes a drop in blood pressure.

alkalosis An acid–base imbalance in which the blood pH is higher than 7.45.

alveolus (AL-vee-OH-lus) 1. A microscopic air sac of the lung. 2. A tooth socket.

amino acids Small organic molecules with an amino group and a carboxyl group; the

monomers of which proteins are composed.

anabolism (ah-NAB-oh-lizm) Any metabolic reactions that consume energy and construct more complex molecules with higher free energy from less complex molecules with lower free energy; for example, the synthesis of proteins from amino acids. *Compare* catabolism.

anaerobic fermentation (AN-err-OH-bic) A reduction reaction independent of oxygen that converts pyruvic acid to lactic acid and enables glycolysis to continue under anaerobic conditions.

anatomical position A reference posture that allows for standardized anatomical terminology. A subject in anatomical position is standing with the feet flat on the floor, arms down to the sides, and the palms and eyes directed forward. (fig. 1.7)

androgen (AN-dro-jen) Testosterone or a related steroid hormone. Stimulates body changes at puberty in both sexes, adult libido in both sexes, development of male anatomy in the fetus and adolescent, and spermatogenesis.

angiotensin II (AN-jee-oh-TEN-sin) A hormone produced from angiotensinogen (a plasma protein) by the kidneys and lungs; raises blood pressure by stimulating vasoconstriction and stimulating the adrenal cortex to secrete aldosterone.

anion (AN-eye-on) An ion with more electrons than protons and consequently a net negative charge.

antebrachium (AN-teh-BRAY-kee-um) The region from elbow to wrist; the forearm.

anterior Pertaining to the front (facial–abdominal aspect) of the body; ventral.

anterior root The branch of a spinal nerve that emerges from the anterior side of the spinal cord and carries efferent (motor) nerve fibers; often called *ventral root.* (fig. 8.17)

antibody A protein of the gamma globulin class that reacts with an antigen and aids in protecting the body from its harmful effect; found in the blood plasma, in other body fluids, and on the surfaces of certain leukocytes and their derivatives.

antidiuretic hormone (ADH) (AN-tee-DYE-you-RET-ic) A hormone released by the posterior lobe of the pituitary gland in response to low blood pressure; promotes water retention by the kidneys. Also known as *vasopressin*.

antigen (AN-tih-jen) Any large molecule capable of binding to an antibody or immune cell and triggering an immune response.

antigen-presenting cell (APC) A cell that phagocytizes an antigen and displays fragments of it on its surface for recognition by other cells of the immune system; chiefly macrophages and B lymphocytes.

apical surface The uppermost surface of an epithelial cell, usually exposed to the lumen of an organ. (fig. 3.1)

apocrine Pertaining to certain sweat glands with large lumens and relatively thick, aromatic secretions, and to similar glands such as the mammary gland.

appendicular (AP-en-DIC-you-lur) Pertaining to the limbs and their supporting skeletal girdles. (fig. 6.10)

areolar tissue (AIR-ee-OH-lur) A fibrous connective tissue with loosely organized, widely spaced fibers and cells and an abundance of fluid-filled space; found under nearly every epithelium, among other places. (fig. 4.14)

arteriole (ar-TEER-ee-ole) A small artery that empties into a metarteriole or capillary.

arteriosclerosis (ar-TEER-ee-oh-sclare-OH-sis) Stiffening of the arteries correlated with age or disease processes, caused primarily by cumulative free radical damage and tissue deterioration. *Compare* atherosclerosis.

articular cartilage A thin layer of hyaline cartilage covering the articular surface of a bone at a synovial joint, serving to reduce friction and ease joint movement. (fig. 6.33)

articulation A skeletal joint; any point at which two bones meet; may or may not be movable.

atherosclerosis (ATH-ur-oh-skleh-ROE-sis) A degenerative disease of the blood vessels characterized by the presence of atheromas, often leading to calcification of the vessel wall and obstruction of coronary, cerebral, or other vital arteries. *Compare* arteriosclerosis.

atrophy (AT-ro-fee) Shrinkage of a tissue due to age, disuse, or disease.

autoantibody An antibody that fails to distinguish the body's own molecules from foreign molecules and thus attacks host tissues, causing autoimmune diseases.

autoimmune disease Any disease in which antibodies fail to distinguish between foreign and self-antigens and attack the body's own tissues; for example, rheumatoid arthritis and type 1 diabetes mellitus

autonomic nervous system (ANS) (AW-toe-NOM-ic) A motor division of the nervous system that innervates glands, smooth muscle, and cardiac muscle; consists of sympathetic and parasympathetic divisions and functions largely without voluntary control. *Compare* somatic nervous system.

axial (AC-see-ul) Pertaining to the head, neck, and trunk; the part of the body excluding the appendicular portion. (fig. 6.10)

axillary (ACK-sih-LERR-ee) Pertaining to the armpit.

axon A fibrous extension of a neuron that transmits action potentials; also called a *nerve fiber*. There is only one axon to a neuron, and it is usually much longer and less branched than the dendrites. (fig. 8.2)

B

baroreceptors (BAR-oh-re-SEP-turz) Pressure sensors located in the heart, aortic arch, and carotid sinuses that trigger autonomic reflexes in response to fluctuations in blood pressure.

base **1.** A chemical that binds protons from solution; a proton acceptor. **2.** Any of the purines or pyrimidines of a nucleic acid (adenine, thymine, guanine, cytosine, or uracil) serving in part to code for protein structure. **3.** The broadest part of a tapered organ such as the uterus or heart, or the inferior aspect of an organ such as the brain.

basophil (BAY-so-fill) A granulocyte with coarse cytoplasmic granules that produces heparin, histamine, and other chemicals involved in inflammation.

B lymphocyte A lymphocyte that functions as an antigen-presenting cell and, in humoral immunity, differentiates into an antibody-producing plasma cell; also called a *B cell.*

body **1.** Part of a cell, such as a neuron, containing the nucleus and most other organelles. **2.** The largest or principal part of an organ such as the stomach or uterus; also called the *corpus.*

brachial (BRAY-kee-ul) Pertaining to the arm proper, the region from shoulder to elbow.

brainstem The stalklike lower portion of the brain, composed of all of the brain except the cerebrum and cerebellum. (Many authorities also exclude the diencephalon and regard only the medulla oblongata, pons, and midbrain as the brainstem.) (fig. 9.5)

bronchiole (BRONK-ee-ole) A pulmonary air passage that is usually 1 mm or less in

diameter and lacks cartilage but has relatively abundant smooth muscle, elastic tissue, and a simple cuboidal, usually ciliated, epithelium.

bronchus (BRONK-us) A relatively large pulmonary air passage with supportive cartilage in the wall; any passage beginning with the main bronchus at the fork in the trachea and ending with segmental bronchi, from which air continues into the bronchioles.

brush border A fringe of microvilli on the apical surface of an epithelial cell, serving to enhance surface area and promote absorption. (fig. 3.4)

C

calorie The amount of thermal energy that will raise the temperature of 1 g of water by 1°C. Also called a *small calorie.*

Calorie *See* kilocalorie.

capillary (CAP-ih-LERR-ee) The narrowest type of vessel in the cardiovascular and lymphatic systems; engages in fluid exchanges with surrounding tissues.

capillary exchange The process of fluid transfer between the bloodstream and tissue fluid.

capsule The fibrous covering of a structure such as the spleen or a synovial joint.

carbohydrate A hydrophilic organic compound composed of carbon and a 2:1 ratio of hydrogen to oxygen; includes sugars, starch, glycogen, and cellulose.

carcinogen (car-SIN-oh-jen) An agent capable of causing cancer, including certain chemicals, viruses, and ionizing radiation.

cardiovascular system An organ system consisting of the heart and blood vessels, serving for the transport of blood. *Compare* circulatory system.

carpal Pertaining to the wrist (carpus).

carrier A protein in a cellular membrane that performs carrier-mediated transport.

carrier-mediated transport Any process of transporting materials through a cellular membrane that involves reversible binding to a transport protein; includes active transport and facilitated diffusion.

catabolism (ca-TAB-oh-lizm) Any metabolic reactions that release energy and break relatively complex molecules with high free energy into less complex molecules with lower free energy; for example, digestion and glycolysis. *Compare* anabolism.

cation (CAT-eye-on) An ion with more protons than electrons and consequently a net positive charge.

caudal (CAW-dul) **1.** Pertaining to a tail or narrow tail-like part of an organ. **2.** Relatively distant from the forehead, especially in reference to structures of the brain and spinal cord; for example, the medulla oblongata is caudal to the pons. *Compare* rostral.

central nervous system (CNS)   The brain and spinal cord.

centriole (SEN-tree-ole)   An organelle composed of a short cylinder of nine triplets of microtubules, usually paired with another centriole perpendicular to it; origin of the mitotic spindle; identical to the basal body of a cilium or flagellum. (fig. 3.10)

cephalic (seh-FAL-ic)   Pertaining to the head.

cerebellum (SERR-eh-BEL-um)   A large portion of the brain posterior to the brainstem and inferior to the cerebrum, responsible for equilibrium, motor coordination, certain perceptual abilities, and memory of learned motor skills. (fig. 9.6)

cerebrospinal fluid (CSF) (SERR-eh-bro-SPY-nul, seh-REE-bro-SPY-nul)   A liquid that fills the ventricles of the brain, the central canal of the spinal cord, and the space between the CNS and dura mater.

cerebrum (SERR-eh-brum, seh-REE-brum)   The largest and most superior part of the brain, divided into two convoluted cerebral hemispheres separated by a deep longitudinal fissure.

cervical (SUR-vih-cul)   Pertaining to the neck or any cervix.

cervix (SUR-vix)   **1.** The neck. **2.** A narrow or necklike part of an organ such as the uterus and gallbladder. (fig. 19.6)

channel protein   A protein in the plasma membrane that has a pore through it for the passage of materials between the cytoplasm and extracellular fluid. (fig. 3.3)

chemoreceptor   An organ or cell specialized to detect chemicals, as in the carotid bodies and taste buds.

cholecystokinin (CCK) (CO-lee-SIS-toe-KY-nin)   A polypeptide employed as a hormone and neurotransmitter, secreted by some brain neurons and cells of the digestive tract.

cholesterol (co-LESS-tur-ol)   A steroid that functions as part of the plasma membrane and as a precursor for all other steroids in the body.

chondrocyte (CON-dro-site)   A cartilage cell. (fig. 4.19)

chronic   **1.** Long-lasting. **2.** Pertaining to a disease that progresses slowly and has a long duration. *Compare* acute.

chronic bronchitis   A chronic obstructive pulmonary disease characterized by damaged and immobilized respiratory cilia, excessive mucus secretion, infection of the lower respiratory tract, and bronchial

inflammation; caused especially by cigarette smoking. *See also* chronic obstructive pulmonary disease.

chronic obstructive pulmonary disease (COPD)   Certain lung diseases (chronic bronchitis and emphysema) that result in long-term obstruction of airflow and substantially reduced pulmonary ventilation; one of the leading causes of death in old age.

cilium (SIL-ee-um)   A hairlike process, with an axoneme, projecting from the apical surface of an epithelial cell; often motile and serving to propel matter across the surface of an epithelium, but sometimes nonmotile and serving a sensory role. (fig. 3.5)

circulatory shock   A state of cardiac output inadequate to meet the metabolic needs of the body.

circulatory system   An organ system consisting of the heart, blood vessels, and blood. *Compare* cardiovascular system.

cisterna (sis-TUR-nuh)   A fluid-filled space or sac, such as the cisterna chyli of the lymphatic system and a cisterna of the endoplasmic reticulum or Golgi complex. (fig. 3.10)

clone   A population of cells that are mitotically descended from the same parent cell and are identical to each other genetically or in other respects.

coagulation (co-AG-you-LAY-shun)   The clotting of blood, lymph, tissue fluid, or semen.

collagen (COLL-uh-jen)   The most abundant protein in the body, forming the fibers of many connective tissues in places such as the dermis, tendons, and bones.

columnar   A cellular shape that is significantly taller than wide. (fig. 4.2)

commissure (COM-ih-shur)   A bundle of nerve fibers that crosses from one side of the brain or spinal cord to the other.

complement   **1.** To complete or enhance the structure or function of something else, as in the coordinated action of two hormones. **2.** A system of plasma proteins involved in defense against pathogens.

concentration gradient   A difference in chemical concentration from one point to another, as on two sides of a plasma membrane.

conception   The fertilization of an egg, producing a zygote.

conceptus   All products of conception, ranging from a fertilized egg to the full-term fetus with its embryonic membranes, placenta, and umbilical cord. *Compare* embryo, fetus, preembryo.

condyle (CON-dile)   An articular surface on a bone, usually in the form of a knob (as on

the mandible), serving to smooth the motion of a joint. (fig. 6.9)

conformation   The three-dimensional structure of a protein that results from interaction among its amino acid side groups, its interactions with water, and the formation of disulfide bonds.

congenital   Present at birth; for example, an anatomical defect, a syphilis infection, or a hereditary disease.

connective tissue   A tissue usually composed of more extracellular than cellular volume and usually with a substantial amount of extracellular fiber; forms supportive frameworks and capsules for organs, binds structures together, holds them in place, stores energy (as in adipose tissue), or transports materials (as in blood).

cooperative effect   Effect in which two hormones, or both divisions of the autonomic nervous system, work together to produce a single overall result.

coronal plane   *See* frontal plane.

coronary circulation   A system of blood vessels that serve the wall of the heart. (fig. 13.5)

corpus   Body or mass; the main part of an organ, as opposed to such regions as a head, tail, or cervix.

cortex (plural, *cortices*)   The outer layer of some organs such as the adrenal gland, cerebrum, lymph node, and ovary; usually covers or encloses tissue called the medulla.

costal (COSS-tul)   Pertaining to the ribs.

cranial (CRAY-nee-ul)   Pertaining to the cranium of the skull.

cranial nerve   Any of 12 pairs of nerves connected to the base of the brain and passing through foramina of the cranium.

cranium   The complex of 8 bones that enclose the brain; together with 14 facial bones, these form the skull.

crista   An anatomical crest or ridge, such as the crista galli of the ethmoid bone or the crista of a mitochondrion.

cross section   A cut perpendicular to the long axis of the body or an organ.

cuboidal (cue-BOY-dul)   A cellular shape that is roughly like a cube or in which the height and width are about equal; typically looks squarish in tissue sections. (fig. 4.2)

current   A moving stream of charged particles such as ions or electrons.

cutaneous (cue-TAY-nee-us)   Pertaining to the skin.

cyanosis (SY-uh-NO-sis)   A bluish color of the skin and mucous membranes due to ischemia or hypoxemia.

cyclic adenosine monophosphate (cAMP)   A cyclic molecule produced from ATP by the removal of two phosphate groups; serves as a second messenger in many hormone and neurotransmitter actions.

cytoplasm The contents of a cell between its plasma membrane and its nuclear envelope, consisting of cytosol, organelles, inclusions, and the cytoskeleton.

cytoskeleton A system of protein microfilaments, intermediate filaments, and microtubules in a cell, serving in physical support, cellular movement, and the routing of molecules and organelles to their destinations within the cell. (fig. 3.9)

cytosol A clear, featureless, gelatinous colloid in which the organelles and other internal structures of a cell are embedded.

cytotoxic T cell A T lymphocyte that directly attacks and destroys infected body cells, cancerous cells, and the cells of transplanted tissues.

D

daughter cells Cells that arise from a parent cell by mitosis or meiosis.

decomposition reaction A chemical reaction in which a larger molecule is broken down into smaller ones. *Compare* synthesis reaction.

decussation (DEE-cuh-SAY-shun) The crossing of nerve fibers from the right side of the central nervous system to the left or vice versa, especially in the spinal cord, medulla oblongata, and optic chiasm.

deep Relatively far from the body surface; opposite of *superficial*. For example, most bones are deep to the skeletal muscles.

denaturation A change in the three-dimensional conformation of a protein that destroys its enzymatic or other functional properties, usually caused by extremes of temperature or pH.

dendrite Extension of a neuron that receives information from other cells or from environmental stimuli and conducts signals to the soma. Dendrites are usually shorter, more branched, and more numerous than the axon and are incapable of producing action potentials. (fig. 8.2)

dendritic cell An antigen-presenting cell of the epidermis and mucous membranes. (fig. 5.2)

dense connective tissue A connective tissue with a high density of fiber, relatively little ground substance, and scanty cells; seen in tendons and the dermis, for example.

depolarization A shift in the electrical potential across a plasma membrane toward a value less negative than the resting membrane potential, associated with excitation of a nerve or muscle cell.

dermis The deeper of the two layers of the skin, underlying the epidermis and composed of fibrous connective tissue.

desmosome (DEZ-mo-some) A patchlike intercellular junction that mechanically links two cells together. (fig. 3.6)

diabetes mellitus (DM) (mel-EYE-tus) A form of diabetes that results from hyposecretion of insulin or from a deficient target cell response to it; signs include hyperglycemia and glycosuria.

diaphysis (dy-AFF-ih-sis) The shaft of a long bone. (fig. 6.4)

diarthrosis (DY-ar-THRO-sis) *See* synovial joint.

diastole (dy-ASS-tuh-lee) A period in which a heart chamber relaxes and fills with blood; especially ventricular relaxation.

differentiation Development of a relatively unspecialized cell or tissue into one with a more specific structure and function.

diffusion Spontaneous net movement of particles from a place of high concentration to a place of low concentration (down a concentration gradient).

diploid (2n) In humans, having 46 chromosomes in 23 homologous pairs; in any organism or cell, having paired chromosomes of maternal and paternal origin.

disaccharide (dy-SAC-uh-ride) A carbohydrate composed of two simple sugars (monosaccharides) covalently bonded together; for example, lactose, sucrose, and maltose. (fig. 2.10b)

distal Relatively distant from a point of origin or attachment; for example, the wrist is distal to the elbow. *Compare* proximal.

disulfide bond A covalent bond that links two cysteine residues through their sulfur atoms (—S—S—), serving to link one peptide chain to another or to hold a single chain in its three-dimensional conformation.

diuretic (DY-you-RET-ic) A chemical that increases urine output.

dorsal Toward the back (spinal) side of the body; in humans, usually synonymous with *posterior*.

dorsal root *See* posterior root.

dorsiflexion (DOR-sih-FLEC-shun) A movement of the ankle that reduces the joint angle and raises the toes. (fig. 6.41)

duodenum (DEW-oh-DEE-num, dew-ODD-eh-num) The first portion of the small intestine extending for about 25 cm from the pyloric valve of the stomach to a sharp bend called the duodenojejunal flexure; receives chyme from the stomach and secretions from the liver and pancreas. (fig. 17.13)

E

edema (eh-DEE-muh) Abnormal accumulation of tissue fluid resulting in swelling of the tissue.

effector A molecule, cell, or organ that carries out a response to a stimulus.

efferent (EFF-ur-ent) Carrying away or out, such as a blood vessel that carries blood away from a tissue or a nerve fiber that conducts signals away from the central nervous system.

elastic fiber A connective tissue fiber, composed of the protein elastin, that stretches under tension and returns to its original length when released; responsible for the resilience of organs such as the skin and lungs.

elasticity The tendency of a stretched structure to return to its original dimensions when tension is released.

electrolyte A salt that ionizes in water and produces a solution that conducts electricity; loosely speaking, any ion that results from the dissociation of such salts, such as sodium, potassium, calcium, chloride, and bicarbonate ions.

electron micrograph A photograph made with a scanning or transmission electron microscope rather than with a light microscope.

elevation A joint movement that raises a body part, as in hunching the shoulders or closing the mouth.

embolism (EM-bo-lizm) The obstruction of a blood vessel by an embolus.

embolus (EM-bo-lus) Any abnormal traveling object in the bloodstream, such as agglutinated bacteria or blood cells, a blood clot, or an air bubble.

embryo A developing individual from the sixteenth day of gestation when the three primary germ layers have formed, through the end of the eighth week when all of the organ systems are present. *Compare* conceptus, fetus, preembryo.

endocrine gland (EN-doe-crin) A ductless gland that secretes hormones into the bloodstream; for example, the thyroid and adrenal glands. *Compare* exocrine gland.

endocytosis (EN-doe-sy-TOE-sis) Any process in which a cell forms vesicles from its plasma membrane and takes in large particles, molecules, or droplets of fluid; for example, phagocytosis and pinocytosis.

endometrium (EN-doe-MEE-tree-um) The mucosa of the uterus; the site of implantation and source of menstrual discharge.

endoplasmic reticulum (ER) (EN-doe-PLAZ-mic reh-TIC-you-lum) An extensive system of interconnected cytoplasmic tubules or channels; classified as rough ER or smooth ER depending on the presence or absence of ribosomes on its membrane. (fig. 3.10)

endothelium (EN-doe-THEEL-ee-um) A simple squamous epithelium that lines the lumens of the blood vessels, heart, and lymphatic vessels.

enteric (en-TERR-ic) Pertaining to the small intestine, as in enteric hormones.

eosinophil (EE-oh-SIN-oh-fill) A granulocyte with a large, often bilobed nucleus and coarse cytoplasmic granules that stain with eosin; phagocytizes antigen–antibody complexes, allergens, and inflammatory chemicals and secretes enzymes that combat parasitic infections.

epidermis A stratified squamous epithelium that constitutes the superficial layer of the skin, overlying the dermis. (fig. 5.1)

epinephrine (EP-ih-NEFF-rin) A catecholamine that functions as a neurotransmitter in the sympathetic nervous system and as a hormone secreted by the adrenal medulla; also called *adrenaline.*

epiphysis (eh-PIF-ih-sis) **1.** The head of a long bone. (fig. 6.4) **2.** The pineal gland (epiphysis cerebri).

epithelium A type of tissue consisting of one or more layers of closely adhering cells with little intercellular material and no blood vessels; forms the coverings and linings of many organs and the secretory tissue and ducts of the glands.

erectile tissue A tissue that functions by swelling with blood, as in the penis and clitoris.

erythrocyte (eh-RITH-ro-site) A red blood cell.

erythropoiesis (eh-RITH-ro-poy-EE-sis) The production of erythrocytes.

erythropoietin (eh-RITH-ro-POY-eh-tin) A hormone that is secreted by the kidneys and liver in response to hypoxemia and stimulates erythropoiesis.

estrogen (ESS-tro-jen) Collective name of three similar steroid sex hormones known especially for producing female secondary sex characteristics and regulating various aspects of the menstrual cycle and pregnancy.

excitability The ability of a cell to respond to stimuli, especially the ability of nerve and muscle cells to produce membrane voltage changes in response to stimuli; irritability.

excitation–contraction coupling Events that link the synaptic stimulation of a muscle cell to the onset of contraction.

excretion The process of eliminating metabolic waste products from a cell or from the body. *Compare* secretion.

exocrine gland (EC-so-crin) A gland that secretes its products into another organ or onto the body surface, usually by way of a duct; for example, salivary and gastric glands. *Compare* endocrine gland.

exocytosis (EC-so-sy-TOE-sis) A process in which a vesicle in the cytoplasm of a cell fuses with the plasma membrane and releases its contents from the cell; used in the elimination of cellular wastes and in the release of gland products and neurotransmitters.

expiration **1.** Exhaling. **2.** Dying.

extension Movement of a joint that increases the angle between articulating bones (straightens the joint). (fig. 6.35) *Compare* flexion.

extracellular fluid (ECF) Any body fluid that is not contained in the cells; for example, blood, lymph, and tissue fluid.

extrinsic (ec-STRIN-sic) **1.** Originating externally, such as extrinsic blood-clotting factors; exogenous. **2.** Not fully contained within an organ but acting on it, such as the extrinsic muscles of the hand and eye. *Compare* intrinsic.

F

facilitated diffusion The process of transporting a chemical through a cellular membrane, down its concentration gradient, with the aid of a carrier that does not consume ATP; enables substances to diffuse through the membrane that would do so poorly, or not at all, without a carrier.

fascia (FASH-ee-uh) A layer of connective tissue between the muscles or separating the muscles from the skin. (fig. 7.14)

fascicle (FASS-ih-cul) A bundle of muscle or nerve fibers ensheathed in connective tissue; multiple fascicles bound together constitute a muscle or nerve as a whole. (figs. 7.14, 8.17)

fat **1.** A triglyceride molecule. **2.** Adipose tissue.

fatty acid An organic molecule composed of a chain of an even number of carbon atoms with a carboxyl group at one end and a methyl group at the other; one of the structural subunits of triglycerides and phospholipids.

fenestrated (FEN-eh-stray-ted) Perforated with holes or slits, as in fenestrated blood capillaries and the elastic sheets of large arteries.

fetus In human development, an individual from the beginning of the ninth week when all of the organ systems are present, through the time of birth. *Compare* conceptus, embryo, preembryo.

fibrin (FY-brin) A sticky fibrous protein formed from fibrinogen in blood, tissue fluid, and lymph; forms the matrix of a blood clot.

fibroblast A connective tissue cell that produces collagen fibers and ground substance; the only type of cell in tendons and ligaments.

fibrosis Replacement of damaged tissue with fibrous scar tissue rather than by the original tissue type; scarring. *Compare* regeneration.

fibrous connective tissue Any connective tissue with a preponderance of fiber, such as areolar, reticular, dense regular, and dense irregular connective tissues.

filtrate A fluid formed by filtration, as at the renal glomerulus and other capillaries.

filtration A process in which hydrostatic pressure forces a fluid through a selectively permeable membrane (especially a capillary wall).

fire To produce an action potential, as in nerve and muscle cells.

fix To hold a structure in place, for example, by fixator muscles that prevent unwanted joint movements.

flexion A joint movement that, in most cases, decreases the angle between two bones. (fig. 6.35) *Compare* extension.

fluid balance A state in which the average daily gain and loss of water are equal and water is properly distributed among the body's fluid compartments.

fluid compartment Any of the major categories of fluid in the body, separated by selectively permeable membranes and differing from each other in chemical composition. Primary examples are the intracellular fluid, tissue fluid, blood, and lymph.

follicle (FOLL-ih-cul) A small space, such as a hair follicle, thyroid follicle, or ovarian follicle.

follicle-stimulating hormone (FSH) A hormone secreted by the anterior pituitary gland that stimulates development of the ovarian follicles and egg cells.

foramen (fo-RAY-men) A hole through a bone or other organ, in many cases providing passage for blood vessels and nerves.

fossa (FOSS-uh) A depression in an organ or tissue, such as the fossa ovalis of the heart or a cranial fossa of the skull.

fovea (FOE-vee-uh) A small pit, such as the fovea capitis of the femur or fovea centralis of the retina.

free radical A particle derived from an atom or molecule, having an unpaired electron that makes it highly reactive and destructive to cells; produced by intrinsic processes such as aerobic respiration and by extrinsic agents such as chemicals and ionizing radiation.

frontal plane An anatomical plane that passes through the body or an organ from right to left and superior to inferior, such as a vertical plane that separates the anterior portion of the chest from the back; also called a *coronal plane*. (fig. 1.7)

fundus The base, the broadest part, or the part farthest from the opening of certain organs such as the stomach and uterus.

fusiform (FEW-zih-form) Spindle-shaped; elongated, thick in the middle, and tapered at both ends, such as the shape of a smooth muscle cell or a muscle spindle.

G

gamete (GAM-eet) An egg or sperm cell.

ganglion (GANG-glee-un) A cluster of nerve cell bodies in the peripheral nervous system, often resembling a knot in a string.

gangrene Tissue necrosis usually resulting from ischemia and often involving infection.

gap junction A junction between two cells consisting of a pore surrounded by a ring of proteins in the plasma membrane of each cell; allows solutes to diffuse from the cytoplasm of one cell to the next, thereby serving for cell-to-cell electrical and chemical communication in tissues such as cardiac and smooth muscle. (fig. 3.6)

gastric Pertaining to the stomach.

gate A protein channel in a cellular membrane that can open or close in response to chemical, electrical, or mechanical stimuli, thus controlling when substances are allowed to pass through the membrane.

gene An information-containing segment of DNA that codes for the production of a molecule of RNA, which in most cases goes on to play a role in the synthesis of one or more proteins.

germ cell An egg or sperm cell (gamete) or any precursor cell destined to become a gamete.

gestation (jess-TAY-shun) Pregnancy.

globulin (GLOB-you-lin) A globular protein such as an enzyme, antibody, or albumin; especially a family of proteins in the blood plasma that includes albumin, antibodies, fibrinogen, and prothrombin.

glucagon (GLUE-ca-gon) A hormone secreted by alpha cells of the pancreatic islets in response to hypoglycemia; promotes glycogenolysis and other effects that raise blood glucose concentration.

glucocorticoid (GLUE-co-COR-tih-coyd) Any hormone of the adrenal cortex that affects carbohydrate, fat, and protein metabolism; chiefly cortisol and corticosterone.

gluconeogenesis (GLUE-co-NEE-oh-JEN-eh-sis) The synthesis of glucose from noncarbohydrates such as fats and amino acids.

glucose A monosaccharide ($C_6H_{12}O_6$) also known as blood sugar; glycogen, starch, cellulose, and maltose are made entirely of glucose, and glucose constitutes half of a sucrose or lactose molecule. The isomer involved in human physiology is also called *dextrose*.

glycocalyx (GLY-co-CAY-licks) A layer of carbohydrate molecules covalently bonded to the phospholipids and proteins of a plasma membrane; forms a surface coat on all human cells.

glycogen (GLY-co-jen) A glucose polymer synthesized by liver, muscle, uterine, and vaginal cells that serves as an energy-storage polysaccharide.

glycogenesis (GLY-co-JEN-eh-sis) The synthesis of glycogen.

glycogenolysis (GLY-co-jeh-NOLL-ih-sis) The hydrolysis of glycogen, releasing glucose.

glycolipid (GLY-co-LIP-id) A phospholipid molecule with a carbohydrate covalently bonded to it, found in the plasma membranes of cells.

glycolysis (gly-COLL-ih-sis) A series of anaerobic oxidation reactions that breaks a glucose molecule into two molecules of pyruvic acid and produces a small amount of ATP.

glycoprotein (GLY-co-PRO-teen) A protein molecule with a smaller carbohydrate covalently bonded to it; found in mucus and the glycocalyx of cells, for example.

goblet cell A mucus-secreting gland cell, shaped somewhat like a wineglass, found in the epithelia of many mucous membranes. (fig. 4.6)

Golgi complex (GOAL-jee) An organelle composed of several parallel cisternae, somewhat like a stack of saucers, that modifies and packages newly synthesized proteins and synthesizes carbohydrates. (fig. 3.12)

Golgi vesicle A membrane-bounded vesicle pinched from the Golgi complex, containing its chemical product; may be retained in the cell as a lysosome or become a secretory vesicle that releases the product by exocytosis.

gonad The ovary or testis.

gonadotropin (go-NAD-oh-TRO-pin) A pituitary hormone that stimulates the gonads; specifically FSH and LH.

gradient A difference or change in any variable, such as pressure or chemical concentration, from one point in space to another; provides a basis for molecular movements such as gas exchange, osmosis, and facilitated diffusion, and for bulk movements such as blood flow and air flow.

gray matter A zone or layer of tissue in the central nervous system where the neuron cell bodies, dendrites, and synapses are found; forms the cerebral cortex and basal nuclei; cerebellar cortex and deep nuclei; nuclei of the brainstem; and core of the spinal cord. (figs. 9.3, 9.4)

gross anatomy Bodily structure that can be observed without magnification.

growth factor A chemical messenger that stimulates mitosis and differentiation of target cells that have receptors for it; important in such processes as fetal development, tissue maintenance and repair, and hemopoiesis; sometimes a contributing factor in cancer.

growth hormone (GH) A hormone of the anterior pituitary gland with multiple effects on many tissues, generally promoting tissue growth.

gyrus (JY-rus) A wrinkle or fold in the cortex of the cerebrum or cerebellum.

H

hair cell Sensory cell of the cochlea, semicircular ducts, utricle, and saccule, with a fringe of surface microvilli that respond to the relative motion of a gelatinous membrane at their tips; responsible for the senses of hearing, body position, and motion.

hair follicle An epithelial pit that contains a hair and extends into the dermis or hypodermis.

haploid (n) In humans, having 23 unpaired chromosomes instead of the usual 46 chromosomes in homologous pairs; in any organism or cell, having half the normal diploid (2n) number of chromosomes for that species. *Compare* diploid.

helper T cell A type of lymphocyte that performs a central coordinating role in humoral and cellular immunity; target of the human immunodeficiency virus (HIV).

hematocrit (he-MAT-oh-crit) The percentage of blood volume that is composed of erythrocytes; also called *packed cell volume.*

hematoma (HE-muh-TOE-muh) A mass of clotted blood in the tissues; forms a bruise when visible through the skin.

heme (heem) The nonprotein, iron-containing component of a hemoglobin or myoglobin molecule; oxygen binds to its iron atom. (fig. 12.4)

hemoglobin (HE-mo-GLO-bin) The red gas-transport pigment of an erythrocyte.

hemopoiesis (HE-mo-poy-EE-sis) Production of any of the formed elements of blood.

hemopoietic stem cell A cell of the red bone marrow that can give rise, through a series of intermediate cells, to leukocytes, erythrocytes, platelets, and various kinds of macrophages.

heparin (HEP-uh-rin) A polysaccharide secreted by basophils and mast cells that inhibits blood clotting.

hepatic (heh-PAT-ic) Pertaining to the liver.

hepatic portal system A network of blood vessels that connect capillaries of the intestines to capillaries (sinusoids) of the liver, thus delivering newly absorbed nutrients directly to the liver.

high-density lipoprotein (HDL) A lipoprotein of the blood plasma that is about 50% lipid and 50% protein; functions to transport phospholipids and cholesterol from other organs to the liver for disposal. A high proportion of HDL to low-density lipoprotein (LDL) is desirable for cardiovascular health.

hilum (HY-lum) A point on the surface of an organ where blood vessels, lymphatic vessels, or nerves enter and leave, usually marked by a depression and slit; the midpoint of the concave surface of any organ that is roughly bean-shaped, such as the lymph nodes, kidneys, and lungs. Also called the *hilus.* (fig. 15.6b)

histamine (HISS-ta-meen) An amino acid derivative secreted by basophils, mast cells, and some neurons; functions as a neurotransmitter or local chemical signal to stimulate effects such as gastric secretion, bronchoconstriction, and vasodilation.

histology **1.** The microscopic structure of tissues and organs. **2.** The study of such structure.

homeostasis (HO-me-oh-STAY-sis) The tendency of a living body to maintain relatively stable internal conditions in spite of changes in its external environment.

hormone A chemical messenger that is secreted by an endocrine gland or isolated gland cell, travels in the bloodstream, and triggers a physiological response in distant cells with receptors for it.

host cell Any cell belonging to the human body, as opposed to foreign cells introduced to it by such causes as infections and tissue transplants.

human chorionic gonadotropin (HCG) (COR-ee-ON-ic) A hormone of pregnancy secreted by the chorion that stimulates continued growth of the corpus luteum and secretion of its hormones. HCG in urine is the basis for pregnancy testing.

human immunodeficiency virus (HIV) A virus that infects human helper T cells and other cells, suppresses immunity, and causes AIDS.

hyaline cartilage (HY-uh-lin) A form of cartilage with a relatively clear matrix and fine collagen fibers but no conspicuous elastic fibers or coarse collagen bundles as in other types of cartilage; the most widespread type of cartilage in the human body.

hydrogen bond A weak attraction between a slightly positive hydrogen atom on one molecule and a slightly negative oxygen or nitrogen atom on another molecule, or between such atoms on different parts of the same molecule; responsible for the cohesion of water and the coiling of protein and DNA molecules, for example.

hydrolysis (hy-DROL-ih-sis) A chemical reaction that breaks a covalent bond in a molecule by adding an —OH group to one side of the bond and —H to the other side, thus consuming a water molecule.

hydrophilic (HY-dro-FILL-ic) Pertaining to molecules that attract water or dissolve in it because of their polar nature.

hydrophobic (HY-dro-FOE-bic) Pertaining to molecules that do not attract water or dissolve in it because of their nonpolar nature; such molecules tend to dissolve in lipids and other nonpolar solvents.

hydrostatic pressure The physical force generated by a liquid such as blood or tissue fluid, as opposed to osmotic and atmospheric pressures.

hypercalcemia (HY-pur-cal-SEE-me-uh) An excess of calcium ions in the blood.

hyperextension A joint movement that increases the angle between two bones beyond 180°. (fig. 6.35)

hyperglycemia (HY-pur-gly-SEE-me-uh) An excess of glucose in the blood.

hyperkalemia (HY-pur-ka-LEE-me-uh) An excess of potassium ions in the blood.

hypernatremia (HY-pur-na-TREE-me-uh) An excess of sodium ions in the blood.

hyperplasia (HY-pur-PLAY-zhuh) The growth of a tissue through cellular multiplication, not cellular enlargement. *Compare* hypertrophy.

hypertension Excessively high blood pressure; criteria vary but it is often considered to be a condition in which resting systolic pressure exceeds 140 mm Hg or diastolic pressure exceeds 90 mm Hg.

hypertonic Having a higher osmotic pressure than human cells or some other reference solution and tending to cause osmotic shrinkage of cells.

hypertrophy (hy-PUR-tro-fee) The growth of a tissue through cellular enlargement, not cellular multiplication; for example, the growth of muscle under the influence of exercise. *Compare* hyperplasia.

hypocalcemia (HY-po-cal-SEE-me-uh) A deficiency of calcium ions in the blood.

hypodermis (HY-po-DUR-miss) A layer of connective tissue deep to the skin; also called *superficial fascia, subcutaneous tissue,* or when it is predominantly adipose, *subcutaneous fat.*

hypoglycemia (HY-po-gly-SEE-me-uh) A deficiency of glucose in the blood.

hypokalemia (HY-po-ka-LEE-me-uh) A deficiency of potassium ions in the blood.

hyponatremia (HY-po-na-TREE-me-uh) A deficiency of sodium ions in the blood.

hypothalamic thermostat (HY-po-thuh-LAM-ic) A nucleus in the hypothalamus that monitors body temperature and sends afferent signals to hypothalamic heat-promoting or heat-losing centers to maintain thermal homeostasis.

hypothalamus (HY-po-THAL-uh-mus) The inferior portion of the diencephalon of the brain, forming the walls and floor of the third ventricle and giving rise to the posterior pituitary gland; controls many fundamental physiological functions such as appetite, thirst, and body temperature, and exerts many of its effects through the endocrine and autonomic nervous systems. (fig. 9.2)

hypothesis An informed conjecture that is capable of being tested and potentially falsified by experimentation or data collection.

hypotonic Having a lower osmotic pressure than human cells or some other reference solution and tending to cause osmotic swelling and lysis of cells.

hypoxemia (HY-pock-SEE-me-uh) A deficiency of oxygen in the bloodstream.

hypoxia (hy-POCK-see-uh) A deficiency of oxygen in any tissue.

I

immunity The ability to ward off a specific infection or disease, usually as a result of prior exposure and the body's production of antibodies or lymphocytes against a pathogen. *Compare* resistance.

immunoglobulin (IM-you-no-GLOB-you-lin) *See* antibody.

implantation The attachment of a conceptus to the endometrium of the uterus.

inclusion Any visible object in the cytoplasm of a cell other than an organelle or cytoskeletal element; usually a foreign body or a stored cell product, such as a virus, dust particle, lipid droplet, glycogen granule, or pigment.

infarction (in-FARK-shun) **1.** The sudden death of tissue from a lack of blood perfusion. **2.** An area of necrotic tissue produced by this process; also called an *infarct*.

inferior Lower than another structure or point of reference from the perspective of anatomical position; for example, the stomach is inferior to the diaphragm.

inflammation (IN-fluh-MAY-shun) A complex of tissue responses to trauma or infection serving to ward off a pathogen and promote tissue repair; recognized by the cardinal signs of redness, heat, swelling, and pain.

inguinal (IN-gwih-nul) Pertaining to the groin.

innervation (IN-ur-VAY-shun) The nerve supply to an organ.

insertion The point at which a muscle attaches to another tissue (usually a bone) and produces movement, opposite from its stationary origin; the origin and insertion of a given muscle sometimes depend on what muscle action is being considered. *Compare* origin. (fig. 7.15)

inspiration **1.** Inhaling. **2.** The stimulus that resulted in this book.

integral protein A protein that extends through a plasma membrane and contacts both the extracellular and intracellular fluid. (fig. 3.2)

intercalated disc (in-TUR-kuh-LAY-ted) A complex of mechanical and electrical junctions that joins two cardiac muscle cells end to end, microscopically visible as a dark line that helps to histologically distinguish this muscle type; functions as a mechanical and electrical link between cells. (fig. 13.6)

intercellular Between cells.

intercostal (IN-tur-COSS-tul) Between the ribs, as in the intercostal muscles, arteries, veins, and nerves.

interdigitate (IN-tur-DIDJ-ih-tate) To fit together like the fingers of two folded hands; for example, at the dermal–epidermal boundary, intercalated discs of the heart, and foot processes of the podocytes in the kidney. (fig. 16.8)

interleukin (IN-tur-LOO-kin) A hormonelike chemical messenger from one leukocyte to another, serving as a means of communication and coordination during immune responses.

interneuron (IN-tur-NEW-ron) A neuron that is contained entirely in the central nervous system and, in the path of signal conduction, lies anywhere between an afferent pathway and an efferent pathway.

interosseous membrane (IN-tur-OSS-ee-us) A fibrous membrane that connects the radius to the ulna and the tibia to the fibula along most of the shaft of each bone. (fig. 6.27)

interphase That part of the cell cycle between one mitotic phase and the next, from the end of cytokinesis to the beginning of the next mitosis.

interstitial (IN-tur-STISH-ul) **1.** Pertaining to the extracellular spaces in a tissue. **2.** Located between other structures, as in the interstitial cells of the testis.

intervertebral disc A cartilaginous pad between the bodies of adjacent vertebrae.

intracellular Within a cell.

intracellular fluid (ICF) The fluid contained in the cells; one of the major fluid compartments.

intravenous (I.V.) **1.** Present or occurring within a vein, such as an intravenous blood clot. **2.** Introduced directly into a vein, such as an intravenous injection or I.V. drip.

intrinsic (in-TRIN-sic) **1.** Arising from within, such as intrinsic blood-clotting factors; endogenous. **2.** Fully contained within an organ, such as the intrinsic muscles of the hand and eye. *Compare* extrinsic.

involuntary Not under conscious control, including tissues such as smooth and cardiac muscle and events such as reflexes.

involution (IN-vo-LOO-shun) Shrinkage of a tissue or organ by autolysis, such as involution of the thymus after childhood and of the uterus after pregnancy.

ion A chemical particle with unequal numbers of electrons or protons and consequently a net negative or positive charge; it may have a single atomic nucleus as in a sodium ion or a few atoms as in a bicarbonate ion, or it may be a large molecule such as a protein.

ionic bond The force that binds a cation to an anion.

ionizing radiation High-energy electromagnetic rays that eject electrons from atoms or molecules and convert them to ions, frequently causing cellular damage; for example, X-rays and gamma rays.

ischemia (iss-KEE-me-uh) Insufficient blood flow to a tissue, typically resulting in metabolite accumulation and sometimes tissue death.

isotonic Having the same osmotic pressure as human cells or some other reference solution.

J

jaundice (JAWN-diss) A yellowish color of the skin, corneas, mucous membranes, and body fluids due to an excessive concentration of bilirubin; usually indicative of a liver disease, obstructed bile secretion, or hemolytic disease.

K

ketone (KEE-tone) Any organic compound with a carbonyl (C = O) group covalently bonded to two other carbons; ketones are produced by fat oxidation and are a cause of diabetic acidosis.

ketonuria (KEE-toe-NEW-ree-uh) The abnormal presence of ketones in the urine; a sign of diabetes mellitus but also occurring in other conditions that entail rapid fat oxidation.

kilocalorie The amount of heat energy needed to raise the temperature of 1 kg of water by 1°C; 1,000 calories. Also called a *Calorie* or *large calorie. See also* calorie.

L

labium (LAY-bee-um) A lip, such as those of the mouth and the labia majora and minora of the vulva.

lactic acid A small organic acid produced as an end product of the anaerobic fermentation of pyruvic acid; a contributing factor in muscle fatigue.

lacuna (la-CUE-nuh) A small cavity or depression in a tissue such as bone, cartilage, and the erectile tissues.

lamella (la-MELL-uh) A little plate, such as the lamellae of bone. (fig. 6.3)

lamina (LAM-ih-nuh)   A thin layer, such as the lamina of a vertebra or the lamina propria of a mucous membrane. (figs. 4.30a, 6.19)

lamina propria (PRO-pree-uh)   A thin layer of areolar tissue immediately deep to the epithelium of a mucous membrane. (fig. 4.30a)

larynx (LAIR-inks)   A cartilaginous chamber in the neck containing the vocal cords; colloquially called the *voice box.*

lateral   Away from the midline of an organ or median plane of the body; toward the side. *Compare* medial.

law   A verbal or mathematical description of a predictable natural phenomenon or of the relationship between variables; for example, the law of complementary base pairing.

leukocyte (LOO-co-site)   A white blood cell.

libido (lih-BEE-do)   Sex drive.

ligament   A cord or band of tough collagenous tissue binding one organ to another, especially one bone to another, and serving to hold organs in place; for example, the cruciate ligaments of the knee and falciform ligament of the liver.

linea (LIN-ee-uh)   An anatomical line, such as the linea aspera of the femur.

lingual (LING-gwul)   Pertaining to the tongue, as in lingual papillae.

lipase (LY-pace)   An enzyme that hydrolyzes a triglyceride into fatty acids and glycerol.

lipid   A hydrophobic organic compound composed mainly of carbon and a high ratio of hydrogen to oxygen; includes fatty acids, fats, phospholipids, steroids, and prostaglandins.

lipoprotein (LIP-oh-PRO-teen)   A protein-coated lipid droplet in the blood plasma or lymph, serving as a means of lipid transport; for example, chylomicrons and high- and low-density lipoproteins.

load   **1.** To pick up a gas for transport in the bloodstream. **2.** The resistance acted upon by a muscle.

lobule (LOB-yool)   A small subdivision of an organ or of a lobe of an organ, especially of a gland.

long bone   A bone such as the femur or humerus that is markedly longer than wide and that generally serves as a lever.

longitudinal   Oriented along the longest dimension of the body or of an organ.

loose connective tissue   *See* areolar tissue.

low-density lipoprotein (LDL)   A blood-borne droplet of about 20% protein and 80% lipid (mainly cholesterol) that transports cholesterol from the liver to other tissues for their use.

lumbar   Pertaining to the lower back and sides, between the thoracic cage and pelvis.

lumen (LOO-men)   The internal space of a hollow organ such as a blood vessel or the esophagus, or a space surrounded by cells as in a gland acinus.

lymph   The fluid contained in lymphatic vessels and lymph nodes, produced by the absorption of tissue fluid.

lymphatic system (lim-FAT-ic)   An organ system consisting of lymphatic vessels, lymph nodes, the tonsils, spleen, and thymus; functions include tissue fluid recovery and immunity.

lymph node   A small organ found along the course of a lymphatic vessel that filters the lymph and contains lymphocytes and macrophages, which respond to antigens in the lymph. (fig. 14.9)

lymphocyte (LIM-foe-site)   A relatively small agranulocyte with numerous types and roles in nonspecific defense, humoral immunity, and cellular immunity.

lysosome (LY-so-some)   A membrane-bounded organelle containing a mixture of enzymes with a variety of intracellular and extracellular roles in digesting foreign matter, pathogens, and expired organelles.

lysozyme (LY-so-zime)   An enzyme found in tears, milk, saliva, mucus, and other body fluids that destroys bacteria by digesting their cell walls.

M

macromolecule   Any molecule of large size and high molecular weight, such as a protein, nucleic acid, polysaccharide, or triglyceride.

macrophage (MAC-ro-faje)   Any cell of the body, other than a leukocyte, that is specialized for phagocytosis; usually derived from a blood monocyte and often functioning as an antigen-presenting cell.

macula (MAC-you-luh)   A patch or spot, such as the *macula lutea* of the retina.

malignant (muh-LIG-nent)   Pertaining to a cell or tumor that is cancerous; capable of metastasis.

mast cell   A connective tissue cell, similar to a basophil, that secretes histamine, heparin, and other chemicals involved in inflammation; often concentrated along the course of blood capillaries.

matrix   **1.** The extracellular material of a tissue. **2.** The fluid within a mitochondrion containing enzymes of the citric acid cycle. **3.** The substance or framework within which other structures are embedded, such as the fibrous matrix of a blood clot. **4.** A mass of epidermal cells from which a hair root or nail root develops.

medial   Toward the midline of an organ or median plane of the body. *Compare* lateral.

median plane   The sagittal plane that divides the body or an organ into equal right and left halves; also called *midsagittal plane.* (fig. 1.7)

mediastinum (ME-dee-ass-TY-num)   The thick median partition of the thoracic cavity that separates one pleural cavity from the other and contains the heart, great blood vessels, esophagus, trachea, and thymus. (fig. 1.12)

medulla (meh-DULE-uh, meh-DULL-uh)   Tissue deep to the cortex of certain two-layered organs such as the adrenal glands, lymph nodes, hairs, and kidneys.

medulla oblongata (OB-long-GAH-ta)   The most caudal part of the brainstem, immediately superior to the foramen magnum of the skull, connecting the spinal cord to the rest of the brain. (figs. 9.2, 9.5)

meiosis (my-OH-sis)   A form of cell division in which a diploid cell divides twice and produces four haploid daughter cells; occurs only in gametogenesis.

melanocyte   A cell of the stratum basale of the epidermis that synthesizes melanin and transfers it to the keratinocytes.

meninges (meh-NIN-jeez) (singular, *meninx*)   Three fibrous membranes between the central nervous system and surrounding bone: the dura mater, arachnoid mater, and pia mater. (fig. 9.3)

merocrine (MERR-oh-crin)   Pertaining to gland cells that release their product by exocytosis.

mesentery (MESS-en-tare-ee)   A serous membrane that binds the intestines together and suspends them from the abdominal wall; the visceral continuation of the peritoneum. (figs. 1.14, 17.18)

mesoderm (MEZ-oh-durm)   The middle layer of the three primary germ layers of an embryo; gives rise to muscle and connective tissue.

metabolism (meh-TAB-oh-lizm)   The sum of all chemical reactions in the body.

metabolite (meh-TAB-oh-lite)   Any chemical produced by metabolism, especially a breakdown product or waste.

microtubule   An intracellular cylinder of protein, forming centrioles, the axonemes of cilia and flagella, and part of the cytoskeleton.

microvillus   An outgrowth of the plasma membrane that increases the surface area of a cell and functions in absorption and some sensory processes; distinguished from cilia and flagella by its smaller size and lack of an axoneme.

milliequivalent   One-thousandth of an equivalent, which is the amount of an electrolyte that would neutralize 1 mole of H⁺ or OH⁻. Electrolyte concentrations are commonly expressed in milliequivalents per liter (mEq/L). *See* appendix C.

mitochondrion (MY-toe-CON-dree-un)   An organelle specialized to synthesize ATP, enclosed in a double unit membrane with infoldings of the inner membrane called cristae.

mitosis (my-TOE-sis)   A form of cell division in which a cell divides once and produces two genetically identical daughter cells; sometimes used to refer only to the division of the genetic material or nucleus and not to include cytokinesis, the subsequent division of the cytoplasm.

molarity   A measure of chemical concentration expressed as moles of solute per liter of solution. *See* appendix C.

mole   The mass of a chemical equal to its molecular weight in grams, containing 6.023×10^{23} molecules. *See* appendix C.

monocyte   An agranulocyte specialized to migrate into the tissues and transform into a macrophage.

monomer (MON-oh-mur)   **1.** One of the identical or similar subunits of a larger molecule in the dimer to polymer range; for example, the glucose monomers of starch, the amino acids of a protein, or the nucleotides of DNA. **2.** One subunit of an antibody molecule, composed of four polypeptides.

monosaccharide (MON-oh-SAC-uh-ride)   A simple sugar, or sugar monomer; chiefly glucose, fructose, and galactose.

motor end plate   *See* neuromuscular junction.

motor neuron   A neuron that transmits signals from the central nervous system to any effector (muscle or gland cell); its axon is an efferent nerve fiber.

motor protein   Any protein that produces movements of a cell or its components owing to its ability to undergo quick repetitive changes in conformation and to bind reversibly to other molecules; for example, myosin.

motor unit   One motor neuron and all the skeletal muscle fibers innervated by it.

mucosa (mew-CO-suh)   A tissue layer that forms the inner lining of an anatomical tract that is open to the exterior (the respiratory, digestive, urinary, and reproductive tracts). Composed of epithelium, connective tissue (lamina propria), and often smooth muscle (muscularis mucosae). (fig. 4.30a)

mucous membrane   A mucosa.

muscle fiber   One skeletal muscle cell.

muscle tone   A state of continual, partial contraction of resting skeletal or smooth muscle.

muscularis externa   The external muscular wall of certain viscera such as the esophagus and small intestine. (fig. 17.2)

muscularis mucosae (MUSS-cue-LERR-iss mew-CO-see)   A layer of smooth muscle immediately deep to the lamina propria of a mucous membrane. (fig. 4.30a)

muscular system   An organ system composed of the skeletal muscles, specialized mainly for maintaining postural support and producing movements of the bones.

mutagen (MEW-tuh-jen)   Any agent that causes a mutation, including viruses, chemicals, and ionizing radiation.

mutation   Any change in the structure of a chromosome or a DNA molecule, often resulting in a change of organismal structure or function.

myelin (MY-eh-lin)   A lipid sheath around a nerve fiber, formed from closely spaced spiral layers of the plasma membrane of a Schwann cell or oligodendrocyte. (fig. 8.6)

myocardium (MY-oh-CAR-dee-um)   The middle, muscular layer of the heart.

myocyte   A muscle cell, especially a cell of cardiac or smooth muscle.

myoepithelial cell   An epithelial cell that has become specialized to contract like a muscle cell; important in dilation of the pupil and ejection of secretions from gland acini.

myofilament   A protein microfilament responsible for the contraction of a muscle cell, composed mainly of myosin or actin. (fig. 7.3)

myoglobin (MY-oh-GLO-bin)   A red oxygen-storage pigment of muscle; supplements hemoglobin in providing oxygen for aerobic muscle metabolism.

myosin   A motor protein that constitutes the thick myofilaments of muscle and has globular, mobile heads that bind to actin molecules.

N

necrosis (neh-CRO-sis)   Pathological tissue death due to such causes as infection, trauma, or hypoxia. *Compare* apoptosis.

negative feedback   A self-corrective mechanism that underlies most homeostasis, in which a bodily change is detected and responses are activated that reverse the change and restore stability and preserve normal body function.

negative feedback inhibition   A mechanism for limiting the secretion of a pituitary hormone. The pituitary hormone stimu-lates another endocrine gland to secrete its own hormone, and that hormone inhibits further release of the pituitary hormone.

neonate (NEE-oh-nate)   An infant up to 6 weeks old.

neoplasia (NEE-oh-PLAY-zee-uh)   Abnormal growth of new tissue, such as a tumor, with no useful function.

nephron   One of approximately 1 million blood-filtering, urine-producing units in each kidney; consists of a glomerulus, glomerular capsule, proximal convoluted tubule, nephron loop, and distal convoluted tubule. (fig. 16.6)

nerve   A cordlike organ of the peripheral nervous system composed of multiple nerve fibers ensheathed in connective tissue.

nerve fiber   The axon of a single neuron.

nerve impulse   A wave of self-propagating action potentials traveling along a nerve fiber.

nervous tissue   A tissue composed of neurons and neuroglia.

neural pool   A group of interconnected neurons of the central nervous system that performs a single collective function; for example, the vasomotor center of the brainstem and speech centers of the cerebral cortex.

neural tube   A dorsal hollow tube in the embryo that develops into the central nervous system. (fig. 20.6)

neuroglia (noo-ROG-lee-uh)   All cells of nervous tissue except neurons; cells that perform various supportive and protective roles for the neurons.

neuromuscular junction   A synapse between a nerve fiber and a muscle fiber; also called a *motor end plate.* (fig. 7.6)

neuron (NOOR-on)   A nerve cell; an electrically excitable cell specialized for producing and transmitting action potentials and secreting chemicals that stimulate adjacent cells.

neurotransmitter   A chemical released at the distal end of an axon that stimulates an adjacent cell; for example, acetylcholine, norepinephrine, or serotonin.

neutrophil (NOO-tro-fill)   A granulocyte, usually with a multilobed nucleus, that serves especially to destroy bacteria by means of phagocytosis, intracellular digestion, and bactericidal secretions.

nitrogenous base (ny-TRODJ-eh-nus)   An organic molecule with a single or double carbon–nitrogen ring that forms one of the building blocks of ATP, other nucleotides, and nucleic acids; the basis of the genetic code. (fig. 2.18)

nitrogenous waste   Any nitrogen-containing substance produced as a metabolic waste and excreted in the urine; chiefly ammonia, urea, uric acid, and creatinine.

nociceptor (NO-sih-SEP-tur)   A nerve ending specialized to detect tissue damage and produce a sensation of pain; pain receptor.

norepinephrine (nor-EP-ih-NEF-rin)   An organic molecule that functions as a neurotransmitter secreted by neurons, especially in the sympathetic nervous system, and as a hormone secreted by the adrenal gland.

nuclear envelope (NEW-clee-ur)   A pair of membranes enclosing the nucleus of a cell, with prominent pores allowing traffic of molecules between the nucleus and cytoplasm. (fig. 3.10)

nucleic acid (new-CLAY-ic)   An acidic polymer of nucleotides found or produced in the nucleus, functioning in heredity and protein synthesis; of two types, DNA and RNA.

nucleotide (NEW-clee-oh-tide)   An organic molecule composed of a nitrogenous base, a monosaccharide, and a phosphate group; the monomer of a nucleic acid.

nucleus (NEW-clee-us)   **1.** A cell organelle containing DNA and surrounded by a double membrane. **2.** A mass of neurons (gray matter) surrounded by white matter of the brain, including the basal nuclei and brainstem nuclei. **3.** The positively charged core of an atom, consisting of protons and neutrons. **4.** A central structure, such as the nucleus pulposus of an intervertebral disc.

O

olfaction (ole-FAC-shun)   The sense of smell.

oocyte (OH-oh-site)   In the development of an egg cell, any haploid stage between meiosis I and fertilization.

oogenesis (OH-oh-JEN-eh-sis)   The production of a fertilizable egg cell through a series of mitotic and meiotic cell divisions; female gametogenesis.

opposition   A movement of the thumb in which it approaches or touches any fingertip of the same hand.

orbit   The eye socket of the skull.

organ   Any anatomical structure that is composed of at least two tissue types, has recognizable structural boundaries, and has a discrete function different from the structures around it. Many organs are microscopic and many organs contain smaller organs, such as the skin containing numerous microscopic sense organs.

organelle   Any structure within a cell that carries out one of its metabolic roles, such as mitochondria, centrioles, endoplasmic reticulum, and the nucleus; an intracellular structure other than the cytoskeleton and inclusions.

origin   The relatively stationary attachment of a skeletal muscle. (fig. 7.15) *Compare* insertion.

osmolarity (OZ-mo-LERR-ih-tee)   The molar concentration of dissolved particles in a solution. *See* appendix C.

osmosis (oz-MO-sis)   The net flow of water through a selectively permeable membrane, resulting from either a chemical concentration difference or a mechanical force across the membrane.

osseous (OSS-ee-us)   Pertaining to bone.

ossification (OSS-ih-fih-CAY-shun)   Bone formation.

osteoblast   Bone-forming cell that arises from an osteogenic cell, deposits bone matrix, and eventually becomes an osteocyte.

osteoclast   Macrophage of the bone surface that dissolves the matrix and returns minerals to the extracellular fluid.

osteocyte   A mature bone cell formed when an osteoblast becomes surrounded by its own matrix and entrapped in a lacuna.

osteon   A structural unit of compact bone consisting of a central canal surrounded by concentric cylindrical layers of matrix. (fig. 6.3)

osteoporosis (OSS-tee-oh-pore-OH-sis)   A degenerative bone disease characterized by a loss of bone mass, increasing susceptibility to spontaneous fractures, and sometimes deformity of the vertebral column; causes include aging, estrogen hyposecretion, and insufficient resistance exercise.

ovulation (OV-you-LAY-shun)   The release of a mature oocyte by the bursting of an ovarian follicle.

ovum   Any stage of the female gamete from the conclusion of meiosis I until fertilization; a primary or secondary oocyte; an egg.

oxidation   A chemical reaction in which one or more electrons are removed from a molecule, lowering its energy content; opposite of reduction and always linked to a reduction reaction.

P

pancreatic islet (PAN-cree-AT-ic EYE-let)   A small cluster of endocrine cells in the pancreas that secretes insulin, glucagon, somatostatin, and other intercellular messengers; also called *islet of Langerhans.* (fig. 11.10)

papilla (pa-PILL-uh)   A conical or nipplelike structure, such as a lingual papilla of the tongue or the papilla of a hair bulb.

papillary (PAP-ih-lerr-ee)   **1.** Pertaining to or shaped like a nipple, such as the papillary muscles of the heart. **2.** Having papillae, such as the papillary layer of the dermis.

parasympathetic nervous system (PERR-uh-SIM-pa-THET-ic)   A division of the autonomic nervous system that issues efferent fibers through the cranial and sacral nerves and exerts cholinergic effects on its target organs.

parathyroid hormone (PTH)   A hormone secreted by the parathyroid glands that raises blood calcium concentration by stimulating bone resorption by osteoclasts, promoting intestinal absorption of calcium, and inhibiting urinary excretion of calcium.

parietal (pa-RY-eh-tul)   **1.** Pertaining to a wall, as in the parietal cells of the gastric glands and parietal bone of the skull. **2.** The outer or more superficial layer of a two-layered membrane such as the pleura, pericardium, or glomerular capsule. *Compare* visceral. (fig. 1.13)

pathogen   Any disease-causing microorganism.

pedicle (PED-ih-cul)   A small footlike process, as in the vertebrae and the renal podocytes; also called a *pedicel.*

pelvic cavity   The space enclosed by the true (lesser) pelvis, containing the urinary bladder, rectum, and internal reproductive organs.

pelvic girdle   A ring of three bones—the sacrum and two hip bones—that encloses the pelvis and links the lower limbs to the vertebral column. (fig. 6.29)

pelvis   **1.** The basinlike enclosure of the inferior abdominopelvic cavity, composed of the pelvic girdle and its associated body-wall ligaments and muscles. **2.** A basinlike structure such as the urine-collecting funnel near the hilum of the kidney. (fig. 16.2)

peptide   Any chain of two or more amino acids. *See also* polypeptide, protein.

peptide bond   A group of four covalently bonded atoms (a —C=O group bonded to an —NH group) that links two amino acids in a protein or other peptide. (fig. 2.15)

perfusion   The amount of blood supplied to a given mass of tissue in a given period of time.

perichondrium (PERR-ih-CON-dree-um)   A layer of fibrous connective tissue covering the surface of hyaline or elastic cartilage.

perineum (PERR-ih-NEE-um)   The region between the thighs bordered by the coccyx, pubic symphysis, and ischial tuberosities; contains the orifices of the urinary, reproductive, and digestive systems. (fig. 19.9)

periosteum (PERR-ee-OSS-tee-um)   A layer of fibrous connective tissue covering the surface of a bone. (fig. 6.3)

peripheral (peh-RIF-eh-rul)   Away from the center of the body or of an organ, as in peripheral vision and peripheral blood vessels.

peripheral nervous system (PNS)   A subdivision of the nervous system composed of all nerves and ganglia; all of the nervous system except the central nervous system.

peristalsis (PERR-ih-STAL-sis)   A wave of constriction traveling along a tubular organ such as the esophagus or ureter, serving to propel its contents.

peritoneum (PERR-ih-toe-NEE-um)   A serous membrane that lines the peritoneal cavity of the abdomen and covers the mesenteries and viscera.

phagocytosis (FAG-oh-sy-TOE-sis)   A form of endocytosis in which a cell surrounds a foreign particle with pseudopods and engulfs it.

pharynx (FAIR-inks)   A muscular passage in the throat at which the respiratory and digestive tracts cross.

phospholipid   An amphipathic molecule composed of two fatty acids and a phosphate-containing group bonded to the three carbons of a glycerol molecule; composes most of the molecules of the plasma membrane and other cellular membranes.

piloerector   A bundle of smooth muscle cells associated with a hair follicle, responsible for erection of the hair; also called a *pilomotor muscle*. (fig. 5.4)

pinocytosis (PIN-oh-sy-TOE-sis)   A form of endocytosis in which the plasma membrane sinks inward and imbibes droplets of extracellular fluid.

plantar (PLAN-tur)   Pertaining to the sole of the foot.

plaque   A small scale or plate of matter, such as dental plaque, the fatty plaques of atherosclerosis, and the amyloid plaques of Alzheimer disease.

plasma   The noncellular portion of the blood.

plasma membrane   The membrane that encloses a cell and controls the traffic of molecules into and out of the cell. (fig. 3.2)

platelet   A formed element of the blood known especially for its role in stopping bleeding, but with additional roles in dissolving blood clots, stimulating inflammation, promoting tissue growth and maintenance, and destroying bacteria.

pleura (PLOOR-uh)   A double-walled serous membrane that encloses each lung.

plexus   A network of blood vessels, lymphatic vessels, or nerves, such as a choroid plexus of the brain or brachial plexus of nerves.

polymer   A molecule that consists of a long chain of identical or similar subunits, such as protein, DNA, or starch.

polypeptide   Any chain of more than 10 or 15 amino acids. *See also* protein.

polysaccharide (POL-ee-SAC-uh-ride)   A polymer of simple sugars; for example, glycogen, starch, and cellulose.

polyuria (POL-ee-YOU-ree-uh)   Excessive output of urine.

popliteal (po-LIT-ee-ul)   Pertaining to the posterior aspect of the knee.

positron emission tomography (PET)   A method of producing a computerized image of the physiological state of a tissue using injected radioisotopes that emit positrons.

posterior   Near or pertaining to the back or spinal side of the body; dorsal.

posterior root   The branch of a spinal nerve that enters the posterior side of the spinal cord and carries afferent (sensory) nerve fibers; often called *dorsal root*. (fig. 8.17)

postganglionic (POST-gang-glee-ON-ic)   Pertaining to a neuron that conducts signals from a ganglion to a more distal target organ.

postsynaptic (POST-sih-NAP-tic)   Pertaining to a neuron or other cell that receives signals from the presynaptic neuron at a synapse. (fig. 8.12)

potential   A difference in electrical charge from one point to another, especially on opposite sides of a plasma membrane; usually measured in millivolts.

preembryo   A developing individual from the time of fertilization to the time, at 16 days, when the three primary germ layers have formed. *Compare* conceptus, embryo, fetus.

preganglionic (PRE-gang-glee-ON-ic)   Pertaining to a neuron that conducts signals from the central nervous system to a ganglion.

presynaptic (PRE-sih-NAP-tic)   Pertaining to a neuron that conducts signals to a synapse. (fig. 8.12)

primary germ layers   The ectoderm, mesoderm, and endoderm; the three tissue layers of an early embryo from which all later tissues and organs arise.

prime mover   The muscle that produces the most force in a given joint action; also called the *agonist*.

programmed cell death   *See* apoptosis.

pronation (pro-NAY-shun)   A rotational movement of the forearm that turns the palm downward or posteriorly. (fig. 6.40)

proprioception (PRO-pree-oh-SEP-shun)   The nonvisual perception, usually subconscious, of the position and movements of the body, resulting from input from proprioceptors and the vestibular apparatus of the inner ear.

proprioceptor (PRO-pree-oh-SEP-tur)   A sensory receptor of the muscles, tendons, and joint capsules that detects muscle contractions and joint movements.

prostaglandin (PROSS-ta-GLAN-din)   A modified fatty acid with a five-sided carbon ring in the middle of a hydrocarbon chain, playing a variety of roles in inflammation, neurotransmission, vasomotion, reproduction, and metabolism.

prostate gland (PROSS-tate)   A male reproductive gland that encircles the urethra immediately inferior to the bladder and contributes to the semen. (fig. 19.1)

protein   A large polypeptide; while criteria for a protein are somewhat subjective and variable, polypeptides over 50 amino acids long are generally classified as proteins.

proximal   Relatively near a point of origin or attachment; for example, the shoulder is proximal to the elbow. *Compare* distal.

pseudopod (SOO-doe-pod)   A temporary cytoplasmic extension of a cell used for locomotion (ameboid movement) and phagocytosis.

pseudostratified columnar   A type of epithelium with tall columnar cells reaching the free surface and shorter basal cells that do not reach the surface, but with all cells resting on the basement membrane; creates a false appearance of stratification. (fig. 4.7)

pulmonary circuit   A route of blood flow that supplies blood to the pulmonary alveoli for gas exchange and then returns it to the heart; all blood vessels between the right ventricle and the left atrium of the heart.

pyrogen (PY-ro-jen)   A fever-producing agent.

R

ramus (RAY-mus)   An anatomical branch, as in a nerve or of the pubis.

receptor   **1.** A cell or organ specialized to detect a stimulus, such as a taste cell or the eye. **2.** A protein molecule that binds and responds to a chemical such as a hormone, neurotransmitter, or odor molecule.

receptor-mediated endocytosis   A process in which certain molecules in the extracellular fluid bind to receptors in the plasma membrane, these receptors gather together, the membrane sinks inward at that point, and the molecules become incorporated into vesicles in the cytoplasm.

reduction **1.** A chemical reaction in which one or more electrons are added to a molecule, raising its energy content; opposite of oxidation and always linked to an oxidation reaction. **2.** Treatment of a fracture by restoring the broken parts of a bone to their proper alignment.

reflex A stereotyped, automatic, involuntary response to a stimulus; includes somatic reflexes, in which the effectors are skeletal muscles, and visceral (autonomic) reflexes, in which the effectors are usually visceral muscle, cardiac muscle, or glands.

reflex arc A simple neural pathway that mediates a reflex; involves a receptor, an afferent nerve fiber, sometimes one or more interneurons, an efferent nerve fiber, and an effector.

refractory period **1.** A period of time after a nerve or muscle cell has responded to a stimulus in which it cannot be reexcited by a threshold stimulus. **2.** A period of time after male orgasm when it is not possible to reattain erection or ejaculation.

regeneration Replacement of damaged tissue with new tissue of the original type. *Compare* fibrosis.

renin (REE-nin) An enzyme secreted by the kidneys in response to hypotension; converts the plasma protein angiotensinogen to angiotensin I, leading indirectly to a rise in blood pressure.

repolarization Reattainment of the resting membrane potential after a nerve or muscle cell has depolarized.

residue Any one of the amino acids in a protein or other peptide.

resistance **1.** A nonspecific ability to ward off infection or disease regardless of whether the body has been previously exposed to it. *Compare* immunity. **2.** A force that opposes the flow of a fluid such as air or blood. **3.** A force, or load, that opposes the action of a muscle or lever.

resting membrane potential (RMP) A stable voltage across the plasma membrane of an unstimulated nerve or muscle cell.

reticular cell (reh-TIC-you-lur) A delicate, branching phagocytic cell found in the reticular connective tissue of the lymphatic organs.

reticular fiber A fine, branching collagen fiber coated with glycoprotein, forming part of the framework of lymphatic organs and some other tissues and organs.

reticular tissue A connective tissue composed of reticular cells and reticular fibers, found in bone marrow, lymphatic organs, and in lesser amounts elsewhere.

ribosome A granule found free in the cytoplasm or attached to the rough endoplasmic reticulum and nuclear envelope, composed of ribosomal RNA and enzymes; specialized to read the nucleotide sequence of messenger RNA and assemble a corresponding sequence of amino acids to make a protein.

risk factor Any environmental factor or characteristic of an individual that increases one's chance of developing a particular disease; includes such intrinsic factors as age, sex, and race and such extrinsic factors as diet, smoking, and occupation.

rostral Relatively close to the forehead, especially in reference to structures of the brain and spinal cord; for example, the frontal lobe is rostral to the parietal lobe. *Compare* caudal.

ruga (ROO-ga) **1.** An internal fold or wrinkle in the mucosa of a hollow organ such as the stomach and urinary bladder; typically present when the organ is empty and relaxed but not when the organ is full and stretched. **2.** Tissue ridges in such locations as the hard palate and vagina. (fig. 17.8)

S

saccule (SAC-yule) A saclike receptor in the inner ear with a vertical patch of hair cells, the macula sacculi; senses the orientation of the head and responds to vertical acceleration, as when riding in an elevator or standing up. (fig. 10.7)

sagittal plane (SADJ-ih-tul) Any plane that extends from anterior to posterior and cephalic to caudal and that divides the body into right and left portions. *Compare* median plane.

sarcomere (SAR-co-meer) In skeletal and cardiac muscle, the portion of a myofibril from one Z disc to the next, constituting one contractile unit. (fig. 7.4)

sarcoplasmic reticulum (SR) The smooth endoplasmic reticulum of a muscle cell, serving as a calcium reservoir. (fig. 7.2)

scanning electron microscope (SEM) A microscope that uses an electron beam in place of light to form high-resolution, three-dimensional images of the surfaces of objects; capable of much higher magnifications than a light microscope.

sclerosis (scleh-RO-sis) Hardening or stiffening of a tissue, as in multiple sclerosis of the central nervous system or atherosclerosis of the blood vessels.

sebum (SEE-bum) An oily secretion of the sebaceous glands that keeps the skin and hair pliable.

secondary active transport A mechanism in which solutes are moved through a plasma membrane by a carrier that does not itself use ATP but depends on a concentration gradient established by an active transport pump elsewhere in the cell.

second messenger A chemical that is produced within a cell (such as cAMP) or that enters a cell (such as calcium ions) in response to the binding of a messenger to a membrane receptor, and that triggers a metabolic reaction in the cell.

secretion **1.** A chemical released by a cell to serve a physiological function, such as a hormone or digestive enzyme. **2.** The process of releasing such a chemical, often by exocytosis. *Compare* excretion.

selectively permeable membrane A membrane that allows some substances to pass through while excluding others; for example, the plasma membrane and dialysis membranes.

semicircular ducts Three ring-shaped, fluid-filled tubes of the inner ear that detect angular acceleration of the head; each is enclosed in a bony passage called the semicircular canal. (fig. 10.9)

semilunar valve A valve that consists of crescent-shaped cusps, including the aortic and pulmonary valves of the heart and valves of the veins and lymphatic vessels. (fig. 13.4)

senescence (seh-NESS-ense) Degenerative changes that occur with age.

sensation Conscious perception of a stimulus; pain, taste, and color, for example, are not stimuli but sensations resulting from stimuli.

sensory nerve fiber An axon that conducts information from a receptor to the central nervous system; an afferent nerve fiber.

serosa (seer-OH-sa) *See* serous membrane.

serous fluid (SEER-us) A watery, low-protein fluid similar to blood serum, formed as a filtrate of the blood or tissue fluid or as a secretion of serous gland cells; moistens the serous membranes.

serous membrane A membrane such as the peritoneum, pleura, or pericardium that lines a body cavity or covers the external surfaces of the viscera; composed of a simple squamous epithelium and a thin layer of areolar connective tissue. (fig. 4.30b)

sex chromosomes The X and Y chromosomes, which determine a person's sex.

shock   In the circulatory sense, a state of cardiac output that is insufficient to meet the body's physiological needs, with consequences ranging from fainting to death.

sign   An objective manifestation of illness that any observer can see, such as cyanosis or edema. *Compare* symptom.

simple epithelium   An epithelium in which all cells rest directly on the basement membrane; includes simple squamous, cuboidal, and columnar types, and pseudostratified columnar. (fig. 4.3)

sinus   1. An air-filled space in the cranium. 2. A modified, relatively dilated vein that lacks smooth muscle and is incapable of vasomotion, such as the dural sinuses of the cerebral circulation and coronary sinus of the heart.   3. A small fluid-filled space in an organ such as the spleen and lymph nodes.   4. Pertaining to the sinoatrial node of the heart, as in *sinus rhythm.*

somatic   1. Pertaining to the body as a whole. 2. Pertaining to the skin, bones, and skeletal muscles as opposed to the viscera. 3. Pertaining to cells other than germ cells.

somatic nervous system   A division of the nervous system that includes afferent fibers mainly from the skin, muscles, and skeleton and efferent fibers to the skeletal muscles. *Compare* autonomic nervous system.

somatosensory   1. Pertaining to widely distributed *general senses* in the skin, muscles, tendons, joint capsules, and viscera, as opposed to the *special senses* found in the head only. 2. Pertaining to the cerebral cortex of the postcentral gyrus, which receives input from such receptors. Also called *somesthetic.*

somite   One segment in a linear series of mesodermal masses that form on each side of the embryonic neural tube and give rise to trunk muscles, vertebrae, and dermis. (fig. 20.7a, b)

spermatogenesis (SPUR-ma-toe-JEN-eh-sis)   The production of sperm cells through a series of mitotic and meiotic cell divisions; male gametogenesis.

spermatozoon (SPUR-ma-toe-ZOE-on)   A sperm cell.

sphincter (SFINK-tur)   A ring of muscle that opens or closes an opening or passageway; found, for example, in the eyelids, around the urinary orifice, and at the beginning of a blood capillary.

spinal nerve   Any of the 31 pairs of nerves that arise from the spinal cord and pass through the intervertebral foramina.

spindle   1. An elongated structure that is thick in the middle and tapered at the ends (fusiform). 2. A football-shaped complex of microtubules that guide the movement of chromosomes in mitosis and meiosis.

(fig. 3.15)   3. A stretch receptor in the skeletal muscles.

spine   1. The vertebral column. 2. A pointed process or sharp ridge on a bone, such as the styloid process of the cranium and spine of the scapula.

squamous (SKWAY-mus)   Having a flat, scaly shape; pertains especially to a class of epithelial cells. (figs. 4.2, 4.8)

stem cell   Any undifferentiated cell that can divide and differentiate into more functionally specific cell types such as blood cells and germ cells.

stenosis (steh-NO-sis)   The narrowing of a passageway such as a heart valve or uterine tube; a permanent, pathological constriction as opposed to normal physiological constriction of a passageway.

steroid (STERR-oyd, STEER-oyd)   A lipid molecule that consists of four interconnected carbon rings; cholesterol and several of its derivatives.

stimulus   A chemical or physical agent in a cell's surroundings that is capable of creating a physiological response in the cell; especially agents detected by sensory cells, such as chemicals, light, and pressure.

stratified epithelium   A type of epithelium in which some cells rest on top of others instead of on the basement membrane; includes stratified squamous, cuboidal, and columnar types, and transitional epithelium. (fig. 4.3)

stress   1. A mechanical force applied to any part of the body; important in stimulating bone growth, for example. 2. A condition in which any environmental influence disturbs the homeostatic equilibrium of the body and stimulates a physiological response, especially involving the increased secretion of certain adrenal hormones.

stroke volume   The volume of blood ejected by one ventricle of the heart in one contraction.

subcutaneous (SUB-cue-TAY-nee-us)   Beneath the skin.

substrate   1. A chemical that is acted upon and changed by an enzyme. 2. A chemical used as a source of energy, such as glucose and fatty acids.

sulcus (SUL-cuss)   A groove in the surface of an organ, as in the cerebrum or heart.

summation   A phenomenon in which multiple muscle twitches occur so closely together that a muscle fiber cannot fully relax between twitches but develops more tension than a single twitch produces. (fig. 7.10)

superficial   Relatively close to the surface; opposite of deep. For example, the ribs are superficial to the lungs.

superior   Higher than another structure or point of reference from the perspective of anatomical position; for example, the lungs are superior to the diaphragm.

supination (SOO-pih-NAY-shun)   A rotational movement of the forearm that turns the palm so that it faces upward or forward. (fig. 6.40)

surfactant (sur-FAC-tent)   A chemical that reduces the surface tension of water and enables it to penetrate other substances more effectively. Examples include pulmonary surfactant and bile acids.

sympathetic nervous system   A division of the autonomic nervous system that issues efferent fibers through the thoracic and lumbar nerves and usually exerts adrenergic effects on its target organs; includes a chain of ganglia adjacent to the vertebral column, and the adrenal medulla.

symphysis (SIM-fih-sis)   A joint in which two bones are held together by fibrocartilage; for example, between bodies of the vertebrae and between the right and left pubic bones.

symptom   A subjective manifestation of illness that only the ill person can sense, such as dizziness or nausea. *Compare* sign.

synapse (SIN-aps)   1. A junction at the end of an axon where it stimulates another cell. 2. A gap junction between two cardiac or smooth muscle cells at which one cell electrically stimulates the other; called an *electrical synapse.*

synaptic cleft (sih-NAP-tic)   A narrow space between the synaptic knob of an axon and the adjacent cell, across which a neurotransmitter diffuses. (fig. 8.12)

synaptic knob   The swollen tip at the distal end of an axon; the site of synaptic vesicles and neurotransmitter release. (fig. 8.12)

synaptic vesicle   A spheroid organelle in a synaptic knob containing neurotransmitter. (fig. 8.12)

syndrome   A suite of related signs and symptoms stemming from a specific pathological cause.

synergist (SIN-ur-jist)   A muscle that works with the prime mover (agonist) to contribute to the same overall action at a joint.

synergistic   An effect in which two agents working together (such as two hormones) exert an effect that is greater than the sum of their separate effects. For example, neither follicle-stimulating hormone nor testosterone alone stimulates significant sperm production, but the two of them together stimulate production of vast numbers of sperm.

synovial fluid (sih-NO-vee-ul) A lubricating fluid similar to egg white in consistency, found in the synovial joint cavities and bursae.

synovial joint A point where two bones are separated by a narrow, encapsulated space filled with lubricating synovial fluid; most such joints are relatively mobile. Also called a *diarthrosis*.

systemic (sis-TEM-ic) Widespread or pertaining to the body as a whole, as in the systemic circulation.

systemic circuit All blood vessels that convey blood from the left ventricle to all organs of the body and back to the right atrium of the heart; all of the cardiovascular system except the heart and pulmonary circuit.

systole (SIS-toe-lee) The contraction of any heart chamber; ventricular contraction unless otherwise specified.

systolic pressure (sis-TOLL-ic) The peak arterial blood pressure measured during ventricular systole.

T

target cell A cell acted upon by a nerve fiber, hormone, or other chemical messenger.

tarsal Pertaining to the ankle (tarsus).

T cell A type of lymphocyte involved in non-specific defense, humoral immunity, and cellular immunity; occurs in several forms including helper, cytotoxic, and regulatory T cells and natural killer cells.

tendon A collagenous band or cord associated with a muscle, usually attaching it to a bone and transferring muscular tension to it.

tetanus **1.** A state of sustained muscle contraction produced by summation as a normal part of contraction; also called *tetany*. **2.** Spastic muscle paralysis produced by the toxin of the bacterium *Clostridium tetani*.

thalamus (THAL-uh-muss) The largest part of the diencephalon, located immediately inferior to the corpus callosum and bulging into each lateral ventricle; a point of synaptic relay of nearly all signals passing from lower levels of the CNS to the cerebrum. (figs. 9.2, 9.5)

theory An explanatory statement, or set of statements, that concisely summarizes the state of knowledge on a phenomenon and provides direction for further study; for example, the fluid-mosaic theory of the plasma membrane and the sliding filament theory of muscle contraction.

thermogenesis The production of heat, for example, by shivering or by the action of thyroid hormones.

thermoreceptor A neuron specialized to respond to heat or cold, found in the skin and mucous membranes, for example.

thermoregulation Homeostatic regulation of the body temperature within a narrow range by adjustments of heat-promoting and heat-losing mechanisms.

thorax A region of the trunk between the neck and the diaphragm; the chest.

threshold The minimum voltage to which the plasma membrane of a nerve or muscle cell must be depolarized before it produces an action potential.

thrombosis (throm-BO-sis) The formation or presence of a thrombus.

thrombus A clot that forms in a blood vessel or heart chamber; may break free and travel in the bloodstream as a thromboembolus. *Compare* embolus.

thyroid hormone Either of two similar hormones, thyroxine and triiodothyronine, synthesized from iodine and tyrosine.

thyroid-stimulating hormone (TSH) A hormone of the anterior pituitary gland that stimulates the thyroid gland; also called *thyrotropin*.

thyroxine (T₄) (thy-ROCK-seen) The thyroid hormone secreted in greatest quantity, with four iodine atoms; also called *tetraiodothyronine*.

tight junction A region in which adjacent cells are bound together by fusion of the outer phospholipid layer of their plasma membranes; forms a zone that encircles each cell near its apical pole and reduces or prevents flow of material between cells. (fig. 3.6)

tissue An aggregation of cells and extracellular materials, usually forming part of an organ and performing some discrete function for it; the four primary classes are epithelial, connective, muscular, and nervous tissue.

trabecula (tra-BEC-you-la) A thin plate or layer of tissue, such as the calcified trabeculae of spongy bone or the fibrous trabeculae that subdivide a gland. (fig. 6.3)

trachea (TRAY-kee-uh) A cartilage-supported tube from the inferior end of the larynx to the origin of the main bronchi; conveys air to and from the lungs; colloquially called the *windpipe*.

translation The process in which a ribosome reads an mRNA molecule and synthesizes the protein specified by the genetic code.

transmission electron microscope (TEM) A microscope that uses an electron beam in place of light to form high-resolution, two-dimensional images of ultrathin slices of cells or tissues; capable of extremely high magnification.

triglyceride (try-GLISS-ur-ide) A lipid composed of three fatty acids joined to a glycerol; also called a *neutral fat*. (fig. 2.11b)

trunk **1.** That part of the body excluding the head, neck, and appendages. **2.** A major blood vessel, lymphatic vessel, or nerve that goes for only a short distance and gives rise to smaller branches; for example, the pulmonary trunk and spinal nerve trunks.

T tubule Transverse tubule; a tubular extension of the plasma membrane of a muscle cell that conducts action potentials into the sarcoplasm and excites the sarcoplasmic reticulum. (fig. 7.2)

tunic (TOO-nic) A layer that encircles or encloses an organ, such as the tunics of a blood vessel or eyeball.

tympanic membrane The eardrum.

U

ultraviolet radiation Invisible, ionizing, electromagnetic radiation with shorter wavelength and higher energy than violet light; causes skin cancer and photoaging of the skin but is required in moderate amounts for the synthesis of vitamin D.

unmyelinated (un-MY-eh-lih-nay-ted) Lacking a myelin sheath.

urea (you-REE-uh) A nitrogenous waste produced from two ammonia molecules and carbon dioxide; the most abundant nitrogenous waste in the blood and urine.

uterine tube A duct that extends from the ovary to the uterus and conveys an egg or conceptus to the uterus; also called *fallopian tube* or *oviduct*.

V

varicose vein A vein that has become permanently distended and convoluted due to a loss of competence of the venous valves; especially common in the lower limb, esophagus, and anal canal (where they are called *hemorrhoids*).

vas (vass) (plural, *vasa*) A vessel or duct.

vascular Pertaining to blood vessels.

vasoconstriction (VAY-zo-con-STRIC-shun) The narrowing of a blood vessel due to muscular constriction of its tunica media.

vasodilation (VAY-zo-dy-LAY-shun) The widening of a blood vessel due to relaxation of the muscle of its tunica media and the outward pressure of the blood exerted against the wall.

vasomotion (VAY-zo-MO-shun) Collective term for vasoconstriction and vasodilation.

vasomotor center A nucleus in the medulla oblongata that transmits efferent signals to the blood vessels and regulates vasomotion.

ventral Pertaining to the front of the body, the regions of the chest and abdomen; anterior.

ventral root *See* anterior root.

ventricle (VEN-trih-cul) A fluid-filled chamber of the brain or heart.

venule (VEN-yool) The smallest type of vein, receiving drainage from capillaries.

vertebra (VUR-teh-bra) One of the bones of the vertebral column.

vertebral column (VUR-teh-brul) A posterior series of usually 33 vertebrae; encloses the spinal cord, supports the skull and thoracic cage, and provides attachment for the limbs and postural muscles. Also called the *spine* or *spinal column.*

vesicle (VESS-ih-cul) A fluid-filled tissue sac or an organelle such as a synaptic or secretory vesicle.

vesicular transport The movement of particles or fluid droplets through the plasma membrane by the process of endocytosis or exocytosis.

viscera (VISS-er-uh) (singular, *viscus*) The organs contained in the body cavities, such as the brain, heart, lungs, stomach, intestines, and kidneys.

visceral (VISS-er-ul) **1.** Pertaining to the viscera. **2.** The inner or deeper layer of a two-layered membrane such as the pleura, pericardium, or glomerular capsule. *Compare* parietal. (fig. 1.13)

visceral muscle A form of smooth muscle found in the walls of blood vessels and the digestive, respiratory, urinary, and reproductive tracts.

viscosity The resistance of a fluid to flow; the thickness or stickiness of a fluid.

voluntary muscle Muscle that is usually under conscious control; skeletal muscle.

vulva The female external genitalia; the mons, labia majora, and all superficial structures between the labia majora.

W

water balance *See* fluid balance.

white matter White myelinated nervous tissue deep to the cortex of the cerebrum and cerebellum and superficial to the gray matter of the spinal cord. (figs. 9.3, 9.4)

X

X chromosome The larger of the two sex chromosomes; males have one X chromosome, and females have two, in each somatic cell.

X-ray **1.** A high-energy, penetrating electromagnetic ray with wavelengths in the range of 0.1 to 10 nm; used in diagnosis and therapy. **2.** A photograph made with X-rays; radiograph.

Y

Y chromosome Smaller of the two sex chromosomes, found only in males and having little genetic function except development of the testis.

yolk sac An embryonic membrane that encloses the yolk in vertebrates that lay eggs; serves in humans as the origin of the first blood and germ cells.

Z

zygomatic arch An arch of bone anterior to the ear, formed by the zygomatic processes of the temporal, frontal, and zygomatic bones; origin of the masseter muscle.

zygote A single-celled, fertilized egg.

Photo Credits

Design Elements

"Before You Go On" (mountain climber): © Alessandro Contadini RF; **"Perspectives on Health" (reviewing x-ray)**: © Ghislain & Marie David de Lossy/The Image Bank/Getty Images; **"Clinical Application" (stethoscope)**: © Nathan Blaney RF; **"Study Guide" (mountaineer at the top of mountain)**: © SimonKr d.o.o; **"Base Camp" (tent/mountains)**: © James Balog

Connective Issues

Circulatory System: Getty Images; **Lymphatic and Immune Systems**: © Russell Kightley/Photo Researchers, Inc.; **Respiratory System**: © McGraw-Hill Companies, Inc.; **Urinary System**: © Medical Body Scans/Photo Researchers, Inc.; **Reproductive System, Digestive System, Nervous System, Muscular System, Integumentary System**: iStockphoto; **Skeletal System**: © U.H.B. Trust/Tony Stone Images/Getty Images

Chapter 1

Opener: © Science Photo Library/Getty Images; **Figure 1.1a**: © U.H.B. Trust/Tony Stone Images/Getty Images; **1.1b**: © Alessandro Contadini RF Custom Medical Stock Photos, Inc.; **1.1c**: © CNRI/Phototake; **1.1d**: © Tony Stone Images/Getty Images; **1.1e**: © Alfred Pasieka/Science Source; **1.2a**: © Alexander Tsiaras/Photo Researchers, Inc.; **1.2b**: © Ken Saladin; **1.7**: © The McGraw-Hill Companies, Inc./Joe DeGrandis, photographer; **1.14**: © MedicImage/Getty Image; p. 27: © ERproductions Ltd/Blend Images LLC RF

Chapter 2

Figure 2.7a–d: © Ken Saladin; p. 59: © Will & Deni McIntyre/Photo Researchers, Inc.

Chapter 3

Opener: © Eye of Science/Photo Researchers, Inc.; **3.4**: Courtesy of Dr. S. Ito, Harvard Medical School; **3.5a**: © Custom Medical Stock Photo, Inc.; **3.5c**: © Don Fawcett/Photo Researchers, Inc.; **3.9b**: © K. G. Murti/Visuals Unlimited; **3.15(1–4)**: © Ed Reschke; **3.16b**: © Biophoto Associates/Photo Researchers, Inc.; p. 93: © iStockphoto

Chapter 4

Opener: © Steve Gschmeissner/Photo Researchers, Inc.; **4.4a, 4.5a**: © The McGraw-Hill Companies, Inc./Dennis Strete, photographer; **4.6a**: © L. V. Bergman, The Bergman Collection; **4.7a**: © The McGraw-Hill Companies, Inc./Dennis Strete, photographer; **4.8**: © Dr. Richard Kessel & Dr. Randy Kardon/Tissues & Organs/Visuals Unlimited, Inc.; **4.9a**: © The McGraw-Hill Companies, Inc./Dennis Strete, photographer; **4.10a**: © Ed Reschke; **4.11a**: © The McGraw-Hill Companies, Inc./Dennis Strete, photographer; **4.12a**: © Johnny R. Howze; **4.13**: © The McGraw-Hill Companies, Inc./Rebecca Gray, photographer; **4.14a**: © The McGraw-Hill Companies, Inc./Dennis Strete, photographer; **4.15a**: © The McGraw-Hill Companies, Inc./Al Telser, photographer; **4.16a–4.18a**: © The McGraw-Hill Companies, Inc./Dennis Strete, photographer; **4.19a, 4.20a**: © Ed Reschke; **4.21a**: © Dr. Alvin Telser; **4.22a**: © The McGraw-Hill Companies, Inc./Dennis Strete, photographer; **4.23a–4.26a**: © Ed Reschke; **4.31**: © Jere Mammino, DO; **p. 125**: © Leca/Photo Researchers, Inc.

Chapter 5

Opener: © Eye of Science/Photo Researchers, Inc.; **5.3a(top)**: © Tom & Dee Ann McCarthy/Corbis; **(bottom)**: © The McGraw-Hill Companies, Inc./Dennis Strete, photographer; **5.3b (top)**: © Creatas/PunchStock; **(bottom)**: © The McGraw-Hill Companies, Inc./Dennis Strete, photographer; **5.4a**: © MNSB/Custom Medical Stock Photo, Inc.; **5.4b**: © James Stevenson/SPL/Photo Researchers, Inc.; **5.5b**: © CBS/Phototake; **5.5c**: © SPL/Photo Researchers, Inc.; **5.7a–c**: © The McGraw-Hill Companies, Inc./Dennis Strete, photographer; **p. 143**: © Lauren Shear/Photo Researchers, Inc.

Chapter 6

Opener: © Dr. P. Marazzi/Photo Researchers, Inc.; **6.1**: © D. W. Fawcett/Visuals Unlimited; **6.3b**: © D. W. Fawcett/Visuals Unlimited; **6.3c**: © Visuals Unlimited; **6.9a**: © Dr. Ken Greer/Getty Images; **6.9b**: © CNRI/Science Photo Library/Photo Researchers, Inc.; **6.29**: © NHS Trust/Tony Stone Images/Getty Images; **6.33d**: © Walter Reiter/Phototake; **6.36a–6.42c**: © The McGraw-Hill Companies, Inc./Timothy L. Vacula, photographer; **p. 198**: © ERproductions Ltd./Blend Images/Corbis

Chapter 7

Opener: © Custom Medical Stock, Inc.; **7.1**: © Ed Reschke; **7.4**: © Visuals Unlimited; **7.6a**: © Victor B. Eichler; **7.12a**: © Ed Reschke; **7.13a**: © The McGraw-Hill Companies, Inc./Dennis Strete, photographer; **7.14b**: © Victor Eroschenko; **p. 245**: © Rolf Hicker/All Canada Photos/Getty Images

Chapter 8

Opener: © Dr. Torsten Wittmann/Photo Researchers, Inc.; **8.2c**: © Ed Reschke; **8.6c**: © The McGraw-Hill Companies, Inc./Dr. Dennis Emery, Dept. of Zoology and Genetics, Iowa State University, photographer; **8.14a**: © Custom Medical Stock Photo; **8.14b**: © Simon Fraser/Photo Researchers, Inc.; **8.16c**: © Sarah Werning; **8.17**: © Biophoto Associates/Photo Researchers, Inc.; **8.18b**: © Dr. Richard Kessel & Dr. Kardon/Tissues & Organs/Visuals Unlimited, Inc.; **p. 284**: © David Grossman/Photo Researchers, Inc.

Chapter 9

Opener: © Living Art Enterprises/Photo Researchers, Inc.; **9.1c**: © The McGraw-Hill Companies, Inc./Rebecca Gray, photographer/Don Kincaid, dissections; **9.2b**: © The McGraw-Hill Companies, Inc./Dennis Strete, photographer; **9.4c**: © The McGraw-Hill Companies, Inc./Rebecca Gray, photographer/Don Kincaid, dissections; **9.13**: © The McGraw-Hill Companies, Inc./Bob Coyle, photographer; **9.19 (left)**: © Stockbyte/PunchStock RF; **(right)**: Copyright © Foodcollection RF; **p. 322**: © Larry Mulvehill/Photo Researchers, Inc.

Chapter 10

Opener: © 3D4Medical/Photo Researchers, Inc.; **10.4c**: © Ed Reschke; **10.11**: © Quest/Science Photo Library/Photo Researchers, Inc.; **10.17a**: © Lisa Klancher; **10.17c**: © Peter Dazeley/Getty Images; **10.19a**: © The McGraw-Hill Companies, Inc./Dennis Strete, photographer; **10.20a**: Courtesy of Beckman Vision Center at UCSF School of Medicine/D. Copenhagen, S. Mittman and M. Maglio; **10.21b**: © Steve Allen/Getty Images RF; **p. 358**: © Huntstock/Getty Images RF

Chapter 11

Opener: © David Mack/Photo Researchers, Inc.; **11.8b**: © Robert Calentine/Visuals Unlimited; **11.10c–11.13b**: © Ed Reschke; **p. 386**: © Rolf Bruderer/Getty Images RF

Chapter 12

Opener: © Dr. Yorgos Nikas/Photo Researchers, Inc.; **12.3b:** © Susumu Nishinaga/Getty Images; **12.3c:** © Dr. Don W. Fawcett/Visuals Unlimited; **12.5:** © Meckes/Ottawa/Photo Researchers, Inc.; **12.8:** © SIU/Photo Researchers, Inc.; **12.9–12.11b:** © Ed Reschke; **12.12a:** © NIBSC/Science Photo Library/Photo Researchers, Inc.; **12.13:** © P. Motta/SPL/Photo Researchers, Inc.; **p. 412:** © Ariel Skelley/Blend Images/Getty Images

Chapter 13

Opener: © SPL/Photo Researchers, Inc.; **13.5c:** © Visuals Unlimited; **13.5d, 13.6a:** © Ed Reschke; **13.11c:** © Biophoto Associates/Photo Researchers, Inc.; **p. 465:** © MCT via Getty Images

Chapter 14

Opener: © Eye of Science/Photo Researchers, Inc.; **14.3 (both):** © Visuals Unlimited; **14.5:** © The McGraw-Hill Companies, Inc./Dennis Strete, photographer; **14.6b:** © Biophoto Associates/Photo Researchers, Inc.; **14.7b:** © The McGraw-Hill Companies, Inc./Dennis Strete, photographer; **14.10:** © David M. Phillips/Photo Researchers, Inc.; **14.18:** © Steve Gschmeissner/Science Source; **14.20a,b:** © Dr. Don W. Fawcett/Visuals Unlimited; **14.24b:** © NIBSC/SPL/Photo Researchers, Inc.; **p. 498:** Getty Images

Chapter 15

Opener: © Anatomical Travelogue/Photo Researchers, Inc.; **15.2a:** © The McGraw-Hill Companies/Joe DeGrandis, photographer;

15.4: © Phototake; **15.5d:** © Custom Medical Stock Photo; **15.7:** © Ralph Hutchings/Visuals Unlimited; **15.8a:** © Dr. Gladden Willis/Visuals Unlimited; **15.8b:** © Visuals Unlimited; **15.12:** © BSIP/Photo Researchers, Inc.; **15.14a:** © The McGraw-Hill Companies, Inc./Dennis Strete, photographer; **15.14b:** © Biophoto Associates/Photo Researchers, Inc.; **p. 524:** © McGraw-Hill Companies, Inc./Rick Brady, photographer

Chapter 16

Opener: © Susumu Nishinaga/Photo Researchers, Inc.; **16.2a:** © Ralph Hutchings/Visuals Unlimited; **16.5b:** © Dr. Richard Kessel & Dr. Randy Kardon/Visuals Unlimited, Inc.; **16.8:** © Don Fawcett/Photo Researchers, Inc.; **p. 558:** © John Birdsall/Visuals Unlimited, Inc.

Chapter 17

Opener: 3d4Medical.com/Corbis; **17.9d:** © Visuals Unlimited; **17.11a–17.11c:** © Eye of Science/Photo Researchers, Inc.; © CNRI/SPL/Photo Researchers, Inc.; **17.15a:** © Meckes/Ottawa/Photo Researchers, Inc.; **17.17a:** © Burke/Triolo Productions/Getty Images RF; **17.17b:** © Pixtal/AGE Fotostock RF; **17.17c:** © MIXA/Getty Images RF; **p. 596:** © MIXA/Getty Images RF

Chapter 18

Opener: © Dr. Donald Fawcett/Visuals Unlimited/Corbis; **p. 627:** © Photo Researchers, Inc.

Chapter 19

Opener: © David M. Phillips/Photo Researchers, Inc.; **19.2a:** © The McGraw-Hill

Companies, Inc./Dennis Strete, photographer; **19.2c:** © Dr. Richard Kessel & Dr. Kardon/Tissues & Organs/Visuals Unlimited, Inc.; **19.2d:** © Ed Reschke; **19.7a:** © Photo Researchers, Inc.; **19.8a,b:** © SPL/Photo Researchers, Inc.; **19.10b:** From Anatomy & Physiology Revealed, © The McGraw-Hill Companies, Inc./The University of Toledo, photography and dissection; **19.10d:** © Dr. Donald Fawcett, T. Nagato/Visuals Unlimited, Inc.; **19.13a:** © Visuals Unlimited; **19.14 (Primordial & Primary):** © Ed Reschke; **(Secondary):** © The McGraw-Hill Companies, Inc./photo by Dr. Alvin Telser; **(Tertiary):** © Ed Reschke; **(mature):** © Dr. Landrum Shettles; **(Corpus luteum):** © The McGraw-Hill Companies, Inc./photo by Dr. Alvin Telser; **19.17 (crowning, expulsion):** © D. Van Rossum/Photo Researchers, Inc.; **(afterbirth):** © Visuals Unlimited; **p. 670:** © Photo Researchers, Inc.

Chapter 20

Opener: Anatomical Travelogue/Photo Researchers/Getty Images; **20.5c,d:** Dr. Kurt Benirschke; **20.7a:** © Dr. Landrum Shettles; **20.7b,c:** © Anatomical Travelogue/Photo Researchers, Inc.; **20.7d:** © Martin Rotker/Phototake, Inc.; **20.7e:** © Photo Researchers, Inc.; **20.7f:** © Dr. Landrum Shettles; **20.9a:** © Photo Researchers, Inc.; **20.9b:** Courtesy Mihaly Bartalos, from Bartalos 1967: Medical Cytogenetics, fig 10.2, pg. 154, Waverly, a division of Williams & Wilkins; **20.10:** © Time & Life Pictures/Getty Images; **20.11:** © Bettmann/Corbis; **p. 703:** MCT via Getty Images

Periodic Table of the Elements

Nineteenth-century chemists discovered that when they arranged the known elements by atomic weight, certain properties reappeared periodically. In 1869, Russian chemist Dmitri Mendeleev published the first modern periodic table of the elements, leaving gaps for those that had not yet been discovered. He accurately predicted properties of the missing elements, which helped other chemists discover and isolate them.

Each row in the table is a *period* and each column is a *group (family)*. Each period has one electron shell more than the period above it, and as we progress from left to right within a period, each element has one more proton and electron than the one before. The dark steplike line from boron (5) to astatine (85) separates the metals to the left of it (except hydrogen) from the nonmetals to the right. Each period begins with a soft, light, highly reactive *alkali metal,* with one valence electron, in family IA. Progressing from left to right,

the metallic properties of the elements become less and less pronounced. Elements in family VIIA are highly reactive gases called *halogens,* with seven valence electrons. Elements in family VIIIA, called *noble (inert) gases,* have a full valence shell of eight electrons, which makes them chemically unreactive.

Ninety-one of the elements occur naturally on earth. Physicists have created elements up to atomic number 118 in the laboratory, but the International Union of Pure and Applied Chemistry has established formal names only through element 109 to date.

The 24 elements with normal roles in human physiology are color-coded according to their relative abundance in the body (see chapter 2). Others, however, may be present as contaminants with very destructive effects (such as arsenic, lead, and radiation poisoning).